Solar Neutrinos

Solar Neutrinos

The First Thirty Years

Edited by

John N. Bahcall
Raymond Davis, Jr.
Peter Parker
Alexei Smirnov
Roger Ulrich

Westview Press
A Member of the Perseus Books Group

ABP *Advanced Book Program*

The Advanced Book Program

Published by Westview Press, a member of the Perseus Books Group.
Find us on the World Wide Web at http://www.westviewpress.com

Cataloging-in-Publication Data is available from the Library of Congress
ISBN 0–8133–4037–3

Westview Press books are available at special discounts for bulk purchases in the U.S. by corporations, institutions, and other organizations. For more information, please contact the Special Markets Department at the Perseus Books Group, 11 Cambridge Center, Cambridge, MA 02142, or call (800) 255–1514 or (617) 252–5298, or e-mail j.mccrary@perseusbooks.com.

First paperback printing, June 2002

4 5 6 7 8 9 10—04 03

Frontiers in Physics

David Pines, Editor

Volumes of the Series published from 1961 to 1973 are not officially numbered. The parenthetical numbers shown are designed to aid librarians and bibliographers to check the completeness of their holdings.
Titles published in this series prior to 1987 appear under either the W. A. Benjamin or the Benjamin/Cummings imprint; titles published since 1986 appear under the Westview Press imprint.

1.	N. Bloembergen	Nuclear Magnetic Relaxation: A Reprint Volume, 1961
2.	G. F. Chew	S-Matrix Theory of Strong Interactions: A Lecture Note and Reprint Volume, 1961
3.	R. P. Feynman	Quantum Electrodynamics: A Lecture Note and Reprint Volume
4.	R. P. Feynman	The Theory of Fundamental Processes: A Lecture Note Volume, 1961
5.	L. Van Hove, N. M. Hugenholtz, L. P. Howland	Problem in Quantum Theory of Many-Particle Systems: A Lecture Note and Reprint Volume, 1961
6.	D. Pines	The Many-Body Problem: A Lecture Note and Reprint Volume, 1961
7.	H. Frauenfelder	The Mössbauer Effect: A Review—With a Collection of Reprints, 1962
8.	L. P. Kadanoff, G. Baym	Quantum Statistical Mechanics: Green's Function Methods in Equilibrium and Nonequilibrium Problems, 1962
9.	G. E. Pake	Paramagnetic Resonance: An Introductory Monograph, 1962 [cr. (42)—2nd edition]G. E. Pake
10.	P. W. Anderson	Concepts in Solids: Lectures on the Theory of Solids, 1963
11.	S. C. Frautschi	Regge Poles and S-Matrix Theory, 1963
12.	R. Hofstadter	Electron Scattering and Nuclear and Nucleon Structure: A Collection of Reprints with an Introduction, 1963
13.	A. M. Lane	Nuclear Theory: Pairing Force Correlations to Collective Motion, 1964
14.	R. Omnès, M. Froissart	Mandelstam Theory and Regge Poles: An Introduction for Experimentalists, 1963
15.	E. J. Squires	Complex Angular Momenta and Particle Physics: A Lecture Note and Reprint Volume, 1963
16.	H. L. Frisch, J. L. Lebowitz	The Equilibrium Theory of Classical Fluids: A Lecture Note and Reprint Volume, 1964
17.	M. Gell-Mann, Y. Ne'eman	The Eightfold Way (A Review—With a Collection of Reprints), 1964
18.	M. Jacob, G. F. Chew	Strong-Interaction Physics: A Lecture Note Volume, 1964
19.	P. Nozières	Theory of Interacting Fermi Systems, 1964

20.	J. R. Schrieffer	Theory of Superconductivity, 1964 (revised 3rd printing, 1983)
21.	N. Bloembergen	Nonlinear Optics: A Lecture Note and Reprint Volume, 1965
22.	R. Brout	Phase Transitions, 1965
23.	I. M. Khalatnikov	An Introduction to the Theory of Superfluidity, 1965
24.	P. G. deGennes	Superconductivity of Metals and Alloys, 1966
25.	W. A. Harrison	Pseudopotentials in the Theory of Metals, 1966
26.	V. Barger D. Cline	Phenomenological Theories of High Energy Scattering: An Experimental Evaluation, 1967
27.	P. Choquàrd	The Anharmonic Crystal, 1967
28.	T. Loucks	Augmented Plane Wave Method: A Guide to Performing.Electronic Structure Calculations—A Lecture Note and Reprint Volume, 1967
29.	Y. Ne'eman	Algebraic Theory of Particle Physics: Hadron Dynamics In Terms of Unitary Spin Current, 1967
30.	S. L. Adler R. F. Dashen	Current Algebras and Applications to Particle Physics, 1968
31.	A. B. Migdal	Nuclear Theory: The Quasiparticle Method, 1968
32.	J. J. J. Kokkedee	The Quark Model, 1969
33.	A. B. Migdal	Approximation Methods in Quantum Mechanics, 1969
34.	R. Z. Sagdeev	Nonlinear Plasma Theory, 1969
35.	J. Schwinger	Quantum Kinematics and Dynamics, 1970
36.	R. P. Feynman	Statistical Mechanics: A Set of Lectures, 1972
37.	R. P. Feynman	Photon-Hadron Interactions, 1972
38.	E. R. Caianiello	Combinatorics and Renormalization in Quantum Field Theory, 1973
39.	G. B. Field H. Arp J. N. Bahcall	The Redshift Controversy, 1973
40.	D. Horn F. Zachariasen	Hadron Physics at Very High Energies, 1973
41.	S. Ichimaru	Basic Principles of Plasma Physics: A Statistical Approach, 1973 (2nd printing, with revisions, 1980)
42.	G. E. Pake T. L. Estle	The Physical Principles of Electron Paramagnetic Resonance, 2nd Edition, completely revised, enlarged, and reset, 1973 [cf. (9)—1st edition

Volumes published from 1974 onward are being numbered as an integral part of the bibliography.

43 .	C. Davidson	Theory of Nonneutral Plasmas, 1974
44.	S. Doniach E. H. Sondheimer	Green's Functions for Solid State Physicists, 1974
45.	P. H. Frampton	Dual Resonance Models, 1974
46.	S. K. Ma	Modern Theory of Critical Phenomena, 1976
47.	D. Forster	Hydrodynamic Fluctuation, Broken Symmetry, and Correlation Functions, 1975
48.	A. B. Migdal	Qualitative Methods in Quantum Theory, 1977
49.	S. W. Lovesey	Condensed Matter Physics: Dynamic Correlations, 1980

50.	L. D. Faddeev A. A. Slavnov	Gauge Fields: Introduction to Quantum Theory, 1980
51.	P. Ramond	Field Theory: A Modern Primer, 1981 [cf. 74—2nd ed.]
52.	R. A. Broglia A. Winther	Heavy Ion Reactions: Lecture Notes Vol. I, Elastic and Inelastic Reactions, 1981
53.	R. A. Broglia A. Winther	Heavy Ion Reactions: Lecture Notes Vol. II, 1990
54.	H. Georgi	Lie Algebras in Particle Physics: From Isospin to Unified Theories, 1982
55.	P. W. Anderson	Basic Notions of Condensed Matter Physics, 1983
56.	C. Quigg	Gauge Theories of the Strong, Weak, and Electromagnetic Interactions, 1983
57.	S. I. Pekar	Crystal Optics and Additional Light Waves, 1983
58.	S. J. Gates M. T. Grisaru M. Rocek W. Siegel	Superspace *or* One Thousand and One Lessons in Supersymmetry, 1983
59.	R. N. Cahn	Semi-Simple Lie Algebras and Their Representations, 1984
60.	G. G. Ross	Grand Unified Theories, 1984
61.	S. W. Lovesey	Condensed Matter Physics: Dynamic Correlations, 2nd Edition, 1986.
62.	P. H. Frampton	Gauge Field Theories, 1986
63.	J. I. Katz	High Energy Astrophysics, 1987
64.	T. J. Ferbel	Experimental Techniques in High Energy Physics, 1987
65.	T. Appelquist A. Chodos P. G. O. Freund	Modern Kaluza-Klein Theories, 1987
66.	G. Parisi	Statistical Field Theory, 1988
67.	R. C. Richardson E. N. Smith	Techniques in Low-Temperature Condensed Matter Physics, 1988
68.	J. W. Negele H. Orland	Quantum Many-Particle Systems, 1987
69.	E. W. Kolb M. S. Turner	The Early Universe, 1990
70.	E. W. Kolb M. S. Turner	The Early Universe: Reprints, 1988
71.	V. Barger R. J. N. Phillips	Collider Physics, 1987
72.	T. Tajima	Computational Plasma Physics, 1989
73.	W. Kruer	The Physics of Laser Plasma Interactions, 1988
74.	P. Ramond	Field Theory: A Modern Primer, 2nd edition, 1989 [cf. 51—1st edition]
75.	B. F. Hatfield	Quantum Field Theory of Point Particles and Strings, 1989
76.	P. Sokolsky	Introduction to Ultrahigh Energy Cosmic Ray Physics, 1989
77.	R. Field	Applications of Perturbative QCD, 1989
80.	J. F. Gunion H. E. Haber G. Kane S. Dawson	The Higgs Hunter's Guide, 1990

81.	R. C. Davidson	Physics of Nonneutral Plasmas, 1990
82.	E. Fradkin	Field Theories of Condensed Matter Systems, 1991
83.	L. D. Faddeev A. A. Slavnov	Gauge Fields, 1990
84.	R. Broglia A. Winther	Heavy Ion Reactions, Parts I and II, 1990
85.	N. Goldenfeld	Lectures on Phase Transitions and the Renormalization Group, 1992
86.	R. D. Hazeltine J. D. Meiss	Plasma Confinement, 1992
87.	S. Ichimaru	Statistical Plasma Physics, Volume I: Basic Principles, 1992
88.	S. Ichimaru	Statistical Plasma Physics, Volume II: Condensed Plasmas, 1994
89.	G. Grüner	Density Waves in Solids, 1994
90.	S. Safran	Statistical Thermodynamics of Surfaces, Interfaces, and Membranes, 1994
91.	B. d'Espagnat	Veiled Reality: An Analysis of Present Day Quantum Mechanical Concepts, 1994
92.	J. Bahcall R. Davis, Jr. P. Parker A. Smirnov R. Ulrich	Solar Neutrinos: The First Thirty Years
93.	R. Feynman F. Morinigo W. Wagner	Feynman Lectures on Gravitation
94.	M. Peskin D. Schroeder	An Introduction to Quantum Field Theory
95.	R. Feynman	Feynman Lectures on Computation
96.	M. Brack R. Bhaduri	Semiclassical Physics
97.	D. Cline	Weak Neutral Currents
98.	T. Tajima K. Shibata	Plasma Astrophysics
99.	J. Rammer R. Hazeltine F. Waelbroeck	Quantum Transport Theory The Framework of Plasma Physics
100.	R. Hazeltine F. Waelbroeck	The Frameworkof Plasma Physics
101	P. Ramond	Journeys Beyond the Standard Moel
102.	P. Philips	Advanced Solid State Physics

Editor's Foreword

The problem of communicating in a coherent fashion recent developments in the most exciting and active fields of physics continues to be with us. The enormous growth in the number of physicists has tended to make the familiar channels of communication considerably less effective. It has become increasingly difficult for experts in a given field to keep up with the current literature; the novice can only be confused. What is needed is both a consistent account of a field and the presentation of a definite "point of view" concerning it. Formal monographs cannot meet such a need in a rapidly developing field, while the review article seems to have fallen into disfavor. Indeed, it would seem that the people who are most actively engaged in developing a given field are the people least likely to write at length about it.

Frontiers in Physics was conceived in 1961 in an effort to improve the situation in several ways. Leading physicists frequently give a series of lectures, a graduate seminar, or a graduate course in their special fields of interest. Such lectures serve to summarize the present status of a rapidly developing field and may well constitute the only coherent account available at the time. One of the principal purposes of the *Frontiers in Physics* series is to make notes on such lectures available to the wider physics community.

A second way to improve communication in very active fields of physics is by the publication of collections of reprints of key articles relevant to our present understanding. Such collections are themselves useful to people working in the field or to students considering entering it. The value of the reprints is, however, considerably enhanced when the collection is accompanied by a commentary which might, in itself, constitute a brief survey of the present status of the field.

The present volume represents just such a collection, with reprints selected and annotated by leading contributors to the subject under review. John Bahcall, Ray Davis, Peter Parker, Alexei Smirnov, and Roger Ulrich have made significant seminal contributions to our present understanding of solar neutrinos; each introduces in these pages a selection of reprints in the subfield of his own particular expertise. Taken together, the introductions and the reprints provide the reader, expert and non-expert alike, with a fine overview of both how much we have learned about solar neutrinos during the past three decades, and of the interesting problems which lie ahead. It gives me pleasure to welcome the authors to *Frontiers in Physics.*

David Pines
Urbana, IL
March, 1995

Contents

Preface to the 2002 Paperback Edition

In the eight years since the first edition of this collection was prepared, the field of solar neutrinos has become widely recognized as one of the most active frontiers of particle physics. The "solar neutrino problem" has been replaced by the "solar neutrino opportunity."

The adiabatic transformation from problem to opportunity has been driven by three important developments: 1). the precise confirmation of solar model parameters by helioseismological measurements; 2). the precise measurement by the Super-Kamiokande collaboration of the average rate, the time-dependences, and the recoil electron energy spectrum of ^{8}B neutrino interactions with ordinary water; and 3). the dramatic demonstration by the SNO collaboration that the electron neutrino flux from ^{8}B beta-decay, measured in heavy water, is less than the combined flux of electron, muon, and tau neutrinos that is detected by Super-Kamiokande in ordinary water.

This volume reproduces not only the original papers, collected in the first edition, which created and clarified the solar neutrino problem, but also includes three additional papers that led to the transformation of the subject. The first of these new papers (March, 1998), entitled "How uncertain are solar neutrino predictions?", showed that helioseismological measurements confirm the standard solar model predictions to a high precision (0.1% for the solar interior sound speeds). This precise agreement between standard solar model predictions and helioseismological measurements the paper demonstrates (see Figure 1) that the long-standing discrepancies between solar model predictions and solar neutrino measurements could not be explained away as inadequacies in the standard solar model.

In the second paper (June 2001), entitled "Solar ^{8}B and hep Neutrino Measurements from 1258 Days of Super-Kamiokande data", the Super-Kamiokande collaboration reported the results of more than 18,000 neutrino-electron scattering events, increasing the number of previously reported solar neutrino events by an order of magnitude. These results provided precise information about the total scattering rate , the recoil electron energy spectrum, and the day-night and other temporal dependences.

The third paper (August 2001), entitled " Measurement of the Rate of ν_e+d → p+p+e$^-$ Interactions Produced by Solar ^{8}B Neutrinos at the Sudbury Neutrino Observatory", describes the results of an epochal measurement of the electron-neutrino flux from solar ^{8}B neutrinos. In this paper, the SNO collaboration showed that the ν_e flux that is measured in their heavy water detector is less than the combined flux of ν_e plus (detected with less efficiency) $\nu_\mu+\nu_\tau$ that is measured in the Super-Kamiokande ordinary water detector. Thus ν_e are converted to $\nu_\mu+\nu_\tau$ on the way to the Earth from the center of the Sun. In addition, the comparison of the SNO and Super-Kamiokande rates implied that the flux of active ^{8}B solar neutrinos is in excellent agreement with the prediction of the standard solar model (given in the first of the three new papers and in many previous papers in this series going back in time almost four decades).

Thus the combined results of the Super-Kamiokande and SNO measurements show that: 1) new physics is required to described the propagation of solar neutrinos, and 2) the standard solar model prediction is verified to high accuracy, provided that the electron neutrino mixes significantly with only the two other known active neutrino flavors, symbolized by ν_μ and ν_τ. These are extraordinary achievements, which are built upon the contributions of thousands of previous workers whose research results are described, directly or indirectly, in the original papers reprinted in this volume.

What are the challenges and the opportunities of the next decade of solar neutrino research? In the realm of particle physics, new experiments (with solar neutrinos using the SNO, BOREXINO, and ICARUS detectors and with reactor anti-neutrinos using the KamLAND detector) are just beginning to come into operation that will help determine accurately—together with the previous solar neutrino data—the propagation characteristics (including masses and mixing angles) of solar neutrinos. One of the principal challenges will be to establish or constrain tightly the role of sterile neutrinos. In the realm of astrophysics, the combined results of all of the solar neutrino experiments will provide more accurate values for the fluxes of ^{8}B and ^{7}Be solar neutrinos, in order to test and refine standard solar model predictions. Further precision laboratory measurements of the nuclear fusion reactions that produce ^{8}B and ^{7}Be isotopes in the Sun are essential (and some are underway) in order to interpret the solar neutrino measurements with their full power.

The focus of solar neutrino research will shift gradually over the next decade from the relatively high-energy ^{8}B neutrinos (maximum energy greater than 14 MeV) to the low-energy, ^{7}Be, p-p, pep, and CNO neutrinos (energies less than or of order 1 MeV). For the standard solar model prediction, more than 98% of the solar neutrino flux lies below 1 MeV. The p-p neutrinos alone are predicted to carry about 91% of the flux, whereas the great science that has been done so far on the ^{8}B neutrinos uses only about 0.01% of the total predicted solar neutrino flux.

The currently favored neutrino oscillation solutions exhibit revealing energy dependencies below 1 MeV. Naturally, all of the currently favored oscillation solutions give similar predictions in the higher energy region about 5 MeV, where the Kamiokande, Super-Kamiokande, and SNO data are best. Future measurements of the low-energy total fluxes, time dependences, energy spectra, flavor composition, and sterile component (if any) will contribute in an important way to understanding the physics of neutrinos.

The standard solar model predictions[1] for the dominant low energy ^{7}Be and p-p neutrinos are relatively accurate (± 1% for the p-p neutrinos and ± 10% for the ^{7}Be neutrinos). Low energy solar neutrino experiments will test these fundamental predictions of the theory of stellar evolution. Improved measurements of the gallium solar neutrino capture rate with the GNO and SAGE experiments will provide important constraints on the combined capture rate of all of the ν_e solar neutrinos above 0.23 MeV, but these radiochemical experiments must be supported by additional experiments that measure the neutrinos from individual neutrino branches.

Redundant experiments measuring the same physical or astrophysical properties in different ways are essential to be sure that we have converged on the correct answer. The implications of the experimental results are too important for physics and for astronomy to depend upon single experiments.

J. N. Bahcall, R. Davis Jr., P. D. Parker,
A. Smirnov, and R. K. Ulrich (March, 2002)

Notes

1. "Solar Models: current epoch and time dependences, neutrinos, and helioseismological properties" (John N. Bahcall, M. H. Pinsonneault, and Sarbani Basu), *ApJ*, **555**, 990–1012 (2001 July 10).

I. Standard Model Expectations

Introduction: John N. Bahcall

The expectations of what should be observed in solar neutrino experiments can be conveniently organized into four categories: A) Neutrino Cross Sections, B) Early Solar Models, C) Sensitivity Studies, and D) Refined Solar Models. The expectations discussed in this section depend upon both the standard electroweak model, which implies that essentially nothing happens to solar neutrinos from the time they are created in the center of the sun until they arrive at detectors on earth (but see Section IV), and the standard solar model, which can be used to predict the fluxes of neutrinos of different energies that are produced in the sun.

In 1964, the experience gained using a chlorine detector near a reactor that is described in Paper 2.A.II was combined with the realization, described in Paper 1.A.I, that a superallowed transition between ^{37}Cl and ^{37}Ar increases by about a factor of 20 the expected event rate due to solar neutrinos. The combined result of these two papers was a suggestion that the expected rate of neutrino events in a large chlorine detector is large enough for a practical experiment. The nuclear-physics prediction of the superallowed transition was confirmed experimentally within months (Papers 4.A.I and 5.A.I) by studies of the beta-decay of ^{37}Ca, the isotopic analogue of ^{37}Cl. Following the proposal of a neutrino-electron scattering experiment (in Paper 3.A.II), the angular distribution of the recoil electrons was calculated (in Paper 4.A.I) and shown to be strongly peaked in the direction of the incident neutrinos, presumably the direction toward the sun. The interaction cross sections relevant to other solar neutrino experiments were also quickly calculated and, when the standard electroweak model was developed, t'Hooft provided (Paper 7.A.I) the remaining essential element–the neutrino-electron scattering cross section including both charged and neutral currents.

The first calculation of solar neutrino fluxes that made use of a detailed solar model was presented in Paper 1.B.I. The first study of the uncertainties in solar neutrino calculations was then given in Paper 2.B.I . The early exploratory studies of solar models lasted about three years, from 1963–1965. Actually, the first claim that a solar neutrino experiment was practical was made in Paper 4.B.I, which was presented by Bahcall and Davis in a conference in New York in the fall of 1963, but was not published until 1965. Paper 4.B.I made preliminary use of the neutrino absorption cross sections calculated in Paper 1.A.I, of the neutrino fluxes calculated in Papers 1.B.I and 2.B.I, and of the estimated sensitivity of a chlorine detector that was described in Paper 2.A.II .

With the publication of the first experimental results from the chlorine detector (Paper 1.B.II), results that differed from the theoretical expectations, the calculations entered a new stage that is illustrated by the papers in section C.I. The principal focus during this period was the determination of the sensitivity of the calculated neutrino fluxes to various input quantities including nuclear cross sections, stellar radiative opacities, and the equation of state. The partial derivatives of the neutrino fluxes with respect to all the major input parameters were calculated in papers that appeared in the years 1968–1971 [1].

Many subsequent investigations have confirmed these early sensitivity studies and a large number of modern papers focus on more detailed evaluations of the effects of a particular input parameter or an assumed piece of physics (e. g., a low-energy nuclear cross section or the radiative opacity or equation of state). Iben (in Paper 2.C.I) appears to have been the first person courageous enough to declare in print that the discrepancy between solar model calculations and observations was not due to inadequate solar models.

In recent years, solar models have become increasingly more precise as more detailed physics has been added to the models and as the input experimental parameters have been measured more accurately (see, e. g., Section II). Some representative papers from this era, which continues today with further improvements, are listed in Section D.I Beginning with Paper 1.D.I, formal methods have been developed to determine the uncertainties in the calculated event rates in different solar neutrino experiments. These developments led to a Monte Carlo study of the theoretical uncertainties using a thousand realizations of the standard solar model (given in Paper 2.D.I and further analyzed in Paper 9.D.II). In addition, the development of helioseismology (see Section V) provided thousands of accurately measured $p-$mode eigenfrequencies that could be used to test the details of the solar model, at least in the outer half of the sun. The fact that the standard solar model cor-

---

[1]The now famous unit of solar neutrino interactions, the *SNU* (10^{-36} interactions per target atom per sec), was introduced as a pun in 1969 in a paper (not reprinted here) entitled: "What Next With Solar Neutrinos?" (J. Bahcall, Phys. Rev. Lett. **23**, 251, 1969). See footnote 10 of this paper for a brief description of the origin and phonetics of the term *SNU*.

rectly predicts the measured $p-$mode eigenfrequencies (to one part in 10^3 or better) is one of the principal reasons for having confidence in the predicted neutrino fluxes.

I. STANDARD MODEL EXPECTATIONS - Bahcall

A. Neutrino Cross Sections

1. "Solar Neutrinos. I. Theoretical," J. N. Bahcall, *Phys. Rev. Lett.* **12**, 300-302 (1964).

2. "Calcium-37," J. C. Hardy and R. I. Verrall, *Phys. Rev. Lett.* **13**, 764-766 (1964).

3. "New Delayed-Proton Emitters: Ti^{41}, Ca^{37}, and Ar^{37}," P. L. Reeder, A. M. Poskanzer, and R. A. Esterlund, *Phys. Rev. Lett.* **13**, 767-769 (1964).

4. "Neutrino Opacity. I. Neutrino-Lepton Scattering," J. N. Bahcall, *Phys. Rev.* **136**, B1164, B1170 (1964).

5. "Absorption of Solar Neutrinos in Deuterium, " F. J. Kelly and H. Uberall, *Phys. Rev. Lett.* **16**, 145-147 (1966).

6. "Solar Neutrinos," J. N. Bahcall, *Phys. Rev. Lett.* **17**, 398-401 (1966).

7. "Prediction for Neutrino-Electron Cross-Sections in Weinberg's Model of Weak Interactions " G. 't Hooft, *Phys. Lett.* **37B**, 195-196 (1971).

8. "Solar Neutrino Experiments," J. N. Bahcall , *Rev. Mod. Phys.* **50**, 881 (1978).

B. Early Solar Models

1. "Solar Neutrino Flux," J. N. Bahcall, W. A. Fowler, I. Iben Jr., and R. L. Sears, *Ap. J.* **137**, L344-L346 (1963).

2. "Helium Content and Neutrino Fluxes in Solar Models," R. L. Sears, *Ap. J.* **140**, 477-484 (1964).

3. "A Study of Solar Evolution," D. Ezer and A. G. W. Cameron, *Canadian J. of Phys.* **43**, 1497 (1965).

4. "On the Problem of Detecting Solar Neutrinos," J. N. Bahcall and R. Davis Jr., in *Stellar Evolution* ed. R. F. Stein and A. G. W. Cameron, Plenum Press, NY, 241-243 (1966).

C. Sensitivity Studies

1. "Present Status of the Theoretical Predictions for the ^{37}Cl Solar-Neurino Experiment," J. N. Bahcall, N. A. Bahcall, and G. Shaviv, *Phys. Rev. Lett.* **20**, 1209-1212 (1968).

2. "Solar Neutrinos and the Solar Helium Abundance," I. Iben Jr., *Phys. Rev. Lett.* **21**, 1208-1212 (1968).

3. "Sensitivity of Solar Neutrino Fluxes," J. N. Bahcall, N. A. Bahcall, and R. K. Ulrich, *Ap. J.* **156**, 559, 564 (1969).

4. "More Solar Models and Neutrino Fluxes," Z. Abraham and I. Iben Jr., *Ap. J.* **170**, 157 (1971).

D. Refined Solar Models

1. "Standard Solar Models and the Uncertainties in Predicted Capture Rates of Solar Neutrinos," J. N. Bahcall, W. F. Huebner, S. H. Lubow, P. D. Parker, and R. K. Ulrich, *Rev. Mod. Phys.* **54**, 767 (1982).

2. "Solar Models, Neutrino Experiments, and Helioseismology," J. N. Bahcall and R. K. Ulrich, *Rev. Mod. Phys.* **60**, 297 (1988).

3. "Our Sun I. The Standard Solar Model: Successes and Failures," I. J. Sackmann, A. I. Boothroyd, and W. A. Fowler, *Ap. J.* **360**, 727 (1990).

4. "Standard Solar Models, With and Without Helium Diffusion, and the Solar Neutrino Problem," J. N. Bahcall and M. H. Pinsonneault, *Rev. Mod. Phys.* **64**, 885 (1992).

5. "On the Depletion of Lithium in the Sun," J. Ahrens, M. Stix, and M. Thorn, *Astron. Astrophys.* **264**, 673 (1992).

6. "Standard Solar Models with CESAM code: neutrinos and helioseismology," J. Berthomie J. Provost, P. Morel, and Y. Lebreton, *Astron. Astrophys.* **268**, 775 (1993).

7. "Toward a Unified Classical Model of the Sun: On the Sensitivity of Neutrino and Helioseismology to the Microscopic Physics," S. Turck-Chieze and J. Lopes, *Ap. J.* **408**, 347 (1993).

8. "Solar Neutrinos and Nuclear Reactions in the Solar Interior," V. Castellani, S. Degl'Innoc and G. Fiorentini, *Astron. Astrophys.* **271**, 601 (1993).

9. "Do Solar Neutrino Experiments Imply New Physics?," J. N. Bahcall and H. A. Bethe, *Phys. Rev.* **D47**, 1298-1301 (1993).

10. "Effects of Heavy Element Settling on Solar Neutrino Fluxes and Interior Structure," C. R. Proffitt, *Ap. J.* **425**, 849-855 (1994).

SOLAR NEUTRINOS. I. THEORETICAL*

John N. Bahcall
California Institute of Technology, Pasadena, California
(Received 6 January 1964)

The principal energy source for main-sequence ars like the sun is believed to be the fusion, in e deep interior of the star, of four protons to rm an alpha particle.[1] The fusion reactions e thought to be initiated by the sequence $^1H(p,$ $^+\nu)^2H(p,\gamma)^3He$ and terminated by the following quences: (i) $^3He(^3He, 2p)^4He$; (ii) $^3He(\alpha,\gamma)^7Be$- $^-\nu)^7Li(p,\alpha)^4He$; and (iii) $^3He(\alpha,\gamma)^7Be(p,\gamma)^8B$- $^+\nu)^8Be^*(\alpha)^4He$. No direct evidence for the istence of nuclear reactions in the interiors of ars has yet been obtained because the mean ee path for photons emitted in the center of a star is typically less than 10^{-10} of the radius of the star. Only neutrinos, with their extremely small interaction cross sections, can enable us to see into the interior of a star and thus verify directly the hypothesis of nuclear energy generation in stars.

The most promising method[2] for detecting solar neutrinos is based upon the endothermic reaction ($Q = -0.81$ MeV) $^{37}Cl(\nu_{solar}, e^-)^{37}Ar$, which was first discussed as a possible means of detecting neutrinos by Pontecorvo[3] and Alvarez.[4] In this note, we predict the number of absorptions of solar neutrinos per terrestrial ^{37}Cl atom by combining results of recent theoretical investigations[5-7] of the solar neutrino fluxes with calculations[8] of the relevant neutrino absorption cross sections on ^{37}Cl. The result of a preliminary experiment by Davis[2] is then used to set an upper limit on the central temperature of the sun and also to give information about the structure of ^{4}Li and its role in the proton-proton chain.

The neutrino fluxes from the hydrogen-burning reactions described in the first paragraph have recently been calculated using detailed models of the sun[5,6] and the effects of uncertainties in nuclear cross sections, as well as solar composition, opacity, and age, have been determined by Sears.[7] The most important predictions are these (uncertainties estimated from the work of Sears[7]): $\varphi_\nu(^7\mathrm{Be}) = (1.2 \pm 0.5) \times 10^{+10}$ neutrinos per cm^2 per sec and $\varphi_\nu(^8\mathrm{B}) = (2.5 \pm 1) \times 10^{+7}$ neutrinos per cm^2 per sec, at the earth's surface.

The cross sections for ^{7}Be and ^{8}B neutrinos to produce transitions from the ground state of ^{37}Cl to the ground state of ^{37}Ar can readily be calculated from known quantities; the results are[8] $\sigma_g(^7\mathrm{Be}) = 1.5\sigma_0$ and $\bar{\sigma}_g(^8\mathrm{B}) = 3.9 \times 10^{+2}\sigma_0$, where $\sigma_0 = 1.91 \times 10^{-46}$ cm^2 is a convenient combination of ground-state parameters and $\bar{\sigma}(^8\mathrm{B})$ has been averaged over the ^{8}B neutrino spectrum. Three excited states[9] in ^{37}Ar also have large matrix elements for neutrino absorption by the ground state of ^{37}Cl (which is a $d_{3/2}{}^3$, $J = 3/2^+$, $T = 3/2$ state); the three excited states of importance in ^{37}Ar are (with their expected energies) (i) $J = 1/2^+$, $T = 1/2$ (1.4 MeV); (ii) $J = 5/2^+$, $T = 1/2$ (1.6 MeV); and (iii) $J = 3/2^+$, $T = 3/2$ (5.1 MeV). The $J = 3/2^+$, $T = 3/2$ excited state of ^{37}Ar is the analog state of the ground state of ^{37}Cl; hence the transition from the ground state of ^{37}Cl to the 5.1-MeV excited state of ^{37}Ar is superallowed and has a large matrix element for neutrino absorption. The calculated absorption cross sections[8] averaged over the ^{8}B neutrino spectrum[10] are, in order of increasing excitation energy, $\bar{\sigma}(^8\mathrm{B})/\sigma_0 = 0.96 \times 10^{+3}$, $1.3 \times 10^{+3}$, and $4.4 \times 10^{+3}$. The net uncertainty in the magnitude of the sum of the above cross sections is estimated to be about 25%.[8,11]

The total predicted number of absorptions per terrestrial ^{37}Cl atom per second, using the above estimates for fluxes and cross sections, is found to be

$$\sum \varphi_\nu(\mathrm{solar})\sigma_{\mathrm{abs}} = (4 \pm 2) \times 10^{-35}\ \mathrm{sec}^{-1}. \quad (1)$$

Only about 10% of the predicted number of absorptions is due to ^{7}Be neutrinos, although the ^{7}Be neutrino flux is predicted to be approximately 500 times the ^{8}B neutrino flux.[12] The solar val[ue] of $\sum \varphi\bar{\sigma}$ given by Eq. (1) is at least several orders of magnitude greater than one would expec[t] from cosmic neutrinos[13] or from neutrinos produced in the earth's atmosphere by the decay of cosmic ray secondaries.[13,14]

The ^{8}B neutrino flux is extremely sensitive[5,12] to the central temperature of the sun because of the large Coulomb barrier, compared to sola[r] thermal energies, for the reaction $^7\mathrm{Be}(p,\gamma)^8\mathrm{B}$ of sequence (iii). An upper limit on the central temperature of the sun can therefore be derived by combining the experimental upper limit already obtained by Davis,[2] on the number of solar neutrinos captured per terrestrial ^{37}Cl atom, with Eq. (1) and the known temperature dependence of the $^7\mathrm{Be}(p,\gamma)^8\mathrm{B}$ reaction. In this way we find that the central temperature of the sun is less than 20 million degrees[5] and that a measurement of the ^{8}B neutrino flux accurate to ±50% would determine the central temperature to better than ±10%.

The role of ^{4}Li in the proton-proton chain has long been recognized as an important astrophysical problem,[1,15] but one that has not yet been solved by direct nuclear physics experiments. The upper limit obtained by Davis[2] on the numbe[r] of solar neutrinos captured per terrestrial ^{37}Cl atom can be used, however, to show that ^{4}Li doe[s] not play a significant role in the proton-proton chain in the sun. The relevant cross section for neutrino absorption (with $q_\nu^{\max} = 20$ MeV) is[8] $\bar{\sigma}(^4\mathrm{Li}) = 2 \times 10^{-42}$ cm^2 and hence $\varphi_\nu(^4\mathrm{Li}) \leq 2 \times 10^{+8}$ neutrinos per cm^2 per sec. The fraction of terminations of the proton-proton chain that occur via ^{4}Li can be calculated[16] as a function of the energy, E_γ, by which the mass of the ground state of ^{4}Li exceeds the mass of ^{3}He plus a proton. One can also calculate an upper limit on th[e] fraction of terminations that occur via $^3\mathrm{He}(p,\gamma)^4\mathrm{Li}(\beta^+\nu)^4\mathrm{He}$ by comparing the above upper limi[t] on $\varphi_\nu(^4\mathrm{Li})$ (multiplied by 17 MeV, the thermal energy release in such a termination) with the observed solar constant ($8.7 \times 10^{+11}$ MeV cm^{-2} sec^{-1}). In this way we find that $E_\gamma \geq 20$ keV[17] and conclude that ^{4}Li participates in at most 0.2[%] of the proton-proton terminations in the sun.

The author is grateful to the entire staff of Kellogg Radiation Laboratory for enthusiastic support and stimulating comments. It is a pleasure to acknowledge many valuable suggestions by R. Davis, Jr.

*Work supported in part by the U. S. Office of Naval Research and in part by the National Aeronautics and Space Administration.

[1]See, for example, W. A. Fowler, Mem. Soc. Roy. Sci. Liege 3, 207 (1960). The CNO cycle is responsible for only a few percent of the energy generation in the sun and the relatively low-energy neutrinos produced by this cycle are unimportant for solar neutrino detection with ^{37}Cl.

[2]R. Davis, Jr., following Letter [Phys. Rev. Letters 12, 303 (1964)]. See also R. Davis, Jr., Phys. Rev. 97, 766 (1955).

[3]B. Pontecorvo, National Research Council of Canada Report No. P.D. 205, 1946 (unpublished), reissued by the U. S. Atomic Energy Commission as document 200-18787.

[4]L. W. Alvarez, University of California Radiation Laboratory Report No. UCRL-328, 1949 (unpublished).

[5]J. N. Bahcall, W. A. Fowler, I. Iben, Jr., and R. L. Sears, Astrophys. J. 137, 344 (1963). The central temperature of the sun for the theoretical model used in this paper (developed by Sears) is 16.2 million degrees. See also R. L. Sears, Mem. Soc. Roy. Sci. Liege 3, 479 (1960).

[6]P. Pochoda and H. Reeves (to be published).

[7]R. L. Sears (to be published).

[8]J. N. Bahcall (to be published). This reference will contain an extensive discussion of neutrino absorption cross sections that are relevant to the detection of solar neutrinos. A variety of experimental tests of the assumptions used to calculate the excited-state neutrino absorption cross sections for ^{37}Cl$(\nu, e^-)^{37}$Ar will also be discussed.

[9]I am grateful to Professor B. R. Mottelson and Professor M. A. Preston for comments that sparked the investigation of excited-state transitions.

[10]The proton-proton ($q_\nu^{max} = 0.42$ MeV) and ^{7}Be electron-capture ($q_\nu = 0.86$ MeV) neutrinos do not have sufficient energy to induce transitions to excited states in ^{37}Ar.

[11]The assumptions made in calculating the ^{37}Cl neutrino absorption cross sections could be directly checked by measuring the ft values for the ^{37}Ca$\to^{37}$K decays, one of whose branches is also superallowed. Two other experiments that would be useful in testing the assumptions made in the cross-section calculations are (i) a measurement of the branching ratios in the ^{37}K$\to^{37}$Ar decay, and (ii) a measurement with improved accuracy of the ft values in the ^{35}Ar$\to^{35}$Cl decay. Predictions for the lifetimes and energies of all branches involved in the above decays are available upon request and will appear in reference 8.

[12]The possible importance of ^{8}B solar neutrinos was first pointed out by W. A. Fowler, Astrophys. J. 127, 551 (1958); A. G. W. Cameron, Ann. Rev. Nucl. Sci. 8, 299 (1958).

[13]See, for example, H. Greisen, Proceedings of International Conference for Instrumentation in High Energy Physics, Berkeley, California, September 1960 (Interscience Publishers, Inc., New York, 1961), p. 209; F. Reines, Ann. Rev. Nucl. Sci. 10, 1 (1960); B. Pontecorvo and Ya. Smorodinskii, Zh. Eksperim. i Teor. Fiz. 41, 239 (1961) [translation: Soviet Phys.–JETP 14, 173 (1962)]. The preliminary experiment of Davis (reference 2) implies that the energy density of 1-MeV cosmic neutrinos is less than 5 MeV/cm^3; however, the galactic energy density of starlight is only about 1 eV/cm^3. Thus the Davis experiment does not furnish a very stringent upper limit on the energy density of low-energy cosmic neutrinos.

[14]G. T. Zatsepin and V. A. Kuz'min, Zh. Eksperim. i Teor. Fiz. 41, 1818 (1961) [translation: Soviet Phys.–JETP 14, 1294 (1962)]; M. A. Markov and I. M. Zheleznykh, Nucl. Phys. 27, 385 (1961); T. D. Lee, H. Robinson, M. Schwartz, and R. Cool, Phys. Rev. 132, 1297 (1963).

[15]H. A. Bethe, Phys. Rev. 55, 434 (1939); H. Reeves, Phys. Rev. Letters 2, 423 (1959); S. Bashkin, R. W. Kavanagh, and P. D. Parker, Phys. Rev. Letters 3, 518 (1959). The possibility of terminating the proton-proton chain through a particle-unstable but thermally populated ground state of ^{4}Li was apparently overlooked.

[16]Details of this calculation will appear in a paper by P. D. Parker, J. N. Bahcall, and W. A. Fowler (to be published). I am especially grateful to Dr. Parker for valuable collaboration on this point. Note that if ^{4}Li were particle stable, all proton-proton terminations in the sun would occur via ^{3}He$(p,\gamma)^4$Li$(\beta^+\nu)^4$He because of the relatively low Coulomb barrier for the ^{3}He$(p,\gamma)^4$Li reaction and the high abundance of protons.

[17]This result implies that there are no $T = 1$ alphaparticle bound states below 19 MeV.

CALCIUM-37†

J. C. Hardy and R. I. Verrall
Foster Radiation Laboratory, McGill University, Montreal, Canada
(Received 13 November 1964)

This Letter reports the observation of the previously unreported ^{37}Ca by measurements on the delayed protons following its β^+ decay.

There has been much recent interest in the detection of solar neutrinos by means of the reaction $^{37}\text{Cl} + \nu_{\text{solar}} \rightarrow {}^{37}\text{Ar} + e^-$. Bahcall[1] has shown that this reaction may proceed either to the ground or to any of three excited states of ^{37}Ar. The total cross section for this reaction was predicted using the measured lifetime for the ground-state inverse transition $^{37}\text{Ar} \rightarrow {}^{37}\text{Cl}$, a calculated ft value for the transition to the first $T = \frac{3}{2}$ state of ^{37}Ar, and estimated ft values for transitions to the two other excited states.

It has been pointed out[2] that a better estimate for branching to the latter two states can be obtained from a knowledge of the lifetime of ^{37}Ca, since the positron decay of $^{37}_{20}\text{Ca}_{17}$, namely $^{37}\text{Ca} \rightarrow {}^{37}\text{K} + e^+ + \nu$, is the mirror reaction to neutrino capture by $^{37}_{17}Cl_{20}$. This lifetime and other properties of the ^{37}Ca decay have now been measured. The nuclide ^{37}Ca is the latest one of the $(4n+1)$ series of delayed-proton precursors studied in this Laboratory,[3] the preceding members being ^{13}O, ^{17}Ne, ^{21}Mg, ^{25}Si, ^{29}S, and ^{33}Ar. The fact that ^{37}Ca has appeared in its logical place in this series lends confidence to the results to be reported here.

Measurements were taken using a surface barrier silicon detector of 200-mm^2 area and 500-μ depletion depth (8 MeV for protons), mounted 6 cm from the thin target foil and inserted on a radial probe into the circulating proton beam of the McGill synchrocyclotron. Counting was performed between repetitive bursts, typically 30 msec long, of normal cyclotron operation. A clock-controlled counting period was initiated following a 100-msec delay which was long enough to allow dissipation of beam storage effects in the cyclotron. This period could be accurately divided into four equal intervals, enabling the spectrum to be stored sequentially in the four quarters of a 256-channel analyzer. Thus, four-point decay curves for the peaks in the spectrum could be obtained.

To calibrate the energy response of the system, the ^{210}Po α line was used in conjunction with pulses from a mercury pulse generator fed onto the input of the preamplifier. Several ten-minute counting periods were alternated with pulser calibrations in order to determine the energy of a peak.

Figure 1 shows the delayed proton spectrum obtained following proton bombardment of a 0.9-mg/cm^2 calcium target, vacuum evaporated on a thin (200 μg/cm^2) gold backing. Similar peaks were obtained from a potassium target (KOH vacuum-evaporated on thin gold), but with a reduced counting rate. Bombardment of a chlorine target (LiCl) produced a proton spectrum which was considered[3] to follow the decay of ^{33}Ar; that spectrum included a peak near channel 47, but its other features indicate that its contribution to the spectrum in Fig. 1 is less than a few percent. Consequently, the nuclide responsible for the present activity must be an isotope of either potassium or calcium.

Yield curves for the prominent peak in Fig. 1 were plotted for both the calcium and potassium targets by measuring the delayed-proton production as a function of the target radius in the cyclotron. Observations near threshold were difficult because of the β background under this peak; also, radial oscillations in the cyclotron beam make the actual bombarding energy at a given target radius less than the nominal cyclotron energy at that radius by an amount not well determined (2 to 5 MeV). Relative threshold measurements, however, could be made to within 2 MeV by comparing the excitation functions over a range of energies above the threshold.

The threshold for production from stable calcium (97% ^{40}Ca) was found to be 7 MeV higher than that from potassium (93% ^{39}K), and was approximately 47 MeV. These results are compatible only with the reactions ^{40}Ca(p, $d2n$)^{37}Ca and ^{39}K($p,3n$)^{37}Ca, whose calculated laboratory energy thresholds are 44.6 and 38.5 MeV. This establishes the activity as following the decay of ^{37}Ca.

From decay curves of the area above background of the main proton peak, each taken over approximately five half-lives, we adopt the value (173 ± 4) msec as the half-life of ^{37}Ca.

The small peak in Fig. 1 appears to have a half-life not significantly different from that of the main peak, but threshold measurements could not be made, and positive identification was not possible. Since its intensity is only about 2 percent of that of the main peak, it is

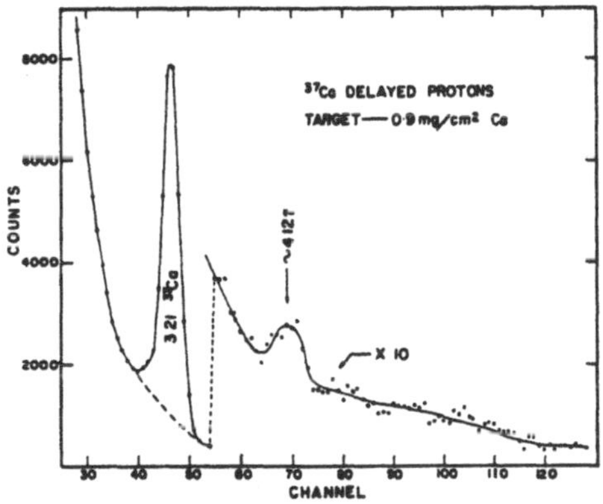

FIG. 1. The spectrum of delayed protons from a calcium target bombarded at 85 MeV in the cyclotron. The two peaks are marked with their center-of-mass energies computed for mass 37, but only the 3.21-MeV peak has been definitely assigned as following the ^{37}Ca decay.

ignored in what follows.

The energies of the peaks are shown, corrected to center of mass, in Fig. 1. The peak at (3.21 ± 0.04) MeV corresponds to the decay of a level at (5.07 ± 0.04) MeV in ^{37}K [using Endt and van der Leun's value[4] of 1.86 MeV for the mass difference between ($^{36}Ar + p$) and ^{37}K], which agrees with the predicted[1] excitation energy (5.12 MeV) of the first $T = \frac{3}{2}$ level in ^{37}K. Such an assignment to this level would account for the fact that this peak appears with relatively high intensity, because the β^+ decay to it from ^{37}Ca would be superallowed.

Figure 2 shows the proposed decay scheme, which includes all relevant levels, and where the bracketed energies have been predicted only. The *ft* value for the electron capture decay ${}^{37}_{18}Ar_{19} - {}^{37}_{17}Cl_{20}$ is known to be 1.14×10^5 sec, and one may assume that this value applies very nearly to the ground-state β^+ decay ${}^{37}_{20}Ca_{17} - {}^{37}_{19}K_{18}$; Bahcall[1] also calculates an *ft* value of 1.9×10^3 sec for decay to the $T = \frac{3}{2}$ level in ^{37}K. Using these assumptions and the measured lifetime, we have calculated approximate branching ratios for the two previously mentioned decays and the total of all other decays. Since decays to the two levels at 1.57 and 1.46 MeV are the only other ones that would contribute significantly, the branching ratios and decays should appear as in Fig. 2.

It is now possible to use these branching values for the mirror neutrino-capture reaction on ^{37}Cl. In reference 1, Bahcall made calculations concerning the cross section for the neutrino capture using estimated branching ratios. Repeating the calculation using the measured half-life and the subsequently modified ratios, a revised estimate for this cross section is obtained. This value predicts the total number of solar neutrino captures to be 3.1×10^{-35}/(^{37}Cl atom)-sec, a value 14% lower than that in reference 1, Eq. (36), but well within the quoted errors.

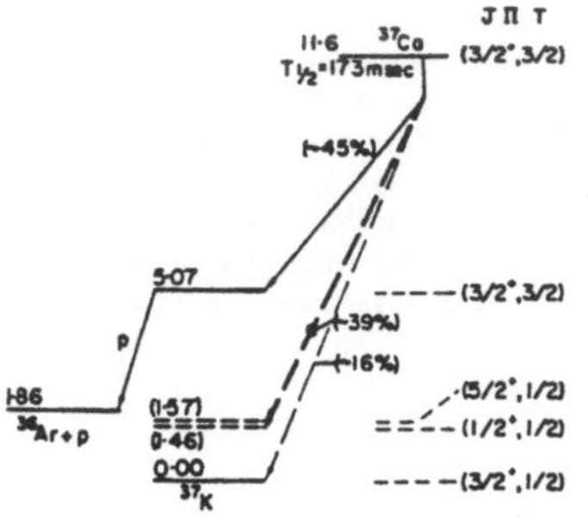

FIG. 2. Proposed decay scheme of ^{37}Ca. Levels and decays that have not been directly observed are shown by dashed lines. Energies that are only estimated are shown in brackets. The β^+ branching ratios have been estimated as described in the text.

We should like to thank Professor R. E. Bell, the director of the Laboratory, and Dr. R. Barton for helpful discussions. We acknowledge, also, scholarships awarded by the National Research Council.

---

†This research was supported by a grant from the Atomic Energy Control Board (Canada).

[1]J. N. Bahcall, Phys. Rev. 135, B137 (1964).

[2]J. N. Bahcall and C. A. Barnes, Phys. Letters 12, 48 (1964).

[3]R. Barton et al., Can. J. Phys. 41, 2007 (1963); R. McPherson, J. C. Hardy, and R. E. Bell, Phys. Letters 11, 65 (1964) (^{17}Ne); R. McPherson and J. C. Hardy, to be published (^{21}Mg and ^{25}Si); J. C. Hardy and R. I. Verrall, to be published (^{29}S); J. C. Hardy and R. I. Verrall, to be published (^{33}Ar).

[4]P. M. Endt and C. van der Leun, Nucl. Phys. 34, 1 (1962).

NEW DELAYED-PROTON EMITTERS: Ti^{41}, Ca^{37}, AND Ar^{33} †

P. L. Reeder, A. M. Poskanzer, and R. A. Esterlund
Chemistry Department, Brookhaven National Laboratory, Upton, New York
(Received 13 November 1964)

Three new nuclides, Ti^{41}, Ca^{37}, and Ar^{33}, have been observed to be delayed proton emitters of the type that undergo beta decay to proton unstable states of daughter nuclei. These nuclides extend to higher mass the series of known delayed proton emitters with even Z and $A = 2Z-3$: S^{29},[1] Si^{25},[2] Mg^{21},[2] Ne^{17},[3,4] and O^{13}.[5] The beta decay of Ca^{37} is of particular interest[6] with regard to the proposed solar neutrino experiment.[7,8] This isotope is the mirror nucleus of Cl^{37}, and thus, knowledge of the half-life of Ca^{37} is of help in estimating[8] the neutrino capture rate in Cl^{37}.

The nuclides were produced by the (He^3, $2n$) reaction in the external beam of the Bookhaven 60-in. cyclotron. The maximum energy of the beam was 31.8 MeV and the average intensity was about 0.1 microampere. The beam was pulsed by means of a mechanical chopper with a frequency that could be varied to suit the half-life of the nuclide being studied. The gaseous targets, Ar^{36} and H_2S for producing Ca^{37} and Ar^{33}, respectively,[9] were contained at $\frac{2}{3}$ of an atmosphere in a brass cell. The beam windows of the cell consisted of 0.25-mil Ni and the proton window parallel to the beam was 0.25 or 1.0-mil Mylar. The solid targets, Ca and S for producing Ti^{41} and Ar^{33}, respectively, were vacuum-deposited to a thickness of 1 mg/cm^2 on 0.1-mil Ni backing foils. The target foils were oriented at 10° to the He^3 beam in order to increase the path of the beam through the target material. Two detectors were mounted side by side, perpendicular to the target. Surface barrier detectors three cm^2 in area and 7000 to 8000 ohm-cm resistivity were used. Proton energy spectra as a function of time were recorded between beam pulses by a two-parameter analyzer normally having 255 energy channels by 16 time channels. Data for many beam pulses were accumulated by means of a repetitive electronic timer which controlled the beam chopper and initiated the analyzer.

Table I. Half-lives of delayed proton emitters.

Target	Product	Half-life (msec)
S	Ar^{33}	182 ± 5
Ar^{36}	Ca^{37}	170 ± 5
Ca	Ti^{41}	90.5 ± 2

The data were processed by a computer, using a least-squares program for fitting decay curves. For the determination of a half-life, the counts in the proton peaks in each energy spectrum were summed before the decay curve analysis was performed. The half-lives obtained for the three new nuclides are presented in Table I. In the case of Ti^{41}, which has several prominent proton peaks, analysis of each peak yielded a half-life which agreed within the errors with that obtained from the integrated spectrum. To obtain an energy spectrum resolved from background, each energy channel of the original data was analyzed with the half-lives of Table I. The resulting proton spectra for Ti^{41} and Ca^{37} are plotted in Figs. 1 and 2, as initial counting rate versus energy channel. The width of the time channel is indicated in the ordinate. A peak resulting from a calibrated 60-cps pulser is shown dashed on the high-energy side of each spectrum. The center-of-mass energies, which have been corrected for energy loss in the target material and in the window of the gas cell, are shown above the proton peaks and in Table II. The proton spectrum for Ar^{33} is similar to that for Ca^{37}

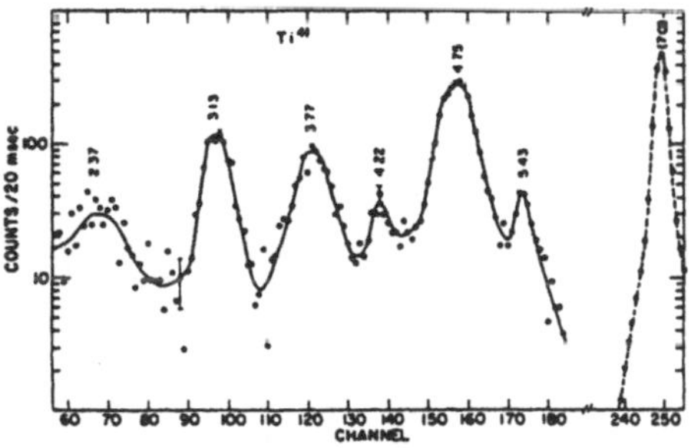

FIG. 1. Proton spectrum following the decay of Ti^{41} taken at 30-V bias. The 5.43-MeV protons penetrated beyond the depletion layer of the detector. The energy indicated for this peak was obtained from other spectra taken at higher bias.

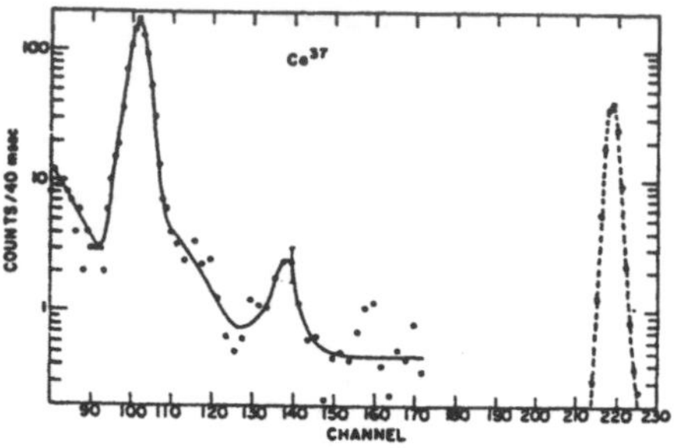

FIG. 2. Proton spectrum following the decay of Ca^{37}.

and is also summarized in Table II. The peaks were shown to be due to protons and not alpha particles from variation of gas pressure and window thickness, and from the narrowness of the peak widths for the target thicknesses used. Also, in the case of Ar^{33} the peak energies derived from the solid and gas targets were the same. In order to relate the center-of-mass energies to energy levels in the daughter nucleus resulting from beta decay, one has to add the proton-separation energy from the daughter nucleus. The values used[10] (in MeV) were 2.285 ± 0.012 for Cl^{33}, 1.876 ± 0.041 for K^{37}, and 1.082 ± 0.012 for Sc^{41}. The resulting energy levels of the daughter nucleus are listed in Table II.

Table II. Proton energies.

Nuclide	$E_{c.m.}$ (MeV)	$E_{c.m.}+S_p$ (MeV)	Relative intensity[a] (%)
Ar^{33}	3.256 ± 0.030	5.54 ± 0.03[b]	98
	3.90 ± 0.10	6.19 ± 0.10	2
Ca^{37}	3.172 ± 0.030	5.05 ± 0.05[b]	99
	4.00 ± 0.10	5.9 ± 0.1	1
Ti^{41}	2.37 ± 0.1	3.45 ± 0.1	8
	3.13 ± 0.03	4.21 ± 0.03	17
	3.77 ± 0.03	4.85 ± 0.03	16
	4.22 ± 0.05	5.30 ± 0.05	4
	4.75 ± 0.03	5.83 ± 0.03[b]	50
	5.43 ± 0.05	6.51 ± 0.05	5

[a] Intensities as percent of protons observed for each isotope.
[b] Energy of the lowest $T = \frac{3}{2}$ state in the daughter nucleus.

Evidence for the identification of these nuclides was obtained from measurements of excitation functions. The excitation function observed for Ca^{37} has a threshold at 20 ± 2 MeV which is consistent with the predicted threshold of 19.4 MeV for the $(He^3, 2n)$ reaction.[6,8] The (He^3, n) reaction is exothermic and the $(He^3, 3n)$ has a threshold at 33 MeV and these reactions are thus inconsistent with the observed excitation function. The energetic threshold for the production of Ar^{33} by the reaction $Ar^{36}(He^3, 2n\alpha)$ is 26.5 MeV. However, inclusion of a Coulomb barrier for the emission of the α particle would raise the effective threshold to about 32 MeV. Thus our results for Ca^{37} have no contribution from the very similar isotope Ar^{33}. The only reactions with thresholds similar to $Ar^{36}(He^3, 2n)$ are the reactions $Ar^{36}(He^3, n\alpha)Ar^{34}$ and $Ar^{36}(He^3, t)K^{36}$. From the masses predicted[11] for these products it can be shown that the amount of energy available for beta decay is insufficient to give rise to the 3-MeV protons which we have observed. The predicted $(He^3, 2n)$ thresholds for producing Ti^{41} and Ar^{33} are quite similar to that for Ca^{37}. The measured excitation functions, although cruder, are consistent with the predicted thresholds. Similar arguments concerning competing reactions apply to these nuclides also.

Each of the proton spectra, particularly in the case of Ca^{37} and Ar^{33}, is dominated by a peak due to a superallowed beta transition to a daughter state which is the analog of the parent nucleus ground state.[2] The excitation energies of these levels in the daughter nuclides, the lowest energy $T = \frac{3}{2}$ states, are indicated by footnote b in Table II. The masses of the parent nuclides may be estimated[11] by adding a Coulomb correction ($1.2Z/A^{1/3}$ MeV) to the energy of the analog state and subtracting the neutron-hydrogen mass difference (0.78 MeV). Using this method we calculate the Q_β values shown in Table III. Using Wilkinson's method[12]

Table III. Calculated values of Q_β in MeV.

Nuclide	From Coulomb correction to analogue state	From Wilkinson
Ti^{41}	12.70	12.58
Ca^{37}	11.46	11.46
Ar^{33}	11.49	11.40

and known ground-state masses[10] of the three other members of the isobaric quartet, we obtain the values shown in the last column of Table III. The good agreement supports our assignment of the analog-state proton peaks.

For Ca^{37}, the energy of the analog state agrees within the errors with that predicted by Bahcall[6,8] while the observed half-life is slightly longer than the 0.13-sec value predicted by him. However, the observation of Ca^{37} with properties so close to those predicted supports the estimation of the neutrino-capture cross sections for the solar-neutrino experiment.

We wish to acknowledge helpful discussions with R. Davis, Jr., G. Friedlander, and R. McPherson, and assistance from R. W. Stoenner.

---

†Research performed under the auspices of the U. S. Atomic Energy Commission.

[1]J. C. Hardy and R. I. Verrall, to be published.

[2]R. McPherson and J. C. Hardy, to be published.

[3]R. McPherson, J. C. Hardy, and R. E. Bell, Phys. Letters 11, 65 (1964).

[4]V. A. Karnaukhov, G. M. Ter-Akopian, L. A. Petrov, and V. G. Subbotin, Zh. Eksperim. i Teor. Fiz. 45, 1280 (1963) [translation: Soviet Phys.—JETP 18, 879 (1964)].

[5]R. Barton, R. McPherson, R. E. Bell, W. R. Frisken, W. T. Link, and R. B. Moore, Can. J. Phys. 41, 2007 (1963).

[6]J. N. Bahcall and C. A. Barnes, Phys. Letters 12, 48 (1964).

[7]J. N. Bahcall, Phys. Rev. Letters 12, 300 (1964); R. Davis, Jr., Phys. Rev. Letters 12, 303 (1964).

[8]J. N. Bahcall, Phys. Rev. 135, B137 (1964).

[9]90 % Ar^{36}, supplied by Oak Ridge National Laboratory.

[10]L. A. König, J. H. E. Mattauch, and A. H. Wapstra, Nucl. Phys. 31, 18 (1962).

[11]A. I. Baz', V. I. Gol'danskii, and Ya. B. Zel'dovich, Usp. Fiz. Nauk 72, 211 (1960) [translation: Soviet Phys.—Usp. 3, 729 (1961)].

[12]D. H. Wilkinson, Phys. Letters 12, 348 (1964).

Neutrino Opacity I. Neutrino-Lepton Scattering*

JOHN N. BAHCALL
California Institute of Technology, Pasadena, California
(Received 24 June 1964)

The contribution of neutrino-lepton scattering to the total neutrino opacity of matter is investigated; it is found that, contrary to previous beliefs, neutrino scattering dominates the neutrino opacity for many astrophysically important conditions. The rates for neutrino-electron scattering and antineutrino-electron scattering are given for a variety of conditions, including both degenerate and nondegenerate gases; the rates for some related reactions are also presented. Formulas are given for the mean scattering angle and the mean energy loss in neutrino and antineutrino scattering. Applications are made to the following problems: (a) the detection of solar neutrinos; (b) the escape of neutrinos from stars; (c) neutrino scattering in cosmology; and (d) energy deposition in supernova explosions.

I. INTRODUCTION

EXPERIMENTS[1,2] designed to detect solar neutrinos will soon provide crucial tests of the theory of stellar energy generation. Other neutrino experiments have been suggested as a test[3] of a possible mechanism for producing the high-energy electrons that are inferred to exist in strong radio sources and as a means[4] for studying the high-energy neutrinos emitted in the decay of cosmic-ray secondaries. From a theoretical standpoint, the production of neutrinos by electron-positron annihilation, or by related processes, has been shown[5,6] to play a pivotal role in the later stages of stellar evolution and in the formation of the elements near the iron peak. Moreover, neutrinos play an important role in a number of cosmological considerations, including the question of the energy density in the universe[7,8] and the problem of distinguishing between cosmological models.[9,10]

The processes by which neutrinos are emitted or absorbed have therefore been extensively discussed by many authors.[5,6,11,12] However, neutrino scattering has only been discussed for the special situation of electrons initially at rest.[13,14]

In this paper, we investigate the contribution of neutrino-lepton scattering to the total neutrino opacity of matter and show, contrary to previous beliefs, that neutrino-lepton scattering dominates the neutrino opacity for many astrophysically important conditions. Here, neutrino opacity is defined, analogously to photon opacity, as the inverse of the neutrino mean free path times the matter density [i.e., $K_\nu = (\lambda\rho)^{-1}$]. In a subsequent paper with Frautschi,[15] the contribution of neutrino-nucleon interactions to neutrino opacity is discussed.

As a basis for astrophysical applications, we give the rates, under a variety of conditions, for neutrino-electron scattering, i.e.,

$$\nu_\beta + e^- \rightarrow \nu_\beta' + e^{-\prime}, \qquad (1)$$

and antineutrino-electron scattering, i.e.,

$$\bar{\nu}_\beta + e^- \rightarrow \bar{\nu}_\beta' + e^{-\prime}. \qquad (2)$$

The formulas we present are derived from the conserved vector current theory.[13]

The cross sections for reactions (1) and (2) are identical, by *PC* invariance, with the cross sections for the interactions among the corresponding antiparticles, i.e.,

$$\bar{\nu}_\beta + e^+ \rightarrow \bar{\nu}_\beta' + e^{+\prime}, \qquad (1')$$

and

$$\nu_\beta + e^+ \rightarrow \nu_\beta' + e^{+\prime}. \qquad (2')$$

Hence, we discuss explicitly only reactions (1) and (2), although (1′) and (2′) also occur in a number of astronomical situations. The cross sections for neutrino-muon scattering can be obtained from the cross sections given in this paper by substituting, in all formulas, the

* Work supported in part by the U. S. Office of Naval Research and the National Aeronautics and Space Administration.

[1] R. Davis, Jr., Phys. Rev. Letters 12, 302 (1964); J. N. Bahcall, *ibid.* 12, 300 (1964).

[2] F. Reines and W. R. Kropp, Phys. Rev. Letters 12, 457 (1964).

[3] J. N. Bahcall and S. C. Frautschi, Phys. Rev. 135, B788 (1964).

[4] K. Greisen, *Proceedings of the International Conference on High Energy Physics* (Interscience Publishers, Inc., New York, 1960), p. 209; T. D. Lee, H. Robinson, M. Schwartz, and R. Cool, Phys. Rev. 132, 1297 (1963).

[5] W. A. Fowler and F. Hoyle, Astrophys. J. Suppl. 91, 1 (1964). This article contains an extensive review of the role of most of the neutrino-emission processes as well as recent work on the abundance of the iron-peak elements.

[6] H. Y. Chiu and P. Morrison, Phys. Rev. Letters 5, 573 (1960); H. Y. Chiu and R. Stabler, Phys. Rev. 122, 1317 (1961). See also H. Y. Chiu, Ann. Phys. (N. Y.) 26, 364 (1964) for recent work and references on the relation between supernovae, neutrinos, and neutron stars.

[7] B. Pontecorvo and Ya. Smorodinskii, Zh. Eksperim. i Teor. Fiz. 41, 239 (1961) [English transl.: Soviet Phys.—JETP 14, 173 (1962)].

[8] G. Marx, Nuovo Cimento 30, 1555 (1963).

[9] S. Weinberg, Nuovo Cimento 25, 15 (1962); S. Weinberg, Phys. Rev. 128, 1457 (1962).

[10] J. V. Narlikar, Proc. Roy. Soc. (London) 270, 553 (1962).

[11] R. N. Euwema, Phys. Rev. 133, B1046 (1964).

[12] J. N. Bahcall, Phys. Rev. 135, B137 (1964).

[13] R. P. Feynman and M. Gell-Mann, Phys. Rev. 109, 193 (1958). See in particular footnote seventeen for neutrino-electron scattering.

[14] Y. Yamaguchi, Progr. Theoret. Phys. (Kyoto) 23, 1117 (1960); S. M. Berman, *International Conference on Theoretical Aspects of Very High Energy Phenomena* (CERN, Geneva, 1961), p. 17.

[15] J. N. Bahcall and S. C. Frautschi, Phys. Rev. 135, B788 (1964).

the recoil electrons are strongly peaked in the direction of the incident neutrinos. In fact, one can show by an elementary calculation using Eq. (8) that the maximum angle, θ_{max}, that a recoil electron makes with respect to the incident neutrino direction is given approximately by:

$$\theta_{max} \cong \left[\frac{2(\omega_{max} - \epsilon_{min}')}{\omega_{max}\epsilon_{min}'}\right]^{+\frac{1}{2}}, \qquad (54)$$

where ω_{max} is the maximum incident neutrino energy and ϵ_{min}' is the minimum energy that a recoil electron must have before it is counted. For the conditions suggested by Reines and Kropp, $\theta_{max} \approx 10°$. Equation (14) shows, in fact, that most of the recoil electrons that are counted will actually lie inside θ_{max}.

Thus the use of reaction (1) to detect neutrinos can in principle enable one to locate the direction (presumably toward the sun) of an extraterrestrial neutrino signal.

For the experiment under consideration,[3] the angular distributions and total cross sections are affected only slightly by atomic binding of the initial electrons (see Sec. IV).

B. Neutrino Escape from Stars

It has usually been assumed that neutrinos escape without interaction from the interiors of stars; we investigate the validity of this assumption. The ratio of the neutrino mean free path to the stellar radius R is given by

$$(\lambda/R) \equiv (\kappa_\omega \rho R)_{average}^{-1}, \qquad (55)$$

where κ_ω is the opacity of the stellar matter to a neutrino of energy ω and ρ is the density of the stellar matter. Recall that, according to the definition given in Sec. I,

$$\kappa_\omega \equiv (\textstyle\sum_i \sigma_i n_i)/\rho, \qquad (56)$$

where σ_i is the neutrino interaction cross section for particles of number density n_i, and ρ is the stellar density. A crude but convenient numerical approximation to formula (56) is given by

$$(\lambda/R) \cong 10^{+9}\mu_e(1+2\omega)\omega^{-2}(R/R_\odot)^2(M_\odot/M), \qquad (57)$$

where μ_e is the mean molecular weight per electron, M is the total stellar mass traversed, $R_\odot \cong 7\times10^{+10}$ cm, and $M_\odot \cong 2\times10^{+33}$ g. For antineutrinos, $(1+2\omega)$ should be replaced by $\omega[1-(1+2\omega)^{-3}]^{-1}$ in formula (57).

In making the transition from Eq. (55) to Eq. (57), we have neglected the effect of neutrino absorption by nucleons, the neutrino red shift, and the statistical effects discussed in Sec. V. Neutrino absorption by nucleons has been considered by Euwema[11] and is rediscussed in Ref. 15; the effect of the gravitational red shift can be estimated with the help of the relation[31] $(\Delta\nu/\nu) = -GM/RC^2$ and is not important for the order of magnitude estimates made here. Statistical effects, which were treated in Sec. V, should, of course, be included in any detailed investigation.

One can easily show with the help of Eq. (57) that the probability is only about 10^{-8} or 10^{-9} that a neutrino emitted from the center of the sun will be scattered before escaping from the surface of the sun. For massive stars in the range $10^{+6}M_\odot$ to $10^{+8}M_\odot$, which have been considered by Fowler and Hoyle,[32] (λ/R) does not equal unity until the massive stars have contracted well inside the Schwarzschild radius. Even for white dwarfs, for which[25,33] $(R/R_\odot) \gtrsim 10^{-2}$ and $(M/M_\odot \sim 1$, (λ/R) is much greater than unity and hence neutrino scattering is not important for white dwarfs.

However, for neutron stars,[34,35] $(R/R_\odot) \sim 10^{-4}$ to 10^{-5} and $(M/M_\odot) \sim 1$. Thus, (λ/R) can be much less than unity for neutron stars. Hence the neutrino opacity of neutron stars (including statistical effects and the mean angle of scattering) should be included in future models of these stars.

C. Neutrino Scattering in Cosmology

Neutrinos play a major role in a number of cosmological speculations.[7–10] It is therefore interesting to note that neutrinos (and antineutrinos) with energies less than a BeV have only negligible interactions with matter that is distributed according to current estimates of the large scale composition of the universe. For example, the mean free path of neutrinos with energies of the order of a few MeV to scattering (or absorption[15]) by matter at cosmological densities ($\sim 10^{-29}$ g) is about 10^{+20} times the "radius of the universe," i.e., the mean free path is about 10^{+48} cm.[36]

D. Neutrino Energy Deposition in Supernova Explosions

Colgate and White[26] suggested that neutrinos emitted from the core of a dense collapsing star can deposit sufficient energy by absorption in the mantle of such a star to produce a supernova explosion and blow off the mantle. The conditions under which neutrino energy deposition might explode the mantle depend critically

[31] In this relation, G is, of course, the gravitational coupling constant, *not* the weak interaction coupling constant.

[32] W. A. Fowler and F. Hoyle, Astrophys. J. 140, 830 (1964); F. Hoyle and W. A. Fowler, Monthly Notices Roy. Astron. Soc. 125, 169 (1963).

[33] M. Schwarzschild, *Structure and Evolution of the Stars* (Princeton University Press, Princeton, New Jersey, 1958).

[34] J. R. Oppenheimer and G. M. Volkoff, Phys. Rev. 55, 374 (1939).

[35] A. G. W. Cameron, Astrophys. J. 130, 884 (1959); V. A. Ambartsumyan and G. S. Saakyan, Astron. Zh. 38, 785 (1961) [English transl.: Soviet Astron.—AJ 5, 601 (1962)]; H. Y. Chiu and E. E. Salpeter, Phys. Rev. Letters 12, 413 (1964).

[36] This estimate also includes the possibility of resonant (see Sec. IIIC) antineutrino scattering by high-energy leptons in the primary cosmic radiation. A probable upper limit on the amount of resonant scattering that occurs can easily be estimated by assuming that the cosmic rays fill all space with the local spectrum and energy density ($\sim$1 eV/cc) and that the high-energy leptons constitute, as they do at lower energies, about 1% of the cosmic-ray energy density. The formulas given in Sec. IIIC then yield a mean free path of approximately 10^{+46} cm.

ABSORPTION OF SOLAR NEUTRINOS IN DEUTERIUM

Francis J. Kelly

Department of Physics, The Catholic University of America, Washington, D. C., and U. S. Naval Ordnance Laboratory, Silver Spring, Maryland

and

H. Überall

Department of Physics, The Catholic University of America, Washington, D. C.

(Received 13 December 1965)

The nuclear reactions occurring in the solar interior which are thought to be responsible for solar energy production[1] may be verified directly, by observing the neutrinos created in the over-all reaction in the cycle

$$4H^1 \to He^4 + 2e^+ + 2\nu_e . \qquad (1)$$

Bahcall[2] has obtained a theoretical spectrum of solar neutrinos, and has suggested the initiation of an experimental program of solar neutrino spectroscopy, based upon an observation of neutrino-induced nuclear reactions with various thresholds. Such a program would also check on details of solar models used in the calculation of the neutrino fluxes and provide a measure for the temperature at the sun's center. Most efficient for inducing nuclear reactions would be the neutrinos from the step

$$B^8 \to Be^{8*} + e^+ + \nu_e \qquad (2)$$

of the p-p cycle, because of their high energy (end point at 14.1 MeV).[2] So far, only experimental upper limits on solar-neutrino fluxes have been set;[4] positive results are expected shortly from two experiments now in progress.[5,6] The experiment of Jenkins[6] proposes to detect the solar neutrinos in the reaction

$$\nu_e + H^2 \to 2H^1 + e^- , \qquad (3)$$

by observation of the Čerenkov radiation of the electrons in a 2000-liter heavy-water target. The cross section and the electron spectrum and angular distribution of Reaction (3) are obtained theoretically in the following.[7] The conclusion is that the Coulomb repulsion will reduce the cross section only by an insignificant amount, so that solar-neutrino detection via (3) appears feasible.

Calling $\vec{\nu}$ = neutrino momentum, T_e (E_e) = electron kinetic (total) energy, $\vec{p}_e$ = electron momentum, $\vec{p}$ = two-proton relative momentum, m_p = proton mass, the kinematics of (3) gives

$$\nu = Q + T_e + (p^2/m_p) \qquad (4)$$

with a threshold of $Q = 1.44$ MeV. We use the conventional nonrelativistic weak-interaction Hamiltonian $G_V + G_A \vec{\sigma}\cdot\vec{\sigma}^N$, where $G_A = -1.20 G_V$ and $G_V = 10^{-5} m_p^{-2}$, and obtain the differential cross section

$$d\sigma = 2^{-2}\pi^{-3} G_A |I|^2$$
$$\times (1 - \tfrac{1}{3}\vec{\nu}\cdot\vec{p}_e/\nu E_e) m_p p \, d\Omega_p p_e E_e dE_e d\Omega_e ; \qquad (5)$$

it is isotropic in $\vec{p}$, thus $d\Omega_p = 2\pi$ (identical particles), and shows a backward electron angular distribution. The following assumptions were made[7]: (1) The electron is treated as a free particle (as justified by its generally high energy). (2) The two protons emerge in a ^1S state; thus only the Gamow-Teller matrix element enters. (3) Retardation is neglected. (4) The effective-range approximation is used.[8,9] Then the matrix element I containing the wave functions u_d, u_{2p} of the relative motion (with the two-proton wave function normalized to 2) becomes

$$I = \int \mu_{2p}^* \mu_d dr \qquad (6a)$$

$$\cong (2/p) \exp[i(\sigma_0 + \delta)] N \sin\delta [J - (r_s + r_t)/4C_0], \qquad (6b)$$

where[9,10] $4\pi N^2 = 150$ MeV, $C_0^2 = 2\pi\eta(e^{2\pi\eta} - 1)^{-1}$, $\eta = (2pR)^{-1}$, $R = 2.88 \times 10^{-12}$ cm, and (see Preston[10])

$$C_0^2 p \cot\delta + R^{-1} h(\eta) = -a_s^{-1} + \tfrac{1}{2} r_s p^2 , \qquad (7)$$

$$a_s = -7.72 \times 10^{-13} \text{ cm}, \quad r_s = 2.72 \times 10^{-13} \text{ cm},$$

$$r_t = 1.71 \times 10^{-13} \text{ cm},$$

and

$$J = \int_0^\infty e^{-\gamma r} [G_0(pr) + F_0(pr) \cot\delta] dr ; \qquad (8)$$

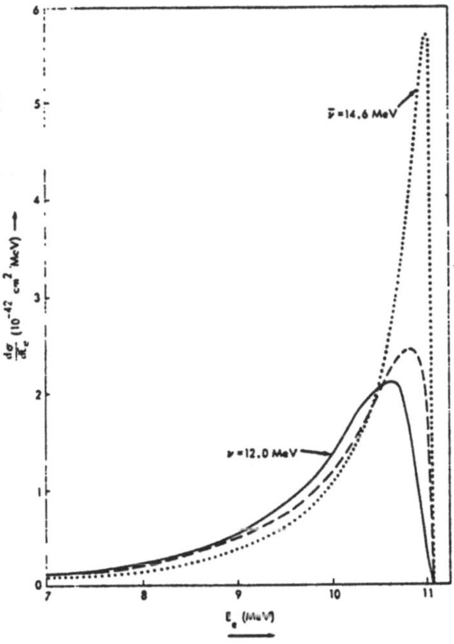

FIG. 1. Electron spectrum of Reaction (3) at a neutrino energy $\nu = 12$ MeV (solid line); for "uncharged protons" (broken line); and for the corresponding antineutrino reaction[7] (dotted line) at the equivalent energy $\bar{\nu} = 14.6$ MeV (giving rise to the same values of E_e, p).

$\gamma = 2.31 \times 10^{12}$ cm^{-1}. Here, F_0 and G_0 are the regular and irregular Coulomb wave functions, with asymptotically unit amplitude, and related to the confluent hypergeometric functions. Integration gives[11]

$$J = \frac{p\,|\Gamma(1-i\eta)|}{(\gamma+ip)^2} e^{-\frac{1}{2}\pi\eta}\left\{\frac{e^{i\delta}}{\sin\delta}\left(\frac{\gamma+ip}{\gamma-ip}\right)^{1-i\eta} + \frac{2ie^{\pi\eta}}{\Gamma(1+i\eta)}\,\frac{F(1-i\eta, 2; 2-i\eta; z)}{\Gamma(2-i\eta)}\right\}, \qquad (9)$$

$z = (\gamma-ip)/(\gamma+ip)$, with F the hypergeometric function.[12] Using all this, we have evaluated numerically the electron spectrum at fixed neutrino energies. Figure 1 presents the spectrum at $\nu = 12$ MeV (solid line), showing a large peak near the upper end, because of the almost two-body kinematics due to the strongly attractive ^1S p-p interaction. The repulsive Coulomb force diminishes this attraction somewhat, as seen by comparison with the broken line for fictitious "uncharged protons"[13]: Its effect is a shift of the peak to smaller electron energies with a broadening (i.e., less resemblance with a two-body kinematics), but practically no decrease of area. Mainly as a consequence of the different scattering lengths, therefore, the total cross section (shown in Fig. 2) is reduced on the order of 10% as compared to Weneser's[7] antineutrino cross section (dotted curve).[14] As mentioned before, detection of solar neutrinos via Reaction (3) in the experiment now carried out by Jenkins[6] therefore seems possible.

It may be of interest to compare counting rates in this experiment, as predicted by our theory, with expected counting rates in other experiments believed to be suitable for solar-neutrino detection. Using the graph for the solar-neutrino flux from Reaction (2) per cm^2 sec MeV at the earth's position as given by Bahcall,[2] we obtain for the integrated product of flux and cross section for Reaction (3), per ^{2}H atom,

$$\int \sigma(\nu_e + {}^2\text{H})\varphi({}^8\text{B})d\nu = 2.3 \times 10^{-35}\ \text{sec}^{-1}. \qquad (10)$$

For the reaction studied by Davis,[5] the corresponding expression is,[15] per ^{37}Cl atom,

$$\int \sigma(\nu_e + {}^{37}\text{Cl})\varphi({}^8\text{B})d\nu = (4 \pm 2) \times 10^{-35}\ \text{sec}^{-1}. \qquad (11)$$

Similarly, for the quantity of importance in Reines and Kropp's experiment,[4] which is based

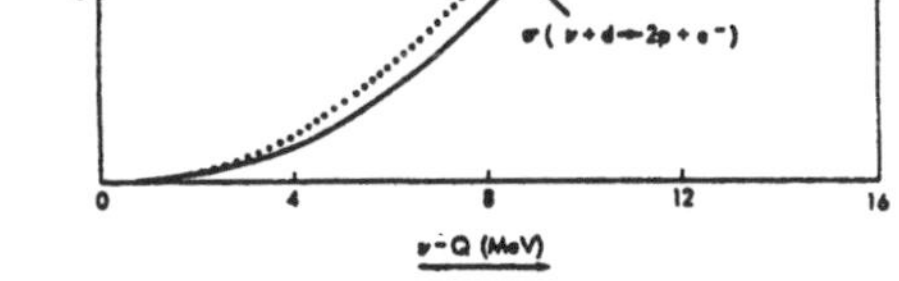

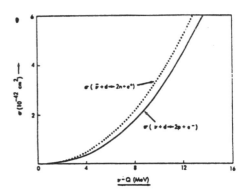

FIG. 2. Total cross section of Reaction (3) plotted versus (ν−1.44) MeV (solid line), and of the corresponding antineutrino reaction[7] plotted versus ($\bar{\nu}$−4.04) MeV (dotted line).

on the hypothetical reaction

$$\nu_e + e^- \to \nu_e + e^-, \qquad (12)$$

we find, using expressions for the total cross section of Reaction (12) given in the literature,[16]

$$\int \sigma(\nu_e + e^-)\varphi(^8\mathrm{B})d\nu = 2.5\times 10^{-36}\ \mathrm{sec}^{-1} \qquad (13)$$

per target electron.

We wish to thank Professor T. L. Jenkins for stimulating discussions and correspondence, and for transmitting the results of an independent numerical computation of our results. We would also like to acknowledge interesting discussions with Dr. Howard Reiss.

*Work supported in part by the National Science Foundation.

[1]H. A. Bethe, Phys. Rev. 55, 434 (1939).

[2]J. N. Bahcall, Science 147, 115 (1965).

[3]V. A. Kuzmin, Phys. Letters 17, 27 (1965), recently mentioned neutrinos from $H^1 + He^3 \to He^4 + e^+ + \nu_e$ with an end point of 18.8 MeV. It is not clear whether their flux is big enough to compete with Reaction (2), which has a flux of 2.5×10^7 cm^{-2} sec^{-1} (Carl Werntz, private communication).

[4]R. Davis, Phys. Rev. Letters 12, 303 (1964); F. Reines and W. R. Kropp, Phys. Rev. Letters 12, 457 (1964).

[5]R. Davis and D. S. Harmer, based on $\nu_e + Cl^{37} \to Ar^{37} + e^-$.

[6]T. L. Jenkins, to be published.

[7]J. Weneser, Phys. Rev. 105, 1335 (1957), gave a theoretical calculation on $\bar{\nu}_e + H^2 \to 2n + e^+$, with a two-neutron final state; the threshold is at 4.04 MeV. With the two-component neutrinos used in the present work Weneser's cross section must be multiplied by 2.

[8]H. A. Bethe and C. Longmire, Phys. Rev. 77, 647 (1950).

[9]See also H. Überall and L. Wolfenstein, Nuovo Cimento 10, 136 (1958).

[10]M. A. Preston, Physics of the Nucleus (Addison-Wesley Publishing Company, Inc., Reading, Massachusetts, 1962).

[11]A. Erdélyi, Higher Transcendental Functions (McGraw-Hill Publishing Company, Inc., New York, 195?

[12]P. M. Morse and H. Feshbach, Methods of Theoretical Physics (McGraw-Hill Publishing Company, Inc., New York, 1953).

[13]Obtained by setting $e \to 0$, i.e., $R^{-1}h(\eta) \to 0$, $\eta \to 0$ and keeping all other parameters unchanged.

[14]We used, like Weneser, the value $a_s = -23.8\times 10^{-13}$ cm (the experimental n-p scattering length) for the n-n scattering length needed in the antineutrino reaction.

[15]J. N. Bahcall, Phys. Rev. Letters 12, 300 (1964).

[16]See, e.g., J. N. Bahcall, Phys. Rev. 136, B1164 (1965).

SOLAR NEUTRINOS*

John N. Bahcall
California Institute of Technology, Pasadena, California
(Received 11 July 1966)

The predicted capture rate in the Cl^{37} experiment for detecting solar neutrinos is calculated using the results of recent nuclear experiments and solar-model investigations. It is shown that additional experiments (e.g., with H^2, Li^7, B^{11}, or ν-e scattering) are necessary to establish the relative contributions of the proton-proton chain and the CNO cycle to solar energy generation.

An experiment is currently under way[1,2] to test directly the theory of nuclear energy generation in stars by detecting neutrinos from the interior of the sun via the reaction

$$\nu_e + Cl^{37} \rightarrow Ar^{37} + e^{-}. \qquad (1)$$

The rate at which solar neutrinos will cause Reaction (1) has previously been calculated[3,4] using a theoretical model of the nuclear mass-37 system and the results of extensive solar-model investigations.[5] The rare mode of the proton-proton chain involving the beta decay of B^8 (maximum neutrino energy ~14 MeV) was shown[3] to be that most important solar neutrino source when Cl^{37} is used as a detector. In this Letter we report the results of more accurate calculations for the cross sections of Reaction (1) leading to the ground state or any of the excited states of Ar^{37}; these calculations were made possible by the large amount of experimental information that has recently become available for the mass-37 system.[6-9] We also discuss the expected solar neutrino fluxes using a revised set of consistent nuclear parameters. The largest remaining experimental uncertainties in predicting the solar-induced rate of Reaction (1) are noted. Finally, we show that experiments with either H^2, Li^7, B^{11}, or ν-e

scattering when combined with the Cl^{37} experiment could furnish a test of the relative contribution of the CNO cycle and proton-proton chain to solar energy generation. Using the absorption cross sections and fluxes calculated in this Letter, the reaction $He^3(H^1, e^+\nu)He^4$ is shown elsewhere[10] to be of minor importance for the Cl^{37} experiment.

The average absorption cross sections in Reaction (1) for neutrinos from the beta decay of B^8 and the reaction[11] $He^3 + H^1 \to He^4 + e^+ + \nu_e$ have been calculated using the detailed results on the beta-decay spectrum of Ca^{37} recently obtained by Poskanzer et al.,[7] the mass-37 level assignments of Kavanagh and Goosman,[8,9] and theoretical formulas previously given.[4] The results are:

$$\langle\sigma\rangle_{B^8} = (1.35 \pm 0.1) \times 10^{-42}\ \mathrm{cm}^2 \qquad (2)$$

and

$$\langle\sigma\rangle_{He^3+H^1} = (4.5 \pm 0.4) \times 10^{-42}\ \mathrm{cm}^2. \qquad (3)$$

The previously calculated[4] absorption cross sections for the other major sources of solar neutrinos (e.g., $e^- + Be^7 \to Li^7 + \nu_e$ and $N^{13} \to C^{13} + e^+ + \nu_e$) are unaffected, because of their much lower decay energies, by the recent experimental results on the mass-37 system. The calculations on which Eqs. (2) and (3) are based include 15 known levels in K^{37} (and therefore with sufficient accuracy[8,9] Ar^{37}) and an estimate of the small contribution of levels of high excitation energy which are inaccessible in the Ca^{37} decay. For both $\langle\sigma\rangle_{B^8}$ and $\langle\sigma\rangle_{He^3+H^1}$, approximately 65% of the total contribution comes from the analog state of Cl^{37} which occurs at an excitation energy of 5.1 MeV in Ar^{37}. The B^8 cross section was averaged over the profile of the Be^{8*} state (or states) using the experimental spectrum[12] for the resulting two alpha particles; this averaging lowers the cross section to the analog level by 14% and that to the ground state by 5%, compared with the cross section calculated for a hypothetical decay to a sharp state at 2.9-MeV excitation energy in Be^{8*}. The quoted uncertainties in Eqs. (2) and (3) include estimated errors in the assignment to particular levels of K^{37} of unidentified transitions in the decay of Ca^{37} and smaller uncertainties[13] in the Gamow-Teller part of the theoretical matrix elements for the analog transition. These results for $\langle\sigma\rangle_{B^8}$ agree with the earlier estimate[3,4] by the present author of this quantity well within the 25% stated uncertainty of the previous estimate, but differ by more than a factor of 2 from some purely theoretical calculations.[14]

Neutrino fluxes from four previously published[15-18] models of the sun have been calculated using detailed information, kindly supplied by the authors, regarding the internal parameters of their models. The results for the $He^3 + H^1$ and B^8 fluxes are given in Table I. The B^8 fluxes for the models listed in rows one, three, and four have been calculated earlier[15,17,18]; our results differ from the previously published fluxes because we have used Parker's[19] more accurate value for the $Be^7 + H^1$ cross-section factor and a consistent set of nuclear-reaction data.[20] The model labels in Table I have the following interpretation: (i) Sears J,[15] a standard solar model; (ii) Weymann and Sears,[16] an improved solar model using more accurate opacities and a nonadiabatic convective envelope, and including radiation pressure; (iii) Ezer and Cameron,[17] a solar model with a convective core, probably the result of using a special opacity law; and (iv) Ezer and Cameron varying-G,[18] a solar model com-

Table I. Solar neutrino fluxes at the earth.

Solar model	$\varphi_\nu(He^3+H^1)^a$ $(10^{+5}\ \mathrm{cm^{-2}\ sec^{-1}})$	$\varphi_\nu(B^8)$ $(10^{+7}\ \mathrm{cm^{-2}\ sec^{-1}})$
Sears J	1.95	2.80
Weymann and Sears	2.00	2.05
Ezer and Cameron	2.05	2.08
Ezer and Cameron (varying G)	1.70	10.2

^{a}The value of this flux is proportional to the low-energy cross-section factor for the reaction $He^3 + H^1$. The numbers in Table I were derived assuming $S_0(He^3 + H^1) = 10^{-18}$ keV b (cf. Ref. 10).

puted with the assumption that the gravitational constant varies with time according to the theory of Brans and Dicke.[21] Note that the varying-*G* model of Ezer and Cameron gives a much higher B^8 flux than any of the other models. All the models quoted assume a primordial solar heavy-element abundance of 2%. This number is uncertain, however, and a heavy-element abundance of 3.4% would probably[15] lead to a B^8 flux for each model about twice that given in Table I. For the purpose of the following discussion, we assume that *G* does not vary and therefore adopt for the B^8 neutrino flux at the earth

$$\varphi_{\nu}(B^8) = (2.1^{+2}_{-1}) \times 10^{+7} \text{ neutrinos cm}^{-2}\text{ sec}^{-1}. \quad (4)$$

All other neutrino fluxes are changed from the values quoted in the original sources[15,17,18] by amounts insignificant for the Cl^{37} experiment.

We find, using Eqs. (2) and (4) and previously published values for the neutrino fluxes[15,17,18] and absorption cross sections[4] for other neutrino sources, the following value for the sum of the fluxes times cross sections for all known sources of solar neutrinos:

$$\sum(\varphi_{\nu}\sigma) = (3.0^{+3}_{-1.5}) \times 10^{-35} \text{ per Cl}^{37}\text{ atom per second} \quad (5)$$

or six captures per day in the experiment, using 10^{+5} gal of C_2Cl_4, that is under way. After B^8, the largest contribution to the predicted counting rate comes from Be^7 neutrinos and is about 7% of the total rate. The value (3.0×10^{-35} sec^{-1}) given in Eq. (5) is only a factor of 5 below the upper limit (16×10^{-35} sec^{-1}) obtained in a preliminary experiment[2] with 10^{+3} gal. The prediction 15×10^{-35} sec^{-1} obtained using the Ezer and Cameron model with a varying gravitational constant[18] and including N^{13}, O^{15}, and Be^7 neutrinos is only 10% below the present upper limit.

The primordial (or surface) composition assumed in computing the solar models represents the largest recognized uncertainty in the predicted capture rate; the errors given in Eq. (5) are no more than guesses for the magnitude of this uncertainty. The assumed composition is based largely on interpretations[15,22] of rocket observations with nuclear emulsions[23] of solar cosmic rays. Further experimental and theoretical work on the determination of the He, C, O, and Ne abundances on the solar surface is essential to an increased understanding of the solar interior.

If the CNO cycle were the dominant mode of energy production in the sun, then the reaction rate for the Cl^{37} experiment would be 3.5×10^{-35} sec^{-1}, independent of the central temperature of the sun.[24] This rate agrees within the recognized uncertainties with the value, given in Eq. (5), based on the assumption that the proton-proton chain is dominant. Thus, the Cl^{37} experiment alone cannot establish whether the sun operates on the proton-proton chain or the CNO cycle. (The solar models used in compiling Table I all have a CNO contribution to the total energy generation of only a few per cent. However, no direct experimental proof is available that the proton-proton chain is dominant in the sun.) Fortunately, a number of other possible detectors of solar neutrinos have been proposed, and the contribution of various solar neutrino sources to the expected counting rates have been calculated.[25] For the four likely detectors recently discussed in this Journal[26-28] (H^2, Li^7, B^{11}, and ν-e scattering), B^8 is again predicted to be the most important solar neutrino source. (In fact, the decay of B^8 provides the only important source[10] of solar neutrinos above the threshold for a B^{11} detector; the proposed experimental conditions[27,28] would preclude the lower-energy neutrinos from the CNO cycle in the Li^7 and ν-e scattering experiments.) If the proton-proton chain is dominant, then the flux inferred from any of the above-mentioned four experiments, by assuming that the source spectrum has the same shape as the B^8 spectrum, should equal the flux inferred from the Cl^{37} experiment.[29] This is an important prediction to check since (1) it is independent of the absolute value of the B^8 flux; (2) it is true only if the spectrum of the primary neutrino source has essentially the same shape as the B^8 spectrum; and therefore, (3) it would not be true if the CNO cycle contributed significantly to the energy production in the sun.

I am grateful to D. Goosman, R. W. Kavanagh, P. D. Parker, and A. M. Poskanzer for supplying me with vital experimental data prior to publication, and to A. G. W. Cameron, D. Ezer, and R. L. Sears for providing detailed information regarding their solar models. It is a pleasure to acknowledge informative conversations with R. Davis, Jr., W. A. Fowler, J. Faulkner, and I. Iben, Jr.

*Work supported in part by the U. S. Office of Naval Research Grant No. Nonr-220(47) and the National Science Foundation Grant No. GP-5391.

[1]R. Davis, Jr., Phys. Rev. Letters 12, 302 (1964).

[2]R. Davis, Jr. and D. S. Harmer, report presented at the Conference on Experimental Neutrino Physics, CERN, Geneva, Switzerland, January, 1965 (to be published); R. Davis, Jr. and D. S. Harmer, paper presented at Second Texas Symposium on Relativistic Astrophysics, Austin, Texas, December, 1964 (to be published).

[3]J. N. Bahcall, Phys. Rev. Letters 12, 300 (1964).

[4]J. N. Bahcall, Phys. Rev. 135, B137 (1964).

[5]R. L. Sears, Astrophys. J. 140, 477 (1964).

[6]P. L. Reeder, A. M. Poskanzer, and R. A. Esterlund, Phys. Rev. Letters 13, 767 (1964); J. C. Hardy and R. I. Verrall, ibid. 13, 764 (1964).

[7]A. M. Poskanzer, R. McPhearson, R. A. Esterlund, and P. L. Reeder, Phys. Rev. (to be published). This beautiful work allows one to extract directly from experiment most of the relevant nuclear matrix elements.

[8]D. R. Goosman and R. W. Kavanagh, to be published.

[9]R. W. Kavanagh and D. R. Goosman, Phys. Letters 12, 229 (1964). See also for Ar^{37}, B. Rosner and E. J. Schneid, Phys. Rev. 139, B66 (1965); J. Cerny, to be published; and J. McNally, to be published; and for K^{37}, see V. E. Storizhko and A. I. Popov, Izv. Akad. Nauk SSSR Ser. Fiz. 28, 1145 (1964) [translation: Bull. Acad. Sci. USSR Phys. Ser. 28, 1048 (1964)]; and A. K. Valter, E. G. Kopanets, A. N. Lvov, and S. P. Tsytko, ibid. 28, 1137 (1964) [translation: ibid. 28, 1040 (1964)].

[10]J. N. Bahcall, J. R. Isper, G. J. Stephenson, Jr., and T. A. Tombrello, to be published.

[11]V. A. Kuzmin, Phys. Letters 17, 27 (1965).

[12]J. Farmer and C. M. Class, Nucl. Phys. 15, 626 (1960). See also D. Alburger, P. F. Donovan, and D. H. Williams, Phys. Rev. 132, 334 (1963).

[13]J. C. Hardy and B. Margolis, Phys. Letters 15, 276

[14]G. A. P. Englebertink and P. J. Brussard, Nucl. Phys. 76, 442 (1965).

[15]R. L. Sears, Astrophys. J. 140, 477 (1964).

[16]R. Weymann and R. L. Sears, Astrophys. J. 142, 174 (1965). The opacities used in this model were taken from A. N. Cox, J. N. Stewart, and D. D. Eilers, Astrophys. J. Suppl Ser. 11, 1 (1965).

[17]D. Ezer and A. G. W. Cameron, Can. J. Phys. 43, 1497 (1965).

[18]D. Ezer and A. G. W. Cameron, Can. J. Phys. 44, 593 (1966).

[19]P. D. Parker, Phys. Rev. (to be published).

[20]P. D. Parker, J. N. Bahcall, and W. A. Fowler, Astrophys. J. 139, 602 (1964).

[21]C. Brans and R. H. Dicke, Phys. Rev. 124, 925 (1961).

[22]S. Biswas, C. E. Fitchel, D. E. Guss, and C. J. Waddington, J. Geophys. Res. 68, 3109 (1963).

[23]J. E. Gaustad, Astrophys. J. 139, 406 (1964).

[24]Each CN cycle liberates 25.05 MeV of useful energy plus one N^{13} and one O^{15} neutrino. Thus the neutrino fluxes ($\varphi_\nu = 3.5 \times 10^{+10}$ cm^{-2} sec^{-1} for both N^{13} and O^{15}) can be found by simply dividing the solar constant by 25 MeV if the CNO cycle is dominant. The relevant cross sections are given in Ref. 4.

[25]J. N. Bahcall, Phys. Letters 13, 332 (1964); J. N. Bahcall, review paper presented at Second Texas Symposium on Relativistic Astrophysics, Austin, Texas, December, 1964 (to be published).

[26]F. J. Kelly and H. Überall, Phys. Rev. Letters 16, 145 (1966); T. L. Jenkins, "A Proposed Experiment for the Detection of Solar Neutrinos" (unpublished).

[27]F. Reines and R. M. Woods, Jr., Phys. Rev. Letters 14, 20 (1965).

[28]F. Reines and W. R. Kropp, Phys. Rev. Letters 12, 457 (1964).

[29]There are small corrections ($\leq 10\%$) due to the contribution from other neutrino sources and uncertainties in the absorption cross sections.

PREDICTION FOR NEUTRINO-ELECTRON CROSS-SECTIONS IN WEINBERG'S MODEL OF WEAK INTERACTIONS

G. 't HOOFT
Institute for Theoretical Physics, University of Utrecht, Utrecht, the Netherlands

Received 27 October 1971

Weinberg's theory of purely leptonic weak interactions can be tested in neutrino-electron scattering experiments. Cross-sections must be measured as a function of the energy of the recoil electron. If Weinberg's theory is correct, then the masses of the intermediate vector bosons can be derived from the measured effective coupling constants.

In an elegant way purely leptonic weak interactions can be cast in the form of a Yang-Mills theory, in which a local SU(2) × U(1) symmetry is broken spontaneously. Weinberg constructed such a model [1] and showed it to be consistent with all known data on leptons. Recently, furthermore, the author has shown that this model is renormalizable [2], which makes it an attractive alternative to many more complicated weak interaction theories. A problem arises if one tries to include hadrons: it is difficult to understand why the neutral intermediate W-boson is not seen. Possibly the hadronic neutral current vanishes due to some mechanism.

Assuming Weinberg's model to be correct for leptons, we get very precise predictions for neutrino-electron elastic scattering. There is one arbitrary parameter, the weak coupling constant g. It is related to the mass of the intermediate vector bosons, $M_\pm$ and M_0, according to

$$g^2 = 4\sqrt{2}\, G M_\pm^2 ; \qquad M_0^2 = M_\pm^2 (1 - e^2/g^2)^{-1}$$

where $e^2/4\pi = 1/137$, and $G = 1.02 \times 10^{-5}\, m_p^{-2}$.

If the neutrino energy is small compared to the intermediate boson mass, the effective Lagrangian pertinent to (e, ν) scattering is

$$\mathcal{L} = \frac{G}{\sqrt{2}} \bar{\nu} \gamma_\mu (1 + \gamma_5) \nu \, \bar{e} \gamma_\mu (g_V + g_A \gamma_5) e \tag{2}$$

It is obtained from the graphs depicted in fig. 1 (to the first of these we applied a Fierz-transformation). The contribution proportional to e^2/g^2 arises from an analogue of the vector dominance principle in strong interactions.

In Weinberg's model we have

for (ν_e, e) and $(\bar{\nu}_e, e)$ scattering: $g_V = \frac{1}{2} + 2e^2/g^2$,

$$g_A = \tfrac{1}{2} ; \tag{3}$$

for (ν_μ, e) and $(\bar{\nu}_\mu, e)$ scattering: $g_V = -\frac{1}{2} + 2e^2/g^2$,

$$g_A = -\tfrac{1}{2} .$$

(In a theory with only charged intermediate bosons these numbers are 1, 1, 0, 0 respectively).

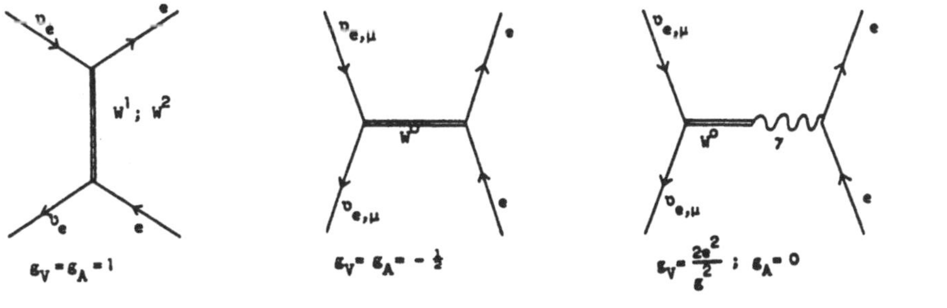

Fig.1.

The number e^2/g^2 must fulfill the inequality

$$0 < e^2/g^2 < 1 \quad . \tag{4}$$

(If $e^2/g^2 \lesssim 1/137$ then the perturbation expansion breaks down and the calculations are invalid, but there arises no inconsistency with what is known about muon decay etc.).

The quantities g_V and g_A can be measured if one knows the energy distribution of the neutrino beam, and measures the energy E_e of the recoil electron. Let the neutrino energy be E_ν. The cross-section for neutrino-electron scattering is

$$\frac{d\sigma}{dE_e} = \frac{G^2 m_e}{2\pi E_\nu^2} \times \tag{5}$$

$$[E_\nu^2(g_V + g_A)^2 + (E_\nu - E_e)^2(g_V - g_A)^2 + m_e E_e(g_A^2 - g_V^2)]$$

The cross-section for antineutrino-electron scattering can be found from eq. (5) by replacing g_A by $-g_A$. Furthermore we have

$$0 < E_e < E_\nu/(1 + m_e/2E_\nu) \quad . \tag{6}$$

If one can measure the angle θ between the recoil electron and the neutrino beams accurately, one finds E_ν directly: we have approximately ($E_{e,\nu} \gg m_e$):

$$E_\nu \approx \frac{2E_e m_e}{2m_e - \theta^2 E_e} \quad , \tag{7}$$

and θ lies in the range

$$\theta \sim \sqrt{\frac{m_e}{E_e}} \quad . \tag{8}$$

We conclude that the proposed experiment could disprove or lend support to the theory. Besides, if Weinberg's theory is correct, then the intermediate boson mass can be found from the number e^2/g^2, following eq. (1).

The theory predicts (cf eqs. (1) and (4))

$$M_\pm > 37\,\text{GeV}, \qquad M_0 \geqslant 74\,\text{GeV}. \tag{9}$$

If g_V in ($\nu_\mu e$) scattering can be determined with at least 20% accuracy then the masses in (9) can be derived if they are lower than ~ 150 GeV.

The author wishes to thank prof. M. Veltman for stimulating discussions and prof. S. Weinberg who brought this model to his notice.

References

[1] S. Weinberg, Phys. Rev. Letters 19 (1967) 1264.
[2] G. 't Hooft, Nucl. Physics, B35 (1971) 167.
Anomalies due to parity non-conservation give rise to complications in the renormalization procedure, but do not make renormalization impossible.

Solar neutrino experiments

John N. Bahcall

Institute for Advanced Study, Princeton, New Jersey 08540
and Department of Nuclear Physics, Weizmann Institute, Rehovot, Israel

New results are presented for absorption cross sections of nine possible detectors of solar neutrinos (^{7}Li, ^{37}Cl, ^{51}V, ^{55}Mn, ^{71}Ga, ^{81}Br, ^{87}Rb, ^{115}In, and ^{205}Tl). Special attention is given to nuclear physics uncertainties. The calculated cross sections are used (with the aid of illustrative solar models and *ad hoc* assumptions about neutrino propagation) to discuss what can be learned about the sun or weak interactions from each of the nine suggested solar neutrino experiments. An experimental program for neutrino spectroscopy of the solar interior is outlined. It is shown in addition that stellar collapses can be detected to typical distances of several kpc (kiloparsecs) by the proposed ^{7}Li, ^{37}Cl, and ^{115}In solar neutrino detectors (provided that electron neutrinos do not decay or oscillate).

CONTENTS

I. INTRODUCTION

A. The problem

Solar neutrino experiments offer a unique opportunity for studying the interior of a star. Conventional information about stars is provided by photons that are emitted from stellar surfaces. The mean free path for photons in stellar interiors, where nuclear fusion occurs, is much less than a centimeter. Neutrinos, unlike photons, interact so weakly with matter that they can escape directly from a stellar interior. Thus neutrinos allow us to look inside a star and to test directly the theoretical predictions for the rates at which certain nuclear reactions occur.

The predicted solar neutrino fluxes make possible well-defined tests of the theory of stellar evolution. We know more about the sun than about any other star. We know its mass, luminosity, radius, surface temperature, surface composition, and age much more accurately than for any other star. The sun is also in what is believed to be the best-understood stage of stellar evolution, the quiescent main sequence phase. If we are to have confidence in the many astronomical and cosmological applications of the theory of stellar evolution, it ought at least to give the right answers for the sun.

The first solar neutrino experiment, which has been performed using ^{37}Cl by R. Davis, Jr. and his associates (Davis, Harmer, and Hoffman 1968; Davis, 1969; Davis and Evans, 1973; Rowley *et al.*, 1977) has revealed a serious discrepancy (see Bahcall and Davis, 1976) between observation and the stardard theory of stellar evolution (using the best estimates for all atomic and nuclear parameters). The origin of this disagreement is unknown. It is not even known for certain that the fault lies in the standard astronomical model of the sun rather than in the conventional physical theory for the propagation of neutrinos.

A number of exotic solutions, modifying either the physics or the astronomy (and in some cases both), have been proposed. Even if one grants that the source of the discrepancy is astronomical, there is no general agreement as to what aspect of the theory is most likely to be incorrect. Many of the proposed solutions of the solar neutrino problem have broad implications for conventional astronomy and cosmology. Some of them would change the theoretical ages of old stars or the inferred primordial element abundances. On the other hand, modified theories of the weak interactions have been proposed in which neutrinos may disappear by mixing or decay in transit from the sun to the earth, but for which there are no terrestrially measurable consequences. It is conceivable that one of these modified theories of the weak interactions is correct and the standard solar model is not in conflict with observations.

The present paper contains discussions of proposed solar neutrino experiments with attention to both the technical questions regarding their sensitivity (i.e., absorption cross sections) and the broader questions of what such experiments can teach us about neutrino physics and astrophysics.

SOLAR NEUTRINO FLUX*

The discovery by Holmgren and Johnston (1958, 1959) of an unexpectedly large cross-section for the $He^3(\alpha, \gamma)Be^7$ reaction led to studies by Fowler (1958) and Cameron (1958) which showed that the proton-proton chain in the present sun is frequently completed by a series of reactions involving Be^7. Fowler and Cameron also discussed the possibility that the decay of B^8, formed by $Be^7(p, \gamma)B^8$ reactions in the interior of the sun, produces a terrestrially measurable flux of high-energy neutrinos ($0 < E_\nu < 14$ Mev). The detection of solar neutrinos is the only experiment that we can think of which could provide *direct* evidence of specific nuclear reactions occurring in the interior of a star.

* This paper presents results of one phase of research carried out at the California Institute of Technology under contract NASW-6 (W0-98001) sponsored by the National Aeronautics and Space Administration and contract Nonr-220(18) sponsored by the Joint Program of the Office of Naval Research and the U.S. Atomic Energy Commission.

We have made use of recently obtained accurate values for the Be^7 electron-capture cross-section (Bahcall 1962) and the Be^7 formation cross-section (Parker and Kavanagh 1962) to make a detailed calculation of the expected B^8 solar neutrino flux. Other relevant nuclear cross-sections have been taken from the report of Fowler (1960). The cross-section constants, corrected for shielding factors, which we have used are, in units of kev-barns, as follows: $S_{11} = 3.5 \times 10^{-22}$, $S_{33} = 1300$, $S_{34} = 0.5$, $S_{17} = 0.03$. The Be^7 decay rate is

$$\lambda_e(Be^7) = 2.12 \times 10^{-9} \rho (1 + x_H) T_6^{-1/2}\ \mathrm{sec}^{-1}\ .$$

The rate of neutrino emission per gram has been integrated over a new model for the interior of the present sun (Iben and Sears 1962); this model has a central temperature of $16.2 \times 10^6\ ^\circ$ K, a central density of 142 gm/cm^3, and a central composition $x_H = 0.333$, $x_{He} = 0.633$, compared with a surface composition of $x_H = 0.630$, $x_{He} = 0.336$. The opacity and energy-generation rates with B^8 reactions included were taken from the work of Iben and Ehrman (1962); we find that 1.0×10^{35} high-energy neutrinos are generated in the sun per second and that the expected neutrino flux at the earth from B^8 decays in the sun is

$$\phi_\nu(B^8) = 3.6 \times 10^7\ \text{neutrinos cm}^{-2}\ \text{sec}^{-1}\ .$$

The neutrino generation corresponds approximately to 1 neutrino for every 1500 proton-proton reactions, and the flux should be compared with the value of 6.4×10^{10} low-energy neutrinos cm^{-2} sec^{-1} from the *pp*-reaction and the Be^7-decays. The flux is a factor of 10 less than could be detected with current experimental techniques using the $Cl^{37}(\nu, e)A^{37}$ reaction and a detector consisting of 10^5 gallons of perchlorethylene (Davis 1962).

However, Davis (1962) has pointed out to us that the more energetic Be^7 neutrinos ($E_\nu = 0.861$ MeV, 88 per cent; 0.383 MeV, 12 per cent) are just above threshold for detection by Cl^{37} absorption ($Q = -0.814$ MeV). The Be^7 solar neutrino flux above the Cl^{37} threshold is

$$\phi_\nu(Be^7; 0.861\ \text{MeV}) = 1.0 \times 10^{+10}\ \text{cm}^{-2}\ \text{sec}^{-1}\ .$$

Since the Cl^{37} neutrino-absorption cross-section for the 0.861-MeV Be^7 neutrinos is about a factor of 200 less than the average absorption cross-section for B^8 neutrinos, about one-half of the detectable solar neutrinos are from B^8 decays, according to the model of Iben and Sears (1962).

The effective nuetrino-generation temperature of a solar model can be defined as the temperature at that radius within which one-half of the high-energy neutrinos is produced; this neutrino temperature can be simply related to the central solar temperature. We find the effective neutrino temperature to be approximately $15 \times 10^6\ ^\circ$ K for the model of Iben and Sears. The number of $Be^7(p, \gamma)B^8$ reactions occurring per unit of time in the sun is an extremely sensitive function of this effective temperature (Fowler 1958), while the competing $Be^7(e, \nu)Li^7$ reaction is a slowly varying function of the temperature as indicated above. A rough calculation shows that an increase in the effective temperature by 20 per cent over the value obtained with the solar model of Iben and Sears (1962) would raise the predicted B^8 neutrino flux by about a factor of 15. Thus an experiment that established even a fairly high upper limit on the solar neutrino flux would nevertheless provide a valuable experimental upper limit on the effective temperature for neutrino generation in the sun.

J. N. Bahcall
William A. Fowler
I. Iben, Jr.
R. L. Sears

REFERENCES

Bahcall, J. N. 1962, *Phys. Rev.*, **128**, 1297.
Cameron, A. G. W. 1958, *Ann. Rev. Nuclear Sci.*, **8**, 299.
Davis, R. D., Jr. 1962, private communications.
Fowler, W. A. 1958, *Ap. J.*, **127**, 551.
———. 1960, *Mem. Soc. R. Sci. Liège*, Ser. 5, **3**, 207.
Holmgren, H. D., and Johnston, R. L. 1958, *Bull. Amer. Phys. Soc.*, Ser. II, **1**, 328.
———. 1959, *Phys. Rev.*, **113**, 1556.
Iben, I., Jr,. and Ehrman, J. R. 1962, *Ap. J.*, **135**, 770.
Iben, I., Jr., and Sears, R. L. 1962, unpublished.
Parker, P. D., and Kavanagh, R. W. 1963, *Bull. Amer. Phys. Soc.*, Ser. II, **8** (in press).

HELIUM CONTENT AND NEUTRINO FLUXES IN SOLAR MODELS*

R. L. SEARS
California Institute of Technology, Pasadena, California
Received March 2, 1964

ABSTRACT

A variety of evolutionary sequences of models for the solar interior has been computed, corresponding to variations in input data, to obtain some idea of the uncertainties involved in predicting a solar neutrino flux. It is concluded that the neutrino flux can be estimated to within a factor of 2, the primary uncertainty being the initial homogeneous solar composition; detailed results are given. With a preferred value of the heavy-element-to-hydrogen ratio $Z/X = 0.028$, the helium content necessary to fit a model to the observed solar luminosity is found to be $Y = 0.27$.

I. INTRODUCTION

Theoretical models of the internal structure of the Sun are no longer at the frontier of the theory of stellar structure and evolution. Since the recognition of the proton-proton chain as the major energy source, the general features of solar structure have become quite well established (see, e.g., Schwarzschild 1958, Sec. 23). A few particular features, however, are of sufficient current interest to warrant detailed calculations of solar models. Theories of convective transport of energy in stellar envelopes encounter observational tests in the well-measured radius of the Sun and in those visible surface phenomena that may be attributed to convection. The helium content of the Sun, observable but not accurately measurable in the solar atmosphere, can be obtained from solar models (Schwarzschild 1946). Neutrinos from various beta-decays in the nuclear energy-production processes in the Sun can apparently be detected at the Earth (Bahcall 1964*a*). The latter two features are the subject of the present report, the principal purpose of which is to attempt to illustrate the uncertainties in calculated helium content and neutrino fluxes as they depend on uncertainties in input data of solar models.

The essential problem in this work is to construct a model of the present Sun. According to the Vogt-Russell theorem, a stellar model is determined by its mass and by the distribution of chemical composition, X, Y, and Z, respectively the mass fractions of hydrogen, helium, and elements heavier than helium. The procedure used here, in principle, to obtain the helium content, Y, involves using the mass of the Sun, the observed spectroscopic Z/X ratio, and a trial value of Y to construct a trial model. If the luminosity of the trial model does not agree with the observed luminosity of the Sun, a new trial value of Y is selected and a new model is constructed, using the same mass and Z/X ratio as before. Eventually a helium content is found for which the resulting model gives the luminosity fit. Then the nuclear-reaction rates may be integrated over the model to obtain the various neutrino fluxes.

II. DETAILS OF MODEL CONSTRUCTION

We have used the classical differential equations of stellar structure (e.g., Schwarzschild 1958, eqs. [12.1]–[12.4]) to construct equilibrium models by fitting numerical integrations from the center and from the surface (Haselgrove and Hoyle 1956; Sears and Brownlee 1964). At the surface we have used the boundary condition $P = KT^{2.5}$, where K is a constant, to treat the solar convective envelope (Schwarzschild, Howard, and Härm 1957; Schwarzschild 1958); this approximation is sufficiently good for the present problem, as shown in Section III. Each point in each model has been tested as to

* Supported in part by the Office of Naval Research and the National Aeronautics and Space Administration.

whether radiative or convective transport is appropriate, according to the usual stability condition of K. Schwarzschild. It may be noted here that no convective *cores* occurred in any of the present models, which is consistent with the results of Iben and Ehrman (1962).

To construct a model of the present Sun requires, in fact, the construction of an evolutionary sequence of models, since the chemical composition is inhomogeneous in the interior as a consequence of hydrogen burning over the past 4.5×10^9 years, the assumed age of the Sun (Patterson 1956; Fowler 1961). To construct an evolutionary sequence we first use the mass and an initial homogeneous composition to obtain a "zero-age" model, including the hydrogen-burning rates in it; the rates multiplied by a time step of 5×10^8 years give a new composition distribution for the next model, which is then obtained. Time steps subsequent to the first are 1×10^9 years, and thus a total of six models comprises the evolutionary sequence, with the last model representing the present Sun. (A check sequence of ten models with time steps all 5×10^8 years gave a luminosity for the last model less than 1 per cent greater than the luminosity for the last model of the six-model sequence, hence the latter has sufficiently short time steps for our purposes.) The six-model sequence requires a computation time of about 10 minutes on an IBM 7090 computer with the program used here; this computation time is fairly reasonable for the iteration procedure to obtain the (initial) helium content as described in the preceding section.

To investigate the various uncertainties in solar-model input data, we first constructed a "basic" model, Model A, and subsequently constructed other models, each with a variation on Model A. The basic data of Model A are as follows: mass, 1.989×10^{33} gm (Allen 1963); Z/X ratio, 0.053; and observed luminosity, 3.90×10^{33} erg sec^{-1} (Allen 1963). The Z/X ratio is derived from the photospheric oxygen/hydrogen ratio determined by Osterbrock and Rogerson (1961) and from the ratios of oxygen to the other elements heavier than helium compiled for the solar system by Aller (1961, Table 8-3). The resulting Z/X ratio is subject to some revision, as will be discussed presently. It is assumed, of course, that the Z/X ratio observed in the photosphere is the same as in the initial homogeneous Sun; this is borne out insofar as the depth of the convective envelope is concerned, since it did not extend to the region of nuclear transmutation in the present models.

Besides these basic data of a model, one needs to specify the constitutive relations for the gas density (equation of state), radiative opacity, and energy-generation rate. The following relations were used in Model A. The density is given by equation (6) of Haselgrove and Hoyle (1956), which allows for partial degeneracy of electrons; radiation pressure, however, is neglected here. The opacity is given by the formula of Iben and Ehrman (1962), which fits the tables of Keller and Meyerott (1955) in the relevant composition range; conduction opacity has been included following Haselgrove and Hoyle (1959). The energy-generation rate includes both gravitational contraction (Schwarzschild 1958, eq. [5.10]) and hydrogen burning, the latter being obtained from a routine prepared and kindly made available by Dr. Iben, based on the *pp*-chain (Fowler 1960; Parker, Bahcall, and Fowler 1964) and on the CN cycle (Fowler 1960). The reaction $N^{15}(p, \gamma)O^{16}$ and its consequences have been neglected, since Caughlan and Fowler (1962) have shown that O^{16} does not begin to burn until $T > 16 \times 10^6$ °K, and the models here reach that temperature only in the later stages and at the center (Tables 1–3).

The main results for Model A are given in the first row of Table 1. The number of significant digits displayed for the initial composition is rounded off to three; in practice, five digits were required to satisfy the conditions that the initial Z/X ratio be 0.053 and that the luminosity of the 4.5-billion-year model be 3.90×10^{33} erg sec^{-1}. The value of K for the adiabatic relation was chosen such that the radius of the model equal approximately the observed solar radius, 0.69598×10^{11} cm (Allen 1963). The neutrino fluxes at

TABLE 1

SOLAR MODELS AND NEUTRINO FLUXES

(Initial Z/X = 0.053; Mass = 1.989×10^{33} gm)

Model and Identification	Initial Composition			Surface Values			Central Quantities			Neutrino Fluxes at Earth (no. cm^{-2} sec^{-1})			
	X	Y	Z	L (erg sec^{-1}) (10^{33})	log K	R (cm) (10^{11})	T_c (° K) (10^6)	ρ_c (gm cm^{-3})	X_c	N_{pp} (10^{10})	N_{Be^7} (10^{10})	N_{B^8} (10^7)	$N_{N^{13}} = N_{O^{15}}$ (10^9)
A. Basic	0.638	0.328	0.034	3.896	−2.6	0.694	16.6	164	0.270	5.2	1.1	4.0	1.8
B. log K increase	.638	.328	.034	3.902	−2.4	.665	16.6	164	.270	5.2	1.1	4.0	1.8
C. log K decrease	.638	.328	.034	3.894	−2.8	.719	16.6	164	.270	5.2	1.1	4.0	1.8
D. Los Alamos opacity	.663	.302	.035	3.897	−2.2	.690	16.3	176	.285	5.4	1.0	3.0	1.4
E. $S_0(3,3)$ = 200 keV-b	.636	.330	.034	3.898	−2.6	.702	16.3	164	.254	4.7	1.7	4.5	1.3
F. Li^4 stable	.655	.310	.035	3.890	−2.6	.730	16.6	182	.257	4.9	$N_{Li^4} = N_{pp}$		1.9
G. Age = 5.5×10^9 yr	.646	.320	.034	3.902	−2.6	.708	16.8	183	.220	5.1	1.2	4.4	2.2
H. $L_\odot$ variation	0.641	0.325	0.034	3.776	−2.6	0.694	16.4	160	0.284	5.2	1.0	3.4	1.5

TABLE 2

SOLAR MODELS AND NEUTRINO FLUXES

(Various Initial Z/X; Mass = 1.989×10^{33} gm)

Model	Identification of Model			Initial Composition			Surface Values			Central Quantities			Neutrino Fluxes at Earth (no. cm^{-2} sec^{-1})			
	Z/X	O/H	Ne/O	X	Y	Z	L (erg sec^{-1}) (10^{33})	log K	R (cm) (10^{11})	T_c (° K) (10^6)	ρ_c (gm cm^{-3})	X_c	N_{pp} (10^{10})	N_{Be^7} (10^{10})	N_{B^8} (10^7)	$N_{N^{13}} = N_{O^{15}}$ (10^9)
A	0.053	0.0014	0.56	0.638	0.328	0.034	3.896	−2.6	0.694	16.6	164	0.270	5.2	1.1	4.0	1.8
I	.0364	.00096	.56	.683	.292	.025	3.898	−2.7	.692	16.0	160	.329	5.6	0.92	2.5	0.79
J	0.028	0.00096	0.1	0.708	0.272	0.020	3.903	−2.8	0.694	15.7	158	0.359	5.8	0.82	1.9	0.48

1 A.U. from the Sun, given in the last four columns, were obtained from integrations of the nuclear-reaction rates over the model, via a routine prepared and kindly made available by Dr. Iben. The relevant neutrino-producing reactions are as follows:

$$H^1(p, \beta^+\nu)D^2 \qquad (N_{pp})$$
$$Be^7(e^-, \nu)Li^7 \qquad (N_{Be^7})$$
$$B^8(\beta^+\nu)Be^{8*} \qquad (N_{B^8})$$
$$N^{13}(\beta^+\nu)C^{13} \qquad (N_{N^{13}})$$
$$O^{15}(\beta^+\nu)N^{15} \qquad (N_{O^{15}}) .$$

TABLE 3

EVOLUTIONARY SEQUENCE OF MODELS WITH MASS = 1.989×10^{33} GM AND INITIAL COMPOSITION $X = 0.708$, $Y = 0.272$, $Z = 0.020$, AND WITH LOG $K = -2.8$

Age (years) (10^9)	L (erg sec^{-1}) (10^{33})	R (cm) (10^{11})	T_c (° K) (10^6)	ρ_c (gm cm^{-3})	X_c
0.	2.78	0.66	13.7	90	0.708
0.5.	2.87	.66	13.8	94	.673
1.5.	3.08	.67	14.2	105	.601
2.5.	3.31	.67	14.6	118	.525
3.5.	3.58	.68	15.1	135	.445
4.5.	3.90	0.69	15.7	158	0.359

MARCH OF PHYSICAL VARIABLES IN MODEL J

M_r/M	ρ (gm cm^{-3})	P (dyn cm^{-2}) (10^{16})	r (cm) (10^{11})	T (° K) (10^6)	L_r (erg sec^{-1}) (10^{33})	X
0.0.	158	252	0	15.7	0	0.359
0.1.	83	133	0.08	12.8	2.13	.584
0.2.	59	87	.10	11.3	3.09	.648
0.3.	43	58	.13	10.1	3.55	.679
0.4.	31	38	.15	9.0	3.77	.694
0.5.	22	25	.17	8.1	3.86	.702
0.6.	15	15	.20	7.1	3.90	.705
0.7.	9.4	7.9	.23	6.2	3.90	.707
0.8.	5.0	3.5	.26	5.1	3.90	.708
0.9.	1.8	0.97	.32	3.9	3.90	.708
1.0.	0	0	0.69	0	3.90	0.708

The neutrino fluxes from the last two beta-decays are of course equal, since it is assumed that the CN cycle is in equilibrium. The neutrinos from B^8, which have the highest maximum energy (14.1 MeV) of the lot, are expected to produce the great majority of detectable processes at the earth, in spite of their comparative rarity, according to the capture cross-sections worked out by Bahcall (1964*a*, *b*) and the experimental arrangement under construction by Davis (1964). (The neutrinos produced by electron captures by H^1, He^3, B^8, N^{13}, and O^{15} have been shown by Bahcall [1964*b*] not to be important for Davis' experiment.)

III. VARIATIONS ON THE BASIC MODEL

Models B and C, in Table 1, were constructed in the same way as was Model A ex-pt that the value of K was increased for B and decreased for C. The luminosity of the tter does not fit exactly our condition, but the discrepancy is negligible for the present ırpose. Now the radii of Models B and C are, respectively, appreciably smaller and rger than the observed solar radius, and it may be expected that the two models acket, in a sense, the true envelope conditions. Specifically, our neglect of hydrogen nization leads to too steep a temperature gradient (Osterbrock 1953), while our neglect the superadiabatic temperature gradient that obtains in realistic convective transport chwarzschild 1958, Sec. 7) leads of course to too shallow a temperature gradient. The oint to be noticed here is that an appreciable range in K hardly affects the deep interior chwarzschild 1958, Sec. 17; Sears 1959), in particular the neutrino fluxes and the erived helium content (via the luminosity). But the well-measured radius of the Sun annot be used as a fitting condition for models—to determine, e.g., the age of the Sun— ntil a definitive theory of convective transport becomes available (cf. Iben 1963).

Model D was constructed with a routine using tables of Los Alamos opacities kindly ıade available by Cox and Stewart (1962; see also Cox [1964]). Opacity tables for four omposition sets were used: $(X, Y, Z) = (0.90, 0.06, 0.04)$, $(0.50, 0.46, 0.04)$, $(0.90, .08, 0.02)$, and $(0.50, 0.48, 0.02)$; the heavy-element distribution was that of Aller (1961, able 8-3). Linear double interpolation of log κ in X and Z was used for the compositions eeded. The value of K needed to give an approximate fit to the solar radius was found be larger than for Model A because the line opacities occurring below $T = 5 \times 0^6$ ° K, not included in the Keller-Meyerott tables, tend to increase the radius. The eutrino fluxes in Model D, compared to those in Model A, differ only slightly, in a sense nderstandable by the reduced central temperature which is due to slightly lower pacities (Los Alamos versus Keller-Meyerott) in the deep interior.

Model E was constructed with a reduced effective cross-section for the $He^3(He^3, p)He^4$ reaction, following an informal suggestion by Bahcall (see Parker, Bahcall, and owler 1964). The resulting B^8 neutrino flux is increased because of the enhancement of $Ie^3(\alpha, \gamma)Be^7$, relative to Model A; the central temperature is reduced, however, and hould the cross-section factor for $He^3(He^3, 2p)He^4$ at stellar energies be even less than he guess used here, the further reduction in T_c would tend to reduce the B^8 neutrino ux.

Model F was constructed on the assumption that Li^4 be stable; i.e., that the proton-roton chain proceed through $H^1(p, \beta^+\nu)D^2(p, \gamma)He^3(p, \gamma)Li^4(\beta^+\nu)He^4$. The neutrinos rom the last reaction have a maximum energy of 18.8 MeV, and the flux from Model F vould have already been detected by Davis (1964) if Li^4 were actually stable, as noted y Bahcall (1964*a*). Bahcall has also shown that it is unlikely that a resonance exists at tellar energies for the $He^3 + p$ system whereby a proton-unstable ground state of Li^4 night serve to terminate the *pp*-chain (Parker, Bahcall, and Fowler 1964).

Model G was constructed in the same way as Model A except that the evolutionary equence covered 5.5×10^9 years. The longer age leads to a reduced central hydrogen ontent, X_c, and thus to an increased central temperature, and hence to an increased alue of N_{B^8}.

Model H is fitted to a formerly popular value for the observed solar luminosity, 3.78×10^{33} erg sec^{-1} (Chandrasekhar 1939; Schwarzschild 1958) to illustrate the sen-itivity of the neutrino fluxes to this observational quantity.

The preceding models have all been constructed with the same initial Z/X ratio, 0.053. Recently, Gaustad (1964) has pointed out advances in solar composition analyses that hange the above ratio significantly, for the present purposes. One advance is a redis-cussion of the photospheric oxygen/hydrogen ratio determined by Osterbrock and Rogerson (1961), which Gaustad now finds to be O/H = 0.00096 (by number). This result is just the mean of the independently derived results by Faulkner and Mugglestone (1962) and by Goldberg, Müller, and Aller (1960), namely, 0.00102 and 0.00091, respectively. Model I, in Table 2, was constructed in the same way as Model A, except with O/H = 0.00096, which seems definitely preferable. This corresponds to $Z/X =$ 0.0364, and the differences between Models A and I may be regarded as consequent on the reduced opacity in the latter.

Gaustad has also emphasized the importance of rocket measurements of the relative abundance of He, C, N, O, and Ne from cosmic rays from solar flares, by Biswas, Fichtel, Guss, and Waddington (1963). These workers (and Gaustad) argue that their results represent the photospheric abundances, one point being that the relative proportions of C, N, and O are essentially the same as those in the spectroscopic analysis by Goldberg, Müller, and Aller (1960). The rocket results for He and Ne are thus of extreme interest, since the abundance of neither of these elements is measurable spectroscopically in the photosphere. The rocket Ne/O ratio quoted by Gaustad is 0.1, which is much less than the value implicit in Model A (0.56), from Aller (1961, Table 8-3), who used the neon abundance from B stars as being the best then available. (Pottasch [1963], from an analysis of the far-ultraviolet lower coronal spectrum, also finds that Ne/O = 0.1.) Model J, in Table 2, was constructed in the same way as Model A, except with O/H = 0.00096 and Ne/O = 0.1, i.e., $Z/X = 0.028$. Model J has a significantly reduced value of Z, the heavy-element content, compared to Model A; and therefore the opacity is reduced, leading to the lower values of Y and of N_{B^8} shown in Table 2. The value of Y in Model J may now be compared with the independently derived helium abundance from the rocket observations. For the latter, Gaustad gives the number ratio He/(C + N + O) = 54 ± 6 (p.e.). The same quantity appropriate to Model J involves He/H from the model, O/(C + N + O) from Aller (1961, Table 8-3), and the preferred value of O/H; these can be combined to give

$$\frac{\mathrm{He}}{\mathrm{C+N+O}} = \frac{\mathrm{He}}{\mathrm{H}} \times \frac{\mathrm{O}}{\mathrm{C+N+O}} \times \frac{\mathrm{H}}{\mathrm{O}}$$

$$= 0.096 \times 0.636 \times \frac{1}{0.00096}$$

$$= 64\,.$$

We do not attempt to assign a probable error to this result, but it is clearly not inconsistent with the rocket result. Model I, which uses the higher neon content and thus has a higher He/H ratio, gives a result of 68 for He/(C + N + O); thus Model J is preferable insofar as its helium and neon contents are concerned vis-à-vis the rocket data.

IV. CONCLUSION

We regard Model J as being most nearly consistent with the available data for the Sun at the present time. This model has been used as a basis for theoretical calculations (Bahcall 1964*a*, *b*) on the solar-neutrino detection experiment currently being undertaken by Davis (1964). Table 3 lists some of the details of Model J—the evolutionary sequence and the march of physical variables.

As regards the uncertainties in predicted neutrino fluxes, the results in Tables 1 and 2 suggest that an estimate of uncertainty of a factor of 2 in N_{B^8}, the most important neutrino flux for the Davis experiment, is not unduly optimistic, in the present state of knowledge. The principal sources of error might be expected to be the opacity, the nuclear cross-sections, and the composition. It is gratifying that the two essentially independent opacity tables (Cox [1964] compares them) used in Models A and D give such similar results. For completeness, a model with the Los Alamos opacities and with

$Z/X = 0.028$ has been constructed: the result for N_{B^8} is 1.8×10^7, which differs little from the result for Model J, which was based on Keller-Meyerott opacities. As for the nuclear cross-sections, the one for the $He^3(He^3, 2p)He^4$ reaction is the most uncertain in the proton-proton chain (Parker, Bahcall, and Fowler 1964); Models A and E represent the results of a reasonable range for the uncertainty. For completeness, a model with the same low cross-section factor as in Model E and with $Z/X = 0.028$ has been constructed: the result for N_{B^8} is 2.7×10^7, which compares in the same way with Model J as Model E does with Model A. Of course the flux of B^8 neutrinos also depends on the cross-section for the $Be^7(p, \gamma)B^8$ reaction: since the rate of this reaction has negligible influence on the structure of the models, N_{B^8} varies linearly with the cross-section, which, according to Parker, Bahcall, and Fowler (1964), has an uncertainty about 30 per cent. As for the effect of the composition (specifically, the initial Z/X ratio), on neutrino fluxes, Table 2 illustrates the rather sensitive relationships. It may be hoped that further observations of far-ultraviolet spectra and of solar cosmic rays will improve the abundance determinations of those elements which contribute the most weight to Z, namely, C, N, O, and Ne.

As regards the uncertainty in the initial helium content, Y, the range of the results in Tables 1 and 2 is not embarrassingly large. The result on the initial composition as represented by Model J may be written as

$$X = 0.71, \qquad Y = 0.27, \qquad Z = 0.02\,.$$

This is hardly different from the conclusion arrived at by Gaustad (1964), viz., $X = 0.72$, $Y = 0.26$, $Z = 0.02$, derived partly but not wholly on the same basis as Model J. A comparison with compositions of other objects is also of interest (cf. Osterbrock and Rogerson 1961; Schmidt-Kaler 1961). Mendez (1963) has recently carried out an extensive spectrophotometric study of the Orion Nebula, and he finds that $X = 0.72$, $Y = 0.26$, $Z = 0.02$! The number ratio He/H has shown a curious uniformity in recent years: in the Sun (Model J) it is 0.096; in the Orion Nebula (Mendez 1963), 0.091; in three other diffuse nebulae, including NGC 604 in M33 (Mathis 1962), 0.102 ± 0.005; in the Small Magellanic Cloud (Aller and Faulkner 1962), 0.11; in the B stars (Aller 1961, Table 5-4), 0.16. The last value is subject to appreciable scatter, as Aller points out. In nine planetary nebulae, O'Dell (1963) reports a range in He/H from 0.09 to 0.19, which he believes is cosmic scatter.

In conclusion, it may be remarked that the solar-neutrino detection experiment of Davis, aside from proving directly the existence of nuclear reactions in stars (Bahcall, Fowler, Iben, and Sears 1963; Bahcall 1964*a*), will provide a most useful parameter for stellar structure.

It is a pleasure to thank Professor W. A. Fowler for his support; Dr. I. Iben, Jr., for his valuable contributions and discussions; Dr. J. N. Bahcall for his suggestions and encouragement; Drs. A. N. Cox and J. A. Stewart for making available their opacity tables; and Dr. J. E. Gaustad for providing a copy of his paper in advance of publication.

REFERENCES

Allen, C. W. 1963, *Astrophysical Quantities* (2d ed.; London: Athlone Press).
Aller, L. H. 1961, *The Abundance of the Elements* (New York: Interscience Publishers, Inc.).
Aller, L. H., and Faulkner, D. J. 1962, *Pub. A.S.P.*, **74**, 219.
Bahcall, J. N. 1964*a*, *Phys. Rev. Letters*, **12**, 300.
———. 1964*b*, *Phys. Rev.* (In press.)
Bahcall, J. N., Fowler, W. A., Iben, I., Jr., and Sears, R. L. 1963, *Ap. J.*, **137**, 344.
Biswas, S., Fichtel, C. E., Guss, D. E., and Waddington, C. J. 1963, *J. Geophys. Res.*, **68**, 3109.
Caughlan, G. R., and Fowler, W. A. 1962, *Ap. J.*, **136**, 453.
Chandrasekhar, S. 1939, *An Introduction to the Study of Stellar Structure* (Chicago: University of Chicago Press).

Cox, A. N. 1964, *Stellar Structure*, ed. L. H. Aller and B. Middlehurst (Chicago: University of Chicago Press). (In press.)
Cox, A. N., and Stewart, J. N. 1962, *A.J.*, **67**, 113.
Davis, R., Jr. 1964, *Phys. Rev. Letters*, **12**, 302.
Faulkner, D. J., and Mugglestone, D. 1962, *M.N.*, **124**, 11.
Fowler, W. A. 1960, *Mém. Soc. R. Sci. Liège*. Ser. 5, **3**, 207.
———. 1961, *Proceedings of the Rutherford Jubilee International Conference*, ed. J. B. Birks (London: Heywood & Co. Ltd.), p. 640.
Gaustad, J. E. 1964, *Ap. J.*, **139**, 406.
Goldberg, L., Müller, E. A., and Aller, L. H. 1960, *Ap. J. Suppl.*, **5** (No. 45) 1.
Haselgrove, C. B., and Hoyle, F. 1956, *M.N.*, **116**, 515.
———. 1959, *ibid.*, **119**, 112.
Iben, I., Jr. 1963, *Ap. J.*, **138**, 452.
Iben, I., Jr., and Ehrman, J. R. 1962, *Ap. J.*, **135**, 770.
Keller, G., and Meyerott, R. E. 1955, *Ap. J.*, **122**, 32.
Mathis, J. S. 1962, *Ap. J.*, **136**, 374.
Mendez, M. E. 1963, thesis, California Institute of Technology.
O'Dell, C. R. 1963, *Ap. J.*, **138**, 1018.
Osterbrock, D. E. 1953, *Ap. J.*, **118**, 529.
Osterbrock, D. E., and Rogerson, J. B., Jr. 1961, *Pub. A.S.P.*, **73**, 129.
Parker, P. D., Bahcall, J. N., and Fowler, W. A. 1964, *Ap. J.*, **139**, 602.
Patterson, C. C. 1956, *Geochim. et Cosmochim. Acta*, **10**, 230.
Pottasch, S. R. 1963, *Ap. J.*, **137**, 945.
Schmidt-Kaler, T. 1961, *Observatory*, **81**, 226.
Schwarzschild, M. 1946, *Ap. J.*, **104**, 203.
———. 1958, *Structure and Evolution of the Stars* (Princeton, N.J.: Princeton University Press).
Schwarzschild, M., Howard, R., and Härm, R. 1957, *Ap. J.*, **125**, 233.
Sears, R. L. 1959, *Ap. J.*, **129**, 489.
Sears, R. L., and Brownlee, R. R. 1964, *Stellar Structure*, ed. L. H. Aller and D. B. McLaughlin (Chicago: University of Chicago Press). (In press.)

A STUDY OF SOLAR EVOLUTION

D. Ezer and A. G. W. Cameron
Institute for Space Studies, Goddard Space Flight Center, NASA, New York, N.Y.
Received April 30, 1965

ABSTRACT

Evolutionary sequences of solar models have been calculated using the Henyey method of model construction. These sequences were started at the threshold of stability, at which the released gravitational potential energy of the sun is just sufficient to supply the thermal, dissociation, and ionization energies of the model. It was found that the present solar characteristics were closely reproduced when a mixing length equal to twc pressure scale heights was used in the convection theory, and when a solar initial helium abundance, based on solar cosmic-ray measurements, was chosen. The sun was again found to have a high luminosity during its contraction phase and to approach the main sequence in only a few million years. Tables of selected characteristics of some of the models are presented.

INTRODUCTION

It has been established that stars undergoing gravitational contraction to the main sequence follow a nearly vertical track in the Hertzsprung–Russell diagram (Hayashi 1961; Ezer and Cameron 1963; Weymann and Moore 1963). This early evolution of the sun plays an important role in studies of the origin, structure, and history of the solar system.

Preceding this early evolutionary phase is the collapse of a primordial gas cloud from the interstellar medium which requires detailed hydrodynamic considerations, including the conservation of angular momentum and the influence of interstellar magnetic fields and external gravitational fields. This collapse will continue until the released gravitational potential energy of the gas cloud becomes larger than the energy required for storage as internal energy, dissociation and ionization of molecules and atoms, and energy loss from the interior. This early collapse phase of the sun has been studied by Cameron (1962).

In the present study, the construction of solar models starts from the point where the collapsing protosun becomes stable against further dynamical collapse. The authors previously studied the early evolution of the sun under the assumption of homologous contraction (Ezer and Cameron 1963). A series of models, each corresponding to a definite radius, was obtained by a classical fitting method using the luminosity L, central pressure P_c, central temperature T_c, and a contraction parameter (defined below) as adjustable parameters. It was found that the collapsing protosun becomes stable, from simple energy principles, at a radius of $57R_{\odot}$. At this radius the gravitational energy,

$$E_{\text{grav}} = -G\int_0^M \frac{M_r \, \mathrm{d}M_r}{r},$$

becomes, in absolute amount, barely larger than the thermal, ionization, and dissociation energies of the model star. During subsequent contraction the

ON THE PROBLEM OF DETECTING SOLAR NEUTRINOS

John N. Bahcall† and Raymond Davis, Jr.‡

The evidence supporting the theory of nuclear-energy generation in stars is indirect, based largely upon observations of electromagnetic radiation emitted from the surface of stars and upon theoretical stellar models that have not been subjected to independent experimental tests. It is interesting, therefore, to try to think of a way of *directly* testing the theory of stellar-energy generation in stars. In order to make such a test, one would like to be able to "see" into the deep interior of a star where the nuclear reactions are believed to occur. Thus, an information carrier with a mean free path of the order of 10^{+11} cm ($\sim R_{\odot}$) is required. In the interior of a star like the sun, light has a mean free path of less than a centimeter. Only neutrinos, which have extremely small interaction cross-sections, can enable us to "see" into the interior of a star. Thus, the observation of solar neutrinos would constitute the most direct test that we can think of for the hypothesis that hydrogen-burning nuclear reactions provide the main energy source for stars like the sun. Moreover, the requirement that a theoretical solar model would have to yield the observed solar neutrino flux would provide an additional, and rather restrictive, condition on acceptable solar models.

Recent theoretical and observational results (Bahcall, 1964a and b, Davis, 1964, and Sears, 1964) have qualitatively changed our ideas concerning the possibility of detecting solar neutrinos. Detailed descriptions of these new results have been given elsewhere; we will summarize here the main conclusions.

The hydrogen-burning fusion reactions in the sun are believed to be initiated by the sequence $H^1(p, e^+ \nu)H^2(p, \gamma)He^3$ and terminated by the sequences

$$\text{(1)}\quad He^3(He^3, 2p)He^4$$

$$\text{(2)}\quad He^3(\alpha, \gamma)Be^7(e^-, \gamma)Li^7(p, \alpha)He^4$$

and

$$\text{(3)}\quad He^3(\alpha, \gamma)Be^7(p, \gamma)B^8(e^+ \nu)Be^{8*}(\alpha)He^4$$

The CNO cycle is believed (Sears, 1964 and 1965) to contribute only a few percent of the energy generation in the sun and a negligible amount of the observable solar neutrino flux. The neutrino fluxes from the hydrogen-burning fusion reactions just listed have been calculated by Sears and others (Sears, 1964 and 1965, Bahcall *et al.*, 1963, and Reeves and Pochoda, 1964) using detailed solar models. When these results are combined with a theoretical discussion of the nuclear physics involved in detecting neutrinos by inverse electron capture, it is found (Bahcall, 1964a and b, and Davis, 1964) that the neutrinos from B^8 decay [sequence (3)] produce 90% of

†Work supported in part by the Office of Naval Research and the National Aeronautics and Space Administration.

‡Work performed under the auspices of the United States Atomic Energy Commission.

the observable reactions, although they constitute less than 0.1% of the total solar neutrino flux.

The method that will be used to detect solar neutrinos makes use of the inverse electron-capture process $Cl^{37}(\nu, e^-)Ar^{37}$ and has been extensively discussed by one of us elsewhere (Davis, 1955, 1958, and 1964). On the basis of experience gained in a preliminary experiment (Bahcall, 1964, and Davis, 1964) involving two 500-gal tanks of perchlorethylene, C_2Cl_4, Davis is undertaking an experiment that utilizes a 100,000-gal tank of perchlorethylene as a detector (roughly equivalent to an Olympic-sized swimming pool of cleaning fluid). The most important features of the detection method being employed are that tiny amounts of neutrino-produced Ar^{37} can be removed from the large volume of liquid detector by the simple procedure of sweeping with helium and that the characteristic decay of Ar^{37} can be observed in a counter with essentially zero background. Background effects in an experiment using chemically pure perchlorethylene in a mine 4500 ft deep are expected to be at least a factor of 10 below the predicted rate for solar neutrinos.

Sears (1964 and 1965) has investigated the uncertainties in the predicted neutrino fluxes due to uncertainties in nuclear cross-sections. as well as solar composition, opacity, and age. There are, in addition, uncertainties (Bahcall. 1964) in the predicted neutrino-absorption cross-sections for the reaction $Cl^{37}(\nu, e^-)Ar^{37}$ due to our incomplete knowledge of the nuclear structure of Cl^{37} and Ar^{37}; experimental studies are currently underway in a number of laboratories in this country that are designed to remove the major gaps in our knowledge of Cl^{37} and Ar^{37}. When the best current estimates of solar and nuclear physics uncertainties were combined, the total predicted number of neutrino captures was found to be $(3.6 \pm 2) \times 10^{-35}$ per Cl^{37} atom per second (Bahcall, 1964a). This rate corresponds to between 4 and 11 predicted solar neutrino captures per day in the 100,000-gal deep-mine experiment.

In order to illustrate how the observation of the solar neutrino flux can be used to determine parameters in the solar interior, we recall that the B^8 neutrino flux is extremely sensitive to the central temperature of the sun (Fowler, 1958). This extreme temperature sensitivity is due to the large Coulomb barrier, compared to solar thermal energies, for the reaction $Be^7(p, \gamma)B^8$ of sequence (3). An experimental upper limit on the central temperature of the sun can therefore be obtained by combining the results of the preliminary experiment (Bahcall, 1964b, and Davis, 1964), which provides an upper limit on the neutrino captures per second per Cl^{37} atom, with the predicted rate and the known temperature dependence of the $Be^7(p, \gamma)B^8$ reaction. Keeping the solar luminosity constant, one finds (Bahcall, 1964b, and Davis, 1964) that the central temperature of the sun must be less than 20 million degrees and that a measurement of the B^8 neutrino flux accurate to $\pm 50\%$ would determine the central temperature of the sun to better than $\pm 10\%$ (Bahcall, 1964b, and Davis, 1964).

It should be noted that a positive result from an experiment of the kind described here would be subject to some ambiguity in interpretation due to the possibility of a galactic source of neutrinos, although present estimates of the probable galactic background indicate it to be negligibly small. A possible method of distinguishing between solar and galactic neutrinos would be to take advantage of the eccentricity of the earth's orbit and measure the 7% difference in solar neutrino intensity between aphelion and perihelion. With a signal as low as 7 captures per day, such an experiment would be marginal, but if a somewhat higher signal is observed a test for the seasonal variation of the neutrino flux will be possible.

POSTSCRIPT

Since the above talk was given, Reines and Kropp (1964) have proposed an experiment to detect solar neutrinos by observing recoil electrons from neutrino–electron scattering. Reines and Kropp point out that such an experiment, which is complementary to the one described in our talk, can in principle give information about the neutrino energy spectrum. In addition, Bahcall (1964c) has shown that under the experimental conditions suggested by Reines and Kropp, all the observed recoil electrons will be confined to a cone of opening angle $\approx 10°$ with respect to the incident neutrino direction. Thus, the observation of electron scattering by neutrinos can in principle enable one to determine the direction (presumably toward the sun) of an extraterrestrial neutrino signal.

REFERENCES

J. N. Bahcall (1964a), *Phys. Rev.* **135**: 137.
J. N. Bahcall (1964b), *Phys. Rev. Letters* **12**: 300.
J. N. Bahcall (1964c), *Phys. Rev.* (to be published).
J. N. Bahcall, W. A. Fowler, I. Iben, Jr., and R. L. Sears (1963), *Astrophys. J.* **137**: 344.
R. Davis, Jr. (1955), *Phys. Rev.* **97**: 766.
R. Davis, Jr. (1958), in: *Radioisotopes in Scientific Research, Vol. 1*, Pergamon Press, New York.
R. Davis, Jr. (1964), *Phys. Rev. Letters* **12**: 302.
W. A. Fowler (1958), *Astrophys. J.* **127**: 551.
H. Reeves and P. Pochoda (1964), To be published.
F. Reines and W. R. Kropp, Jr. (1964), *Phys. Rev. Letters* **12**: 457.
R. L. Sears (1964), *Astrophys. J.* **140**: 477.
R. L. Sears (1965), "Solar Models and Neutrino Fluxes," this volume, p. 245.

PRESENT STATUS OF THE THEORETICAL PREDICTIONS FOR THE ${}^{36}Cl$ SOLAR-NEUTRINO EXPERIMENT*

John N. Bahcall† and Neta A. Bahcall‡
California Institute of Technology, Pasadena, California

and

Giora Shaviv§
Cornell University, Ithaca, New York
(Received 8 April 1968)

The theoretical predictions for the ${}^{37}Cl$ solar-neutrino experiment are summarized and compared with the experimental results of Davis, Harmer, and Hoffman. Three important conclusions about the sun are shown to follow.

The experiment of Davis, Harmer, and Hoffman,[1,2] designed to detect solar neutrinos with a ${}^{37}Cl$ target, has prompted a continuing investigation[3-7] of the accuracy with which the flux of neutrinos produced by nuclear reactions in the sun's interior can be predicted. We report here calculations of the solar-neutrino fluxes made using the more accurate rate for the proton-proton reaction recently derived by Bahcall and May[8] and the improved determination of the abundance ratio of heavy elements to hydrogen recently obtained by Lambert and Warner.[9] We also discuss some of the important, recognized uncertainties that influence the predictions of the solar-neutrino fluxes and conclude that the present results of Davis, Harmer, and Hoffman[1] are not in obvious conflict with the theory of stellar structure. We show, however, that a counting rate of less than $0.03 \times 10^{-35}/{}^{37}Cl$ atom sec would cast serious doubt on the correctness of current ideas con-

Table I. Some important quantities for five solar models.

Model	S_{11} (10^{-25} MeV b)	X	Y	Z	T_c (10^6 °K)	ρ_c (10^2 g cm^3)
A	3.36	0.715	0.258	0.027	15.7	1.7
B	3.36	0.768	0.217	0.015	15.2	1.6
C	3.78	0.764	0.221	0.015	14.9	1.5
D	3.93	0.800	0.190	0.010	14.5	1.4
E	3.63	0.740	0.240	0.020	15.2	1.6

cerning the way nuclear fusion reactions produce the sun's luminosity. We then enumerate some of the most important experiments that are necessary to limit the uncertainties in the theoretical predictions. Finally, we show that the experiment of Davis, Harmer, and Hoffman implies the following: (1) that the sun does not derive most of its radiated energy from the CNO cycle, (2) the heavy-element mass fraction in the sun is probably less than 2%, and (3) the primordial helium content was of the order of 22% by mass. The latter two inferences depend upon the validity of current theoretical models for the solar interior.

In Table I we list some important quantities derived from five evolutionary models for the sun that were obtained by numerically integrating the relevant equations of stellar structure[10] as described in Ref. 7. In Table II we give the neutrino fluxes and predicted counting rates for the experiment of Davis, Harmer, and Hoffman[1,2] that were calculated from the same solar models. The quantities X, Y, Z, T_C, and ρ_C of Table I are, respectively, the primordial hydrogen mass fraction, the primordial helium mass fraction, the heavy-element (atomic number greater than four) mass fraction, the central temperature, and the central density. It is assumed that the heavy-element abundance observed on the surface of the sun is the same as the primordial (and present) heavy-element abundance in the center of the sun. This assumption requires further theoretical investigation, but is supported by the agreement between our inferred helium abundance [cf. conclusion (3)] and rocket measurements of the helium abundance in solar cosmic rays (cf. Ref. 9). The neutrino fluxes from the various neutrino-emitting isotopes[11,12] are given in columns two through six of Table II: the neutrinos from the reaction $^1H + {}^1H \rightarrow {}^2D + e^- + \nu$ are represented by the flux $\varphi_\nu(^1H + {}^1H)$ and those from the reaction $^1H + {}^1H + e^- \rightarrow {}^2D + \nu$ are represented by the flux $\varphi_\nu(^1H + e^- + {}^1H)$. The quantities $\sum_{\text{all}}(\varphi_\nu \sigma_\nu)$ and $\sum_{\text{all but B}}(\varphi_\nu \sigma_\nu)$ are the predicted capture rates per ^{37}Cl atom. The cross sections are taken from the work of Bahcall.[5,12] All of the models listed in Tables I and II have a luminosity, after 4.7×10^9 yr of nuclear burning, that equals the solar luminosity[13] of 3.83×10^{33} erg/sec within ±0.2%; all of the nuclear parameters, with the exception of the rate of the proton-proton reaction, are taken from the recent review by Fowler, Caughlan, and Zimmerman.[14]

Model *A* was constructed for a heavy-element mass fraction of $Z = 0.027$ and a low-energy cross-section factor[14] for the proton-proton reaction of $S_{11} = 3.36 \times 10^{-25}$ MeV b. A similar model was regarded as their most probable one by Bahcall and Shaviv[7] and has been used by Davis, Harmer, and Hoffman[1] in discussing the results of their experiment. The present model *A*, and

Table II. Neutrino fluxes and counting rates from five solar models.

Model	$10^{-7}\varphi_\nu(^8B)$ (cm^{-2} sec^{-1})	$10^{-9}\varphi_\nu(^7Be)$ (cm^{-2} sec^{-1})	$10^{-9}\varphi_\nu(^{13}N)$ (cm^{-2} sec^{-1})	$10^{-10}\varphi_\nu(^1H + {}^1H)$ (cm^{-2} sec^{-1})	$\varphi_\nu(^1H + e^- + {}^1H)$ (10^8 cm^{-2} sec^{-1})	$\sum_{\text{all}}(\varphi_\nu \sigma_\nu)$ (10^{-35} sec^{-1})	$\sum_{\text{all but B}}(\varphi_\nu \sigma_\nu)$ (10^{-35} sec^{-1})
A	1.35	4.7	1.1	6.0	1.6	2.1	0.27
B	0.69	3.4	0.3	6.2	1.7	1.1	0.16
C	0.47	2.9	0.2	6.4	1.7	0.77	0.13
D	0.25	2.1	0.1	6.5	1.7	0.44	0.10
E	0.70	3.7	0.4	6.3	1.6	1.1	0.17

all other models discussed in this Letter, differ from the one selected as most probable by Bahcall and Shaviv[7] in that three rather small effects not previously included have been taken account of in the present work. These effects are the Debye-Hückel correction to the equation of state,[15] the contributions of electron conduction to the opacity,[16] and partial conversion of ^{16}O to ^{14}N via the reactions $^{16}O(^1H,\gamma)^{17}F(\beta^+\nu)^{17}O(^1H,\alpha)^{14}N$. The net result of the inclusion of these effects has been to increase the predicted counting rate calculated from model A by about 15% compared with the most probable model of Ref. 7.

Since the work of Bahcall and Shaviv was completed, two important experimental data have become available. The two data are the improved measurement of the mass ratio of heavy elements to hydrogen on the surface of the sun[9] and the redetermination of the neutron lifetime.[17] Model B was constructed using the mass ratio of heavy elements to hydrogen of 0.019 obtained by Lambert and Warner,[9] and the traditional value[14] for the proton cross-section factor, $S_{11}=3.36\times10^{-25}$ MeV b. Note that $\sum_{\text{all}}(\varphi\sigma)$ is lowered by about a factor of 2 when the newer composition is used. Model C was constructed using the values of the low-energy proton cross-section factor $S_{11}=3.78\times10^{-25}$ MeV b and its logarithmic derivative $(d\ln S_{11}/dE)_{E=0}=11.2\ \text{MeV}^{-1}$, derived recently by Bahcall and May.[8] The result quoted above differs from the previous value for S_{11} mainly because Bahcall and May used the newer lifetime measurement for the neutron[17]; small changes were also introduced because of their more accurate calculations of the nuclear matrix element and beta-decay phase-space factors, and their treatment of radiative corrections. Note that the 12.5% increase in S_{11} from model B to model C decreased the predicted counting rate by 32%.

Model C yields our most probable theoretical results. We find that[18]

$$\sum_{\text{all}}(\varphi\sigma)\Big|_{\text{most probable}} = (0.75\pm0.3)\times10^{-35}\ \text{sec}^{-1}\frac{S_{17}}{0.043\ \text{keV b}}. \quad (1)$$

The quantity S_{17} is the low-energy cross-section factor for the reaction $^7Be(^1H,\gamma)^8B$. If we use in Eq. (1), as we have throughout Table II, the value of 0.043 keV b obtained for S_{17} by Parker,[19] the most probable predicted counting rate is about a factor of 2 larger than the probable upper limit set by Davis, Harmer, and Hoffman.[1] However, the preliminary results of Vaughn et al.[20] suggest that Parker's value may require revision downward. The error estimate in Eq. (1) was made by constructing models D and E in which Z and S_{11} were chosen equal to their probable extreme values.[8,9] The opacities used in all of the above-described calculations were obtained in the usual way[7] by interpolation within published tables of Cox, Stewart, and Eilers.[21] As an additional check, J. N. Stewart and A. N. Cox kindly supplied us with opacity tables for precisely the solar composition of heavy elements that was obtained by Lambert and Warner.[9] A recalculation of model C using this more direct approximation to the solar opacity yielded values for the most important quantities that were within a few percent of the values listed in Tables I and II.

It is apparent from Eq. (1) that there is no irreconcilable discrepancy between our predictions and the experiment of Davis, Harmer, and Hoffman[1] when the uncertainties in the various parameters that enter the calculation are taken into account.[22]

The neutrino flux from the reaction $^1H+{}^1H+e^-\rightarrow{}^2D+\nu$ is very nearly model independent as may be seen in Table II. Hence we can predict a lower limit on the counting rate that is consistent with current ideas about the way nuclear fusion reactions produce the sun's liminosity. We find (cf. Ref. 12) that

$$(\varphi\sigma)_{\text{only }^1H+^1H+e^-\rightarrow^2D+\nu} = 0.03\times10^{-35}\ \text{sec}^{-1}. \quad (2)$$

It is important to measure accurately several crucial quantities in order that the relationship between the observed and predicted counting rates may more clearly reveal the adequacy or inadequacy of the current theory of stellar interiors. The quantities of most importance are (1) the neutron lifetime from which the axial-vector coupling constant, and hence the rate of the proton-proton reaction, are determined,[8] (2) the low-energy cross section for $^7Be(^1H,\gamma)^8B$ to which the predicted counting rate is directly proportional, and (3) the heavy-element abundance on the surface of the sun.

We now list several conclusions that can be drawn from the results of the experiment of Davis, Harmer, and Hoffman. First, the sun does not derive most of its radiated energy from the CNO cycle since this implies, independent of the theory of stellar models (cf. Ref. 5), a counting rate of 3.5×10^{-35} sec$^{-1}/^{37}$Cl atom. Second, if the usual theory of stellar interiors is correct, then the heavy-element abundance Z must be less than 2% by mass in order for the predicted neutrino-capture rate not to exceed the observed value. Third, assuming the measured value[9] of $Z/X \cong 0.019$, we can deduce the primordial helium abundance of the sun by requiring that the calculated luminosity of our solar models equals, after 4.7×10^9 yr of nuclear burning, the observed solar luminosity. We find $Y=0.22\pm0.03$, where the uncertainty in Y reflects the uncertainties in the parameters that characterize various solar models.

We are grateful to R. Davis, Jr., R. P. Feynman, William A. Fowler, P. Goldreich, D. L. Lambert, R. M. May, and F. Reines for stimulating and informative conversations, and to J. N. Stewart and A. N. Cox for supplying us with opacity tables for the Lambert-Warner solar composition.

---

*Supported in part by the National Science Foundation [GP-7976, formerly GP-5391] and the Office of Naval Research [Nonr-220(47)].

†Alfred P. Sloan Foundation Fellow.

‡Also at Tel Aviv University, Tel Aviv, Israel.

§Also supported by the National Science Foundation [GP-6928] with Cornell University.

[1] R. Davis, Jr., D. S. Harmer, and K. C. Hoffman, preceding Letter [Phys. Rev. Letters 20, 1205 (1968)].

[2] R. Davis, Jr., Phys. Rev. Letters 12, 303 (1964).

[3] J. N. Bahcall, W. A. Fowler, I. Iben, Jr., and R. L. Sears, Astrophys. J. 137, 344 (1963).

[4] R. L. Sears, Astrophys. J. 140, 477 (1964).

[5] J. N. Bahcall, Phys. Rev. Letters 17, 398 (1966).

[6] J. N. Bahcall, M. Cooper, and P. Demarque, Astrophys. J. 150, 723 (1967); G. Shaviv, J. N. Bahcall, and W. A. Fowler, Astrophys. J. 150, 725 (1967); J. N. Bahcall, N. A. Bahcall, W. A. Fowler, and G. Shaviv, Phys. Letters 26B, 359 (1968).

[7] J. N Bahcall and G. Shaviv, to be published.

[8] J. N. Bahcall and R. M. May, Astrophys. J. Letters 152, 37 (1968).

[9] D. L. Lambert, Nature 215, 43 (1967), and Monthly Notices Roy. Astron. Soc. 138, 143 (1967), and Observatory 87, 228 (1968); D. L. Lambert and B. Warner, Monthly Notices Roy. Astron. Soc. 138, 181, 213 (1968); B. Warner, Monthly Notices Roy. Astron. Soc. 138, 219 (1968); D. L. Lambert, private communications.

[10] M. Schwarzschild, Structure and Evolution of the Stars (Princeton University Press, Princeton, N. J., 1958).

[11] J. N. Bahcall, Science 147, 115 (1965).

[12] J. N. Bahcall, Phys. Rev. 135, B137 (1964), and Phys. Rev. Letters 12, 300 (1964).

[13] R. Stair, W. R. Waters, and H. T. Ellis, spring meeting of the Optical Society of America, 1967, Abstract THG15 (unpublished); R. Stair and H. T. Ellis, "The Solar Constant Based on New Spectral Irradiance Data from 3100 to 5300 Angstroms" (to be published); A. J. Drummond, J. R. Hickey, W. J. Sholes, and E. G. Lane, paper presented at the Fifth Aerospace Sciences Meeting, American Institute of Aeronautics and Astrophysics, New York, January, 1967 (unpublished); H. Neckel, Z. Astrophys. 65, 133 (1967).

[14] W. A. Fowler, G. R. Caughlan, and B. A. Zimmerman, Ann. Rev. Astronomy Astrophys. 5, 525 (1967).

[15] See, for example, L. D. Landau and E. M. Lifshitz, Statistical Physics (Pergamon Press, London, England, 1958) p. 232. The corrected pressure can be written in the form $P \approx P_0[1-4.4\times10^{-2}(3-X)^{3/2}\rho^{1/2}/(5X+3)T_6^{3/2}]$, where P_0 is the pressure of a perfect gas, ρ is the density in g cm^3, and T_6 is the temperature in units of 10^6 °K. We have assumed in the above formula that $Z \ll X$. For consistency, the Debye-Hückel correction must also be taken into account in calculating the entropy.

[16] We use the analytic approximation of C. B. Haselgrove and F. Hoyle, Monthly Notices Roy. Astron. Soc. 119, 112 (1959), to the results of L. Mestel, Proc. Cambridge Phil. Soc. 46, 331 (1950).

[17] C. J. Christensen, A. J. Nielsen, A. Bahnsen, W. K. Brown, and B. M. Rustach, Phys. Letters 26B, 11, (1967).

[18] Only the ^{8}B contribution to the counting rate is proportional to S_{17}, but this is almost all of the total predicted counting rate for the most likely models (cf. Table II).

[19] P. D. Parker, Astrophys. J. 145, 960 (1966), and Phys. Rev. 150, 851 (1966). Cf. the results of R. W. Kavanagh, Nucl. Phys. 15, 411 (1960).

[20] F. J. Vaughn, R. A. Chalmers, D. A. Kohler, and L. F. Chase, Jr., Bull. Am. Phys. Soc. 12, 1177 (1967).

[21] A. N. Cox and J. N. Stewart, Astron. J. 67, 113 (1962), and Astrophys. J. Suppl. 11, 22 (1966); A. N. Cox, J. N. Stewart, and D. D. Eilers, Astrophys. J. Suppl. 11, 1 (1966).

[22] A fuller discussion of these uncertainties is given in Refs. 6 and 7.

SOLAR NEUTRINOS AND THE SOLAR HELIUM ABUNDANCE*

Icko Iben, Jr.
Massachusetts Institute of Technology, Cambridge, Massachusetts
(Recieved 19, July 1968)

The upper limit on the solar neutrino flux set by Davis, Harmer, and Hoffman places an upper limit on the sun's initial helium abundance that is small compared with that estimated for other galactic objects. Adopting current estimates of low-energy nuclear cross-section factors, the upper limit is essentially equal to a lower bound set by demanding that the sum is at least $4\frac{1}{2} \times 10^9$ yr old.

The preliminary upper limit on the solar neutrino flux set recently by Davis, Harmer, and Hoffman[1] is an order of magnitude smaller than the flux that had been expected on the basis of solar model calculations prepared prior to the establishment of this limit. The Davis, Harmer, and Hoffman result has therefore forced a rethinking of the standard assumptions concerning both the input parameters and the input physics that are necessary for the construction of solar models.[2-4]

In an effort to contribute to a better understanding of the implications of the Davis, Harmer, and Hoffman limit, I have prepared an extensive analysis of the relationship between the neutrino flux derived from solar models and several solar input parameters. Many of my results are consistent with those already in the literature.[2-7] However, several new results have emerged and several conclusions are at variance with inferences drawn in two recent papers.[2,8] In this communication, a statement of my basic conclusions will be offered first, followed by a summary of the supporting evidence. A more complete discussion will appear elsewhere.

(1) With the standard choice of solar input parameters, the Davis, Harmer, and Hoffman limit implies an upper limit on the sun's initial helium abundance that is small compared with the helium abundance estimated for other galactic objects. The upper limit on Y (initial He^4 abundance by mass) required for consistency with the Davis, Harmer, and Hoffman limit is $Y_0 \cong 0.16$-0.17. On the other hand, almost every attempt to estimate Y for galactic objects other than the sun has led to values in the range 0.2-0.4, the most probable values clustering about 0.25-0.30. The evidence for a possibly universal, high value for Y has been amply catalogued.[5,9]

Bahcall, Bahcall, and Shaviv[2] claim that a solar $Y = 0.22 \pm 0.03$ (~0.22 with standard assumptions) is consistent with the Davis, Harmer, and Hoffman limit. Despite this claim, the quantitative results in the Bahcall, Bahcall, and Shaviv paper clearly indicate that consistency with the Davis, Harmer, and Hoffman upper bound can be achieved only with $Y \leq Y_0 \sim 0.16$ (with standard assumptions), in agreement with the limit presented here.

(2) With the standard assumptions, the upper

limit on Y is essentially equivalent to a lower limit on Y ($Y_{lower} \sim 0.15$-0.18) set by demanding that the sun's age is at least $4\frac{1}{2} \times 10^9$ yr. If, therefore, the eventual upper limit on the counting rate determined by the Davis, Harmer, and Hoffman experiment is reduced much below the preliminary limit, a clear internal discrepancy will be established, regardless of outside arguments for a larger helium abundance.

(3) By varying the relevant nuclear cross-section parameters as far as possible within quoted limits in directions most favorable for increasing the upper bound on Y, it is possible to escape comfortably, for the present, the embarrassment of a lower bound that exceeds an upper bound. The resultant upper bound of $Y_0 \cong 0.20$ is still small compared with the most probable Y estimated for other galactic objects. Only by varying several nuclear cross-section factors considerably beyond quoted limits is it possible to obtain a Y_0 on the order of 0.25. It is suggested that such large variations are not out of the question. Experimental cross sections can be measured only at energies large compared with energies relevant in the solar interior; it is quite possible that an extrapolation from known to unknown regions may hold surprises.

(4) Ezer and Cameron[3] have suggested that, because of mixing currents, the helium produced at any point in the solar interior need not remain at the site of formation over the sun's $4\frac{1}{2} \times 10^9$-yr lifetime. They estimate that, in the case of complete mixing, the solar neutrino flux is significantly reduced relative to the case of no mixing. A working out of the Ezer and Cameron suggestion reveals that the upper bound on Y increases strongly with the assumed degree of mixing. When complete mixing is permitted, but all other standard assumptions are retained, the upper bound on Y implied by comparison with the Davis, Harmer, and Hoffman limit is $Y_0 \cong 0.24$. A very minor variation of any one of several nuclear cross-section factors (within quoted limits) permits one to achieve a Y_0 in the range 0.25-0.30. Whatever the merits of the mixing assumption may be, this conclusion is in conflict with the conclusion of Bahcall, Bahcall, and Shaviv,[9] who state that "The primordial composition necessary to obtain a solar model···is···almost completely independent of the amount of mixing."

(5) The relationship between calculated neutrino fluxes and the mean interior Y is, to first order, independent of the choice of opacity. It is therefore, to first order, independent of the relationship between Y and the opacity parameter Z. Estimates of mean interior Z, which is approximated roughly by the total abundance of elements heavier than He^4, are normally obtained on the basis of spectroscopic estimates of abundances near the solar surface. Surface abundances, even if they were known exactly, are not necessarily identical to abundances in the deep interior. To exhibit but a few examples, C^{12} has been converted almost completely into N^{14} over the inner half of the sun's mass,[10] while spallation reactions and selective diffusion have possibly affected surface abundances relative to interior abundances. Finally, even if the interior heavy-element abundances were known exactly, there remain many known sources of large errors in the opacity. A Z estimated from spectroscopic data is therefore not necessarily the appropriate choice for the opacity parameter Z.

The value of $Y \cong 0.22$ quoted by Bahcall, Bahcall, and Shaviv[2] is the result of a specific choice of Z and of a particular opacity law. $Y = 0.22$ is not consistent with the Davis, Harmer, and Hoffman limit. An insistence on consistency with this limit, rather than an insistence on a particular choice for Z, leads instead to an upper limit $Y_0 \cong 0.16$-0.17.

(6) The only uncertainty in the equation of state that appears capable of producing a significant increase in derived upper bounds on Y is that associated with the presence or absence of large-scale internal magnetic fields. An average field strength which drops off (according to the two-thirds power of the density) from 10^9 G at the center to 200 G near the surface would lead to an increase of about 0.04 in all upper limits on Y. Whether or not such a field is possible has not been explored.

The supporting evidence for the above conclusions will now be summarized.

Theoretical estimates of $\sum \sigma_i \varphi_i$ (where σ_i is the effective cross section for the absorption of neutrinos that impinge on the earth with a flux φ_i, distributed in an energy spectrum of type i) are obtained by weighting the neutrino fluxes from solar models with theoretically calculated neutrino absorption cross sections.[11]

The canonical choices for the relevant center-of-mass cross-section factors[12] used here are (in keV b) $S_{11}{}^0 = 3.5 \times 10^{-22}$ ($p + p \rightarrow d + e^+ + \nu$), $S_{33}{}^0 = 6.5 \times 10^3$ ($2He^3 \rightarrow He^4 + 2p$), $S_{34}{}^0 = 0.6$ ($He^3 + He^4 \rightarrow Be^7 + \gamma$), $S_{17}{}^0 = 0.03$ ($Be^7 + p \rightarrow B^8 + \gamma$), $S_{114}{}^0 = 3$ ($N^{14} + p \rightarrow O^{15} + \gamma$). With the exception of $S_{33}{}^0$, these cross-section factors are identical to those used by the author in previous investigations.[10]

In Fig. 1, the ratio (call it R) of the counting rate associated with any given solar model to the current preliminary upper limit of Davis, Harmer, and Hoffman ($\sum \sigma_i \varphi_i \leq 3 \times 10^{-36}$ sec^{-1} per Cl^{37} nucleus) is plotted as a function of assumed initial solar helium abundance Y for several different choices of cross-section factors. Beside each curve is that cross-section factor which differs from the canonical set. For the curve labeled $S_{ij}{}^1$'s, $S_{11}{}^1 = S_{11}{}^0 \times (4/3.5)$, $S_{34}{}^1 = \frac{1}{2} S_{34}{}^0$, $S_{17}{}^1 = \frac{1}{2} S_{17}{}^0$, and all other $S_{ij}{}^1 = S_{ij}{}^0$. All curves in Fig. 1 pertain to solar models that have evolved for $4\frac{1}{2} \times 10^9$ yr with no mixing (the He^4 produced at any point in the interior remains near the site of formation).

It is clear that, for any given choice of cross-section factors, the specification of an experimental upper limit on $\sum \sigma_i \varphi_i$ establishes an upper limit on the initial solar Y. The entries in the second column of Table I give this upper limit on Y as a function of cross section factors—if the Davis, Harmer, and Hoffman limit is a hard upper limit, i.e., if $R \leq 1$. When the statistics are improved, the Davis, Harmer, and Hoffman limit could either decrease (the preliminary result is consistent with $R = 0$) or increase. Anticipating the latter possibility, upper limits on Y have been entered into successive columns of Table I for $R \leq 2$, 3, 5, and 10.

For every value of maximum R ($t = 4\frac{1}{2}$ and no mixing), a "penultimate" maximum to Y may be defined by varying all cross-section factors to

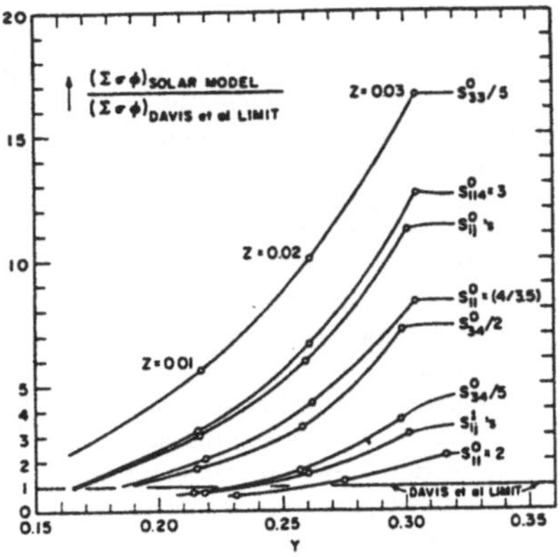

FIG. 1. The ratio of calculated values of $\sum \sigma_i \varphi_i$ to the Davis, Harmer, and Hoffman upper limit.

Table I. Upper and lower limits to the initial solar helium abundance.

S_{ij}'s	R = 1	R = 2	R = 3	R = 5	R = 10	Y_{min}
S^0_{ij}'s	0.166*	0.193	0.216	0.249	0.294	0.174
$S^0_{114} \times 3$	.165*	.191	.212	.243	.288	.174
$S^0_{114} \times 10$	.164*	.190	.210	.239	.283	.175
$S^0_{11} \times (4/3.5)$	.185	.217	.241	.271	.317	.177
$S^0_{11} \times 2$	.263	.310	.344			.187
$S^0_{34}/2$	.189	.226	.252	.280	.316	.172
$S^0_{34}/5$	.231	.267	.288	.318		.171
$S^0_{33}/5$	.140*	.158	.177	.209	.261	.175
S^1_{ij}'s	.238	.278	.301	.337		.177
$(\frac{1}{u}) = -0.01$	.165*	.191	.213	.244	.288	.167
$t = 6$, S^0_{ij}'s	.148*	.169	.186	.215	.262	.158
$t = 3$, S^0_{ij}'s	.179*	.224	.254	.285	.326	.191
$t = 0$, S^0_{ij}'s	.276	.315	.342	.382		.231
$B^2S(AB), t = 4.7$	.160*	.185	.205	.236	.282	.163
$B^2S(CDE), t = 4.7$	0.165	0.206	0.231	0.263	0.310	0.148

the edge of their stated limits in a direction most favorable for increasing Y. Setting $S_{11} = 4.1 \times 10^{-22}$ keV b, $S_{17} = 0.03$ keV b, $S_{34} = 0.38$ keV b, and $S_{33} = 7 \times 10^{3}$ keV b, inspection of Table I yields $Y(1) = 0.20$ as the maximum value of Y permissible if $R \leq 1$ and if cross-section factors are held within quoted limits. In a similar fashion, $Y(2) = 0.24$ and $Y(3) = 0.27$ for $R \leq 2$ and $R \leq 3$, respectively.

Cross-section factors stated as $A \pm B$ are not unique interpretations of experimental results. A stated value of A is an extrapolation from cross sections obtained at energies considerably above those of relevance in the solar interior. A re-examination of the experimental data suggests that the following "ultimate" limits may not be out of the question: $S_{17} \geq 0.015$ keV b, $S_{34} \geq 0.30$ keV b. An "ultimate" limit of $S_{11} \leq 4.3 \times 10^{-22}$ is also not out of the question. Adopting these "ultimate" limits, in addition to $S_{33} = 7 \times 10^{3}$ keV b, "ultimate" upper limits on Y may be determined from the information in Table I. These limits are $\bar{Y}(1) = 0.25$, $\bar{Y}(2) = 0.28$, and $\bar{Y}(3) = 0.32$ for $R \leq 1$, 2, and 3, respectively.

The extent to which uncertainties in several input parameters affect the limits on Y is exhibited in Fig. 2. Curves labeled $t = 4\frac{1}{2}, 3, 2, 1, 0$ are for different assumed solar ages but all with $S_{ij} = S_{ij}^{0}$. Partial results of two other investigations[2,5] are also shown. When normalized to the same set of input parameters, the relationships $\sum \sigma_i \varphi_i$ vs Y given by Sears[5] and by Bahcall, Bahcall, and Shaviv[2] are essentially identical to those presented here.

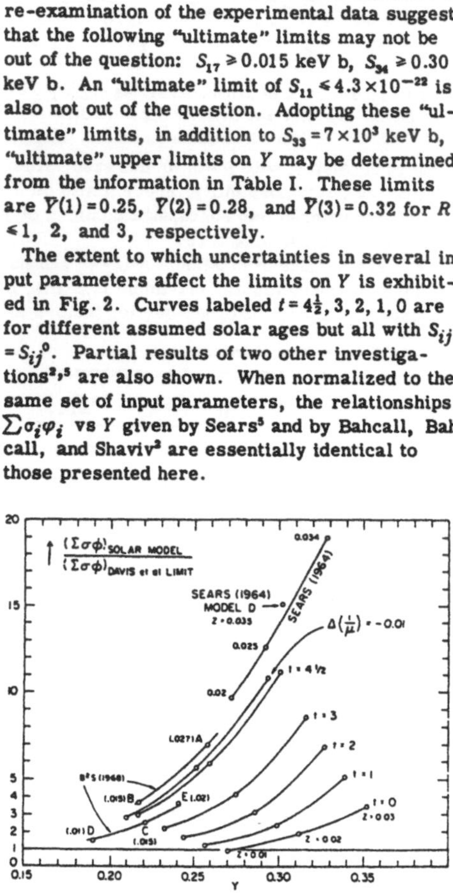

FIG. 2. The dependence of $\sum \sigma_i \varphi_i$ on solar age, opacity, and equation of state. B²S = Bahcall, Bahcall, and Shaviv.

Inspection of the Sears result shows that, although the relationship between Y and Z is fairly sensitive to the choice of opacity, the relationship $\sum \sigma_i \varphi_i$ vs Y is, to first order, independent of this choice. The Sears model D differs from other Sears models only in the choice of opacity.

The curve in Fig. 2 labeled $\Delta(1/\mu) = -0.01$ and row 11 in Table I exhibit the effect of small variations in the equation of state. Here μ is the average molecular weight (in atomic mass units) and the variation chosen is roughly equivalent to neglecting electron degeneracy in the equation of state. Large-scale magnetic fields in the solar interior would act analogously to an increase in μ^{-1}.

The sun's age is presumably well established at about $4\frac{1}{2} \times 10^{9}$ yr. The relationships $\sum \sigma_i \varphi_i$ vs Y shown in Fig. 2 for considerably more youthful suns are nevertheless of interest in the light of Ezer and Cameron's[3] suggestion that large-scale currents associated with a spin-down mechanism may have maintained chemical homogeneity throughout the sun's interior. The effect of mixing between regions where nuclear transformations are occurring and the rest of the sun is equivalent to choosing a smaller solar age. The limit of complete mixing during an assumed $4\frac{1}{2} \times 10^{9}$-yr solar lifetime is equivalent to the limit of a "zero-age" sun, with a slight adjustment in Y because Y has increased by about 0.035 in the $4\frac{1}{2} \times 10^{9}$-yr old, fully mixed sun. The entries in Table I labeled $t = 0$ may, therefore, be interpreted either as upper limits on Y for a "zero-age" sun or, when 0.035 is subtracted, as upper limits on Y for a fully mixed, $4\frac{1}{2} \times 10^{9}$-yr old sun. It is highly unlikely that the sun is younger than $4\frac{1}{2} \times 10^{9}$ yr, but it is not out of the question that a certain amount of mixing may have taken place.

A lower limit on the sun's initial Y is set by the requirement that the solar model reach the sun's present luminosity in $4\frac{1}{2} \times 10^{9}$ yr. The last column in Table I contains extrapolated estimates of this lower limit. Note that, in several instances (distinguished by an asterisk), an extrapolated lower limit exceeds the appropriate upper limit. This means that the parameters describing the spurious upper limit do not form a valid combination. For example, with all $S_{ij} = S_{ij}^{0}$, there exists no Y such that $R \leq 1$.

In summary, if one accepts (1) current best estimates for nuclear cross-section factors, (2) a $4\frac{1}{2} \times 10^{9}$-yr old sun, (3) the absence of significant mixing currents and magnetic fields in the sun's interior, and (4) the Davis, Harmer, and Hoffman interpretation that $\sum \sigma_i \varphi_i \leq 3 \times 10^{-36}$ sec^{-1} per Cl^{37} atom, then an upper limit to the initial solar helium abundance lies in the range $Y_0 \cong 0.16$-0.17. This limit is uncomfortably close to a lower limit of $Y_{\text{lower}} \cong 0.15$-$0.18$ set by demanding that the solar model have the sun's luminosity after $4\frac{1}{2} \times 10^9$ yr. If all cross-section factors are varied in a direction favorable for decreasing $\sum \sigma_i \varphi_i$, but not beyond limits customarily quoted, then a "penultimate" upper limit on Y is $Y(1) \cong 0.20$. Finally, if cross-section factors are varied beyond conventional limits by extrapolating low-energy cross-section measurements differently, an "ultimate" upper limit on Y is $\overline{Y}(1) \cong 0.25$. Thus, if conventional assumptions about the sun are maintained, a close similarity between model-determined values for solar Y and estimates of Y for other galactic objects can be achieved only by adopting cross-section factors outside commonly accepted limits. This conclusion could, of course, be avoided if the Davis, Harmer, and Hoffman limit were an underestimate.

*Work supported in part by the National Science Foundation Contract No. GP-8060 and in part by the National Aeronautics and Space Administration Contract No. NsG-496.

[1]R. Davis, Jr., D. S. Harmer, and K. C. Hoffman, Phys. Rev. Letters 20, 1205 (1968).

[2]J. N. Bahcall, N. A. Bahcall, and G. Shaviv, Phys. Rev. Letters 20, 1209 (1968).

[3]D. Ezer and A. G. W. Cameron, Astrophys. Letters 1, 177 (1968).

[4]S. Torres-Peimbert, R. K. Ulrich, and E. Simpson, to be published.

[5]R. L. Sears, Astrophys. J. 140, 477 (1964).

[6]J. N. Bahcall and G. Shaviv, Astrophys. J. 153, 117 (1968).

[7]Additional papers containing results of solar-model calculations are listed in Refs. 1 and 2.

[8]J. N. Bahcall, N. A. Bahcall, and R. K. Ulrich, to be published.

[9]See, e.g., R. J. Taylor, Observatory 87, 193 (1967); see also other contributions to the Proceedings of the Herstmonceaux Conference on Helium, Observatory 87, 193 ff (1967).

[10]See, e.g., I. Iben, Jr., Astrophys. J. 141, 993 (1965), and 147, 624 (1967).

[11]J. N. Bahcall, Phys. Rev. Letters 12, 300 (1964), and 17, 398 (1966), and Phys. Rev. 135, B137 (1964).

[12]A discussion of cross-section factors may be found in the compilation by W. A. Fowler, G. A. Caughlin, and B. A. Zimmerman, Ann. Rev. Astronomy Astrophys. 5, 525 (1967).

SENSITIVITY OF THE SOLAR-NEUTRINO FLUXES*

JOHN N. BAHCALL,† NETA A. BAHCALL,‡ AND ROGER K. ULRICH
California Institute of Technology, Pasadena
Received October 7, 1968

ABSTRACT

The sensitivity of the solar-neutrino fluxes to localized changes in the opacity and the equation of state is calculated. The sensitivity to changes in the equation of state reaches a maximum near 8×10^6 °K and is in general much larger than the sensitivity to opacity changes. The effect of various parameters on the predicted rate of neutrino capture in the ^{37}Cl experiment is summarized in equation (8). The characteristic numerical properties of a standard solar model are then presented. Several general conclusions are justified on the basis of the variations considered. In particular, an assessment is given of the relation between the theory of solar models and the experimental results of Davis, Harmer, and Hoffman.

I. INTRODUCTION

Davis, Harmer, and Hoffman (1968) have reported the first results of an experiment designed (Davis 1964) to detect neutrinos produced (Bahcall 1964*a*) by nuclear-fusion reactions in the solar interior. The probable upper limit of Davis *et al.* (1968) on the capture rate by ^{37}Cl is about a factor of 2 smaller than the current theoretical estimates (Bahcall, Bahcall, and Shaviv 1968). We attempt to answer in this paper several questions regarding solar models and neutrino fluxes. First, are there particularly sensitive regions of the Sun where small changes in the opacity or the equation of state can make large changes in the calculated neutrino fluxes? Second, how does the predicted capture rate in the ^{37}Cl experiment depend on the important parameters such as nuclear cross-sections, solar age, and heavy-element abundance? Third, to what extent do the experimental results of Davis *et al.* (1968) represent a conflict with the theory of stellar structure?

Even if the current agreement between theory and experiment were good, it would be important to know how sensitive the predicted capture rate is to the various quantities entering the theory. Only after the sensitivity of the fluxes is established can one decide what it is that is being measured (or tested) in solar-neutrino experiments. Moreover, a knowledge of the general level of sensitivity of the predicted capture rate to various quantities can indicate whether it is important to include small improvements in, for example, the equation of state or the opacity.

In § II we outline the calculational procedures, and list the standard parameters, that we have used. We describe in § III the calculated sensitivity of the rate of neutrino capture by ^{37}Cl to localized changes in opacity and pressure. Our results, many of which are summarized in Figures 1 and 2, reveal a much higher sensitivity to changes in the equation of state than to changes in opacity. We have found, unexpectedly, that the counting rate in the ^{37}Cl experiment is especially sensitive to the assumed form of the equation of state at a temperature of about 8×10^6 ° K. We then summarize in equations (8) and (9) of § IV the results of an investigation of the effect of various parameters on the predicted rate of neutrino capture. This part of our work is an extension and refinement of previous studies (cf. § IV for references). We present in § V and in Tables 1

* Supported in part by the National Science Foundation [GP-7976 and GP-9114] and the Office of Naval Research [Nonr-220(47)].

† Alfred P. Sloan Foundation Fellow.

‡ Also at Tel Aviv University.

using the parameters described in § II, that

$$\sum_{\text{all}}(\phi\sigma) = 1.35 \times 10^{-36}\ \text{sec}^{-1}\ \text{per}\ {}^{37}\text{Cl atom} \times \left(\frac{S_{11}}{S_{11}(0)}\right)^{-2.5}\left(\frac{S_{33}}{S_{33}(0)}\right)^{-0.37}$$

$$\times \left(\frac{S_{34}}{S_{34}(0)}\right)^{+0.8}\left[1 + 3.47\left(\frac{S_{17}}{S_{17}(0)}\right)^{+1.0}\left(\frac{\lambda_{e7}(0)}{\lambda_{e7}}\right)^{+1.0}\right] \qquad (8)$$

$$\left(\frac{t_{\text{age}}}{4.7 \times 10^{9}\ \text{yr}}\right)^{+1.4}\left(\frac{Z}{0.015}\right)^{+1.1}.$$

Each exponent in equation (8) is based on the results from at least three different evolutionary sequences of solar models. The quantities $S_{11}(0)$, $S_{33}(0)$, $S_{34}(0)$, and $S_{17}(0)$ are, respectively, the low-energy nuclear cross-section factors for the reactions $p + p \rightarrow {}^2D + e^+ + \nu$, ${}^3He + {}^3He \rightarrow \alpha + 2\,{}^1H$, ${}^3He + {}^4He \rightarrow {}^7Be + \gamma$, and ${}^7Be + {}^1H \rightarrow {}^8B + \gamma$. The actual values of the $S_{ij}(0)$ used in determining the constants in equation (8) are

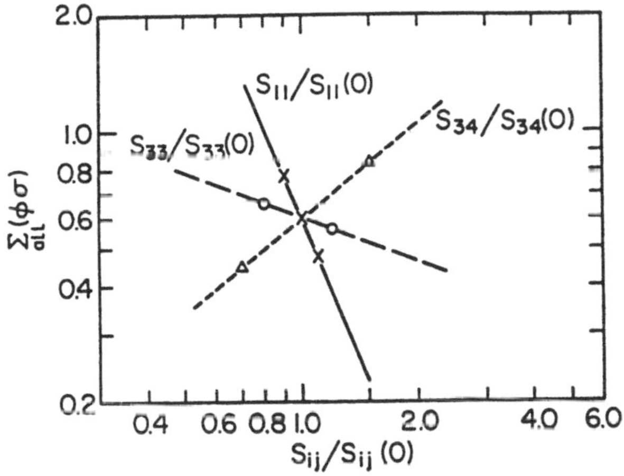

FIG. 3.—Effect of some changes in the nuclear reaction rates on the total capture rate per sec per ^{37}Cl atom $\Sigma_{\text{all}}(\phi\sigma)$ in units of 10^{-35}.

given in the references listed in § II; the quantity $\lambda_{e7}(0)$ is the rate of the reaction ${}^7Be + e^- \rightarrow {}^7Li + \nu$ as computed by Bahcall and Moeller (1969). Also, t_{age} and Z are, respectively, the time the Sun has been burning nuclear fuel and the heavy-element abundance by mass. The dependence on S_{17} and λ_{e7} given in equation (8) is almost exact. The changed counting rate when any of the other variables is altered from its standard value by ~25 per cent or less should be given to an accuracy of the order of 10 per cent by equation (8). A convenient graphical representation is given in Figure 3 of some of the results that led to equation (8).

The capture rate obtained from equation (8) by setting all quantities equal to their most likely values is 6.0×10^{-36} sec^{-1} per ^{37}Cl atom. This rate is smaller than the rate, 7.5×10^{-36} sec^{-1}, estimated by Bahcall, Bahcall, and Shaviv (1968) because we have used Parker's (1968) revised value for S_{17} and the improved estimate of $\lambda_{e7}(0)$ by Bahcall and Moeller (1969).

The quantity $\Sigma_{\text{all but B}}\,(\phi\sigma)$ is not particularly sensitive to parameter changes. We find

MORE SOLAR MODELS AND NEUTRINO FLUXES*†

ZULEMA ABRAHAM AND ICKO IBEN, JR.
Massachusetts Institute of Technology
Received 1971 June 1; revised 1971 June 30

ABSTRACT

We present neutrino fluxes from a sequence of solar models that differ from one another in regard to opacity, equation of state, and nuclear cross-section factors. Using current estimates of the relevant input parameters, we obtain capture rates that range between 3 and 10 times the most recent result of the Davis ^{37}Cl neutrino-capture experiment. The contribution to a theoretical capture rate due to neutrinos from all reactions other than ^8B decay ranges from 0.5 to 1.5 times the latest observational result. Comparison with results of other solar model calculations indicates reasonable agreement when results are normalized to the same input parameters.

I. MOTIVATION AND SUMMARY

During the past three years, numerous suggestions have been made for reducing the ^8B neutrino flux from calculated models of the Sun in an effort to account for the first result of the Davis experiment (Davis, Harmer, and Hoffman 1968)—a capture rate *less than* 3×10^{-36} per $^{37}Cl\ s^{-1}$. No serious suggestion that has thus far been made is capable, by itself, of reducing the ^8B flux from solar models to be consistent with the most recent result of the Davis experiment (Davis, Rogers, and Radeka 1971), a capture rate $\Sigma_{DRR} = (1.5 \pm 1) \times 10^{-36}$ per $^{37}Cl\ s^{-1}$.

We have come to feel that it is not a basic, overall deficiency in the solar models themselves that is responsible for what is now generally recognized as a rather clear discrepancy between theory and observation, and that standard models (age $\sim$4.5 billion years, negligible large-scale mixing, slow interior rotation, negligible interior magnetic fields) are in principle as reliable as they were thought to be prior to the Davis experiment.

In this paper we present a few results of standard model calculations that we have performed over the past two years using (1) several choices for the opacity, (2) several choices for nuclear cross-section parameters, and (3) a modification in the equation of state that partially mimics the effect that large interior magnetic fields might have. Our objective is to provide neutrino flux values for a range of composition parameters for the solar *interior* wide enough to correspond to the actual uncertainties. In common with all recent work of a similar nature, our results contain no surprises. On comparison with other careful solar model work, we find that, after differences in input physics have been reduced to a minimum, no major discrepancies remain between model results.

Our principal results are these. (1) Choosing the same opacity tables, the same equation of state, and nearly the same nuclear cross-sections as used by Bahcall and Ulrich (1970) for their standard model with high iron, we obtain an initial helium abundance $Y = 0.253$ and a total counting rate $\Sigma = 9.9 \times 10^{-36}$ per $^{37}Cl\ s^{-1}$, a value 6.6 times larger than the value $\Sigma^0_{DRR} = 1.5 \times 10^{-36}$ per $^{37}Cl\ s^{-1}$ estimated by Davis *et al.* (1971). This Σ is only 27 percent larger than that found by Bahcall and Ulrich. (2) Choosing the parameter $(Z/X) = 0.019$ and "low" iron, we estimate $\Sigma/\Sigma^0_{DRR} \sim (5 \rightarrow 7)$ for several different choices of opacity. (3) When we alter the crucial cross-

* Based on a thesis presented by one of us (Z. A.) in partial fulfillment of the requirements for a doctor of philosophy in physics at M.I.T., May 1971.

† Supported in part by the National Science Foundation (GP-11277) and in part by the National Aeronautics and Space Administration (NGL 22-009-019).

Standard solar models and the uncertainties in predicted capture rates of solar neutrinos

John N. Bahcall
Institute for Advanced Study, Princeton, New Jersey 08540

Walter F. Huebner
Los Alamos National Laboratory, Los Alamos, New Mexico 87545

Stephen H. Lubow
Hewlett-Packard Corporation, Cupertino, California 95014

Peter D. Parker
Yale University, New Haven, Connecticut 06520

Roger K. Ulrich
University of California at Los Angeles, Los Angeles, California 90024

The uncertainties that affect the prediction of solar neutrino fluxes are evaluated with the aid of standard solar models. The uncertainties are determined from available data for all measured quantities that are known to affect significantly the neutrino fluxes; these include nuclear reaction rates, the solar constant, and the primordial surface composition of the sun. Uncertainties in theoretical quantities (such as the stellar opacity, the equation of state, and the rate of the proton-proton reaction) are estimated from the range of values in published state-of-the-art calculations. The uncertainty in each neutrino flux that is caused by a specified uncertainty in any of the parameters is evaluated with the aid of a series of standard solar models that were constructed for this purpose; the results are expressed in terms of the logarithmic partial derivative of each flux with respect to each parameter. The effects on the neutrino fluxes of changing individual parameters by large amounts can usually be estimated to satisfactory accuracy by making use of the tabulated partial derivatives. An overall "effective 3σ level of uncertainty" is defined using the requirement that the true value should lie within the estimated range unless someone has made a mistake. Effective 3σ levels of uncertainty, as well as best estimates, are determined for the following possible detectors of solar neutrinos: ^{2}H, ^{7}Li, ^{37}Cl, ^{71}Ga, ^{79}Br, ^{81}Br, ^{97}Mo, ^{98}Mo, ^{115}In, and electron-neutrino scattering. The most important sources of uncertainty in the predicted capture rates are identified and discussed for each detector separately. For the ^{37}Cl detector, the predicted capture rate is 7.6$\pm$3.3 (effective 3σ errors) SNU. The measured production rate is (Cleveland, Davis, and Rowley, 1981) 2.1$\pm$0.3 SNU (1σ error). For a ^{71}Ga detector, the expected capture rate is 106$(1^{+0.12}_{-0.08})$ SNU (also effective 3σ errors). The relatively small uncertainty quoted for the ^{71}Ga detector is a direct result of the fact that ^{71}Ga is primarily sensitive to neutrinos from the basic proton-proton reaction, the rate of which is determined largely by the observed solar luminosity. The Caltech and Munster measured values for the cross-section factor for the reaction ^{3}He$(\alpha,\gamma)^{7}$Be are inconsistent with each other. The capture rates quoted above were obtained using the Caltech value for the cross-section factor. If the Munster value is used instead, then the predicted capture rate for the ^{37}Cl experiment is 4.95 ± 2.1 SNU (effective 3σ errors) and, for the ^{71}Ga experiment, 96.7 $(1^{+0.12}_{-0.08})$ SNU (effective 3σ errors). In order for the best-estimate value to agree with the observation of Davis (1978) of 2 SNU for the ^{37}Cl experiment, the cross-section factor $S_{34}(0)$ would have to be reduced by about 15σ to less than the Caltech value, i.e. to 7σ less than the Munster value. The characteristics of the standard solar model, constructed with the best available nuclear parameters, solar opacity, and equation of state, are presented in detail. The computational methods by which this and similar models were obtained are also described briefly. The primordial helium abundance inferred with the aid of standard solar models is $Y = 0.25\pm0.01$. The complementary relation between observations of solar neutrinos and of the normal modes of oscillation of the sun is examined. It is shown that the splitting of the observed large-n, small-l, p-mode (five minute) oscillations of the sun primarily originates in the outer ten percent of the solar mass, while the neutrinos from ^{8}B beta decay originate primarily in the inner five percent of the solar mass. The solar luminosity, and the flux of neutrinos from the proton-proton reaction, come mostly from an intermediate region.

Solar models, neutrino experiments, and helioseismology

John N. Bahcall

Institute for Advanced Study, Princeton, New Jersey 08540

Roger K. Ulrich

University of California at Los Angeles, Los Angeles, California 90024

The event rates and their recognized uncertainties are calculated for eleven solar neutrino experiments using accurate solar models. The same solar models are used to evaluate the frequency spectrum of the p and g oscillation modes of the sun and to compare with existing observations. A numerical table of the characteristics of the standard solar model is presented. Improved values have been calculated for all of the neutrino absorption cross sections evaluating the uncertainties for each neutrino source and detector as well as the best estimates. The neutrino capture rate calculated from the standard solar model for the ^{37}Cl experiment is $7.9(1\pm0.33)$ SNU, which spans the total theoretical range; the rate observed by Davis and his associates is (2.0 ± 0.3) SNU. The ratio of the observed to the predicted flux at Earth of neutrinos from ^{8}B decay lies in the range $0\leq[\varphi(^8\mathrm{B})_{\mathrm{observed}}/\varphi(^8\mathrm{B})_{\mathrm{predicted}}]\leq0.5$. The recent results from the Kamiokande II electron scattering experiment confirm this conclusion. This discrepancy between calculation and observation is the solar neutrino problem. Measurements of the energy spectrum of solar neutrinos can discriminate between suggested solutions of the solar neutrino problem. Nonstandard solar models, many examples of which are also calculated in this paper, preserve the shape of the energy spectrum from individual neutrino sources, whereas most proposed weak-interaction explanations imply altered neutrino energy spectra. Detailed energy spectra of individual neutrino sources are presented as well as a composite solar neutrino spectrum. hep neutrinos from the $^3\mathrm{He}+p$ reaction, probe a different region of the solar interior than do ^{8}B neutrinos. Measurements of the very rare but highest-energy hep neutrinos are possible in proposed experiments using electron scattering, ^{2}H, and ^{40}Ar detectors. The standard solar model predicts p-mode oscillation frequencies that agree to within about 0.5% with the measured frequencies and reproduce well the overall dispersion relation of the modes. However, there are several small but significant discrepancies between the measured and observed frequencies. The complementarity of helioseismology and solar neutrino experiments is demonstrated by constructing a solar model with a drastically altered nuclear energy generation that eliminates entirely the important high-energy ^{8}B and ^{7}Be neutrinos, but which affects by less than 0.01% the calculated p-mode oscillation frequencies.

OUR SUN. I. THE STANDARD MODEL: SUCCESSES AND FAILURES[1]

I.-JULIANA SACKMANN, ARNOLD I. BOOTHROYD,[2] AND WILLIAM A. FOWLER
W. K. Kellogg Radiation Laboratory, California Institute of Technology
Received 1989 December 15; accepted 1990 March 9

ABSTRACT

We have computed a number of standard solar models. Our best Standard Sun has $Y = 0.278$, $Z = 0.0194$, and $\alpha = 2.1$, using $L_\odot \equiv 3.86 \times 10^{33}$ ergs s^{-1}, $R_\odot \equiv 6.96 \times 10^{10}$ cm, $t_\odot = 4.54$ Gyr, and the value of $Z/X = 0.02766$ published by Grevesse in 1984; we used LAOL opacities, including molecular opacities, nuclear rates published by Caughlin and Fowler in 1988, and neutrino capture cross sections published by Bahcall and Ulrich in 1988. We predicted a ^{37}Cl neutrino capture rate of 7.7 SNUs, which would be observed if all solar neutrinos reach Earth (i.e., in the absence of such effects as the MSW neutrino oscillation effect which could reduce the flux of electron neutrinos). This is in agreement with results of other authors, but a factor of 4 larger than the average observed rate. We predict neutrino capture rates for other targets: 26 SNUs for ^{81}Br, 17 SNUs for ^{98}Mo, 125 SNUs for ^{71}Ga, 615 SNUs for ^{115}In, and 47 SNUs for ^{7}Li.

We have investigated the sensitivity of the standard solar model to uncertainties in the solar luminosity, solar age, and observed Z/X ratio as well as to changes in molecular opacities, pressure ionization effects in the envelope, and nonequilibrium ^{3}He energy contributions. Of these, only the uncertainty in Z/X has a significant effect on the solar Y value and the ^{37}Cl neutrino capture rate: use of the older value published by Ross and Aller in 1976 of $Z/X = 0.02282$ decreases Y by 0.014 and decreases the ^{37}Cl rate by 1.5 SNUs. While the 1989 work of Guenther, Jaffe, and Demarque has shown that Y can be significantly affected by the choice of stellar interior opacities, we find that even large changes in the *low-temperature molecular opacities* have *no* effect on Y, nor even on conditions at the *base* of the convective envelope. The large molecular opacities *do* cause a large increase in the mixing-length parameter α but do *not* cause the convective envelope to reach deeper. The temperature remains too low for lithium burning, and there is no surface lithium depletion, let alone the observed depletion of a factor of 100: the lithium problem of the standard solar model remains.

Subject headings: neutrinos — nucleosynthesis — Sun: interior

I. INTRODUCTION

The initial impetus for creating standard solar models came from our interest in creating nonstandard solar models. Guzik, Willson, and Brunish (1987) have produced very interesting nonstandard solar models: they considered the possibility of large amounts of mass loss during the early main-sequence stage. We wished to look at this possibility in more detail (see Boothroyd, Sackmann, and Fowler 1990, hereafter Paper II). A standard model was clearly required for comparison purposes. It is also of considerable interest in its own right. Therefore we made an effort to use the current (observed) solar composition values, as well as up-to-date input physics, nuclear reaction rates, and opacities.

The ratio Z/X of solar metallicity to solar hydrogen abundance is fairly well determined by observations. However, the solar helium abundance is not constrained very tightly. Another important quantity, which is not directly observable at all, is the parameter $\alpha \equiv l/H_p$, the ratio of the convective mixing length to the pressure scale height. This parameter determines the depth of the outer convective envelope and therefore strongly affects the radius of a stellar model. Fortunately, two key boundary conditions allow us to determine these quantities: the luminosity $L_\odot$ and radius $R_\odot$ of the Sun at the present age, which are relatively well determined by observations.

It has only recently been demonstrated conclusively (Guenther, Jaffe, and Demarque 1989) that the value obtained for the presolar helium abundance changes significantly (from $Y \approx 0.24$ to $Y \approx 0.28$) if one uses interior opacities from the Los Alamos Opacity Library (LAOL) rather than the older opacities of Cox and Stewart (1970). (This had been suspected earlier, from a comparison of solar models of different authors: see, e.g., Lebreton and Maeder 1986.) The molecular opacities at temperatures below about 10^4 K are still very uncertain; a number of recent authors have included molecular opacities, from a number of different sources (see discussion in § III). We wanted to investigate whether this uncertainty in the molecular opacity would also have a large effect on the solar models. There is some uncertainty in the solar composition, age, and luminosity, so we tested the effect of varying these also. We also tested the effect of adding pressure ionization effects ("depression of the continuum") to our model, and the effect of nonequilibrium ^{3}He burning on the nuclear energy production.

We constructed evolutionary tracks from the zero-age main sequence (ZAMS) up to the solar age. Each track requires an input value for the initial helium abundance Y and metallicity Z, as well as for the value of the mixing-length parameter α. Holding either Z or Z/X fixed, we varied Y and α until our tracks had the solar luminosity and effective temperature (and thus radius) at the solar age. This determined a unique value for Y, Z, and α for our Standard Sun. We could not resist exploring the Sun's future fate: for our best Standard Sun, we carried the evolution up through the red giant stage and core helium burning stage to the asymptotic giant branch stage.

[1] This work was supported in part by a grant from the National Science Foundation PHY-8817296.

[2] Now at the Canadian Institute for Theoretical Astrophysics, University of Toronto.

Standard solar models, with and without helium diffusion, and the solor neutrino problem

J. N. Bahcall

School of Natural Sciences, Institute for Advanced Study, Princeton, New Jersey 08540

M. H. Pinsonneault

Department of Astronomy, Yale University, New Haven, Connecticut 06511

We first show that, with the same input parameters, the standard solar models of Bahcall and Ulrich; of Sienkiewicz, Bahcall, and Paczyński; of Turck-Chièze, Cahen, Cassé, and Doom; and of the current Yale code all predict event rates for the chlorine experiment that are the same within ±0.1 SNU (solar neutrino units), i.e., approximately 1% of the total calculated rate. We then construct new standard solar models using the Yale stellar evolution computer code supplemented with a more accurate (exportable) nuclear energy generation routine, an improved equation of state, recent determinations of element abundances, and the new Livermore (OPAL) opacity calculations. We evaluate the individual effects of different improvements by calculating a series of precise models, changing only one aspect of the solar model at a time. We next add a new subroutine that calculates the diffusion of helium with respect to hydrogen with the aid of the Bahcall-Loeb formalism. Finally, we compare the neutrino fluxes computed from our best solar models constructed with and without helium diffusion. We find that helium diffusion increases the predicted event rates by about 0.8 SNU, or 11% of the total rate, in the chlorine experiment; by about 3.5 SNU, or 3%, in the gallium experiments; and by about 12% in the Kamiokande and SNO experiments. The best standard solar model including helium diffusion and the most accurate nuclear parameters, element abundances, radiative opacity, and equation of state predicts a value of 8.0±3.0 SNU for the ^{37}Cl experiment and 132^{+21}_{-17} SNU for the ^{71}Ga experiment. The quoted errors represent the total theoretical range and include the effects on the model predictions of 3σ errors in measured input parameters. All 15 calculations since 1968 of the predicted rate in the chlorine experiment given in this series of papers are consistent with both the range estimated in the present work and the 1968 best-estimate value of 7.5±2.3 SNU. Including the effects of helium diffusion and the other improvements in the description of the solar interior that are implemented in this paper, the inferred primordial solar helium abundance is $Y=0.273$. The calculated depth of the convective zone is $R=0.707R_{\odot}$, in agreement with the value of $0.713R_{\odot}$ inferred by Christensen-Dalsgaard, Gough, and Thompson from a recent analysis of the observed p-mode oscillation frequencies. Including helium diffusion increases the calculated present-day hydrogen surface abundance by about 4%, decreases the helium abundance by approximately 11%, and increases the calculated heavy-element abundance by about 4%. In the Appendix, we present detailed numerical tables of our best standard solar models computed both with and without including helium diffusion. In the context of the MSW (Mikheyev-Smirnov-Wolfenstein) or other weak-interaction solutions of the solar neutrino problem, the numerical models can be used to compute the influence of the matter in the sun on the observed neutrino fluxes.

On the depletion of lithium in the Sun

B. Ahrens, M. Stix, and M. Thorn
Kiepenheuer-Institut für Sonnenphysik, Schöneckstr. 6, W-7800 Freiburg, Federal Republic of Germany

Received April 15, accepted June 26, 1992

Abstract. We describe evolutionary sequences of solar models starting in a fully convective phase at the Hayashi line. An overshoot layer of $\simeq 0.30$ pressure scale heights is sufficient for the depletion of lithium by a factor 100. The destruction of Li occurs almost entirely during the pre-main-sequence evolution, within a period of $1.5 \cdot 10^7$ years when the star develops a radiative core and departs from the Hayashi line. For the present Sun, the same overshoot layer leads to a convection zone of $\simeq 200\,000$ km total depth, in good agreement with the helioseismic evidence and with the needs of the solar dynamo.

Key words: solar evolution — lithium abundance — convective overshoot

1. Introduction

Greenstein & Richardson (1951) showed that lithium is about one hundred times less abundant in the solar atmosphere than on Earth or in meteorites. They also discussed the destruction of lithium in the interior of the Sun by nuclear reactions, a possibility mentioned earlier (McMillan 1933) in view of the abundance ratio of the lithium isotopes. For such destruction to take place within a period of order 10^9 years a temperature in excess of $2.5 \cdot 10^6$ K is required, and Greenstein & Richardson also discussed the inferences of their result to the mixing of the Sun's outer layers. Beryllium, in contrast to lithium, is only slightly depleted in the solar atmosphere (Greenstein & Tandberg-Hanssen 1954). As the destruction of Be proceeds above, say, $3.6 \cdot 10^6$ K (depending on the time the material spends at such temperature), the observed Be abundance as well as the fact that there is still some lithium left on the Sun set a limit to the depth of the mixed layers. The whole problem has already been described by Schwarzschild (1958); a recent review, including the stellar evidence, has been presented by Michaud & Charbonneau (1991), and a comprehensive compilation of abundance determinations has been published by Anders & Grevesse (1989).

For a detailed calculation of the destruction of Li in the Sun it appears necessary to include the hydrostatic pre-main-sequence evolution, because convective mixing down to a sufficiently high temperature most likely occurs during the period when the star departs from the Hayashi line in the Hertzsprung-Russell diagram. This was shown by Ezer & Cameron (1963), Weymann & Moore (1963), and Bodenheimer (1965), although these early studies did not achieve the desired large depletion of Li. It also appears necessary to include, in addition to the mixing in the outer convection zone of the standard solar model, a layer of extra mixing, as first suggested by Böhm (1963). Straus et al. (1976) have demonstrated the efficacy of a layer of convective overshoot in depleting the Sun's lithium. However, their calculations were started at the zero-age main sequence, as were those of Lebreton & Maeder (1987) who found that another mechanism, namely turbulent diffusion within the solar core, alone would not suffice to explain the observed Li depletion. On the other hand, both pre-main-sequence evolution and extra mixing were taken into account by D'Antona & Mazzitelli (1984) in their evolutionary sequences of stars between 0.6 and 1.2 solar masses. For the Sun, depletion of Li by a factor of 100 was obtained if the layer of extra mixing had a thickness between 0.7 and 1.4 pressure scale heights, depending on the helium and metal abundances used, and on the mixing-length parameter; the depletion was essentially completed during the pre-main-sequence evolution. The results of Pinsonneault et al. (1989) are similar, although they attribute the extra mixing to the instabilities arising from the redistribution of angular momentum, and although their lithium depletion occurs partly during the pre-main-sequence phase and partly at the main sequence itself.

The results reported in the present contribution confirm the necessity of both pre-main-sequence evolution and extra mixing. But we wish to point out that according to our calculations the thickness of the layer of extra mixing should be only about 0.3 pressure scale heights. We interpret this layer as a layer of convective overshoot. In view of the large efficiency of the convective heat transport, we shall assume that the temperature gradient in the overshoot layer is adiabatic, with a sharp transition to the radiative gradient at its base.

In the two subsequent sections we shall describe the ingredients of our solar model in general and of our calculations of the evolution of Li and Be in particular. In the final section we discuss the result which, as we believe, is in good agreement with a recent helioseismological estimate of the total depth of the solar convection zone, and also with the requirements of the storage of sufficient magnetic flux for the solar dynamo process.

2. The solar model

We have calculated our model of the Sun with the stellar evolution program originally described by Kippenhahn et al. (1967), and in an updated version by Stix (1989). Further updatings will be mentioned subsequently. The model is a standard solar model, spherically symmetric, with magnetic and rotational effects ne-

Standard solar models with CESAM code: neutrinos and helioseismology

G. Berthomieu[1], J. Provost[1], P. Morel[1], and Y. Lebreton[2]

[1] Département Cassini, URA CNRS 1362, Observatoire de la Côte d'Azur, BP 229, F-06304 Nice Cédex 4, France
[2] DASGAL, URA CNRS 335, Observatoire de Paris, F-92195 Meudon Principal Cédex, France

Received June 3, accepted October 15, 1992

Abstract. A new code for stellar evolution, named CESAM, has been constructed at Nice by P. Morel. Standard solar models have been computed using this code. Their global characteristics, predicted capture rates of neutrinos for the chlorine and gallium experiments and their seismological properties are given and compared to the observational constraints. The emphasis is put on the effect of the recent opacities of Livermore for different mixtures, corresponding to recent abundances determination, on the neutrino predictions and the solar oscillations.

Key words: Sun: interior – Sun: oscillations – stars: evolution

1. Introduction

Present solar models have to be calculated with high accuracy in order to be discussed in the framework of the very strong observational constraints of helioseismology. The new stellar evolution code CESAM[1] developed at Nice Observatory by P. Morel was adjusted and updated for that purpose (Morel 1992a,b). The comparison with other codes in the frame of the Solar Model Comparison initiated by GONG (Global Oscillations Network Group) has shown its ability to integrate the set of differential equations describing the structure and the evolution of a one solar mass star with high accuracy (Christensen - Dalsgaard 1991; Gabriel 1991).

Moreover great efforts are being recently made by the community to improve the physics that describes the stellar interiors. New opacities have been calculated by the Livermore group (Iglesias & Rogers 1991; Rogers & Iglesias 1992; Iglesias et al. 1992) for different mixtures and the results of the Opacity Project group will be soon available (Seaton 1992). We have calculated several solar models using the CESAM code and different opacity tables, equation of state, abundances... One of these models, calculated with the more recent set of physical data is presented as our reference standard model. Its properties, global parameters, neutrino fluxes and helioseismological characteristics, are given and compared to previous standard models (Bahcall & Ulrich 1988, Lebreton & Däppen 1988, Turck-Chièze et al. 1988, Sackmann et al. 1990, Dziembowski et al. 1992, Guenther et al. 1992, Schaller et al. 1992) in Sect. 2 and 3. In a second step we analyse the sensitivity of the solar model to the different physical inputs and examine how the resulting models satisfy the observational constraints of neutrinos and helioseismology (Sect. 4).

[1] Code d'Evolution Stellaire Adaptatif et Modulaire

2. Reference solar model computation

The solar models discussed in this paper have been computed with the code CESAM. A brief description of the numerical procedures which are involved (Morel et al. 1990; Morel 1992a,b) is given in Sect. 2.2. We assume a spherically symmetrical model which is in hydrostatic equilibrium with no macroscopic motion except convection, no mass loss, no rotation and no magnetic field.

2.1. Input physics

2.1.1. Thermonuclear reactions

The evolution of the chemical abundances is due to the nuclear reactions (we do not take into account microscopic diffusion). The nuclear reaction networks include the proton-proton chain and the CNO cycles which are implemented following Clayton (1968). The models where the numerical integration of the evolution in time of all the elements is performed do not differ from the models where ^{7}Li and ^{7}Be are taken at equilibrium (less than 10^{-4} for global quantities and less than 1% for the neutrino fluxes) so that in most cases these simplifying assumptions have been made.

The thermonuclear reaction rates are taken from the tables of Caughlan & Fowler (1988). The relation between the S factors for the p-p, ^{7}Be(p,γ)^{8}B and ^{3}He(^{3}He,2p)^{4}He reactions corresponding to these tables and those adopted by Bahcall & Ulrich

TOWARD A UNIFIED CLASSICAL MODEL OF THE SUN: ON THE SENSITIVITY OF NEUTRINOS AND HELIOSEISMOLOGY TO THE MICROSCOPIC PHYSICS

SYLVAINE TURCK-CHIÈZE AND ILIDIO LOPES[1]
Service d'Astrophysique, DAPNIA, Centre d'Etudes de Saclay, 91191 Gif-sur-Yvette, France
Received 1992 August 3; accepted 1992 October 29

ABSTRACT

This paper focuses mainly on the neutrino puzzle and discusses the point of view that neutrinos and helioseismology are two complementary probes of the solar interior. We first analyze the physical differences noticed between already published solar models and their consequences for neutrino predictions. Including improvements achieved in microscopic physics these last 3 years, we propose new results on the solar neutrino predictions and acoustic mode frequencies for $l = 0$–150, in the classical framework of stellar evolution. Doing so, we quantify the influence of precise composition, nuclear reaction rates, screening effect, and opacity calculations on both neutrino and acoustic mode frequency predictions. Our present predictions are 6.4 ± 1.4 SNU for the chlorine experiment, $4.4 \pm 1.1 \times 10^6$ cm^{-2} s^{-1} for the water detector, and 122.5 ± 7 SNU for the gallium detector. Considering that the present experimental situation may support the hypothesis that neutrinos and helioseismology are both representative of the Sun, we then try to derive from the sound speed behavior an estimate of how the neutrino predictions may progress, including phenomena not yet simulated at present; we suggest that a correct description of the region located between the nuclear region and the convective zone justifies an increase of the ^{8}B neutrino prediction by no more than 10%–15%, and that the nuclear reaction rates of the *p-p* chain should be revisited.

Subject headings: elementary particles — nuclear reactions, nucleosynthesis, abundances — Sun: interior

1. INTRODUCTION

The Sun is a unique star because of the quality of its observations (luminosity, mass, radius, and composition). Moreover, its evolutionary stage corresponds to the first stage of hydrogen burning, which is the best known and the longest one. So the astrophysical community is extremely motivated to study this stage in detail in order to put some constraints on the helium initial content and on the age of low-mass stars. In the early 1980s, people working on stellar evolution produced a reference solar model along the framework of the classical stellar evolution (Bahcall et al. 1982; Christensen-Dalsgaard 1982; VandenBerg 1983; Noels, Scuflaire, & Gabriel 1984; Bahcall et al. 1985; Lebreton & Maeder 1986). This activity was encouraged by the detection and interpretation of a large number of acoustic modes. On the neutrino side, the surprise was twofold: the predictions were in disagreement with the first neutrino experiment, the chlorine experiment developed by Davis and collaborators (Davis 1964; Rowley, Cleveland, & Davis 1984), and the predictions differed significantly among various authors, even with rather similar hypotheses. With the development of helioseismology, more stratified information has been offered, and the sensitivity of such models has pushed large and complementary communities to work together.

At the end of the 1980s, more and more astrophysical groups became conscious of the importance of a precise determination of a reference solar model. In fact, the astrophysical community would like to establish some constraints deduced from specific solar observations or solar neutrino detections, on the phenomena which are present in stars but which are still too difficult to include in a stellar calculation: rotation, convection (correctly treated), and mixing (Schatzman et al. 1981; Pinsonneault et al. 1989). A detailed comparison between models of different groups has convinced us (Turck-Chièze et al. 1988a) that the differences were not due to numerical inaccuracy or errors in the computation but came, in fact, from the adopted microscopic description of the Sun. Moreover, the multiplication of the results—Kamiokande neutrinos (Hirata et al. 1989, 1990, 1991), gallium experiments, the detection of about 2700 acoustic modes (Libbrecht, Woodard, & Kaufman 1990)—has stimulated atomic and nuclear communities to improve both the experimental and the theoretical aspects of the solar problem.

The purpose of this paper is to return once more to this first stage of the calculation, that is, in the classical framework of stellar evolution, and extensively study the effect of microscopic physics on the predicted observables: neutrinos and accoustic modes.

We first focus on the interpretation of the differences between the already published solar models (§§ 2 and 3) which reveal the importance of some ingredients and allow us to extend previous similar analysis (Bahcall & Pinsonneault 1992). In order to perform a comparative analysis of neutrinos and accoustic modes, we then introduce the recent improvements noticed in the last 3 years, completing some other recent works (Guenther et al. 1992; Bahcall & Pinsonneault 1992; Berthomieu et al. 1992) by a discussion on screening and photospheric abundances (§ 4). Section 5 presents the effect of these improvements on the structure of the Sun and on the neutrino flux predictions and compares with the recent experimental results. Sections 6 and 7 will be devoted to the helioseismology signature of such improvements. Section 8 is speculative and suggests possible astrophysical consequences of what we have learned from the general experimental situation.

[1] Also Centro de Astrofisica, Universidade do Porto, Portugal.

Solar neutrinos and nuclear reactions in the solar interior

V. Castellani[1,2], S. Degl'Innocenti[3], and G. Fiorentini[3]

[1] Dipartimento di Fisica dell'Università di Pisa, Piazza Torricelli 2, I-56100 Pisa, Italy
[2] Osservatorio Astronomico di Collurania, I-64100 Teramo, Italy
[3] Dipartimento di Fisica dell'Università di Ferrara and Istituto Nazionale di Fisica Nucleare, Sezione di Ferrara, I-44100 Ferrara, Italy

Received May 13, accepted October 20, 1992

Abstract. We discuss the results of Homestake and Kamioka experiments, showing that – if the results of these experiments are taken at their face values – one way to save "conventional neutrinos" is to look for a nuclear solution decreasing both ^{7}Be and ^{8}B neutrino fluxes with respect to the predictions of the standard solar models. Recent GALLEX results appear in agreement with such a conclusion. We discuss the sensitivity of the ^{8}B and ^{7}Be neutrino fluxes to the behaviour of the low energy ^{3}He + ^{3}He and ^{3}He + ^{4}He cross sections. We derive analytically the dependence of the neutrino fluxes on the low energy nuclear cross sections. This analytical approach has been supported by numerical experiments based on a new Standard Solar Model. In the non-resonant case, reduction of the neutrino fluxes to about 1/3 of the Standard Solar Model could be obtained if the true value of $S_{34}(0)$ is three times smaller than the presently accepted extrapolated value. Alternatively, one should have $S_{33}(0)$ wrong by a factor nine. A resonance in the ^{3}He + ^{3}He channel could yield a sufficient reduction of ^{8}B neutrinos and, furthermore, a suppression of ^{7}Be neutrinos larger than that of ^{8}B neutrinos provided that $E_R \leq 21.4$ keV, an energy region so far almost unexplored experimentally. We show that future experiments in underground laboratories should be able to explore the region down to $E_R = 10$ keV with a significant sensitivity. We also compare our Standard Solar Model with the results of previous calculations.

Key words: Sun: interior – Sun: particle emission – nuclear reactions – elementary particles

1. Foreword

The status of the solar neutrino problem, with the Homestake and Kamioka experiments – (see Davis 1990; Hirata et al. 1990, 1991) – reporting a neutrino signal significantly smaller than the theoretical estimate (Bahcall 1989), seems to point towards unconventional neutrino properties, as flavour oscillations (Pontecorvo 1968; Gribov & Pontecorvo 1969; Wolfenstein 1978, 1979; Mikheyev & Smirnov 1986a–c), magnetic moment transitions (Voloshin et al. 1986a, b), neutrino decay (Bahcall et al. 1972; Berezhiani et al. 1987). However, preliminary results of the Soviet American Gallium experiment (Abazov et al. 1991) supporting this conclusion have been recently challenged by the result of GALLEX collaboration (1992), which gave no clear evidence against conventional neutrinos. In such a situation, since the solar neutrino problem is now extremely important, we believe that one has to be extremely critical and not to dismiss any possibility. Thus it appears worth to investigate the room left to possible errors in the theoretical estimate of solar neutrino fluxes.

In this respect, it is worth recalling that the values of the nuclear cross sections adopted in the calculations do not correspond to directly measured quantities. So far, they are always obtained by extrapolating experimental data which are measured at significantly higher energies than those relevant to the solar interior (see Rolfs & Rodney 1988 for a review on the argument). More in general, one has to observe that the theoretical values of the neutrino fluxes are based on stellar evolutionary codes which are rather complex from both the physical and numerical point of view. It is thus extremely desirable to investigate the model dependence of solar neutrino fluxes, looking for information which is independent, or very weakly dependent, on the details of the solar model calculations.

In this spirit, the main purpose of this paper is to examine which space is still left for a "nuclear physics solution" of the solar neutrino problem. We will try to get information directly from both Homestake and Kamioka data, without making assumptions on nuclear cross sections and by using as little as possible a Standard Solar Model (SSM). Next we shall discuss how the present situation can be clarified by experiments which are now in progress or in preparation. We refer of course to the Gallium detectors, which are now producing their first

Send offprint requests to: V. Castellani (first address above)

Do solar-neutrino experiments imply new physics?

John N. Bahcall
Institute for Advanced Study, Princeton, New Jersey 08540

H. A. Bethe
Newman Laboratory of Nuclear Studies, Cornell University, Ithaca, New York 14853

(Received 24 August 1992)

None of the 1000 solar models in a full Monte Carlo simulation is consistent with the results of the chlorine or the Kamiokande experiments. Even if the solar models are forced artificially to have a ^{8}B neutrino flux in agreement with the Kamiokande experiment, none of the fudged models agrees with the chlorine observations. The GALLEX and SAGE experiments, which currently have large statistical uncertainties, differ from the predictions of the standard solar model by 2σ and 3σ, respectively.

PACS number(s): 96.60.Kx, 12.15.Ff, 14.60.Gh

Four solar-neutrino experiments [1–4] yield results different from the combined predictions of the standard solar model and the standard electroweak model with zero neutrino masses. The question physicists most often ask each other about these results is: Do these experiments require new physics beyond the standard electroweak model? We provide a quantitative answer using the results of a detailed Monte Carlo study of the predictions of the standard solar model.

The basis for our investigation is a collection of 1000 precise solar models [5] in which each input parameter (the principal nuclear reaction rates, the solar composition, the solar age, and the radiative opacity) for each model was drawn randomly from a normal distribution with the mean and standard deviation appropriate to that variable. In the calculations described in this paper, the uncertainties in the neutrino cross sections [5] for chlorine and for gallium were included by assuming a normal distribution for each of the absorption cross sections with its estimated mean and error. Since for gallium the estimated uncertainties in the neutrino absorption cross sections are not symmetric with respect to the best-estimate absorption cross sections, two different normal distributions were used to simulate the detection uncertainties for each neutrino flux to which gallium is sensitive. This Monte Carlo study automatically takes account of the nonlinear relations among the different neutrino fluxes that are imposed by the coupled partial differential equations of stellar structure and by the boundary conditions of matching the observed solar luminosity, heavy element to hydrogen ratio, and effective temperature at the present solar age.

This is the first Monte Carlo study that uses large numbers of standard solar models that satisfy the equations of stellar evolution and that is designed to determine if physics beyond the standard electroweak theory is required.

Related investigations have been carried out assuming [6–8] that each solar model could be represented by a single parameter, its central temperature. The flux of neutrinos from each nuclear source is represented in these studies by a power law in the central temperature. This simplification, while providing a semiquantitative understanding of some of the important relationships, leads to serious errors in some cases. For example, the parametrization in terms of central temperature predicts a ^{8}B flux for the maximum rate model that is too low by more than a factor of 4 [9]. Just as detailed Monte Carlo calculations are necessary in order to understand the relative and absolute sensitivities of complicated laboratory experiments, a full Monte Carlo calculation is required to determine the interrelations and absolute values of the different solar neutrino fluxes. The Sun is as complicated as a laboratory accelerator or a laboratory detector, for which we know by painful experience that detailed simulations are necessary. For example, the fact that the ^{8}B flux may be crudely described as $\phi(^8\mathrm{B}) \propto T_{\mathrm{central}}^{18}$ and $\phi(^7\mathrm{Be}) \propto T_{\mathrm{central}}^{8}$ does not specify whether the two fluxes increase and decrease together or whether their changes are out of phase with each other. The actual variations of the calculated neutrino fluxes are determined by the coupled partial differential equations of stellar evolution and the boundary conditions, especially the constraint that the model luminosity at the present epoch be equal to the observed solar luminosity. Fortunately, the simplified method and the full Monte Carlo calculation yield similar results for the probability that new physics is required, although the full Monte Carlo calculation yields a more accurate numerical statement.

Figures 1(a)–1(c) show the number of solar models with different predicted event rates for the chlorine solar neutrino experiment, the Kamiokande (neutrino-electron scattering) experiment, and the two gallium experiments (GALLEX and SAGE). For the chlorine experiment, which is sensitive to neutrinos above 0.8 MeV, the solar model with the best input parameters predicts [5] an event rate of about 8 solar neutrino units (SNU). None of the 1000 calculated solar models yields a capture rate below 5.8 SNU, while the observed rate is [1]

$$\langle \phi\sigma \rangle_{\mathrm{Cl\,exp}} = (2.2 \pm 0.2)\ \mathrm{SNU}, \quad 1\sigma\ \mathrm{error}\,. \qquad (1)$$

The discrepancy that is apparent in Fig. 1(a) was for two

decades the entire "solar neutrino problem." Figure 1(a) implies that something is wrong with either the standard solar model or the standard electroweak description of the neutrino.

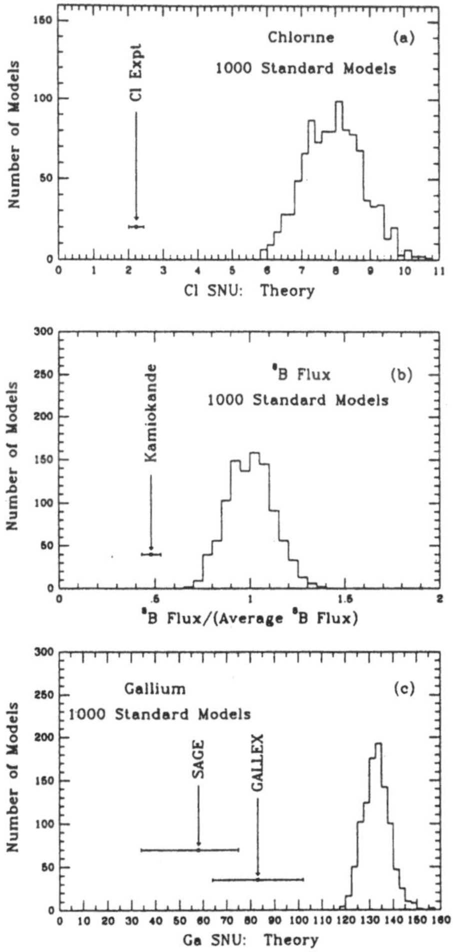

FIG. 1. 1000 solar models vs experiments. The number of precisely calculated solar models that predict different solar neutrino event rates are shown for the chlorine (a), Kamiokande (b), and gallium (c) experiments. The solar models from which the fluxes were derived satisfy the equations of stellar evolution including the boundary conditions that the model luminosity, chemical composition, and effective temperature at the current solar age be equal to the observed values [5]. Each input parameter in each solar model was drawn independently from a normal distribution having the mean and the standard deviation appropriate to that parameter. The experimental error bars include only statistical errors (1σ).

Figure 1(b) shows the number of solar models with different ^{8}B neutrino fluxes. For convenience, we have divided each ^{8}B flux by the average ^{8}B flux so that the distribution is peaked near 1.0. The rate measured for neutrinos with energies above 7.5 MeV by the Kamiokande II and III experiments is [2]

$$\langle\phi(^8\text{B})\rangle=[0.48\pm0.05(1\sigma)\pm0.06(\text{syst})]\langle\phi(^8\text{B})\rangle_{\text{average}} \quad (2)$$

for recoil electrons with energies greater than 7.5 MeV. Here $\langle\phi(^8\text{B})\rangle_{\text{average}}$ is the best-estimate theoretical prediction. None of the 1000 standard solar models lie below $0.65\langle\phi(^8\text{B})\rangle_{\text{average}}$. If one takes account of the Kamiokande measurement uncertainty (±0.08) in the Monte Carlo simulation, one still finds that none of the solar models is consistent with the observed event rate. These results provide independent support for the existence of a solar neutrino problem.

Figure 1(c) shows the number of solar models with different predicted event rates for gallium detectors and the recent measurements by the SAGE [3,10] [$58^{+17}_{-24}\pm14$(syst) SNU] and GALLEX [$83\pm19(1\sigma)$ ±8(syst) SNU] Collaborations [4]. With the current large statistical errors, the results differ from the best-estimate theoretical value [5] of 132 SNU by approximately 2σ (GALLEX) and 3.5σ (SAGE). The gallium results provide modest support for the existence of a solar neutrino problem, but by themselves do not constitute a strong conflict with standard theory.

Can the discrepancies between observation and calculation that are summarized in Fig. 1 be resolved by changing some aspect of the solar model? We have argued previously [11] that this is difficult to do because the energy spectrum of any specific neutrino source is unchanged by the solar environment [12] and because the uncertainties in all of the important sources except ^{8}B are relatively small. A comparison of Figs. 1(a) and 1(b) shows that the discrepancy with theory appears to be energy dependent. The larger discrepancy occurs for the chlorine experiment, which is sensitive to lower neutrino energies than is the Kamiokande experiment. If one normalizes the ^{8}B flux by the best-estimate measurement from Kamiokande [see Eq. (2)], then the implied rate in the chlorine experiment from ^{8}B alone is 3.0 SNU. We also argued that the other neutrino fluxes would most likely yield at least another 1 SNU, implying a minimum total rate of about 4 SNU. On this basis, we concluded that the two experiments, chlorine and Kamiokande, are inconsistent with the combined standard electroweak and solar models. However, our argument did not take into account in a well-defined way the errors in the predictions and in the measurements. We remedy this shortcoming in the following discussion using the previously described Monte Carlo simulation.

Figure 2 provides a quantitative expression of the difficulty in reconciling the Kamiokande and chlorine experiments by changing solar physics. We constructed Fig. 2 using the same 1000 solar models as were used in constructing Fig. 1, but for Fig. 2 we artificially replaced the ^{8}B flux for each standard model by a value drawn ran-

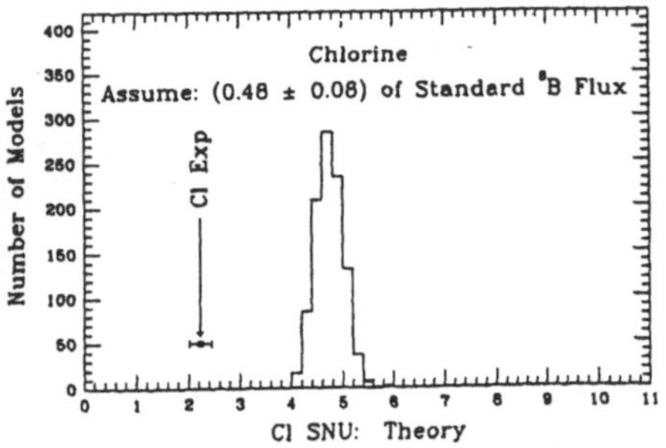

FIG. 2. 1000 artificially modified fluxes. The ^{8}B neutrino fluxes computed for the 1000 accurate solar models were replaced in the figure shown by values drawn randomly for each model from a normal distribution with the mean and the standard deviation measured by the Kamiokande experiment [2].

domly for that model from a normal distribution with the mean and the standard deviation measured by Kamiokande [see Eq. (2)]. This *ad hoc* replacement is motivated by the fact that the ^{7}Be$(p,\gamma)^8$B cross section is the least accurately measured of all the relevant nuclear fusion cross sections and by the remark that the ^{8}B neutrino flux is more sensitive to solar interior conditions than any of the other neutrino fluxes. The peak of the resulting distribution is moved to 4.7 SNU (from 8 SNU) and the full width of the peak is decreased by about a factor of 3. The peak is displaced because the measured (i.e., Kamiokande) value of the ^{8}B flux is smaller than the calculated value. The width of the distribution is decreased because the error in the Kamiokande measurement is less than the estimated theoretical uncertainty ($\approx 12.5\%$) and because ^{8}B neutrinos constitute a smaller fraction of each displaced rate than of the corresponding standard rate.

Figure 2 was constructed by assuming that something is seriously wrong with the standard solar model, something that is sufficient to cause the ^{8}B flux to be reduced to the value measured in the Kamiokande experiment. Nevertheless, there is no overlap between the distribution of fudged standard model rates and the measured chlorine rate. None of the 1000 fudged models lie within 3σ (chlorine measurement errors) of the experimental result.

The results presented in Figs. 1 and 2 suggest that new physics is required beyond the standard electroweak theory if the existing solar neutrino experiments are correct within their quoted uncertainties. Even if one abuses the solar models by artificially imposing consistency with the Kamiokande experiment, the resulting predictions of all 1000 of the "fudged" solar models are inconsistent with the result of the chlorine experiment (see Fig. 2).

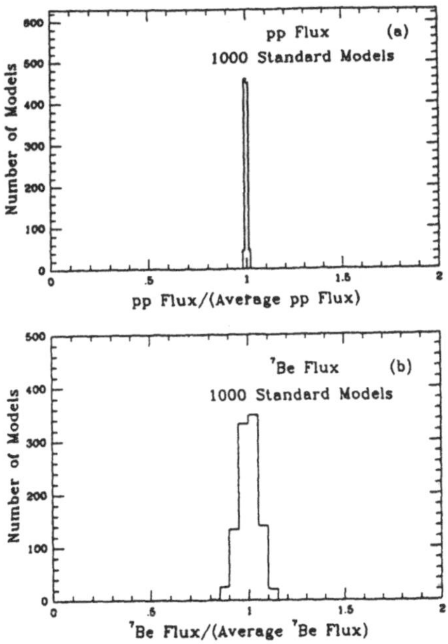

FIG. 3. The *pp* and ^{7}Be neutrino fluxes. The histogram of the number of 1000 precisely calculated solar models that predict different *pp* neutrino fluxes is shown in (a) and the number that predict different ^{7}Be neutrino fluxes is shown in (b). The individual neutrino fluxes are divided by their respective average values.

Figure 3 shows the relatively high precision with which the *pp* and ^{7}Be neutrino fluxes can be calculated. Since the 1σ theoretical uncertainty in the flux of ^{7}Be neutrinos is only $\simeq 5\%$ [5,9], an accurate measurement of this quantity in the proposed Borexino experiment [13] will constitute another important test of the standard model.

All of the arguments in this paper depend to some extent on our understanding of the solar interior. In the future, it will be possible to use solar neutrinos to test electroweak theory independent of solar models by measuring the energy spectrum of the ^{8}B neutrinos with the Super-Kamiokande [14], the Sudbury Neutrino Observatory (SNO) [15], and the Imaging of Cosmic and Rare Underground Signals (ICARUS) experiments [16] and by measuring the ratio of charged to neutral currents with the SNO experiment [15].

This work was supported in part by the NSF via Grant No. PHY-92 45317 at IAS and PHY-87-15272 at Cornell.

[1] R. Davis, Jr., in *Seventh Workshop on Grand Unification*, Proceedings, Toyama, Japan, 1986, edited by J. Arafune (World Scientific, Singapore, 1987), p. 237; R. Davis, Jr., K. Lande, C. K. Lee, P. Wildenhain, A. Weinberger, T. Daily, B. Cleveland, and J. Ullman, in *Proceedings of the 21st International Cosmic Ray Conference*, Adelaide, Australia, 1990, edited by R. J. Protheroe (Graphic Services, Northfield, South Australia, 1990); J. K. Rowley, B. T. Cleveland, and R. Davis, Jr., in *Solar Neutrinos and Neutrino Astronomy*, Proceedings of the Conference, Lead, South Dakota, 1984, edited by M. L. Cherry, W. A. Fowler, and K. Lande, AIP Conf. Proc. No. 126 (AIP, New York, 1985), p. 1.

[2] K. S. Hirata *et al.*, Phys. Rev. Lett. **63**, 16 (1989); **65**, 1297 (1990).

[3] A. I. Abazov *et al.*, Phys. Rev. Lett. **67**, 332 (1991).

[4] P. Anselmann *et al.*, Phys. Lett. B **285**, 376 (1992).

[5] J. N. Bahcall and R. K. Ulrich, Rev. Mod. Phys. **60**, 297 (1988); J. N. Bahcall, *Neutrino Astrophysics* (Cambridge University Press, Cambridge, England, 1989).

[6] S. A. Bludman, D. C. Kennedy, and P. G. Langacker, Nucl. Phys. **B374**, 373 (1992).

[7] P. Anselmann *et al.*, Phys. Lett. B **285**, 390 (1992).

[8] S. A. Bludman, N. Hata, D. C. Kennedy, and P. G. Langacker, Phys. Rev. D (to be published).

[9] J. N. Bahcall and M. H. Pinsonneault, Rev. Mod. Phys. **64**, 885 (1992).

[10] V. N. Gavrin *et al.*, in Proceedings of the XXVI International Conference on High Energy Physics, Dallas, Texas, 1992 (unpublished).

[11] J. N. Bahcall and H. A. Bethe, Phys. Rev. Lett. **65**, 2233 (1990).

[12] J. N. Bahcall, Phys. Rev. D **44**, 1644 (1991).

[13] R. S. Raghavan, in *Proceedings of the XXVth International Conference on High Energy Physics*, Singapore, 1990, edited by K. K. Phua and Y. Yamaguchi (World Scientific, Singapore, 1990), Vol. 1, p. 482; C. Arpasella *et al.*, in "Borexino at Gran Sasso: Proposal for a real-time detector for low energy solar neutrinos," Vols. I and II, University of Milan, INFN report (unpublished).

[14] Y. Totsuka, in *Proceedings of the International Symposium on Underground Physics Experiments*, edited by K. Nakamura (ICRR, University of Tokyo, 1990), p. 129.

[15] G. Aardsma *et al.*, Phys. Lett. B **194**, 321 (1987).

[16] J. N. Bahcall, M. Baldo-Ceolin, D. Cline, and C. Rubbia, Phys. Lett. B **178**, 324 (1986).

EFFECTS OF HEAVY-ELEMENT SETTLING ON SOLAR NEUTRINO FLUXES AND INTERIOR STRUCTURE

CHARLES R. PROFFITT
Science Programs, Computer Sciences Corporation and *IUE* Observatory, Code 684.9, Goddard Space Flight Center, Greenbelt, MD 20771
Received 1993 August 9; accepted 1993 October 22

ABSTRACT

We consider the effects of gravitational settling of both He and heavier elements on the predicted solar neutrino fluxes and interior sound speed and density profiles. We find that while the structural changes that result from the inclusion of both He and heavy-element settling are only slightly larger than the changes resulting from the inclusion of He settling alone, the additional increases in expected neutrino fluxes are of comparable size. Our preferred model with both He and heavy-element settling has neutrino count rates of 9.0 SNU for ^{37}Cl detectors and 137 SNU for ^{71}Ga detectors, as compared to 7.1 and 127 SNU for a comparable model without any diffusive separation, or 8.0 and 132 SNU for a model that includes He settling alone. We suggest that the correction factors by which the predicted neutrino fluxes of solar models calculated without including the effects of diffusion should be multiplied are 1.25 ± 0.08 for Cl detectors, 1.07 ± 0.02 for Ga detectors, and 1.28 ± 0.09 for the ^{8}B flux (1 σ errors).

Comparison of internal sound speed and density profiles strongly suggests that the additional changes in calculated *p*-mode oscillation frequencies due to the inclusion of heavy-element settling will be small compared to the changes that result from He settling alone, especially for the higher degree modes. All models with diffusive separation give much better agreement with the observed depth of the convection zone than do non-diffusive models. The model that includes both He and heavy-element settling requires an initial He mass fraction $Y = 0.280$ and has a surface He abundance of $Y = 0.251$ at the solar age.

Subject headings: diffusion — elementary particles — Sun: abundances — Sun: Interior — Sun: oscillations

1. INTRODUCTION

In radiative regions of stellar interiors atomic diffusion acts to cause chemical separation of the elements. Noerdlinger (1977) first included the effects of He diffusion on the evolution of the Sun, while Cox, Guzik, & Kidman (1989) were the first to include the settling of heavy elements (see also Guzik & Cox 1993). Proffitt & Michaud (1991b) reviewed the different diffusion calculations done up to that time and explained most of the apparent discrepancies between them.

However, all of the above calculations had more or less subtle calibration problems that rendered them less than perfect for isolating the effects of diffusion alone, and the different calculations used varying opacity calculations, equations of state, and energy generation rates, making detailed comparisons difficult. Bahcall & Pinsonneault (1992; hereafter BP) argued that calibrated solar models should be carefully matched to all have the same age, luminosity, radius, and surface Z/X to a high degree of precision. The last criteria was the one that has been most ignored. Most workers simply choose a fixed initial value of Z (usually 0.02). The choice of Z is, however, especially critical for the precise calculation of neutrino fluxes. BP calculated carefully calibrated models that included He diffusion according to the prescription of Bahcall & Loeb (1990). (While Bahcall & Loeb proposed a prescription for the calculation of heavy-element settling, BP did not include this or any other approximation to heavy-element settling in their calculations.) They found that including He diffusion in solar models increases the predicted ^{37}Cl detector neutrino count rate by 12% and the ^{71}Ga detector rate by 3%.

In this paper we will calculate carefully calibrated models which include both He and heavy-element diffusion and discuss the resulting changes in the calculated neutrino fluxes, as well as changes in structural parameters (especially sound speed) relevant for calculation of oscillation frequencies. This paper will concentrate on the differential effects of including heavy-element diffusion. Actual calculation of oscillation frequencies will be deferred to a later paper.

1.1. *Constraints from Helioseismology*

Until recently, only a few parameters (age, mass, luminosity, radius, and surface heavy-element abundances) were measured for the Sun. Helioseismic observations have recently provided a great deal of information about the solar interior and place new constraints on the models. Solar *p*-modes are closely related to simple acoustic waves and as such their frequencies depend strongly on the sound speed as a function of radius in the interior. In the limit that certain asymptotic approximations are made, the frequencies of high *l*-modes depends only on the sound speed (cf. Deubner & Gough 1984). This allows the observed *p*-mode frequencies to be inverted to estimate the solar interior sound speed (e.g., Christensen-Dalsgaard et al. 1985). The density profile can also be determined using inversion techniques, but this requires numerical inversions and cannot take advantage of the asymptotic properties of the modes (cf. Däppen et al. 1991). It is also sensitive to a smaller number of modes, and the results may be less robust than is the case for sound speed inversions.

Once the sound speed profile in the solar interior is known, a variety of other information can be inferred. The transition between the nearly adiabatic gradient within the solar convection zone, and the subadiabatic gradient below it, leads to a distinct feature in the sound speed profile which Christensen-Dalsgaard, Gough, & Thompson (1991) used to determine that the base of the adiabatically stratified region in the Sun is at a radius $R_{con} = 0.713 \pm 0.003\ R_\odot$. The variation of the adiabatic exponent in the second He ionization zone also leads to a feature in the sound speed that can in principle be used to determine the He abundance of the solar convection zone. Kosovichev et al. (1992) have recently reviewed the efforts made in this direction and concluded that a He mass fraction of 0.232 ± 0.006 best agrees with the observed *p*-mode frequencies for their preferred equation of state. However, they also note that it is not possible at this time to put firm limits on how large uncertainties in the equation of state for the He ionization zone are (cf. Däppen 1993), and this might lead to an unknown systematic error in the derived He abundance.

Christensen-Dalsgaard, Profitt, & Thompson (1993) (CPT) and Guzik & Cox (1993) have made detailed studies of the effects of diffusion on *p*-mode oscillation frequencies. CPT considered only H-He diffusion, while Guzik & Cox included an approximate calculation of heavy-element settling. Both papers considered only models with an initial heavy-element abundance $Z = 0.02$. CPT used an asymptotic inversion technique to estimate the difference in internal sound speed between their models and the Sun, while Guzik & Cox used direct comparison between observed and calculated frequencies. Both papers found that including diffusion results in improved agreement with observations and that attempting to smooth out the He abundance gradient below the surface convection zone with slow turbulent mixing worsens the agreement with observations.

2. INPUT PHYSICS

We have adopted the same nuclear reaction rates, energy values, screening calculations, and neutrino cross sections of BP. In order to simplify comparisons, we have incorporated their subroutine for the calculation of these quantities into our stellar evolution code, and it is used for all of our calculations in this paper. We follow the abundances of ^{1}H, ^{3}He, ^{4}He, ^{12}C, ^{13}C, ^{14}N, and ^{16}O. The equation of state used here is the Eggleton, Faulkner, & Flannery (1973) equation of state with some modifications. The pressure ionization terms have been altered as discussed in Proffitt & Michaud (1989) in order to force full ionization of H and He everywhere below the surface convection zone. A simplified approximation to Coulomb corrections has been added (Proffitt 1993). This approximation is essentially exact in the interior where H and He are fully ionized, but is less accurate in partially ionized regions near the stellar surface. The approach used here should be adequate for calculating stellar neutrino fluxes and for differential comparisons of the structure between different models, but for direct comparisons with solar oscillation data and the inferred sound speed in the solar convection zone a more sophisticated equation of state would be required.

2.1. *Calculation of Atomic Diffusion*

The physics assumed in our diffusion calculations is the same as that used by Proffitt & Michaud. Diffusion coefficients are calculated using the numerical results of Paquette et al. (1986), and the equations for multispecies ionic diffusion proposed by Burgers (1969). The coefficients of Paquette et al. assume that interparticle interactions can be described as elastic binary collisions with a screened Debye-Huckel interparticle potential. A discussion of the applicability of these equations to the solar interior, as well as a quantitative comparison with the approach of Bahcall & Loeb, may be found in Michaud & Profitt (1993). They found that for H-He diffusion our approach gives diffusion velocities that are qualitatively similar but somewhat larger than Bahcall & Loeb. For heavier elements the differences are more substantial.

In this paper we calculate the diffusion for a mixture of H, He, and electrons, and treat the diffusion of the heavy elements as if they were trace elements in that H-He background. Calculations of multicomponent plasmas show that the inaccuracies introduced by treating the heavy elements as trace constituents are much smaller than the other uncertainties discussed below. The solution procedures for diffusion, settling and nuclear reactions are iterated to give a consistent abundance solution.

Assuming complete ionization below the Sun's surface convection zone may be a good approximation for H, He, and the CNO elements (i.e., Ulrich 1982), but is not very good for heavier elements such as Fe. Less than fully ionized elements are expected to diffuse downward more quickly. For conditions near the base of the solar convection zone, we find that $^{56}Fe^{+17}$ should have a diffusion velocity about 40% larger than $^{56}Fe^{+26}$, while $^{22}Ne^{+8}$ will have a velocity about 20% larger than $^{22}Ne^{+10}$. Our approach will be to assume fully ionized CNO elements, and to approximate the remainder of the heavy elements as a single species. For our baseline model, we will choose to calculate the diffusion for all elements heavier than O as if they were fully ionized $^{56}Fe^{+26}$.

If the average opacity per atom for a given element is large enough, it is possible that the radiative force on that species might contribute a significant upward component to the velocity. A crude estimate of the possible radiative effects on Fe by Turcotte, Proffitt, & Michaud (1993), based on Rosseland mean opacity tables with varying Fe abundances computed by Iglesias, Rogers, & Wilson (1992), suggests that the radiative force on Fe is likely to be largest at the base of the solar convection zone, where it may be as large as 30% of the force of gravity. If correct, this might largely offset the increased settling velocity expected due to the incomplete ionization of Fe. Radiative forces on the CNO elements are probably negligible.

The question of the proper uncertainty to assign to theoretically calculated values of solar neutrino fluxes has been discussed extensively in the literature (e.g., Bahcall 1993; BP; Bahcall & Ulrich 1988; Turck-Chieze & Lopes 1993 [T-C&L]). There is some disagreement as to how to best estimate the uncertainty for theoretically calculated quantities, and as to which error estimates for some experimentally determined quantities are most reliable. As it is not the purpose of this paper to reanalyze the question of the uncertainties in theoretically determined neutrino fluxes, we will discuss errors quantitatively only with regard to diffusion. For errors in other quantities we will adopt the results of BP.

It is, however, difficult to assign a quantitative value for the uncertainty of the calculated diffusion velocities. The derivation of the diffusion coefficients by Paquette et al. (1986) and others assumes that the kinetic energies of colliding particles are much larger than their Coulomb interaction energies. The "Coulomb logarithm" is often used to measure the strength of this plasma coupling for collisions between particles of type *i* and *j*;

$$\ln \Lambda_{ij} = \ln\left(\text{constant}\,\frac{2kT\lambda_d}{Z_i Z_j e^2}\right). \tag{1}$$

Here the Debye screening length is given by

$$\lambda_d = \left(\frac{kT}{4\pi e^2 \sum_a n_a Z_a^2 \theta_a}\right)^{1/2}, \tag{2}$$

with Z_i the charge and n_i the number density of particle type i. The sum in equation (2) is over all particle types in the plasma, not just the two particle types for which the interaction is being considered. When $\ln \Lambda \gg 0$ the plasma is weakly coupled and the calculations of Paquette et al. should be essentially exact. Taking the constant $= \frac{2}{3}$ in equation (1) gives the best correspondence with the numerical results of Paquette et al., and this gives for conditions below the solar surface convection zone $\ln \Lambda \approx 2.8 - \ln Z_i Z_j$ for collisions between particles with charges Z_i and Z_j (Michaud & Proffitt).

If we simply guess that the errors in the diffusion coefficients are of order $1/\Lambda$, than the expected uncertainty in the solar interior H-He diffusion coefficients is about 15%. The same guess suggests uncertainties of order 50% for H-O collision coefficients and of order 100% for He-ones. Fortunately, the diffusion *velocities* of heavy elements in the solar interior are less sensitive to uncertainties in the heavy-element diffusion coefficients than might be naively expected, since the relative diffusion of the H-He background is only weakly affected by changes in their interactions with heavier elements. It is especially difficult to decrease the expected heavy-element settling velocities by large amounts, as making the collisions with He too strong simply causes the heavy element to be dragged down along with the sinking He. Numerical experiments in which we solve Burgers' system of equations while arbitrarily varying the resistance coefficients show that, under conditions appropriate to the solar interior, even if we increase the strength of collisions with H by 50% above the Paquette et al. values, and decrease the strength of collisions with He to a very small value, the net settling velocity of various heavy elements decreases by less than 35% as compared to our normal scenario. Further, such numerical experiments suggest that the values we have guessed for the uncertainties in the diffusion coefficients imply uncertainties of $(+50; -35\%)$ for the heavy-element diffusion velocities. Recently Thoul, Bahcall, & Loeb (1993) also considered the diffusion velocities of H, He, and heavy elements in the solar interior and, while making somewhat different choices for the diffusion coefficients, still found velocities within 15% of those found by our approach. As discussed above, allowing heavy elements to be partially ionized may further increase the allowed settling velocity.

2.2. *Opacities*

We take into account the changes in opacity due to changes in heavy-element abundances by using the full set of tables from Rogers & Iglesias (1993), and for each point performing a four-point Lagrangian interpolation in each dimension [X, $\log T$, $\log(\rho/T^3)$, and Z] of the table. However, the heavy-element mass fraction Z also changes due to CNO processing, increasing as C is converted to N, and decreasing slightly due to ON cycling. As the number abundance of CNO elements is conserved in this process, it is not appropriate to change the local opacity as if the abundance of all heavy elements had changed uniformly. Indeed, Iglesias & Rogers (1991) find that opacity changes in the solar interior due to the CNO cycle are quite small. We therefore calculate a quantity Z_κ, in which we subtract out the part of the change in the heavy-element mass fraction, Z, due to changes in the relative CNO abundances, but keep the part of the increase caused by the any change in the total number abundance of CNO atoms, i.e.,

$$Z_\kappa = Z - \sum_{\text{cno}} X_i + \left(\frac{\sum_{\text{cno}} X_i/A_i}{\sum_{\text{cno}} X_i'/A_i}\right) \sum_{\text{cno}} X_i', \qquad (3)$$

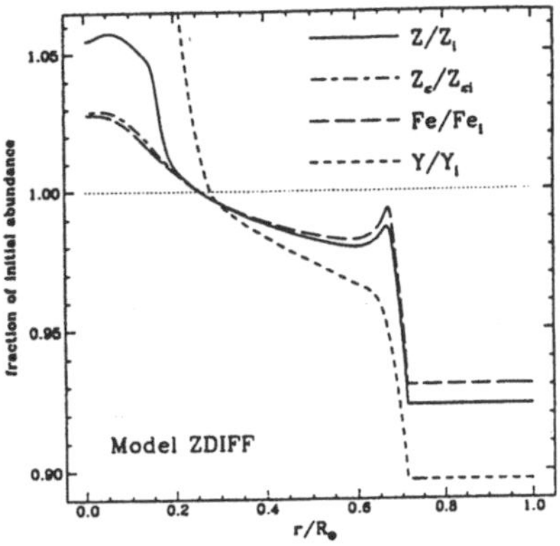

FIG. 1.—Internal abundance profiles as a function of radius for model ZDIFF at 4.63 Gyr normalized to initial abundances. The quantity Z_κ is the mass fraction of heavy elements after subtracting out changes due to redistribution of relative CNO abundances.

where the primed abundances inside the square brackets match the relative CNO mass fractions assumed in the opacity table, and the unprimed abundances refer to the current abundances at the point of interest. This quantity Z_κ is then used in place of the total Z for interpolation in the opacity tables (see also Fig. 1). In models without heavy-element setting Z_κ will always be equal to the initial value of Z. This allows direct comparison to models calculated using opacity tables for a single metal abundance.

Rather than simply using an average Z to determine the opacity, it would be better if each element that makes a significant contribution to the opacity were followed in detail with the diffusion velocity calculated for the correct distribution of ionization states and for the correct radiative forces, and the opacity then adjusted at each point for the abundance change in that particular element. This would require not only the complete partition function for each element, but also $d \log \kappa/dX_i$ at each point in the table, for each element i of interest. Without such information, the value of more carefully following the transport of each element in detail is questionable.

Low-temperature opacities are from Kurucz (1993). Photospheric pressures are taken from a grid of model atmospheres by Bell (1993).

3. MODEL CALIBRATION

We will take as our default values for the solar radius and luminosity to be $L_\odot = 3.846 \times 10^{33}$ ergs s^{-1}, and $R = 6.9599 \times 10^{10}$ cm. Our models are started in a fully convective configuration on the Hayashi track near 40 $L_\odot$. Most other workers start with a homogeneous model in thermal equilibrium near the zero-age main sequence (ZAMS). Our models reach a comparable configuration at an age of about 0.03 Gyr, and therefore our preferred age of 4.63 Gyr is equivalent to 4.60 Gyr models of these workers. This age was chosen to be consistent with that used by BP and CPT, but an age about 0.1 Gyr smaller might be a better choice for direct comparisons to the Sun (cf. Guenther 1989).

TABLE 1

COMPARISON WITH BAHCALL & PINSONNEAULT NONDIFFUSION MODELS, $L = 3.86 \times 10^{33}$ ergs s^{-1} $R = 6.96 \times 10^{10}$ cm, $Z/X = 0.02668$

Quantity	BP	This Work
Z	0.01895	0.01890
^4He_i	0.2716	0.2725
^4He_c	0.6270	0.6299
T_c (10^7 K)	1.559	1.558
ρ_c (g cm^{-3})	151.3	152.7
$R_{conv}/R_{\odot}$	0.721	0.726
$M_{conv}/M_{\odot}$	0.0216	0.0205
$\phi(pp)$ (cm^{-2} s^{-1})	6.04×10^{10}	6.03×10^{10}
$\phi(pep)$ (cm^{-2} s^{-1})	1.43×10^8	1.43×10^8
$\phi(hep)$ (cm^{-2} s^{-1})	1.25×10^3	1.25×10^3
$\phi(^7Be)$ (cm^{-2} s^{-1})	4.61×10^9	4.60×10^9
$\phi(^8B)$ (cm^{-2} s^{-1})	5.06×10^6	5.01×10^6
$\phi(^{13}N)$ (cm^{-2} s^{-1})	4.35×10^8	4.50×10^8
$\phi(^{15}O)$ (cm^{-2} s^{-1})	3.72×10^8	3.81×10^8
$\phi(^{17}F)$ (cm^{-2} s^{-1})	4.67×10^6	4.58×10^6
$\sum(\phi\sigma)_{Cl}$	7.18	7.14
$\sum(\phi\sigma)_{Ga}$	127.5	127.3

As noted above, using the Rogers & Iglesias (1993) opacity tables which include the slightly altered CN abundances, as well as the lower (meteoritic) Fe abundance, gives an observed surface solar $Z/X = 0.02694$. Recently it has been suggested that the O abundance also be lowered by about 15% (Biémont et al. 1991), but this change is not incorporated into the opacity tables that we use, and so we do not take it into account in our value for Z/X. We will, however, discuss the sensitivity of our results to uncertainties in Z/X.

For each set of assumed input physics, the initial Z and Y and the mixing length parameter α were varied until the radius, luminosity, and Z/X at the solar age matched the desired values to within one part in 10^6. This is, of course, far more precise than the accuracy with which any of these values are measured, and far more precise than the absolute numerical accuracy of the calculation, but such matching is useful for comparing models with slightly different physics (though for the purposes of this paper, matching to a few parts in 10^5 would have been sufficient). Typically, four models were calculated to determine the dependency of the final results on the initial parameters, and another two to three models were needed to converge to the desired parameters.

At each time step Newton-Raphson procedures were used to separately converge the structure and composition equations. The solutions of the structure and composition equations were iterated until they were mutually consistent with each having final corrections of less than one part in 10^6. To eliminate interpolation errors a fixed Langrangian grid of 1156 mesh points was used. The evolution between 0.3 Gyr and the solar age was divided into 201 equal time steps. The small time steps and fine zoning ensure acceptable accuracy despite our use of first order accurate, forward time centered implicit difference equations for both the structure and the composition equations. A test calculation using the same input parameters with time steps 2.6 times as large gave a final model that differed by less than 0.05% in temperature and density throughout the interior and by less than 0.4% in the predicted neutrino count rates.

3.1. *Detailed Comparison to Bahcall & Pinsonneault (1992)*

Bahcall & Pinsonneault used slightly different values for the assumed input quantities. They used $L_{\odot} = 3.86 \times 10^{33}$ ergs s^{-1}, and $R = 6.96 \times 10^{10}$ cm. As they used the Anders & Grevesse (1989) mixture with the meteoritic Fe abundance, their assumed $Z/X = 0.02668$. We computed a standard (nondiffusive) model with these parameters, but with our preferred relative heavy-element abundance mixture and opacity tables. BP's nondiffusion results with 1 σ errors are 7.2 ± 0.9 SNU for the Cl experiment, 127(+6, −5) SNU for the Ga experiment, and $5.1 \pm 0.7 \times 10^6$ (cm^{-2} s^{-1}) for the ^8B flux. A more detailed comparison with BP is shown in in Table 1. T-C&L, using somewhat different nuclear parameters and the lower O abundance discussed above, predict lower neutrino fluxes than BP. T-C&L's preferred model predicts rates of 6.4 ± 1.4 SNU for the chlorine experiment, $4.4 \pm 1.1 \times 10^6$ (cm^{-2} s^{-1}) for the ^8B flux, and 122.5 ± 7 SNU for gallium detectors.

The differences between our nondiffusion model and BP's are very small. The initial He mass fraction we find is about 0.001 larger, and we find the convection zone to be slightly shallower (0.726 $R_{\odot}$ vs. 0.721 $R_{\odot}$ for BP). Comparing our fluxes for the individual neutrino sources with those of BP ($\sum \delta\phi_i \sigma_i$) shows BP's predicted count rate for ^{37}Cl detectors to be 0.04 SNU (0.6%) larger than ours, while for ^{71}Ga detectors, their value is 0.14 SNU (0.1%) larger. To some extent this excellent agreement is the result of partial cancellation between our finding slightly smaller *pp*-cycle fluxes and slightly larger ^{13}N and ^{15}O fluxes. Our predicted ^8B flux is ≈ 1% smaller, and the CNO neutrino rates differ by between 2% and 3.5%. BP reported neutrino flux differences from their preferred calculations of similar magnitude when they used the opacity tables and abundances of Rogers & Iglesias (1993) that our calculation also uses. Indeed, when we compare the opacity values produced with our interpolation procedure to those in Table 6 of BP, we find that for the range of parameters relevant for the solar interior, our opacities are systematically lower by 0.1%–1.5%.

We believe these differences are large enough to account for the small differences between our results and BP's, but we cannot rule out numerical or other physics differences of comparable magnitude. In any case the differences in predicted neutrino fluxes are quite small compared to the theoretical uncertainties in the assumed physics.

4. RESULTS

4.1. *Predicted Neutrino Fluxes*

Results with our preferred parameters are shown in Table 2. For consistency with CPT we chose slightly different values for $L_{\odot}$ and $R_{\odot}$ than did BP. The models discussed in this section use the same solar age, 4.63 Gyr, as our models in § 3. The slight differences from BP luminosity and Z/X result in a nondiffusion model with only very slightly increased Y_i and Z_i, and slightly smaller neutrino flux rates. We have computed three models with this set of parameters; a model without any diffusion (labeled NOND), a model following diffusion for H and ^4He only (YDIFF), and one which also follows diffusion for heavier elements in the approximation discussed above (ZDIFF). For comparison purposes we have computed another model (LZDIFF), which uses the same physics as ZDIFF, but with the initial Z fixed to be the same as in the nondiffusion model NOND, even though this gives much lower value for the final Z/X. Because they differ in the values for both their initial Z and final Z/X, direct comparisons

TABLE 2

EFFECTS OF DIFFUSION MODELS WITH $L = 3.846 \times 10^{33}$ ergs s^{-1}, $R = 6.9599 \times 10^{10}$ cm, AND $^3He = 5 \times 10^{-5}$

QUANTITY	MODEL NOND	YDIFF	ZDIFF	LZDIFF
Diffusion	None	H, ^{4}He	H, ^{4}He, Z	H, ^{4}He, Z
Z/X	0.02694	0.02694	0.02694	0.01962
Z initial	0.01907	0.01979	0.02127	0.01907
Z surface	0.01907	0.01979	0.01964	0.01758
Z central	0.01957	0.02030	0.02243	0.02013
Z_{κ} central	0.01907	0.01979	0.02188	0.01962
^{4}He initial	0.2729	0.2740	0.2803	0.2707
^{4}He surface	0.2729	0.2456	0.2514	0.2422
^{4}He central	0.6292	0.6412	0.6488	0.6370
α	1.574	1.729	1.711	1.677
$T_c(10^7$ K)	1.577	1.567	1.581	1.566
ρ_c(g cm^{-3})	152.3	155.7	155.9	154.7
$R_{conv}/R_{\odot}$	0.7254	0.7103	0.7115	0.7150
$M_{conv}/M_{\odot}$	0.0206	0.0248	0.0248	0.0234
$\phi(pp)$ (cm^{-2} s^{-1})	6.01×10^{10}	5.96×10^{10}	5.91×10^{10}	5.98×10^{10}
$\phi(pep)$ (cm^{-2} s^{-1})	1.43×10^8	1.42×10^8	1.39×10^8	1.42×10^8
$\phi(hep)$ (cm^{-2} s^{-1})	1.25×10^3	1.22×10^3	1.20×10^3	1.23×10^3
$\phi(^7Be)$ (cm^{-2} s^{-1})	4.57×10^9	4.91×10^9	5.18×10^9	4.79×10^9
$\phi(^8B)$ (cm^{-2} s^{-1})	4.95×10^6	5.70×10^6	6.48×10^6	5.46×10^6
$\phi(^{13}N)$ (cm^{-2} s^{-1})	4.50×10^8	5.21×10^8	6.40×10^8	5.02×10^8
$\phi(^{15}O)$ (cm^{-2} s^{-1})	3.81×10^8	4.48×10^8	5.57×10^8	4.27×10^8
$\phi(^{17}F)$ (cm^{-2} s^{-1})	4.57×10^6	5.43×10^6	6.79×10^6	5.16×10^6
$\sum(\phi\sigma)_{Cl}$(SNU)	7.06	8.02	9.02	7.71
$\sum(\phi\sigma)_{Ga}$(SNU)	126.7	131.6	136.9	130.1

between models YDIFF and LZDIFF are not particularly meaningful.

Interior abundance profiles at 4.63 Gyr for the model ZDIFF are shown in Figure 1. The difference in the inner 20% of the radius between the total Z and the quantity Z_{κ} used to calculate the opacity is due to CN cycling, and the small central dip in Z is from a small amount of ON cycling.

Despite the differences between our procedure for calculating H-He separation and Bahcall & Loeb's prescription (see the discussion in Michaud & Proffitt), the changes we find from including He settling are only slightly larger than those found by BP. For instance, we find a 13.6% increase in the Cl neutrino signal between models YDIFF and NOND, while BP found a 10.5% increase for their comparable models. We find that for model YDIFF the final surface Y is 10.3% lower than its initial value, while BP found a decrease of 9.6%.

The inclusion of heavy-element settling causes additional increases in the predicted neutrino fluxes comparable in size to the increases caused by including He settling. The ^8B neutrino flux calculated for model ZDIFF is 31% larger than for model NOND. The predicted signal for the Cl detector is 28% larger, while the flux for Ga detector is 8% larger.

4.2. *Differential Effect on Interior Sound Speed*

Models that include He settling (with or without heavy-element settling) have a convection zone depth that is in substantially better agreement with the helioseismologically determined depth of $0.713 \pm 0.003\ R_{\odot}$ found by Christensen-Dalsgaard et al. (1991) than are the nondiffusion models. The surface He abundances found in diffusion models are also in better agreement with the low surface He values found by many seismological studies, but as was discussed above, possible systematic errors in the equation of state limit the value of this comparison.

Figure 2*a* shows the difference in interior sound speed between each of the models that include settling and the non-diffusion model NOND. While the inclusion of He diffusion has a substantial effect on the interior sound speed profile everywhere below the base of the convection zone, the additional changes to the sound speed caused by the inclusion of heavy-element settling are rather small in the outer 80% of the radius. As the asymptotic properties of high-degree p-modes are mostly determined by the sound speed profile, the changes due to heavy-element settling will be expected to have little additional effect on modes for which the asymptotic approximations are valid. We therefore do not expect that including heavy-element settling would have substantially changed the results of CPT's asymptotic inversion for the determination of the solar interior sound speed (see Fig. 3 of CPT).[1] The *additional* changes due to heavy-element settling (i.e., beyond those due to He settling alone) are largest in the inner 20% of the radius and may have a substantial effect on the small frequency spacings (e.g., $\nu_{nl} - \nu_{n-1,l+2}$ for small angular degree l) that are sensitive to the gradient of sound speed near the solar center.

Figure 2*b* shows a similar comparison for the interior density. The density profile in the convection zone is more sensitive to the details of the model than is the sound speed in the same region, and in so far as this quantity can be reliably determined from helioseismological studies it will provide a powerful diagnostic of the structure of the surface convection zone.

Model LZDIFF, which uses the same initial Z as model NOND, illustrates the importance of adjusting the initial heavy-element abundance in order to match the current Z/X

[1] Although CPT used a fixed value of $Z = 0.02$, their model 2 with He diffusion has a final $Z/X = 0.02738$, which is only 1.6% larger than the value we have adopted, and their model 2 is therefore a reasonable comparison model to the Sun.

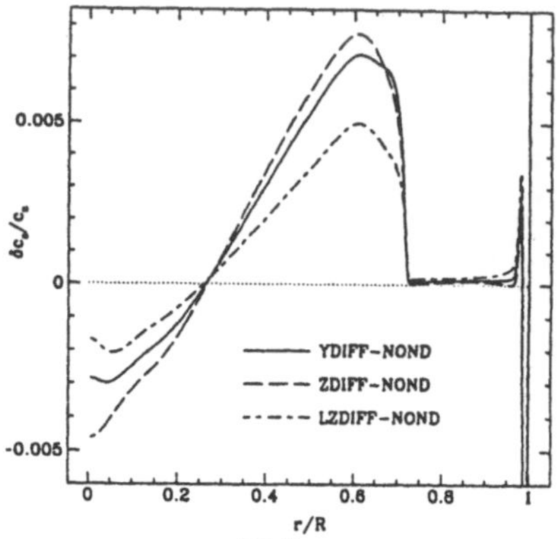

FIG. 2a

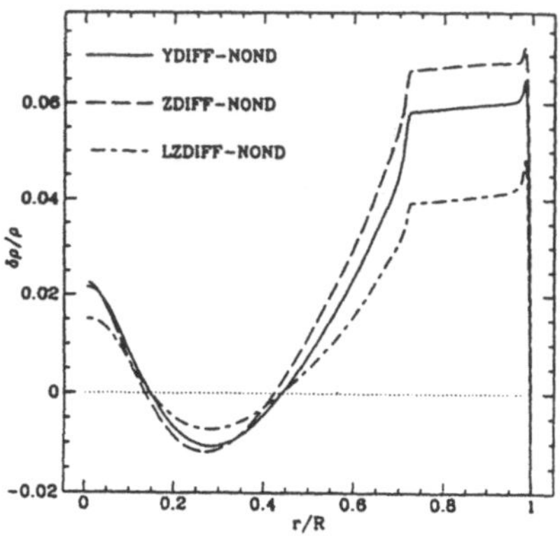

FIG. 2b

FIG. 2.—(a) Fractional difference in sound speed as a function of radius between models with and without diffusion. (b) As in (a), except for the fraction difference in density.

ratio. In model LZDIFF, the changes in the neutrino fluxes relative to the nondiffusion model NOND are only one-third as large as in model ZDIFF. The initial Y abundance of LZDIFF is 0.002 *lower* than NOND, while for ZDIFF it is 0.007 *higher* than NOND. The changes in the interior sound speed and density profiles (see Figs. 2a and 2b) are also substantially smaller when a fixed initial Z is used.[2] Although the difference in Z/X between LZDIFF and the other models is at least 3 times as large as the likely uncertainty in Z/X (cf. Anders & Grevesse 1989), this model still gives a good indication of how the uncertainty in this quantity might affect the calculated neutrino fluxes and interior structure.

5. DISCUSSION

We can apply our estimates of the uncertainties in the diffusion velocities (§ 2.1) to modify existing estimates for the uncertainties in solar neutrino fluxes. We will assume that the uncertainty in the H-He diffusion rate causes uncertainties in neutrino fluxes equal to 15% of the difference in the fluxes of models NOND and YDIFF, and that the uncertainties in the heavy-element diffusion rates cause uncertainties equal to 50% of the flux differences between models YDIFF and ZDIFF. To be conservative, we will assume that the errors caused by uncertainties in He and heavy-element diffusion are always correlated. We will consider these uncertainties to be equivalent to 1 σ errors. Since our predicted neutrino fluxes are very similar to those of BP, we will take the errors in neutrino fluxes that are not related to diffusion to be the same as the errors that BP estimate for their nondiffusion model (note that the uncertainties listed in BP are 3 σ errors, while here we quote 1 σ errors).

In the above discussion we have assumed that any mixing below the base of the surface convection zone can be ignored when computing the effects of diffusion. The well-known depletion of surface solar Li clearly indicates some mixing has occurred below the solar convection zone. Proffitt & Michaud (1991b) argued that realistic scenarios for mixing that are consistent with observed Li abundances in young clusters can reduce the amount of surface settling by at most one-third. The results of CPT suggest that the inhibition of settling by turbulence may be even less than this. Mixing outside the nuclear-burning regions will have negligible effects on the required initial He abundance and on the effects of the He settling on the predicted neutrino fluxes. However, since the initial heavy-element abundance is adjusted to match the observed Z/X ratio, any reduction of the surface settling due to turbulence will reduce the needed change in Z, and reduce the increase in the neutrino flux caused by heavy-element settling. To take this into account we will reduce our final estimates for the neutrino fluxes below the values given for model ZDIFF by an amount equal to $\frac{1}{6} \pm \frac{1}{6}$ of the difference in flux between models YDIFF and ZDIFF.

With the above assumptions we can derive correction factors (and associated uncertainties) by which the predicted nuetrino fluxes of our nondiffusion model needs to be multiplied to derive the best estimate for the predicted neutrino fluxes. Very similar correction factors should be applicable to any nondiffusion model of the Sun, even one assuming somewhat different opacities or nuclear reaction rates. For the Cl experiment we find a correction factor $F_D(\mathrm{Cl}) = 1.25 \pm 0.08$, while for Ga experiments $F_D(\mathrm{Ga}) = 1.07 \pm 0.02$, and for the ^{8}B flux $F_D(^8\mathrm{B}) = 1.28 \pm 0.09$. Table 3 shows our preferred estimates for the predicted neutrino fluxes and uncertainties. As noted above, the neutrino flux uncertainties not associated with diffusion are simply taken to be one-third of BP's 3 σ errors for their best nondiffusion model. Also shown are the results of applying our correction factors F_D to the nondiffusion results of T-C&L.

The net result of including both He and heavy-element settling is to increase the predicted neutrino fluxes by about 1–1.5 σ above the results for models that ignore diffusion. The additional increases in the neutrino fluxes due to the inclusion of heavy-element settling are comparable to the increases due to

[2] Had CPT adjusted both their diffusion and nondiffusion to match the solar Z/X, they would have found a larger difference in sound speed between these two models, and between their nondiffusion model and the Sun.

TABLE 3

ADJUSTED NEUTRINO FLUXES WITH 1 σ ERROR ESTIMATES

Flux	This Paper or Adjusted BP	Adjusted T-C&L	Observed
Cl (SNU)	8.9 ± 1.1	8.0 ± 1.5	2.23 ± 0.22[a]
^{8}B (10^6 cm^{-2} s^{-1})	6.4 ± 0.9	5.6 ± 1.2	$2.67 \pm 0.28 \pm 0.34$[b]
Ga (SNU)	136^{+7}_{-6}	131.5 ± 8	$83 \pm 19 \pm 8$[c]
			$85^{+22}_{-32} \pm 20$[d]

NOTES.—Where two error values are quoted for observed values, the first is the statistical error, and the second is the estimated systematic error.

[a] Davis et al. 1990.

[b] GALLEX experiment; Anselmann et al. 1992.

[c] SAGE experiment; Gavrin et al. 1992.

[d] Kamiokondo; Hirata et al. 1991.

He settling. About two-thirds of the total change in neutrino fluxes is due to the larger initial heavy-element abundance needed in models with settling.

Even without including the effects of settling, it has not been possible to find a standard or nonstandard stellar model that explains the observed neutrino fluxes for all existing experiments (cf. the discussion in Bahcall & Bethe 1993). The increased neutrino fluxes that we have found for solar models that include settling will make it even more difficult to find a solar structure solution to the solar neutrino problem. The implications for particle physics solutions to the solar neutrino problem, such as the MSW theory (Mikheyev & Smirnov 1985; Wolfenstein 1978; Rosen 1991), are not dramatic, and will require only slight shifts in the theory parameters to match the observations.

CPT showed that including He settling substantially improves the agreement between the calculated and observed solar oscillation frequencies. Comparison of the interior sound speed profiles of the models in this paper suggests that including heavy-element settling in solar models will have a substantially smaller effect on high degree p-modes than does He settling. However, there will likely be detectable effects on p-modes of low angular degree.

We would like to thank John Bahcall for providing a copy of the Bahcall & Pinsonneault (1992) subroutine for calculating nuclear reaction rates and neutrino fluxes. Thanks are also due to J. Christensen-Dalsgaard, G. Michaud, J. Guzik, J. Bahcall, and J. Richer for reviewing and commenting on earlier drafts of this paper. This research project was supported in part by a grant from the Ultraviolet, Visible, and Gravitational Astrophysics Branch of NASA's Astrophysics Division, as well as by the Computer Sciences Corporation and the Space Telescope Science Institute.

REFERENCES

Anders, E., & Grevesse, M. 1989, Geochim. Cosmochim. Acta, 53, 197
Anselmann, P., et al. 1992, Phys. Lett. B, 285, 390
Bahcall, J. N. 1993, in Proc. 4th Recontres de Blois (France, June 15–20, 1992), (Particle Astrophysics, Editions Frontières), in press
Bahcall, J. N., & Bethe, H. A. 1993, Phys. Rev. D, 47, 1298
Bahcall, J. N., & Loeb, A. 1990, ApJ, 360, 274
Bahcall, J. N., & Pinsonneault, M. H. 1992, Rev. Mod. Phys., 64, 885 (BP)
Bahcall, J. N., & Ulrich, R. K. 1988, Rev. Mod. Phys., 60, 297
Bell, R. A. 1993, private communication
Biémont, E., Hibbert, A., Godefroid, M., Vaeck, N., & Fawcett, B. C. 1991, ApJ, 375
Burgers, J. M. 1969, Flow Equations for Composite Gases (New York: Academic)
Christensen-Dalsgaard, J., Duvall, T. L., Gough, D. O., Harvey, J. W., & Rhodes, E. J., Jr. 1985, Nature, 315, 378
Christensen-Dalsgaard, J., Gough, D. O., & Thompson, M. J. 1991, ApJ, 378, 413
Christensen-Dalsgaard, J., Proffitt, C. R., & Thompson, M. J. 1993, ApJ, 403, L75 (CPT)
Cox, A. N., Guzik, J. A., & Kidman, R. B. 1989, ApJ, 342, 1187
Däppen, W. 1993, in IAU Coll. 137, Inside the Stars, ed. W. W. Weiss & A. Baglin (San Francisco: Astronomical Society of the Pacific), 208
Däppen, W., Gough, D. O., Kosovichev, A. G., & Thompson, M. J. 1991, in Challenges to Theories of the Structure of Moderate-Mass Stars (Lecture Notes in Physics, 288), ed. D. O. Gough & J. Toomre (Heidelberg: Springer-Verlag), 111
Davis, R., Jr. 1990, in Proc. 21st Intern. Cosmic Ray Conf., Adelaide, 134
Deubner, F.-L., & Gough, D. O. 1984, ARA&A, 22, 593
Eggleton, P. P., Faulkner, J., & Flannery, B. P. 1973, A&A, 23, 325
Gavrin, V. N., et al. 1991, Nucl. Phys. B, Proc. Suppl., 19, 84
Guenther, P. B. 1989, ApJ, 339, 1156
Guzik, J. A., & Cox, A. N. 1993, ApJ, 411, 394
Hirata, K. S., et al. 1991, Phys. Rev. D, 44, 2241
Iglesias, C. A., & Rogers, F. J. 1991, ApJ, 371, 408
Iglesias, C. A., Rogers, F. J., & Wilson, B. G. 1992, ApJ, 397, 717
Kosovichev, A. G., Christensen-Dalsgaard, J., Däppen, W., Dziembowski, W. A., Gough, D. O., & Thompson, M. J. 1992, MNRAS, 259, 536
Kurucz, R. L. 1993, private communication
Michaud, G., & Proffitt, C. R. 1993, in IAU Coll. 137, Inside the Stars, ed. W. W. Weiss & A. Baglin (San Francisco: Astronomical Society of the Pacific), 246
Mikheyev, S., & Smirnov, A. Yu. 1985, Soviet J. Nucl. Phys., 42, 913
Noerdlinger, P. D. 1979, A&A, 57, 407
Paquette, C., Pelletier, C., Fontaine, G., & Michaud, G. 1986, ApJS, 61, 1977
Proffitt, C. R. 1993, ApJ, submitted
Proffitt, C. R., & Michaud, G. 1989, ApJ, 346, 976
———. 1991a, ApJ, 371, 584
———. 1991b, ApJ, 380, 238
Rogers, F. J., & Iglesias, C. A. 1993, private communication
Rosen, S. P. 1991, in Solar Atmosphere and Interior ed. A. N. Cox, W. C. Livingston, & M. S. Matthews (Tucson: Univ. Arizona Press), 86
Thoul, A. A., Bahcall, J. N., & Loeb, A. 1993, ApJ, 421, 828
Turcotte, S., Proffitt, C. R., & Michaud, G. 1993, in preparation
Turck-Chieze, S., & Lopes, I. 1993, ApJ, 408, 347 (T-C&L)
Ulrich, R. K. 1992, ApJ, 258, 404
Wolfenstein, L. 1978, Phys. Rev., 17D, 2369

II. Solar Neutrino Experiments

Introduction: Raymond Davis Jr.

The Beginning of Solar Neutrino Studies

The concept of studying the neutrino radiation from the sun as a technique for exploring the deep interior, and verifying that the sun's energy source is indeed the thermal fusion of the light nuclei into helium was rarely expressed. The early papers postulating detailed mechanisms for the fusion processes in stars did not include the neutrinos in the beta decay processes, even though the neutrino was conceptually associated with nuclear beta decay. Perhaps it was too obvious to express, or the neutrino was a mysterious particle that was nearly impossible to detect. The principal goal at the time was to account for the lifetime of main sequence stars by thermonuclear processes. Experimentalists in the late 1940's suggested using the sun and nuclear reactors as intense sources of neutrinos useful for experimental detection of the free neutrino (Paper 1.A.II). The views on the neutrino as a useful particle for observing the sun, the earth's heat sources and supernovae changed dramatically after the unique experiment of Reines and Cowan[1] the first detection of the antineutrinos from a reactor. After this event a number of detectors were built in the world's deepest mines to detect cosmic ray produced neutrinos[2, 3] and the high energy physicists built neutrino beams at accelerators to study neutrino physics at high energy[4].

It was during this period of high interest in neutrino physics that Holmgren and Johnston discovered that the ^{3}He(^{4}He, γ)^{7}Be reaction was a direct capture reaction and consequently had a high cross section, over 100 times higher than expected (Paper 1.A.III). This reaction had a sufficiently high cross section to compete with the ^{3}He(^{3}He, $2p$)^{4}He reaction for burning ^{3}He and terminating the p-p chain. It is interesting to note that the ^{3}He–^{4}He reaction was considered by Bethe in his famous paper[5]. It was suggested as a means for terminating the p-p chain with the production of ^{4}He. The mechanism involved the decay of ^{7}Be to ^{7}Li that subsequently captured a proton forming two ^{4}He nuclei, or ^{7}Be could capture a proton to form ^{8}B which promptly decays to ^{8}Be* and two ^{4}He nuclei.

The new result of Holmgren and Johnston was noted by W. A. Fowler and by A. G. W. Cameron (Papers 2.A.III and 3.A.III). They raised again the possibility that ^{7}Be could either decay by electron capture emitting monoenergetic neutrinos (0.862; 0.383 MeV) or capture a proton forming ^{8}B that beta decays yielding a neutrino spectrum extending to 14 MeV. Their suggestion of two competing branches was not only an important clarification of the p-p chain, it led to a potential major change in the neutrino energy spectrum of the sun. Before this discovery the only hope of observing solar neutrinos was the remote possibility that the sun was generating its energy by the CNO cycle alone, in which case the capture rate in ^{37}Cl would be approximately 36×10^{-36}cm^{-2}s^{-1} per ^{37}Cl[6] or 36 SNU, a readily measurable rate. However, it was then a generally accepted conclusion that less than 5 percent of the sun's energy was being produced by the CNO cycle[7]. Thus, the prospects were poor for observing solar neutrinos.

The author received letters from both Fowler and Cameron (January, 1958) pointing out that the neutrinos from the newly proposed branches of the p-p cycle could perhaps be observed by the ^{37}Cl–^{37}Ar method of Pontecorvo (Paper 1.A.II), a method that was being applied at this time in an experiment at the Savannah River Laboratory to study the possibility that the neutrino and the antineutrino were identical[8]. Following their incentive letters a pilot experiment was built in the Pittsburgh Plate Glass Company Mine at Barberton, Ohio (ref. [4], page 201). Following the discovery by Bahcall (Paper 1.A.I) of the large increase in the theoretical rate due to analog transitions, the result of this background study was the basis of our suggestion to build a 100 times larger experiment to observe solar neutrinos (Paper 2.A.II). However, before embarking on a full scale experiment designed to measure the total solar neutrino capture rate in ^{37}Cl, it was essential to have a reasonable estimate of the flux and energy of the neutrinos from the sun. Particularly, it was necessary to know the cross sections of all the pertinent reactions at energies relevant to the solar core temperatures, i.e., $^4\mathrm{He}(^4\mathrm{He},\gamma)^7\mathrm{Be}$, $^3\mathrm{He}(^3\mathrm{He},2p)^4\mathrm{He}$, $^7\mathrm{Be}(p,\gamma)^8\mathrm{B}$, to calculate the electron capture rate of ^{7}Be in the solar interior, and to incorporate this nuclear information in a solar model calculation to determine the neutrino fluxes. Most of this effort was carried out at the California Institute of Technology under the leadership of W. A. Fowler.

Unfortunately, the first calculation of neutrino fluxes from a detailed solar model (Paper 1.B.I, Bahcall, Fowler, Iben, and Sears) gave at the end of 1962 disappointingly low fluxes! Nevertheless, these results were of great importance, giving the experimentalists guidance in planning experiments. Therefore, the results from this paper were often quoted; they were: $^8\mathrm{B} = 2.5 \times 10^7$, $^7\mathrm{Be} = 1.0 \times 10^{10}\mathrm{cm}^{-2}\mathrm{s}^{-1}$. Two follow-up papers appeared in the next year and a half, one by R. Sears (Paper 2.B.I), and one by Pochoda and Reeves[9], both of which gave many details about the solar models, These papers yielded fluxes that were in good agreement with the first calculation.

The Case Western Reserve Experiments

The calculated ^{8}B flux values were sufficiently large to encourage three proposals for pilot experiments from F. Reines and his associates at the Case Western Reserve University. This group was already heavily engaged in neutrino physics experiments at the Savannah River Reactor, the Rand Gold Mine in South Africa, and in a local salt mine in Ohio. Paper 3.A.II describes a 200 liter scintillation counter that Reines and Kropp had been operating in the Morton Salt Company mine at Fairport Harbor, Ohio to study proton decay. The results of this experiment were interpreted using the Feynman and Gell-Mann and Marshak and Sundarshan $V - A$ theory of elastic scattering of neutrinos with electrons to set a limit on the flux of ^{8}B neutrinos. They obtained a limit on the flux of $< 10^9 \text{cm}^{-2}\text{s}^{-1}$, compared with $2.5 \times 10^7 \text{cm}^{-2}\text{s}^{-1}$ of Bahcall, Fowler, Iben, and Sears (Paper 1.B.I).

Encouraged by these results, a much larger experiment was built in the Rand Mine with the hope of observing ^{8}B decay solar neutrinos. The new experiment was built at the great depth of 3,200 meters, sufficiently deep to eliminate all of the cosmic ray muon background. The detector contained 4,000 liters of liquid scintillator in a cylindrical tank with interior baffles. The events were recorded in five separate regions by 58 5-inch photomultiplier tubes. The installation was shielded with 40 tons of paraffin and boron to reduce the flux of fast neutrons from the surrounding rock wall. The predicted elastic scattering event rate of ^{8}B decay neutrinos above 6 MeV was 24 per year. Unfortunately, the detector was plagued with a high background from gammas with energies 8 to 20 MeV arising from spontaneous fission of ^{238}U in the rock wall[10]. The results of this experiment are of great significance in evaluating the expected background rate in presently proposed neutrino detectors sensitive to gamma rays. Plans to improve this experiment were abandoned when the first results from the Homestake experiment were announced.

Reines and Woods[11] designed a lithium (or boron) detector that used the concept of slabs of the target material immersed in liquid scintillator. Electrons produced by neutrino capture in ^{7}Li or ^{11}B escaping the slabs could be observed by photomultiplier tubes viewing the scintillator liquid between the slabs. A total of 570 kg of metallic lithium in the form of 14 slabs was planned. This detector anticipated approximately 40 events per year from a flux of ^{8}B decay solar neutrinos of $2.5 \times 10^7 \text{cm}^{-2}\text{s}^{-1}$. The authors pointed out that the method in principle could identify the sun as the source by studying the angular distribution of events. The apparatus was ready for final assembly when the results from the Homestake detector were announced. The detector was then modified to measure the stopped muon intensity[12] at the depth of the Morton Salt Mine at Fairport Harbor, Ohio (1,440 m.w.e.).

A third Case Western Reserve University solar neutrino detector was built by T. L. Jenkins and his graduate student F. Dix[13]. They built a 2,000 liter heavy water Cerenkov counter viewed by 55 5-inch photomultiplier tubes. The heavy water was furnished by the U.S. Atomic Energy Commission. Solar neutrino events were considered from both neutral and charged current reactions. The Cerenkov counter was surrounded by a 30 cm thick anti-coincidence liquid scintillator detector. The counting rate expected by solar neutrinos was 96 events per year, assuming as before a ^{8}B solar neutrino flux of $2.5 \times 10^7 \text{cm}^{-2}\text{s}^{-1}$. The sensitivity of the detector was limited by a high background counting rate. The results of this experiment are given in the thesis of F. Dix (1970). He concluded the flux of ^{8}B neutrinos is less than

$9 \times 10^8 \mathrm{cm}^{-2}\mathrm{s}^{-1}$. This impressive early effort is a forerunner of the Sudbu Neutrino Observatory (Paper 12.A.II).

To end this discussion of the contributions of Fred Reines and his ass ciates I would like to point out that Fred also invented (ref.[4], p. 213) directional Cerenkov light neutrino telescope, based upon the forward ang lar distribution of recoil electrons calculated by Bahcall (Paper 4.A.I) th follows from the Feynman and Gell-Mann elastic neutrino scattering th ory. The experiments discussed above and others, are described in an ear conference sponsored by the Royal Society, reference[14].

The Homestake Experiment

A chlorine solar neutrino experiment, based upon the neutrino captu reaction $^{37}\mathrm{Cl}(\nu, e^-)^{37}\mathrm{Ar}$, was built by Brookhaven National laboratory in th Homestake Gold Mine at Lead, South Dakota during the period 1965–196 The facility was sponsored by the U.S.Atomic Energy Commission, Chemic Research Division. The experiment depends upon the fact that argon is a ra gas and ^{37}Ar produced by neutrino interactions are easily removed by purgin with helium. Radioactive ^{37}Ar decays by electron capture with a 35 da half-life emitting 3–4 Auger electrons with a total energy of 2.82 keV. Th radiation is totally absorbed in the gas of a miniature proportional counte There are two features of radiochemical experiments that are responsible fo their high sensitivity: 1) they are very insensitive to gamma radiation, an 2) the product is a radioactive isotope that can be removed quantitativel and measured in a very small detector. There are background processes tha must be carefully evaluated. These attractive features were suggested b Bruno Pontecorvo in 1946 (Paper 1.A.II). Luis Alvarez made a proposal t carry out a reactor experiment using 40,000 liters of carbon tetrachloride[15 In the appendix of the proposal L. Shiff calculated the average cross sectio for fission product anti-neutrinos, presuming that the transition $^{37}\mathrm{Cl} - {}^{37}\mathrm{A}$ is allowed. Shiff stressed the importance of excited states in evaluating th total cross section. This was indeed a well conceived proposal. Howeve neither Pontecorvo nor Alvarez carried out the experiment.

There were three additional advantages of the radiochemical chlorin method that Pontecorvo and Alvarez could not have realized in 1948.

1. A $^{37}\mathrm{Cl} - {}^{37}\mathrm{Ar}$ detector is only sensitive to electron neutrinos, a clea advantage for solar neutrino observations.

2. The contribution from excited states, particularly the isobaric analogu state in ^{37}Ar at an excitation energy of 5.0 MeV, greatly increases th neutrino capture cross section for energetic neutrinos (Paper 1.A.I Furthermore, Bahcall suggested that by studying the beta decay o ^{37}Ca to ^{37}K that feeds various excited states in ^{37}K one can determin the ft-values for the individual transitions. These beta decay processe

are the mirror of the neutrino absorption reaction in ^{37}Cl to these isospin equivalent states in ^{37}Ar. Bahcall (Paper 1.A.I and reference [16]) made an analysis of the then sparse nuclear data, before the first measurement of the ^{37}Ca decay, that gave a cross section for ^{8}B decay neutrinos of $1.27(1 \pm 0.25) \times 10^{-42}\text{cm}^{-2}$, a factor of 17 higher than the ground state cross section. For the most recent evaluation of the cross section, $1.11(1 \pm 0.04) \times 10^{-42}\text{cm}^{-2}$, see reference [17]. This development increased enormously the sensitivity of the Homestake experiment for observing the ^{8}B solar neutrino flux, a factor that was crucial in obtaining funding for the project.

However, the sensitivity to the ^{8}B decay neutrinos means that the Homestake experiment is very dependent on the neutrino flux of a source that itself depends upon details of the parameters used in the solar modeling.

3. Pulse rise-time measurements for proportional counters were developed in 1968. The application of this technique greatly increases the discrimination for resolving ^{37}Ar-like events from beta rays and Compton electrons, the most prevalent source of counter background. This later technique permitted the Homestake experiment to observe a clear signal from solar neutrinos (Paper 2.B.I).

The early results from the Homestake experiment were published in Paper 1.B.II. The design and operation of the experiment is described; also described were the tests performed to insure that ^{37}Ar would be recovered by the procedures used. The results were given from two exposures in the form of limit on the total neutrino capture rate in ^{37}Cl of $0.3 \times 10^{-35}\text{s}^{-1}$ per ^{37}Cl atom or 3 SNU. A paper presented at the Moscow conference on neutrino astrophysics gives more information on background processes[18]. During this conference I discovered there was considerable effort devoted to the design of a Cl–Ar experiment in the USSR. More information about this important program will be given later on in this introduction.

The publication of the initial results was accompanied by a set of new theoretical model calculations (Paper 1.C.I) that greatly reduced the predicted solar neutrino capture rate for ^{37}Cl to (7.5 ± 3) SNU. One important change in this solar model was the adoption of a significantly lower heavy element composition, $z = 0.015$ for their model C, compared to 0.034 used by Bahcall, Fowler, Iben and Sears (Paper 1.B.I).

Some additional experiments were performed with the Homestake detector to verify the low results, and to carry out various tests to measure the extraction efficiency of ^{37}Ar. It was clear that the sensitivity of the experiment must be improved considerably to search for a neutrino signal from the sun. The early results were obtained using pulse-height analysis to distinguish ^{37}Ar decay events from counter background in the miniatu proportional counters that were used.

I learned from Gordon Garmire, an X-ray astronomer from the Californi Institute of Technology, of using pulse-rise-time measurements for recognizin X-ray and Auger events in gas proportional counters. An electronic syste was developed at Brookhaven National Laboratory by Radeka, Chase, an Rogers that could distinguish between ^{37}Ar decays and background even from beta rays and Compton electrons. The method they developed is calle an amplitude of the differentiated pulse (ADP) versus energy (E) or ADP- method[19, 20, 21]. This timely development gave the Homestake experime a new life.

Fortunately, after introducing pulse rise-time analysis a clear ^{37}Ar sign was observed. Some of the early experiments using the method were given i the review article (Paper 1.B.II), and[22]. In these articles a simplified deca analysis of the time distribution of the pulses in the ^{37}Ar region (defined i energy and ADP) into a decaying component with a half-life of 35 days and constant background rate. A maximum likelihood method was developed at later time to analyze the counting data. This method was developed by Bruc Cleveland[23] specifically for analyzing the data from the rise-time syste (ADP-E). This system was used to analyze all Homestake data from th time that the ADP-E method was used. The first experiment that employe the ADP-E system was run 18 in 1970.

To end this discussion on the Homestake Cl–Ar radiochemical experime I would like to give a set of references and some comments. The results of th Homestake experiment were not readily accepted by the scientific communit nor were the solar model calculations and the nuclear data that was use There followed a long period of searching for explanations.

John Bahcall and I wrote an account of the development of solar neutrin research in a volume dedicated to Professor W. A. Fowler on the occasio of his seventieth birthday[24]. There were also two conferences devoted t a detailed discussion of all topics related to solar neutrino research. Ever word was recorded in both of these conferences and transcribed in the report The first was organized by Fred Reines at the University of California, Irvin in February 1972[25]. There was a lively discussion on the details of th Homestake experiment and how it could be tested, solar modeling, nuclea reaction measurements, and possible new experiments.

The second conference was organized by Gerhart Friedlander at Brookha National Laboratory, January 5–7, 1978[26]. The same topics were dis cussed, but there was more emphasis on possible new experiments: ^{7}Li, ^{71}Ga ^{81}Br, ^{87}Rb, and ^{205}Tl radiochemical, directional Cerenkov and fine graine scintillators, and direct counting ^{115}In. At this conference Heidelberg an Brookhaven formed a collaboration to develop a gallium radiochemical de tector. These two conferences defined the future direction in solar neutrin research.

Neutrino Research in Russia

A vigorous program in cosmic rays and neutrino astrophysics was developed in the Soviet Union in the early 1960's under the direction of M. A. Markov, G. T. Zatsepin and others. The new developments in solar neutrino research became a major part of their program. Initially they planned a large chlorine experiment. However, deep mines were not available in the Soviet Union, and it was necessary for them to build an underground laboratory for cosmic ray and neutrino experiments. They chose to build their laboratory in Baksan Valley in the North Caucasus Mountains by drilling a 4 km long horizontal adit under Mt. Andyrchi, and building chambers for experimental facilities as they penetrated deeper into the mountain.

The suggestion of using the neutrino capture reaction $^{71}Ga(\nu, e^-) \geq 71$ to observe the low energy p-p neutrinos was made in 1963 (Paper 5.A.II). The gallium experiment was part of a plan to use three separate radiochemical detectors to resolve the principal neutrino sources in the sun; the p-p reaction, ^7Be decay and ^8B decay using ^{71}Ga, ^7Li, and ^{37}Cl as target elements[27]. A similar concept was published by Bahcall (Paper 4.A.II). Observing the solar neutrino spectrum by using three radiochemical experiments has been an overall goal of their program.

A 60 ton gallium experiment became a first step in this program. This experiment is now in operation in a chamber at 4 km from the entrance. They chose to use metallic gallium as the target material. Although, using a metal target involves using an unusual chemical procedure to extract ≥ 71, their chemical process was well tested. After the experiment was built two groups, from Los Alamos National Laboratory and the University of Pennsylvania, were invited to join the Soviet gallium experiment (SAGE). Their primary role was to assist in improving counting procedures and process chemistry. The initial results from 30 tons of gallium was published in 1992 (Paper 5.B.II). The results of the initial observations were much lower than the predictions of the standard solar model, only an upper limit of 55 SNU was given. The latest results and details of their chemical procedures are described in a paper published in 1994 (Paper 8.B.II). In the new paper a higher rate was obtained with a full 60 ton experiment, 73 ± 20 SNU. The expected rate predicted by the standard model given in Paper 2.D.I is 132 SNU.

A one ton lithium detector is now planned as a pilot experiment to test their chemical and counting procedures[28]. The lithium radiochemical detector will use molten metallic lithium as the target material. Using a metal target requires an unusual chemical extraction procedure for recovering ^7Be. Liquid lithium was chosen because it greatly reduces the production of ^7Be by background processes from cosmic ray muons. The lithium-beryllium experiment, based upon the mirror transition ^7Li to ^7Be, has the highest neutrino capture rate per ton of target element of all radiochemical experiments. However, there are two major difficulties to be overcome to make it into a practical experiment, background effects and a practical, efficient method for counting ^{7}Be decays. The Russian group has promised solutions for both these problems[28].

For many years a 3,000 ton chlorine experiment has been planned and a considerable effort has been devoted to this project. However, a very large chamber will be required to house this experiment. At the present time, it seems unlikely that the mining effort needed to excavate the chamber will be supported. One tank to hold 1,000 tons of perchloroethylene has already been fabricated, and is ready for assembly. The necessary extraction equipment has been built. We must wait for better days to complete this project.

The GALLEX Experiment

The gallium solar neutrino experiment became a practical method for observing the p-p reaction neutrinos when industry began extracting ton quantities of gallium as a by-product from the manufacture of aluminum. Producing gallium in quantity was motivated by the need for gallium to produce various electronic devices. In 1974 research on chemical procedures for extracting germanium from gallium began at Brookhaven National Laboratory in collaboration with the University of Pennsylvania. A similar effort was started in the Soviet Union. Two practical chemical procedures were developed. One procedure used gallium metal as a target material, and the second procedure used an acid solution of gallium chloride. The procedures are outlined in Paper 6.A.II. This paper served as an announcement that the gallium experiment, proposed by Kuzmin many years earlier, was a practical means of observing the low energy neutrinos from the basic proton-proton fusion reaction. The theoretical calculations indicated that the flux of p-p neutrinos was expected to be the dominant signal observed by a gallium detector. The flux of p-p neutrinos is accurately calculated from solar models, it is essentially independent of many factors that influence the calculations of the ^{8}B neutrino flux. A measurement of the solar p-p flux was regarded as a critical test of our knowledge of neutrino physics and the fusion processes in the interior of the sun.

The theory of matter oscillations was published in 1978 by Wolfenstein (Paper 1.C.IV). This concept was considered to be a factor in understanding neutrino transport in the sun. The resonance character of matter oscillations discovered by Mikeyev and Smirnov in 1984 (Paper 4.C.IV) greatly amplified the effects expected from matter oscillations. Measuring the p-p reaction neutrinos by a gallium experiment was clearly an important means of testing these new concepts in neutrino physics.

A collaboration was formed in 1978 between Brookhaven National laboratory, the University of Pennsylvania, the Institute of Advanced Study and the Max-Planck Institute at Heidelberg. A pilot gallium experiment was carried out using 500 kg of gallium in the form of an acid solution of gallium chloride. However, the full-scale project was not funded in the U.S.

The Max-Planck Institute formed a new collaboration in Europe. This collaboration (GALLEX) built a 30 ton gallium experiment in the Gran Sasso Underground laboratory. Their first observations are described in Paper .B.II, and a paper published in 1994 summarizes the first 30 experimental runs (Paper 7.B.II). The combined results for 30 experiments is 79 $\pm$ 13 SNU, compared to a standard model value (Paper 4.D.I) of 132 SNU (Paper 2.D.I). One of the difficulties with comparing the results with theoretical expectations is that the contribution of excited states in ≥ 71 is not well determined. Excited state cross sections that were used in predicting the standard model rate were derived from (p,n) reaction studies made in 1985[29]. There have been improvements in this technology since 1985, and it would be very important to make a new set of (p,n) reaction studies to measure again the Gamow-Teller strengths of excited states in ≥ 71.

However, one can tentatively conclude from the results of the two gallium experiments that the p-p reaction is occurring in the sun at the expected rate. This is indeed a gratifying result.

The Kamiokande II Detector

The Kamiokande detector is an imaging water Cerenkov detector designed originally to observe proton decay. the detector was modified to study the ^{8}B neutrino flux from the sun by adding an anti-coincidence shield, purifying the water to remove the natural radioactivities, and adding a new electronic system to provide recording of the time distribution of the photomultiplier tube pulses. After making these changes they had the good fortune to record the anti-neutrinos from Supernova 1987A! Neutrinos from the sun are observed by neutrino-electron elastic scattering from electrons. The energy and direction of the recoiling electron is observed by the emitted Cerenkov radiation by photomultipliers on the walls of the detector tank. The neutrino scattering process is directional (Paper 4.A.I) and allows a determination of the direction of the neutrino that produces the recoil electron, modified by electron scattering in the water. These scattering recoil electrons are not directly related to the energy of the neutrino that is scattered. The recoil electron spectrum from a monoenergetic source of neutrinos has a flat energy spectrum, all recoil energies are equally probable up to the energy of the neutrino. These dynamic effects must be taken into account in analyzing the recorded events. The Kamiokande II detector can give important directional information on the neutrino that cannot be obtained by radiochemical detectors, or direct counting detectors based upon an inverse beta process.

The modified detector, designated Kamiokande II, began observing the ^{8}B decay neutrino flux in January 1987. Their first report was published in 1989. The Paper 4.B.II describes their first 1,000 live days of observations. A more detailed paper for the first 1,000 days of operation is listed as reference[30]. Their recorded recoil electron rates are given as a function of the cosine of the angle toward the sun. A rise in the signal rate in the solar direction is evidence that the observed rate can be ascribed to a flux of ^{8}B decay neutrin from the sun. On the same plot the rate expected from the standard mod of Bahcall and Ulrich (Paper 2.D.I) is shown. The final result is given i terms of the percentage of the expected signal for two energy thresholds the recoil electron. They observe a ^{8}B neutrino flux of 2.5×10^6 compare to a standard model value of 5.6×10^6 $\mathrm{cm^{-2}s^{-1}}$. They make a compariso with the Homestake results that shows there is general agreement within th errors of the two experiments[30, 20, 21]. Observations are being continue with the detector, now designated as Kamiokande III.

A new and much larger experiment is under construction in the same loca tion, Superkamiokande. The new experiment will be able to search for longe lifetime of the proton, will serve as a higher rate solar neutrino detector, an will be a sensitive monitor for collapsing stars (Paper 11.A.II).

The Sudbury Neutrino Observatory (SNO)

Herbert Chen discussed the early work on neutrino interactions with deu terium and suggested a large scale heavy water Cerenkov detector could serv as an important instrument to observe the ^{8}B solar neutrino flux and searc for muon or tau neutrinos from the sun by the neutral current dissociation the deuteron (Paper 7.A.II). A proposal was made to the Canadian Atomi Energy Commission for the use of 1,000 tons of heavy water. A collaboratio was formed to design and build the observatory in the International Nick Company mine at Sudbury (Paper 12.A.II).

There are many advantages of a heavy water solar neutrino detector; 1 the neutrino capture reaction, $D + \nu \rightarrow 2p + e^-$, will be used to measure th shape of the ^{8}B neutrino spectrum, 2) the neutral current process, $D + \nu_x$ – $p + n + \nu_x$, will be used to search for solar muon or tauon neutrinos, 3) th experiment will have a high detection rate of ^{8}B solar neutrinos that ca be applied to a search for variations in the solar neutrino flux and neutrin production in solar flares, and 4) be a sensitive monitor for neutrinos fron supernova events. The Sudbury detector is currently being built at a ver great depth underground, 6,000 m.w.e. The observatory will be operationa in 1995–1996.

A interesting set of articles on various aspects of the Sudbury Observator are published in the March 1992 issue of the journal *Physics in Canada*, bulletin of the Canadian Association of Physicists.

Direct Counting Detectors Under Development

Many years ago R. S. Raghaven proposed using indium as a target fo determining the low energy neutrinos from the p-p reaction in the sun (Pape 2.C.II). Neutrino detection depends upon the absorption of a neutrino b ^{115}In (abundance 95.7%) to produce ^{115}Sn in an isomeric state that decays i 3.26 microseconds to the ground state emitting two successive gammas. Th process allows one to use a delayed coincidence method to achieve neutrino detection. The cross section for the transition is estimated theoretically and by (p,n) reactions. A major difficulty with this method is that ^{115}In is a long lived beta emitter itself (4.4×10^{14} years). Therefore, a 4 ton indium detector needed for an event rate of one per day would have a high internal background counting rate. By making many small individual detectors this experimental problem can be solved in principle, but in practice it is quite difficult. The threshold for the process is 128 keV, similar to a gallium detector, 238 keV. A large effort has been devoted to the development of an indium-115 detector, but the problem of devising a practical system that resolves the neutrino signal from background has not been solved[31]. The most recent development has been carried out at Oxford by Norman Booth and his associates[32]; their approach used many tunnel junction detectors at superconducting temperatures.

The Borexino experiment began as an experiment designed to observe nuclear excitation in boron and other elements following the theory expressed in Paper 9.A.II using a very pure ethyl borate solution. This is a useful delayed coincidence technique that could observe the ^{8}B neutrino spectrum above 4 MeV. However, the 1991 proposal for support for the Borexino detector uses a pure hydrocarbon scintillation liquid. The collaboration proposes to observe the ^{7}Be neutrino line by electron scattering, and a number of other observations related to solar neutrinos[33]. A major question concerning this detector is whether a sufficiently low background will be achieved to allow measuring a clear solar neutrino signal.

Superfluid Helium experiment was proposed in Paper 5.C.II. The operation of the detector depends upon neutrino excitation of rotons and phonons induced in helium below 0.1 mK. These waves interact in the liquid and induce a helium atom to evaporate and be detected calormetrically. The development of the technology has achieved success in observing stopped 5 MeV alphas in 2 liters of helium at 0.1 mK[34].

Radiochemical Experiments Under Development

The Bromine experiment was conceived as an experiment to follow the Homestake experiment. Neutrino capture in ^{81}Br produces ^{81}Kr in an isomeric state that rapidly decays to the long lived ground state of ^{81}Kr (half-life 2×10^5 years. The technology for removing krypton from a very large volume of ethylene dibromide is identical to that used in the Homestake experiment. The number of product ^{81}Kr can be counted by a single atom counting technique developed by G. S. Hurst and his associates. A proposal to build a bromine experiment is described in Paper 4.C.II. The neutrino capture cross section to form the isomeric state in ^{81}Kr was determined, and the sensitivity to high energy neutrinos was measured by (p,n) reaction studies. A bromine detector would have a higher sensitivity to ^{7}Be solar neutrinos than the Homestake experiment. The bromine experiment remains an attractive and technically feasible experiment.

The Iodine experiment described in Paper 10.A.II was proposed by Haxton as a prospective high rate radiochemical experiment for observing ^8B solar neutrinos. The cross sections were estimated, and means of measuring them suggested. The technology of recovering and counting ^{131}I again is easily accomplished. A 100 ton experiment is approved and will be built in the Homestake mine in the next two years, 1994–1996. Experiments have been performed to measure the neutrino capture cross section at the Los Alamos Meson Facility using a stopped positive muon decay neutrino spectrum. If a suitable megacurrie source of ^{37}Ar can be made, the cross section to the first excited state can be measured; a state that would be reached by ^7Be neutrinos. The experiment is being carried out at the University of Pennsylvania under the direction of Kenneth Lande.

Finally two geological experiments will be discussed. These experiments were motivated by the thought that the neutrino luminosity of the sun could vary over long periods of time and thus explain the poor agreement of current neutrino measurements and the solar model. An experiment was proposed by M. S. Friedman and associates to use the solar neutrino capture rate in ^{205}Tl to yield ^{205}Pb in a low level excited state, a reaction that has a very low energy threshold of 46 keV (Paper 1.C.II). A specific thallium mineral lorandite was chosen as one suitable for the measurement. This mineral occurs in an ancient mine in Yugoslavia. Unfortunately, after the paper was written, it was shown[35] that the cross section estimate was incorrect, and there is to date no feasible means of measuring the cross section. The experiment still has its supporters. The thallium experiment was discussed in detail in a conference in Dubrovnik, Yugoslavia in 1986[36].

A second geological experiment was proposed by Cowan and Haxton that takes advantage of inverse beta reactions on molybdenum isotopes to produce two long lived isotopes of technetium ^{97}Tc and ^{98}Tc (Paper 3.C.II). The experiment makes use of the Henderson Molybdenum Mine at Climax, Colorado, and their mineral separation and preparation of molybdenum metal. A complete large scale experiment was carried out by Cowan and Wolfsberg[37]. The results were unsatisfactory because of a high background of technetium isotopes.

References

[1] F. Reines and C. L. Cowan, Jr., Phys. Rev., 92, 830 (1953); F. Reines, C. L. Cowan, Jr., F. B. Harrison, A. D. McGuire, and H. W. Kruse, Phys. Rev., 117, 159 (1960).

[2] F. Reines, Ann. Rev. Nuclear Science, 10, 1 (1960); B. S. Meyer, J. P. F. Sellschop, M. Crouch, W. R. Kropp, H. W. Sobel, H. S. Gurr, J. Lathrup, and F. Reines, Phys. Rev. D, 1, 2229 (1970).

[3] M. G. Menon, et al., Proc. Royal Soc. A, 301, 137 (1967).

[4] Proceedings of Informal Conference on Experimental Physics, CERN (January 20–22, 1965), CERN 65-32, ed. C. Franzinetti.

[5] H. A. Bethe, Phys. Rev., 55, 434 (1939).

[6] R. Davis, Jr., Phys. Rev., 97, 766 (1957); J. N. Bahcall, Phys. Rev. Lett. 17, 398 (1966).

[7] E. M. Burbidge, G. R. Burbidge, W. A. Fowler, and F. Hoyle, Rev. Mod. Phys., 29, 547 (1957).

[8] R. Davis, Jr., *Radioisotopes in Scientific Research*, Proc. of the First (UNESCO) Int. Conf., Paris, 1957, 1, p. 728, ed. R. C. Extermann; for analysis of results from an 18.4 m-ton experiment, see J. N. Bahcall and H. Primakoff, Phys. D, 18, 3463 (1978).

[9] P. Pochoda, and H. Reeves, Planetary Space Sci., 12, 199 (1964).

[10] H. W. Sobel, A. A. Hruschka, W. R. Kropp, J. Lathrop, F. Reines, M. F. Crouch, B. S. Meyer, and J. P. F. Sellschop, Phys. Rev. C, 7, 1564 (1973).

[11] F. Reines, and R. M. Woods, Phys. Rev. Lett., 14, 20 (1965).

[12] W. K. Kropp, Jr., F. Reines, and R. M. Woods, Phys. Rev. Lett., 20, 1451 (1968).

[13] T. L. Jenkins, A Proposed Experiment for Detection of Solar Neutrinos, Case Western Reserve Univ. Report COO-818-62 (1962).

[14] A Discussion on Neutrinos, organized by C. F. Powell and G. D. Rochester for the Royal Society, Proc. Roy. Soc. A, 301, 107 (1967).

[15] L. Alvarez, Proposed Test of the Neutrino Theory, Univ. of California Radiation Laboratory Report UCRL 1949.

[16] J. N. Bahcall, Phys. Rev. B, 135, 137 (1964).

[17] M. B. Aufderheide, S. D. Bloom, D. A. Restler, and C. D. Goodman, Phys. Rev. C, 49, 678 (1994).

[18] Conference on Neutrino Astrophysics, Moscow, September 9–12, 1968, p. 99–132, or Brookhaven National laboratory Report No. 12981; see also ref.[4], p. 201.

[19] R. Davis, Jr., J. C. Evans, V. Radeka, and L. C. Rogers, Neutrino '72 Europhysics Conference, Balatonfured, Hungary, June 11–17, 1972, Vol. 1, p. 5, ed. A. Frenkel and G. Marx, OMKDA-YECHNOINFORM.

[20] R. Davis, Jr., Prog. Part. Nucl. Phys., 32, 13 (1994).

[21] R. Davis, Jr., 6th International Workshop on Neutrino Telescopes, Venezia, Italy, February 22–24, 1994, ed. Milla Baldo Ceolin.

[22] R. Davis, Jr., and John C. Evans, 13th International Cosmic Ray Conference, Denver, Colorado, August 17–31, 1973.

[23] B. T. Cleveland, Nucl. Inst. and Methods, 214, 451 (1983).

[24] J. N. Bahcall, and R. Davis, Jr., Essays in Nuclear Astrophysics, p. 243, ed. C. A. Barnes, D. D. Clayton and D. N. Schramm (Cambridge University Press, 1982). This article is reprinted in Neutrino Astrophysics by J. N. Bahcall (Cambridge University Press, 1989).

[25] Proc. Solar Neutrino Conf., Univ. California, Irvine, February 25–26, 1972, ed. F. Reines and V. Trimble; summary in Rev. Mod. Phys., 45, 1 (1973).

[26] Proc. Informal Conf. on Status and Future of Solar Neutrino Research, Brookhaven National Laboratory, Upton, NY, January 5–7, 1978, BNL Report 50879, 2 Vols., ed. G. Friedlander; summary in Comments on Astrophysics, 8, 47 (1978).

[27] V. A. Kuzmin, and C. T. Zatsepin, Proc. Int. Conf. on Cosmic Rays, Jaipur, India, 1965, paper Mu-Nu 36, page 1023.

[28] A. Kopylov, Frontiers of Neutrino Astrophysics, ed. by Y. Suzuki and K. Nakamura (Universal Academy Press, Inc., Tokyo, Japan, 1993).

[29] D. Krofcheck, et al., Phys. Rev. Lett., 55, 1051 (1985).

[30] K. S. Hirata and the Kamiokande Collaboration, Phys. Rev. D, 44, 2241 (1991).

31] R. Raghavan, and M. Deutsch, Technical Memorandum submitted to Scientific Panel on Solar Neutrinos (DOE/NSF Nuclear Science Advisory Committee), R. Vandenbosch, Chairman, June 12, 1985.

32] N. Booth, Frontiers of Neutrino Astrophysics, ed. by Y. Suzuki and K. Nakamura (Universal Academy Press, Inc., Tokyo, Japan, 1993).

33] The Borexino Collaboration, Borexino at Gran Sasso, August 1991, ed. G. Bellini, M. Campanella, D. Giugni, and R. Raghavan.

34] S. R. Bandler, R. E. Lanou, and others, Proc. IV Int. Workshop on Low Temperature Detection of Neutrinos and Dark Matter, Gif-sur-Yvette, France, 1991, p. 421, ed. N. E. Booth and G. L. Salmon (Editions Frontiéres, 1992).

35] J. N. Bahcall, Rev. Mod. Phys., 50, 881 (1978).

36] Proc. Int. Conf. on Solar Neutrino Detection with Thalium-205 and Related Topics, Dubrovnik, Yugoslavia, September 29–October 3, 1986, Nuclear Instruments and Methods in Physics Research, A271 (no. 2), August 15, 1988.

37] G. E. Kocharov, and M. Wolfsberg in The Sun in Time, ed. C. P. Sonnet et al. (Univ. of Arizona Press 1991).

II. SOLAR NEUTRINO EXPERIMENTS - Davis

A. First Proposals of Experiments Now Being Performed

1. "Inverse Beta Processes," B. Pontecorvo, Chalk River Laboratory Report PD-205 (1946).

2. "Solar Neutrinos. II. Experimental," R. Davis, Jr. *Phys. Rev. Lett.* **12**, 303-305 (1964).

3. "Limits on Solar Neutrino Flux and Elastic Scattering," F. Reines and W. R. Kropp, *Phys. Rev. Lett.* **12**, 457-459 (1964).

4. "Neutrino-Spectroscopy of the Solar Interior," J. N. Bahcall *Phys. Lett.* **13**, 332–333 (1964).

5. "Detection of Solar Neutrinos by Means of the $^{71}Ga(\nu, e^-)^{71}Be$ Reaction," V. A. Kuzmin, *Soviet Phys. JETP* **22**, 1051 (1966).

6. "A Proposed Solar-Neutrino Experiment Using ^{71}Ga," J. N. Bahcall, B. Cleveland, R. Davis, Jr., I. Dostrovsky, J. C. Evans, Jr. W. Frati, G. Friedlander, K. Lande, K. Rowley, R. W. Stoenner, and J. Weneser, *Phys. Rev. Lett.* **40**, 1351–1354 (1978).

7. "Direct Approach to Resolve the Solar Neutrino Problem," H. H. Chen, *Phys. Rev. Lett.* **55**, 1534-1536 (1985).

8. "Predictions for A Liquid Argon Solar Neutrino Detector," J. N. Bahcall, M. Baldo-Ceollin, D. Cline, and C. Rubbia *Phys. Lett. B* **178**, 324–328 (1986).

9. "New Tools for Solving the Solar Neutrino Problem," R. S. Raghavan, S. Pakvasa, and B. A. Brown, *Phys. Rev. Lett.* **57**, 1801 (1986).

10. "Radiochemical Neutrino Detection via $^{127}\mathrm{I}(\nu_e, e^-)^{127}\mathrm{Xe}$," W. C. Haxton, *Phys. Rev. Lett.* **60**, 768 (1988).

11. "The Superkamiokande," M. Takita, in *Frontiers of Neutrino Astrophysics* edited by Y. Suzuki and K. Nakamura (Universal Academy Press, Tokyo), 135-137 (1993).

12. "Sudbury Neutrino Observatory," G. T. Ewan, in *Frontiers of Neutrino Astrophysics* edited by Y. Suzuki and K. Nakamura (Universal Academy Press, Tokyo), 147-149 (1993).

B. Initial Experimental Results

1. "Search for Neutrinos from the Sun," R. Davis, Jr., D. S. Harmer, and K. C. Hoffman, *Phys. Rev. Lett.* **20**, 1205-1209 (1968).

2. "Solar Neutrinos: A Scientific Puzzle" J. N. Bahcall and R. Davis, Jr., *Science* **191**, 264–267 (1976).

3. "The Chlorine Solar Neutrino Experiment," J. K. Rowley, B. T. Cleveland, B. T. and R. Davis, Jr., in *Solar Neutrinos and Neutrino Astronomy*, AIP Conference Proceedings No. 126, Ed. Cherry, M.L., Fowler, W. A., and Lande, K., 1 (1985).

4. "Results from One Thousand Days of Real-Time, Directional Solar-Neutrino Data," K. S. Hirata, *et al.*, *Phys. Rev. Lett.* **65**, 1297-1300 (1990).

5. "The Baksan Neutrino Observatory, Soviet-American Gallium Solar Neutrino Experiment," A. I. Abazov, *et al.*, *Phys. Rev. Lett.* **67**, 3332 (1991).

6. "Solar Neutrinos Observed by GALLEX at Gran Sasso," P. Anselmann, *et al.*, *Phys. Lett.* **B285**, 376-389 (1992).

7. "GALLEX results from the first 30 solar neutrino runs," P. Anselmann, *et al.*, *Phys. Lett.* **B327**, 377-378 (1994).

8. "Results from SAGE, (The Soviet-American Gallium Solar Neutrino Experiment)," J. N. Abdurashitov, *et al.*, *Phys. Lett.* **B328**, 234 (1994).

C. **Other Proposed Experiments**

1. "Solar Neutrinos: Proposal for a New Test," M. S. Freedman, *et al.* *Science* **193**, 1117-1118 (1976).

2. "Inverse β Decay of ^{115}In $\rightarrow$ ^{115}Sn*: A New Possibility for Detecting Solar Neutrinos from the Proton-Proton Reaction," R. S. Raghavan, *Phys. Rev. Lett.* **37**, 259 (1976).

3. "Solar Neutrino Production in ^{97}Tc and ^{98}Tc," G. A. Cowan, and W. C. Haxton, *Science* **216**, 51 (1982).

4. "Feasibility of a ^{81}Br $(\nu, e^-)^{81}$Kr Solar Neutrino Experiment," G. S. Hurst, *et al.*, *Phys. Rev. Lett.* **53**, 1116 (1984).

5. "Detection of Solar Neutrinos in Superfluid Helium," R. E. Lanou, H. J. Maris, and G. M. Seidel, *Phys. Rev. Lett.* **58**, 2498 (1987).

National Research Council of Canada

Division of Atomic Energy

Inverse β Process

P.D. - 205

A Lecture

by

B. Pontecorvo

Chalk River, Ontario

20 November, 1946

INVERSE β PROCESS

Introduction

The Fermi theory of the β disintegration is not yet in a final stage; not only detail problems are to be solved, but also the fundamental assumption — the neutrine hypothesis — has not yet been definitely proven. I will recall briefly the main experimental facts which have lead Pauli to propose the neutrine hypothesis.

1. In a β disintegration, the atomic nucleus Z changes by one unit, while the mass number does not change.
2. The β spectrum is continuous, while the parent and the daughter states correspond to well defined energy values of the nuclei Z and Z $\pm$ 1.
3. The difference in energy between the initial and final states involved in a β transition is equal to the upper limit of the continuous spectrum.

We see that the fundamental facts can be reconciled only with one of the following alternative assumptions:

i. The law of the conservation of the energy does not hold in a single β process.

ii. The law of the conservation of the energy is valid, but a new hypothetical particle, undetectable in any colorimetric measurement - the neutrino - is emitted together with a β particle in a β transition, in such a way that the energy available in such transition is shared between the electron and the neutrino. This suggestion was made by Pauli and on this basis Fermi has built a consistent quantitative theory of the β disintegration. In addition to the difficulties already mentioned, the assumption ii removes some difficulties connected with the conservation of the spin and of the type of statistics of which we cannot speak here.

The main neutrino properties follow "by definition" and are: zero charge, spin $\frac{1}{2}$ and Fermi's statistic.

The problem of the β disintegration has been attacked experimentally in many ways:

(a) β spectroscopy, i.e., study of:- the form of the spectrum, the relationship between the energy released and the probability of disintegration, the ratio of positron to electron emission in cases where both electrons and positrons can be emitted, the ratio of the number of the K capture transitions to positron transitions.

(b) Neutron Decay. This fundamental β transition, the transformation of a free neutron into a proton, has not yet been detected. Plans for its detection, as well as for the study of the angular distribution of the proton and electron emitted, have been made in several laboratories in the U.S.A. and in the Chalk River Laboratory.

(c) Experiments on the recoils of nuclei in a β-ray disintegration. Several authors have attempted experiments of this type. The common feature of all these experiments is that the magnitude of the recoil energy of the nucleus having undergone a β decay process is examined in the light of the laws of the conservation of energy and momentum. The most significant results were obtained by Allen, who studied the recoil of a nucleus having undergone a K electron capture, and by Jacobsen and Kofoed-Hansen, who deduced from their experiments that neutrinos and electrons are emitted prevalently in the same direction. It should be noticed that experiments of this type, while of fundamental significance in the understanding of the β process, cannot bring decisive direct evidence on the basic assumption of the existence of the neutrino. This statement can be understood if we keep in mind that recoil experiments are

> interpreted on the basis of the laws of the conservation of the energy and momentum in individual β processes, i.e., on the basis of the alternative ii, which in effect corresponds essentially to the assumption of the existence of the neutrino.

Direct proof of the existence of the neutrino, consequently, must be based on experiments, the interpretation of which does not require the law of conservation of energy, i.e., on experiments in which some characteristic process produced by free neutrinos, (a process produced by neutrinos after they have been omitted in a β disintegration) is observed.

Inverse β Process

It is clear that inverse β transformations produced by neutrinos are processes of this type and certainly can be produced by neutrino's, if neutrino's exist at all. They consist of the concomitant absorption of a neutrino and emission of a β-particle (positron or negatron), by a nucleus. It is obvious, on thermodinomical grounds, that such process must have an extreme low yield since their inverse, the β process, is so unlikely. It has been currently stated in the literature than an inverse β process produced by neutrino's cannot be observed, due to the low yield. As it will be shown below, this statement seems to be too drastic. The object of this note is to show that the experimental observation of an inverse β process produced by neutrino's is not out of the question with the modern experimental facilities, and to suggest a method which might make an experimental observation feasible.

For completeness, we will mention also some inverse β processes produced by other particles than a neutrino; an inverse β process, more generally, can be defined as the transformation of a neutron into a proton, or vice-versa, produced artificially by bombardment with neutrino's, electrons, or γ-rays. These processes are:

(a) Absorption of a negative β particle ($\bar{\beta}$) with emission of a neutrino. (ν) $\bar{\beta} + Z \longrightarrow \nu + (Z - 1)$

(b) Absorption of a neutrino with emission of a particle:-

$$\nu + Z \longrightarrow \bar{\beta} + (Z + 1)$$

$$\nu + Z \longrightarrow \overset{+}{\beta} + (Z - 1)$$

(c) Absorption of a neutrino accompanied by a K electron capture

$$\nu + Z + (\bar{K}) \longrightarrow Z - 1$$

(d) processes induced by γ radiation

$$\gamma + Z \longrightarrow \nu + \bar{\beta} + (Z + 1)$$

$$\gamma + Z \longrightarrow \nu + \overset{+}{\beta} + (Z - 1)$$

Proposed Method

It is true that the actual β transition involved, i.e., the actual emission of a β particle in processes (b) and (d), and the emission of X radiation in process (c) is certainly not detectable in practice. However, the nucleus of charge Z $\pm$ 1, which is produced in any of the reactions indicated above, may be (and generally will be) radio-active, with a decay period well known, (see, for example, Seaborg's Table of Radioelements). Consequently, the radio-activity of the produced nucleus may be looked for as a proof of the inverse β process.

The essential point, in this method, is that radio-active atoms produced by an inverse β-ray process have different chemical properties from the irradiated atoms. Consequently, it may be possible to concentrate the radio-active atoms of known period from a very large irradiated volume. In the case of electron irradiation, the effective volume irradiated may be of the order of cubic centimeters; in the case of γ-ray irradiation, the volume may be of the order of a liter and for neutrino irradiation, the volume is limited only by practical considerations and may be as high as 1 cubic meter. Elements to be considered for irradiation must be selected according to a compromise between their desirable properties, which are:-

1. The material to be irradiated must not be too expensive, since large volumes are involved.

2. The nucleus produced in inverse β transformation must be radio-active with a period of at least one day, because of the long time involved in the separation.

3. The separation of the radio-active atoms from the irradiated material must be relatively simple. If a chemical separation is involved, it is necessary that the addition of only a few grams of a non-isotopic carrier, per hundred liters of material treated, gives an efficient separation. Isotopic carriers must be used only in the last phase of the separation. An electro-chemical separation is another possibility, presenting some advantage because of the absence of carriers. If the nucleus formed in the inverse β process is a rare gas, the separation can be obtained by physical methods, again without carrier, for example, by boiling the material irradiated. This is the most promising method, according to Dr. Frisch and the writer.

4. The maximum energy of the β-rays emitted by the radio element produced must be very small, i.e., the difference in mass of the elements Z and Z $\pm$ 1 must be small. This is so because the probability of an inverse β process increase rapidly with the energy of the particle emitted, as it will be explained below. Of course, the requirement that the mass of Z is close to the mass of Z $\pm$ 1 is not important if the bombarding particles have an energy much higher than the difference in the masses of Z and Z $\pm$ 1. While γ-rays or electrons produced by betatrons or sincotrons may easily satisfy this condition, strong sources of high energy neutrinos are not

available, so that the requirement is of importance in a neutrino experiment.

5. The background, (i.e., the production of element Z $\pm$ 1 by other causes than the inverse β process), must be as small as possible.

<u>An Example</u>

There are several elements which can be used for neutrino radiation in the suggested investigation. Clorine and Bromine, for example, fulfill reasonably well the desired conditions. The reactions of interest would be:-

$$\nu + Cl^{37} \longrightarrow \bar{\beta} + A^{37}$$

$$A^{37} \longrightarrow Cl^{37} \quad \text{(34 days K capture)}$$

and

$$\nu + Br^{79,81} \longrightarrow \bar{\beta} + Kr^{79,81}$$

$$Kr^{79,81} \longrightarrow Br^{79,81}$$

(34 hrs.; emission of positrons of .4 MEV.)

The experiment with Clorine, for example, would consist in irradiating with neutrinos a large volume of Chlorine or Carbon Tetra-Chloride, for a time of the order of one month, and extracting the radio-active A^{37} from such volume by boiling. The radio-active argon would be introduced inside a small counter; the count efficiency is close to 100%, because of the high Auger electron yield. Conditions 1, 2, 3, 4, are reasonably fulfilled in this example. It can be shown also that condition 5, implying a relatively low background, is fulfilled.

Causes other than inverse β processes capable of producing the radio element looked for are:-

(a) (np) processes and Nuclear Explosions. The production of background by (np) process against the nucleus bombarded is zero, if the particular inverse β process selected involves the emission of a negatron rather than the emission of a positron.

This is the case in the inverse β process which would produce A^{37} from Cl^{37}. Similar arguments show that "cosmic ray stars" cannot produce a direct background of A^{37} from Cl^{37}. As for (np) processes in impurities, the fact that K^{37} does not exist in nature rules out this possibility.

(b) (n,γ) Process. This effect can produce background only through impurities. In principle at least, it can be reduced by addition of neutrons absorbing material. In the case considered, A^{37} could be produced by absorption of neutrons in A^{36} present to an extent of .3% in natural argon still present as contamination. It is estimated than (n,2n) effects, again through impurities, would not produce high background.

(c) (p,n) effects. These effects are estimated to be very small. They would arise from cosmic rays, and are consequently independent of the neutrino strength used. They could be investigated in a blank experiment.

Cross-Sections

If W is the mass difference between the two atoms involved in the inverse transition, Ep is the energy of the impinging particle, E is the energy of the emitted particle, we have $E = Ep - W$. We will see that the cross-section $\sigma_{inv.}$ for the inverse β process increases rapidly with E, so that there is advantage in having a small W, at least for an energy of the primary particle smaller than 10 MEV.

Fierz and Bethe first gave a theoretical value for the cross-sections of an inverse β process. A general dimensional argument given by Bethe and Peierls will be given here. This argument permits the estimate of the order of magnitude of $\sigma_{inv.}$ by using only the empirical knowledge of the β-ray lifetimes.

On thermodinomical grounds, the cross-section $\sigma_{inv.}$ of an inverse β process produced by neutrino must be given by a formula of the type $\sigma = \frac{K}{\tau}$ cm^2, where $\frac{1}{\tau}$ is the probability per unit time of a β disintegration involving energy E and K is a constant of proportionality having the dimensions of Cm^2 X sec.. The largest possible length involved is the wave length of the impinging neutrino and the longest time involved is that length divided by c. We can write then, the above formula in the form

$$\sigma_{inv.} \leq \lambda^2 \times \frac{\lambda}{c} \times \frac{1}{\tau}$$

which has a quite clear physical meaning. From the above formula, we can recognize immediately that the cross-section will increase with the energy of the impinging particle, if $\frac{1}{\tau}$ increases with a power of E bigger than E^3. Now $\frac{1}{\tau}$ according to our knowledge of the β disintegration, increases about as E^5 for energy of the order of 1 MEV. For energies very high, the dependence of $\frac{1}{\tau}$ on the energy is not known. It might be considerably higher. The Konopinski and Uhlenbeck modification of Fermi theory would give a dependence $\propto E^7$. We can conclude that the cross-section for an inverse β process produced by neutrinos with emission of a β particle, increases with a high power of the energy of the bombarding neutrino.

For E = 5 MEV., τ might be as small as .1 sec., λ^2 and $\frac{\lambda}{c}$ are respectively of the order of 10^{-21} cm^2 and 10^{-22} sec., so that $\sigma_{inv.}$ for neutrinos of 5 MEV. may be of the order of 10^{-42} cm^2. The evaluation is more complicated when many levels participate to the process, because of the uncertain dependence of the matrix elements on the energy excitation.

Assuming, for example, that 1 cu. meter of CCl^4 is used for the experiment, the number of nuclei of Cl^{37} is about 10^{28}, and the number of disintegrations N per second of A^{37} produced at saturation in such volume is,

N = neutrino flux X $\sigma_{inv.}$ X $10^{28} \sim$ neutrino flux X 10^{-14}

The effect might be detected if N is of the order of 1, requiring a neutrino flux of the order of 10^{14} neutrinos per cm^2/sec. Such value of the neutrino flux, though extremely high, is not too far from what could be obtained with present day facilities.

Sources

The neutrino flux from the sun is of the order of 10^{10} neutrinos/ cm^2/sec. The neutrinos emitted by the sun, however, are not very energetic. The use of high intensity piles permits two possible strong neutrino sources:

1. The neutrino source is the pile itself, during operation. In this case, neutrinos must be utilized beyond the usual pile shield. The advantage of such an arrangement is the possibility of using high energy neutrinos emitted by all the very short period fission fragments. Probably this is the most convenient neutrino source.
2. The neutrino source is the "hot" uranium metal extracted from a pile, or the fission fragment concentrate from "hot" uranium metal. In this case, neutrinos can be utilized near to the surface of the source, but the high energy neutrinos emitted by the short period fragments are not present.

In the case of the investigation of inverse β processes produced by electrons or γ-rays of high energy, the best source is a betatron or a sincotron.

SOLAR NEUTRINOS. II. EXPERIMENTAL*

Raymond Davis, Jr.
Chemistry Department, Brookhaven National Laboratory, Upton, New York
(Received 6 January 1964)

The prospect of observing solar neutrinos by means of the inverse beta process $^{37}Cl(\nu, e^-)^{37}Ar$ induced us to place the apparatus previously described[1] in a mine and make a preliminary search. This experiment served to place an upper limit on the flux of extraterrestrial neutrinos. These results will be reported, and a discussion will be given of the possibility of extending the sensitivity of the method to a degree capable of measuring the solar neutrino flux calculated by Bahcall in the preceding paper.[2]

The apparatus consists of two 500-gallon tanks of perchlorethylene, C_2Cl_4, equipped with agitators and an auxiliary system for purging with helium. It is located in a limestone mine 2300 feet below the surface[3] (1800 meters of water equivalent shielding, m.w.e.). Initially the tanks were swept completely free of air argon by purging the tanks with a stream of helium gas. ^{36}Ar carrier (0.10 cm^3) was introduced and the tanks exposed for periods of four months or more to allow the 35-d ^{37}Ar activity to reach nearly the saturation value. Carrier argon along with any ^{37}Ar produced were removed from the tanks by sweeping them in series with 5000 liters of helium. Argon was extracted from the helium gas stream with activated charcoal at 78°K. Finally the argon was desorbed from the charcoal, purified and counted. The over-all efficiency of the processing was determined by ^{36}Ar isotopic analysis of the recovered argon. The recovery of carrier argon was always greater than 95%. The entire argon sample was placed in a small proportional counter 1.2 cm long and 0.3 cm in diameter to measure the ^{37}Ar activity. Pulse-height analysis was used, and counters were recorded in anticoincidence with a ring of proportional counters, and an enveloping NaI crystal. The counter was provided with an end window to permit exposure of the counting volume to ^{55}Fe x rays for energy calibration and determination of the resolution of the counter. The resolution, full width at half-height for the 2.8-keV Auger electrons from the ^{37}Ar decay, was 26%. The over-all counter efficiency for ^{37}Ar in the full peak was 46%. The counting rate with the sample was 3 counts in 18 days and this is the same as the background rate for the counter filled with air argon. Therefore the observed counting rate of 3 counts in 18 days is probably entirely due to the background activity. However, if one assumes that this rate corresponds to real events and uses the efficiencies mentioned, the upper limit of the neutrino capture rate in 1000 gallons of C_2Cl_4 is ≤ 0.5 per day or $\varphi\bar{\sigma} \leq 3\times10^{-34}$ sec^{-1} (^{37}Cl atom)$^{-1}$. From this value, Bahcall[2] has set an upper limit on the central temperature of the sun and other relevant information.

On the other hand, if one wants to measure the solar neutrino flux by this method one must use a much larger amount of C_2Cl_4, so that the expected ^{37}Ar production rate is well above the background of the counter, 0.2 count per day. Using Bahcall's expression,

$$\sum\varphi_\nu(\text{solar})\,\sigma_{\text{abs}} = (4\pm2)\times10^{-35}\ \text{sec}^{-1}\ (^{37}\text{Cl atom})^{-1},$$

then the expected solar neutrino captures in 100 000 gallons of C_2Cl_4 will be 4 to 11 per day, which is an order of magnitude larger than the counter background. On the basis of experience obtained with the present experiment, an increase in the volume of liquid to 100 000 gallons would not present any insuperable difficulties. The result of such an experiment would provide a valid test for the present theory of the solar energy generation process. The important features of the method are that small amounts of ^{37}Ar can be removed efficiently from large volumes of liquid by the simple procedure of sweeping with helium and that the characteristic decay of ^{37}Ar can be observed in a counter with an essentially zero background. There are, however, a number of other processes that could produce ^{37}Ar at these low levels in a tank of perchlorethylene in an underground mine; these other effects constitute an undesirable background. Alvarez[4] made a thorough analysis of these unwanted effects in his original proposal some years ago. In general, background effects may arise from cosmic-ray muons, from fast neutrons from the surrounding rock wall, and from nuclear reactions arising from internal contaminations in the liquid.

Cosmic-ray background effects underground arise by the $^{37}Cl(p, n)^{37}Ar$ reaction from the protons produced in muon interactions. The magni-

Table I. ^{37}Ar production rates in C_2Cl_4 by cosmic-ray muons underground.

Depth below surface (m.w.e.)	Muon intensity (cm^{-2} sec^{-1} sr^{-1})	Muon star production cross section (cm^2/nucleon)	^{37}Ar production rate per day for 10^5 gallons C_2Cl_4
25	2×10^{-3}	3×10^{-30}	6500 (measured)
1800	2×10^{-7}	17×10^{-30}	3.5
4000	6×10^{-9}	22×10^{-30}	0.14

tude of this background effect can be calculated from measurements made at a depth of 25 m.w.e.[5] where the nucleonic component is essentially eliminated, and the ^{37}Ar is produced by muons. At this depth a ^{37}Ar production rate of 210 atoms per day was observed 3000 gallons of CCl_4. Below this depth the ^{37}Ar production should decrease with the product of the muon intensity and the cross section for star production by muons. Table I lists the muon intensities[6] and cross sections[7] used to calculate the ^{37}Ar production by muons at 1800 and 4000 m.w.e.

It may be noted from Table I that the calculated rate at 1800 m.w.e. is below the limit set by the present 1000-gallon experiment. However, it is clear that a large-scale experiment would have to be performed at a much greater depth. If the proposed experiment were conducted in a mine approximately 4500 feet deep (4000 m.w.e.) the muon-produced ^{37}Ar would be a factor of 30 below the expected rate of 4 to 11 per day from solar neutrinos.

^{37}Ar may also be produced in the liquid by energetic neutrons. Neutrons having an energy above 0.97 MeV will produce protons by the exothermic $^{35}Cl(n,p)^{35}S$ reaction with sufficient energy to produce ^{37}Ar by the $^{37}Cl(p,n)^{37}Ar$ reaction. This effect was evaluated by irradiating the liquid with a Pu-Be neutron source. These measurements gave a yield of one ^{37}Ar atom per 1.4×10^{8} neutrons absorbed. Fast neutrons from the surrounding rock could produce one ^{37}Ar atom per day if the neutron flux on the surface of the 100 000-gallon tank (26 ft diameter × 26 ft high) were 4×10^{-4} neutron cm^{-2} min^{-1}. The fast neutron flux may be kept below this value by a water shield, the thickness depending on the uranium and thorium content of the rock wall.

Internal contaminations leading to ^{37}Ar production in the materials of the tank or the liquid itself cannot be shielded out, and would serve as an inherent background that could not be separated from a neutrino signal. We have, however, found that the thorium and uranium content of perchlorethylene was less than 2×10^{-9} g per gram. At this level internal neutron production is unimportant; less than 0.01 ^{37}Ar would be produced per day by these neutrons. However, even at this uranium and thorium level the sulfur content must be below 0.5% to reduce the ^{37}Ar produced by the $^{34}S(\alpha,n)^{37}Ar$ reaction to less than one per day.

We may conclude from the above considerations that an experiment using 100 000 gallons of pure perchlorethylene in a mine 4500 feet deep, properly shielded from fast neutrons, would have a background ^{37}Ar production rate at least a factor of ten below the expected rate from solar neutrinos. It should be noted that if a positive result were obtained from such an experiment there would remain a small ambiguity in interpretation because of the possibility of a galactic source of neutrinos. A possible method of distinguishing between solar and galactic neutrinos would be to take advantage of the eccentricity of the earth's orbit and measure the 7% difference in solar neutrino intensity between aphelion and perihelion. With a signal as low as 7 per day (a total of 350 ^{37}Ar atoms) such an experiment would be marginal, but if a somewhat higher signal was observed such a test would be possible.

Dr. John Bahcall, Dr. R. L. Sears, and Professor W. A. Fowler of the California Institute of Technology have provided much of the theoretical and experimental information that makes this experiment meaningful. I would like to thank them for keeping me informed of these developments. Also, I would like to acknowledge the assistance of Robert L. Chase for designing the low-background counting system.

---

*Research performed under the auspices of the U. S. Atomic Energy Commission.

[1]R. Davis, Jr., in Radioisotopes in Scientific Research (Pergamon Press, New York, 1958), Vol. 1.

[2]J. Bahcall, preceding Letter [Phys. Rev. Letters 12, 300 (1964)]. If only ground-state transitions are considered, a solar neutrino capture rate in 10^5 gallons of C_2Cl_4 from ^{7}Be and ^{8}B neutrinos would be 0.6 and

0.3 per day. An additional contribution to the rate of 5.7 per day would be expected from ^{8}B neutrinos captured to form excited states in ^{37}Ar (approximately 50% error in flux). It is the contribution from excited states that produces an amount of ^{37}Ar in 10^{5} gallons of C_2Cl_4 well above that expected from background effects.

[3]The chemical division of the Pittsburgh Plate Glass Co. kindly allowed us to use their limestone mine at Barberton, Ohio, for this experiment. A more complete report will be published.

[4]L. W. Alvarez, University of California Radiation Laboratory Report No. UCRL-328, 1949 (unpublished).

[5]R. Davis, Jr., and D. S. Harmer (to be published).

[6]P. H. Barrett, L. M. Bollinger, G. Cocconi, Y. Eisenberg, and K. Greisen, Rev. Mod. Phys. 24, 133 (1952); J. Pine, R. J. Davisson, and K. Greisen, Nuovo Cimento 14, 1181 (1959).

[7]G. N. Fowler and A. W. Wolfendale, Progr. Elem. Particle Cosmic Ray Phys. 4, 105 (1958).

LIMITS ON SOLAR NEUTRINO FLUX AND ELASTIC SCATTERING*

F. Reines and W. R. Kropp
Department of Physics, Case Institute of Technology, Cleveland, Ohio
(Received 17 February 1964)

The calculations of Bahcall, Fowler, Iben and Sears[1] on the flux at the earth of neutrinos from the decay of ^{8}B produced in the interior of the sun predict a value of $(2.5\pm1)\times10^7$ ν_e/cm^2 sec. Since the ^{8}B decay neutrinos have energies up to 13.7 MeV, i. e., well in excess of natural radioactivity, we are able to interpret results of an underground experiment performed in another connection[2] in terms of an upper limit on the flux of solar neutrinos providing we assume an elastic scatter of ν_e by electrons as predicted by Feynman and Gell-Mann[3] and Marshak and Sudarshan[4]:

$$\nu_e + e^- \rightarrow \nu_e + e^-.$$

Our data can also be interpreted as an upper limit on the elastic scattering process[5] assuming the flux calculations cited by Bahcall to be correct. We describe these arguments in this Letter following a discussion of the complementary relationship between the present approach and that of Davis.[1] Davis proposes to detect solar ν_e by using the inverse beta decay of ^{37}Cl,

$$^{37}\mathrm{Cl} + \nu_e \rightarrow {}^{37}\mathrm{Ar} + \beta^-.$$

The cross section for this reaction is predictable so that the only unknown factor is the ^{8}B ν_e flux, f. Therefore, given the results of the projected experiment of Davis, f can be deduced, assuming of course that the reaction which is responsible is known from other considerations—the projected Davis experiment is expected to give not f but the product of f times the cross section integrated over the relevant neutrino spectrum.

The direct counting experiment utilizing the elastic scattering process has the additional feature that it is in principle capable of giving information regarding the neutrino energy distribution, so helping to identify the responsible reaction. A negative result here in the face of a positive result from the Davis experiment could be interpreted either as due to the absence of the elastic scattering process or as an indication that the ^{8}B reaction is not responsible for the anticipated Davis results and does not occur in the sun as predicted.

A modest experiment was performed which enabled us to set some limits and helps assess the full-scale effort. It consisted of looking for unaccompanied counts in a 200-liter liquid scintillation detector (5×10^{28} electrons) which was surrounded by a large Cherenkov anticoincidence detector and located 2000 feet underground in a salt mine. The experimental details will be published elsewhere.[2] In a counting time of 4500 hours only three events were observed in the energy range 9 to 15 MeV, unaccompanied by pulses in the anticoincidence guard. If we ascribe these events to recoil electrons produced by the elastic scattering of solar neutrinos from the ^{8}B or ^{4}Li decays we obtain the conservative upper limits

$$f(^8\mathrm{B}) < 10^9\ \nu_e/\mathrm{cm}^2\ \mathrm{sec},\ (<2\times10^8\ \nu_e/\mathrm{cm}^2\ \mathrm{sec}).$$

The number in the brackets is the corresponding flux limit set by Bahcall and Davis. Table I summarizes the calculation of limits assuming the ^{8}B reaction. The results are seen to depend on the lower limit of the recoil electron energy E, which is considered. Figure 1 shows the calculated cross-section, σ_e per incident ^{8}B ν_e for

Table I. Elastic scattering of ^{8}B neutrinos.

E MeV	$\sigma_e(>E)$ ($\times10^{45}$ cm^2)	Calculated rate[a] R_1 ($\times10^4$ day^{-1})	Present experimental limit R_2 ($\times10^2$ day^{-1})	R_2/R_1	Upper limit on solar ν_e flux (ν_e/cm^2 sec)
8.0	4.7	6.3			
8.5	4.2	4.6	1.6	35	9×10^8
9.0	3.1	3.4	1.6	50	1.2×10^9
9.5	2.1	2.3	1.1	50	1.2×10^9
10.0	1.4	1.5	1.1	75	

[a] $R_1=\sigma_e(>E)Nf$, $N=5\times10^{28}$ target electrons, $f=2.5\times10^7$ ν_e/cm^2 sec.

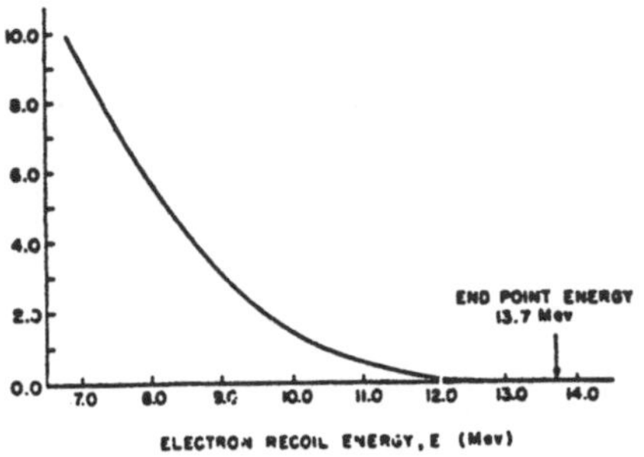

FIG. 1. Cross section, $\sigma_e(>E)$, per target electron, per incident ^{8}B neutrino, for the production of recoil electrons with energy in excess of E.

the production of recoil electrons above the energy E. On this basis, the elastic scattering cross section is seen from the table to be <35 times the expected value, a better limit by ~10 than that previously set for $\bar{\nu}_e$'s from a fission reactor.[8]

It is interesting to inquire how large a detector would be required to see the predicted ^{8}B solar neutrino and then to consider the question of backgrounds. If the flux is taken to be 2.5×10^7 ν_e/cm^2 sec and the lower limit on electron recoil energy is set at 8 MeV, then we predict a rate of 1.5 events per year in a metric ton of scintillator. Since it is within our current experience with large detectors to consider a sensitive volume of 10^4 gallons we could build a system with a predicted rate ~50/yr. The background which appears most serious is that resulting from the "pileup" of pulses due to natural radioactivity. The most troublesome natural gamma emitter is ThC" which[7] produces a cascade of gammas of energy totalling from 2.62 to 3.96 MeV. Assuming a threefold pileup is required to give an apparent pulse >8 MeV and a resolving time of 10^{-7} sec (modest) in a detector made up of $N(=24)$ identical elements, the singles rate per element, n sec^{-1}, consistent with a total background rate, R, of 10/yr is given by

$$R = 2n^3\tau^2 \times 3.1\times10^7 N.$$

Inserting numbers we find

$$n = 88/\text{sec per element}.$$

If the resolving time is reduced a correspondingly larger value of n becomes acceptable.

As an example of attainable backgrounds[8] a scintillation detector of ~10 m^2 surface area and located 2000 feet underground in a salt mine exhibits an integral count rate above 3 MeV of only 6 sec^{-1}, an order of magnitude below the required limit per detector element estimated above.

The twofold pileup background in the complete detector array, using the integral rate above 4 MeV in the 10-m^2 test detector to make an estimate, is ~50/yr, again assuming a resolving time of 10^{-7} sec. It therefore seems reasonable that the background problem due to natural radioactivity can be reduced to acceptable levels.

A second source of background can result from the passage of cosmic rays through the detector. It is improbable that a detector located near the earth's surface can be sufficiently well shielded by a charged-particle anticoincidence detector from cosmic rays which deposit energy in the range, 8-18 MeV, of interest here. It therefore appears necessary to locate such a detector deep underground where the cosmic radiation is penetrating muons and their secondaries. If, to assess the situation, we assume a detector of 25 m^2 surface area as seen by the cosmic rays, then at a location 2000 feet underground, allowing an anticoincidence factor[9] >10^4, the residual rate due to all cosmic rays is estimated to be <300/yr. Considering only those cosmic rays which deposit between 7 and 18 MeV it appears possible that the rate would be reduced by as much as an order of magnitude or more. There always remains the possibility of going deeper underground.

We wish to thank the Morton Salt Company for continued hospitality in their Fairport Harbor Mine and Dr. M. Crouch and Dr. T. L. Jenkins for interesting discussions.

Note added in proof. – It has been called to our attention by L. Heller that Azimov and Schekhter[10] and Heller[11] have recalculated the elastic scattering process according to the conserved vector current theory and find a cross section twice as large as previously quoted. On this basis our limits on the elastic scattering process are closer by a factor of two to prediction and a detector smaller by a factor of two would suffice to detect solar neutrinos.

*Work supported in part by the United States Atomic Energy Commission.

[1]J. N. Bahcall, C. W. Fowler, I. Iben, Jr., and R. L. Sears, Astrophys. J. 137, 344 (1963). We are indebted to Dr. R. Davis, Jr., and Dr. J. N. Bahcall for calling our attention to the latest estimates of the ^{8}B rates as well as preprints of their publications on

the subject of solar neutrinos.

[2]A paper by the present authors is in preparation. It will contain a discussion of the various limits which can be set on nucleon stability and neutrino fluxes by a consideration of the unaccompanied counts. The earlier stages of this work are described in a paper by C. C. Giamati and F. Reines, Phys. Rev. 126, 2178 (1962).

[3]R. P. Feynman and M. Gell-Mann, Phys. Rev. 109, 193 (1958).

[4]R. E. Marshak and E. C. G. Sudarshan, Phys. Rev. 109, 1860 (1958); Proceedings of the Padua-Venice Conference on Mesons and Newly Discovered Particles, September, 1957 (Società Italiana di Fisica, Padua-Venice, 1958).

[5]Our results can also be used to set an upper limit on the product of the elastic scattering cross section averaged over the upper end of the ^{4}Li decay spectrum (on the unlikely assumption that it is particle stable) times the solar ^{4}Li-produced ν_e flux at the earth. The flux limit so obtained is $<2\times10^3$ ν_e/cm^2 sec. This is to be compared with the Bahcall-Davis limit of $<1\times10^8$ ν_e/cm^2 sec.

[6]C. L. Cowan, Jr., and F. Reines, Phys. Rev. 107, 528 (1957). This experiment was interpreted in terms of an upper limit on the neutrino magnetic moment. We here reinterpret these data in terms of the conserved vector current predictions.

[7]D. Strominger, J. M. Hollander, and G. T. Seaborg, Rev. Modern Phys. 30, 585 (1958).

[8]L. V. East, T. L. Jenkins, and F. Reines (unpublished).

[9]M. K. Moe, T. L. Jenkins, and F. Reines, Rev. Sci. Instr. 35, 370 (1964).

[10]Ya. I. Azimov and V. M. Shekhter, Zh. Eksperim. i Teor. Fiz. 41, 592 (1961) [translation: Soviet Phys.—JETP 14, 424 (1962)].

[11]L. Heller, Los Alamos Scientific Laboratory Report LAMS-3013, 1964 (unpublished).

NEUTRINO-SPECTROSCOPY OF THE SOLAR INTERIOR *

J. N. BAHCALL
California Institute of Technology, Pasadena, California

Received 20 November 1964

An experiment is currently underway [1] to test directly the theory of stellar energy generation by observing the relatively high-energy neutrinos (14 MeV maximum energy) emitted from the interior of the sun in the rare mode of the proton-proton chain that involves B^8[2]. As a follow-up to this experiment, the author has proposed [3] that a program of neutrino-spectroscopy of the solar interior be carried out to determine quantitatively the conditions in the interior of the sun by using a variety of neutrino-absorbers having, for example, different absorption thresholds. It would be particularly desirable to try to observe the numerous low-energy neutrinos from the basic reaction, $H^1(p, e^+\nu_e)H^1$, of the proton-proton chain (0.43 MeV maximum energy) and from the frequently occurring $Be^7(e^-, \nu)Li^7$ reaction (0.86 MeV maximum energy). The ratio of B^8 neutrinos to low-energy neutrinos would provide a crucial and stringent test of current theories of the interior of main sequence stars. If no neutrinos are detected in the Davis-Harmer experiment [1] (0.81 MeV threshold energy), it will be even more desirable to try to observe the low-energy neutrinos.

In this note we present some theoretical results concerning the following neutrino absorbers: H^3, Li^7, B^{11} and Rb^{87}. Two of these absorbers (H^3 and Rb^{87}) can detect primarily the low-energy solar neutrinos ** and the other two *** (Li^7 and B^{11}) are primarily sensitive to the B^8 neutrinos. In addition, Li^7 and B^{11} can be used to establish the direction of the neutrino source (presumably toward the sun) by observing the direction in which the created electrons are produced †. The ability of targets made from Li^7 or B^{11} to distinguish the direction of the neutrino source is especially valuable since it will probably be difficult to demonstrate by the method of Davis and Harmer that neutrinos observed in their experiment come from the sun rather than a galactic background [1].

The cross sections for absorption of neutrinos from the most important sources in the sun are given in columns two through six of table 1 for each of the four targets mentioned above. These cross sections include [6] †† corrections due to excited-state transitions and have been averaged over the appropriate incident neutrino spectra. In column seven of table 1, the asymmetry parameter α is given for each of the isotopes

* Supported in part by the Office of Naval Research [Nonr-220(47)] and the National Aeronautics and Space Administration [NGR-05-002-028].

** The use of Sr^{87} as a possible detector of low-energy neutrinos was originally proposed by Sunyar and Goldhaber [4].

*** The suitability of B^{11} as a target for detecting B^8-neutrinos was suggested to the author by F. Reines. Reines also suggested that the reaction $B^{11}(\nu_e, e^-)C^{11}$ could be used to establish the direction of the neutrino source by observing the direction of the created electron.

† The idea of using the direction of the created electrons to establish the direction of the neutrino source was originally proposed by Marx and Menyhard [5].

†† The parameters needed in the present study can all be calculated from known experimental data except for the $\langle\sigma\rangle^2$ for excited-state transitions in the reaction $B^{11}(\nu, e^-)C^{11*}$; these matrix elements can, however, be estimated with sufficient accuracy from a sum rule for Gamow-Teller matrix elements [7].

Table 1

Cross sections and asymmetry parameters. The subscripts such as p-p or Be^7 indicate the neutrino source in the sun (refs. 2 and 7). The cross sections for Rb^{87} refer to the low-lying metastable state of Sr^{87} (ref. 4).

Target	σ_{p-p}	σ_{Be^7}	σ_{B^8}	$\sigma_{N^{13}}$	$\sigma_{O^{15}}$	α
	($\times 10^{45}$ cm²)					
H^3	48	170	6700	150	230	-0.1
Li^7	0	0	4600	5	23	-0.1
B^{11}	0	0	1900	0	0	+0.35 ± 0.05
Rb^{87}	7.1	25	970	22	35	-0.3

Table 2

Predicted number of neutrino-induced reactions per target particle per second. The fluxes, φ, used here are adapted from ref. 10.

Target	$(\varphi\sigma)_{p-p}$	$(\varphi\sigma)_{Be^7}$	$(\varphi\sigma)_{B^8}$	$(\varphi\sigma)_{N^{13}+O^{15}}$
	($\times 10^{35}$ per sec)			
H^3	250 ± 30	200 ± 100	17 ± 8	38 ± 20
Li^7	0	0	11 ± 6	3 ± 1.5
B^{11}	0	0	4.7 ± 2.5	0
Rb^{87}	38 ± 4	30 ± 15	2.4 ± 1.3	6 ± 3

listed *. The parameter α determines the extent to which the outgoing electron indicates the incident neutrino direction in a reaction such as:

$$\nu + Li^7 \rightarrow e^- + Be^7. \quad (1)$$

The angular distribution for a reaction such as (1) is of the form:

$$\frac{d\sigma}{d\Omega_e} = \frac{\sigma_{total}}{4\pi}\left[1 + \alpha\,\frac{v_e}{c}\cdot\hat{q}\right], \quad (2)$$

where v_e is the outgoing electron's velocity and $\hat{q}$ is a unit vector in the direction of the incident neutrino's momentum. For the V-A theory **,

$$\alpha = \frac{\langle 1\rangle^2 - \frac{1}{3}C_A^2\langle\sigma\rangle^2}{\langle 1\rangle^2 + C_A^2\langle\sigma\rangle^2}, \quad (3)$$

where $\langle 1\rangle$, $\langle\sigma\rangle$ are reduced matrix elements with the usual meaning [8].

In order to convert the cross sections given in table 1 into expected numbers of reactions per second per target particle, one must multiply by the predicted solar neutrino fluxes. Table 2 gives the predicted numbers of neutrino-induced reactions per second per target particle from the most important solar-neutrino sources; the fluxes are taken from the work of Sears [9] and the uncertainties shown represent the present author's estimates [6] of uncertainties in the flux predictions.

An experiment that uses either Li^7 or B^{11} as a target to detect neutrinos from B^8 decay and to establish the direction of the neutrino source appears feasible with current technology + [9]. Experiments designed to detect the low-energy neutrinos are more difficult. For example, ten kilograms of tritium are required in order to obtain a counting rate of 300 events per year induced by solar neutrinos. In addition, the rare events induced by solar neutrinos (which primarily produce electrons with energies of the order of a few hundred keV) must be distinguished from the much more numerous events arising from the normal radioactive decay of tritium (maximum energy of electrons ~ 18 keV), presumably by counting only electrons with energies in excess of 18 keV ++.

* The asymmetry parameter α given in table 1 was calculated by averaging over all relevant states weighted according to their relative population by solar-neutrino absorption. If one only counts electrons with energies in excess of some minimum energy W_{min}, α (for Li^7 and B^{11}) is a function of W_{min}. The asymptotic value of α for Li^7 (B^{11}) is + 0.1 (+ 0.5) for $W_{min} \gtrsim 13.6$ MeV ($\gtrsim 12$ MeV).

** See footnote † on page 332.

References

1. R.Davis Jr., Phys. Rev. Letters 12 (1964) 303; R.Davis Jr. and D.S.Harmer (private communication).
2. W.A.Fowler, Astrophys. J. 127 (1958) 551; P.D.Parker, J.N.Bahcall and W.A.Fowler, Astrophys. J. 139 (1964) 602.
3. J.N.Bahcall, Science (to be published, 1964).
4. A.W.Sunyar and M.Goldhaber, Phys.Rev. 120 (1960) 871.
5. G.Marx and N.Menyhard, Science 131 (1960) 299.
6. J.N.Bahcall, Phys. Rev. Letters 12 (1964) 300; Phys. Rev. 135 (1964) B137.
7. J.N.Bahcall and H.A.Weidenmüller, to be published.
8. E.J.Konopinski, Ann. Rev. Nucl. Sci. 9 (1958) 99.
9. R.L.Sears, Astrophys. J. 140 (1964) 477; J.N.Bahcall, W.A.Fowler, I.Iben Jr. and R.L. Sears, Astrophys. J. 137 (1963) 344.
10. F.Reines and R.M.Woods Jr., to be published.

\+ See footnote *** on page 332.

++ A possible method for distinguishing between tritium-decay electrons and neutrino-produced electrons might be to cover the tritium target with a thin absorber sufficient to stop the decay electrons without preventing the neutrino-induced electrons from reaching a surrounding scintillator.

DETECTION OF SOLAR NEUTRINOS BY MEANS OF THE $Ga^{71}(\nu,e^-)Ge^{71}$ REACTION

V. A. KUZ'MIN

P. N. Lebedev Physics Institute, Academy of Sciences, U.S.S.R.

Submitted to JETP editor June 14, 1965

J. Exptl. Theoret. Phys. (U.S.S.R.) 49, 1532-1534 (November, 1965)

The possibilities are examined of detecting solar neutrinos by means of the low threshold reaction $Ga^{71}(\nu,e^-)Ge^{71}$, which can be efficiently employed for recording neutrinos from the $H^1(p,e^+\nu)H^2$ and $Be^7(e^-,\nu)Li^7$ reactions. The merits of the radiochemical method based on this reaction are compared with those of other reactions. The cross section for absorption of solar neutrinos by the Ga^{71} nucleus and the magnitude of the effects on the earth due to neutrinos from various reactions occurring in the interior of the sun are presented. It is pointed out that transitions to the excited states of the Ge^{71} nucleus may be important for neutrino absorption.

1. In connection with the extensively discussed problems of detection of solar neutrinos and studies of the internal structure of the sun[1-10], we wish to call attention in this communication to unique possibilities connected with the development of a radiochemical method based on the reaction $Ga^{71}(\nu, e^-)Ge^{71}$. This reaction was pointed out in [1, 5, 9] as being the most convenient for the detection of solar neutrinos from the reactions $H^1(p, e^+, \nu)H^2$ and $Be^7(e^-, \nu)Li^7$ (and Li^{7*}).

2. We wish to point out here the advantages of the method, based on the reaction $Ga^{71}(\nu, e^-)Ge^{71}$, which make it especially attractive compared with other possible realizations on the idea of the radiochemical method proposed by Pontecorvo.[11]

1) The rather low energy threshold for neutrino absorption, $E_\nu^{thr} = 0.237$ MeV, makes it possible to register neutrinos from practically all reactions in the interior of the sun and, most importantly, the neutrinos with the greatest flux density,[6] from the reactions $H^1(p, e^+\nu)H^2$ and $Be^7(e^-, \nu)Li^7$ (Li^{7*}). The only neutrinos that cannot be registered are those from $He^3(e^-, \nu)H^3$, with energy ~18 keV. We note that in accordance with the Pontecorvo-Davis method,[11, 4] which is based on the reaction $Cl^{37}(\nu, e^-)Ar^{37}$, the threshold amounts to $E_\nu^{thr} = 0.816$ MeV.

2) The cross section for the absorption of a neutrino by the Ga^{71} nucleus is relatively large even for the transition to the ground state of Ge^{71}, log ft ≈4.3. This ensures, by the same token, a very high sensitivity of such a detector to low-energy neutrinos (we note by way of comparison, for example, that Mn^{55}, with a low threshold $E_\nu^{thr} = 0.0220$ MeV, has log ft = 6.0, while Cl^{37} has log ft = 5.0).

3) The Ge^{71} half-life, $t_{1/2} = 11.4$ days,[12] which is very convenient for purposes of chemical extraction of Ge^{71} atoms from a large quantity of Ga (or its compounds), makes it possible to assume that the efficiency of the method can be made sufficiently high if the extraction procedure lasts about five days. The method of [10], connected with the use of $Rb^{87}(\nu, e^-)Sr^{87m}$, has from this point of view the shortcoming that the half-life of Sr^{87m} is small, ~ 2.9 hours, and the extraction procedure can hardly be carried out with sufficient speed. In the case of Fe^{55}, the half-life, to the contrary, is too long, $t_{1/2} = 2.6$ years.

4) The energy released by K-capture in Ge^{71}, ~ 12 keV, is noticeably larger than for K-capture in Ar^{37} (~2.8 keV), giving grounds for hoping to be able to discriminate more readily the false background in the counter.

5) The considerable abundance of the Ge^{71} isotope, ~40%, makes it possible to obtain a noticeable effect per unit mass of the natural isotope mixture.

3. Finally, we wish to call attention to the transitions that are possible under the influence of neutrinos and excited states of Ge^{71}, and can greatly increase the total cross section for the absorption of solar neutrinos. The scheme and identification[12] of the Ge^{71} levels allows us to think that the transitions to the level E = 1.75 MeV ($5/2^-$) will play a noticeable role in neutrino absorption. The role of other states is not clear. In estimating the influence of this transition on the total cross section and the change in the selective sensitivity of the detector, we have assumed that the matrix element of the transition to the 0.175 MeV level in one spin state is equal to the corresponding matrix

Proposed Solar-Neutrino Experiment Using ^{71}Ga

J. N. Bahcall, B. T. Cleveland, R. Davis, Jr., I. Dostrovsky, J. C. Evans, Jr., W. Frati, G. Friedlander, K. Lande, J. K. Rowley, R. W. Stoenner, and J. Weneser
Institute for Advanced Study, Princeton, New Jersey 08540, and Brookhaven National Laboratory, Upton, New York 11973, and University of Pennsylvania, Philadelphia, Pennsylvania 19104, and Battelle Pacific Northwest Laboratories, Richland, Washington 99352, and Weizmann Institute, Rehovot, Israel
(Received 20 March 1978)

A solar-neutrino experiment that uses ^{71}Ga as a detector can distinguish between broad classes of explanations for the discrepancy between prediction and observation in the ^{37}Cl experiment. A radiochemical experiment with the required amount of ^{71}Ga is feasible.

The results of the ^{37}Cl solar-neutrino experiment are in disagreement with the predictions made using a standard model of the solar interior.[1] Many authors have argued that this discrepancy shows that the standard theory of stellar evolution is wrong in some basic aspect and have proposed conceivable ways of modifying the conventional assumptions that are used in stellar model calculations. Other authors have suggested that neutrinos produced in the interior of the sun do not reach the earth, at least not in the form or quantity in which they are emitted. We show in this Letter that these two broad classes of explanation can be distinguished by a feasible experiment involving the reaction

$$\nu_{e,\mathrm{solar}} + {}^{71}\mathrm{Ga} \to e^- + {}^{71}\mathrm{Ge} \qquad (1)$$

first suggested by Kuzmin.[2] The proposed experiment will require about 50 tons of Ga.

The neutrino capture cross sections for solar neutrinos incident on ^{71}Ga are given in Table I for ground-state to ground-state transitions. We present cross sections for all the important sources of solar neutrinos and also for ^{65}Zn (which can be used as a terrestrial neutrino source[3] to verify the overall validity of the proposed experiment). The cross sections have been calculated accurately by the standard methods.[4-7]

The predicted capture rates for various extreme assumptions about the solar interior or neutrino propagation are shown in Table II. The first five models listed in Table II are concerned only with aspects of the solar interior and they are as follows: (1) a standard solar model[8]; (2) a model in which the only neutrinos come from the basic, low-energy reactions ($p+p \to {}^2\mathrm{H}+e^+ + \nu_e$ and $p+e^- + p \to {}^2\mathrm{H}+\nu_e$); (3) a model in which the solar interior is depleted of heavy elements[8]; (4) a model in which the composition of the sun is completely homogenized for its entire lifetime[9]; and (5) an extreme model in which the central temperature of the sun is so high that all of the nuclear energy is produced by the CNO cycle. The standard solar model and the CNO model are inconsistent with the ^{37}Cl experiment. The p-p, low–heavy-elements, and complete-mixing models are consistent with the ^{37}Cl experiment, but are apparently inconsistent with other aspects of the theory or observation of stellar evolution. The adoption of any of the nonstandard models would have important implications for many branches of astronomy and cosmology via the dating of old stars or the inferred helium abundance.

The first four models shown in Table II all give about the same capture rate [80 ± 10 SNU (solar neutrino units)] since they are primarily sensitive to the low-energy p-p (and pep) neutrinos (≥ 70% of the capture rate would come from these basic reactions). The p-p (and pep) neutrino fluxes are practically invariant to parameter

TABLE I. Neutrino absorption cross sections[a] for ^{71}Ga. Also given are the products of flux and cross section for a standard solar model in solar neutrino units (1 SNU = 10^{-36} captures per target atom per sec).

Source	Cross section[b] (10^{-46} cm^2)	Flux × (cross section) (Standard model; SNU)
p-p	10.7	65
pep	157	2
^{7}Be	64	21
^{8}B	3×10^3	1
^{13}N	53	1
^{15}O	92	2
^{65}Zn	67	...

[a]These are cross sections for transitions to the ^{71}Ge ground state only; they have uncertainties of ≤ 5%. The contributions of transitions to excited states are small (cf. Ref. 5).

[b]The cross sections are given per reaction (for p-p and pep) or per decay (for the radioactive sources).

TABLE II. Predicted capture rates for some extreme hypotheses.

Model	Predicted capture rate[a] (SNU)	Tons of Ga for 1 capture/day
Standard sun	92	37
Only *p-p* and *pep*	71	47
Heavy elements depleted	79	43
Homogenized sun	82	41
CNO	487	7
Neutrino oscillations	≤ 31	≥ 110
Neutrino decay	0	$\sim \infty$

[a]The uncertainties in these rates due to uncertainties in the cross sections are $\leq$ 10% except for the pure CNO model (cf. Ref. 5).

changes since, if the sun is currently supplying its luminosity by nuclear fusion via the *p-p* cycle of reactions, then the *p-p* (and *pep*) fluxes are a direct measure of the sun's luminosity.

The last two rows of Table II refer to situations in which the standard theory of weak interactions has been modified to include either neutrino oscillations[10] or neutrino decay.[11] They give much lower predicted rates than any of the listed astronomical hypotheses. The effect of neutrino oscillations, when averaged over the solar neutrino spectrum,[12] yields a reduction factor, R_{osc}, that is the same for all experiments. Comparison of the predictions for the ^{37}Cl experiment with observations[1] yields the value of $R_{osc} \leq \frac{1}{3}$ used in row 6 of Table II. Note, in connection with row 7, that if the higher-energy neutrinos that are most important in the ^{37}Cl experiment decay on their way to the earth, then certainly the lower-energy *p-p* (and *pep*) neutrinos also decay. A counting rate below 70 SNU could also arise, in principle, if the sun is now in an abnormal phase in which its nuclear energy generation rate is much less than its surface luminosity.[13] However, for most of the models that are in the literature,[14] the reduction in the counting rate would not be nearly as great as for either the oscillation or decay hypotheses. Moreover, these latter two processes give specific predictions for the gallium experiment when combined with the ^{37}Cl experiment.

We conclude that, if a ^{71}Ga experiment gave a result in the range of 70–90 SNU, neutrino decay or oscillations over a distance of 1 AU could be ruled out, putting the burden of explaining the low result for ^{37}Cl squarely on the astrophysicists; on the other hand, a ^{71}Ga result at about one-third that level or lower would be evidence for neutrino oscillations, and a zero result would indicate neutrino decay.

We note, from the last column of Table II, that an experiment with enough sensitivity to detect the *p-p* and *pep* neutrinos would require about 50 tons of gallium.[15] The half-life of ^{71}Ge is 11.8 days, so that production of one ^{71}Ge atom per day leads to a steady-state amount of about seventeen ^{71}Ge atoms. Thus a procedure for quantitative and reliable isolation of a few atoms of ^{71}Ge from 50 tons of gallium and for their subsequent purification and counting is required. We have developed suitable procedures for the separation of Ge from two different forms of target: an acidified 8*M* $GaCl_3$ solution and gallium metal.

From a $GaCl_3$ solution the germanium can be swept out as $GeCl_4$ by a He purge at about 60°C. From a metallic Ga target the germanium, even when carrier-free, can be separated by contacting the liquid Ga with an acid solution that contains an oxidizing agent such as H_2O_2. Since a heterogeneous system is involved, a large surface area must be provided. Conditions were found for forming, for a definite and controlled time, a finely divided dispersion of liquid gallium metal in weakly acidic aqueous phase. At the end of the allocated interval of time the dispersion breaks down spontaneously and cleanly into two phases which can be separated readily, and the germanium, along with a small amount ($\leq$0.2%) of the gallium, is found in the aqueous phase. The whole process can be adjusted to take a few minutes. After addition of HCl, the germanium can be swept out of the aqueous solution as $GeCl_4$ by a stream of He. The ensuing treatment in both procedures is the same: $GeCl_4$ is trapped in an alkaline solution and then reduced to germane (GeH_4) with sodium borohydride. Finally, germane is purified by gas chromatography and introduced with argon into a small proportional counter. The techniques developed for ^{37}Ar measurements,[1] including pulse-height and rise-time discrimination, are directly applicable to ^{71}Ge counting. The chemical yield of the procedure is determined by measuring the volume of GeH_4 obtained in relation to the amount of Ge carrier added initially (about 1 mg).

The chemical separation using a $GaCl_3$ target is most appealing because of its simplicity, elegance, and well-understood chemistry; however, since practically all gallium is produced and commercially used in the form of the metal, economic considerations favor a metallic Ga target. We have tested both of the chemical schemes many times on amounts of gallium up to 20 kg with germanium recoveries of 90% or better, and we anticipate no difficulties in scaling each up to full scale. Quantitative recovery of carrier-free trace quantities of $^{69,71}Ge$ introduced into Ga metal by (p, n) reactions has been demonstrated.

An extractor for 200 kg is being designed as a pilot model for the ultimate extractor of 1–2-ton capacity. The procedure for the extraction of Ge from a large experiment (50 ton Ga) is envisaged to involve semicontinuous batch processing of 1–2 tons at a time. The whole operation from the extraction to the counter filling should take about a day.

An interfering reaction in all radiochemical neutrino detectors is the (p, n) reaction on the target material which gives the same product as neutrino capture. The major sources of protons in a deep-mine experiment are muon interactions, (α, p) reactions from natural α emitters, and (n, p) reactions from fast neutrons originating in the rock wall. This last source can be eliminated by water shielding as is done in the ^{37}Cl experiment. A unique advantage in the Ga system is that ^{69}Ge ($t_{1/2} = 39$ h) is also produced by these interfering (p, n) reactions but not by low-energy neutrinos. The activity of ^{69}Ge observed may thus be used to monitor the effectiveness of the various measures taken to eliminate background reactions. Furthermore, ^{72}As (26 h) and ^{74}As (17.8 day) are produced in Ga by α particles originating from Th and U via direct (α, n) reactions and are, therefore, made in much higher yield than the secondary product ^{71}Ge. The Ga system possesses, therefore, the unique feature of being self-monitoring and providing its own corrections independent of any other measurements if ^{69}Ge, ^{72}As, and ^{74}As yields are measured along with ^{71}Ge. Fortunately the chemical procedure described results in the removal of As along with Ge. After volatilization of GeH_4 and AsH_3, the two are separated by gas chromatography.

We have measured the cross sections for ^{69}Ge and ^{71}Ge production in Ga and in 8*M* $GaCl_3$ and fo[r] ^{37}Ar production in C_2Cl_4 with 225-GeV muons at Fermilab. They are, respectively, 30 ± 3 and 8 ± 1 μb per Ga atom in 8*M* $GaCl_3$, 64 ± 5 and 16 ± 2 μb in Ga metal, and 5.3 ± 0.5 μb per Cl atom in C_2Cl_4. Using the cosmic-ray–induced ^{37}Ar production rate[16] of 0.080 ± 0.024 atoms/day in 610 tons of C_2Cl_4 in the Homestake Mine experiment [depth 4400 hg/cm² (1 hg = 10^2 g), average muon energy 320 GeV], we estimate the production rates for ^{69}Ge and ^{71}Ge in 50 tons of Ga at the same depth to be 0.022 ± 0.007 and 0.006 ± 0.002 atoms/day in 8*M* $GaCl_3$ and 0.047 ± 0.016 and 0.012 ± 0.004 atoms/day in Ga metal. Thus the muon-induced ^{71}Ge will be negligible in the proposed experiment; furthermore, the larger ^{69}Ge production makes it possible, at least in principle, to distinguish between muon and neutrino signals.

We have also measured yield curves for the production, in gallium, of ^{69}Ge, ^{71}Ge, ^{72}As, and ^{74}As by α particles in the range of interest (4–11 MeV). The results show that 0.7 g of Th in equilibrium with its daughters or 0.05 μg of ^{226}Ra will produce 0.1 ^{71}Ge atoms per day in a metallic Ga target. The tank walls and Ga metal must therefore be free of α emitters down to about these levels. However, verification that these impurity specifications have been met can be obtained from the monitoring reactions mentioned; e.g., the α particles of the ^{232}Th chain will produce about 35 times as many atoms of ^{74}As as of ^{71}Ge, so that ^{74}As will be readily detectable via its electron-capture branch at any α level that requires a significant ^{71}Ge correction. Although ^{71}Ge can be produced from Zn impurity by (α, n) reaction on ^{68}Zn, the effect of Zn was shown to be entirely negligible for all reasonable levels of this impurity. Furthermore, the production of ^{69}Ge by $^{66}Zn(\alpha, n)$ again serves as an internal monitor.

The conclusion from all these considerations is that a ^{71}Ga detector for low-energy solar neutrinos is feasible and desirable.

We are grateful to many colleagues for advice, suggestions, and constructive criticism. This work was supported by the Office of Energy Research of the U. S. Department of Energy and by the National Science Foundation.

[1] J. N. Bahcall and R. Davis, Jr., Science 191, 164

(1976); J. K. Rowley, B. T. Cleveland, R. Davis, Jr., and J. C. Evans, Jr., in *Proceedings of the Neutrino—77 Conference, June 1977, Baksan Valley, U. S. S. R.*, edited by M. A. Markov (Nauka, Moscow, 1978), Vol. I, p. 15.

[2]V. A. Kuzmin, Zh. Eksp. Teor. Fiz. 49, 1532 (1965) [Sov. Phys. JETP 22, 1051 (1966)].

[3]L. Alvarez, Lawrence Berkeley Laboratory Report No. LBL 767 (unpublished).

[4]J. N. Bahcall, Phys. Rev. 135, B137 (1964).

[5]J. N. Bahcall, to be published. Three excited states of ^{71}Ge at excitation energies of 0.175 MeV ($J = \frac{5}{2}^-$), 0.50 MeV ($J = \frac{3}{2}^-$), and 0.71 MeV ($J = \frac{5}{2}^-$) must be considered in determining solar-neutrino absorption cross sections. From experimental information on seventeen β-decay transitions with approximately the same shell-model description as the transitions of interest, one can estimate the maximum effect of these excited states by making the extreme assumption that each of the transitions to excited states of ^{71}Ge occurs at the *fastest* rate exhibited by an analogous transition in the available experimental data. This assumption leads to capture rates < 7% above those in Table II for all the cases considered except for the very high-counting example of CNO (where the increase is 28%).

[6]Transitions to the analog state of ^{71}Ge do not contribute to the formation of ^{71}Ge(g.s.) since the analog state is particle unstable. The average cross section for the reaction $\nu_e + {}^{71}\text{Ga} \rightarrow {}^{70}\text{Ge}\ ({}^{70}\text{Ga}) + n\ (p) + e^-$ is, for ^{8}B neutrinos, 4×10^{-43} cm^2.

[7]The results given here, and in more detail in Ref. 5, are in good agreement with the independent results of G. Domogatski, Yad. Fiz. 25, 1125 (1977). His results are typically 5% larger than those given here (presumably because of less numerical precision in the calculation of Fermi functions and the ft value) except for ^{8}B (where our value is 25% larger than his). Domogatski did not report the ^{65}Zn cross section nor did he consider the effect of excited states.

[8]J. N. Bahcall, W. F. Huebner, N. H. McGee, A. L. Merts, and R. K. Ulrich, Astrophys. J. 184, 1 (1973).

[9]J. N. Bahcall, N. A. Bahcall, and R. K. Ulrich, Astrophys. Lett. 2, 91 (1968).

[10]B. Pontecorvo, Zh. Eksp. Teor. Fiz. 53, 1717 (1967) [Sov. Phys. JETP 26, 984 (1968)]; V. Gribov and B. Pontecorvo, Phys. Lett. 28B, 493 (1969).

[11]J. N. Bahcall, N. Cabbibo, and Y. Yahil, Phys. Rev. Lett. 28, 316 (1972).

[12]J. N. Bahcall and S. C. Frautschi, Phys. Lett. 29B, 623 (1969).

[13]See, for example, W. A. Fowler; Nature (London) 238, 24 (1972); F. W. W. Dilke and D. O. Gough, Nature (London) 240, 262 (1972); W. R. Sheldon, Nature (London) 221, 650 (1969); and E. E. Salpeter, Comments Nucl. Part. Phys. 2, 97 (1968).

[14]R. T. Rood, Nature (London) 240, 179 (1972): D. Ezer and A. G. W. Cameron, Nature (London) 240, 181 (1972); R. K. Ulrich and R. T. Rood, Nature (London) 241, 111 (1972). These authors do not give p-p or pep neutrino fluxes, but the appropriate fluxes can be inferred approximately from their results for the theoretical solar luminosity since the p-p (and pep) fluxes are approximately proportional to the computed luminosity.

[15]Since the gallium will undoubtedly be acquired over a period of several years, exposures will first be made with smaller quantities. We note that a 7-ton experiment will serve the useful purpose of verifying the conclusion, already indicated by the ^{37}Cl experiment, that the sun is not currently generating its luminosity by the CNO cycle of reactions.

[16]A. W. Wolfendale, E. C. M. Young, and R. Davis, Jr., Nature (London) PS 238, 130 (1972); G. Cassidy, in *Proceedings of the Thirteenth International Conference on Cosmic Rays, Denver, Colorado, 1973* (Univ. of Denver, Denver, Colo., 1973), Vol. 3, p. 1958.

Direct Approach to Resolve the Solar-Neutrino Problem

Herbert H. Chen
Department of Physics, University of California, Irvine, California 92717
(Received 27 June 1985)

A direct approach to resolve the solar-neutrino problem would be to observe neutrinos by use of both neutral-current and charged-current reactions. Then, the total neutrino flux and the electron-neutrino flux would be separately determined to provide independent tests of the neutrino-oscillation hypothesis and the standard solar model. A large heavy-water Cherenkov detector, sensitive to neutrinos from ^{8}B decay via the neutral-curent reaction $\nu+d\to\nu+p+n$ and the charged-current reaction $\nu_e+d\to e^-+p+p$, is suggested for this purpose.

PACS numbers: 96.60.Kx, 14.60.Gh

The solar-neutrino problem, i.e., fewer neutrinos are assigned to the sun in the chlorine-argon radiochemical experiment of Davis and co-workers[1] than predicted by the standard solar model,[2] has prompted a variety of possible solutions ranging from neutrino oscillations[3] to a very large number of nonstandard solar models.[4] The neutrino-oscillation hypothesis postulates interesting new properties of the neutrino in order to decrease the electron-neutrino flux, but this hypothesis cannot be fully tested in experiments using terrestial neutrino sources. The nonstandard solar models were developed primarily to decrease the central temperature of the sun in order to suppress ^{8}B production. Then, the smaller number of ^{8}B-decay neutrinos would reduce the anticipated signal in the chlorine-argon radiochemical experiment. These possibilities have been discussed widely over the past decade, and the discussions continue. In the absence of further experimental information, there will not be a resolution to this problem.

The new radiochemical experiments—^{71}Ga,[5-7] sensitive to neutrinos from the *pp* reaction; ^{81}Br,[8] sensitive to the ^{7}Be-decay neutrino—and the geochemical experiment—^{98}Mo,[9] sensitive to the ^{8}B-decay neutrino flux averaged over the past several million years—have been widely discussed, and they will add to our knowledge when completed. However, these experiments address the problem indirectly because they detect neutrinos via the charged-current (CC) reaction, and thus have a sensitivity only to electron neutrinos.

An experiment which directly addresses the solar-neutrino problem should be sensitive to all neutrino species equally. Such a measurement could determine the total solar-neutrino flux *even if neutrinos oscillate.* At the low energies relevant for solar neutrinos, however, the only possible reactions are neutral-current (NC) reactions with nuclei since these NC cross sections are independent of the neutrino type. Note that the (ν,e^-) scattering reactions are not appropriate because the (ν_e,e^-) reaction, in the standard electroweak theory, has both CC and NC contributions that make its cross section about 6 times larger[10] than the other (ν,e^-) reactions.[11,12] Thus, a measurement of the CC and NC rates on a nucleus fixes the ν_e flux and the total neutrino flux, respectively. Measurement of the total neutrino flux tests the standard solar model independent of the neutrino-oscillation hypothesis and measurement of the ratio—of the electron-neutrino flux to the total neutrino flux tests the neutrino-oscillation hypothesis[13] independent of the standard solar model.

The NC reaction on a nucleus is difficult to detect since the outgoing neutrino carries most of the available energy, especially at the energies relevant here. But one NC reaction, the neutrino disintegration of the deuteron, was observed some time ago with use of reactor $\bar{\nu}_e$'s by detecting the product neutron.[14] The neutron was seen by ^{3}He-gas proportional counters immersed in a tank of heavy water which provided the target deuterons.

Recently, the possibility of observing ^{8}B-decay solar neutrinos in a large heavy-water Cherenkov detector (1000 to 1500 metric tons) by use of (ν,e^-) scattering and the CC ν_e-*d* reaction was raised.[15] That this experiment can be seriously considered is a result of the successful operation of large light-water Cherenkov detectors built deep underground to search for proton decay.[16,17] The existence of many reactors with heavy water as a moderator, e.g., the Canadian deuterium uranium power reactors, provides encouragement that a large volume of heavy water could be made available for this purpose.

The CC reaction on the deuteron is relevant in the present discussion. The event rate can be calculated by use of the upper limit for the ^{8}B ν_e flux allowed by Davis and co-workers[1] and the theoretical cross section.[18] This reaction has been observed,[19] albeit with large errors. The expected rate is

$$R(\mathrm{CC})=F(\nu_e)\sigma_{\mathrm{CC}}N_d=(2\times10^6\ \mathrm{cm}^{-2}\ \mathrm{sec}^{-1})\times(1.2\times10^{-42}\ \mathrm{cm}^2)\times(6\times10^{31})=12/\mathrm{kt\text{-}d}\ (\text{kiloton-day}),$$

where $F(\nu_e)$ is the ^{8}B electron-neutrino flux allowed by Davis, σ_{CC} is the average CC ν_e-d cross section, and N_d is the number of target deuterons per kiloton of heavy water. These CC events produce an electron which carries essentially all of the energy of the incident neutrino (minus the threshold energy). Thus about 70% of these events would be above 7 MeV.

The detection of such low-energy electrons using Cherenkov light in water is made feasible[20,21] by the development of large (20-in.-diam) photomultiplier tubes (PMT's)[22] for the Kamioka proton-decay experiment. These allow coverage of a large fraction of the surfaces surrounding the detector volume by photocathodes in order to maximally collect the small amount of Cherenkov light.

Considerations of backgrounds for this experiment led to a problem which is unique to heavy water, i.e., photodisintegration of the deuteron followed by neutron capture on a deuteron to produce a 6.25-MeV γ ray. These γ's would undergo either Compton scattering to produce an electron of up to 6 MeV or pair production to generate a background event. This mechanism for the production of a relatively high-energy γ from low-energy γ's (and from thermal neutrons) places a most severe constraint on the allowable radioactivity in detector materials.

On the assumption that such backgrounds can be reduced to the required level, then neutrino disintegration of the deuteron may be detected either via the same 6.25-MeV γ from neutron capture on the deuteron or by loading the heavy water. In order to detect these γ's, the heavy-water Cherenkov detector being considered will have to be more highly instrumented since several Compton electrons and/or an electron-positron pair would be produced and each electron (positron and γ) with less than 0.26 MeV kinetic energy would be below the Cherenkov threshold in water. It is also possible, if necessary, to enhance Cherenkov light detection by loading the heavy water with an appropriate wavelength shifter.[19] Besides increasing the number of photoelectrons produced in the PMT's of the detector, the isotropic light distribution from this wave-shifted light improves event reconstruction which, at these low energies, depends only on PMT position and event timing.

The neutrino disintegration rate of deuterons by ^{8}B-decay neutrinos in the standard solar model is

$$R(\mathrm{NC}) = F(^8\mathrm{B}) \times \sigma_{NC} \times N_d = (4.6\times10^6\ \mathrm{cm^{-2}\ sec^{-1}}) \times (0.6\times10^{-42}\ \mathrm{cm^2}) \times (6\times10^{31}) = 14/\mathrm{kt\text{-}d},$$

where $F(^8\mathrm{B})$ is the expected ^{8}B neutrino flux from the standard solar model,[2] and σ_{NC} is the average NC deuteron disintegration cross section.[23] The fourteen neutrons will produce about eleven γ's of energy 6.25 MeV from deuteron capture with the remainder mostly captured[24] by ^{16}O or ^{17}O to produce a low-energy γ or an α, respectively. A small residual fraction of light water will further decrease conversion of the neutron to this high-energy (6.25 MeV) γ ray. However, appropriate loading of the heavy water could improve this detection efficiency back towards 14/kt-d. Alternatively, it is also possible to decrease the neutron detection efficiency in order to have less backgrounds for the CC ν_e-d reaction by appropriate loading of the heavy water, e.g., by ^{10}B.

Atmospheric-neutrino–generated backgrounds cannot be avoided. This flux has been measured by the proton-decay experiments with use of CC reactions[16,17] and these results agree well with calculations.[25] Thus, the atmospheric-neutrino flux is known and it is substantially lower than the solar-neutrino flux though much higher in energy. We note that the total atmospheric-neutrino CC rate is about 100/kt-d, i.e., much lower than the ^{8}B rates quoted above which are about 4000/kt-d.

Cosmic-ray– and radioactivity-induced backgrounds and many other questions relevant to a large heavy-water detector in an experiment to detect ^{8}B solar neutrinos via the (ν,e^-) and CC ν_e-d reactions are being fully addressed in a collaborative effort which has been underway since early this year.[26] If the NC ν-d reaction suggested here is to be used to detect ^{8}B solar neutrinos, a further increase in light detection sensitivity and in background reduction beyond that considered so far would be required.

However, detailed considerations suggest that there are no insurmountable problems, in principle, that would prevent such an experiment using this direct approach from resolving the solar-neutrino problem. In practice, the radioactivity-background problem is the most severe, though the use of water which can be continuously purified[27] and of acrylic which is made from highly processed polymers provides encouragement that the extremely low levels of radioactivity, which are required within the detector, can be achieved and maintained.

Useful discussions with many people have had an influence on the possibilities presented here. Members of the Sudbury Neutrino Observatory Collaboration, the Irvine-Michigan-Brookhaven Collaboration, the Kamioka Collaboration, the University of California–Irvine Neutrino Group, and others have contributed useful suggestions. Support of this work by the United States Department of Energy and the National Science Foundation is gratefully acknowledged.

[1]R. Davis, Jr., D. S. Harmer, and K. C. Hoffman, Phys.

Rev. Lett. **20**, 1205 (1968); J. N. Bahcall and R. Davis, Jr., Science **191**, 264 (1976); J. K. Rowley, B. T. Cleveland, and R. Davis, Jr., in *Solar Neutrinos and Neutrino Astronomy (Homestake, 1984)*, edited by N. L. Cherry, W. P. Fowler, and K. Lande, AIP Conference Proceedings No. 126 (American Insitute of Physics, New York, 1985), p. 1.

[2]J. N. Bahcall, W. F. Huebner, S. H. Lubow, P. D. Parker, and R. K. Ulrich, Rev. Mod. Phys. **54**, 767 (1982).

[3]B. Pontecorvo, Zh. Eksp. Teor. Fiz. **53**, 1717 (1967) [Sov. Phys. JETP **26**, 984 (1968)]; V. Gribov and B. Pontecorvo, Phys. Lett. **28B**, 495 (1969); S. M. Bilenky and B. Pontecorvo, Phys. Rep. **41**, 226 (1978).

[4]R. T. Rood, in Proceedings of an Informal Conference on the Status and Future of Solar Neutrino Research, Upton, N.Y., 1978, edited by G. Friedlander, BNL Report No. 50879, 1978 (unpublished), Vol. 1, p. 175; J. N. Bahcall and R. Davis, Jr., in *Essays in Nuclear Astrophysics*, edited by C. A. Barnes, D. D. Clayton, and D. N. Schramm (Cambridge Univ. Press, Cambridge, England, 1982), p. 243.

[5]J. N. Bahcall, B. T. Cleveland, R. Davis, Jr., I. Dostrovsky, J. C. Evans, Jr., W. Frati, G. Friedlander, K. Lande, J. K. Rowley, W. Stoenner, and W. Weneser, Phys. Rev. Lett. **40**, 1351 (1978); W. Hampel *et al.*, in *Science Underground (Los Alamos, 1982)*, edited by M. M. Nieto *et al.*, AIP Conference Proceedings No. 96 (American Institute of Physics, New York, 1983), p. 88.

[6]W. Hampel, in *Solar Neutrinos and Neutrino Astronomy (Homestake, 1984)*, edited by N. L. Cherry, W. P. Fowler, and K. Lando, AIP Conference Proceedings No. 126 (American Institute of Physics, New York, 1985), p. 162.

[7]I. R. Barabanov, E. P. Veretenkin, V. N. Gavrin, S. N. Danshin, L. A. Eroshkina, G. T. Zatsepin, Yu. I. Zakharov, S. A. Klimova, Yu. B. Klimov, T. V. Knodel, A. V. Kopylov, I. V. Orekhov, A. A. Tikhonov, and M. I. Churmaeva, in Ref. 6, p. 175.

[8]G. S. Hurst, C. H. Chen, S. D. Kramer, M. G. Payne, and R. D. Willis, in *Science Underground (Los Alamos, 1982)*, edited by M. M. Nieto *et al.*, AIP Conference Proceedings No. 96 (American Institute of Physics, New York, 1983), p. 96; G. S. Hurst, C. H. Chen, S. D. Kramer, B. T. Cleveland, R. Davis, Jr., R. K. Rowley, F. Gabbard, and F. J. Schima, Phys. Rev. Lett. **53**, 1116 (1984).

[9]G. A. Cowan and W. C. Haxton, Science **216**, 51 (1982); K. Wolfsberg, G. A. Cowan, E. A. Bryant, K. S. Daniels, S. W. Downey, W. C. Haxton, V. G. Niesen, N. S. Nogar, C. M. Miller, and D. J. Rokop, in Ref. 6, p. 196.

[10]R. C. Allen, V. Bharadwaj, G. A. Brooks, H. H. Chen, P. J. Doe, R. Hausammann, H. J. Mahler, A. M. Rushton, K. C. Wang, T. J. Bowles, R. L. Burman, R. D. Carlini, D. R. F. Cochran, J. S. Frank, E. Piasetzky, V. D. Sandberg, D. A. Krakauer, and R. C. Talaga, in *Neutrino '84, Proceedings of the International Conference on Neutrino Physics and Astrophysics, Dortmund, Federal Republic of Germany, 1984*, edited by K. Klienkinecht and E. A. Paschos (World Scientific, Singapore, 1985), p. 322.

[11]F. Bergsma *et al.*, Phys. Lett. **117B**, 272 (1982).

[12]L. A. Ahrens *et al.*, Phys. Rev. Lett. **51**, 1514 (1983), and **54**, 18 (1985).

[13]F. Reines, H. W. Sobel, and E. Pasierb, Phys. Rev. Lett. **45**, 1307 (1980).

[14]E. Pasierb, H. S. Gurr, J. F. Lathrop, F. Reines, and H. W. Sobel, Phys. Rev. Lett. **43**, 96 (1979).

[15]H. H. Chen, University of California–Irvine Neutrino Group Report No. 120, 1984 (unpublished), and in Ref. 6, p. 249.

[16]R. M. Bionta *et al.*, Phys. Rev. Lett. **51**, 27 (1983); H. S. Park *et al.*, Phys. Rev. Lett. **54**, 22 (1985).

[17]M. Koshiba, in *Proceedings of the Twenty-Second International Conference on High Energy Physics, Leipzig, East Germany, 1984*, edited by A. Meyer and E. Wieczore (Akademie der Wissenschaften der DDR, Zeuth, East Germany, 1984), Vol. 2, p. 67, and *ibid.*, Vol. 1, p. 250; also, see T. Suda, in *Neutrino '81, Proceedings of the International Conference on Neutrino Physics and Astrophysics, Maui, Hawaii, 1981*, edited by R. G. Cence, E. Ma, and A. Roberts (Univ. Of Hawaii Press, Honolulu, 1981), Vol. 1, p. 224.

[18]F. J. Kelly and H. Uberall, Phys. Rev. Lett. **16**, 145 (1966); S. D. Ellis and J. N. Bahcall, Nucl. Phys. **A114**, 636 (1968).

[19]S. E. Willis *et al.*, Phys. Rev. Lett. **44**, 522 (1980), and **45**, 1370(E) (1980).

[20]T. W. Jones and J. van der Velde, unpublished.

[21]M. Koshiba, in *Proceedings of the International Colloquium on Baryon Nonconservation, Salt Lake City, January 1984*, edited by D. Cline (Univ. of Wisconsin Press, Madison, 1984); A. K. Mann, in *Proceedings of the Conference on the Intersections of Particle and Nuclear Physics, Steamboat Springs, Colorado, 1984*, edited by Richard E. Mischke, AIP Conference Proceedings No. 123 (American Institute of Physics, New York, 1984).

[22]H. Kume, S. Sawaki, M. Ito, K. Arisaka, T. Kajita, A. Nishimura, and A. Suzuki, Nucl. Instrum. Methods **205**, 443 (1983).

[23]A. Ali and C. A. Dominguez, Phys. Rev. D **12**, 3673 (1975).

[24]The abundance of ^{17}O is increased by less than a factor of 2 in heavy water. E. D. Earle (Atomic Energy of Canada Limited, Chalk River), private communication.

[25]T. K. Gaisser, T. Stanev, S. A. Bludman, and H. Lee, Phys. Rev. Lett. **51**, 223 (1983); T. K. Gaisser and T. Stanev, in Ref. 6, p. 277.

[26]D. Sinclair, A. L. Carter, D. Kessler, E. D. Earle, P. Jagam, J. J. Simpson, R. C. Allen, H. H. Chen, P. J. Doe, E. D. Hallman, W. F. Davidson, R. S. Storey, A. B. McDonald, G. T. Ewan, H.-B. Mak, and B. C. Robertson (Sudbury Neutrino Observatory Collaboration), talk presented at the Conference on Underground Physics, Aosta, Italy, April 1985 (unpublished).

[27]The purification process would likely remove whatever is added to increase or decrease the neutron detection efficiency and/or improve the Cherenkov light-collection efficiency. Thus, these would have to be continuously replaced. In order not to increase radioactivity-related backgrounds, such additives would have to be quite pure.

PREDICTIONS FOR A LIQUID ARGON SOLAR NEUTRINO DETECTOR

J.N. BAHCALL
Institute for Advanced Study, Princeton, NJ 08540, USA

M. BALDO-CEOLIN
Università di Padova and I.N.F.N., Sezione di Padova, Padua, Italy

D.B. CLINE
University of Wisconsin, Madison, WI 53706, USA

and

C. RUBBIA
HEPL, Harvard University, Cambridge, MA 02138, USA

Received 3 July 1986

A liquid argon time projection chamber can be used to make separate measurements of electron neutrinos and neutrinos of different flavors, to determine the incident neutrino spectrum, and to verify that neutrinos come from the direction of the sun. Resonant neutrino oscillation has a dramatic effect on the shape of the predicted spectrum of recoil electrons.

1. Introduction. The solar neutrino problem is well known: observation differs from expectation [1]. The results of the ^{37}Cl experiment are in disagreement with theoretical calculations based upon standard models of how the sun shines. This disagreement is of importance both for physics and for astronomy. For physics, the sun provides a beam of collimated low-energy (MeV) neutrinos that traverse a large distance before they are detected. Hence, solar neutrino experiments provide a means for exploring, for large mixing angles, neutrino mass matrix elements as small as 10^{-12} eV^2 and, for small mixing angles, more modest neutrino masses ($\lesssim 10^{-4}$ eV^2). For astronomy, solar neutrino experiments constitute rigorous tests of our understanding of how stars generate energy and evolve. The long-standing discrepancy between calculation and observation in the ^{37}Cl experiment could be caused by new weak interaction physics or by our lack of understanding of the simplest stage of stellar evolution.

A liquid argon detector is especially suited for resolving this problem. Measurement of the rate of absorption will determine the incident flux of ^{8}B neutrinos. Measurement of the rate of neutrino–electron scattering will establish the extent to which missing ^{8}B neutrinos are converted to different flavors. The energy spectra of the electrons produced by absorption and by scattering provide details concerning the mixing angles and mass differences.

In this letter, we present predictions for a liquid argon solar neutrino experiment and give specific rates for a detector, Icarus [2], that will be built in the Gran Sasso Laboratory. One of the goals of the Icarus Collaboration is to study solar neutrionos. The fundamental requirement of a liquid argon imaging detector is the ability to drift ionization electrons over large distances, which has been achieved by the work of several groups [2,3]. The unique feature of the detector is the recording of any event occurring within its volume. The detector will form a three-dimensional electronic image by drifting electrons in a homogeneous electric field onto a readout plane, where the charge is recorded. The arrival positions of the drifting electrons on the readout plane give two spatial coordinates of the points (separated by about 2 mm) on the original track. The drift time determines the third coordinate. The detector will have both energy and angular resolution, and real-time analysis capability.

We present in section 2 the theoretical predictions for a liquid argon detector and describe in section 3 how the experiment can be used in practice to solve the solar neutrino problem. The detector is ideally suited to study resonant amplification of neutrino oscillations.

2. Predictions. Neutrinos from the decay of ^8B in the solar interior can produce both absorption (^{40}Ar goes to ^{40}K) and neutrino–electron scattering in the liquid argon detector. We present for both processes the calculated total cross sections, energy spectra and information about the angular distributions of the recoil electrons.

Neutrino *absorption* by ^{40}Ar will be dominated by transitions to the isotopic analogue state [‡1] ($T = 2$, $J = 0^+$) at 4.38 MeV above the ground state of ^{40}K. This analogue state decays by emitting a 2.09 MeV gamma ray (branching ratio 76%) or a 1.65 MeV gamma, followed by characteristic lower energy gammas. The difference in the nuclear ground state energies between ^{40}K and ^{40}Ar is 1.505 MeV; the threshold for neutrino absorption to the analogue state is therefore 5.885 MeV. The angular distribution of the electrons that are produced is *slowly varying*, proportional to [5] $1 + (v_e/c)\cos\theta_{\text{scattering}}$.

Fig. 1a shows the energy spectrum of recoil electrons that are produced by neutrino absorption to the isotopic analogue state. *This recoil spectrum reflects the assumed spectrum of solar neutrino energies and will be greatly changed if resonant neutrino oscillation occurs (see below).*

The total cross sections (averaged over the incident neutrino spectrum) for excitation of the analogue state are given in table 1 as a function of W_{cutoff}, the minimum recoil energy an electron must have to be counted. The absorption cross section decreases relatively slowly for $W_{\text{cutoff}} < 5$ MeV, but decreases rather rapidly for larger cutoff energies. The absorption cross sections given here include all the usual atomic and nuclear physics corrections [6]. In addition, the cross sections were calculated using a recently derived ^8B neutrino spectrum [7] that accurately takes account of the shape of the broad excited state to which ^8B decays and includes forbidden weak interaction corrections. *The absorption cross section to the analogue state can be calculated to an accuracy of a few percent or better since the nuclear matrix element can be determined from isotopic spin invariance.* Contributions from transitions to other states in ^{40}K are expected to be numerically unimportant (a claim which should be checked by (p, n) experiments and detailed shell-model calculations) and will not be in-coincidence with the characteristic gamma decay of the analogue state (see below).

The *scattering* cross sections have been computed [8] in the standard electroweak theory [9] also using the recently derived spectrum of ^8B neutrinos and different values of W_{cutoff}. At low recoil energies, the event rate in any practical detector is dominated by background events. At very large values of W_{cutoff}, the event rate becomes too low to be measured. A reasonable compromise for a liquid argon detector is reached when the threshold energy for detection of recoil electrons is set at 5 MeV. For $W_{\text{cutoff}} = 5$ MeV, the scattering cross sections per electron are [8] (for $\sin^2\theta_W = 0.22$):

$$\sigma_{\nu_e-e^-} = 2.1 \times 10^{-44}\ \text{cm}^2\,, \tag{1a}$$

$$\sigma_{\nu_\mu-e^-} = 3.45 \times 10^{-45}\ \text{cm}^2\,. \tag{1b}$$

In the standard theory [9], the scattering by electrons is the same for muon neutrinos and for tau neutrinos; we denote for simplicity all "converted" neutrinos by ν_μ. Above about 9 MeV electron recoil energy, essentially all the expected solar-induced events are from electron–neutrino scattering (cf. table 1). For $W_{\text{cutoff}} = 9$ MeV, the scattering cross sections are [8]

$$\sigma_{\nu_e-e^-} = 3.4 \times 10^{-45}\ \text{cm}^2\,, \tag{2a}$$

$$\sigma_{\nu_\mu-e^-} = 5.4 \times 10^{-46}\ \text{cm}^2\,. \tag{2b}$$

Figs. 1b and 1c show the calculated energy spectra of electrons that are scattered by electron and muon neutrinos.

The forward-peaked angular distribution of the re-

‡1 The importance of this transition was emphasized in a prescient discussion in ref. [4].

Table 1
The absorption cross sections for ^8B solar neutrinos incident on ^{40}A are given for different values of the minimum accepted total energy, W_{cutoff}, for the recoil electron. The neutrinos are assumed to be pure electron neutrinos (ν_e) when they reach the earth. The absorption cross sections refer to transitions from the ground state of ^{40}A to the $T = 2$ isotopic analogue state at 4.38 MeV excitation in ^{40}K.

W_{cutoff} (MeV)	$\sigma_{absorption}$ ($\times 10^{-46}$ cm^2)
0.511	7.84×10^3
1.0	7.81×10^3
2.0	7.50×10^3
3.0	6.72×10^3
4.0	5.43×10^3
5.0	3.79×10^3
6.0	2.15×10^3
7.0	8.76×10^2
8.0	1.97×10^2
8.5	6.53×10^1
9.0	1.64×10^1
9.5	3.17
10.0	0.36

coil electrons can be used to demonstrate experimentally that the detected neutrinos are coming from the direction of the sun. For $W_{cutoff} = 5$ MeV, 90% (50%) of the recoil electrons are confined to a cone with an opening angle of 16° (10°) about the earth–sun direction; for $W_{cutoff} = 9$ MeV, 90% (50%) of the recoil electrons are contained within a cone of 9° (5°). The cumulative probabilities for the scattering of ν_μ are the same as for ν_e to the numerical accuracy quoted above. Details of the calculations are given in ref. [8].

3. Experimental possibilities. The planned [2] underground detector for the Gran Sasso Laboratory will contain about 6.5 kilotons of argon. If the standard solar model [10,1] is correct and electron neutrinos are unaffected as they traverse the solar interior

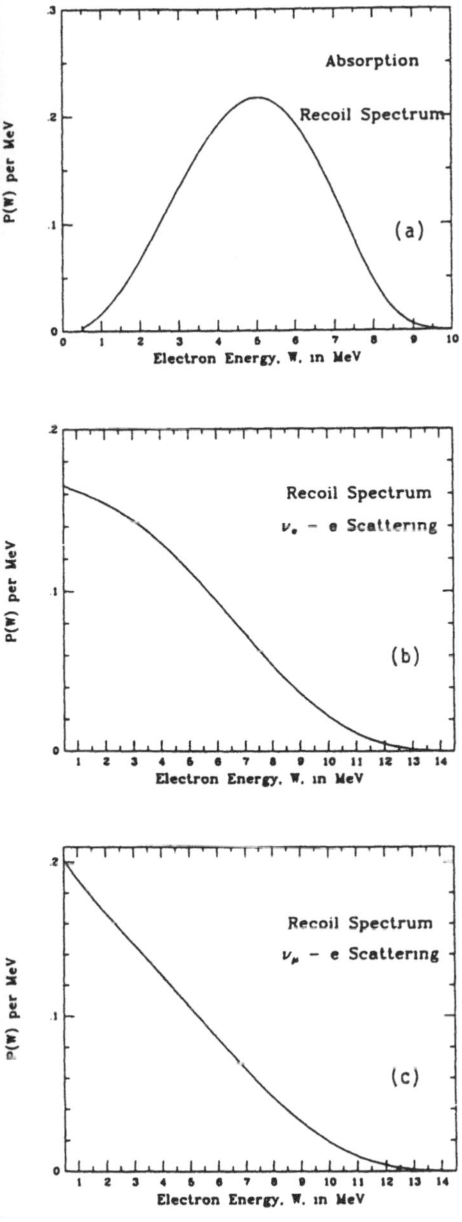

Fig. 1. Energy spectra of recoil electrons. (a) The electron energy spectrum resulting from absorption of neutrinos by ^{40}A leading to the analogue state at 4.38 MeV excitation energy in ^{40}K. (b) Electron energy spectrum from ν_e–e^- scattering. (c) Electron energy spectrum from ν_μ–e^- scattering, assuming the incident neutrinos are fully converted. For (a), the incident neutrino energy is equal to 5.885 MeV plus the electron energy. In all cases, a ^8B neutrino spectrum, *unaltered by oscillations*, is assumed.

and travel to earth, then this detector should experience 4.7×10^3 absorption events per year that produce recoil electrons above 5 MeV with a broad angular distribution. There should be an approximately equal number of forward-peaked scattering events with $W_{cutoff} = 5$ MeV. If the background can be kept sufficiently small, then in *one day* this detector should be able to measure, from absorption events alone, the flux of ^8B solar neutrinos predicted by the standard solar model to a 1σ accuracy of 30%!

The high event rate expected in a liquid argon detector will make possible a test of the theory of stellar evolution at a flux level which tests even the most speculative stellar model "solutions" of the solar neutrino problem. The ^{37}Cl experiment [11] has shown that the flux of electron neutrinos from ^8B decay in the sun is less than 2×10^6 cm^{-2} s^{-1} at the earth. In one year of operation, a 6.5 kiloton argon detector should detect with 10% accuracy (1σ) a flux of 9×10^4 cm^{-2} s^{-1}, *23 times less* than the current limit. None of the so-far proposed "non-standard" solar models give ^8B fluxes this low [‡2].

The proposed experiment has several consistency checks. First, gamma rays from the decay of the 4.83 MeV excited state in ^{40}K will be associated in space and time with the recoil electrons from neutrino absorption. Estimates of the background rates in liquid argon suggest that the probability of accidental γ–e correlations with the required energies is negligible. *The characteristic gamma rays from the decay of the ^{40}K analogue state will provide the "smoking gun" proving that ν_e's have been absorbed by ^{40}A*. This characteristic signal should permit the operation of the detector with W_{cutoff} as low as 2 MeV for absorption events (total track energy >8 MeV), doubling the event rate cited above. Second, the event rate should be independent of time except for the seasonal variation caused by the eccentricity (1.7%) of the earth's orbit or by a variation with the solar cycle caused by an interaction between the neutrino magnetic moment and the solar magnetic field [14]. The eccentricity should be discernible in data that are accumulated over a period of a few years. Third, the direction from which the neutrinos come can be inferred from the angular distribution of the scattered electrons and must point to the sun. The angular resolution, which is determined mainly by multiple scattering, is inversely proportional to electron energy. For 10 MeV electrons, a projected angular resolution of order 12° appears possible provided only that one can reconstruct the initial 1 cm of electron track. Fourth, methane can be used instead of argon, allowing a "beam-off" test for the absorption process. The (forward-peaked) electron scattering will be the same per electron of methane or argon. By contrast, the cross section for absorption of ^8B neutrinos by normal methane is negligibly small [‡3].

The hypothesis of adiabatic resonant neutrino oscillations [15], the MSW effect, predicts a dramatically different spectrum for the recoil electrons from absorption, different from all other explanations of the solar neutrino problem. *The standard solar model, non-standard solar models and non-resonant neutrino oscillations all predict that the recoil spectrum should have the shape shown in fig. 1a. MSW predicts that the spectrum will be greatly distorted.* Both of the possible solution branches of the MSW effect produce large changes in the predicted recoil spectrum, either depleting all the high-energy electron neutrinos or preferentially removing the low-energy electron neutrinos. Simulations of the energy resolution in Icarus suggest that an accuracy of order 5% (FWHM) at 10 MeV is possible. Either of the MSW solutions will show up clearly in the detector by measuring the energy spectrum of the recoil electrons.

We are grateful to E. Aprile, K.L. Giboni, and G. Puglierin for valuable contributions to the experimental program. It is a pleasure to thank E.S. Phinney and M. Schwarzschild for stimulating conversations. This work was supported in part by the National Science Foundation.

‡2 For non-standard models see refs. [1,6,12]. For a recent suggestion, see ref. [13].

‡3 The threshold for neutrino absorption by ^{12}C, 17.3 MeV, is beyond the nominal endpoint of ^8B neutrinos. The absorption cross section calculated with the spectrum of ref. [7] (which includes the width of the ^8Be final state) is negligibly small ($< 10^{-47}$ cm^2). The cross section for absorption of ^8B neutrinos by the 1.1% contamination of ^{13}C is 7×10^{-43} cm^2 for $W_{cutoff} = 5$ MeV.

References

[1] J.N. Bahcall and R. Davis, Science 191 (1976) 264, in: Solar neutrinos and neutrino astronomy (Homestake, 1984), AIP Conf. Proc. No. 126, eds. M.L. Cherry, W.A. Fowler and K. Lande (AIP, New York, 1985).
[2] CERN–Harvard–Milano–Padova–Rome–Tokyo–Wisconsin Collab., ICARUS, A proposal for the Gran Sasso Laboratory, preprint INFN-AE-85-7 (September 1985) (available upon request);
C. Rubbia, CERN Internal Report 77-78;
W.A. Huffman, J.M. Lo Secco and C. Rubbia, IEEE Trans. Nucl. Sci. NS-26 (1979) 64;
E. Aprile, K. Giboni and C. Rubbia, Nucl. Instrum. Methods 241 (1985) 62;
M. Baldo-Ceolin, Symp. on Photon and lepton interactions (Kyoto, 1985) p. 432.
[3] H.H. Chen and J.F. Lathrop, Nucl. Instrum. Methods 150 (1978) 585;
P.J. Doe et al., Nucl. Instrum. Methods 199 (1982) 639;
H.J. Mahler et al., IEEE Trans. Nucl. Sci. NS-30 (1983) 86;
H.H. Chen et al., Science Underground (1982) AIP Conf. Proc. No. 96 (AIP, New York) p. 182.
[4] R.S. Raghavan, Liquid argon detectors: a new approach to observation of neutrinos from meson factories and collapsing stars, Bell Laboratories, Technical Memorandum 79-1131-31 (1979), unpublished.
[5] J.N. Bahcall, Phys. Lett. 12 (1964) 332.
[6] J.N. Bahcall, Rev. Mod. Phys. 50 (1978) 881.
[7] J.N. Bahcall and B.R. Holstein, submitted to Phys. Rev. D (1986).
[8] J.N. Bahcall, Astrophys. J. (1986), to be submitted.
[9] S.L. Glashow, Nucl. Phys. 22 (1961) 579;
S. Weinberg, Phys. Rev. Lett. 19 (1967) 1264;
A. Salam, in: Elementary particle theory, ed. N. Svartholm (Almqvist and Wiksell, Stockholm, 1968) p. 367.
[10] J.N. Bahcall, W.R. Huebner, S.H. Lubow, P.D. Parker and R.K. Ulrich, Rev. Mod. Phys. 54 (1982) 767;
J.N. Bahcall et al., Astrophys. J. Lett. 292 (1985) L79.
[11] R. Davis, in: Proc. Informal Conf. on Status and future of solar neutrino research ed. G. Friedlander, BNL Report 40879 (1978), Vol. 1, p. 1.
[12] J.N. Bahcall and R. Davis, in: Essays in Nuclear Astrophysics, eds. C.A. Barnes, D.D. Clayton and D. Schramm (Cambridge U.P., Cambridge, 1982) p. 243.
[13] D.N. Spergel and W.H. Press, Astrophys. J. 294 (1985) 663;
J. Faulkner and R.L. Gilliland, Astrophys. J. 299 (1985) 994;
L.M. Krauss et al., Astrophys. J. 299 (1985) 1001.
[14] L.B. Okun, M.B. Voloshin and M.I. Vysotsky, ITEP preprint (1986).
[15] S.P. Mikheyev and A.Yu. Smirnov, 10th Intern. Workshop (Savonlinna, Finland, June 1985);
L. Wolfenstein, Phys. Rev. D20 (1979) 2634;
H.A. Bethe, Phys. Rev. Lett. 56 (1986) 1305;
P.S. Rosen and J.M. Gelb, Phys. Rev. D, submitted.

New Tools for Solving the Solar-Neutrino Problem

R. S. Raghavan
AT&T Bell Laboratories, Murray Hill, New Jersey 07974

Sandip Pakvasa
Department of Physics and Astronomy, University of Hawaii at Manoa, Honolulu, Hawaii 96822

and

B. A. Brown
Cyclotron Laboratory, Michigan State University, East Lansing, Michigan 48824
(Received 26 June 1986)

Neutrino (ν) excitation of nuclear levels (NUEX) via neutral currents is shown to be crucially important to the solar-ν problem as a method of detecting ν's regardless of flavor. We examine ^{11}B, ^{40}Ar, and ^{35}Cl as NUEX targets. The ^{11}B system offers means for concurrent solar-ν spectroscopy by NUEX and by charged-current ν_e capture with a set of differing thresholds. It promises a self-contained solution to the solar-ν problem as well as key probes for revealing the structure of resonant flavor and nonflavor ν oscillations in the sun. This approach may be feasible in an ICARUS-type underground experiment.

PACS numbers: 96.60.Kx, 13.15.Hq, 25.30.Pt, 96.40.Qr

The central dilemma of the solar-neutrino (ν) problem is that it is impossible to decide if ν oscillations or the shortcomings of the standard solar model are at the root of the disparity between predictions and experiment. Either possibility is of fundamental importance to astrophysics or particle physics. The concept of resonant enhancement of ν oscillations in solar matter[1,2] implies that even for small flavor-mixing angles, serious distortions at all energies of the standard solar-ν spectrum are possible, setting us the difficult task of acquiring very precise, wide-ranging spectral data.[3] A direct attack on the problem is possible by the devising of a method to observe ν's *regardless of flavor* in the same experiment designed to detect only electron neutrinos (ν_e). A difference in the solar-ν flux as measured by the two detection modes would be *specific* evidence for ν oscillations while the flux seen by the "all-flavor" mode should be *the true ν flux from the sun.* We show that this strategy can be achieved by flavor-independent ν *excitation of nuclear levels* (NUEX) via neutral currents.

NUEX is a process of general applicability to solar-ν detectors. It can be observed in practice by the signatures of a subsequent deexcitation mode or product ($\nu + A \rightarrow A^* + \nu' \rightarrow A + \gamma$ or $A' + p,n,\alpha$) in close analogy to γ excitation. The choice of the NUEX mode is guided by our desire for comparative operation of NUEX and a flavor-dependent charged-current (CC) ν_e-detection mode. Known schemes for CC solar-ν_e detection may thus be examined first for NUEX capability. The NUEX target could be the original CC target, an isotopic or other constituent, or a new target species mixed into the detector. The viability of the NUEX mode depends on sensitivity to solar ν's in terms of threshold energies and excitation strengths specific to the target, a signature to identify NUEX events, and compatibility of the NUEX mode with the technique of CC ν_e detection. A particular case of this method is deuteron breakup by ν or $\bar{\nu}$ combined with ν_e or $\bar{\nu}_e$ capture by deuterons, as employed by Reines, Sobel, Pasierb[4] or as proposed by Chen for solar-ν detection.[5] Following the new avenues opened by the NUEX mode of ν detection, we consider three solar-ν targets: ^{11}B and ^{40}Ar (direct counting) and ^{35}Cl (in the Homestake chemical detector).

The ^{11}B system is uniquely tailored to the NUEX-CC method. The salient features are as follows: safely predictable NUEX ($^{11}B + \nu \rightarrow \nu' + {}^{11}B + \gamma$) strengths to a set of levels in ^{11}B (Fig. 1); a well-defined set of

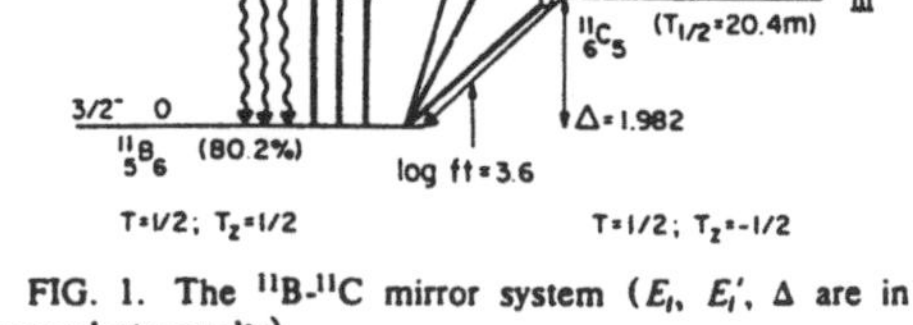

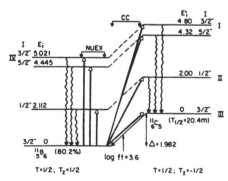

FIG. 1. The ^{11}B-^{11}C mirror system (E_i, E_i', Δ are in megaelectronvolts).

Radiochemical Neutrino Detection via $^{127}I(\nu_e, e^-)^{127}Xe$

W. C. Haxton
Institute for Nuclear Theory, Department of Physics, University of Washington, Seattle, Washington 98195
(Received 6 October 1987)

Solar or supernova neutrinos incident on an iodine-bearing liquid will produce the noble gas ^{127}Xe ($\tau_{1/2} = 36.4$ d), which can be recovered and counted as in the present ^{37}Cl experiment. The rate of neutrino reactions per unit volume of detector could be more than an order of magnitude greater than in perchloroethylene. I discuss the new physics that might be learned from such an experiment.

PACS numbers: 96.40.Qr, 25.30.Pt, 96.60.Kx, 97.60.Bw

I will argue below that a radiochemical detector employing an iodine-bearing liquid could serve as a sensitive observatory for astrophysical neutrinos. The operative reaction, $^{127}I(\nu_e,e)^{127}Xe$ ($\tau_{1/2} = 36.4$ d), is strikingly similar[1] to the reaction $^{37}Cl(\nu_e,e^-)^{37}Ar$ ($\tau_{1/2} = 35.0$ d) employed by Davis and co-workers in the current Homestake experiment.[2] It appears that the recovery and counting of the noble-gas product ^{127}Xe can be performed with existing techniques, and that the capture rate per unit detector volume could exceed that in perchloroethylene by more than an order of magnitude. I argue that calibration experiments should be mounted to determine whether ^{127}I is primarily a ^7Be or an ^8B solar-neutrino detector. I also discuss the sensitivity of such a detector to neutrinos from galactic supernovae.

Some of the relevant nuclear structure is illustrated in Fig. 1. Neutrinos with $\epsilon_\nu \gtrsim 664$ keV can interact with ^{127}I (abundance = 100%) to produce ^{127}Xe, which then captures an electron as it decays back to the parent nucleus with a half-life of 36.4 d. As the $\frac{5}{2}^+ \to \frac{1}{2}^-$ transition to the ^{127}Xe ground state is forbidden, the cross section for capturing ^7Be neutrinos ($\epsilon_\nu = 862$ keV) should depend only on the strength of the Gamow-Teller[3] $\frac{5}{2}^+ \to \frac{3}{2}^+$ transition to the 125-keV state. The high-energy ($\simeq 14$-MeV end point) ^8B solar neutrinos can produce ^{127}Xe by exciting many transitions to states below the threshold for neutron breakup, 7.23 MeV above the ground state.

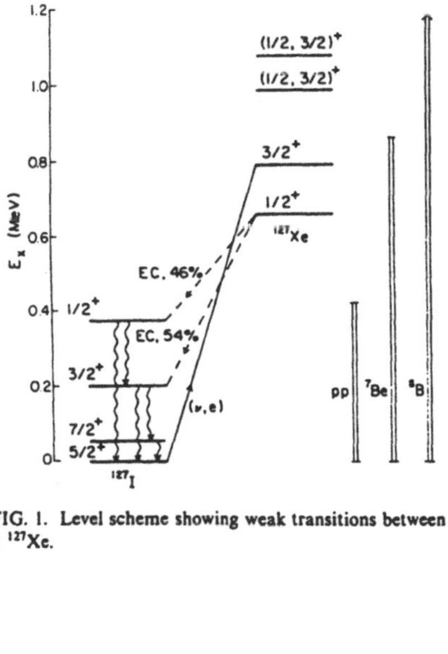

FIG. 1. Level scheme showing weak transitions between ^{127}I and ^{127}Xe.

The strengths of the relevant Gamow-Teller (GT) transitions are not presently known. In recent years it has been demonstrated empirically that the forward-angle (p,n) reaction at medium energies can be used to calibrate GT strengths. This technique provided an estimate of the excited-state contributions to the ^{71}Ga neutrino-capture cross section,[4] and yielded a ^{37}Cl capture cross section in good agreement with other determinations.[5] Though the accuracy of this technique is difficult to quantify, $\pm 20\%$ may be a reasonable estimate for the anticipated ^{127}I ^8B cross-section uncertainty.[6] For the present discussion I have made a rough estimate of the ^{127}Xe GT distribution between threshold and neutron breakup by scaling the measured ^{71}Ga[4] and ^{98}Mo[7] GT distributions by the appropriate ratios of N-Z factors, in accordance with the naive sum rule. On folding with the proper phase-space factors, one obtains cross sections averaged over the ^8B neutrino spectrum of 7.2×10^{-42} cm^2 and 8.9×10^{-42} cm^2, respectively. I adopt the smaller value, with the caution that the uncertainty in this estimate is large. The large cross section is due in part to Coulomb effects, which enhance the phase space by a factor of 2 in going from $Z = 31$ (Ga) to $Z = 53$. For comparison, the ^{37}Cl ^8B cross section is 1.12×10^{-42} cm^2.[5]

The strength of the ^7Be-neutrino $\frac{5}{2}^+ \to \frac{3}{2}^+$ (125 keV) transition could be determined from (p,n) measurements, with the assumption that sufficient resolution is achieved, or directly calibrated in a neutrino-source experiment. Neighboring states at 321 and 412 keV have been identified only as $J^\pi = (\frac{1}{2}, \frac{3}{2})^+$. If the 321-keV state has $J = \frac{3}{2}$, a resolution of approximately 200 keV is required to separate this transition from that to the 125-keV level. This is comparable to the best

The Superkamiokande

Masato Takita
Department of Physics
Osaka University
Machikaneyama-cho 1-1, Toyonaka, Osaka, 560,
Japan

ABSTRACT

The salient detector performance of the Superkamiokande experiment are briefly described. Then, selected physics goals, solar neutrinos, supernova neutrinos, the atmospheric neutrino problem, nucleon decays, are discussed.

1. Introduction

The Superkamiokande experiment [Y. Totsuka(1990)] is a natural extension of the current Kamiokande experiment with an upgraded performance as well as a by far better sensitivity. The physics goal of Superkamiokande is the precise measurement of cosmic or atmospheric neutrinos as well as the observation of possible baryon number violating processes. The Superkamiokande project was funded in 1991 and the construction job started in 1991 and will last until April 1996. We need 3 years for the production of 11200 20-inch-in-diameter photomultipier tubes (PMTs), 3 years for the excavation of a huge cavity of 60000 tons, 1.5 year for a 50000-m^3 water tank construction and 1 year for the installation. The detector construction period competes with that of a typical accelerator experiment, although the Superkamiokande experiment are categorized as a non-accelerator experiment. The whole detector is composed of a huge water Čerenkov inner detector having a photosensitive mass of 32000 tons viewed by 11200 PMTs as dense as 2 PMTs/m^2, and of an outer 4-π solid-angle anticounter layer (2 m thick) which is also of water Čerenkov type with a mass of 18000 tons. The anticounter surrounding the photosensitive volume of the detector is essential to eliminate background (γ rays and neutrons) of external

Table 1: Detector Performance of Superkamiokande and of Kamiokande-II

Parameters	Super-Kamiokande	Kamiokande-II	Remarks
Total size	41mh×39mϕ	16mh×19mϕ	
Total mass	50000t	4500t	
Fiducial mass	32000t	2140t	supernova ν burst
	22000t	1040t	proton decay
	22000t	680t	solar ν
Thickness of anti-counter	2m	1.2m~1.5m	
Number of PMTs	11200	948	
Photosensitive coverage	40%	20%	
PMT timing resolution	2.5nsec	4nsec	at 1 P.E.
Energy	$2.6\%/\sqrt{E}$	$3.6\%/\sqrt{E}$	e of E(GeV)
	2.5%	4%	$\mu(\lesssim 1$ GeV)
resolution	$16\%/\sqrt{E(10MeV)}$	$20\%/\sqrt{E(10MeV)}$	e($\lesssim$20 MeV)
Position	50 cm	110 cm	10 MeV e
resolution	~10 cm	15 cm	$p \rightarrow e^+\pi^0$
Angular	28°	28°	10 MeV e
resolution	~1°	2.7°	Thru-going μ
E_{th}(trigger)	4~5 MeV	5.2 MeV	
E_{th}(analysis)	5 MeV	7.5 MeV	Solar ν
e/μ separation	99 %	98±1 %	$0.03<p_e<1.33$GeV/c $0.2<p_\mu<1.5$GeV/c
$\epsilon_{\mu^+ \rightarrow e^+\nu\nu}$	95 %	87±1 %	

origin. The anticounter also facilitates various analysis, nsince it provides us a clear hardware signal of incoming or outgoing ocharged partilcles. The expected performance of the Superkamiokande idetector is summarized in Table 1. The vertex position resolution and energy resolution are shown in Fig. 1 and Fig. 2, respectively, as a function of electron energy. One doubts that the water transparency (60 m in the case of Kamiokande-II) might affect energy resolution of the detector. As you see in Fig. 2, however, the water transparency can be properly corrected according to the Monte Carlo simulation. The water transparency will be measured with an accuracy of 1 m in the Superkamiokande detector by means of cosmic-ray muons and laser light. The error 1m affects the detector gain only by [exp(-20m/60m)/exp(-20m/59m)]=0.6 % which corresponds to 2.4 % uncertainty in the ^{8}B solar neutrino flux.

The physics goal of the Superkamiokande experiment is itemized below:

1. precise measurement of ^{8}B solar neutrinos

2. precise measurement of real-time supernova neutrinos
3. measurement of diffuse supernova neutrinos
4. high-energy neutrinos from astronomical point sources
5. neutrinos from dark matter annihilaition in the Sun
6. neutrinos from the Rubakov effect (nucleon decays catalyzed by magnetic monopoles) in the Sun
7. nucleon decays
8. neutron-antineutron oscillations
9. precise measurement of atmospheric neutrinos
10. precise measurement of upward through-going muons
11. precise measurement of upward stopping muons

Among various topics, here, I will discuss ^{8}B solar neutrinos, real-time supernova neutrinos, atmospheric neutrinos, and nucleon decays.

2. ^{8}B Solar Neutrinos

Kamiokande-II succeeded in the observation of ^{8}B solar neutrinos. The most recent measurement [K. S. Hirata *et al.*(1991)] of ^{8}B solar neutrino flux gives 46 % of the expected value from the standard solar model [J. N. Bahcall and R. K. Ulrich(1988) and see also S. Turck-Chieze *et al.*(1988)]. Then, Superkamiokande will observe 8400 solar neutrino events per year (taking into account dead time=28 %, E_e>5 MeV, 46 % of the SSM). Here, one point should be stressed: it is not obvious now to lower the analysis threshold down to 5 MeV because of overwhelming background due presumably to ^{222}Rn atoms sneaking into the detector through incomplete air-tightness. To achieve a good S/N ratio at 5 MeV, it is necessary to reduce the background by a factor of 100. We guess (or hopefully) that these ^{222}Rn are of external origin since the mine air contains at least 100 times as much ^{222}Rn as the air outside. With a moderate cost, we can do this job by introducing radon-free air (5 m^3/min) from outside via a long duct and filling the detector site with the radon-free air. However, we are able to lower the analysis threshold down to 6.5 MeV without any serious problem, keeping a good S/N ratio (2∼3), thanks to the better energy resolution than in the current Kamiokande detector. The Superkamiokande experiment will thus accumulate 100 times higher statistics than the current Kamiokande experiment. Using this high statistics, We can make precise measurement of spectral shape, time variation, day/night effect of the ^{8}B solar neutrino flux.

Reference [P. Anselmann *et al.*(1992)] shows a combined allowed region at 90 % C.L. by Homestake (^{37}Cl)[R. Davis Jr., D.S. Harmer and K.C. Hoffman(1968), R. Davis Jr(1990)], Kamiokande-II and Gallex (^{71}Ga) experiments. The allowed region is classified into 2 types: non-adiabatic region and quasi-vacuum region. As is seen in Fig. 3, the Superkamiokande experiment can distinguish between these two cases by means of spectral shape of recoil elec-

Sudbury Neutrino Observatory

G.T. EWAN for the SNO Collaboration[1]
Department of Physics
Queen's University
Kingston, Ontario
CANADA K7L 3N6

Abstract

The Sudbury Neutrino Observatory (SNO) detector is a 1000 tonne heavy water Čerenkov detector designed to study the intensity, energy and direction of neutrinos from the sun and other astrophysical sources such as supernovae. The unique feature is the use of heavy water which enables both electron neutrinos and all types of neutrinos to be measured. These measurements will provide sensitive tests of the standard solar model, vacuum oscillations and MSW effects. It will detect with high sensitivity all types of neutrinos from a supernova in our galaxy.

1. Introduction

The Sudbury Neutrino Observatory is designed to study neutrinos from the sun and other astrophysical sources such as supernovae. The neutrinos are detected in a heavy water Čerenkov detector using 1000 tonnes of heavy water on loan from Atomic Energy of Canada Limited. The detector is presently under construction in a very low background laboratory 2000 m underground in the Creighton mine near Sudbury, Ontario, Canada. Construction is scheduled to be completed in 1995.

The most important and unique feature of the SNO detector is the use of heavy water as a detection medium. Neutrinos interact with the deuterons in the heavy water in three ways discussed below and this enables measurement of both electron neutrinos and all types of neutrinos that reach the detector. If an electron neutrino produced at the center of the sun changes into another type of neutrino this process will be detected and neutrino oscillations observed. In addition to neutrino oscillations in vacuum, the SNO experiment can detect matter enhanced oscillations, the MSW effect, by observing the shape of the energy spectrum of electron neutrinos reaching the earth. The detection of both electron neutrinos and all types of neutrinos from a supernova would give new information on the mechanism of stellar collapse.

2. Neutrino reactions with heavy water

The experimental programme with the SNO detector involves the use of several reactions to measure neutrinos. The ν_e flux and energy spectrum can be studied by two different reactions, the total neutrino flux by a third reaction and the $\overline{\nu}_e$ flux and energy spectrum by a fourth reaction. Three of these reactions are with the deuterium in the heavy water and one with the electrons in heavy water similar to the reaction used in the ordinary water Kamiokande detector. The relatively high cross section of deuterium for solar neutrinos will lead to high counting rates compared to existing experiments and enable possible time variations to be studied with high statistical accuracy.

The reactions in the heavy water are

(1) The charged current (CC) reaction

$$\nu_e + \mathrm{d} \rightarrow \mathrm{p} + \mathrm{p} + \mathrm{e}^- \qquad (\mathrm{Q} = -1.44\ \mathrm{MeV})$$

The Čerenkov light from the electron is measured in the detector. This reaction of the electron neutrino on the deuteron is unique to the SNO detector. It offers excellent spectral information, thereby providing a sensitivity to the MSW effect. The large cross section will make the detector some forty times more sensitive than existing experiments. This reaction would also identify electron neutrinos from the initial burst in the collapse of a supernova.

(2) The neutral current (NC) reaction

$$\nu_x + \mathrm{d} \rightarrow \nu_x + \mathrm{p} + \mathrm{n} \qquad (\mathrm{Q} = -2.2\ \mathrm{MeV}).$$

This reaction is also unique to the SNO detector. It is observed by the detection of the neutron released using the capture of the neutron as a signature. This reaction is equally sensitive to all neutrinos and would be used to measure the total flux of neutrinos with a counting rate of about 10 per day above the threshold of about 2.2 MeV. This will give a direct measure of the total solar ^{8}B neutrino production independent of neutrino oscillation effects.

(3) The neutrino-electron elastic scattering (ES) reaction,

$$\nu_x + \mathrm{e}^- \rightarrow \nu_x + \mathrm{e}^-$$

This is the primary detection mechanism for light water detectors. It is sensitive to all neutrino types, but is dominated by the electron neutrino. The predicted count rate is $\sim 1/10$ of the charged current (CC) rate. The ES reaction is strongly forward peaked and gives excellent directional information.

(4) The anti-neutrino charged current reaction

$$\overline{\nu}_e + \mathrm{d} \rightarrow \mathrm{n} + \mathrm{n} + \mathrm{e}^+ \qquad (\mathrm{Q} = -4.03\ \mathrm{MeV})$$

The signature for this reaction will be the positron signal followed by the delayed signal arising from the capture of a neutron. The signature for $\overline{\nu}_e$ is very different from that of ν_e and if ν_e-$\overline{\nu}_e$ oscillations occur they should be easily detected.

3. The SNO detector

A simplified schematic drawing of the SNO detector is shown in Figure 1. This figure shows the basic components but no details of the support structures for the acrylic vessel and for the PMTs. The neutrinos are detected in 1000 tonnes of 99.85% enriched ultrapure D_2O contained in a spherical thin-walled (5 cm. thick) transparent acrylic vessel of 12 m. diameter which itself is immersed in 7300 tonnes of ultrapure H_2O shield.

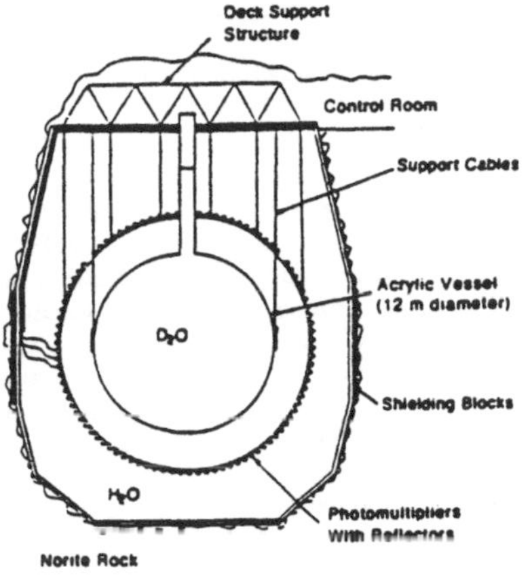

Figure 1. Simplified schematic diagram of the SNO detector

The Čerenkov light from neutrino interactions in the heavy water is detected in an array of 9600 20 cm diameter photomultiplier tubes on a support structure 2.5 m. from the surface of the acrylic vessel. The water attenuates the radiation from the PMTs which are made using special low radioactivity glass. A reflector is mounted in front of each PMT to increase the light collection and the effective photocathode coverage is approximately 60%.

The host cavity, which is barrel-shaped and of 22 m. diameter and 30 m. height, is under excavation at a depth of 2070 m. below surface at the (operating) Creighton Mine of INCO Ltd., near Sudbury, Canada. In a waist region the cavity is lined with low background concrete. A liner is placed inside the cavity enclosing the light water shield. After completion of the cavity the PMT support structure and acrylic vessel will be assembled and installed underground. Great care must be taken to ensure the cleanliness of all components during and after assembly.

Neutrino interactions in the detector produce either relativistic electrons or free neutrons. The neutrons thermalize in the water and are subsequently captured, generating γ-rays which produce relativistic electrons primarily through Compton scattering. The electrons from neutrino interactions or neutron captures will produce Čerenkov photons which pass through the D_2O, acrylic, and H_2O to be detected by the PMTs. The signals from the PMTs are interpreted to give the location, energy, and direction of the electron based on the time of arrival of the photons at the PMTs, the number of PMTs triggered, and the locations of the particular PMTs hit.

SEARCH FOR NEUTRINOS FROM THE SUN*

Raymond Davis, Jr., Don S. Harmer,† and Kenneth C. Hoffman
Brookhaven National Laboratory, Upton, New York 11973
(Received 16 April 1968)

A search was made for solar neutrinos with a detector based upon the reaction $Cl^{37}(\nu, e^-)Ar^{37}$. The upper limit of the product of the neutrino flux and the cross sections for all sources of neutrinos was 3×10^{-36} sec^{-1} per Cl^{37} atom. It was concluded specifically that the flux of neutrinos from B^8 decay in the sun was equal to or less than 2×10^6 cm^{-2} sec^{-1} at the earth, and that less than 9% of the sun's energy is produced by the carbon-nitrogen cycle.

Recent solar-model calculations have indicated that the sun is emitting a measurable flux of neutrinos from decay of B^8 in the interior.[1-8] The possibility of observing these energetic neutrinos has stimulated the construction of four separate neutrino detectors.[9] This paper will present the results of initial measurements with a detection system based upon the neutrino capture reaction $Cl^{37}(\nu, e^-)Ar^{37}$. It was pointed out by Bahcall[10] that the energetic neutrinos from B^8 would feed the analog state of Ar^{37} (a superallowed transition) that lies 5.15 MeV above the ground state. The importance of the contribution of the B^8 neutrino flux is readily seen from the neutrino-capture cross sections and the solar neutrino fluxes given in Table I. The tabulated fluxes were taken from the calculations of Bahcall and Shaviv,[8] who studied the effect of errors in the parameters—solar composition, luminosity, opacity, and nuclear reaction cross sections. These authors have placed a probable error of 60% on the calculated B^8 flux. Their predicted B^8 flux for mean values of the various parameters agrees well with the independent calculations of Ezer and Cameron.[5] On the basis of these predictions, the total solar-neutrino-capture rate in 520 metric tons of chlorine would be in the range of 2 to 7 per day.

The detector design.—A detection system that contains 390 000 liters (520 tons chlorine) of liquid tetrachloroethylene, C_2Cl_4, in a horizontal cylindrical tank was built along the lines proposed earlier.[11] The system is located 4850 ft underground [4400 m (w.e.)] in the Homestake gold mine at Lead, South Dakota. It is essential to place the detector underground to reduce the production of Ar^{37} from (p,n) reactions by protons formed in cosmic-ray muon interactions. The rate of Ar^{37} production in the liquid by cosmic-ray muons at this location is estimated to be 0.1 Ar^{37} atom per day.[11] Background effects from internal α contaminations and fast neutrons from the surrounding rock wall are low. The total Ar^{37} production from all background processes is less than 0.2 Ar^{37} atom per day, which is well below the rate expected from solar neutrinos.

Neutrino detection depends upon removing the Ar^{37} from a large volume of liquid contained in a sealed tank, and observing the decay of Ar^{37} (35-day half-life) in a small proportional counter (0.5 cm^3). It is therefore necessary to have an efficient method of removing a fraction of a cubic centimeter of argon from 390 000 liters of C_2Cl_4. The Ar^{37} activity is removed by purging with helium gas. Liquid is pumped uniformly

Table I. Solar neutrino fluxes and cross sections for the reaction $Cl^{37}(\nu,e^-)Ar^{37}$.

Neutrino source	Cross section[a,b] (cm^2)	Neutrino flux[c] at the earth (cm^{-2} sec^{-1})	$10^{35}\sigma\varphi$ (sec^{-1})
$H+H+e^- \to D+\nu$	1.72×10^{-45}	1.7×10^8	0.03
Be^7 decay	2.9×10^{-46}	3.9×10^9	0.11
B^8 decay	1.35×10^{-42}	$1.3(1=0.6)\times10^7$	$1.8(1\pm0.6)$
N^{13} decay	2.1×10^{-46}	1.0×10^9	0.02
O^{15} decay	7.8×10^{-46}	1.0×10^9	0.08
			$\sum\varphi\sigma=2.0(1\pm0.6)\times10^{-35}$ sec^{-1}

[a]Ref. 4. [b]Ref. 10. [c]Ref. 8.

from the bottom of the tank and returned to the tank through a series of 40 eductors arranged along two horizontal header pipes inside the tank. The eductors aspirate the helium from the gas space (2000 liters) above the liquid, and mix it as small bubbles with the liquid in the tank. The pump and eductor system passes helium through the liquid at a total rate of 9000 liters per minute maintaining an effective equilibrium between the argon dissolved in the liquid and the argon in the gas phase.

Argon is extracted by circulating the helium from the tank through an argon extraction system. Gas flow is again achieved by a pair of eductors in the tank system, and they maintain a flow rate of 310 liters per minute through the argon extraction system. The tetrachloroethylene vapor is removed by a condenser at −40°C followed by a bed of molecular sieve adsorber at room temperature. The helium then passes through a charcoal bed at 77°K to adsorb the argon, and is finally returned to the tank. This arrangement is shown schematically in Fig. 1. The apparatus is located in three separate rooms in the mine as indicated in the diagram.

The argon sample adsorbed on the charcoal trap is removed by warming the charcoal while a current of helium is passed through it. The argon and other rare gases from the effluent gas stream are collected on a small liquid-nitrogen-cooled charcoal trap (1 cm diam by 10 cm long). Finally, the gases from this trap are desorbed and heated over titanium metal at 1000°C to remove all traces of chemically reactive gases. The resulting rare gas contains krypton and xenon in addition to argon. These higher rare gases were dissolved from the atomosphere during exposure of the liquid during the various manufacturing, storage, and transfer operations. Krypton and xenon are much more soluble in tetrachloroethylene than argon, and, therefore, they are more slowly removed from the liquid by sweeping with helium. Since the volume of krypton and xenon in an experimental run is comparable with or exceeds the volume of argon, it is necessary to remove these higher rare gases from the sample. A more important consideration is that atmospheric krypton contains the

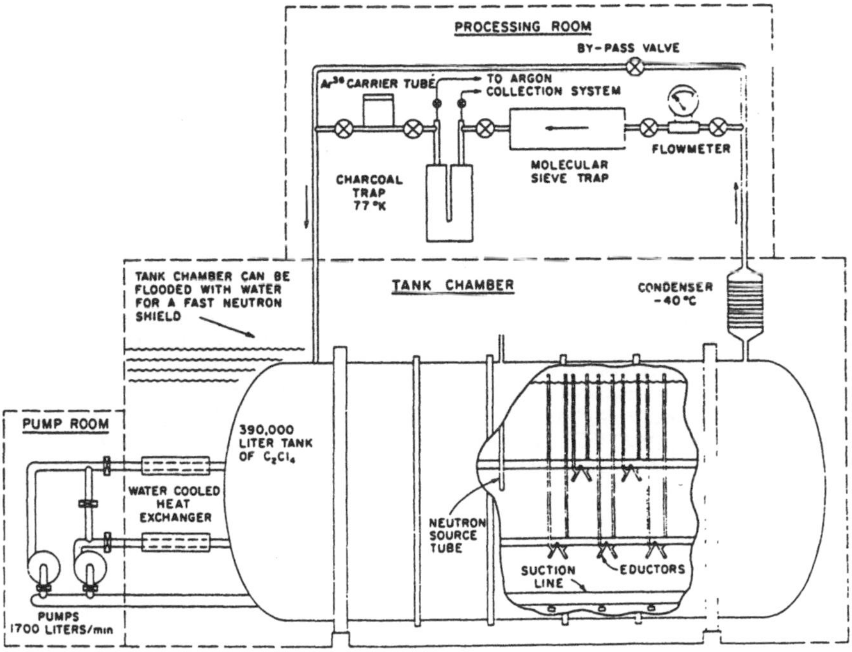

FIG. 1. Schematic arrangement of the Brookhaven solar neutrino dector.

10.8-yr fission product Kr^{85}. The rare gases recovered from the tank are therefore separated by gas chromatography. To insure complete removal of krypton from the argon sample; a second gas chromatographic separation is made of the argon fraction. Experience has shown that these two successive chromatographic separations reduce krypton concentration in the argon sample to less than 10^{-8} parts per volume. The entire purified argon sample is counted in a small proportional counter that will be described later.

Argon recovery tests.—After the air and air argon had been removed from the system by prolonged sweeping with helium, the argon recovery efficiency of the system was measured by an isotope dilution method. A measured volume of 99.9% Ar^{36} was introduced into the tank and dissolved in the liquid with the eductor system. It was then recovered by six separate purging operations. The Ar^{36} recovered from each purge was determined by a volumetric and argon mass-ratio measurement. It was found that the volume of Ar^{36} in the tank dropped exponentially with the volume of helium circulated according to

$$v(\mathrm{Ar})/v_0(\mathrm{Ar}) = e^{-7.21\times10^{-6}V(\mathrm{He})}\ .$$

where $v_0(\mathrm{Ar})$ is the initial volume of Ar^{36} and $v(\mathrm{Ar})$ is the volume remaining after $V(\mathrm{He})$ liters of helium have passed through the extraction system. This test showed that a 95% recovery of argon from the tank can be achieved by circulating 0.42 million liters of helium through the extraction system, which requires a period of 22 h.

Another test of the argon recovery from the tank was performed with Ar^{37} activity produced in the tank by a fast-neutron irradiation. A Ra-Be neutron source with a total neutron emission rate of 7.38×10^{4} neutrons sec^{-1} was inserted in a re-entrant iron pipe that reaches to the center of the tank. The liquid was irradiated with this source for 0.703 days producing Ar^{37} in the liquid by the reaction $Cl^{37}(p,n)Ar^{37}$ from the protons produced in the liquid principally by the reaction $Cl^{35}(n,p)S^{35}$. Carrier Ar^{36} was introduced (1.18 std cc) and the tank was swept three successive times with helium in which the volumes passed were, respectively, 0.35, 0.26, and 0.34 millions of liters of helium. The recovered argon was purified and counted following the procedures given below. The Ar^{37} activities in the three separate purges were found to be 63.4 $\pm$3.6, 2.3$\pm$1.1, and 0.7$\pm$0.5 disintegrations per day at the end of the neutron irradiation. The total Ar^{37} production rate observed in this experiment was $(7.5\pm0.4)\times10^{-7}$ Ar^{37} atom per neutron. This production rate compared favorably with similar measurements in containers of smaller diameters (29 and 120 cm) which gave yields of 3.0×10^{-7} and 6.4×10^{-7} Ar^{37} atom per neutron, respectively. The Ar^{36} recoveries from each of the three successive purges were 90.6, 6.2, and 0.7%, matching closely the Ar^{37} recoveries.

One might question whether Ar^{37} produced by the (ν, e^-) reaction would also be removed efficiently, since it would initially have a lower recoil energy than Ar^{37} produced by the (p,n) reaction. The Ar^{37} recoil energy resulting from neutrino capture ranges from 11 to over 1000 eV for neutrino energies of 1 to 10 MeV. These recoil energies are sufficient to assure that the Ar^{37} ion formed would be free of the parent molecule, and, therefore, it would be expected to behave chemically similarly to an Ar^{37} atom produced by the (p,n) reaction. Once an Ar^{37} atom exists as a free atom it will mix with the carrier Ar^{36} present in the liquid (10^{10} atoms cm^{-3}) and be removed by the helium purge.

Counting.—The argon sample is counted in a small proportional counter with an active volume 3 cm long and 0.5 cm in diameter. A small amount of methane is added to the argon to improve the counting characteristics of the gas. The counter cathode was constructed of zone-refined iron and the exterior envelope is made of silica glass. A thin window in the envelope is located at the end of the counter to facilitate energy calibration of the counter with Fe^{55} x rays. The counter is shielded from external radiations by a cylindrical iron shield 30 cm thick lined with a ring of 5-cm-diam proportional counters for registering cosmic-ray muons. The argon counter is held in the well of a 12.5- by 12.5-cm sodium-iodide scintillation counter located inside the ring counters. Events in anticoincidence with both the ring counters and the scintillation counter are recorded on a 100-channel pulse-height analyzer.[12] The pulse-height and time distribution of the events are recorded on paper tape. Each anticoincidence pulse is displayed on a storage oscilloscope and photographed to allow examination of each pulse shape to insure that it has a proper shape and is not caused by electrical noise. The counter had a 28% resolution (full width at half-maximum) for the 2.8-keV Auger electrons from the Ar^{37} decay. The operating voltage and amplifier gain are adjust-

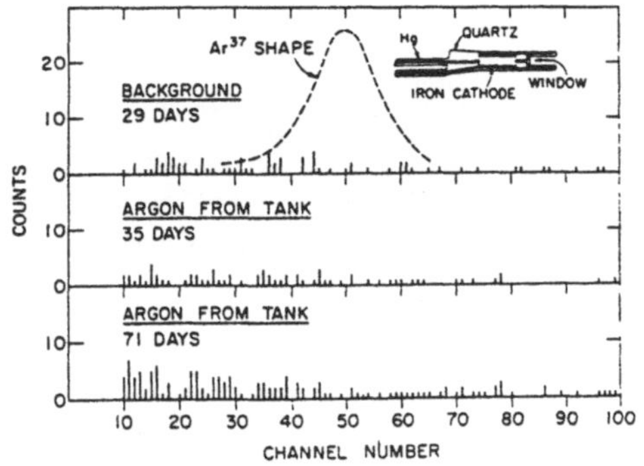

FIG. 2. Pulse-height spectra.

ed to place the center of the 2.8-keV Ar^{37} peak at channel 50 in the spectrum. The background counting rate in the 14 channels centered around channel 50 is 0.3 count per day (see Fig. 2). The efficiency of the counter was determined by filling with argon containing a known amount of Ar^{37}. Its efficiency for Ar^{37} is 51 % for the 14 channels centered about channel 50.

Results and discussion.—Two experimental runs have been performed. In both experiments a measured volume of Ar^{36} was introduced into the tank at the start of the period of exposure, and mixed into the liquid for a period of approximately two hours with the eductor system. During the period of exposure the pumps were not operated. A positive pressure of helium of approximately 250 mm of Hg exists in the tank at all times.

The first exposure was 48 days. The tank was purged with 0.50 million liters of helium. A volume of 1.27 std cc of argon was recovered from the tank, and this volume contained 94 % of the carrier Ar^{36} introduced at the start of the exposure. It was counted for 39 days and the total number of counts observed in the Ar^{37} peak position (full width at half-maximum) in the pulse-height spectrum was 22 counts. This rate is to be compared with a background rate of 31 ± 10 counts for this period. The neutrino-capture rate in the tank deduced from the exposure, counter efficiency, and argon recovery from this experiment was (-1.1 ± 1.4) per day.

A second exposure was made for 110 days from 23 June to 11 October 1967. The tank was purged with 0.53 million liters of helium yielding 0.62 cm^3 of argon with a 95% recovery of the added carrier Ar^{36}. The pulse-height spectra are shown in Fig. 2 for the first 35 days of counting and also for a total period of 71 days. This rate can be compared with the background rate for the counter filled with Ar^{36} purified in an identical manner (shown in Fig. 2). It may be seen from the pulse-height spectrum for the first 35 days of counting that 11 ± 3 counts were observed in the 14 channels where Ar^{37} should appear. The counter background for this period of time corresponded to 12 ± 4 counts. Thus, there is no increase in counts from the sample over that expected from background counting rate of the counter. One would deduce from these rates that the neutrino-capture rate in 610 tons of tetrachloroethylene was equal to or less than 0.5 per day based upon one standard deviation. A similar limit can be obtained if one examines the shape of the pulse-height spectrum for extra counts in the 14 channels centered about channel 50 in the first 35-day count.

This limit, expressed as

$$\sum \varphi\sigma \leq 0.3 \times 10^{-35}\ \text{sec}^{-1} \text{ per } Cl^{37} \text{ atom},$$

can be compared with the predicted value of $(2.0 \pm 1.2) \times 10^{-35}$ sec^{-1} per Cl^{37} atom (Table I). It may be seen that this limit is approximately a factor of 7 below that expected from these solar-model calculations. From this limit and the cross section for B^8 neutrinos given in Table I, it may be concluded that the flux of B^8 neutrinos at the earth is equal to or less than 2×10^6 cm^{-2} sec^{-1}. It may be pointed out that if one accepted all of the 11 counts in the spectrum for the 35-day count as real events, making no allowance for background, then the flux-cross-section product limit would be 0.6×10^{-35} sec^{-1} per Cl^{37} atom.

The solar-model calculation of the flux of B^8 neutrinos is dependent upon the nuclear cross sections, solar composition, solar age and luminosity, and the opacity of solar material. The effect of each of these parameters has been studied, and the present results show that the solar B^8 neutrino flux is outside the present error limits if the uncertainties are treated as probable errors.[5-8] In the following article[13] Bahcall, Bahcall, and Shaviv have re-evaluated the solar neutrino fluxes taking into account a new value for the heavy element composition of the sun, and a new rate for the reaction $H(H, e^+\nu)D$.

Since this experiment is the first one with sufficient sensitivity to detect solar neutrinos from the carbon-nitrogen cycle, it is interesting to draw a conclusion about this energy cycle. Bahcall[4] has calculated the total flux-cross-section product for the carbon-nitrogen cycle to be 3.5×10^{-35} sec^{-1} per Cl^{37} atom, based on this cycle being the only source of the sun's energy. With the limit given above one can conclude that less than 9% of the sun's energy is produced by the carbon-nitrogen cycle.

It is possible to improve the sensitivity of the present experiment by reducing the background of the counter. However, background effects from cosmic-ray muons will eventually limit the detection sensitivity of the experiment at its present location. Detailed studies of the cosmic-ray background are in progress.

The authors would like to thank Professor W. A. Fowler and Professor John N. Bahcall for their initial and continual encouragement in planning this experiment. We would like to acknowledge Professor A. G. W. Cameron's constant interest extending over many years. We are indebted to the Homestake Mining Company for allowing us to build the experiment in their mine, and for their generous assistance in solving many technical problems in the construction of the apparatus. We would like to acknowledge the many useful suggestions and direct assistance from the members of the staff of Brookhaven National Laboratory.

*Research performed under the auspices of the U. S. Atomic Energy Commission.

†Permanent address: Georgia Institute of Technology, Atlanta, Georgia.

[1]J. N. Bahcall, W. A. Fowler, I. Iben, Jr., and R. L. Sears, Astrophys. J. 137, 344 (1963).

[2]R. L. Sears, Astrophys. J. 140, 153 (1964).

[3]P. Pochoda and H. Reeves, Planetary Space Sci. 12, 119 (1964).

[4]J. N. Bahcall, Phys. Rev. Letters 12, 300 (1964); 17, 398 (1966).

[5]D. Ezer and A. G. W. Cameron, Can. J. Phys. 43, 1497 (1965), and 44, 593 (1966); and private communication.

[6]J. N. Bahcall, N. Cooper, and P. Demarque, Astrophys. J. 150, 723 (1967); G. Shaviv, J. N. Bahcall, and W. A. Fowler, Astrophys. J. 150, 725 (1967).

[7]J. N. Bahcall, N. Bahcall, W. A. Fowler, and G. Shaviv, Phys. Letters 26B, 359 (1968).

[8]J. N. Bahcall and G. Shaviv, to be published.

[9]For recent summary see F. Reines, Proc. Roy. Soc. (London) 310A, 104 (1967).

[10]J. N. Bahcall, Phys. Rev. 135, B137 (1964).

[11]R. Davis, Jr., Phys. Rev. Letters 12, 303 (1964); R. Davis, Jr., and D. S. Harmer, CERN Report No. CERN 65-32, 1965 (unpublished).

[12]The circuit used in this work was designed by Mr. R. L. Chase and Mr. Lee Rogers of Brookhaven National Laboratory.

[13]J. N. Bahcall, N. A. Bahcall, and G. Shaviv, following Letter [Phys. Rev. Letters 20, 1209 (1968)].

Solar Neutrinos: A Scientific Puzzle

John N. Bahcall and Raymond Davis, Jr.

For the past 15 years we have tried, in collaboration with many colleagues in astronomy, chemistry, and physics, to understand and test the theory of how the sun produces its radiant energy (observed on the earth as sunlight). All of us have been surprised by the results: there is a large, unexplained disagreement between observation and the supposedly well established theory. This discrepancy has led to a crisis in the theory of stellar evolution; many authors are openly questioning some of the basic principles and approximations in this supposedly dry (and solved) subject.

One may well ask, Why devote so much effort in trying to understand a backyard problem like the sun's thermonuclear furnace when there are so many exciting and exotic discoveries occurring in astronomy? Most natural scientists believe that we understand the process by which the sun's heat is produced—that is, in thermonuclear reactions that fuse light elements into heavier ones, thus converting mass into energy. However, no one has found an easy way to test the extent of our understanding because the sun's thermonuclear furnace is deep in the interior, where it is hidden by an enormous mass of cooler material. Hence conventional astronomical instruments can only record the photons emitted by the outermost layers of the sun (and other stars). The theory of solar energy generation is sufficiently important to the general understanding of stellar evolution that one would like to find a more definitive test.

There is a way to directly and quantitatively test the theory of nuclear energy generation in stars like the sun. Of the particles released by the assumed thermonuclear reactions in the solar interior, only one has the ability to penetrate from the center of the sun to the surface and escape into space: the neutrino. Thus neutrinos offer us a unique possibility of "looking" into the solar interior. Moreover, the theory of stellar aging by thermonuclear burning is widely used in interpreting many kinds of astronomical information and is a necessary link in establishing such basic data as the ages of the stars and the abundances of the elements. The parameters of the sun (its age, mass, luminosity, and chemical composition) are better known than those of any other star, and it is in the simplest and best understood stage of stellar evolution, the quiescent main sequence stage. Thus an experiment designed to capture neutrinos produced by solar thermonuclear reactions is a crucial one for the theory of stellar evolution. We also hoped originally that the application of a new observing technique would provide added insight and detailed information. It is for all of these reasons (a unique opportunity to see inside a star, a well-posed prediction of a widely used theory, and the hope for new insights) that so much effort has been devoted to the solar neutrino problem.

John Bahcall is professor of natural science at the Institute for Advanced Study, Princeton, New Jersey 08540, and Raymond Davis is senior chemist at Brookhaven National Laboratory, Upton, Long Island, New York 11973.

Nuclear Fusion in the Sun

We shall now outline briefly the conventional wisdom (*1*) regarding nuclear fusion as the energy source for main sequence stars like the sun. It is assumed that the sun shines because of fusion reactions similar to those envisioned for terrestrial fusion reactors. The basic solar process is the fusion of four protons to form an alpha particle, two positrons (e^+), and two neutrinos (ν); that is, $4p \rightarrow \alpha + 2e^+ + 2\nu$. The principal reactions are shown in Table 1 with a column indicating in what percentage of the solar terminations of the proton-proton chain each reaction occurs. The rate for the initiating proton-proton (PP) reaction, number 1 in Table 1, is largely determined by the total luminosity of the sun. Unfortunately, these neutrinos are below the threshold, which is 0.81 Mev, for the ^{37}Cl experiment that has been performed to detect solar neutrinos (*2*). The PEP reaction (number 2), which is the same as the familiar PP reaction except for having the electron in the initial state, is detectable in the ^{37}Cl experiment. The ratio of PEP to PP neutrinos is approximately independent of which model (see below) one uses for the solar properties (*3*). Two other reactions in Table 1 are of special interest. The capture of electrons by ^7Be (reaction 6) produces detectable neutrinos in the ^{37}Cl experiment. The ^8B beta decay, reaction 9, was expected (*4*) to be the main source of neutrinos for the ^{37}Cl experiment because of their relatively high energy (14 Mev), although it is a rare reaction in the sun (see Table 1). There are also some less important neutrino-producing reactions from the carbon-nitrogen-oxygen (CNO) cycle, but we shall not discuss them in detail since the CNO cycle is believed to play a rather small role in the energy-production budget of the sun.

In order to calculate how often the various nuclear reactions occur, one must make a detailed model of the interior of the sun. The techniques for constructing such models are standard (*1, 5*) (although greater precision is required for the solar neutrino problem than for most other problems in stellar evolution). The physics involved is elementary. One imposes at each point in the star the condition of hydrostatic equilibrium; that is, the condition that the pressure gradient balances the gravitational attraction. Both radiative and kinetic contributions to the pressure are included. The energy generation is given by an integral over the derived temperature-density distribution, using the calculated rates of all the nuclear reactions. Energy transport in the solar interior is largely by radiation and hence depends inversely on the radiative opacity. It is conventional to assume a primordial chemical composition that is homogeneous throughout the sun and equal to the presently observed surface chemical composition. One makes a sequence of successive solar models, requiring that the calculated luminosity equal the observed solar luminosity at a model age of 4.7×10^9 years, the age of the solar system.

Brookhaven Solar Neutrino Experiment

Neutrinos can be captured in nuclei by a reaction called the inverse beta process, so called because it is the inverse of the beta decay process in which neutrinos are created. The Brookhaven solar neutrino detector is based on the neutrino capture reaction (*2, 6*)

$$\nu + {}^{37}Cl \underset{\text{decay}}{\overset{\text{capture}}{\rightleftharpoons}} {}^{37}Ar + e^-$$

which is the inverse of the electron capture decay of ^{37}Ar. The radioactive decay occurs with a half-life of 35 days. This reaction was chosen for the Brookhaven solar neutrino experiment because of its unique combination of physical and chemical characteristics, which were favorable for building a large-scale solar neutrino detector. Neutrino capture to form ^{37}Ar in the ground state has a relatively low energy threshold (0.81 Mev) and a favorable cross section, nuclear properties that are important for observing neutrinos from ^{7}Be, ^{13}N, and ^{15}O decay and the PEP reaction. If neutrinos are energetic enough, ^{37}Ar can be formed in one of its many excited states (*4, 7*). Neutrinos from ^{8}B decay have sufficient energy to feed these excited states and have a much higher capture cross section than those from the lower energy neutrino sources mentioned above. Because of this, the capture rate was expected to be due primarily to the ^{8}B neutrinos. The nuclear properties of ^{37}Ar, ^{37}K, and ^{37}Ca have been determined in various laboratory measurements, providing a solid experimental basis for the original theoretical calculations of the neutrino capture cross section of ^{37}Cl (*7*). The sensitivity of the detector for neutrinos from the various solar processes depends on these calculated cross sections.

The ^{37}Cl reaction is very favorable from a chemical point of view (*2, 6, 8*). Chlorine is abundant and inexpensive enough that one can afford the many hundreds of tons needed to observe solar neutrinos. The most suitable chemical compound is perchloroethylene, C_2Cl_4, a pure liquid, which is manufactured on a large scale for cleaning clothes. The product, ^{37}Ar, is a noble gas, which should ultimately exist in the liquid as dissolved atoms. The neutrino capture process produces an ^{37}Ar atom with sufficient recoil energy to break free of the parent perchloroethylene molecule and penetrate the surrounding liquid, where it reaches thermal equilibrium. Initially the recoiling argon atom is ionized. As it slows down, it will extract electrons from a neighboring molecule and become a neutral argon atom. A neutral argon atom behaves as dissolved argon, which can be removed easily from the liquid by purging with helium gas. These chemical processes are of crucial importance to the operation of the detector.

The Brookhaven ^{37}Cl detector (see Fig. 1) was built deep underground to avoid the production of ^{37}Ar in the detector by cosmic rays. This was done with the cooperation of the Homestake Gold Mining Company (Lead, South Dakota), who excavated a large cavity in their mine (~1500 m below the surface) to house the experiment. The final detector system consists of an ~ 400,000-liter tank of perchloroethylene, a pair of pumps to circulate helium through the liquid, and a small building to house the extraction equipment, as shown in Fig. 1.

The chemical processing is relatively simple (*2, 8*). A small amount of isotopically pure ^{36}Ar (or ^{38}Ar) carrier gas is placed in the tank and stirred into the liquid to insure that it dissolves. The tank is then allowed to stand about 100 days, permitting the ^{37}Ar activity to grow to nearly the saturation value. After this period the pumps are turned on, circulating helium through the tank. Helium from the tank passes through a chemical extraction system consisting sequentially of a condenser, an absorber for perchloroethylene vapor, and a charcoal trap at liquid nitrogen temperature, which collects the argon gas. The gas is removed from the charcoal absorber, purified, and placed in a miniature proportional counter to observe ^{37}Ar decay events. Recovery of argon from the tank is very high, at least 90 percent, and is determined in each experiment by measuring the amount of ^{36}Ar recovered compared to the amount introduced initially. If the standard solar model were correct, one would expect about 50 ^{37}Ar atoms in the 400,000 liters of liquid at the time it is purged. These few atoms of ^{37}Ar behave chemically in the same way as the 3×10^{19} atoms of ^{36}Ar introduced as a carrier gas. Therefore, a direct determination of the ^{36}Ar recovered is a reliable measure of the ^{37}Ar atom recovery.

Two additional tests have been performed to ensure that ^{37}Ar produced in the large tank is indeed recovered efficiently. In one, a small neutron source was placed in the center of the tank through a reentrant tube. Neutrons produce ^{37}Ar in the liquid by a series of nuclear reactions, and one verifies that the ^{37}Ar is recovered along with the carrier gas. The second test was to introduce a measured number of ^{37}Ar

Table 1. The proton-proton chain in the sun.

Number	Reaction	Solar terminations (%)	Maximum neutrino energy (Mev)
1	$p + p \rightarrow {}^{2}H + e^{+} + \nu$	99.75	0.420
	or		
2	$p + e^{-} + p \rightarrow {}^{2}H + \nu$	0.25	1.44 (monoenergetic)
3	${}^{2}H + p \rightarrow {}^{3}He + \gamma$		
4	${}^{3}He + {}^{3}He \rightarrow {}^{4}He + 2p$	86	
	or		
5	${}^{3}He + {}^{4}He \rightarrow {}^{7}Be + \gamma$		
6	${}^{7}Be + e^{-} \rightarrow {}^{7}Li + \nu$		0.861 (90%), 0.383 (10%) (both monoenergetic)
7	${}^{7}Li + p \rightarrow 2{}^{4}He$	14	
	or		
8	${}^{7}Be + p \rightarrow {}^{8}B + \gamma$		
9	${}^{8}B \rightarrow {}^{8}Be^{*} + e^{+} + \nu$		14.06
10	${}^{8}Be^{*} \rightarrow 2{}^{4}He$	0.02	

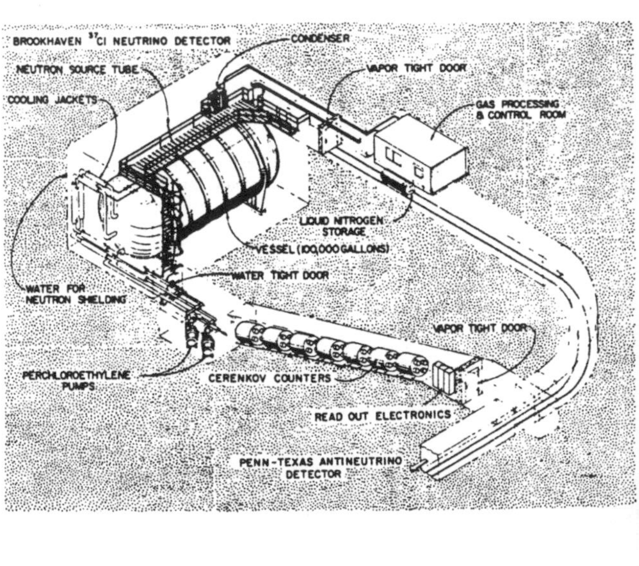

atoms (500) into the detector and then remove them, measuring the overall recovery and counting efficiency. Both of these tests show that ^{37}Ar is efficiently removed by the procedures used.

The entire gas sample from the tank is placed in a small proportional counter with an internal volume of less than a cubic centimeter. The decay of ^{37}Ar is characterized by the energy deposited in the counter by the Auger electrons following the electron capture process. These Auger electrons have a very short range in the counter and produce a characteristic pulse shape, which can be distinguished electronically. In a typical experiment only about two to four events are observed in the counter that have the proper characteristics for an ^{37}Ar decay.

We might add that Jacobs (*9*) has raised the question of whether the chemical tests are valid, and has suggested that ^{37}Ar produced by neutrino capture either does not become a neutral dissolved argon atom or that once formed it is trapped in a molecular cage or compound. The possible formation of an argon molecule ion with perchloroethylene was tested by an experiment of Leventhal and Friedman (*10*), and they showed that the charge exchange process is at least 100 times more likely. The formation of molecular cages or rare gas compounds in perchloroethylene is very unlikely. Even if they were formed, an argon atom would not be retained, as evidenced by the diffusion of rare gases in plastics. One could test the chemical fate of argon (*11*) by studying the formation of ^{36}Ar from molecules of ^{36}Cl-labeled perchloroethylene, $C_2Cl_3{}^{36}Cl$. The recoil dynamics and ultimate chemical behavior of the resulting ^{36}Ar ion produced in the beta

Table 2. Significance of counting rates in the ^{37}Cl experiment. One solar neutrino unit (SNU) = 10^{-36} captures per target particle per second.

Counting rate (SNU)	Significance of counting rate
35	Expected if the CNO cycle produces the solar luminosity
~6 ± 2	Predictions of current models
1.5	Expected as a lower limit consistent with standard ideas of stellar evolution
0.3	Expected from the PEP reaction, hence a test of the basic idea of nuclear fusion as the energy source for main sequence stars

decay of ^{36}Cl (half-life, 308,000 years) are identical to those of the neutrino capture process. Because of the intense ^{36}Cl source needed, this experiment is not an easy one to perform, but it is now being undertaken. There is little question about the chemical fate of ^{37}Ar produced by neutrino capture, and we feel certain that recovery of ^{37}Ar from the 400,000-liter detector is accurately measured by the ^{36}Ar (or ^{38}Ar) recovery measurements described above.

Observational Results

A set of ten experimental runs carried out in the Brookhaven ^{37}Cl experiment over the last 3 years show that the ^{37}Ar production rate in the tank is 0.13 ± 0.13 ^{37}Ar atoms per day (*12*). Even though the tank is nearly a mile underground, a small amount of ^{37}Ar is produced by cosmic rays. An evaluation of data obtained by exposing 7500 liters of C_2Cl_4 at various depths underground indicates that the cosmic-ray production rate in the detector is 0.09 ± 0.03 ^{37}Ar atoms per day (*13*). Thus, the observed rate in the detector is essentially the same as the extrapolated cosmic-ray background, and there is no evidence (at the 90 percent confidence level) for a solar neutrino capture rate of 1.5 solar neutrino units (SNU; 1 SNU = 10^{-36} captures per target particle per second). The individual experiments and the average rate are illustrated in Fig. 2, including some more recent runs.

Even though the average ^{37}Ar production rate shown in Fig. 2 is very low, there are occasional high runs. These may be due to statistical fluctuations in the data or to rare cosmic-ray events. It is unlikely that variations on the time scale of months are due to changes in the solar neutrino flux, since solar thermal time scales are tens of thousands of years or longer.

Is there any hope of improving the sensitivity of the present ^{37}Cl detector? There are two limitations: the background counting rate of the counters used to measure the ^{37}Ar activity, and the cosmic-ray background effect. Attempts are being made to decrease the counter background, which is at present 1 to 2 counts per month, to less than 0.5 per month. With this improvement a search could be made for a solar neutrino flux in the range of 0.5 to 1 SNU.

The ^{37}Cl experiment tests theoretical ideas at different levels of meaning, depending on the counting rate being discussed. The various counting rates and their significance are summarized in Table 2. It is obvious from a comparison of Table 2 with the experimental results given above (and in Fig. 2) that the value (*7*) of 35 SNU's based on the CNO cycle is ruled out. More surprisingly, the best current models (*14*) based on standard theory, which imply ~ 5.5 SNU's, are also inconsistent with the observations. This disagreement between standard theory and observation has led to many speculative suggestions of what might be wrong. One such suggestion (*15*), that in the solar interior the heavy element abundance is at least a factor of 10 less than the observed surface abundance, leads to an expected counting rate of 1.5 SNU's (see Table 2), which is about as low a prediction as one can obtain from solar models without seriously changing current ideas about the physics of the solar interior. We note that present and future versions of the ^{37}Cl experiment are not likely to reach a sensitivity as low as 0.3 SNU, the minimum counting rate (from reaction 2 of Table 1) that can be expected if the basic idea of nuclear fusion as the energy source for main sequence stars is correct.

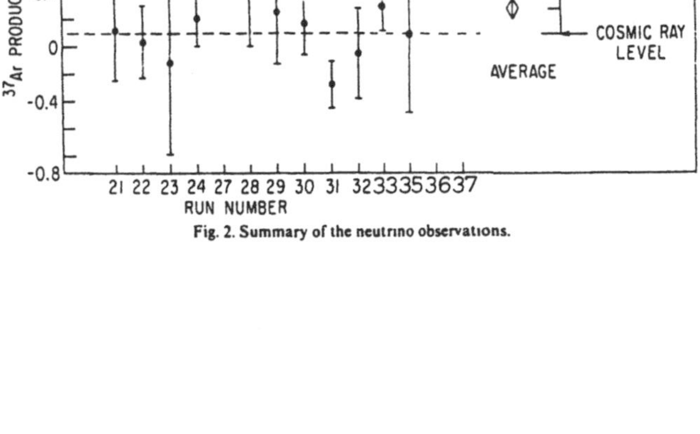

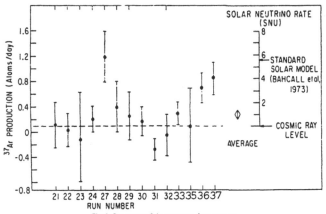

Fig. 2. Summary of the neutrino observations.

Speculations

The conflict between observation and standard theory has led to many speculations about the solar interior that were advanced because their proponents believed that the subject is in a state of crisis. For example, it has been suggested that the sun contains a black hole in its center, which is currently supplying more than half the observed solar luminosity through energy radiated when the black hole accretes mass from the surrounding gas (*16*). It has also been suggested that the sun is in a transient phase during which the interior luminosity produced by nuclear reactions is much less than the observed luminosity, which results from photons slowly diffusing out from the interior to the surface (*17*). These suggestions have not been widely accepted because they require the sun to be in a special state during the ^{37}Cl neutrino observations and also because there is no evidence from theoretical calculations that the dynamical behavior of the sun would be as required by these speculations. Other radical ad hoc assumptions about the solar interior that have recently been put forward include a departure from the Maxwellian velocity distribution at energies large compared to the thermal energy (*18*), the existence of very large central magnetic fields in the solar interior (*19*), and a critical temperature below which hydrogen and helium are immiscible (*20*). One imaginative cosmologist has even suggested that the exterior half of the sun's mass was added, with an entirely different composition from the interior half, about 5 billion years ago (*21*).

In addition to the many speculations about radical changes in the theory of stellar evolution, it has also been suggested (*22*) that the neutrino may behave differently in traversing astronomical distances (10^{13} cm) than has been inferred from laboratory measurements over distances of less than 10^3 cm. It has been proposed (*23*) that the neutrino has a tiny rest mass and decays into a (presently unknown) lower mass boson. The latter suggestion has not been taken very seriously by most physicists because there is no independent evidence for the postulated decay product and because weak interaction theory has a more elegant structure if neutrinos have zero rest mass.

The attitude of many physicists toward the present discrepancy is that astronomers never really understand astronomical systems as well as they think they do, and the failure of the standard theory in this simple case just proves that physicists are correct in being skeptical of the astronomers' claims. Many astronomers believe, on the other hand, that the present conflict between theory and observation is so large and elementary that it must be due to an error in the basic physics, not in our astrophysical understanding of stellar evolution.

The Next Experiment

Another experiment is required to settle the issue of whether our astronomy or our physics is at fault. Fortunately, one can make a testable distinction. The flux of low energy neutrinos from the PP and PEP reactions (numbers 1 and 2 in Table 1) is almost entirely independent of astronomical uncertainties and can be calculated from the observed solar luminosity, provided only that the basic physical ideas of nuclear fusion as the energy source for the sun and of stable neutrinos are correct. If these low energy solar neutrinos are detected in a future experiment, we will know that the present crisis is caused by a lack of astronomical understanding. If the low energy neutrinos are shown not to reach the earth, then even many physicists would be inclined to suspect their physics.

The radiochemical approach, even with its inherent backgrounds and indiscriminate signal, appears to be the only method with sufficient sensitivity to make possible another solar neutrino experiment. If one examines all possible inverse beta processes with low threshold energies, satisfactory neutrino capture cross sections, and suitable product lifetimes, and also considers the availability of the target element and the ease of separation of the target, only one reasonably good candidate has been found capable of observing the abundant flux of PP neutrinos. This reaction is the capture of neutrinos by ^{71}Ga to produce ^{71}Ge, an isotope with an 11-day half-life. The threshold, 233 kev, is ideal for observing neutrinos from the PP reaction. Gallium is very expensive and is now used for making light-emitting diodes for minicomputer readout displays. About 20 tons of gallium are needed for a solar neutrino detector.

Another approach is to use the capture of neutrinos by ^{7}Li to form ^{7}Be, an isotope with a 53-day half-life. Although this reaction has a high threshold, 861 kev, the target, ^{7}Li, has a neutrino capture cross section for PEP neutrinos 34 times higher than that of ^{37}Cl, and the reaction is favorable for observing PEP neutrinos. The neutrino capture cross sections for this target are accurately known (*1*) and could, in principle, permit one to determine the relative frequency of capture of neutrinos from the various reactions listed in Table 1. Development work has started on chemical separation methods and counting techniques. The experiment might involve about 200,000 liters of nearly saturated aqueous lithium chloride solution, from which about 30 atoms of ^{7}Be must be separated. Needless to say, such an experiment is not easy.

References and Notes

1. J. N. Bahcall and R. L. Sears, *Annu. Rev. Astron. Astrophys.* 10, 25 (1972).
2. R. Davis, Jr., *Phys. Rev. Lett.* 12, 303 (1964); *Proc. Int. Conf. Neutrino Phys. Astrophys. (Moscow)* 2, 99 (1969).
3. J. N. Bahcall, N. A. Bahcall, G. Shaviv, *Phys. Rev. Lett.* 20, 1209 (1968).
4. J. N. Bahcall, *ibid.* 22, 300 (1969).
5. M. Schwarzschild, *Structure and Evolution of the Stars* (Princeton Univ. Press, Princeton, N.J., 1958).
6. B. Pontecorvo, *Chalk River Lab. Rep. PD-205* (1946); L. W. Alvarez, *Univ. Calif. Radiat. Lab. Rep. UCRL-328* (1949).
7. J. N. Bahcall, *Phys. Rev. Lett.* 17, 398 (1966).
8. R. Davis, Jr., D. C. Harmer, F. H. Neely, in *Quasars and High-Energy Astronomy*, K. N. Douglas *et al.*, Eds. (Gordon & Breach, New York, 1969), p. 287.
9. K. C. Jacobs, *Nature (London)* 256, 560 (1975).
10. J. J. Leventhal and L. Friedman, *Phys. Rev. D* 6, 3338 (1972).
11. F. Reines and V. Trimble, Eds. *Proceedings of the Solar Neutrino Conference* (University of California, Irvine, 1972); V. Trimble and F. Reines, *Rev. Mod. Phys.* 45, 1 (1973).
12. R. Davis, Jr., and J. M. Evans, *Proc. 13th Int. Cosmic Ray Conf.* 3, 2001 (1973).
13. A. W. Wolfendale, E. C. M. Young, R. Davis, *Nature (London) Phys. Sci.* 238, 130 (1972).
14. J. N. Bahcall *et al.*, *Astrophys. J.* 184, 1 (1973).
15. J. N. Bahcall and R. K. Ulrich, *ibid.* 170, 593 (1971).
16. M. J. Newman, D. D. Clayton, R. J. Talbot, *ibid.*, in press.
17. E. E. Salpeter, *Comments Nucl. Part. Phys.* 2, 97 (1968); W. A. Fowler, *Colloquium on Cosmic Ray Studies in Relation to Recent Developments in Astronomy and Astrophysics* (Tata Institute of Fundamental Research, Bombay, 1968), p. 245; W. R. Sheldon, *Nature (London)* 221, 650 (1969); F. W. W. Dilke and D. O. Gough, *ibid.* 240, 262 (1972); W. A. Fowler, *ibid.* 238, 24 (1972).
18. D. D. Clayton, E. Dwek, M. J. Newman, J. Talbot, *Astrophys. J.* 199, 494 (1975).
19. D. Bartenwerfer, *Astron. Astrophys.* 25, 455 (1973); S. M. Chitre, D. Ezer, R. Stothers, *Astrophys. Lett.* 14, 37 (1973).
20. J. C. Wheeler and A. G. W. Cameron, *Astrophys. J.* 196, 601 (1975).
21. F. Hoyle, *Astrophys. J. Lett.* 197, L127 (1975).
22. B. Pontecorvo, *Sov. Phys. JETP* 26, 984 (1964); V. Gribov and B. Pontecorvo, *Phys. Lett.* 28B, 493 (1969); J. N. Bahcall and S. C. Frautschi, *ibid.* 29B, 623 (1969).
23. J. N. Bahcall, N. Cabibbo, A. Yahil, *Phys. Rev. Lett.* 28, 316 (1972).
24. Research sponsored in part by the National Science Foundation (grant NSF GP-40768X) and in part by the Energy Research and Development Administration.

THE CHLORINE SOLAR NEUTRINO EXPERIMENT

J. K. Rowley, B. T. Cleveland, and R. Davis, Jr.
Brookhaven National Laboratory, Upton, New York 11973

ABSTRACT

The chlorine solar neutrino experiment in the Homestake Gold Mine is described and the results obtained with the chlorine detector over the last fourteen years are summarized and discussed. Background processes producing ^{37}Ar and the question of the constancy of the production rate of ^{37}Ar are given special emphasis.

INTRODUCTION

The first underground experiment in the Homestake Gold Mine was the chlorine solar neutrino experiment, which was initiated nearly 20 years ago. The so-called solar neutrino problem is the discrepancy between the results of this experiment and the result predicted by solar model calculations using the best available input physics, i.e., by the standard solar model. The neutrino capture rate in the chlorine detector calculated using the standard solar model has changed with time as new data have become available. However, since 1969, in spite of great effort producing many new and improved measurements of nuclear reaction cross-sections, new opacity calculations etc., the capture rate predicted by the standard solar model has not changed in a major way. The chlorine detector has been operating regularly since 1970 with 61 experimental runs completed at the present time. The purpose of this paper is to give a brief description of the chlorine solar neutrino experiment and to discuss at greater length the results of these experimental runs.

The standard solar model is based upon the set of nuclear reactions shown in Table I.

Table I The proton-proton reaction chains

	Reaction	Neutrino Energy in MeV
PPI	$H + H \rightarrow D + e^+ + \nu$ (99.75%)	0-0.42 spectrum
	or	
	$H + H + e^- \rightarrow D + \nu$ (0.25%)	1.44 line
	$D + H \rightarrow {}^3He + \gamma$	
	${}^3He + {}^3He \rightarrow 2H + {}^4He$ (87%)	
PPII	${}^3He + {}^4He \rightarrow {}^7Be + \gamma$ (13%)	
	${}^7Be + e^- \rightarrow {}^7Li + \nu$	0.861 (90%) line 0.383 (10%) line
	${}^7Li + H \rightarrow \gamma + {}^8Be \rightarrow 2\,{}^4He$	
PPIII	${}^7Be + H \rightarrow {}^8B + \gamma$ (0.017%)	
	${}^8B \rightarrow {}^8Be^* + e^+ + \nu$	0-14.1

Results from One Thousand Days of Real-Time, Directional Solar-Neutrino Data

K. S. Hirata, K. Inoue, T. Kajita, T. Kifune, K. Kihara, M. Nakahata, K. Nakamura, S. Ohara, N. Sato, Y. Suzuki, Y. Totsuka, and Y. Yaginuma
Institute for Cosmic Ray Research, University of Tokyo, Tanashi, Tokyo 188, Japan

M. Mori, Y. Oyama, A. Suzuki, K. Takahashi, and M. Yamada
National Laboratory for High Energy Physics (KEK), Tsukuba, Ibaraki 305, Japan

M. Koshiba
Tokai University, Shibuya, Tokyo 151, Japan

T. Suda and T. Tajima
Department of Physics, Kobe University, Kobe, Hyogo 657, Japan

K. Miyano, H. Miyata, and H. Takei
Niigata University, Niigata, Niigata 950-21, Japan

Y. Fukuda, E. Kodera, Y. Nagashima, and M. Takita
Department of Physics, Osaka University, Toyonaka, Osaka 560, Japan

K. Kaneyuki and T. Tanimori
Department of Physics, Tokyo Institute of Technology, Meguro, Tokyo 152, Japan

E. W. Beier, L. R. Feldscher, E. D. Frank, W. Frati, S. B. Kim, A. K. Mann, F. M. Newcomer, R. Van Berg, and W. Zhang[(a)]
Department of Physics, University of Pennsylvania, Philadelphia, Pennsylvania 19104
(Received 5 June 1990)

A data sample of 1040 days from the Kamiokande II detector, consisting of subsamples of 450 days at electron-energy threshold $E_e \geq 9.3$ MeV and 590 days at $E_e \geq 7.5$ MeV, yields a clear directional correlation of the solar-neutrino-induced electron events with respect to the Sun and a measurement of the differential electron-energy distribution. These provide unequivocal evidence for the production of ^{8}B by fusion in the Sun. The measured flux of ^{8}B solar neutrinos from the two subsamples relative to a prediction of the standard solar model is $0.46 \pm 0.05(\text{stat}) \pm 0.06(\text{syst})$. The total data sample is tested for short-term time variation; within the statistical error, no significant variation is observed.

PACS numbers: 96.60.Kx, 95.85.Qx, 96.40.Tv

A primary motivation of the study of solar neutrinos is the prospect that it will directly reveal the inner structure of the Sun. At the same time, it may also reveal as yet undetected intrinsic properties of neutrinos, owing to the wide range of matter density, the very long distance from the Sun to the Earth, and the relatively high magnetic field traversed by low-energy solar neutrinos in their passage from the center of the Sun to a detector on Earth.

In a previous paper,[1] we describe the observation of ^{8}B solar neutrinos during an exposure of 450 live days of the detector Kamiokande II in the time period January 1987 through May 1988. The principal result of that observation was a measurement of the value of the ^{8}B solar-neutrino flux relative to that predicted by the standard solar model (SSM),[2] to wit, $0.46 \pm 0.13(\text{stat}) \pm 0.08(\text{syst})$, for observed recoil-electron total energies of $E_e \geq 9.3$ MeV. The neutrino signal was correlated with the direction from the Sun, and the shape of the differential total-energy spectrum was consistent with that predicted from the product of the ^{8}B decay spectrum[3] and the cross section $\sigma(\nu_e e \rightarrow \nu_e e)$, which is the reaction for detecting low-energy ν_e in Kamiokande II.

Here we report the data from an additional 590 live detector days in the period June 1988 through April 1990, obtained with a lower background level due to improved detector performance and reduction of the radioisotopes (primarily ^{222}Rn) present in the detector water. These improvements permitted a lower electron-energy threshold of 7.5 MeV to be used. The data so obtained, in conjunction with the earlier 450-day data sample, are of particular interest, because they extend over a 1040-day period in which the sunspot activity[4] has risen steeply from a minimum value at the end of solar magnetic cycle 21 to a maximum value approximately 15 times larger at the present peak of solar cycle 22. Accordingly, it is possible to test for a correlation of the B solar-neutrino yield with the sunspot activity[5] by comparing the ^{8}B flux value obtained from the later sample with that from the earlier sample and by further subdividing the total data sample into five time intervals, each of approximately 200 live detector days. We also present the resultant relative flux value and differential electron-energy distribution from the combined samples.

The tests described here bear on the possible influence of the solar magnetic field on the ^{8}B solar-neutrino flux;[6] they are essentially empirical and do not rely on quantitative comparison with theory. In particular, the absolute value of the ^{8}B solar-neutrino flux predicted by the SSM is not necessary for these tests.

The Kamiokande II detector has been described previously,[7] as has the method of extracting the solar-neutrino signal from the data.[1,8] In June 1988, however, the gain of the photomultiplier tubes (PMT) was increased by a factor of 2 to improve the single-photoelectron response of the detector. The effect of the PMT-gain increase was to increase the number of hit PMT at a given energy, thus obtaining better event reconstruction as well as improved energy resolution. The energy-resolution improvement is summarized quantitatively by the fact that the gain change, 30 hit (effective[7]) PMT correspond to 10 MeV, whereas before the gain change, the ratio was 26 hit (effective) PMT per 10 MeV. In addition, successful efforts to seal the system of the main detector tank and water circulation equipment against the abundant radon in the mine air resulted in a lower radon content in the system and, therefore, a lower raw trigger rate. Together, the modifications led to a detector trigger threshold of 6.1 MeV with 50% efficiency, and an electron-energy threshold in the analysis of 7.5 MeV. The analysis was also improved to further reduce the background,[9] though the method is essentially the same as before.[1] The previous 450-day data were accordingly reanalyzed, and the resultant flux value was consistent with the value already published[1] with a statistical error improved by 30%.

Two independent analyses were performed on the data. Each analysis obtained the final sample using totally independent programs for event reconstruction and applying different cuts. The results of the two analyses were carefully compared in many ways after each cut. The agreement between the two analyses on the measured signal and the distributions shown below is excellent.

Figure 1 shows the distribution in $\cos\theta_{Sun}$ with $E_e \geq 7.5$ MeV of the 590 days of data taken in the period June 1988 through April 1990. The number of events in the broad peak at $\cos\theta_{Sun} = 1$ is 164, giving a signal-to-background ratio of approximately 0.5 in the signal region ($\cos\theta_{Sun} \gtrsim 0.8$). The resultant value of the ratio data/SSM from the sample in Fig. 1 is 0.45 ± 0.06(stat) ± 0.06(syst). The statistical error is significantly smaller than that given for the earlier, 450-day-sample because of the lower-electron-energy threshold and the later lower background level described above. The systematic error is also reduced by the improved analysis and energy resolution. Sources of the systematic error are uncertainties in energy calibration (0.053), in angular resolution (0.031), and in dead time of various cuts (0.014). The overall systematic error is thus $[(0.053)^2+(0.031)^2+(0.014)^2]^{1/2} = 0.06$.

We plot in Fig. 2 the ratio data/"SSM" as a function of time. We emphasize that the relative positions of the five data points in Fig. 2 do not depend on an absolute value of the ^{8}B flux predicted by the SSM. (We use

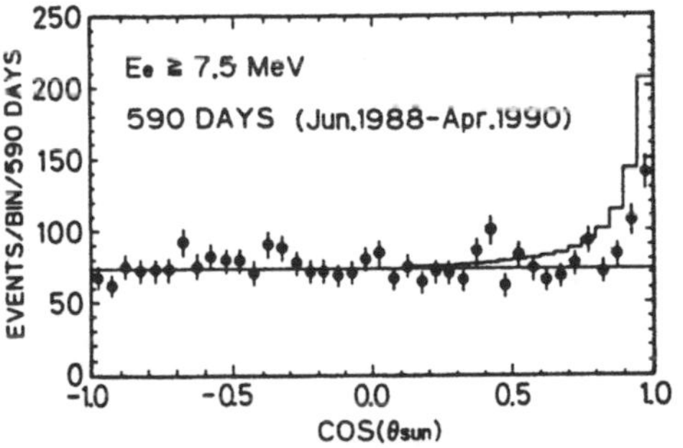

FIG. 1. Distribution in $\cos\theta_{Sun}$ of the 590-day sample for $E_e \geq 7.5$ MeV. θ_{Sun} is the angle between the momentum vector of an electron observed at a given time and the direction from the Sun relative to the detector at that time. The isotropic background (roughly 0.1 event/day bin) is due to spallation products induced by cosmic-ray muons, γ rays from outside the detector, and radioactivity in the detector water. The histogram is the calculated signal distribution based on the full value of the SSM and includes multiple scattering and the angular resolution of the detector.

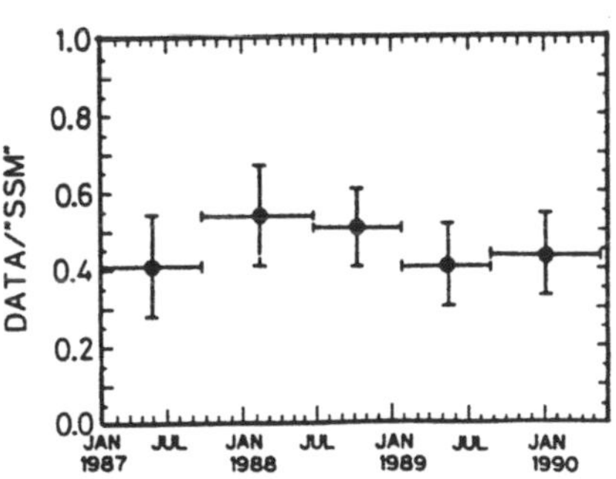

FIG. 2. Plot showing the time variation of the ^{8}B solar-neutrino signal in the Kamiokande II detector. Threshold for the two earlier points is $E_e \geq 9.3$ MeV, while for the three later points $E_e \geq 7.5$ MeV.

"SSM" to remind the reader of that fact.) Nor are those points subject to a significant systematic error arising from the aforementioned gain change because the Monte Carlo calculation that yields the value in the denominator of the ratio data/"SSM" has applied to it the same event criteria as applied to the observed events which enter the numerator. Accordingly, a difference in the energy scales before and after the gain change tends to cancel in the ratio data/"SSM". Furthermore, the distribution in Fig. 2 is not appreciably changed if a common energy threshold, e.g., $E_e \gtrsim 9.3$ MeV, is used for all data, which additionally reduces any effect arising from different energy thresholds in the two data samples. The point-to-point systematic error is estimated using γ rays from Ni(n,γ)Ni reactions and spallation products induced by cosmic-ray muons and it is within $\pm 8\%$.

The $\cos\theta_{Sun}$ distribution of the combined 1040-day data sample for $E_e \gtrsim 9.3$ MeV is shown in Fig. 3. (The choice of $E_e \gtrsim 9.3$ MeV allows the two samples to be combined directly into a single $\cos\theta_{Sun}$ distribution.) The signal in the peak near $\cos\theta_{Sun} = 1$ is 128 events, yielding a signal-to-background ratio of approximately 0.9 in that region. The statistical significance of the directional correlation of the solar-neutrino signal with respect to the Sun is excellent: 36% C.L. for an isotropic distribution plus a peak versus 2×10^{-4}% C.L. for an isotropic distribution only, where the Kolmogorov-Smirnov test was used in the calculation of the C.L. The peak is characterized such that the peak width and height are determined by the experimental angular resolution and the measured total flux value, respectively.

The energy distribution of the scattered electrons from the 1040-day sample is given in Fig. 4. The points in Fig. 4 were obtained by a maximum-likelihood fit to each $\cos\theta_{Sun}$ distribution corresponding to a given electron-energy bin with the fits involving an isotropic background plus an expected angular distribution of the signal. The $\cos\theta_{Sun}$ distribution of each energy bin shows clear enhancement in the direction of the Sun. The shape of the energy spectrum in Fig. 4 is different from that of the background; the probability that the signal and background shapes are the same is less than 3%. Above $E_e = 7.5$ MeV, the shape of the distribution is seen to be consistent with the shape of the Monte Carlo distribution based on the known β-decay spectrum of ^{8}B, the cross section for the reaction $\nu_e e \rightarrow \nu_e e$, and the measured energy resolution and calibration of the detector.

The representative flux value for the combined 450- and 590-day data samples is

$$\text{data/SSM} = 0.46 \pm 0.05(\text{stat}) \pm 0.06(\text{syst}) .$$

In summary, there is no significant difference between the relative flux values of the 450- and 590-day data samples, which are $0.48 \pm 0.09(\text{stat}) \pm 0.08(\text{syst})$ ($E_e \gtrsim 9.3$ MeV; January 1987–May 1988),[9] and $0.45 \pm 0.06(\text{stat}) \pm 0.06(\text{syst})$ ($E_e \gtrsim 7.5$ MeV; June 1988–April 1990), respectively, essentially independent of electron-energy threshold. The finer-binned time-variation plot in Fig. 2 is also consistent statistically with a flux of ^{8}B solar neutrinos constant in time, but a possible time variation related to sunspot activity cannot be definitively ruled out by these data.

The totality of the ^{8}B solar-neutrino data from Kamiokande II provides a clear two-part evidence for a neutrino signal from ^{8}B production and decay in the Sun: namely, the directional correlation of the neutrino signal with the Sun, and the consistency of the differential electron-energy distribution of the signal in

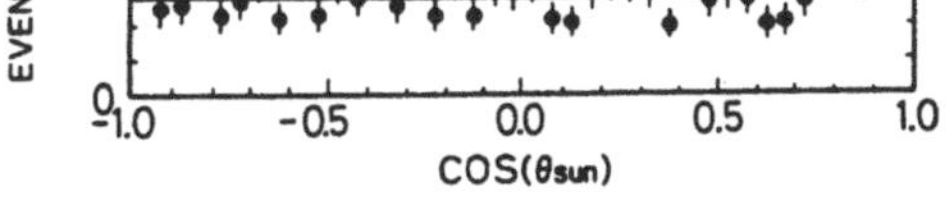

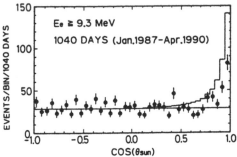

FIG. 3. Distribution in $\cos\theta_{Sun}$ of the combined 1040-day sample for $E_e \gtrsim 9.3$ MeV. The value of the ratio data/SSM from this figure is 0.43 ± 0.06.

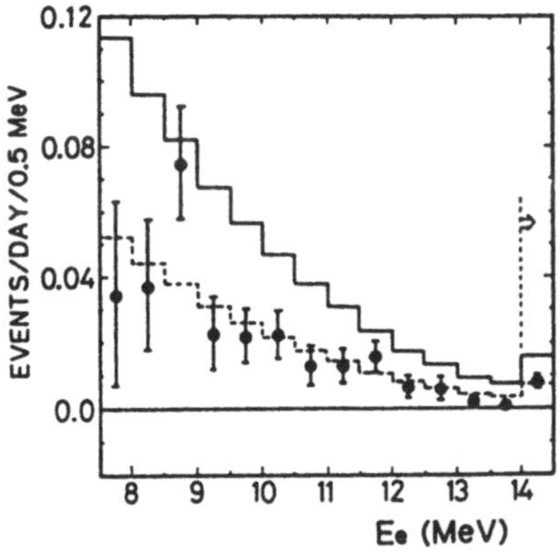

FIG. 4. Differential energy distribution of the recoil electrons from $\nu_e e \rightarrow \nu_e e$ from the 1040-day sample. The last bin corresponds to $E_e = 14$–20 MeV. The solid histogram shows the area and the shape of the distribution predicted by the SSM. The dashed histogram is the best fit ($0.46\times$SSM) of the expected Monte Carlo–calculated distribution to the data.

shape and energy scale with that expected from ^{8}B decay. Accordingly, the mechanism of energy generation in the Sun based on the fusion reactions which give rise to ^{8}B as a by-product would appear to be unequivocally confirmed by the detection of neutrinos which could only have originated in the core of the Sun.

We gratefully acknowledge the cooperation of the Kamioka Mining and Smelting Company. We thank K. Arisaka and B. G. Cortez for their contributions to the early stage of this experiment. This work was supported by the Japanese Ministry of Education, Science and Culture, by the U.S. Department of Energy, and by the University of Pennsylvania Research Fund. Part of this analysis was carried out on the FACOM M780 and M380 at the computer facilities of the Institute for Nuclear Study, University of Tokyo, and part at the David Rittenhouse Computing Facility of the University of Pennsylvania.

---

[a]Now at Los Alamos National Laboratory, Los Alamos, NM 87545.

[1]K. S. Hirata *et al.*, Phys. Rev. Lett. **63**, 16 (1989).

[2]J. N. Bahcall and R. K. Ulrich, Rev. Mod. Phys. **60**, 297 (1988); J. N. Bahcall, *Neutrino Astrophysics* (Cambridge Univ. Press, Cambridge, 1989); S. Turck-Chièze *et al.*, Astrophys. J. **335**, 415 (1988). The SSM-predicted value of the ^{8}B solar-neutrino flux used in Ref. 1 and here, 5.8×10^6/cm^{2}sec, is taken from the review article of Bahcall and Ulrich.

[3]J. N. Bahcall and B. R. Holstein, Phys. Rev. C **33**, 2121 (1986).

[4]Sunspot Number Plots (Daily, Monthly, Smoothed Monthly, and Yearly) National Geophysical Data Center, Boulder, CO 80303.

[5]See R. Davis, Jr., in *Proceedings of the Thirteenth International Conference on Neutrino Physics and Astrophysics, "Neutrino '88,"* edited by J. Schneps *et al.* (World Scientific, Singapore, 1989), p. 518, and references therein.

[6]M. B. Voloshin, M. I. Vysotskiĭ and L. B. Okun, Yad. Fiz. **44**, 677 (1986) [Sov. J. Nucl. Phys. **44**, 440 (1986)]; E. Kh. Akhmedov, Phys. Lett. B **213**, 64 (1988).

[7]K. S. Hirata *et al.*, Phys. Rev. D **38**, 448 (1988).

[8]M. Nakahata, Ph.D. thesis, University of Tokyo, Report No. UT-ICEPP-88-01, 1988 (unpublished); S. B. Kim, Ph.D. thesis, University of Pennsylvania, Report No. UPR-0174E, 1989 (unpublished).

[9]K. S. Hirata *et al.*, in Proceedings of the Fourteenth International Symposium on Lepton and Photon Interactions, Stanford, California, 1989 (to be published). The relative flux value from the reanalyzed 450-day sample is 0.48 ± 0.09(stat) ±0.08(syst) ($E_e\geq9.3$ MeV).

Search for Neutrinos from the Sun Using the Reaction $^{71}Ga(\nu_e, e^-)\,^{71}Ge$

A. I. Abazov, O. L. Anosov, E. L. Faizov, V. N. Gavrin, A. V. Kalikhov, T. V. Knodel, I. I. Knyshenko, V. N. Kornoukhov, S. A. Mezentseva, I. N. Mirmov, A. V. Ostrinsky, A. M. Pshukov, N. E. Revzin, A. A. Shikhin, P. V. Timofeyev, E. P. Veretenkin, V. M. Vermul, and G. T. Zatsepin
Institute for Nuclear Research, Academy of Sciences, U.S.S.R., Moscow 117312, U.S.S.R.

T. J. Bowles, B. T. Cleveland, S. R. Elliott, H. A. O'Brien, D. L. Wark,[a] and J. F. Wilkerson
Los Alamos National Laboratory, Los Alamos, New Mexico 87545

R. Davis, Jr., and K. Lande
University of Pennsylvania, Philadelphia, Pennsylvania 19104

M. L. Cherry
Louisiana State University, Baton Rouge, Louisiana 70803

R. T. Kouzes
Princeton University, Princeton, New Jersey 08544
(Received 7 October 1991)

A radiochemical ^{71}Ga-^{71}Ge experiment using 30 tons of gallium to determine the primary flux of neutrinos from the Sun has begun operation at the Baksan Neutrino Observatory in the U.S.S.R. Assuming that the extraction efficiency for ^{71}Ge atoms produced by solar neutrinos is the same as from natural Ge carrier, we observed the capture rate to be 20^{+15}_{-20}(stat) $\pm$ 32(syst) solar neutrino units (SNU), resulting in a limit of less than 79 SNU (90% C.L.). This is to be compared with 132 SNU predicted by the standard solar model.

PACS numbers: 96.60.Kx, 14.60.Gh, 95.85.Qx

A fundamental problem during the last two decades has been the large deficit of the solar neutrino flux observed in the radiochemical chlorine experiment [1] compared with the theoretical predictions [2,3] based on the standard solar model (SSM). Recent results of the Kamiokande II water Cherenkov experiment [4] have confirmed this deficit. These results may be explained by deficiencies in the solar model in predicting the ^8B neutrino flux or may indicate the possible existence of new properties of the neutrino [5–9]. The role new neutrino properties may play in the suppression of the high-energy solar neutrino flux, as possibly indicated in the chlorine and Kamiokande II experiments, can be determined by a radiochemical gallium experiment. An experiment using ^{71}Ga as the capture material [10] provides the only feasible means at present to measure low-energy solar neutrinos produced in the proton-proton (*p-p*) reaction. Exotic hypotheses aside, the rate of the *p-p* reaction is directly related to the solar luminosity and is insensitive to alterations in the solar models. An observation in a gallium experiment of a strong suppression of the low-energy solar neutrino flux requires the invocation of new neutrino properties.

In this paper we present the first results of the measurement of the solar neutrino flux by the Soviet-American gallium solar neutrino experiment (SAGE). SAGE uses the Gallium-Germanium Neutrino Telescope situated in an underground laboratory at the Baksan Neutrino Observatory of the Institute for Nuclear Research of the Academy of Sciences of the U.S.S.R. in the Northern Caucasus of the U.S.S.R. The main chamber is located 3.5 km from the entrance of a horizontal adit driven into the side of Mount Andyrchi, and has an overhead shielding of approximately 4700 m water equivalent. The 30 tons of liquid gallium used in the present measurement is contained in four Teflon-lined chemical reactors, each holding about 7 tons of gallium.

The chemical extraction of germanium from liquid metallic gallium was first tested on a small scale in the U.S. [11] and later developed and tested at a 7.5-ton pilot installation in the U.S.S.R. [12]. The experimental layout as well as the chemical and counting procedures have been described previously [13] and are only briefly outlined here.

Each measurement of the solar neutrino flux begins by adding approximately 160 μg of natural Ge carrier to each of the four reactors holding the gallium. After a typical exposure interval of 3 to 4 weeks, the Ge carrier and any ^{71}Ge atoms that have been produced by neutrino capture are chemically extracted from the gallium using the following procedure. A weak hydrochloric acid solution is mixed with the gallium metal in the presence of hydrogen peroxide, which results in the extraction of germanium into the aqueous phase. The extracted solutions from the four separate reactors are combined and reduced in volume by vacuum evaporation. Additional HCl is then added and an argon purge is initiated which sweeps the Ge as $GeCl_4$ from the acid solution into 1.2 liters of H_2O. The Ge is then extracted into CCl_4 and back extracted into 0.1 liter of low-tritium H_2O. The counting gas GeH_4 (germane) is then synthesized and purified by gas chromatography. The efficiency of extrac-

Solar neutrinos observed by GALLEX at Gran Sasso

GALLEX Collaboration [1,2,3,4,5]

P. Anselmann, W. Hampel, G. Heusser, J. Kiko, T. Kirsten, E. Pernicka, R. Plaga [6], U. Rönn, M. Sann, C. Schlosser, R. Wink, M. Wójcik [7]
Max-Planck-Institut für Kernphysik (MPIK), Postfach 103980, W-6900 Heidelberg, FRG

R. von Ammon, K.H. Ebert, T. Fritsch, K. Hellriegel, E. Henrich, L. Stieglitz, F. Weyrich
Institut für Heiße Chemie, Kernforschungszentrum Karlsruhe (KFK), Postfach 3640, W-7500 Karlsruhe, FRG

M. Balata, E. Bellotti, N. Ferrari, H. Lalla, T. Stolarczyk
INFN – Laboratori Nazionali del Gran Sasso (LNGS), S.S. 17/bis Km18+910, I-67010 L'Aquila, Italy [3]

C. Cattadori, O. Cremonesi, E. Fiorini, S. Pezzoni, L. Zanotti
Dipartimento di Fisica, Università di Milano e INFI – Sezione di Milano, Via Celoria 16, I-20133 Milan, Italy [3]

F. von Feilitzsch, R. Mößbauer, U. Schanda
Physik Department E15, Technische Universität München (TUM), James-Franck Straße, W-8046 Garching bei München, FRG

G. Berthomieu, E. Schatzman
Observatoire de la Côte d'Azur, Département Cassini, B.P. 229, F-06004 Nice Cedex 4, France and DASGAL, Bâtiment Copernic, Observatoire de Paris, 5, place Jules Janssen, F-92195 Meudon Principal, France

I. Carmi, I. Dostrovsky
Department of Environmental and Energy Research, Weizmann Institute of Science (WI), P.O. Box 26, 76100 Rehovot, Israel

C. Bacci [8], P. Belli, R. Bernabei, S. d'Angelo, L. Paoluzi
Dipartimento di Fisica II Università di Roma "Tor Vergata" and INFN – Sezione di Roma 2, via della Ricerca Scientifica, I-00133 Rome, Italy [3]

S. Charbit, M. Cribier, G. Dupont, L. Gosset, J. Rich, M. Spiro, C. Tao, D. Vignaud
DAPNIA/Service de Physique des Particules, CE Saclay, F-91191 Gif-sur-Yvette Cedex, France [4]

R.L. Hahn, F.X. Hartmann, J.K. Rowley, R.W. Stoenner and J. Weneser
Brookhaven National Laboratory (BNL), Upton, NY 11973, USA [5]

We have measured the rate of production of ^{71}Ge from ^{71}Ga by solar neutrinos. The target consists of 30.3 t of gallium in the form of 8.13 M aqueous gallium chloride solution (101 t), shielded by ≈ 3300 m water equivalent of standard rock in the Gran Sasso Underground Laboratory (Italy). In nearly one year of operation, 14 measurements of the production rate of ^{71}Ge were carried out to give, after corrections for side reactions and other backgrounds, an average value of 83 + 19 (stat.) ± 8 (syst.) SNU (1σ) due to solar neutrinos. This conclusion constitutes the first observation of solar pp neutrinos. Our result is consistent with the presence of the full pp neutrino flux expected according to the "standard solar model" together with a reduced flux of $^8B + ^7Be$ neutrinos as observed in the Homestake and Kamiokande experiments. Astrophysical reasons remain as a possible explanation of the solar neutrino problem. On the other hand, if the result is to be interpreted in terms of the MSW effect, it would fix neutrino masses and mixing angles within a very restricted range.

1. Introduction

Fusion reactions in the solar interior give rise to three major neutrino emissions:

(i) pp neutrinos from $p+p \to d+e^+ + \nu_e$; E_{max}= 420 keV,

(ii) ^7Be neutrinos from $^7Be + e^- \to {}^7Li + \nu_e$; E= (mainly) 860 keV.

(iii) ^8B neutrinos from $^8B \to {}^8Be + e^+ + \nu_e$; E_{max}= 14 MeV.

Star models (e.g. refs. [1–5]) predict their respective fluxes. The relative values are $\phi_{pp} : \phi_{^7Be} : \phi_{^8B} \approx$ 1:0.08:0.0001. Detection of these neutrinos may test solar models [2–5] and/or neutrino properties (see e.g. refs. [6–11]). The radiochemical Chlorine Detector [12] is sensitive both to ^8B and ^7Be neutrinos, while the real-time Kamiokande Cherenkov detector [13,14] responds mainly to ^8B neutrinos. Both have reported fluxes less than half of the theoretical expectations. This discrepancy has become the so-called "solar neutrino problem". Various explanations have been suggested; among them are overestimation of the central temperature in the solar model calculations, or electron-neutrino modifications between Sun and detector, a manifestation of neutrino mass.

The decision between these alternatives would be least ambiguous if a shortage of pp neutrinos were also observed. Given that the present Sun is producing fusion energy in equilibrium with its luminosity, a substantial reduction of pp neutrinos could only be due to some form of electron-neutrino disappearance. Flavor oscillations [7] resonantly enhanced in the dense solar interior, by the MSW effect [6], are a natural consequence of neutrino mass taken together with mixing of mass states in the physical neutrinos. The energy dependence of this effect permits determination of the neutrino-mass and mixing angle parameters.

In any case, the detection of pp neutrinos would be the first *experimental* proof of energy production by fusion of hydrogen inside the Sun. With respect to massive neutrinos we could exclude certain situations free of assumptions, but evaluate the regions of allowed parameter space of neutrino masses and mixing angles only with some ambiguity, since the effect need not be causal for the solar neutrino problem altogether.

A suitable reaction is $^{71}Ga\ (\nu_e, e^-)^{71}Ge$. It was first proposed by Kuzmin [15] and will remain the only practical choice for a long time to come [16]. Its low threshold of 233 keV [17] allows the detection of pp neutrinos. The product ^{71}Ge has a convenient half-life (11.43 d) [18] and can be isolated by radiochemical techniques. Two collaborations have set out to perform gallium experiments in underground lab-

[1] Credit is given to valuable contributions of former GALLEX members which have helped to develop the experiment into its present state: B. Pichard [deceased] (Saclay); M. Breitenbach, G. Eymann, M. Hübner, A. Lenzing, G. Monninger, R. Schlotz, K. Schneider, M. Schneller, K. Zuber (all MPIK); J. Römer, J. Unk (KFK); A. Urban (TUM); G. Friedlander (BNL).

[2] This work has been supported by the German Federal Minister for Research and Technology (BMFT) under the contract number 06HD5541. This work has been generously supported by the ALFRIED KRUPP von BOHLEN und HALBACH-Foundation, Germany.

[3] This work has been supported by Istituto Nazionale di Fisica Nucleare (INFN), Italy.

[4] This work has been supported by Commissariat à l'Energie Atomique, France.

[5] This work has been supported by the Office of Nuclear Physics of the US Department of Energy.

[6] Present address: Max-Planck-Institut für Physik, Föhringer Ring 6, Postfach 401212, W-8000 Munich 40, FRG.

[7] Permanent address: Instytut Fizyki, Uniwersytet Jagiellonski, ul. Reymonta 4, PL-30-059 Cracow, Poland.

[8] Permanent address: Dipartimento di Fisica, Università "La Sapienza", Piazzale A. Moro 2, I-00185 Rome, Italy.

oratories, SAGE [19] in Baksan (Caucasus) and GALLEX at Gran Sasso (Central Italy).

In this letter we report the first neutrino data obtained with GALLEX in 14 runs during the period 14 May 1991–31 March 1992 ("GALLEX I") [#1], covering 295 d of exposure.

The history and development of GALLEX as well as many experimental aspects have been given in previous reports, see e.g. refs. [20–29]. Here we describe the experimental aspects (section 2); side reactions and background considerations (section 3); and the runs performed (section 4). In section 5 we sketch the data treatment and present the preliminary results. Implications of these results are discussed in the accompanying Letter [30]. The future experimental program is outlined in section 6 together with our conclusions.

[#1] Phase "GALLEX I" extends until 29 April 1992, but run B 50 came too late to be included in the present data analysis (see table 1).

2. The GALLEX experiment

2.1. General description

For the inverse beta-decay reaction $^{71}Ga(\nu_e, e^-)^{71}Ge$, standard solar models (SSM) predict production rates ranging from 124 SNU [4] to 132 SNU [3] [#2]. The value from ref. [3], (132^{+21}_{-17}) SNU (3σ) is composed of 74 SNU from pp (and pep) neutrinos, 34 SNU from ^7Be, 14 SNU from ^8B and 10 SNU

[#2] SNU = solar neutrino unit, 10^{-36} captures atom^{-1} s^{-1}.

Table 1
Gallex I runs B 29–B 50.

Type/ctd. # [a]	Run #	Time period	Duration (d)	Carrier [b]	(%)Ge yield [c]	Position [d]	End of counting
SR 1	B 29	14.05.91–04.06.91	20.96	74	83.7	a	07.01.92
[e]	B 30	04.06.91–05.06.91	1.27	72	88.0	a	-
SR 2	B 31	05.06.91–26.06.91	20.77	76	84.4	a	07.01.92
SR 3	B 32	26.06.91–17.07.91	21.00	74	85.0	a	ongoing
SR 4	B 33	17.07.91–07.08.91	21.00	72	87.0	a	18.02.92
SR 5	B 34	07.08.91–28.08.91	21.00	76	85.2	a	10.03.92
SR 6	B 35	28.08.91–19.09.91	22.25	74	92.6 [f]	a	30.03.92
SR 7	B 36	19.09.91–09.10.91	19.70	72	93.5	p	ongoing
BL 1	B 37	09.10.91–10.10.91	1.17	76	92.8	p	30.03.92
SR 8	B 38	10.10.91–30.10.91	19.87	74	88.8	p	ongoing
SR 9	B 39	30.10.91–20.11.91	20.96	72	92.4	p	ongoing
BL 2	B 40	20.11.91–21.11.91	1.17	76	92.8	p	ongoing
SR 10	B 41	21.11.91–11.12.91	19.88	74	90.9	p	ongoing
SR 11	B 42	11.12.91–08.01.92	27.96	72	90.6	p	ongoing
BL 3	B 43	08.01.92–09.01.92	1.17	76	94.2	p	28.04.92
[g]	B 44	09.01.92–29.01.92	19.88	74	-	p	-
SR 12	B 45	29.01.92–19.02.92	20.96	72	92.8	p	ongoing
BL 4	B 46	19.02.92–20.02.92	1.21	76	93.8	p	ongoing
SR 13	B 47	20.02.92–11.03.92	19.81	74	86.1	p	ongoing
BL 5	B 48	11.03.92–12.03.92	1.19	72	92.8	p	ongoing
SR 14	B 49	12.03.92–31.03.92	18.83	76	93.2	p	ongoing
SR 15 [h]	B 50	31.03.92–29.04.92	29.00	74	89.1	p	ongoing

[a] SR = solar neutrino run, BL = short exposure blank run (full scale, involving target).
[b] 72, 74, 76 indicate the use of carrier solutions enriched in ^{72}Ge, ^{74}Ge, ^{76}Ge, respectively.
[c] Integral tank to counter yield, errors are ±1.7%.
[d] a = active (NaI) counting position; p = passive counting position (see text).
[e] Not used because of malfunction of HV-supply while counting.
[f] Since B 35, yields are higher due to mist acidification applied in the desorption steps.
[g] Run lost in preparation.
[h] Too late to be included in the present data analysis.

from ($^{13}N+^{15}O$). Our target is 30.3 t of gallium containing 12 t of ^{71}Ga in the form of an 8.13 M aqueous solution acidified to 2 M in HCl. For our target, the 132 SNU value of ref. [3] would yield an expected production rate of 1.18 atoms d^{-1} of ^{71}Ge or approximately 14 atoms of ^{71}Ge present in the target solution after ≈ three weeks of exposure.

To minimize interfering ^{71}Ge production through (p, n) reactions, the target must be free of radioactive impurities and effectively shielded from cosmic rays. Also, the local fast neutron flux must be low to avoid the creation of secondary protons.

The high concentration of chloride and the acidity of the solution ensure that the Ge will be in the form of the tetrachloride. The volatility of $GeCl_4$ allows it to be separated from the non-volatile $GaCl_3$ by bubbling an inert gas through the solution. Extensive experiments at several of our laboratories and on various scales showed that this separation can be carried out quantitatively, even without the addition of carriers [31–34]. In practice, a small known amount of non-radioactive germanium (about 1 mg) is added to the solution in each run to determine the actual yield of the process and to monitor its efficiency.

The $GeCl_4$ swept out of the solution by the stream of nitrogen is reabsorbed in water and after several stages of volume reduction is chemically converted to the gas, germane (GeH_4). After purification, the germane is introduced into a special miniaturized very-low-background proportional counter where the EC decay of ^{71}Ge is detected. The radiations actually observed are Auger electrons and stopped X-rays (keV range). To achieve the very low and constant backgrounds required for a reliable detection of the few ^{71}Ge decays expected during the first two months of counting, ultimate low-level techniques must be applied in a very stable experimental environment. Four years of running time are scheduled to achieve the desired statistical significance

Special tests using samples of actual B-tank solution (see below) were carried out to seek any "hot atom" effects that might influence the Ge recovery:

In one such test, ^{69}Ge was produced in situ via $^{69}Ga(p, n)^{69}Ge$ with ≈ 10 MeV protons and subsequently purged. The extraction efficiency of > 99% was directly determined from the ratio of the 1106 keV γ-lines of ^{69}Ge measured with a Ge detector before and after desorption in the unchanged irradia tion vessel.

In another series of tests, ^{71}Ge was produced in sit by EC decay of ^{71}As. This decay approximately re produces the kinematics of neutrino capture in ^{71}Ga. The results from counting of ^{71}Ge showed that thes daughter atoms behaved identically chemically to th 1 mg of germanium carrier that had been added t the gallium chloride solution. Recovery of the ^{71}G was quantitative, (99.8 ± 3.7)%.

In summary, there were no Ge "hot atom-chemi try" effects which would effectively influence our re sults. Yet we are aware that these conditions are nc *fully* identical to the capture of solar neutrinos in th gallium chloride target. Thus, a full-scale test of th GALLEX experiment is planned for 1993, when w will produce a $> 5\times10^{16}$ Bq ^{51}Cr neutrino source tha will be inserted into the target tank.

2.2. Experimental setup

This description is condensed. We refer to an illus trated version in ref. [23] and reports as quoted.

The GALLEX underground facilities are located i hall A of the Laboratori Nazionali del Gran Sass (LNGS) [35,36], easily accessible by a 6.3 km high way from the tunnel entrance. One building accom modates two 7 m high 70 m^3 target tanks, the Ge ex traction facilities (absorber columns and automati control system), duplicate GeH_4 synthesis an counter-filling stations, and a calcium-nitrate neu tron monitor. Nearby, another building houses th main counting station enclosed in a Faraday cage, an the computer and electronics laboratory. Abov ground, near the tunnel entrance, is the Technica Outside Facility ("TOF"), where the target solutio was received and preconditioned.

We have installed two similar tanks, "A" and "B" for redundancy and safety [31]; only one is used at a time. All experiments reported here were done in tan B, runs B 29–B 50. The tanks are made from vinyles ter resin reinforced by special glass fibers selected fo low U/Th content (Corning "S2"). Inside, they ar lined with PVDF. This teflon-like material is also use for the sparger pipes within the tanks. These pipe are perforated to distribute the purging gas through out the solution. The gas used is nitrogen, generate by the evaporation of liquid nitrogen stored in a vessel nearby.

The target tank A is equipped with a central elongated thimble designed to contain the ^{51}Cr neutrino source, as well as to accommodate a 470 L vessel containing the calcium-nitrate solution to monitor the fast-neutron background (see section 3).

In many years of proportional counter development, we have continuously refined and optimized the counter design and operating conditions for performance, stability, efficiency, and background [37–40]. The final model, "HD2-Fe/Si" is made from hyperpure Suprasil quartz. It is miniaturized and features a solid Fe or Si cathode (made from single silicon crystals with impurities $\leqslant 2$ ppt ^{238}U; $\leqslant 0.2$ ppt ^{232}Th; $\leqslant 0.1$ ppb ^{40}K) and a 13 μm tungsten anode wire directly sealed into quartz. Active volumes are typically 87%. At present, 22 low-background counters of this type are in use. Special quartz-blowing technology is used to standardize dimensions (6.4 mm diameter, 32 mm length, ≈ 1 cm^3 total volume) so that all counters are similar to each other. The counters are inserted into form-fitted Cu/low activity-Pb housings with void space minimized for radon suppression. The counters operate within a large shield tank inside a Faraday cage. This tank is fitted with 8.6 t of low radioactivity lead. Two sliding-end doors, also filled with lead, are attached to the tank. Entry is via flexible glove-boxes designed to allow handling of counters without venting the low-radon atmosphere inside. At present we can simultaneously operate eight "active" counter lines positioned in the well of a NaI-pair spectrometer at the front end of the shield tank [41] and sixteen "passive" lines positioned in a low radioactivity copper block at the opposite end. The choice of which positions to use is a trade-off between the lower counter backgrounds on the passive side and the diagnostic power of the NaI-spectrometer (see table 1).

The pulses from the proportional counters are amplified with fast charge-sensitive preamplifiers, wideband low noise amplifiers for pulse shape analysis, and shaping amplifiers for pulse-height analysis. Pulse heights are measured in 12 bit ADCs. The rise time can be determined by measuring the amplitude of the differentiated pulse (ADP, time constant 10 ns). Moreover, in order to discriminate later between the desired events and background, the whole pulse shape is recorded with transient digitizers on different time scales (smallest bin width 0.4 ns). We prefer this superior technique and use it exclusively. We define the rise time as the time elapsing between 10% and 70% of the pulse height, registered with the fast transient digitizer. Multiple lines are served by a self-switching multiplexer system connecting to a single Tektronix 7912 AD fast transient digitizer. The overall bandwidth is 100 MHz. The amplitudes of the pulses in the NaI-pair crystal are measured with a 13-bit buffer ADC and a slow transient digitizer. All modules are read out over a CAMAC bus linked with optical fiber cable to a MicroVAX II outside the Faraday cage [42].

2.3. Experimental procedures

A new run starts with the addition of ≈ 1 mg of a stable germanium carrier to the target solution after the end of the previous extraction. The carrier solution is added during 10 min through a pipe at ≈ 1m below the liquid surface. During and after the addition, the solution is stirred by a gentle stream of nitrogen for ≈ 9 h in a closed loop. To monitor carryover from one run to the next by means of a subsequent mass-spectrometric analysis of the counting gas, we successively use carrier solutions of stable enriched isotopes: ^{72}Ge, ^{74}Ge and ^{76}Ge. The Ge concentrations of the stock solutions are determined by atomic absorption spectroscopy and cross checked by other methods. During two years of operation, no measurable concentration changes have been found. At the end of the exposure time, usually three weeks, the Ge in the solution is recovered by purging with 1900 m^3 of nitrogen during 20 h at 20°C. Typically 99% of the carrier is extracted. Proper corrections are made for carryover from previous runs.

The volatile germanium tetrachloride is recovered by scrubbing the gas with 50 L of counter-flowing water at 12°C in three large absorber columns (3 m height, 0.3 m diameter), packed with glass helices. The resulting aqueous solution of Ge is further concentrated by acidifying it with HCl gas to 30% hydrochloric acid, purging the solution again with a stream of nitrogen and reabsorbing the volatilized $GeCl_4$ in about 1 L of water. Further volume reduction is effected by acidifying this solution and extracting the Ge in 500 mL of carbon tetrachloride. The $GeCl_4$ is then back extracted into 50 mL of tritium-free water, and this final solution is used to generate germanium hydride (GeH_4). At pH of 11.4, the Ge solution is mixed with a solution of 0.3 g of low-tritium sodium borohydride in tritium-free water and gently heated. The germane produced is entrained in a stream of helium and passed through a gas-chromatographic column for purification. The volume of the purified germane is measured with high precision using a McLeod gauge. To optimize the counting performance [39], the gas mixture introduced into the low-background counters at a slight overpressure is germane (30%), old xenon (70%).

The integral yield of the germanium recovery is defined as that fraction of the carrier added to the $GaCl_3$ solution in the tank that ends up in the counter. The integral yields for the individual runs are given in table 1.

After the counter is filled, it is imbedded in its lead mold, fitted into the preamplifier box and transported into the Faraday cage. The counter box is inserted by way of the air lock. After connection to the counting system, it is calibrated with the Gd(Ce) source (see below) inside the tank shield and then inserted into its final counting position. The operating voltage is ≈ 1600 V. The period from the end of extraction to the start of counting is typically 14 h.

To insert the newly filled counter into its operating position, all counting has to be stopped. In routine operation, such interventions occur approximately every three weeks, and we have adopted the procedure of calibrating half of the counters at each interruption. Thus, every counter is calibrated every six weeks. To obtain accurate backgrounds, each sample is counted for > six months, a period that is very long compared to the 11.43 d half-life of ^{71}Ge.

The Ge-decay (gallium) peaks occur at 1.17 keV (L-peak) and 10.37 keV (K-peak). For energy calibration, we use fluorescent X-rays from the xenon in the counting gas. These are produced by ≈ 35 keV Ce X-rays from the excitation of cerium by europium X-rays from the EC decay of a ^{153}Gd source (half-life 242 d). This method has the great advantage that the entire active counter volume is rather homogeneously illuminated, and that three lines are available at energies of 1.03, 5.09 and 9.75 keV [43]. From such calibrations, one also derives the energy resolution of the counters.

Energy resolutions (FWHM) of our counters are typically 26% for the K-peak and 43% for the L-peak. The K-peak resolution is not as good as one would expect from the value for the L-peak because we choose to operate at a rather high gas-amplification for a better signal-to-noise ratio for superior pulse shape analysis. This is particularly beneficial for L-pulses (see below). For this, we accept a slightly nonlinear energy scale above 5 keV.

The energy resolution and the absolute counting efficiencies for both peaks are obtained from a comparison with Ce source calibrations of counters filled with high ^{71}Ge activities. This was done in Heidelberg using a counting system similar to that at Gran Sasso. The Ce source obviates the need to bring high-activity ^{71}Ge sources to the Gran Sasso Laboratory with the attendant risk of contamination. Average counting efficiencies (2FWHM) are $\varepsilon_L = 30.6\%$ and $\varepsilon_K = 35.2\%$; hence the overall efficiency is 65.8%. Individual values depend mainly on the ratio of the active volume to the total volume of a counter. The total relative error of the efficiencies is $\pm 4\%$. The error sources are: (i) determination of L- and K-peak positions and of energy resolutions by means of Ce calibrations, (ii) absolute ^{71}Ge counting efficiency calibrations at MPIK, away from Gran Sasso, and (iii) application of these values to the counters used in the actual solar neutrino runs.

The stability of the counting system is very good; the energy scale and the energy resolution have not changed significantly during the counting times relevant in our experiment.

Pulse shape analysis distinguishes between genuine *fast* rising pulses due to point-like ^{71}Ge decays and *slow* pulses of extended ionization tracks from Compton-like background events where the primary ionization occurs at various radial distances from the anode wire. Since this analysis is the basis for the recognition of ^{71}Ge counts, and defines the pulse shape acceptance cuts, the pulse shape stability over extended time periods is crucial. It is checked at each calibration. In addition, there is routine monitoring of the shapes of the test pulses that are regularly fired at each line, currently once every two hours of counting.

During regular counting, the environment inside the Faraday cage remains very stable, the room is never entered and the temperature and the radon concentration inside and outside the shield tank are continuously monitored.

The final acts in an individual run are the quantitative recovery of the counting gas after the end of counting, and the mass-spectrometric isotope analysis to check carryover, as described above.

3. Side reactions and background considerations

^{71}Ge production from non-neutrino sources ("side reactions") is largely dominated by the reaction ^{71}Ga(p, n)^{71}Ge, with the threshold energy of 1.02 MeV. The protons may be secondaries induced by cosmic-ray muons, fast neutrons, or residual radioactivities in the target, e.g. (α, p), (n, p) or knock-on protons. The measured contributions from different side reactions are subtracted from the measured ^{71}Ge rate to give the net rate induced by solar neutrinos.

Concerning penetrating muons, we have redetermined the relevant ^{71}Ge production rate in $GaCl_3$ solution at the CERN muon beam [44]. Together with the measured muon flux at the underground site ($(27\pm3)\,m^{-2}d^{-1}$) [45], this led to an estimate of the cosmic-ray induced ^{71}Ge production rate in the target geometry equivalent to 3.7 ± 1.1 SNU [44]. ^{71}Ge production from atmospheric high energy neutrinos is negligible (see e.g. ref. [46]).

The target solution was conditioned to a Cl/Ga ratio of 3.24, suitable for Ge extractions according to our extraction pilot experiments. The levels of impurities specified for the gallium chloride solution, which were met by the manufacturers, guaranteed that their contributions were negligible [#3] (see table 2). The purity specifications are based on the rate of production of ^{71}Ge by α-particles and fast neutrons as measured in experiments with concentrated $GaCl_3$ solutions irradiated with α-particles at various energies and with fast neutrons from a Pu–Be source [32]. The levels of critical impurities were checked by neutron activation analysis (for U, Th, Fe, As, Se) and by radon determinations using proportional counting (table 2) [47]. Note the remarkably low value for the critical ^{226}Ra concentration. We check the purity of the target solution at approximately yearly intervals and so far it has not changed. Quantitatively, the contributions from internal U, Th and Ra are limited to less than the equivalent of 0.2 SNU.

[#3] We acknowledge the personal concern of Dr. H. Pfeiffer (Gebr. Sulzer) and of Dr. Y. Calvez (Rhône-Poulenc) to meet our quality requirements for the target solution.

Table 2
Key properties of the target solution. Colour is water clear with a slight yellowish shade.

	Measured	Specification
Ga (g/L)	567 ±2 [a]	565 ±10
Cl (g/L)	937 ±4 [a]	
Cl/Ga (mol/mol)	3.24 ±0.01	>3.15
density (g/mL)	1.887	
^{226}Ra (pCi/kg)	≤0.04	<0.50 [b]
U (ppb)	<0.04	<25 [b]
Th (ppb)	<0.04	<2 [b]
Zn (ppb)	≈20	
Se (ppb)	0.2	
As (ppb)	0.6	
Ge (ppb)	<0.5	

[a] Statistical error of multiple analyses.
[b] These specifications would lead to a 1% contribution to the SSM ^{71}Ge production rate for each individual background source.

The impurity levels are specified for the natural radioisotopes. Contamination with transuranic nuclides is not impossible in today's environment, and therefore we have checked for the presence of transuranium elements by means of hypersensitive α-spectrometry. The levels found for these nuclides are negligible in terms of their ^{71}Ge production.

The environmental fast neutron flux at the GALLEX site is measured to be $\phi_n(\geqslant 2.5\,MeV) = 0.23\times10^{-6}\,cm^{-2}\,s^{-1}$ [48,49]. This flux, together with neutrons from the tank walls, was effectively monitored with our 470 L $Ca(NO_3)_2$ solution [44]. By measuring the ^{37}Ar produced in the ^{40}Ca(n, α)^{37}Ar reaction, the ^{71}Ge production rate can be scaled. The ^{71}Ge production rate due to fast neutrons at the experimental site is estimated to be equivalent to 0.15 ± 0.10 SNU [44].

In the early testing phase of GALLEX, we have encountered an unexpected, and temporary, background effect caused not by a side reaction but by residual cosmogenic long-lived ^{68}Ge (half-life 288 d) [26,27,29,33]. This isotope is produced in $GaCl_3$ solutions exposed to cosmic rays at the surface; we estimate that $\approx(1.7\pm0.2)\times10^7$ ^{68}Ge atoms were produced in the target before it was brought underground

[26]. To eliminate this isotope, we purged the solutions received from the manufacturers in the TOF immediately before transfer of the material into the underground laboratory. Purging was continued after the target solution was in tank B (see refs. [26,29]); however, residues of this initially high activity proved hard to remove. It seems that a small fraction of the ^{68}Ge activity was picked up by some trace impurity in the material – most likely polysilicic acids – which released it only slowly. Heat treatment and repeated purging in 28 extraction runs reduced this extraneous activity due to ^{68}Ge to the level of our very low general counter backgrounds (see section 5). The long life of this isotope and its characteristic decay (positrons from the 68 min ^{68}Ga daughter) make it distinguishable from ^{71}Ge. It should be emphasized that this effect is of such a small magnitude as to make it undetectable in terms of recovery yields. If it were not for the very high initial level of ^{68}Ge in the solution, the effect would not have been detected.

^{69}Ge is not formed in appreciable yield by solar neutrinos, but can be produced from ^{69}Ga by (p, n) secondary reactions. Hence, it can serve as a useful monitor of various side reactions caused by impurities, such as the various α-induced reactions discussed above. In none of the runs employing the NaI spectrometer have we seen a single decay with the specific EC/γ signature of this short-lived isotope (half-life of 39 h), but we have not yet fine tuned this check to its full potential sensitivity by tailoring exposure time in devoted runs.

Another possible source of background is the tritium content of the hydrogen atoms of the germane in the counting gas. We have reduced this factor by using low tritium material in all the relevant steps in the synthesis of germane. We estimate that the tritium contribution to the background is negligible (< 0.01 cpd in the acceptance windows).

^{222}Rn (half-life 3.82 d) and its daughters are ubiquitous constituents of laboratory air, especially underground, and at our level of operation are great nuisances. Great pains have been taken to avoid even traces of radon inside and outside our counters [50]. Nevertheless, we have not yet succeeded in eliminating this interference entirely. On the average, the level of this impurity per run introduced with the counting gas is about three ^{222}Rn atoms, which produce about one count in the ^{71}Ge acceptance windows. Fortunately it is possible to recognize the signature of the decay of Rn and its daughters and thus minimize their interference. The radon decay chain includes three α- and two β-decays within less than 3 h. The signatures are overflow counts produced by the energetic α-particles and a very characteristic delayed coincidence of ^{214}Bi β- and ^{214}Po α-decays ("BiPo" events). The delay due to the 164 μs ^{214}Po half-life is registered with a slow transient digitizer. We have adopted respective cuts to eliminate radon-related counts from the data (see section 5). Their efficiency is 92%; the dead time imposed during the first month of counting is 4% in typical runs. From the tagged Rn events and from their known influence on the final result if they were falsely attributed to ^{71}Ge decays, we remain with a 2 ± 1 SNU equivalent subtraction for unidentified Rn events.

4. GALLEX 1: run schedule from B 29–B 50

Since 14 May 1991 we have monitored the solar neutrino signal continuously, at the very low background level for which the experiment was designed. In table 1 we list the dates and the chemical yields of the individual runs from B 29–B 50. These runs define the data taking period "GALLEX 1", terminated on 29 April 1992. During these 50 weeks we have performed 16 solar neutrino runs, 14×3 weeks and 2×4 weeks . One of them (B 44) was lost in preparation; the other fifteen are consecutively labeled [#4] SR 1–SR 15; see table 1 for their B number equivalent. In addition, five blank runs (BL 1–BL 5) have been carried out with minimal exposure times.

The duty factor of 94% for the preparative part of the experiment is well matched by the 97% running time of the counting system. This time splits into 94.5% for sample counting, 2.2% for calibration runs and 0.2% for test runs.

In order to determine the counter background rates properly, it is necessary to count much longer than a few ^{71}Ge life-times. This Letter is based on data taken until 18 May 1992. As of that date, 13 out of the 20 SR and BL runs of GALLEX I were still counting; only seven experiments have really been completed (see table 1). For this reason and for the limited

[#4] SR stands for solar neutrino run.

number of runs done, the data reported in this paper should be considered preliminary. They will be re-evaluated in due time with the complete experimental information then available and probably treated with more sophisticated pulse shape criteria than at present. We do not expect, however, any *substantial* changes in the preliminary data presented here.

5. Data treatment and results

5.1. Data treatment

The first phase of data treatment is to eliminate obvious background events.

(i) *NaI-veto* cancels counter pulses which are time coincident with a signal above a defined noise level in either half of the NaI anticoincidence pair-spectrometer. This cut applies only for runs in the active positions, SR 1–SR 6 (see table 1). Nuclides other than ^{71}Ge can be observed in the coincidence mode (e.g., ^{69}Ge, ^{214}Bi).

(ii) *Radon daughters outside the counter* can stick on the outside surfaces of the counter during handling. Because they are short lived, we cancel the first three hours after start of counting. Since SR 5, this disturbance has been eliminated by a minor modification of the counter filling procedure.

(iii) *Radon inside the counter* is eliminated by the following cuts: We ignore all events occurring up to 3 h after and 15 min before an unvetoed overflow event ($\gtrsim 16$ keV). This covers all cases where at least one of the initial two α-particles of the radon decay chain is seen. In addition, we ignore all counts occurring up to 3 h before a BiPo event.

Having removed these background events, we plot the rise time versus energy for the remaining events. Fig. 1 shows such data combined, for all but two [#5] solar neutrino runs collected during the first 16.49 days (mean life of ^{71}Ge) respectively.

Groupings of fast pulses are seen in the relevant energy regions around 1 and 10 keV. The energy spectrum of the "fast" events (for all 14 SR runs) is shown in fig. 2a for the first mean life (0τ–1τ) of ^{71}Ge. The L- and K-peaks of ^{71}Ge are clearly seen. Similar plots for the time periods (1τ–2τ) and (2τ–4τ), respectively, are shown in figs. 2b and 2c. The decay of the peaks over the characteristic ^{71}Ge lifetime is evident. For comparison, fig. 2d shows the energy spectrum of a ^{71}Ge calibration sample.

[#5] For runs SR 10, SR 11, the rise-time information is incomplete.

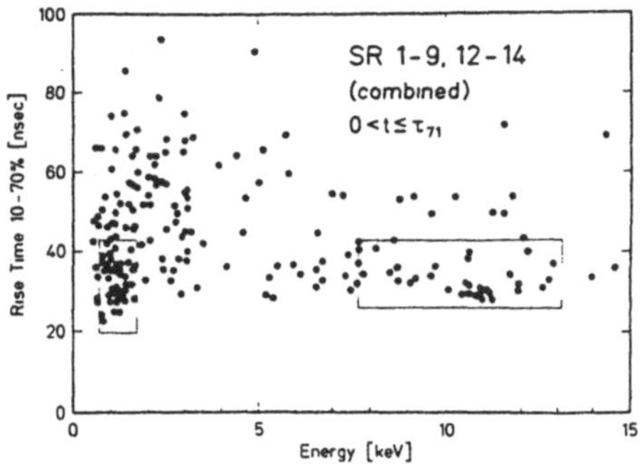

Fig. 1. A plot of all events occurring in 12 solar runs during the first 16.49 days (one mean life of ^{71}Ge) after the respective ends of extraction. The L- and K-energy windows are expanded to cover *all* individual runs included in this illustration. Runs SR 10 and SR 11 are not included since complete rise-time information is not available for them.

To quantify the magnitude of the signal, we select the events that are candidates for ^{71}Ge decays by applying the following cuts (fig. 1):

(i) *Energy* acceptance windows for L- and K-peaks are determined for each counter individually using all available calibrations for that run as described in section 2 (2FWHM acceptance).

(ii) *Pulse shape* acceptance windows are based on short rise times. These rise-time cuts encompass 97.7% of L-pulses ("L-fast") and 95% of K-pulses ("K-fast"). For runs SR 10, SR 11, and BL 3 the rise-time windows were completely opened since there was a fluctuation (the energy remained stable). ADP values are also available for eventual cross-checks.

The event list obtained is used for a maximum likelihood analysis, as conveniently summarized for this type of application by Cleveland [51]. The initial number of ^{71}Ge atoms is obtained by fitting the data to the assumption of a component decaying with 16.49 d mean life plus a constant background, determined for each individual measurement.

^{68}Ge can not be identified on an individual basis but only on its long time effects on the counter backgrounds. With the described coincidence method we

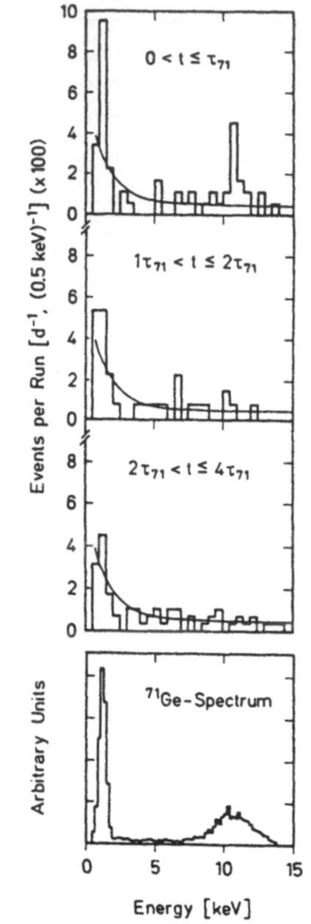

Fig. 2. Energy spectra for the fast events of 14 solar neutrino runs during different counting periods after the end of extraction: (a) $(0\tau-1\tau)$ ($\tau = {}^{71}$Ge mean life); (b) $(1\tau-2\tau)$; (c) $(2\tau-4\tau)$; (d) measured spectrum from a ^{71}Ge calibration sample. The decay of ^{71}Ge during subsequent mean lives is evident. Identical curves are inserted in (a) – (c) to facilitate cross comparisons.

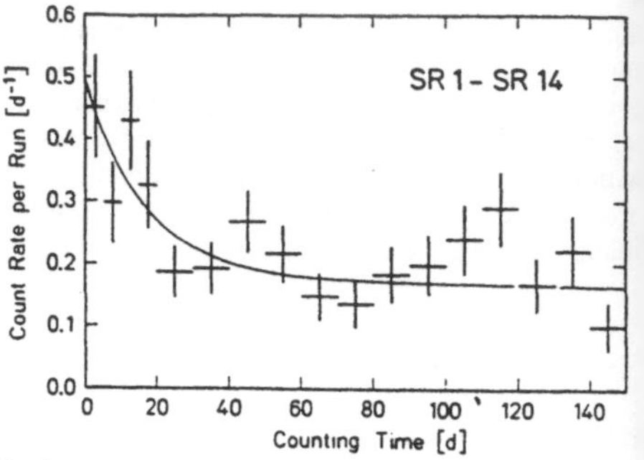

Fig. 3. Count rate of fast L- and K-events averaged over all 14 solar neutrino runs as a function of counting time. The solid line represents the best fit to the data obtained with the maximum likelihood method allowing: (i) ^{71}Ge, (ii) counter backgrounds, and (iii) a small fixed contribution from ^{68}Ge (see text).

have identified in six runs during their total counting time altogether eight ^{68}Ge decays. Using this information, we assume in the maximum likelihood analysis also a known background component decaying with 288 d half-life.

The result of the likelihood analysis is plotted in fig. 3 (solid line) as a function of counting time along with the count rates of fast L- and K-events averaged over all 14 solar neutrino runs. Although individual data points deviate from the fitted curve due to statistical fluctuations, the signal attributed to the decay of solar neutrino produced ^{71}Ge can clearly be seen in the data.

We do analyses (i) for the individual runs, (ii) for all SR runs combined, and (iii) for all BL (blank) runs combined. Furthermore, we analyse the K- and L-windows separately, as well as the K- and L-windows combined.

The resulting decay rates in atoms per day are converted into production rates by taking into account the extraction yields, the individual counter volume efficiencies, and the exposure and delay times respectively of the individual runs. Common to all runs and contributing to the systematic errors are the uncertainties of quantities such as the target size, the general extraction efficiency, the normalized counting efficiency when using the standard cuts, and the side-reaction corrections. The correction for ^{68}Ge beforehand taken care of in the likelihood treatment (see above) has a magnitude equivalent to 9 SNU and introduces an error equivalent to ± 5.5 SNU.

5.2. Results

The ^{71}Ge production rates measured for the individual solar neutrino runs are plotted versus the date of the respective exposures in fig. 4, left-hand scale. Monte Carlo simulations have shown that their distribution is fully consistent with statistical fluctua-

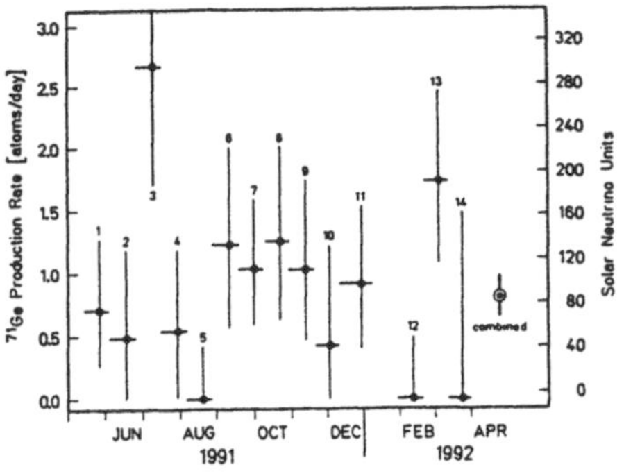

Fig. 4. Results of the first year of solar neutrino recording for runs SR 1–SR14. The left-hand scale is in units of the measured production rate, the right-hand scale, in units of the net solar production rate. Error bars shown are 1σ. The individual data are consistent with statistical fluctuations only.

tions. For the combined SR runs, the analysis gives a mean production rate of $(90 \pm 19) \times 10^{-36}$ ^{71}Ge atoms (target atom)$^{-1}$ s^{-1} (1σ) before corrections for side reactions. The answers obtained if L- or K-peaks are analyzed separately do not differ significantly from the combined result.

Next, we subtract the known contributions which are not due to solar neutrinos (see section 3):
muon induced production: (3.7 ± 1.1) SNU,
actinides in the target: <0.2 SNU,
fast neutrons: 0.15 ± 0.10 SNU,
radon cut scaled to 100% efficiency: 2 ± 1 SNU,
^{69}Ge from muons and ^{8}B neutrinos falsely attributed to ^{71}Ge: 1 ± 1 SNU.

This yields a net rate associated with solar neutrinos of 83 ± 19 (stat.) ± 8 (syst.) SNU (1σ) (fig. 4, right-hand scale).

Apart from the side reactions and the ^{68}Ge correction mentioned above, additional contributions to the final systematic error (expressed in SNU equivalents) come from uncertainties in the chemical yield (± 1.5), the counting efficiency (± 3.6), and the rise-time cut (± 1.8).

The systematic errors were estimated from the figures quoted under the additional assumption that: (i) no unidentified background reactions exist and (ii) that the extraction yield is not much different from the monitored efficiency due to hot atom effects. These conjectures seem justified to us in view of the results from the relevant investigations (see sections 2 and 3).

The statistical errors are consistent with the results of respective Monte Carlo simulations.

The final results for the individual SR runs are given explicitly in table 3, last column. For comparison, we also list the results obtained if K- and L-windows are analyzed separately by maximum likelihood.

The counter backgrounds (without sample) are typically 0.10 cpd (L-fast) and 0.05 cpd (K-fast) in the "active" counting positions and 0.07 cpd (K+L fast) in the "passive" lines [41]. Our combined analysis for B runs since B 29 reveals that the additional ^{68}Ge contribution per run contributes about 0.03 cpd (L-fast) and 0.04 cpd (K-fast). This component stems from the target tank where it is decaying with 288 d half-life. It corresponds to 2.5 ± 1.5 ^{68}Ge atoms released per day of standing time and is taken into account as described.

In order to check the reliability of the result of the experiment, we have performed a number of tests which corroborate the findings:

(i) The overall result is notably stable to changes in the energy and rise-time cuts, even to complete opening of the rise-time windows.

Table 3
Results for individual solar neutrino runs [a)].

Run #	L-only (SNU)	K-only (SNU)	L+K combined (SNU)
SR 1	97	49	72^{+64}_{-50}
SR 2	69	33	48^{+79}_{-48}
SR 3	339	255	291^{+121}_{-106}
SR 4	10	94	54^{+72}_{-54}
SR 5	0	0	0^{+47}_{-0}
SR 6	67	177	131^{+87}_{-74}
SR 7	41	179	109^{+62}_{-50}
SR 8	176	102	134^{+85}_{-71}
SR 9	200	48	108^{+80}_{-63}
SR 10	0	83	41^{+89}_{-41}
SR 11	156	62	96^{+70}_{-58}
SR 12	24	0	0^{+55}_{-0}
SR 13	317	78	187^{+81}_{-74}
SR 14	0	0	0^{+167}_{-0}

[a)] Negative production rates or errors occur for individual runs due to statistics and are truncated at the end since they are unphysical. This effect does not apply to the combined analysis which yeidls the mean production rate.

(ii) A comparison of the SR runs with the BL (blank) runs should reveal effects which do not depend on exposure time. With only five blanks at hand, their combined result, equivalent to 8^{+16}_{-8} SNU, is not yet statistically very revealing but consistent with the absence of a signal in the blanks.

(iii) In a maximum likelihood analysis in which the mean life of the decaying component is entered as a free parameter, the fit yields a mean life of $\tau = (13.5^{+5.2}_{-3.3})$ d (1σ), compatible within statistics with 16.49 d for ^{71}Ge.

The main result of 83 ± 19 (stat.) ± 8 (syst.) SNU (1σ) is to be compared with the outcome of the SAGE experiment: 20^{+15}_{-20}(stat.) ± 32 (syst.) SNU (1σ) [19] and with the standard solar model prediction range 132^{+21}_{-17}) SNU ("3σ") [3]. For 1σ-comparability we have, in SNU, 83 ± 21 (GALLEX), 20^{+35}_{-20}(SAGE) and 132 ± 7 (SSM [3]). We have no explanation of the discrepancy between our results and those of the SAGE Collaboration [19].

Our result implies the first detection ever of the pp neutrinos associated with the bulk of the energy production inside the Sun by nuclear fusion. Further astrophysical and particle physics consequences of this result are discussed in the following Letter [30].

6. Outlook and conclusions

6.1. Next steps

GALLEX is one of the few low-rate experiments to claim a positive signal rather than a limit. To improve the quantitative implications of the result already obtained, more measuring time is needed to reduce the *statistical* error. Conceptually, GALLEX is designed for four years of solar neutrino recording to lower the statistical error to approximately 10 SNU. Our first year of taking data permits only a preliminary result. Based on our experience to date, we fully agree with the original concept of running for several years in order to achieve the needed accuracy. This in turn will allow firmer conclusions to be drawn in comparisons with theory.

In radiochemical solar neutrino experiments it is necessary to ascertain, beyond any reasonable doubt: (i) the complete extraction of the product (^{71}Ge) from the target and (ii) the absence of sources that might mimic a solar neutrino signal. Having observed a *positive* signal, we concentrated on the exclusion of side reactions as being of more immediate importance than the proof of complete germanium extraction. We have therefore devoted considerable effort to reduce the possibilities of side reactions, as discussed above. In this respect our situation is quite opposite to that in which virtually *no* signal is seen [19].

As a further test of the whole experimental procedure, we are planning a calibration experiment in which the target will be irradiated with a man-made neutrino source. The latter will be in the form of $> 5 \times 10^{16}$ Bq ^{51}Cr (half-life 28 d) made by neutron irradiation of isotopically enriched ^{50}Cr (see refs. [52,53] for details). The source is expected to produce a signal at least seven-fold that of the Sun (SSM prediction). The execution of the Cr source experiment is well advanced [52]; the first irradiation at the Grenoble Siloe reactor is scheduled for fall 1993.

To prepare for this, the gallium target solution was transferred from tank B to tank A just after B 50 on30 April 1992, thus terminating the B series of runs ("GALLEX I"). According to our planning, in the fall of 1992 we shall start another year of solar neutrino measurements. ending with the arrival of the chromium source at LNGS. After the source exposures, solar neutrino runs will continue for about two more years. Afterwards, it seems likely that new physical insights from the results of upcoming solar neutrino experiments as well as from GALLEX itself will require the continuation of this experiment for many years to come.

6.2. Conclusions

The GALLEX experiment can truly claim to have observed, for the first time, the primary pp neutrinos.

Our central value, 83 SNU, is to be compared with the 74 SNU attributed to the pp and pep cycles in the SSM. Further, a model in which all of the solar luminosity is attributed only to this cycle gives the somewhat larger value of 79 SNU. Within the 2σ range of the present statistics, the GALLEX result is even compatible with the full SSM prediction. In this sense, neutrinos that have come through unaltered are allowed. However, to explain the GALLEX results plus those of ^{37}Cl and Kamiokande in this way requires stretching the solar models and the data to their error limits; yet the possibility remains open. On the other hand, the results are interpretable in terms of a MSW mechanism, with the neutrino mass and mixing angle parameters fixed within narrow ranges. These conclusions are more fully discussed in our accompanying Letter [30].

Acknowledgements

Invaluable help, inspiration, and encouragement by N. Cabibbo were instrumental for the good fortune of GALLEX. The experiment would not have been possible without the continuous support of B. Povh and H. Völk. Special credit for substantial help is given to W. Klose (KFK), H. Leussink (Alexander Krupp von Bohlen und Halbach Foundation), G. Preiß (MPG), J. Rembser (BMFT), H.A. Staab (MPG), and G. Wilke (MPG). Important motivation and support came from R. Aymar, D. Clayton, D. Cribier, R. Davis, H. Elsässer, J. Horowitz, N.P. Samios and V. Soergel. We wish to thank F. Baumgärtner (TU Munich) and A. Warshawsky (WIS) for their help and advice in dealing with specific radiochemical problems. We greatly acknowledge the excellent support of the technical staff of the Laboratori Nazionali del Gran Sasso and of the participating institutes. In particular, we appreciate outstandingly skillful contributions in the technical realization of counters and counting by P. Mögel, E. Burkert, D. Dörflinger and H. Richter (all MPIK) and creative assistance by G. Pagnozzi (LNGS) and J.P. Soirat (Saclay). Collegial discussions with many members of the SAGE Collaboration and warm hospitality during various visits to Baksan are acknowledged.

References

[1] J.N. Bahcall and R. Ulrich, Rev. Mod. Phys. 60 (1988) 297.
[2] J.N. Bahcall, Neutrino astrophysics (Cambridge U.P., Cambridge, 1989).
[3] J.N. Bahcall and M.H. Pinsonneault, preprint IASSNS-AST 92/10, submitted to Rev. Mod. Phys. (1992).
[4] S. Turck-Chièze, S. Cahen, M. Casse and C. Doom, Astrophys. J. 335 (1988) 415.
[5] G. Berthomieu, J. Provost, P. Morel and Y. Lebreton, Astron. Astrophys. (1992), to appear.
[6] S.P. Mikheyev and A. Yu. Smirnov, Nuovo Cimento 9C (1986) 17.
[7] L. Wolfenstein, Phys. Rev. D 17 (1978) 2369.
[8] J. Bouchez et al., Z. Phys. C 32 (1986) 499.
[9] M. Cribier, W. Hampel, J. Rich and D. Vignaud, Phys. Lett. B 183 (1986) 89.
[10] A.J. Baltz and J. Weneser, Phys. Rev. D 35 (1987) 528; D 37 (1988) 3364.
[11] L. Wolfenstein, in: Neutrino physics, ed. K. Winter (Cambridge U.P., Cambridge, 1991) p. 605.
[12] R. Davis et al., Proc. 21th Intern. Cosmic ray Conf., ed. R.J. Protheroe, Vol. 12 (University of Adelaide Press, Adelaide, Australia, 1990) p. 143.
[13] K. Hirata et al., Phys. Rev. Lett. 65 (1990) 1297, 1301.
[14] K. Hirata et al., Phys. Rev. D 44 (1991) 2241.
[15] V.A. Kuzmin, Sov. Phys. JETP 22 (1966) 1050.
[16] T. Kirsten, in: Neutrino physics, ed. K. Winter (Cambridge U.P., Cambridge, 1991) p. 585.
[17] W. Hampel and R. Schlotz, Proc. AMCO-7 (Darmstadt-Seeheim, FRG, 1984), ed. O. Klepper, p. 89.
[18] W. Hampel and L.P. Remsberg, Phys. Rev. C 31 (1985) 677.
[19] SAGE Collab., A.I. Abazov et al., Phys. Rev. Lett. 67 (1991) 3332.
[20] T. Kirsten, Inst. Phys. Conf. Ser. 71 (1984) 251.
[21] W. Hampel, Neutrino 88, ed. J. Schneps (World Scientific, Singapore, 1989) p. 311.
[22] W. Hampel, Proc. TAUP 1989 (L'Aquila, Italy, 1989), eds. A. Bottino and P. Monacelli (Editions Frontieres, Gif-sur-Yvette, 1990) p. 189.
[23] T. Kirsten, in: Inside the Sun, Proc. IAU Colloq., No. 121 (Versailles, France), eds. G. Berthomieu and M. Cribier (Kluwer Academic, Dordrecht, 1990) p. 187.
[24] M. Cribier, Proc. 25 HEP Conf. (Singapore), ed. K. Phua, SE Asian Theor. Phys. Assoc. Publ. 1 (1990) 689.
[25] T. Kirsten, Neutrino 90, Nucl. Phys. B (Proc. Suppl.) 19 (1991) 77.
[26] R. Plaga, in: Proc. PASCOS 91 (Boston, MA, 1991), ed. P. Nath (World Scientific, Singapore, 1992).
[27] T. Kirsten, Proc. 2nd TAUP Workshop (Toledo, Spain, September 1991), ed. A. Morales, Nucl. Phys. B (Proc. Suppl.) 28 A (1992), to appear.
[28] R. Wink, in: Proc. Particles and cosmology (Baksan, Caucasus), ed. V. Matveev (World Scientific, Singapore, 1992).
[29] T. Kirsten, in: Proc. Particles and fields 91 (Vancouver, Canada, 1992), eds. D. Axen et al., Vol. 2 (World Scientific, Singapore, 1992) p. 942.
[30] GALLEX Collab., P. Anselmann et al., Phys. Lett. B 285 (1992) 390.
[31] E. Henrich, R. von Ammon, K. Ebert, T. Fritsch, K. Hellriegel and M. Balata, KFK-Nachr. 22 (1990) 125.
[32] BNL, Proposal to DOE for a Gallium Solar Neutrino Experiment, BNL report BNL-37081 (1985).
[33] U. Schanda, Master Thesis, Technische Universität München (1989).

[34] I. Dostrovsky, Proc. Conf. on Status and future of solar neutrino research, ed. G. Friedlander, BNL report BNL-50879, Vol. 1 (BNL, Upton, NY, 1978) p. 231;
J.N. Bahcall et al., Phys. Rev. Lett. 40 (1978) 1351.

[35] L. Paoluzi, Nucl. Instrum. Methods A 279 (1989) 133.

[36] A. Zichichi, Proc. ICOMAN '83 (Frascati, Italy, 1983);
E. Bellotti, Nucl. Instrum. Methods A 264 (1988) 1.

[37] R. Wink, Thesis, Universität Heidelberg (1988).

[38] W. Hampel et al., Proc. 19th Intern. Cosmic ray Conf. (La Jolla, CA), Vol. 5 (1985) p. 422.

[39] R. Plaga, Thesis, Universität Heidelberg (1989).

[40] R. Plaga and T. Kirsten, Nucl. Instrum. Methods A 309 (1991) 560.

[41] G. Heusser, Background characteristics of the GALLEX spectrometer, Proc. 2nd Conf. on Trends in Astroparticle physics (Aachen, FRG, 1992), ed. P.C. Bosetti.

[42] H. Lalla, Master Thesis, Universität Heidelberg (1989).

[43] A. Urban, Thesis, TU München (1989).

[44] T. Stolarczyk, Thesis, Université de Paris-Sud (1990).

[45] S.P. Ahlen et al., Phys. Lett. B 249 (1990) 149.

[46] M. Rudzskii and Z.F. Seidov, Sov. J. Nucl. Phys. 39 (1979) 553.

[47] A. Lenzing, Thesis, Universität Heidelberg (1989).

[48] P. Belli et al., Nuovo Cimento 101A (1989) 959.

[49] R. Aleksan et al., Nucl. Instrum. Methods A 274 (1989) 203.

[50] G. Heusser and M. Wójcik, Appl. Radiat. Isot. 43 (1992) 9.

[51] B. Cleveland, Nucl. Instrum. Methods 214 (1983) 451.

[52] M. Cribier et al., Nucl. Instrum. Methods A 265 (1988) 574.

[53] F.X. Hartman and R.L. Hahn, BNL Formal Report BNL-52281 (1991).

GALLEX results from the first 30 solar neutrino runs

GALLEX Collaboration [1,2,3,4]

P. Anselmann, W. Hampel, G. Heusser, J. Kiko, T. Kirsten, M. Laubenstein, E. Pernicka, S. Pezzoni, U. Rönn, M. Sann, C. Schlosser, R. Wink, M. Wojcik [6]

Max-Planck-Institut für Kernphysik (MPIK), Postfach 103980, D-69029 Heidelberg, Germany [1]

R. v. Ammon, K.H. Ebert, T. Fritsch, K. Hellriegel, E. Henrich, L. Stieglitz, F. Weirich

Institut für Heiße Chemie, Kernforschungszentrum Karlsruhe (KFK), Postfach 3640, D-76021 Karlsruhe, Germany

M. Balata, N. Ferrari, H. Lalla

Laboratori Nazionali del Gran Sasso (LNGS), S.S. 17/bis Km 18 + 910, I-67010 L'Aquila, Italy [2]

E. Bellotti, C. Cattadori, O. Cremonesi, E. Fiorini, L. Zanotti

Dipartimento di Fisica, Università di Milano e INFN, Via Celoria 16, I-20133 Milano, Italy [2]

M. Altmann, F. v. Feilitzsch, R. Mößbauer, U. Schanda [5]

Physik Department E15, Technische Universität München (TUM), James-Franck Straße, D-85748 Garching b. München, Germany

G. Berthomieu, E. Schatzman [7]

Observatoire de la Côte d'Azur, Département Cassini, B.P. 229, 06004 Nice Cedex 4, France

I. Carmi, I. Dostrovsky

Department of Environmental and Energy Research, The Weizmann Institute of Science (WI), P.O. Box 26, 76100 Rehovot, Israel

C. Bacci [8], P. Belli, R. Bernabei, S. d'Angelo, L. Paoluzi

Dipartimento di Fisica, II Università di Roma 'Tor Vergata' e INFN, Sezione di Roma 2, Via della Ricerca Scientifica, I-00133 Roma, Italy [2]

A. Bevilacqua, S. Charbit, M. Cribier, L. Gosset, J. Rich, M. Spiro, T. Stolarczyk, C. Tao, D. Vignaud

DAPNIA/Service de Physique des Particules, CE Saclay, F-91191 Gif-sur-Yvette Cedex, France [3]

R.L. Hahn, F.X. Hartmann, J.K. Rowley, R.W. Stoenner, J. Weneser

Brookhaven National Laboratory (BNL), Upton, NY 11973, USA [4]

Abstract

We report new GALLEX solar neutrino results from 15 runs covering 406 days (live time) within the exposure period 19 August 1992–13 October 1993 ("GALLEX II"). With counting data considered until 4 January 1994, the new result is [78 ± 13 (stat.) ± 5 (syst.)] SNU (1σ). It confirms our previous result for the 15 initial runs ("GALLEX I") of [81 ± 17 (stat.) ± 9 (syst.)] SNU. After two years of recording the solar neutrino flux with the GALLEX detector, the combined result from 30 solar runs (GALLEX I + GALLEX II) is [79 ± 10 (stat.) ± 6 (syst.)] SNU (1σ). In addition, 19 "blank" runs gave the expected null result. GALLEX neutrino experiments are continuing.

1. Introduction

The radiochemical GALLEX detector at Gran Sasso monitors the solar neutrino flux with energies above 233 keV. Recording started on 14 May 1991 and is ongoing. The first evidence for neutrino detection in gallium and the claim for the observation of pp-neutrinos was based on 15 solar neutrino runs, (SR1–SR15), performed in tank "B" of the GALLEX instrumentation (GALLEX I). The data were published [1] together with the basic description of the experimental procedures and conditions.

The GALLEX II series of runs started on 19 August 1992, after the gallium chloride target solution had been transferred into tank "A". Preliminary data on the first 6 GALLEX II runs were given together with the final results for GALLEX I and with the combined result for the first 21 solar neutrino runs in [2]. In summary, this updated result of 87 ± 16 SNU was (66–71)% ± 13% (1σ) of the Standard Solar Model predictions that range from 123 SNU [3] to 132 SNU [4]. Ref. [2] also contains further details on data and error evaluation and the description of various significance tests which corroborated the nature of the signal observed.

The GALLEX II exposure period will last until June 1994, with counting continuing until the end of 1994. Because of the large interest in the timely release of our data, we present this condensed update at a stage where data are available from GALLEX II for 15 solar runs, a number equal to the total number of runs in GALLEX I. These GALLEX II solar runs (SR16–SR30) cover the exposure period 19 August 1992–13 October 1993. Our required six months' counting period for each of these runs has been completed, except for the last four runs, SR27–SR30, for which > 2.5 months' of counting data have been accumulated, sufficient to have them included in the present analysis.

We report the results here without repeating the detailed descriptions of experimental procedures and of data evaluation methods that were given in [1,2,5], but we do discuss additions and refinements. This applies, in particular, to the presentation of the results of the short exposure runs (quasi-"blanks") where virtually no solar neutrino signal is expected.

So far, 19 blanks have been evaluated in GALLEX I + II, compared to the 30 solar runs. Even though the experimental effort in doing these blanks is large, they are worthwhile since they further assure us that we are interpreting the solar runs correctly.

[1] This work has been supported by the German Federal Minister for Research and Technology (BMFT). This work has been generously supported by the ALFRIED KRUPP von BOHLEN und HALBACH-Foundation, Germany.

[2] This work has been supported by Istituto Nazionale di Fisica Nucleare (INFN), Italy.

[3] This work has been supported by the Commissariat à l'énergie atomique (CEA), France.

[4] This work has been supported by the Office of High Energy and Nuclear Physics of the U.S. Department of Energy, United States.

[5] Present address: Max-Planck-Institut für Physik, Föhringer Ring 6, Postfach 401212, D-80805 München, Germany.

[6] Permanent address: Instytut Fizyki, Uniwersytet Jagiellonski, ul. Reymonta 4, PL-30059 Kraków, Poland.

[7] Present address: DASGAL, Bâtiment Copernic, Observatoire de Paris, 5 place Jules Janssen, F-92195 Meudon Principal, France.

[8] Permanent address: Dipartimento di Fisica, III Università di Roma, Via C. Segre 2, 00100 Roma, Italy.

Results from SAGE
(The Russian–American Gallium solar neutrino Experiment)

J.N. Abdurashitov [a], E.L. Faizov [a], V.N. Gavrin [a], A.O. Gusev [a], A.V. Kalikhov [a], T.V. Knodel [a], I.I. Knyshenko [a], V.N. Kornoukhov [a], I.N. Mirmov [a], A.M. Pshukov [a], A.M. Shalagin [a], A.A. Shikhin [a], P.V. Timofeyev [a], E.P. Veretenkin [a], V.M. Vermul [a], G.T. Zatsepin [a], T.J. Bowles [b], J.S. Nico [b], W.A. Teasdale [b], D.L. Wark [b,1], J.F. Wilkerson [b], B.T. Cleveland [c], T. Daily [c], R. Davis Jr. [c], K. Lande [c], C.K. Lee [c], P.W. Wildenhain [c], S.R. Elliott [d], M.L. Cherry [e], R.T. Kouzes [f,2]

[a] *Institute for Nuclear Research, Russian Academy of Sciences, Moscow 117312, Russia*
[b] *Los Alamos National Laboratory, Los Alamos, NM 87545, USA*
[c] *University of Pennsylvania, Philadelphia, PA 19104, USA*
[d] *Lawrence Livermore National Laboratory, Livermore, CA 94550, USA*
[e] *Louisiana State University, Baton Rouge, LA 70803, USA*
[f] *Princeton University, Princeton, NJ 08544, USA*

Received 7 April 1994
Editor: K. Winter

Abstract

Fifteen measurements of the solar neutrino flux have been made in a radiochemical ^{71}Ga–^{71}Ge experiment employing initially 30 t and later 57 t of liquid metallic gallium at the Baksan Neutrino Observatory between January 1990 and May 1992. This provides an integral measurement of the flux of solar neutrinos and in particular is sensitive to the dominant, low-energy p–p solar neutrinos. SAGE observed the capture rate to be 73^{+18}_{-16} (stat.) $^{+5}_{-7}$ (syst.) SNU. This represents only 56%–60% of the capture rate predicted by different Standard Solar Models.

1. Introduction

The discrepancy between the solar neutrino capture rate predicted by Standard Solar Model (SSM) calculations and the ^{37}Ar rate measured by the chlorine experiment of Davis et al. has persisted for more than twenty years. Recent calculations of this capture rate give 8.0 ± 1.0 SNU (1 Solar Neutrino Unit $= 10^{-36}$ captures/target atom/s) in the Bahcall–Pinsonneault (B–P) SSM [1] and 6.4 ± 1.4 SNU in the Turck-Chieze and Lopes (T-C-L) SSM [2]. This is compared with the chlorine measurement [3] of 2.3 ± 0.3 SNU. The Kamiokande water Cherenkov experiment [4] observes only $0.50 \pm 0.04 \pm 0.06$ ($0.64 \pm 0.05 \pm 0.08$) of the ^8B flux predicted by the B–P (T-C-L) SSM. The initial SAGE results from the 1990 data [5] reported an upper limit on the capture

[1] Present address: Department of Particle and Nuclear Physics, Oxford University, Keble Road, Oxford OX1 3RH, UK.
[2] Present address: Batelle Pacific Northwest Laboratories, P.O. Box 999, Richland, WA 99352, USA.

Solar Neutrinos: Proposal for a New Test

Abstract. *The predicted flux on the earth of solar neutrinos has eluded detection, confounding current ideas of solar energy production by nuclear fusion. The dominant low-energy component of that flux can be detected by mass-spectrometric assay of the induced tiny concentration of 1.6 × 10⁷ year lead-205 in old thallium minerals. Comments are solicited from those in all relevant disciplines.*

Detection and measurement of the flux on the earth of solar neutrinos (6×10^{10} cm^{-2} sec^{-1}) appears to be the only definitive test of current models of solar energy production, which postulate thermal fusion of light nuclei (hydrogen burning) in the solar core. A recent review by Bahcall and Davis (*1*) summarizes the essentially negative results of the experiment of Davis and co-workers: for the last 5 years the average yield of ^{37}Ar from neutrino capture in ^{37}Cl has been only about one-sixth of the rate predicted for this nucleus by almost all solar models, and within twice the experimental uncertainty the rate is not certainly above the estimated background. However, because of the rather high neutrino energy threshold for capture in ^{37}Cl, 0.82 Mev, the predicted capture rate excludes capture of about 95 percent of the solar neutrino spectrum. In fact, about 80 percent of the neutrinos expected to be captured in ^{37}Cl are those of very low relative intensity (~ 10^{-4} of the total solar neutrinos emitted) and very high energy (E_{max} = 14 Mev) that come from the rare decays of ^{8}B formed in a weak branch of the solar fusion chain. In perhaps the least radical (*2, 3*) of many proposed revisions (*1–6*) of the solar model or of the relevant nuclear or neutrino physics it is predicted that only this small flux component is overestimated in the solar model calculations, which lowers the predicted rate into near agreement with the observed value. Even if such revisions were tenable, the ^{37}Cl value is so low that it challenges the basic ideas (*1*) that solar neutrinos do reach the earth, or leave the sun, or are now being produced. Clearly, progress toward understanding the solar energy mechanism is now blocked pending definitive evidence as to the presence of the expected bulk of *low-energy* solar neutrinos.

This interest has stimulated renewed inquiry into alternate neutrino capture reactions with low energy thresholds, and Davis and co-workers have advanced the feasibility of experiments with ^{7}Li and ^{71}Ga targets in pilot tests (*1*). These experiments should yield of the order of one neutrino capture product atom per month per ton of target of a short-lived radioactive species, which can be extracted chemically and detected with good efficiency by following its decay, as for ^{37}Ar. Usable sensitivity requires massive targets—for example, 40 tons of Ga.

The availability of other methods of detecting modest numbers of neutrino capture product atoms, such as mass spectrometry or laser-excited resonance fluorescence, brings into consideration many capture reactions whose products are too long-lived for decay counting. With a longer half-life of the product atom, a larger concentration of product atoms can build up in the target mass. Thus a much higher overall sensitivity in terms of reduced target mass requirements can be realized, compared to the "decay counting" experiments, despite the lower detection efficiency of these other methods.

We have examined all possible candidate neutrino capture (inverse electron capture) reactions, imposing the requirements of (i) low energy threshold, < 0.4 Mev, to gain sensitivity to the intense neutrinos produced in the proton-proton ($p + p$) reaction (*1*); (ii) adequate and predictable cross section; (iii) very long product half-life, so that the product would have accumulated and have been retained in a suitable target mineral over geologic time, but less than 10^{8} years so that it would not have survived from primordial generation; and (iv) tolerable yields of competing reactions giving the same product in situ. These requirements are satisfied only by the reaction

$$^{205}\mathrm{Tl} + \nu\,(E_\nu \geq 46\ \mathrm{kev}) \rightarrow$$
$$^{205m}\mathrm{Pb}\,(E_{exc} = 2.3\ \mathrm{kev}) + e^-$$

where ν is a neutrino, E_ν is the neutrino energy, E_{exc} is the energy of the excited ^{205m}Pb nucleus, and e^- is an electron. The neutrino capture to the fortuitously low-lying excited state, ^{205m}Pb, is followed by its fast isomeric decay to the ^{205}Pb ground state, whose 16-million-year half-life for $L + M + N$ . . . electron capture effectively traps the neutrino capture event. Direct neutrino capture to the ^{205}Pb ground state is relatively insignificant (~ 10^{-5} of that to the ^{205m}Pb excited state).

The cross-section calculations (*7*) for neutrino capture are based on the log *ft* value (5.3 ± 0.1) for the inverse reaction, orbital electron capture decay of the ^{205m}Pb state. (Log *ft*, where *f* is a function of decay energy and *t* is the half-life, is a measure of the probability of electron capture, and, hence, of neutrino capture.) This decay has an unmeasurably small branching ratio (10^{-13}); its log *ft* value is reliably estimated by analogy with all three neighboring transitions connecting the same single particle states with log *ft* values of 5.1 or 5.2. The predicted neutrino capture rate is 430 ± 100 solar neutrino units [SNU (*1*)] (~ 85 percent arising from the low-energy $p + p$ neutrinos) compared to 5.6 SNU predicted for ^{37}Cl.

In the most available of the very rare occurrences of known thallium crystalline minerals, lorandite ($TlAsS_2$), of estimated age ~ 10 million years, we predict a trapped ^{205}Pb concentration of 132 atoms per gram of lorandite, arising from solar neutrino capture. We are engaged in a detailed study of the feasibility of a solar neutrino flux measurement using a few kilograms of lorandite per sample. Our current considerations on procedure follow.

The methods considered for assay of the ^{205}Pb impose severe tolerance limits on the natural lead abundance in the sample, inasmuch as the tiny amount of ^{205}Pb must be distinguishable from the macroscopic amounts of the stable lead isotopes, masses 204, 206, 207, and 208, that will surely be present. The best anticipated ability of a mass spectrometer to detect mass 205 in the presence of the scattering tail from ~ 10^{10} times as much mass 206 translates into a maximum allowable lead impurity in the mineral of a few parts per million (ppm); even this limit requires inclusion in the Pb-Tl separation procedure of a stage of isotopic enrichment of mass 205 by 10^{4}, at a cost of 90 percent loss of the ^{205}Pb.

Thallium minerals are very rare; only lorandite appears to be available in amounts exceeding a few grams, and that only at the site of its original discovery (*8*) as a minor constituent in an arsenic sulfide ore deposit in southern Macedonia, Yugoslavia. There it is believed to exist in adequate abundance at a depth of about 120 m (*9*). We have measured the Pb contamination of a museum specimen (*10*) from this source as ~ 3 ppm, usable for the neutrino experiment. The estimated age (*9*) of the ore is 10 million to 15 million years.

A serious question exists as to whether the depth of the ore is adequate to provide sufficient shielding against the high energy cosmic-ray muons. These may produce protons in deep rock which can participate in the ^{205}Tl(p,n)^{205}Pb reaction in the lorandite, giving the major contribution of background ^{205}Pb, just as the muon-generated ^{37}Cl(p,n)^{37}Ar reaction gives the principal background in Davis's experiment even at his target depth of 1480 m. We have crudely estimated the depth of burial of lorandite at which the muon-generated background would be tolerable as 300 m. (We are now measuring these cross sections with Davis in the muon beam at Fermilab.) The estimated 300-m depth requirement is much greater than the 120-m depth of the mine shaft in which the Macedonian deposit was found, and it is not yet known if that deposit extends much deeper. However, we have been advised by consultant geologists (*11*) that the rate of erosion of rock overburden at the site of the deposit is estimated to have been at least 200 to 300 m within at most the last million years, so that lorandite now 120 m deep was 300 to 400 m deep a million years ago, and perhaps even deeper earlier. Thus the known deposit was probably fairly well shielded over most of the last mean life of ^{205}Pb, and so is useful for the experiment. It should be possible to refine the depth requirements with the Fermilab experimental results and studies of past erosion rates at the Macedonian site. Lorandite has also been identified in milligram amounts at mine sites in western states (*12*). We are, of course, interested to learn of other sources of crystalline thallium minerals.

The separation and enrichment procedure for lorandite would proceed roughly as follows. Each 3-kg sample is subjected to a sequence of stages of initial chemical separation in which Tl is removed to a residue of ~ 1 mg; the original Pb content, ~ 10 mg, remains. Electromagnetic separation is then used to enrich mass 205, and finally a series of ultrapure liquid-liquid chromatographic separation stages should provide ~ 60-ng samples each containing $\sim 5 \times 10^4$ atoms of ^{205}Pb, $\sim 2 \times 10^{14}$ atoms of ^{204}Pb, and a ratio ^{205}Tl/^{205}Pb ≈ 10 corresponding to a Tl reduction factor of 10^{17}. If the ^{205}Pb assay can employ laser-excited resonance fluorescence source ionization and/or ion-beam detection in the mass spectrometry, the requirements for Tl stripping are much less severe, perhaps by 10^4. There is a backlog of experience at Argonne in such techniques in transplutonic element chemistry at the level of a few atoms from massive targets (*13*).

The samples are then assayed for absolute ^{205}Pb content to 10 to 20 percent accuracy by mass spectrometry using the isotope dilution method. The Argonne 100-inch double focusing mass spectrometer (*14*) attains more than 90 percent transmission efficiency at a resolution of 5000, sufficient to discriminate against the organic mass-205 peak. With an added second magnetic stage and ion retarding lens the scattered background tail from ^{206}Pb at mass 205 will not exceed 10^{-10} of the peak intensity of mass 206. We need 1 to 10 percent ionization efficiency, which we have attained with a surface ionization filament or gas magnetron source. With the former we can tolerate a ^{205}Tl/^{205}Pb ratio of > 10 because of differential volatility. If we can produce Pb ions at 1 percent efficiency using two-stage laser excited resonance fluorescence, the tolerance level for Tl in the mass spectrometer sample will be raised by many orders of magnitude, with a corresponding easing of the burden on the chemical purification.

Corrections are then made for various sources of "background"—that is, for the estimated ^{205}Pb content of lorandite arising from nuclear reactions of Tl and of Pb, Hg, and Bi impurities induced by alpha particle, proton, and neutron irradiations from natural radioactivities in the mineral's environment and from cosmic-ray muons. In view of the low yield from neutrino capture, each such other reaction is clearly a serious hazard, and each must separately be proved negligible. Preliminary estimates show that each is tolerable.

The age of lorandite mineralization is determined by dating techniques: on the lorandite directly by counting spontaneous-fission tracks, if possible, or by the K-Ar method; on associated minerals in the deposit by these methods and/or by U-Pb measurements; and by standard geological stratification analysis.

From the corrected ^{205}Pb concentration observed, and the measured age of the Tl mineral, one obtains the neutrino flux averaged over that age, or, for older minerals, over a time span of the order of 23 million years, the mean life of ^{205}Pb. A result in reasonable agreement with current predictions for the major low-energy part of the solar neutrino spectrum would be especially interesting to compare with the data from the proposed ^{71}Ga experiment (*1*), which is also sensitive to the low-energy neutrinos, but which gives an essentially current view of the flux because of the 11-day life of the ^{71}Ge product of neutrino capture. This comparison could speak to the validity of recent solar model studies (*4, 5*) that suggest possible gross fluctuations or cyclic behavior in the thermal fusion (and hence neutrino emission) rate, with periods of the order of 1 million to 2 million years. Such oscillations would be severely damped out in respect to light radiation from the photosphere, so evidence of their occurrence would be suppressed. If the long-time average solar energy production is indeed due to hydrogen burning, and if the emitted neutrinos reach the earth, the ^{205}Tl detector should show an average neutrino flux value equal to that of the steady-state models, for the several cycles of possible fluctuation over which it integrates, whereas the ^{71}Ga and ^{37}Cl detectors may now be looking at a trough in the neutrino rate.

Estimation of the uncertainties accruing to this measurement—those of the neutrino absorption cross section, the ^{205}Pb background, and the analysis and age determination—as they appear at present yields an overall uncertainty of perhaps 30 to 40 percent for the neutrino flux evaluation. While this can hardly be regarded as precisely testing the current solar model, what is now sorely needed is a resolution of the dilemma as to whether solar neutrinos are indeed reaching the earth, and with an intensity consistent with the model predictions. Should the answer given by this measurement be affirmative, further refinements can yield significantly improved accuracy.

MELVIN S. FREEDMAN
CHARLES M. STEVENS
E. PHILIP HORWITZ
LOUIS H. FUCHS
JEROME L. LERNER
LEONARD S. GOODMAN
WILLIAM J. CHILDS
JAN HESSLER

Argonne National Laboratory, Argonne, Illinois 60439

References and Notes

1. J. N. Bahcall and R. Davis, Jr., *Science* 191, 264 (1976); R. Davis, Jr., personal communication.
2. F. Hoyle, *Astrophys. J. Lett.* 197, 127 (1975).
3. J. N. Bahcall and R. K. Ulrich, *Astrophys. J.* 170, 593 (1971).
4. R. K. Ulrich, *Science* 190, 619 (1975).
5. A. G. W. Cameron, *Rev. Geophys. Space Phys.* 11, 505 (1973).
6. V. Trimble and F. Reines, *Rev. Mod. Phys.* 45, 1 (1973).
7. J. N. Bahcall, *Phys. Rev. B* 135, 137 (1964). We thank J. N. Bahcall for the precision verification of our calculations with his computer programs.
8. J. D. Dana, C. Palache, H. Berman, C. Frondel, *Dana's System of Mineralogy* (Wiley, New York, ed. 7, 1944), vol. 1.
9. T. Ivanov, personal communication.
10. We thank E. Olsen of Chicago's Field Museum of Natural History for the use of these specimens.
11. We are indebted to Professor R. Garrels of Northwestern University for the analysis of topographic features leading to the erosion rate estimate.
12. A. S. Radtke, C. M. Taylor, R. C. Erd, F. W. Dickson, *Econ. Geol.* 69, 121 (1974).
13. E. P. Horwitz and C. A. A. Bloomquist, *J. Inorg. Nucl. Chem.* 37, 425 (1974).
14. C. M. Stevens, J. Terandy, G. Lobell, J. Wolfe, R. Lewis, N. Beyer, in *Advances in Mass Spectrometry*, R. M. Elliott, Ed. (Pergamon, Oxford, 1963), pp. 198–205.

28 June 1976

Inverse β Decay of $^{115}In \rightarrow ^{115}Sn^*$: A New Possibility for Detecting Solar Neutrinos from the Proton-Proton Reaction

R. S. Raghavan
Bell Laboratories, Murray Hill, New Jersey 07974
(Received 8 April 1976)

The basis for a low-threshold, high-efficiency, direct-counting detector for solar neutrinos from the p-p fusion reaction is proposed. The inverse β decay of ^{115}In to the 614-keV excited state of ^{115}Sn ($T_{1/2}=3.26$ μsec) provides a unique delayed-coincidence signature and is estimated to yield a solar-neutrino capture rate of ~750 solar neutrino units, ~85 of which is due to $pp+pep$ neutrinos.

The persisting disparity between theoretical expectations of the solar-neutrino flux and the observational results of the Davis, Harmer, and Hoffman experiment[1] based on the inverse β decay of $^{37}Cl \rightarrow ^{37}Ar$ has intensified a re-examination of questions on the current models of the solar interior as well as those concerning basic physical theory.[2] The predictions of low-energy (~0.4 MeV) solar-neutrino flux from the proton-proton reaction is considered to be independent of astronomical uncertainties and requires only the principle that nuclear fusion is the basic energy-producing mechanism in the sun and that neutrinos are stable.[3] Since the ^{37}Cl experiment is dominantly sensitive only to high-energy (5–10 MeV) neutrinos, predictions of which are more critically dependent on models of the solar interior, the necessity of an experiment which clearly establishes the validity of the physical basis, namely, one which is sensitive mostly to the low-energy pp neutrinos, is becoming increasingly urgent. Although many inverse-β-decay candidates have been considered from this point of view over the years, at present only radiochemical experiments on the cases of $^{71}Ga \rightarrow ^{71}Ge$ or $^{7}Li \rightarrow ^{7}Be$ are judged to be even hopeful.[2] The purpose of this Letter is to propose a new candidate: the inverse β decay of ^{115}In to the excited state at 614 keV in ^{115}Sn ($^{15}Sn^*$) having $T_{1/2}=3.26$ μsec and de-exciting by the emission of two γ rays of energies 116 and 498 keV. The present calculations show that this case is sensitive to the low-energy pp neutrinos and has a very large capture rate, $\varphi\sigma \sim 750$ solar neutrino units (SNU's) [1 SNU

Solar Neutrino Production of Technetium-97 and Technetium-98

G. A. Cowan and W. C. Haxton

Solar Neutrino Production of Technetium-97 and Technetium-98

Abstract. *It may be possible to determine the boron-8 solar neutrino flux, averaged over the past several million years, from the concentration of technetium-98 in molybdenite. The mass spectrometry of this system is greatly simplified by the absence of stable technetium isotopes, and the presence of the fission product technetium-99 provides a monitor of uranium-induced backgrounds. This geochemical experiment could provide the first test of nonstandard solar models that suggest a relation between the chlorine-37 solar neutrino puzzle and the recent ice age.*

A serious discrepancy exists between the solar neutrino capture rate measured in the ^{37}Cl experiment (*1*) and that predicted by the standard solar and weak interaction models (*2*). If, as many have suggested, this discrepancy is due to a misunderstanding of the physics of the solar interior, the implications for present theories of stellar evolution could be profound (*3*). Alternatively, if the sun does produce the expected neutrino flux, then some mechanism must be altering the character of these neutrinos before they reach the earth. This suggestion now seems particularly interesting in view of recent evidence for massive neutrinos (*4*) and neutrino oscillations (*5*).

These two classes of solutions to the ^{37}Cl puzzle can be distinguished. Proposed modifications of the standard solar model to accommodate the ^{37}Cl capture rate result primarily in a reduced flux of high-energy ^{8}B neutrinos, whose production depends most critically on the central temperature of the sun. Neutrino oscillations or decay would, except under unusual circumstances, affect all components of the solar neutrino flux equally. Thus there has been great interest in mounting new experiments to complete the spectroscopy of the neutrino sources shown in Table 1. Among those attracting serious attention are radiochemical experiments employing ^{71}Ga (*6*) and ^{81}Br (*7*) targets and a counter experiment with ^{115}In (*8*).

Several interesting proposals for geochemical experiments have also been made. These suggest measuring the concentrations of certain long-lived isotopes produced in natural ore bodies and salt deposits by solar neutrinos. The advantage of this approach over the laboratory experiments is that much larger concentrations of the daughter atoms can be produced over geologic times. The disadvantage is that one must be able to tolerate the backgrounds found in nature. Three isotopes that have been discussed are ^{205}Pb (half-life, $\tau_{1/2}$ = 14 million years) (*9*), ^{81}Kr ($\tau_{1/2}$ = 0.21 million years) (*7*), and ^{41}Ca ($\tau_{1/2}$ = 0.10 million years) (*10*), which are produced by the neutrino reactions on ^{205}Tl, ^{81}Br, and ^{41}K, respectively. At the moment, unresolved problems complicate each of these proposals. The capture cross section for ^{205}Tl is uncertain (*11*), and no ^{81}Br source has been found for which backgrounds from natural radioactivity and cosmic rays are acceptable (*12*). The technology for discriminating ^{41}Ca from ^{40}Ca at the necessary sensitivity has not yet been demonstrated (*13*) and the background problem is very troublesome.

In this report we discuss another geochemical possibility, the production of ^{98}Tc ($\tau_{1/2}$ = 4.2 million years) from ^{98}Mo (24.1 percent). This technetium isotope is of great interest not only in the context of conventional solar neutrino measurements, but also as a unique test of the steady-state sun hypothesis implicit in the standard solar model. The isotope ^{97}Tc ($\tau_{1/2}$ = 2.6 million years) is also produced by neutrinos but will be masked by the neutron capture product in ^{96}Ru unless the ruthenium content in the molybdenum ore is extremely low.

The threshold for the ^{98}Mo(ν,e)^{98}Tc reaction is 1.68 MeV. However, transitions from the 0$^+$ ground state of ^{98}Mo to the (6)$^+$ ground state and (4,5)$^+$ (23 keV) first excited state in ^{98}Tc are strongly hindered. Thus the effective threshold for neutrino excitation is $\gtrsim$1.74 MeV and only high-energy ^{8}B neutrinos can produce ^{98}Tc. The ^{97}Mo(ν,e)^{97}Tc reaction has a lower threshold, 0.32 MeV, and so in principle can be induced by the main component of the solar neutrino flux, the pp neutrinos. However, the ground state transition (5/2$^+$ → 9/2$^+$) is again strongly hindered [log(ft) ≈ 13]. The second most plentiful neutrinos, from electron capture on ^{7}Be, can induce the Gamow-Teller (GT) transitions to the 7/2$^+$ (216 keV) and 5/2$^+$ (324 keV) excited states in ^{97}Tc, and of course the ^{8}B neutrinos can excite many levels.

Quantitative estimates of capture rates can be made by folding the GT and Fermi strength distributions in ^{97}Tc and ^{98}Tc with the various solar reaction neutrino spectra of Table 1. The model-independent Fermi contribution to the capture rate is relatively weak, as only those few ^{8}B neutrinos with energies $\gtrsim$ 11.4 MeV can excite the analog states. As these states decay by isospin-forbidden neutron emission (*14*) and as ^{99}Mo is unstable, only the Fermi transition in ^{98}Mo yields a product of interest, ^{97}Tc. The production rate, based on the standard model ^{8}B neutrino flux of Bahcall *et al.* (*2*), is 0.5 solar neutrino unit (1 SNU = 10^{-36} capture per target atom per second). However, we bear in mind that the ^{37}Cl experiment proves that today the ^{8}B flux is considerably weaker

Table 1. Neutrino sources and fluxes. Reactions (1) to (4) produce solar neutrinos with continuous distributions, while reactions (5) and (6) are line sources. E_ν^{max} is the maximum energy of the neutrinos for all reactions except (4), where it has been computed with respect to the center of the broad 2.9-MeV ^{8}Be resonance populated in the beta decay of ^{8}B. Fluxes are taken from the standard solar model calculation of Bahcall *et al.* (*2*).

Reaction	E_ν^{max} (MeV)	Flux (10^{10} cm^{-2} sec^{-1})
(1) $p + p \rightarrow {}^2H + e^+ + \nu$	0.420	6.1
(2) ${}^{13}N \rightarrow {}^{13}C + e^+ + \nu$	1.199	4.6×10^{-2}
(3) ${}^{15}O \rightarrow {}^{15}N + e^+ + \nu$	1.732	3.7×10^{-2}
(4) ${}^{8}B \rightarrow {}^{8}Be + e^+ + \nu$	14.02	5.85×10^{-4}
(5) ${}^{7}Be + e^- \rightarrow {}^{7}Li + \nu$	0.862 (89.6 percent) 0.384 (10.4 percent)	4.1×10^{-1}
(6) $p + e^- + p \rightarrow {}^2H + \nu$	1.442	1.5×10^{-2}

Feasibility of a $^{81}Br(\nu,e^-)^{81}Kr$ Solar Neutrino Experiment

G. S. Hurst, C. H. Chen, and S. D. Kramer
Chemical Physics Section, Oak Ridge National Laboratory, Oak Ridge, Tennessee 37830

and

B. T. Cleveland, R. Davis, Jr., and R. K. Rowley
Brookhaven National Laboratory, Upton, New York 11973

and

Fletcher Gabbard
Department of Physics, University of Kentucky, Lexington, Kentucky 40506

and

F. J. Schima
National Bureau of Standards, Washington, D.C. 20234
(Received 4 June 1984)

A solar neutrino experiment utilizing the interaction of $^{81}Br(\nu,e^-)^{81}Kr$ to study the ^{7}Be neutrino source in the interior of the sun is shown to be feasible. Resonance ionization spectroscopy was used to count less than 1000 atoms of 2×10^5-yr ^{81}Kr, making the bromine experiment possible. Except for the method of counting product atoms, the bromine experiment would be very similar to the successful chlorine detector $^{37}Cl(\nu,e^-)^{37}Ar$, and thus it is a natural sequel to the only solar neutrino experiment to date.

PACS numbers: 96.60.Kx, 07.75.+h, 14.60.Gh

The flux of neutrinos on the Earth is believed to be due to several sources in the interior of the sun. Fluxes predicted by the standard solar model of Bahcall *et al.*[1] are given in column 2 of Table I. Mechanisms of energy generation within the sun could be clarified by measuring each of these neutrino flux components.

The only solar neutrino experiment ever undertaken, based on the reaction $^{37}Cl(\nu,e^-)^{37}Ar$ and using a volume of 380 m^3 of C_2Cl_4 at a depth of 410 kg/cm^2 in the Homestake Mine, is primarily sensitive to the weak ^{8}B neutrino source. The capture rate (the product of flux and cross section[2]) predicted by the standard solar model is 7.6 solar neutrino units (SNU), in disagreement with the value from the chlorine experiment[3] of 1.9 ± 0.3 SNU above estimated backgrounds. The proposed ^{71}Ga experiment[4] is most sensitive to the *p-p* neutrinos. This Letter considers use of the reaction $^{81}Br(\nu,e^-)^{81}Kr$, with an energy threshold of 470

TABLE I. Calculated ^{81}Br neutrino capture rates for the major solar nuclear reactions. 1 SNU (solar neutrino unit) equals 10^{-36} capture per target atom per second.

Source and energy (MeV)	Neutrino flux ($10^6/cm^2\cdot s$) Standard model	Consistent model	^{81}Br cross section (10^{-46} cm^2) Bahcall	Haxton	^{81}Br capture rate (SNU) Standard model	Consistent model
p-p (0–0.42)	61 000	64 000	0	0	0	0
pep (1.44)	150	150	78	85	1.3	1.3
^{7}Be (0.862)	4300	1800	25	27	11.8	4.9
^{8}B (0–14)	5.6	1.1	1700	5400	3.0	0.6
^{13}N (0–1.20)	500	160	19	20	1.0	0.3
^{16}O (0–1.73)	400	130	41	44	1.7	0.6
					18.8	7.7

Detection of Solar Neutrinos in Superfluid Helium

R. E. Lanou, H. J. Maris, and G. M. Seidel
Department of Physics, Brown University, Providence, Rhode Island 02912
(Received 4 March 1987)

A new method for detecting solar neutrinos and other weakly interacting particles is proposed and described. The detector consists of a large mass of superfluid helium at low temperatures (20 mK). When a neutrino is scattered off an electron, the recoil energy of the electron (10^{-6} to 10^{-7} erg) is deposited in the helium. This small amount of energy can be detected because of the unusual kinetics of rotons at low temperatures. It should be possible to construct a detector of sufficiently low background and large size to measure solar neutrino spectra.

PACS numbers: 96.60.Kx, 14.60.Gh, 29.40.−n, 67.40.Pm

The experiments of Davis and co-workers[1,2] have shown that neutrinos from the sun with energies greater than 0.814 MeV are three times fewer than predicted by the standard solar model (SSM).[3] This discrepancy between theory and experiment is commonly referred to as the solar neutrino problem (SNP). If both the experiment of Davis *et al.* and the standard solar model are correct, then what are the possible explanations of the solar neutrino problem? One interesting mechanism has been proposed by Mikheyev and Smirnov,[4] based upon earlier work by Wolfenstein[5] (MSW effect). They showed that an electron neutrino ν_e generated in the interior of the sun may be adiabatically converted into a muon neutrino ν_μ through interaction with the electron density of the sun via the charged weak current. The MSW effect can account for the observations of Davis *et al.* over a wide range of neutrino masses and mixing parameters.[6] Other possible explanations include neutrino vacuum oscillations,[7] decay of massive neutrinos,[8] the effect of a neutrino magnetic moment,[9] and modification of the solar model due to the capture of weakly interacting dark matter in the sun.[10]

Experiments using neutrinos produced on earth are at present unable to measure the neutrino parameters which enter into the different possible explanations of the SNP. More detailed measurements of the solar neutrino spectrum are needed to resolve the situation. Ideally, one would like to measure the energy spectrum, both for ν_e and ν_μ, either with a very low background, or with means to discriminate neutrino events from other processes. It could be important to have a detector that operates in real time. For example, if the MSW effect occurs, there may be a night-day variation[11] in the neutrino fluxes.

No detector proposed so far has all of these desired properties; rather, a number of different complementary detectors will be needed. Several detection methods have been discussed[12-17] based on a wide variety of techniques, and their utility is being actively pursued. Of these, only one, the radiochemical GALLEX experiment,[12] is under full construction. While each of the proposed detectors has unique and special features, most possess the common difficulty of background from residual radioactivity in the detecting medium itself or in its internal components.

The major part of the ν_e flux comes from the *pp* reaction

$$p+p\rightarrow d+e^{+}+\nu_e. \qquad (1)$$

These ν_e have a continuous spectrum up to 420 keV, and an expected flux at the surface of the earth of 6.1×10^{10} $\mathrm{cm}^{-2}\,\mathrm{sec}^{-1}$. The reaction by which we plan to detect these neutrinos is the elastic-scattering process

$$\nu+e^{-}\rightarrow\nu+e^{-}. \qquad (2)$$

For reaction (1) the maximum recoil energy of the e^- is 260 keV. We propose to use a detector consisting of a large mass of liquid helium-4. The neutrino events will be detected by means of a measurement of the energy deposited in the helium by the recoil electron. As far as radioactive background is concerned, liquid helium is the ideal detector material. At low temperatures ($T<1$ K) all impurities freeze out on the walls of the container. However, a helium detector has the difficulty that the specific heat per unit mass is very large at any reasonable temperature, and so a conventional calorimetric detection of the recoil energy of the electron is not possible. For example, below 0.5 K the specific heat of the helium is $\sim10^5$ times larger than that of crystalline silicon at the same temperature.

We have devised an approach which avoids the problem of the large specific heat. The experiment is shown schematically in Fig. 1. The large mass of superfluid helium is held at a temperature of $\simeq 20$ mK. In neutrino scattering off an atomic electron in the helium, a fraction f of the electron recoil energy E_{rec} is converted within a short time into low-energy elementary excitations of the helium, i.e., phonons and rotons.[18] The dispersion curve for phonons and rotons is shown in Fig. 2. The volume element of phase space goes as p^2dp. Because of the much larger phase space for rotons compared to phonons (see Fig. 2), we will make the approximation that the energy appears exclusively as rotons. The number of ro-

III. Nuclear Fusion Reactions

Introduction: Peter Parker

The role of nuclear reactions and decays in the generation of energy in stars, including our sun, was recognized as early as the 1920's (e.g., [1]) and was developed quantitatively in the work of Gamow[2] on barrier tunneling, followed by the independent work of Weizsächer[3, 4] and Bethe and Critchfield[5] and Bethe[6] working out the energetics and the rates of the p-p chain and the CN cycle.

Given the long age and the large luminosity of the sun, the initial suggestions that nuclear reactions and decays were required to explain the observed energy output were based largely on the inadequacy of classical energy sources. The experimental measurement of the cross section for the $^3He(\alpha,\gamma)^7Be$ reaction by Holmgren and Johnston ([7] and Paper 1.A.III) demonstrated the possibility of terminating the p-p chain via ^7Be, thereby producing high energy neutrinos in the solar core which could be detected in the C_2Cl_4 tanks of Ray Davis (e.g., [8]). This experiment by Holmgren and Johnston, together with its immediate interpretations by Fowler (Paper 2.A.III) and by Cameron (Paper 3.A.III), marks the beginning of the laboratory-nuclear-physics aspects of the solar neutrino problem.

Subsequent experimental and theoretical studies in the 1960s further defined the nuclear-physics aspects of the production and detection of solar neutrinos and supported the proposal (Papers 1.A.I and 2.A.II) to build a large C_2Cl_4 detector. Those studies included refined measurements of the competing $^3He(^3He,2p)^4He$ and $^3He(\alpha,\gamma)^7Be$ reactions (Papers 3.B.III and 8.B.III), the initial measurements of the $^7Be(p,\gamma)^8B$ reaction [9, 10], theoretical studies of the mechanism of direct capture in order to provide accurate extrapolation of the measured cross sections (Paper 4.A.III), and the the realization of the role of the isobaric analog state in ^{37}Ar in providing a superallowed (ν,e^-) transition for high energy neutrinos in the ^{37}Cl detector[11], together with the subsequent experimental observation of ^{37}Cl and its decay to the isospin mirror state in ^{37}K (Papers 2.A.I and 3.A.I).

In the subsequent 30 years, such measurements have been continuously refined, and their analyses and interpretations have steadily improved; examples of some of the most recent and precise experiments are presented in Section C.III. As recently as 15 years ago, the uncertainty in the neutron lifetime (which is used in determining the rate of the $p+p\to d+e^++\nu_e$ reaction) was a major source of error in calculating the solar neutrino fluxes;

however, more recent measurements (both direct and indirect) of the neutron lifetime have now reduced its uncertainty to $< \pm 1\%$ (e.g. [12]), which is no longer significant in determining the solar neutrino flux. The most significant uncertainty in the determination of the flux of high energy neutrinos produced in the sun is still the rate of the $^7Be(p,\gamma)^8B$ reaction, which has an overall uncertainty of ($\pm 15\%$).

The status of these steadily-improving nuclear-physics studies has been regularly reviewed and summarized, beginning with Papers 1.B.I and 4.B.III.

References

[1] A. S. Eddington, Nature, 106, 14 (1920).

[2] G. Gamow, Z. Phys., 52, 510 (1928).

[3] C. F. v. Weizsäcker, Physik. Z., 38, 176 (1937).

[4] C. F. v. Weizsäcker, Physik. Z., 39, 663 (1938).

[5] H. A. Bethe, and C. L. Critchfield, Phys. Rev., 54, 248 (1938).

[6] H. A. Bethe, Phys. Rev., 55, 434 (1939).

[7] H. D. Holmgren, and R. L. Johnston, B.A.P.S., 3, 26 (1958).

[8] R. Davis, Jr., Phys. Rev., 97, 766 (1955).

[9] R. W. Kavanagh, Nucl. Phys., 15, 411 (1960).

[10] P. D. Parker, Phys. Rev., 150, 851 (1966) and ApJ, 153, L85 (1968).

[11] J. N. Bahcall, Phys. Rev., 135, B137 (1964).

[12] K. Hikasa, et al., Phys. Rev. D, 45, S1 (1992).

III. NUCLEAR FUSION REACTIONS - Parker

A. Exploratory Studies of Alternate Terminations

1. "$^3H(\alpha,\gamma)^7$ Li and $^3He(\alpha,\gamma)^7$ Be Reactions," H. D. Holmgren and R. L. Johnston, *Phys. Rev.* **113**, 1556-1559 (1959).

2. "Completion of the Proton-Proton Reaction Chain and the Possibility of Energetic Neutrino Emission by Hot Stars." W. A. Fowler, *Ap. J.* **127**, 551-556 (1958).

3. "Modification of the Proton-Proton Chain," A. G. W. Cameron, *BAPS* **3**, 227 (1958).

4. "γ-Rays from an Extranuclear Direct-Capture Process," R. F. Christy and I. Duck, *Nucl. Phys.* **24**, 89-90, 99-100 (1961).

B. Early Systematic Studies:

1. "Electron Capture and Nuclear Matrix Elements of ^{7}Be," J. N. Bahcall, *Phys. Rev.* **128**, 1297-1301 (1962).

2. "The Reaction D(pγ)^{3}He Below 50 keV," G. M. Griffiths, M. Lal, and C. D. Scarfe, *Can. J. Phys.* **41**, 724, 731-733 (1963).

3. "^{3}He(α,γ)^{7}Be Reaction," P. D. Parker and R. W. Kavanagh, *Phys. Rev.* **131**, 2578-2582 (1963).

4. "Termination of the Proton-Proton Chain in Stellar Interior," P. D. Parker, J. N. Bahcall, and W. A. Fowler, *Ap. J.* **139**, 602, 605 (1964).

5. "The Effect of ^{7}Be Electron Capture on the Solar Neutrino Flux" I. Iben, Jr., K. Kalata and J. Schwartz, *Ap. J.* **150**, 1001-1004 (1967).

6. "The Rate of the Proton-Proton Reaction," J. N. Bahcall and R. M. May, *Ap. J.* **152**, L17-L20 (1968).

7. "The ^{7}Be Electron-Capture Rate," J. N. Bahcall and C. P. Moeller, *Ap. J.* **155**, 511(1969).

8. "^{3}He(^{3}He,2p)^{4}He Total Cross-Section Measurements Below the Coulomb Barrier," M. R. Dwarakanath and H. C. Winkler, *Phys. Rev.* **C4**, 1532-1540 (1971).

9. "Reaction Rates in the Proton-Proton Chain," R. W. Kavanaugh, in *Cosmology, Fusion and Other Matters: George Gamow Memorial Volume*," ed. F. Reines (1972) p. 169, 180-182.

C. Recent Precision Studies:

1. "Low Energy ^{3}He(α,γ)^{7}Be Cross Section Measurements," J. L. Osborne, C. A. Barnes, R. W. Kavanagh, R. M. Kremer, G. J. Mathews, J. L. Zyskind, P. D. Parker, and A. J. Howard, *Phys. Rev. Lett.* **48**, 1664-1666 (1982).

2. "Proton Capture Cross Section of ^{7}Be and the Flux of High Energy Solar Neutrinos," B. W. Fillippone, A. J. Elwyn, C. N. Davids and D. D. Koetke, *Phys. Rev.* **C28**, 2222-2229 (1983).

3. "Astrophysical S(E) Factor on $^{3}He(^{3}He,2p)^{4}He$ at Solar Energies," A. Krauss, H. W. Becker, H. P. Trautvetter and C. Rolfs, *Nucl. Phys.* **A467**, 273-275, 285, 288-290 (1987).

4. "The Fate of ^{7}Be in the Sun," C. W. Johnson, E. Kolbe, S. E. Koonin and K. Langanke, *Ap. J.* **392**, 320-327 (1992).

$H^3(\alpha,\gamma)Li^7$ and $He^3(\alpha,\gamma)Be^7$ Reactions

H. D. HOLMGREN AND R. L. JOHNSTON
Nucleonics Division, United States Naval Research Laboratory, Washington, D. C.
(Received October 14, 1958)

The cross sections for the $H^3(\alpha,\gamma)Li^7$ and $He^3(\alpha,\gamma)Be^7$ reactions have been measured at bombarding energies of 480, 720, 940, 1130, and 1320 kev. The cross sections at the lower bombarding energies may be fitted within the experimental uncertainties by the expressions $\sigma = 0.12(1-0.00051E_\alpha)E_\alpha^{-1}\exp(-125E_\alpha^{-\frac{1}{2}})$ b for the H^3 reaction and $\sigma = 2.8(1-0.00055E_\alpha)E_\alpha^{-1}\exp(-250E_\alpha^{-\frac{1}{2}})$ b for the He^3 reaction, with E_α in kev.

INTRODUCTION

THE $H^3(\alpha,\gamma)Li^7$ and the $He^3(\alpha,\gamma)Be^7$ reactions are of interest in astrophysics and nucleogenesis.[1–3] On the basis of the calculations of Cameron[1] and Fowler[2] the second reaction may be very important in the energy production of many stars during certain periods of their lifetimes. Our sun appears to be in just such a period. In addition they represent a means of bypassing the mass-5 barrier in the production of elements and may account for the anomalous over-abundance of certain light elements in some stars.[2]

The *pp*-reaction chains in hydrogen-burning stars which contain sufficient quantities of He^4 may proceed in the following ways[1,2]:

$$H^1(p,\beta^+\nu)H^2(p,\gamma)$$

$$He^3(He^3,2p)He^4, \quad (1)$$

$$\searrow (\alpha,\gamma)Be^7(e^-,\nu)Li^7(p,\alpha)He^4, \quad (2)$$

$$\searrow (p,\gamma)B^8(\beta^+\nu)Be^{8*}(\alpha)He^4, \quad (3)$$

where the following amounts of energy are released on the average, exclusive of the neutrino energy losses, in the conversion of 4 protons into an α particle: (1) 26.2 Mev, (2) 25.6 Mev, and (3) 19.1 Mev. The relative importance of each chain will depend upon the temperature, the densities of the constituents, the ratio of the cross section for the $He^3(He^3,2p)He^4$ reaction to that for the $He^3(\alpha,\gamma)Be^7$ reaction, and the ratio of the $Be^7(e^-,\nu)Li^7$ cross section to the $Be^7(p,\gamma)B^8$ cross section.

Due to the astrophysical interest an attempt was made to measure the cross sections for these reactions at as low a bombarding energy as possible. The reactions were found to proceed by nonresonant capture, thus making it possible to extrapolate the cross sections to very low energies with reasonable accuracy.

[1] A. G. W. Cameron, Bull. Am. Phys. Soc. Ser. II, 3, 227 (1958), and private communication.
[2] W. A. Fowler, Astrophys. J. 127, 551 (1958), and private communication.
[3] A. G. W. Cameron, Atomic Energy of Canada, Limited, Report AECL 454 (unpublished).

PROCEDURE

One of the 2-Mv Van de Graaff accelerators of the Naval Research Laboratory was used as a source of singly-ionized α particles[4] for these studies.

The H^3 and He^3 gases were contained in a small gas-target chamber, shown in Fig. 1. The beam of α particles entered the target chamber through a 25-μin. nickel window and were stopped at the other end of the chamber by a gold disk. The target chamber was filled with 94% He^3 to a pressure of 85 mm of Hg in order to study the $He^3(\alpha,\gamma)Be^7$ reaction and with 99% H^3 to a pressure of 20 mm of Hg for the $H^3(\alpha,\gamma)Li^7$ reaction. The number of α particles incident upon the gas was determined by measuring the charge collected in the chamber.

At the highest bombarding energy, 1320 kev, the yield of γ rays from these reactions was measured with a 3-in. diam by 3-in. long NaI (Tl) crystal placed at 90° to the axis of the beam and at a distance of 1 in. from the axis of the target chamber. The absolute cross sections were determined from the calculated efficiencies[5,6] for NaI crystals under these conditions. Since the yields decrease very rapidly with bombarding energies, it was necessary to use the 5-in. diam by 3-in. NaI (Tl) well crystal, shown in Fig. 1, in order to determine the yield at lower energies.

Because the yields from these reactions are very low, it was necessary to determine the background under conditions as similar as possible to those under which the actual runs were made. This was accomplished by first taking a background run with the chamber filled with He^4; then a second run was taken for the same length of time and for the same amount of charge with the chamber filled with H^3 or He^3 to the same pressure as the background run. The H^3 or He^3 gas was then recovered; the chamber was again filled with He^4 and another identical background run was made. Two examples of the γ-ray spectra obtained with a 20-channel differential pulse-height analyzer in the above manner are shown in Figs. 2 and 3. Each run took about

[4] α particles were chosen as the bombarding particles rather than He^3 and H^3 ions because they produce fewer γ rays when they strike most other nuclei.
[5] Lazar, Davis, and Bell, Nucleonics 14, 52 (1956).
[6] Wolicki, Jastrow, and Brooks, Naval Research Laboratory Report NRL-4833 (unpublished).

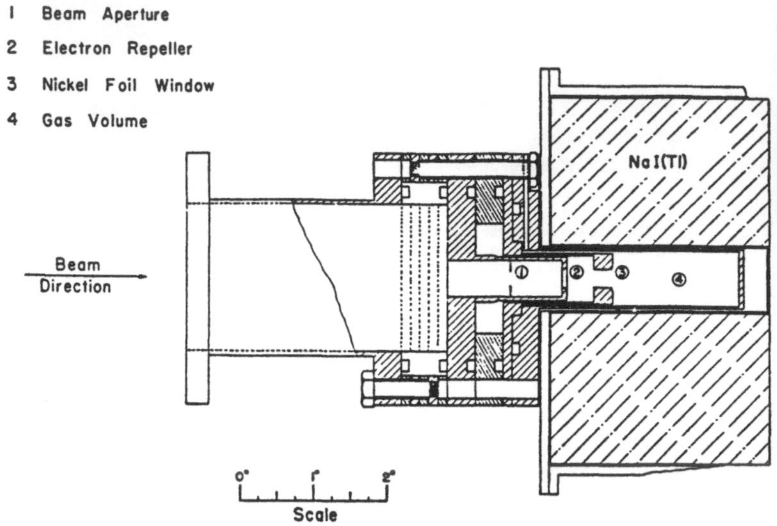

FIG. 1. Schematic diagram of the experimental apparatus.

one hour. The beam currents were limited to 0.2 or 0.3 μa because of the fragility of the nickel window.

In order to determine if the difference in charge of the H^3 and He^3 nuclei would produce any effect on the background due to differences in the Coulomb scattering, the chamber was filled with H^1. It may be seen in Fig. 2 that the background with the chamber filled with H^1 was essentially the same as with the chamber filled with He^4. The background runs were normally taken using He^4 instead of H^1 throughout the H^3 experiment in order to reduce the dilution of the H^3. The backgrounds obtained with He^4 in the chamber were not significantly different from that obtained with the beam off except at the highest bombarding energy. This increase in background was probably due to the interactions in the NaI crystal of neutrons from $C^{13}(\alpha,n)O^{16}$ reactions.[7]

The energy of the γ rays corresponding to the ground-state transition was found to increase continuously with the bombarding energy in agreement with the

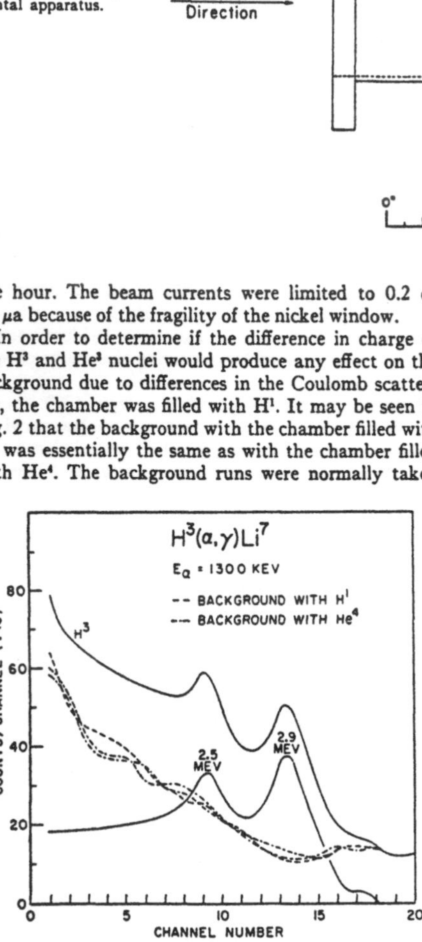

FIG. 2. γ-ray spectra observed with H^3, H^1, and He^4 as target gases. Lower solid curve is with the average background subtracted from the gross spectrum. Of importance is the similarity of background with H^1 and He^4 as target gases. Data points are omitted for clarity.

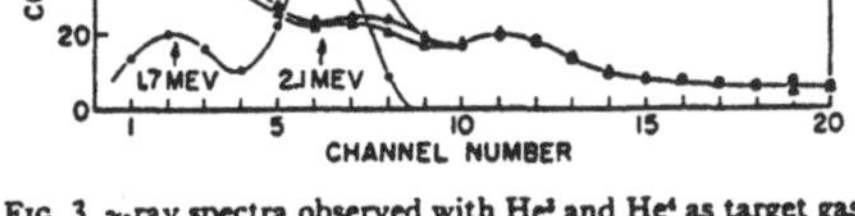

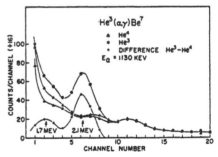

FIG. 3. γ-ray spectra observed with He^3 and He^4 as target gases. Lower curve is with the average background subtracted from spectrum with He^3 as target gas.

[7] The 1% of C^{13} in the very thin layers of normal C which accumulated upon the collimating apertures, window, and beam stopper would probably be sufficient to account for this observed increase in the background. Since this background was subtracted out in the procedure described above, it did not affect the data.

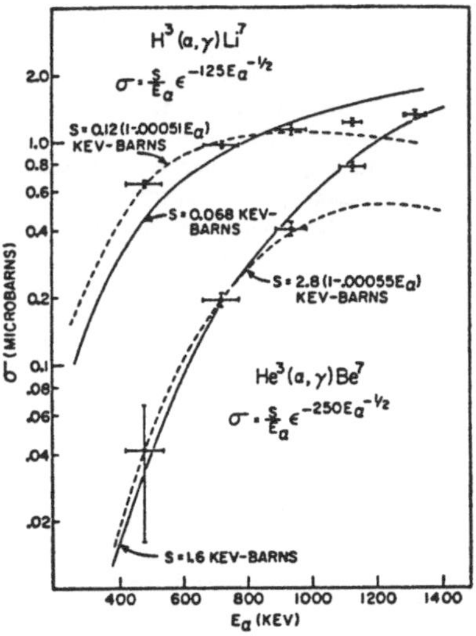

FIG. 4. The total cross sections for the $H^3(\alpha,\gamma)Li^7$ and $He^3(\alpha,\gamma)Be^7$ reactions as a function of energy. The solid curves are for S, a constant in the expression. The dashed curves are for S energy-dependent, as indicated.

following relationship:

$$E_\gamma = Q + (3/7)E_\alpha,$$

where Q is 1.583 Mev for the $He^3(\alpha,\gamma)Be^7$ reaction and 2.465 Mev for the $H^3(\alpha,\gamma)Li^7$ reaction. This continuous change in γ-ray energy confirmed the identification of these γ rays with the reactions studied.

In order to determine the mean energy of the α particles in the reaction chamber, a thick Li^7 target was placed at the center of the chamber. The chamber was then filled to the normal pressure of He^4 used in taking the background runs. (The stopping powers of He and H do not differ sufficiently so that a correction need be made for this effect.) The shifts in the positions of the 400-, 820-, and 960-kev $Li^7(\alpha,\gamma)B^{11}$ resonances were then measured, as well as their apparent widths. These three resonances provided an energy calibration curve by which the mean energy of the α particles in the chamber could be determined to an accuracy of about 20 kev. In addition, the observed width of the resonances gave an indication of the straggling of the beam of α particles. The α particles were found to lose on an average of 720 kev before reaching the center of the chamber (most of this loss was due to the Ni window) and the beam energy spread was about 120 kev at lowest bombarding energy (1200 kev incident on the window). The energy spread was only about 70 kev at the higher resonances.

Due to this observed large straggling at low bombarding energy, the measurements were limited to a lower

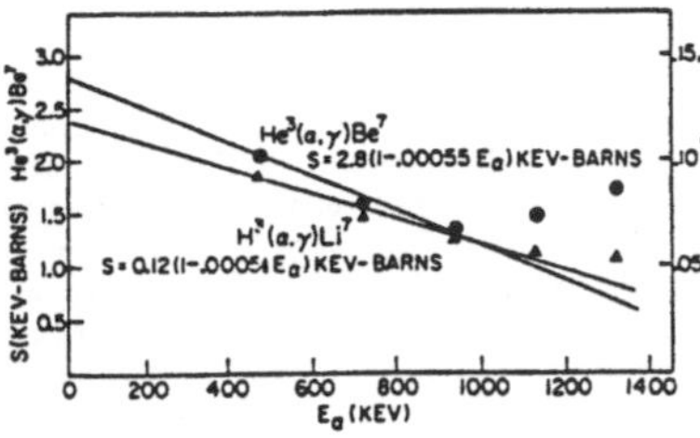

FIG. 5. The experimental values of the cross-section factor, S, as a function of the bombarding energy for the $H^3(\alpha,\gamma)Li^7$ and $He^3(\alpha,\gamma)Be^7$ reactions.

bombarding energy of 480 kev. In addition, at this bombarding energy it was difficult to identify the γ rays originating from the reaction due to the rapid decrease in the cross section with energy.

RESULTS

The cross sections for the $H^3(\alpha,\gamma)Li^7$ and $He^3(\alpha,\gamma)Be^7$ reactions are shown in Fig. 4. The uncertainties indicated in the values of the cross sections are due primarily to the statistical uncertainties and thus indicate only the relative uncertainties of the data. The absolute uncertainties of the data which arise from the uncertainties in the gas pressure, gas purity, solid angle, and crystal efficiency are estimated to be of the order of 30%.

The horizontal bars indicate the spread of the beam energy in the target chamber due to the straggling caused by the nickel window and the energy loss in the gas.

The curves in Fig. 4 represent two attempts to fit each set of data with an expression of the form[8-10]

$$\sigma = \frac{S}{E_\alpha}\exp\left(\frac{2\pi e^2 Z_1 Z_2}{\hbar v}\right),$$

where v is the velocity of relative motion of the two particles (E_α is measured in kev throughout this paper). This expression is based on an approximation to the barrier penetration calculation for a nonresonant capture cross section. A theoretical expression may be obtained for S of the following form:

$$S = \tfrac{1}{2}\pi R^2\left(\frac{M_1+M_2}{M_2}\right)\Gamma \exp[(32R/a)^{\frac{1}{2}}],$$

where

$$a = \hbar^2/(MZ_1Z_2e^2) = 4.25\times10^{-13}\text{ cm},$$

R is the radius of interaction, and Γ is the radiation width for γ rays. Assuming $R \approx 4\times10^{-13}$ cm, we find $S \approx 160\Gamma$ (kev-barns) for the $He^3(\alpha,\gamma)Be^7$ reaction.

---

[8] Burbidge, Fowler, and Hoyle, Revs. Modern Phys. 29, 560 (1957).
[9] H. A. Bethe, Revs. Modern Phys. 2, 186 (1957).
[10] H. A. Bethe, Phys. Rev. 55, 436 (1939).

Salpeter[11] has assumed that Γ is about 0.1 ev, which gives a value of $S=0.016$ kev-barn for this reaction. This theoretical estimate appears to be too small by about a factor of 100. From the present experiment the best experimental estimate of S (for S a constant) is 1.6 kev-barns. On the basis of a more refined expression[8] for S, one expects S to vary with the bombarding energy according to $S=S_0(1-\alpha E_\alpha)$. In Fig. 5, S has been plotted as a function of the bombarding energy (note different ordinates for the two reactions). It may be seen that for both reactions S appears to vary in a linear manner for low bombarding energies and departs from this line at higher energies. In the above expression it has been assumed that only s-wave α particles contribute to the reaction; thus this departure of the cross-section factor from a straight line at higher bombarding energies may be due to the onset of p-wave α-particle contributions. A reasonable fit to the experimental data is obtained when S_0 equals 0.12 kev-barn and $\alpha=0.00051$ kev^{-1} for the $H^3(\alpha,\gamma)Li^7$ reaction, and when $S_0=2.8$ kev-barns and $\alpha=0.00055$ kev^{-1} for the $He^3(\alpha,\gamma)Be^7$ reaction.

On the basis of these estimates of S for the $He^3(\alpha,\gamma)Be^7$ reaction and the calculations of Cameron[1,3] and Fowler,[2] it appears that chain (2) or (3) could compete very favorably with chain (1) at temperatures above 10×10^6 °K.

[11] E. E. Salpeter, Phys. Rev. 88, 552 (1952).

From the spectra of γ rays taken with the 3-in. diam by 3-in. NaI crystal it is estimated that at a bombarding energy of 1320 kev about 50% of the time both reactions proceed through the first excited states, that is the 478 kev state of Li^7 and 431 kev state of Be^7. In the present experiment it was difficult to estimate how the branching ratio changed with energy; since with the 5-in. diam by 3-in. NaI crystal used to measure the yield curves there was a large probability of stopping both γ rays from the cascade.

Riley[12] has found a cross section of 0.2 μb/sterad at 90° for the $H^3(\alpha,\gamma)Li^7$ reaction at 1640 kev and a branching ratio of 5 to 2 for the ground state to first excited state transitions. On the basis of the combined uncertainties this is in reasonable agreement with the present work.

ACKNOWLEDGMENTS

The authors wish to thank Dr. W. A. Fowler of the California Institute of Technology and Dr. A. G. W. Cameron of the Chalk River Project for their helpful and encouraging correspondence. In addition the authors wish to express their gratitude to Dr. M. M. Shapiro for his encouragement of this experiment.

[12] Riley, Warren, and Griffiths, Bull. Am. Phys. Soc. Ser. II, 3, 330 (1958).

COMPLETION OF THE PROTON-PROTON REACTION CHAIN AND THE POSSIBILITY OF ENERGETIC NEUTRINO EMISSION BY HOT STARS*

William A. Fowler
Kellogg Radiation Laboratory, California Institute of Technology
Received February 24, 1958

ABSTRACT

The relatively large value for the cross-section of the $He^3(\alpha, \gamma)Be^7$ reaction found experimentally by Holmgren and Johnston (1958) indicates that this reaction will complete the proton-proton reaction chain in stars which contain helium in comparable amounts to hydrogen and which operate at effective temperatures in excess of 13×10^6 degrees. Modifications to the rates of helium production and of energy generation by the proton-proton chain given by Burbidge, Burbidge, Fowler, and Hoyle (1957) are necessitated by this mode of completion of the chain. Correction factors have been calculated for several values of x_{He}/x_H over an appropriate range of temperatures. The Be^7 produced is consumed by electron capture and by proton capture, and an estimate of the relative rates of these two processes is discussed. In hot stars operating at $>20 \times 10^6$ degrees, the proton capture forming B^8 will probably predominate, and the B^8 decay will result in the emission of energetic neutrinos up to 14.1 Mev in energy. If the proton-capture cross-section for Be^7 is relatively large, it may even be that a substantial flux of such neutrinos is emitted by the sun. The flux at the earth will be at most $\sim 2 \times 10^{+10}$ neutrinos/cm² sec. If the flux is not too small compared to this maximum value, it may be detectable through observations on $Cl^{37}(\nu, \beta^-)A^{37}$, using the techniques developed by Davis (1955).

Holmgren and Johnston (1958), of the Naval Research Laboratory, have successfully detected the reaction $He^3(\alpha,\gamma)Be^7$ and have measured the cross-section σ over the laboratory energy interval from $E_1 = 0.47$ to $E_1 = 1.32$ Mev for alpha particles incident on He^3 gas. Their results yield a low-energy cross-section factor defined by $S_0 \equiv \frac{3}{7} \sigma E_1 \exp(+7.92\, E_1^{-1/2})$, with the value $S_0 \simeq 1.5$ kev-barn in the center-of-mass system when corrections for finite nuclear radius and electron shielding are taken into account. This value is twenty-five hundred times the value, 0.6 ev, given by Cameron (1957) on the basis of estimates for the mean reaction lifetime of He^3 in stars by Salpeter (1952). However, the result is consistent with the large values for S_0 recently found in the non-resonant direct radiative capture of protons by O^{16} in the measurements of Tanner and Pixley (1957) and by Ne^{20} in the measurements of Pixley, Hester, and Lamb (1957).

The large cross-section found for the $He^3(\alpha,\gamma)Be^7$ capture process means that this process will compete successfully with $He^3(He^3,2p)He^4$ in the operation of the *pp*-reaction chain in hydrogen-burning stars which contain sufficient amounts of He^4 and which operate at high enough temperatures. The Be^7 produced in the reaction will be destroyed either by electron or by proton capture, so that the *pp*-chain for converting hydrogen into helium ($4H^1 \rightarrow He^4$) can now be written as follows:

$$H^1(p, \beta^+\nu)D^2(p, \gamma)He^3(He^3, 2p)He^4 \qquad \bar{Q} = 26.2 \text{ Mev (2 per cent loss)}$$

or

$$He^3(\alpha, \gamma)Be^7(e^-, \nu)Li^7(p, \alpha)He^4 \qquad \bar{Q} = 25.6 \text{ Mev (4 per cent loss)}$$

or

$$Be^7(p, \gamma)B^8(\beta^+\nu)Be^{8*}(\alpha)He^4 \qquad \bar{Q} = 19.1 \text{ Mev (29 per cent loss)}.$$

The $\bar{Q}$-values listed are the over-all average energy release in the respective branches exclusive of neutrino energy losses. The total energy release is $Q(4H^1 \rightarrow He^4) = 26.7$ Mev. At low temperatures the *pp*-chain terminates at He^3, and the average energy

* Supported in part by the joint program of the Office of Naval Research and the U.S. Atomic Energy Commission.

release is $\bar{Q}(3H^1 \rightarrow He^3) = 6.68$ Mev. When scaled to the energy release for four protons, this becomes $\frac{4}{3} \times 6.68 = 8.91$ Mev.

In this chain of reactions we have neglected $H^1(e^-,\nu)n(p,\gamma)D^2$, $He^3(e^-,\nu)T^3(p,\gamma)He^4$, and $He^3(p,\beta^+\nu)He^4$. The first two electron-capture processes are important only at very high densities. The last process is the only He^3 reaction at low temperatures, where, however, it is not rapid enough to produce equilibrium with the *pp*-reaction.

As first pointed out by Lauritsen (1951) and Schatzman (1951) and as emphasized by Burbidge, Burbidge, Fowler, and Hoyle (1957),[1] it is the completion of the *pp*-chain through $He^3(He^3,2p)He^4$ which is of key importance in the conversion of hydrogen into helium in the theory of stellar nucleogenesis. Only in this way can a star consisting originally of pure hydrogen produce helium through thermonuclear reactions and thus bring about the first step in nucleogenesis in stars. In cool stars this process will still be the most important, primarily because of the smaller reduced mass and higher relative velocity for $He^3 + He^3$ compared to $He^3 + He^4$ at a given stellar temperature. However, this will no longer necessarily be the case for the sun and hotter stars which contain He^4, either initially or as a result of hydrogen-burning in the star itself.

The comments to follow will serve to supplement the discussion of the *pp*-reaction chain recently given by B²FH(1957). At high temperatures, all of the reaction products except He^4, the end-product, are in equilibrium with the burning hydrogen. This is assured for He^3 by the $He^3(He^3,2p)$ process alone for temperatures greater than 8×10^6 degrees. For lower temperatures the He^3 does not come into equilibrium with the H^1 except in times longer than 4.5×10^9 years, which is taken as a representative time for the age of the solar system. Equilibrium for Be^7 is guaranteed by the electron capture alone under all but exceptional circumstances.

1. Salpeter (1952) and Bosman-Crespin, Fowler, and Humblet (1954) noted that the $He^3(He^3,2p)$ process requires two $H^1(p,\beta^+\nu)$ reactions per He^4 production. The $He^3(\alpha,\gamma)$ leads to He^4 production through only one $H^1(p,\beta^+\nu)$ reaction. Thus the rate of He^4 production is doubled if the $He^3(\alpha,\gamma)$ reaction is very rapid compared to $He^3(He^3, 2p)$. Assuming equilibrium, the reaction rate given by B²FH(1957) must be multiplied, in general, by the temperature-dependent factor,

$$\phi(\alpha) = 1 - \alpha + \alpha\left(1 + \frac{2}{\alpha}\right)^{1/2},$$

where the critical parameter, α, is given by

$$\alpha = \frac{\lambda_{34}^2}{\lambda_{11}\lambda_{33}}\left(\frac{x_{He}}{4x_H}\right)^2 = \frac{S_{34}^2}{S_{11}S_{33}}\left(\frac{x_{He}}{4x_H}\right)^2 \exp(-100T_6^{-1/3})$$

and where T_6 is the temperature in 10^6 degrees Kelvin. The λ's are defined in expressions of the form

$$\frac{1}{A_i}\frac{dx_i}{dt} = \frac{1}{A_j}\frac{dx_j}{dt} = -\lambda_{ij}\rho\frac{x_i}{A_i}\frac{x_j}{A_j},$$

where the x_i's are concentrations by mass. In numerical calculations, the following values for the cross-section factors at low energy have been used: $S_{11} = S_0[H^1(p,\beta^+\nu)] = 3 \times 10^{-22}$ kev-barn; $S_{33} = S_0[He^3(He^3,2p)] = 1500$ kev-barns; $S_{34} = S_0[He^3(\alpha,\gamma)] = 1.5$ kev-barns. Corrections for electron shielding and finite nuclear radius have been included. It will be noted that the new results for S_{34} change α by a factor of the order of 10^7!

The variation of $\phi(\alpha)$ with the interaction temperature above $T = 8 \times 10^6$ degrees is shown as the solid curves labeled ***He⁴ production rate factor*** in the lower part of Figure 1. Since $\phi(\alpha)$ also depends on x_{He}/x_H, two cases have been illustrated—one typical of

[1] This reference will be designated "B²FH(1957)" in what follows.

the interior of the "present sun," for which, according to Schwarzschild, Howard, and Härm (1957), $x_{He}/x_H \approx 1$, and the other typical of the "primeval sun," for which $x_{He}/x_H \approx \frac{1}{4}$ throughout. For inhomogeneous models of the sun, x_{He}/x_H increases with increasing temperature so that the production rate factor actually increases more rapidly than shown in the region from $T = 10$ to 15×10^6 degrees.

In order to find the rate of the *pp*-chain in the sun relative to the rate given in B²FH (1957), it is necessary to average $\phi(\alpha)$ over the range of temperatures at which the chain effectively operates. Because of the larger mass of interacting material at lower tempera-

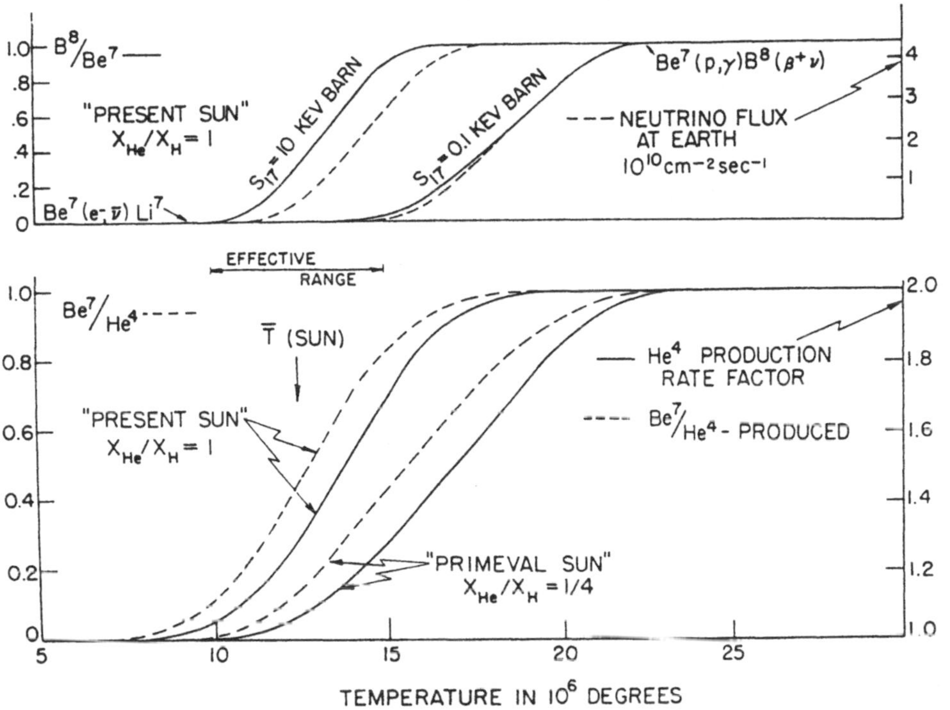

FIG. 1.—*Lower solid curves:* The factor for the increase in the rate of He⁴ production in the *pp*-chain over that given by B²FH(1957) due to new cross-section measurements on He³ (α, γ) Be⁷. This factor is plotted versus temperature for $x_{He}/x = 1$, which is characteristic of the central regions of the "present sun" and for $x_{He}/x_H = \frac{1}{4}$, which was characteristic of the "primeval sun." *Lower dashed curves:* The fraction of Be⁷ nuclei produced by He³ (α, γ) Be⁷ per total number of He⁴ produced in the *pp*-chain. *Upper solid curves:* The fraction of B⁸ nuclei produced by Be⁷ (p, γ Be⁸) per Be⁷ produced in the *pp*-chain. Curves for $S_{17} = S_0$[Be⁷ (p, γ)] = 0.1 and 10 kev-barn are illustrated. *Upper dashed curves:* The energetic ($E\nu \geq 14.1$ Mev) neutrino flux at the earth from B⁸ ($\beta^+\nu$) decay in the sun as a function of temperature. For these curves the temperature is the effective operating temperature assumed for the *pp*-chain.

tures, this effective range extends from ~10 to 15×10^6 degrees, and an effective average temperature is estimated to be $\bar{T} \sim 12.5 \times 10^6$ degrees. Thus, from Figure 1, the He⁴ production rate in the "present" sun is ~1.3 times that given by B²FH(1957). In the "primeval" sun this factor is ~1.1.

At low temperatures the He³ builds up and interacts slowly, and the He⁴ production is very small. The rate of He⁴ production in a star depends on the age of the star. This is illustrated in the solid curve in the lower part of Figure 2. For the curves below 8×10^6 degrees in Figure 2 we have taken 4.5×10^9 years for the age of the "present" sun.

2. Several interesting points depend on the relative amounts of proton and electron capture by Be^7. In electron capture by Be^7 the average neutrino energy loss is only 0.8 Mev. The average loss in $H^1(p,\beta^+\nu)D^2$ is 0.3 Mev; hence the total average loss in the chain becomes 1.1 Mev, and the effective energy release through Be^7 becomes 26.7 − 1.1 = 25.6 Mev. However, in the proton capture by Be^7, the high-energy positron-neutrino emitter B^8 is formed. This nucleus has been shown by Alvarez (1950) and by Gilbert (1954) to decay to the 2.9-Mev excited state of Be^8, which then disintegrates into two alpha particles. Using the mass of B^8 determined by Dunning, Butler, and Bondelid (1956) from the Q-value of $Li^6(He^3,n)B^8$ it is found that the kinetic energy release is 14.1 Mev. Of this amount, 7.3 Mev is taken, on the average, by the neutrino.

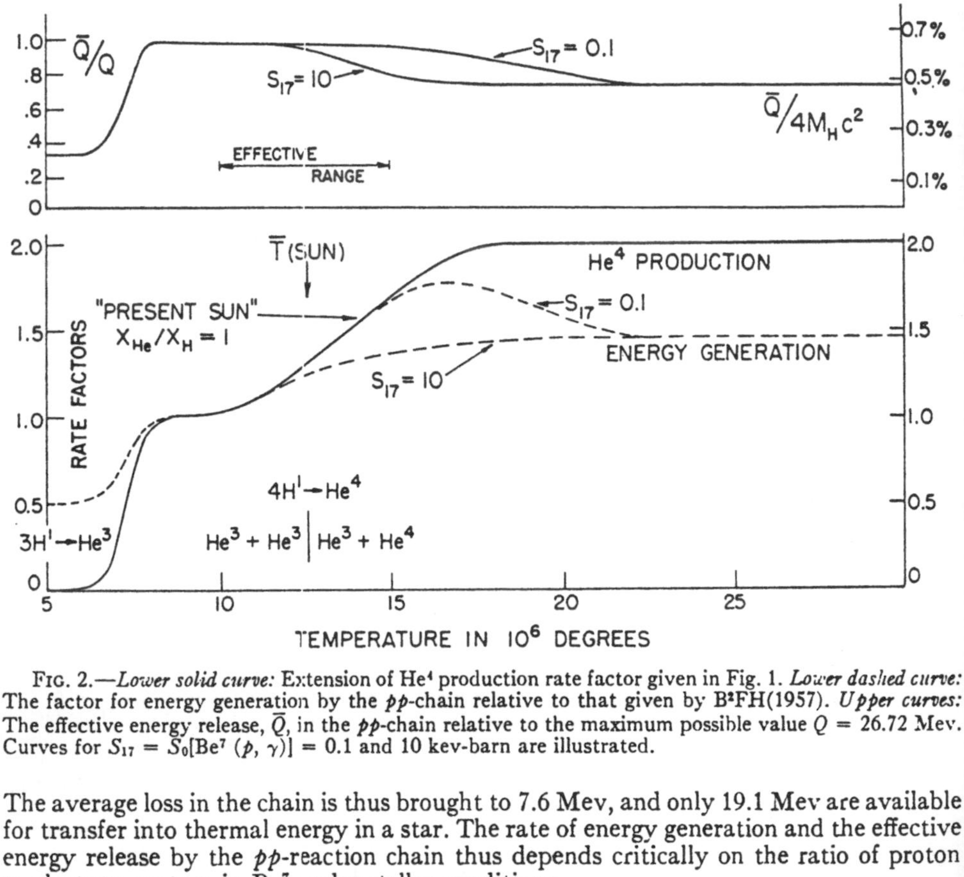

FIG. 2.—*Lower solid curve:* Extension of He^4 production rate factor given in Fig. 1. *Lower dashed curve:* The factor for energy generation by the *pp*-chain relative to that given by $B^2FH(1957)$. *Upper curves:* The effective energy release, $\bar{Q}$, in the *pp*-chain relative to the maximum possible value Q = 26.72 Mev. Curves for $S_{17} = S_0[Be^7(p,\gamma)]$ = 0.1 and 10 kev-barn are illustrated.

The average loss in the chain is thus brought to 7.6 Mev, and only 19.1 Mev are available for transfer into thermal energy in a star. The rate of energy generation and the effective energy release by the *pp*-reaction chain thus depends critically on the ratio of proton to electron capture in Be^7 under stellar conditions.

Bethe (1939) showed that the electron-capture mean lifetime for Be^7 should be about 14 months in the sun. Using present-day values for the K-capture lifetime for Be^7, his results reduce to

$$\tau_e\,(Be^7) = \frac{7 \times 10^4}{\rho\,(1 + x_H)}\ \text{days}\,,$$

where τ_e is the mean lifetime for completely ionized Be^7 in an electron density characteristic of a medium of density ρ and hydrogen concentration x_H. This expression uses the average electron density and neglects the increase over the average density at the Be^7 nucleus due to its positive charge. However, the lifetime calculated by the above expression will be high by at most a factor of 2. The proton-capture lifetime is given approximately by

$$\tau_p\,(Be^7) = \frac{5 \times 10^5}{\rho x_H S_{17}} \left(\frac{12.5}{T_6}\right)^{14} \text{days},$$

where $S_{17} = S_0[Be^7(p,\gamma)]$ is the low-energy cross-section factor for $Be^7(p,\gamma)$. Proton capture by Be^7 has not been measured to date, and in what follows we assume $S_{17} = 0.1$ and 10 kev-barns as representative values covering the range in which this factor may lie. Preliminary estimates made by Christy and Duck (1958) indicate that S_{17} is about 0.2 kev-barn for direct non-resonant capture. The fraction of B^8 nuclei produced per Be^7 nucleus is given by $\tau_e/(\tau_e + \tau_p)$, and this quantity for the present sun with $x_H = 0.5$ is shown as the upper curve in Figure 1. For $S_{17} > 10$ kev-barns, B^8 will be produced a good fraction of the time in the sun, while, for $S_{17} < 0.1$ kev-barn, it will not. For, considerably hotter stars, B^8 will probably always be produced, while for cooler stars the *pp*-chain will proceed through Li^7.

The production of Be^7 and B^7 leads to modifications of the energy-generation rates given by B^2FH(1957). The factor $\epsilon_{pp}/\epsilon_{pp}$ (B^2FH) by which the energy-generation rate given by B^2FH (1957) must be multiplied is included in the lower curves of Figure 2, using the $\bar{Q}$-values given above. It is doubtful whether this factor ever reaches 2, and at high temperature it most probably will approach $2 \times 19.1/26.2 = 1.46$. A rough estimate for an average value over the life-history of the sun would be $\sim$1.2. This factor almost compensates for the 29 per cent decrease from the value given by Salpeter (1952) in the rate of the *pp*-reaction recommended by B^2FH(1957) on the basis of recent determinations of the Gamow-Teller coupling constant in beta decay.

Below 10×10^6 degrees the energy-generation factor decreases because the $He^3(He^3, 2p)He^4$ reaction rate decreases rapidly. The factor will depend on the time scale assumed, just as does the He^4 production rate factor discussed under point 1 above. At very low temperatures the *pp*-chain terminates at He^3, and the factor approaches 0.51, since this is the fraction (6.68/13.1) of the total available energy per *pp*-reaction up to this point.

The effective energy release $\bar{Q}/Q$ in the *pp*-reaction chain is shown as a function of temperature for the "present sun" in the upper curve of Figure 2. The right-hand scale gives $\bar{Q}/4M_Hc^2$, which is the ratio of the energy released to the rest mass energy of four hydrogen atoms. At low temperatures, $\bar{Q}/Q$ is approximately unity for completion of the *pp*-chain through $He^3 + He^3$ or through Li^7, but only 0.7 for completion through B^8. This is important mainly in the calculation of the lifetime of giant systems from the point in the Russell-Hertzsprung diagram where departure from the main sequence is indicated. For stars burning on the main sequence with effective temperatures of 15–20×10^6 degrees, the lifetime on the main sequence will be only 70 per cent of the values previously calculated if $S_{17} \geq 10$ kev-barns. At higher temperatures, the energy generation in general will be due to the CNO-cycle, and these considerations will not be relevant. The temperature at which the CNO-cycle takes over will depend on the carbon-nitrogen-oxygen abundance and on the time scale assumed.

3. If energetic neutrinos ($E_\nu \leq 14.1$ Mev, $\bar{E}_\nu \approx 7.3$ Mev) are released in B^8 decay in the sun, they may be observable at the surface of the earth. The flux at the earth's surface for various effective solar temperatures has been calculated and is shown in the dashed curves in the upper part of Figure 1. The right-hand ordinate should be used. For $S_{17} > 10$ kev-barns, the flux will have about half its maximum possible value, namely, 2×10^{10} neutrinos/cm² sec, while, for $S_{17} < 10$ kev-barns, it will be negligible. In the former case it may be possible to detect this neutrino flux through observations on $Cl^{37}(\nu,\beta^-)A^{37}$, using the techniques developed by Davis (1955, 1956, 1958) at Brook-

haven and Savannah River. The probability that S_{17} exceeds 10 kev-barns is rather low. On the other hand, it is fairly probable that stars hotter than the sun do emit large amounts of energy in the form of energetic neutrinos. The fraction of their total energy release in this form will depend on the ratio of the CNO-cycle rate to that of the *pp*-chain.

4. The new rate for the $He^3(\alpha,\gamma)Be^7$ reaction may make feasible the suggestion by Cameron (1955) that the Li^7 observed in carbon-rich giant stars is produced through the production and decay of Be^7. As will be seen from Figure 1, there is a narrow range of temperature over which Be^7 is produced and not destroyed by proton capture, and this range may well occur in the outer hydrogen-burning regions in giant stars. The Be^7/H ratio by number has a maximum value $\sim 10^{-10}$ over the temperature range 12–18 $\times$ 10^6 degrees. Very rapid convection of the Be^7 to the giant surface with a time scale of not more than 1 year will, of course, still be necessary, as originally emphasized by Cameron.

Many of the above considerations depend on the still unmeasured cross-section for $Be^7(p,\gamma)$ at low energies. An attempt to measure this cross-section is now under way in the Kellogg Radiation Laboratory of the California Institute of Technology. A choice among the possible alternatives discussed above will be possible when these measurements are completed.

I am grateful to H. D. Holmgren for informing me of the results of the $He^3(\alpha,\gamma)$ measurements before publication and to R. Davis, Jr., for information on the status of his experiments on $Cl^{37}(\nu,\beta^-)A^{37}$. It is a pleasure to acknowledge conversations with C. A. Barnes, A. G. W. Cameron, R. F. Christy, J. L. Greenstein, F. Hoyle, R. W. Kavanagh, and D. E. Osterbrock. Modification of the *pp*-reaction chain has also been discussed by Cameron (1958).

REFERENCES

Alvarez, L. 1950, *Phys. Rev.*, **80**, 519.
Bethe, H. A. 1939, *Phys. Rev.*, **55**, 434.
Bosman-Crespin, D., Fowler, W. A., and Humblet, J. 1954, *Bull. Soc. R. Sci., Liège*, **9**, 327.
Burbidge, E. M., Burbidge, G. R., Fowler, W. A., and Hoyle, F. 1957, *Rev. Mod. Phys.*, **29**, 547.
Cameron, A. G. W., 1955, *Ap. J.*, **121**, 144.
———. 1957, Report No. AECL/454.
———. 1958, *Bull. Am. Phys. Soc.*, Ser. II, **3**, 227.
Christy, R. F., and Duck, I. M. 1958, private communication.
Davis, R., Jr. 1955, *Phys. Rev.*, **97**, 766.
———. 1956, *Bull. Am. Phys. Soc.*, Ser. II, **1**, 219.
———. 1958, private communication.
Dunning, K. L., Butler, J. W., and Bondelid, R. O. 1956, *Bull. Am. Phys. Soc.*, Ser. II, **1**, 328.
Gilbert, F. C. 1954, *Phys. Rev.*, **93**, 499.
Holmgren, H. D., and Johnston, R. L. 1958, *Bull. Am. Phys. Soc.*, Ser. II, **3**, 26, and private communication.
Lauritsen, C. C. 1951, quoted by W. A. Fowler, *Phys. Rev.*, **81**, 655 (1951).
Pixley, R. E., Hester, R. E., and Lamb, W. A. S. 1957, *Bull. Am. Phys. Soc.*, Ser. II, **2**, 377.
Salpeter, E. E. 1952, *Phys. Rev.*, **88**, 547.
Schatzman, E. 1951, *C.R.*, **232**, 1740.
Schwarzschild, M., Howard, R., and Härm, R. 1957, *Ap. J.*, **125**, 233.
Tanner, N., and Pixley, R. E. 1957, private communication.

SESSION XA

Saturday Afternoon at 2:00

Shoreham, Main Ballroom

(A. Zucker presiding)

Nuclear Reactions IV, Including Fission

XA1. Modifications of the Proton-Proton Chain. A. G. W. Cameron, *Atomic Energy of Canada Limited*.—The unexpectedly large cross section measured[1] for the $He^3(\alpha,\gamma)Be^7$ reaction implies that this reaction will complete the proton-proton chain in stars with higher temperatures and helium concentrations, supplanting the $He^3(He^3,2p)He^4$ reaction. The energy generation rate is thus doubled because of the more efficient conversion of hydrogen into helium. From recent solar model calculations[2] it appears that the $He^3(\alpha,\gamma)Be^7$ process did not take place in the primordial sun, but now predominates within about the central seven percent of the radius. Be^7 is ionized in the solar interior. The free electron capture rate has been estimated by replacing the wave function for a K electron at the nuclear surface by that for a free electron. This rate appears to be slightly smaller at the solar center than the rate for the $Be^7(p,\gamma)B^8$ process, which has been computed for direct nonresonant capture using a square well. If this is correct, then the very energetic B^8 neutrinos should have a flux at the earth of about 4×10^9 neutrinos cm^{-2} sec^{-1}, which can probably be detected using a 10 000 gal tank of CCl_4 in a mine (R. Davis, private communication). The modified proton-proton chain may produce the large lithium abundances in certain giant stars. Thanks are due W. A. Fowler for bringing the work of reference 1 to the writer's attention before publication.

[1] H. D. Holmgren and R. L. Johnston, Bull. Am. Phys. Soc. Ser. II, 3, 26 (1958), and private communication.
[2] R. L. Sears (private communication).

XA2. Reactions Involving Medium-Light Nuclei.* H. Reeves and E. E. Salpeter, *Cornell University*.—Thermonuclear reactions in very hot stars, consisting mainly of C^{12}, O^{16}, and Ne^{20} are studied. One important reaction is the compound nucleus formation of Mg^{24} in the collision between two C^{12} nuclei, which will occur at temperatures of about 5 to 7×10^8°K. The calculated rates of such reactions depend very strongly on the assumed effective interaction radius for the process. This radius was estimated by extrapolation to low energies of an analysis of laboratory experiments on similar reactions. Calculations were also carried out for $C^{12}+O^{16}$, $O^{16}+O^{16}$, etc. Further experiments are suggested which would greatly improve the accuracy of the calculated rates. Other types of reaction chains, involving successive (α,γ) and (γ,α) reactions are also studied theoretically. Experimental information on the spin and parity of the 5.63-Mev level in Ne^{20} are urgently required for this purpose.

* Supported in part by a joint contract of the U. S. Atomic Energy Commission and the Office of Naval Research.

XA3. Proton Groups from the Reactions $Li^7(Li^7,p)B^{13}$ and $Be^9(Li^6,p)C^{14}$.* C. S. Littlejohn and G. C. Morrison, *University of Chicago* (introduced by S. K. Allison).—The reactions $Li^7(Li^7,p)B^{13}$ and $Be^9(Li^6,p)C^{14}$ are exothermic, with Q values of 5.97 and 15.12-Mev, respectively. By searching for proton groups at 90° in the laboratory system arising from bombardment with 2-Mev Li^+ ions, energy levels in B^{13} and C^{14} may be detected. B^{13} may be explored up to 7.8 Mev of excitation, and C^{14} up to 16.3 Mev. The protons are detected in a scintillation counter gated by a proportional counter and the pulses are displayed on a pulse-height analyzer. Range difference determinations, using standard proton groups for calibration, are used for energy measurements. In the case of B^{13}, no proton groups in the kinetic energy range from 6.40 Mev (arising from the ground state of B^{13}) down to 3.5 Mev are observed, meaning no detectable levels in B^{13} below 2.9 Mev of excitation. Lower energy proton groups would be interfered with by deuterons and tritons from competing reactions and this region is under investigation. In the case of C^{14}, the ground-state group and groups corresponding to the known levels between 6.1 and 8.5 Mev of excitation have been seen. The work of McGruer, Warburton, and Bender on $C^{13}(d,p)C^{14}$ has shown that at their deuteron energy of 14.8 Mev the stripping mechanism predominates, whereas by the lithium beam method the reaction cannot proceed by a single nucleon transfer.

* Research supported by a joint program of the Office of Naval Research and the U. S. Atomic Energy Commission.

XA4. Elastic Scattering of C^{12} from Gold.* H. L. Reynolds and E. Goldberg, *University of California Radiation Laboratory, Livermore*.—The angular distribution of C^{12} ions elastically scattered by gold has been measured at the following laboratory energies: 118, 101, 79.4, and 73.6 Mev. The heavy ions from the Berkeley HILAC entered the scattering chamber through a collimator and were scattered by a gold foil 0.75 mg/cm² thick. Two Ilford El plates recorded scattered ions from 19° to 159°. In all cases the differential cross sections exhibit a Coulomb-like behavior at small angles, a rise above Coulomb of about twenty-five percent as the scattering angle increases, and then a rapid drop below Coulomb in much the same manner as alpha particles scattered from heavy elements in the 20–40 Mev range. Application of the Blair[1] "sharp cutoff" model gives an interaction radius of $11.8\pm0.2\times10^{-13}$ cm at all four energies.

* This work done under the auspices of the U. S. Atomic Energy Commission.
[1] J. S. Blair, Phys. Rev. 95, 1218 (1954).

XA5. Nuclear Reaction Cross Section from the Nitrogen Bombardment of Sulfur. David E. Fisher,* *Oak Ridge National Laboratory*† (introduced by A. Zucker).—Thick targets of ZnS, containing natural sulfur, were bombarded with 28-Mev nitrogen ions in the ORNL 63-inch cyclotron. The compound nucleus resulting from bombardment of sulfur-32 with 28-Mev nitrogen ions is V^{46}, with an excitation energy of 33.3 Mev. Nickel foils of varying thickness were used to degrade the beam energy. Each target was chemically processed to separate the different radioactive reaction products, and the various fractions were beta counted. The following nuclear reactions were detected: (1) $S^{32}(N^{14},p)Ti^{45}$, (2) $S^{32}(N^{14},2p)Sc^{44}$, (3) $S^{32}(N^{14},2pn)Sc^{43}$, (4) $S^{32}(N^{14},2\alpha)K^{38}$, and (5) $S^{32}(N^{14},N^{13})S^{33}$. The yields as a function of incident energy for these reactions were differentiated to obtain excitation functions. Reaction (5) is identified as a transfer reaction. Its cross sections agree well with the systematics of previously studied nitrogen-induced reactions of this type. The cross sections for reactions (1), (2), (3), and (4) are compared with calculations based on the statistical theory of compound nucleus decay.

* ORINS fellow from The University of Florida.
† Operated for U. S. Atomic Energy Commission by Union Carbide Corporation.

XA6. Variation of the Ternary U^{235} Fission Cross Section with Neutron Energy.* E. J. Seppi, *Hanford Laboratories*.—Large differences exist in the results of various experimenters[1] for the value of the ratio fo the neutron induced ternary to binary fission cross section of uranium-235. It has been

γ RAYS FROM AN EXTRANUCLEAR DIRECT CAPTURE PROCESS

R. F. CHRISTY

Institute for Advanced Study, Princeton, New Jersey

and

IAN DUCK

Kellogg Radiation Laboratory, California Institute of Technology, Pasadena, California

Received 2 December 1960

Abstract: Direct electric dipole capture γ-ray transitions are calculated for a number of cases of charged particle capture in nuclei. It is found that when the γ-ray energy is sufficiently low – below about 2 MeV – the capture matrix element is determined by regions external to the usual "nuclear radius". A number of cases of this type are discussed and the calculations compared with experiment. The calculations are extended to the keV region in those cases when the process is of astrophysical interest.

1. Introduction

In many of the light nuclei, the cross section for the (p, γ) reaction shows a smooth variation with proton energy on which are superposed the resonances for the reaction in question. This smooth "background" cross section, the magnitude of which is characteristically $\approx 10^{-30}$ cm^2 at 1 MeV, is the subject of this note. In a number of cases, the angular distribution of the radiation follows the simple $\sin^2\theta$ law which is characteristic of a dipole oscillator oriented along the beam direction. This angular distribution results from an electric dipole p to s transition with no spin flip (i.e. the spin is uncoupled from the orbit). This simple angular distribution, combined with the smooth non-resonant excitation function, has led to the description of the process as direct capture [1]). The process apparently involves a transition from an initial state which does not involve any nuclear interaction to a final bound state. The absence of nuclear interaction in the initial state accounts for the absence of resonance and for the constancy of the spin which is not affected by the electric dipole radiation interaction but only by the nuclear force.

We have found that in a number of cases of this type of transition where the γ-ray energy is less than a few MeV, a straightforward calculation of the matrix element of the electric dipole moment, between an initial (undistorted) Coulomb wave function and a final bound state function of the residual nucleus, shows that the integrand rises to a maximum value at a radius which is in some cases many times the normal nuclear radius. Except at resonance, the irregular Coulomb function, since it strongly interacts with the nucleus, is comparable to the regular function at the nuclear radius. It is therefore negligibly small at such large radii, which justifies its neglect in the calculation. These cases then, when the undisturbed regular Coulomb function dominates the matrix element, will have just those properties ascribed to the direct capture process.

These cases, when the dipole moment is determined entirely by those parts of the initial and final state wave functions which are exterior to what we normally think of as the nucleus, permit a more careful and detailed discussion of the value of the matrix element and of the (p, γ) cross section than is customary since the other nuclear particles and nuclear forces do not enter in such an explicit way.

In what follows we develop detailed expressions for this cross section. Our interest is twofold: On the one hand we have been interested in predicting the cross sections for a number of reactions at stellar energies, on the other hand, we are interested in the light thrown on the nucleus and its surface structure by measurement of these processes coupled with calculations.

2. Theoretical Basis

2.1. GENERAL EXPRESSIONS POR THE CAPTURE CROSS SECTION

We consider the capture of a particle of mass M_1, and charge Z_1 by a target of mass M_2 and charge Z_2, to form a final state of the combined system. If the γ-ray energy is $E_\gamma = \hbar\omega$, the capture cross section for emission of electric dipole and electric quadrupole radiation of polarization m is, respectively,

$$\sigma_{1m} = \frac{16\pi}{9}\left(\frac{E_\gamma}{\hbar c}\right)^3 \frac{1}{\hbar v}|Q_{1m}|^2, \qquad (1)$$

$$\sigma_{2m} = \frac{4\pi}{75}\left(\frac{E_\gamma}{\hbar c}\right)^5 \frac{1}{\hbar v}|Q_{2m}|^2. \qquad (2)$$

The electric moments Q are given by

$$Q_{1m} = \frac{eM_1M_2}{M_1+M_2}\left(\frac{Z_1}{M_1} - \frac{Z_2}{M_2}\right)\int rY^*_{1m}(\theta, \varphi)\Phi_f{}^*(r, \theta, \varphi)\Phi_1(r, \theta, \varphi)\mathrm{d}\tau, \qquad (3)$$

$$Q_{2m} = e\left(\frac{M_1M_2}{M_1+M_2}\right)^2\left(\frac{Z_1}{M_1{}^2} + \frac{Z_2}{M_2{}^2}\right)\int r^2 Y^*_{2m}(\theta, \varphi)\Phi_f{}^*(r, \theta, \varphi)\Phi_1(r, \theta, \varphi)\mathrm{d}\tau, \qquad (4)$$

where r, θ, φ are the relative coordinates of the (centres) of particles 1 and 2. We treat the nuclei involved as single particles since we attempt to evaluate the matrix elements only in the external region where they are separate.

For Φ_1 we use a wave function expressed in channel spin states Ψ_{Im_I} and in Coulomb functions $F_l(kr)$ and $G_l(kr)$ and including the nuclear phase shifts:

$$\begin{aligned}\Phi_1(r, \theta, \varphi) = &\sum_{l_1=0}^{\infty} i^{l_1}\sqrt{4\pi(2l_1+1)}\,\frac{e^{i\eta_{l_1}}}{kr}\sum_{J_1 m_{J_1}}\langle Im_I l_1 0|J_1 m_{J_1}\rangle \\ &\times[F_{l_1}(kr)+(G_{l_1}(kr)+iF_{l_1}(kr))e^{i\delta_{J_1}}\sin\delta_{J_1}]\sum_{m_{l_1} m'_I}\langle l_1 m_{l_1} Im'_I|J_1 m_{J_1}\rangle \qquad (5)\\ &\times Y_{l_1 m_{l_1}}(\theta, \varphi)\Psi_{Im'_I}.\end{aligned}$$

the off resonance intensity is similarly given by $[\int (F \cos \delta_0 + G \sin \delta_0)]^2$ which equals $(\int F)^2$ in the case δ_0 is negligible. In any explicit example, of course, we would refer to the appropriate contributions from the two J values as indicated in sect. 2.

In some cases, as for example $C^{12}+p$, it has been of interest to compare the γ-ray width of the mirror nucleus $C^{13*} \rightarrow C^{13}+\gamma$ with that indicated by the resonance at 1.7 MeV in $C^{12}+p \rightarrow N^{13*} \rightarrow N^{13*}+\gamma$ observed by Woodbury [10]. Now only that part of the matrix element which is largely unaffected by Coulomb effects could legitimately be expected to be similar in the two cases. That is, the G^2 part of the matrix element is the part that may be equal in the two cases. This means that whenever F gives a large contribution, as it does in the examples being discussed, it will be necessary to extract just that part associated with G^2 for the purposes of mirror nucleus comparison. Such an analysis is in principle possible through the fact that $G+iF$ determines the phase of the resonant contribution compared to the non-resonant one when the potential phase shift is zero or is known from the scattering data. It can also be determined by comparing the resonance strength with the non-resonant strength at any particular angle. In this example, $N^{13*} \rightarrow N^{13*}+\gamma$, such an analysis indicates that only about half of the γ-ray width, i.e. about 30 ± 10 MeV out of about 50 ± 10 MeV, can be attributed to G^2 for purposes of comparison with the mirror.

In another example $O^{16}+p \rightarrow F^{17*} \rightarrow F^{17}+\gamma$, the non-resonant γ ray has been observed [11] in the region of the p-wave resonance at 2.66 MeV. This means that not both the matrix elements of F and G vanish. As a consequence we can be sure that the resonance intensity F^2+G^2 cannot vanish. This prediction is in contradiction with the observation of ref. [11] that this transition is not resonant.

4. Discussion

As a result of these calculations, we have found that the exterior contribution to the matrix element in a p–γ reaction will dominate for sufficiently small γ-ray energies and may already be important for γ-ray energies below two MeV. When this is the case, it is possible to calculate the capture cross section in terms of a single unknown parameter which we may call the reduced width of the final state. This unknown is, in fact, just the amplitude for the capture particle in the final state to be found far out in the barrier away from the target nucleus in its ground state.

In certain cases when, either from scattering or other experiments or from theory, we have some a priori knowledge of this reduced width, we can make an absolute calculation of the capture cross section. In other cases we may only be able to predict accurately the energy dependence of the cross section. The results of such absolute calculations are seen, in the results given in sect. 3, to be quite reasonable.

In particular in the case of the capture by O^{16} to form the $S_{\frac{1}{2}}$ state of F^{17}, we would expect that a good calculation would be possible since this state is presumed to be well described by O^{16}+a 2s proton. For such a case, where we can set $\int u^2(r)dr = 1$ as a normalization, we find that $\theta^2_{sp} \approx \frac{1}{2}$. Also, our approximate formulae should be improved when precise calculation is desired. A numerical integration at about 400 keV in this case shows that a correct result is about 1/1.6 times our approximate expression. Consequently the actual agreement with experiment is here very good. This agreement is, of course, somewhat sensitive to the radius r_n where the connection between the exterior function W and the interior function u is made. The calculated cross section is approximately proportional to the square of this radius in this case. We have intentionally chosen a radius significantly larger than the mean nuclear radius of electron scattering experiments in order to arrive more nearly at the radius where the nuclear forces have so decreased that the Coulomb forces become dominant. This radius is not only greater because of the range of the nuclear force but also because of the thickness of the nuclear surface. The value which we have chosen was taken to be $r_n = \frac{1}{2}(A_1^{\frac{1}{3}} + A_2^{\frac{1}{3}})e^2/mc^2 = 1.77\, r_e = 4.98 \times 10^{-13}$ cm $= 1.97(16)^{\frac{1}{3}} \times 10^{-13}$ cm. This is by a factor 1.8 greater than the mean nuclear radius.

At this point, it is convenient to reverse the argument. Once we are convinced that the matrix element is "extranuclear" we may use the cross section measurement at low energy as a direct measurement of the amplitude for finding a proton exterior to the nucleus in the final state. We find in these cases that the Coulomb barrier to the leaking out of such a proton is much smaller than would be obtained with a conventional nuclear radius and can be estimated correctly with a rather enlarged radius as we have done. These measurements can thus be used as a sensitive test of our understanding of the region of the nucleus where the nuclear attractive force changes into the Coulomb repulsion. The most direct way of approaching this problem through these results would be to compute directly from the cross section, the amplitude for finding the proton outside the nucleus. This can then be compared with any detailed theory or measurement of the nuclear surface.

One of us (R. F. C.) would like to thank Professor J. R. Oppenheimer for the hospitality of the Institute for Advanced Study, and to acknowledge the support of the National Science Foundation, which made it possible to write up this work. Support (of I. D.) by an Imperial Oil of Canada Fellowship is also acknowledged.

Electron Capture and Nuclear Matrix Elements of Be^7†

JOHN N. BAHCALL*
Indiana University, Bloomington, Indiana
(Received June 28, 1962)

We calculate the Be^7 electron capture rate from continuum orbits as a function of electron temperature and concentration and consider the effect on the decay rate, for some typical stellar conditions, of the statistical distribution of electrons, the electron-nucleus Coulomb interaction, relativistic and nuclear size corrections, the imperfect overlap between initial and final atomic states, and electron screening in bound decay. We also analyze the experimental information on Be^7 and Li^7 to obtain accurate experimental Gamow-Teller matrix elements for both the ground- and excited-state transitions and compare these matrix elements with the predictions of some simple nuclear models. This comparison supports the conclusion of other authors that near the beginning of the $1p$ nuclear shell L-S coupling is better satisfied than j-j coupling.

I. INTRODUCTION

THE large value for the $He^3(\alpha,\gamma)Be^7$ cross section found experimentally by Holmgren and Johnston[1] led to studies by Fowler[2] and Cameron[3] of the possibility of completing the proton-proton chain with reactions involving Be^7. The latter authors[2,3] showed that the proton-proton chain in the sun is completed more frequently through the $He^3(\alpha,\gamma)Be^7$ reaction than through the $He^3(He^3,2p)He^4$ reaction, in regions of high temperature and high helium-to-hydrogen ratio. They emphasized that the rate of energy generation and the effective energy release by the proton-proton chain depend critically on the ratio of proton-capture lifetime to electron-capture lifetime in Be^7. Fowler[2] and Cameron[3] also showed that the possibility of detecting on earth neutrinos emitted by the sun or other stars depends sensitively on the ratio of the Be^7 proton-capture lifetime to the electron-capture lifetime. Davis and his collaborators[4] are currently performing an experiment to determine if the solar neutrino flux is detectable by the inverse electron-capture process $Cl^{37}(\nu,e^-)Ar^{37}$.

The present work provides an accurate formula for the rate of capture of continuum electrons by Be^7 as a function of electron density and temperature.[5] In deriving this formula, we consider the statistical distribution of the electrons, the electron-nucleus Coulomb interaction, relativistic and nuclear size corrections, the imperfect overlap between initial and final atomic states, and electron screening in bound decay. For temperatures of the order of 2×10^7°K, our results are in good agreement (14%) with a previous calculation by Cameron,[3] but at higher temperatures it is necessary to consider more accurately the electron-nucleus Coulomb interaction.

Using Hartree-Fock atomic wave functions and accurately determined experimental parameters, we analyze existing information to obtain the experimental Gamow-Teller nuclear matrix elements that govern the Be^7 decay. We estimate that the experimental value thus determined for the ground-state (excited-state) transition matrix element is accurate to 10% (15%). We calculate the Gamow-Teller matrix elements on the basis of some simple nuclear models and compare with the experimental values. The results support the conclusion by other authors[6] that near the beginning of the $1p$ nuclear shell L-S coupling is better satisfied than j-j coupling.

II. ELECTRON CAPTURE FROM CONTINUUM ORBITS

The allowed continuum electron capture rate for a single nucleus of charge Z, nuclear energy release W_0, corresponding to one electron in a volume V is[7,8]

$$\lambda=\frac{G_V^2(W_0+W)^2\xi F(Z,W)}{2\pi V},\tag{1}$$

where W is the total energy of the electron and ξ is the usual allowed combination of nuclear matrix elements,[9]

$$\xi\equiv\langle 1\rangle^2+(C_A^2/C_V^2)\langle\sigma\rangle^2.\tag{2}$$

The function $F(Z,W)$ is the ratio of the electron density at the nucleus calculated with a Coulomb distorted wave to the density calculated with a plane wave; G_V is the usual beta-decay coupling constant.

Assuming that the electrons in a star constitute a Fermi gas in which interactions among the electrons can

† Supported by the National Science Foundation.
* Now at California Institute of Technology, Pasadena, California.
1 H. D. Holmgren and R. L. Johnston, Phys. Rev. 113, 1556 (1959).
2 W. A. Fowler, Astrophys. J. 127, 551 (1958).
3 A. G. W. Cameron, Atomic Energy of Canada, Limited Report, CRL-41, 1958 (unpublished), 2nd ed. See also A. G. W. Cameron, Ann. Rev. Nuclear Sci. 8, 299 (1958).
4 R. Davis, Jr. (private communication).
5 At sufficiently low stellar temperatures, not all Be^7 nuclei will be completely ionized and electron capture from bound orbits must be considered. See reference seven for the combined capture rate formula and for some indication of the temperatures at which bound capture becomes important.
6 A. M. Lane, Proc. Phys. Soc. (London) A66, 977 (1953); D. Kurath, Phys. Rev. 101, 216 (1956). See also J. P. Elliott and A. M. Lane, in *Encyclopedia of Physics*, edited by S. Flügge (Springer-Verlag, Berlin, 1957), Vol. 39, p. 241.
7 J. N. Bahcall, Phys. Rev. 126, 1143 (1962).
8 We use units throughout this paper in which $\hbar=m=c=1$.
9 All symbols have their usual meaning. For definitions, see E. J. Konopinski, Ann. Rev. Nuclear Sci. 9, 99 (1959). We use the definition of reduced matrix elements suggested by Konopinski,

$$\langle I'(M')|S_{J_m}{}^+|I(M)\rangle=\langle I'(M')J(m)|I(M)\rangle\langle S_J\rangle.$$

be neglected, the rate of capture of electrons from continuum orbits is[7]

$$\lambda = G_V^2 \xi K / 2\pi^3, \quad (3)$$

where the dimensionless quantity K is given by

$$K \equiv \int_0^\infty dp\, p^2 \frac{F(Z,W)(W_0+W)^2}{1+\exp(-\nu+W/kT)}. \quad (4)$$

If Boltzmann statistics are valid, and if $2\pi\alpha Z\langle v^{-1}\rangle_{av} \gg 1$, $\alpha^2 Z^2 \ll 1$, and $kT \ll mc$,[2] the statistical factor K can be evaluated approximately; the result is[7,10]

$$K_B \cong \pi\alpha Z n_e (2\pi)^{3/2} (kT)^{-1/2} q^2 (1-(15/8)kT) \times[(1+2\mu_0+2\mu_0^2) + (1+x_m/q)^2 (2x_m)^{5/4} e^{-f(x_m)}/(3\alpha Zk^2T^2)^{1/2}], \quad (5a)$$

$$x_m \equiv (2^{-1})[2\pi\alpha ZkT]^{2/3}, \quad (5b)$$

$$f(x) = (x/kT) + \pi\alpha Z(2/x)^{1/2}, \quad (5c)$$

and

$$\mu_0 = kT/q. \quad (5d)$$

The quantity q is the energy of the neutrino emitted in the decay process and is defined by the relation

$$q = W_0 + W \cong W_0 + 1. \quad (6)$$

Equations (5) differ from the corresponding Eqs. (17) in reference 7; everywhere Eqs. (17) in reference 7 contain W_0, Eqs. (5) contain q. Moreover, the error function that appears in reference 7 has been replaced by its asymptotic value for small temperatures and/or large nuclear charge.

Equations (5) were obtained by making use of the approximation[7]

$$F(Z,W) \cong 2\pi\eta[1+\exp(-2\pi\eta)], \quad (7)$$

where

$$\eta \equiv \alpha Z v^{-1}. \quad (8)$$

The term in Eq. (5a) involving x_m arose from the term $\exp(-2\pi\eta)$ in Eq. (7). Thus the term involving x_m is of the order of[11]

$$\exp(-2\pi\alpha Z v^{-1}) \cong \exp[-2\pi\alpha Z(2/\pi kT)^{1/2}]. \quad (9)$$

The central temperature in the sun is less than or of the order of 1.5×10^7 °K,[12] and thus the correction term (9) is about 5% for Be^7 electron capture in the sun.[13] Hence we can omit for Be^7 the term involving x_m in Eq. (5a), although this term must be included in order to obtain accurate results for the electron-capture lifetime of hydrogen or helium, Equation (5a) can be further simplified by observing that kT is less than or of the order of 1 keV for the sun while q is 863(385) keV for the ground-state (excited-state) decay of Be^7. Thus we may neglect kT everywhere with respect to q in Eq. (5a). The dimensionless quantity K can then be written in the form

$$K_B \cong \pi\alpha Z n_e (2\pi)^{3/2} (kT)^{-1/2} q^2, \quad (10)$$

and the continuum electron-capture rate is given simply by

$$\lambda = (2/\pi kT)^{1/2} G_V^2 \alpha Z n_e q^2 \xi. \quad (11)$$

Expression (11) for the allowed electron-capture rate in a nondegenerate Fermi gas of electrons can be obtained heuristically in a simple way by substituting in Eq. (1)

$$n_e = 1/V, \quad (12a)$$

$$F(Z,W) \cong 2\pi\alpha Z\langle v^{-1}\rangle, \quad (12b)$$

and[11]

$$\langle v^{-1}\rangle = (2/\pi kT)^{1/2}. \quad (12c)$$

Formula (11) is then obtained immediately.

III. BOUND ELECTRON CAPTURE

The allowed capture rate for an electron bound in an atomic orbit with quantum numbers[14] n, κ, μ is

$$\lambda_{n,\kappa,\mu} = G_V^2 q^2 \xi |\psi_{n,\kappa}(0)|^2 / 2\pi, \quad (13)$$

where[14,15]

$$|\psi_{n,-1}(0)|^2 = (4\pi)^{-1} g_{n,-1}^2(0), \quad (13a)$$

$$|\psi_{n,+1}(0)|^2 = (4\pi)^{-1} f_{n,-1}^2(0), \quad (13b)$$

The quantity $\Delta b_{n,\kappa}$ is the change in the atomic binding energy when an electron with quantum numbers n, κ is captured; this binding energy change is negligible for Be^7. Only electrons with κ equal to plus or minus one undergo allowed decay; electrons with κ different from one possess orbital angular momentum and hence have their decay rates retarded by centrifugal repulsion. The form of the functions f and g is well known.[14]

Equation (13) neglects the difference between initial and final atomic states due to the change of nuclear charge by one unit. This effect has been studied for Be^7 by Benoist-Gueutal[16] who concludes that the Be^7 capture rate might be decreased by as much as 34% compared to the value given in Eq. (13), due to the imperfect overlap of initial and final atomic states. The estimate of Benoist-Gueutal only indicates the need for further study, since no accurate atomic wave functions were available for the excited states of Li^7. The Li^7

[10] Equations (5) are valid for low temperatures and/or large nuclear charge, since the formulas were obtained by using Eq. (7). The term involving x_m is not given very accurately by Eq. (5); if this term is appreciable, then numerical integration is probably necessary in order to obtain an accurate value for K.

[11] Formula (12c) is a general result for systems obeying Maxwell-Boltzmann statistics.

[12] R. L. Sears, Mém. soc. roy. sci. Liege 3, 479 (1960).

[13] Equation (5a) yields a similar estimate for the correction term.

[14] We use the notation of M. E. Rose, *Relativistic Electron Theory* (John Wiley & Sons Inc., New York, 1961) for relativistic electron wave functions and quantum numbers.

[15] J. N. Bahcall, Phys. Rev. 124, 495 (1961).

[16] P. Benoist-Gueutal, Ann. Phys. (New York) 8, 593 (1953).

atom is usually left in the $1s2s^2$ excited state following Be^7 electron capture.

The present writer has reinvestigated the effect on weak interaction decay rates of a different nuclear charge in initial and final atomic states. Our method differs from the one used by Benoist-Gueutal and has been briefly described in a previous publication.[15] Preliminary unpublished results by the present writer suggest that the imperfect overlap between individual atomic states does not inhibit the total Be^7 decay rate by nearly as much as the 34% upper bound estimated by Benoist-Gueutal.[17] The numerical results presented in this paper ignore the imperfect overlap between initial and final atomic states since the present writer's estimate for the smallness of this effect is not contradicted by the experimental comparisons described in Sec. V of this paper.

In terrestrial experiments, Be^7 decays by allowed electron capture to both the ground and the first excited states of Li^7. We denote by q (q^*), ξ (ξ^*), and λ (λ^*), the ground- (excited-) state neutrino energy, nuclear matrix elements, and transition probability respectively. The laboratory transition probability can then be written,

$$\begin{aligned}\lambda_{\rm lab}&=\lambda+\lambda^*\\&=G_V^2A\,(q^2\xi+q^{*2}\xi^*),\end{aligned}\tag{14}$$

where

$$A=(4\pi^2)^{-1}[g_{1,-1}^2(0)+g_{2,-1}^2(0)].\tag{15}$$

IV. NUMERICAL RESULTS

In this section, we present some numerical results for the Be^7 electron capture rate from continuum orbits under typical stellar conditions. The stellar capture rate can be written as a sum of two terms, analogous to Eq. (14) for the laboratory decay rate, by making use of Eq. (10). Forming the ratio of $\lambda_{\rm star}$ and $\lambda_{\rm lab}$, we obtain

$$\begin{aligned}\lambda_{\rm star}/\lambda_{\rm lab}&=\tau_{\rm lab}/\tau_{\rm star}\\&=A^{-1}(2/\pi kT)^{1/2}\alpha Zn_e.\end{aligned}\tag{16}$$

Since the laboratory half-life, $\tau_{\rm lab}$, is accurately known, the atomic factor A is the only quantity that must still be determined before Eq. (16) can be used. Brysk and Rose[18] have shown that nuclear size effects on the wave functions (13a) and (13b) are negligible for nuclei as small as Be^7. Relativistic effects can easily be estimated by examining the form of g appropriate to a pure Coulomb field. The Dirac $g_{1,-1}$, correct to terms of order α^2Z^2, has the value[19] (at the nuclear radius R)

$$g_{1,-1}(R)\cong2(\alpha Z)^{3/2}e^{-x}\{1+\alpha^2Z^2[(5/4)-\gamma-\ln 2x]\},\tag{17}$$

where

$$x=\alpha ZR,$$

and γ is the Euler-Mascheroni constant. The dependence of $g_{1,-1}(R)$ on the nuclear radius is weak and relativistic corrections, represented by the terms proportional to α^2Z^2, are much less than one percent. Similar results are valid for $g_{2,-1}(R)$.

Electronic screening causes the beryllium atomic wave function to differ appreciably at the origin from the value computed using a pure Coulomb field. Hartree and Hartree[20] computed accurate beryllium radial wave functions using the method of the self-consistent field with exchange; they found (atomic units):

$$g_{1,-1}(0)=14.67,\tag{18a}$$

$$g_{2,-1}(0)=\ 2.67.\tag{18b}$$

The value of $g_{1,-1}(0)$ given above is probably accurate to better than 1%. It differs by much less than 1% from the $g_{1,-1}(0)$ computed by Hartree and Hartree[21] using a self-consistent field without exchange and also agrees to better than 1% with their[21] $g_{1,-1}(0)$ computed, ignoring exchange, for Be^{++}. The values of $g_{1,-1}(0)$ computed by Hartree and Hartree for neutral Be, including exchange, and for Be^{++}, ignoring exchange, agree to within a few tenths of a percent with the value determined by Pekeris[22] for Be^{++} using an elaborate numerical technique.

The electron density calculated from Eq. (18a) is 16% less than the pure Coulomb value,

$$\begin{aligned}[g_{1,-1}^2(0)]_{\rm coulomb}&\equiv4Z^3\\&=256.\end{aligned}\tag{19}$$

The fact that $g_{1,-1}(0)$ is essentially the same for Be and for Be^{++} shows that the 16% decrease in the self-consistent field decay rate compared to the pure Coulomb decay rate is due to the mutual screening of the two $1s$ electrons. The ratio of L to K capture determined solely from Eqs. (18) is 3.3×10^{-2}; this is much less than the pure Coulomb ratio of 12.5×10^{-2}, although it has frequently been stated that for light atoms the L to K capture ratio is given accurately by pure Coulomb wave functions.

The value[8] of A computed from Eqs. (18) is 2.19×10^{-6}; this estimate of A is probably accurate to one percent. The half-life $\tau_{\rm lab}$ is $(4.61\pm0.01)\times10^6$ sec.[23] Substituting these numbers in Eq. (16), we find:

$$\begin{aligned}\tau_{\rm star}^{-1}&\equiv\lambda_{\rm star}/\ln 2\\&=1.02n_eT_6^{-1/2}\times10^{-22}\ {\rm sec}^{-1},\end{aligned}\tag{20}$$

---

[17] It is hoped that a complete account of this work will be available in the near future.

[18] H. Brysk and M. E. Rose, Revs. Modern Phys. **30**, 1169 (1958).

[19] D. Layzer and J. Bahcall, Ann. Phys. (New York) **17**, 177 (1962).

[20] D. R. Hartree and W. Hartree, Proc. Roy. Soc. (London) **A150**, 9 (1935).

[21] D. R. Hartree and W. Hartree, Proc. Roy. Soc. (London) **A149**, 210 (1935).

[22] C. L. Pekeris, Phys. Rev. **112**, 1649 (1958).

[23] F. Ajzenberg-Selove and T. Lauritsen, Nuclear Phys. **11**, 1 (1959).

TABLE I. Experimental parameters.

Quantity	Value[a]	Quantity	Value[a]
q^2	2.85	G_V^2	$(8.82\pm0.12)\times10^{-24}$
$(q^*)^2$	0.568	C_A^2/C_V^2	1.41 ± 0.05
λ^*/λ	0.130 ± 0.013	A	2.19×10^{-6}
$\tau_{1/2}$	$3.58\times10^{+27}$		

[a] We use units in which $\hbar=m=c=1$. Except where shown explicitly, experimental errors are assumed negligible.

where τ_{star} is the half-life for capture of an electron by a Be^7 nucleus, n_e is the number of electrons per cm^3, and T_6 is the temperature in units of 10^6 °K.

Let

$$n_e \equiv \rho/(\mu_e m_p), \tag{21}$$

where ρ is the density of the stellar matter in g/cm^3 and m_p is the mass of the proton. Then,

$$\tau_{star}^{-1} = 6.12\times10^{-9}\rho/\mu_e T_6^{+1/2}\ \text{sec}^{-1}. \tag{22}$$

Equations (20) and (22) are accurate to 1 or 2% if we ignore higher order correction terms in the expansion[24] of $F(Z,W)$ and the imperfect overlap of initial and final atomic states.

Equation (22) is in good agreement with a heuristic estimate given by Cameron.[3] His result differs by only 14% from the value given by Eq. (22), the difference being due almost entirely to the neglect of electron screening in his calculation of A. For temperatures of the order of the central temperature in the sun, i.e., T_6 about 15, Eq. (29) yields a lifetime three times shorter than the order-of-magnitude estimate derived by Bethe and used by Fowler[2] in his estimates of the solar neutrino flux. As a first approximation, Bethe assumed that $F(Z,W)$ was equal to one; it is about three for solar temperatures.

V. NUCLEAR MATRIX ELEMENTS

We can extract experimental values for the nuclear matrix elements involved in Be^7 electron capture by making use of the parameters listed in Table I. The values for q, q^*, λ^*/λ, and $\tau_{1/2}$ were taken from the compilation of Ajzenberg-Selove and Lauritsen.[23] The value of G_V^2 has recently been determined very accurately[25]; our knowledge of C_A/C_V is much less precise.[26] The value of A adopted here has been discussed in Sec. III. Unless explicitly shown in Table 1, it is assumed that experimental uncertainties are unimportant in our considerations.

The experimental nuclear matrix elements can be computed from the following version of formula (13),

$$\begin{aligned}\xi &= \langle 1\rangle^2 + (C_A^2/C_V^2)\langle\sigma\rangle^2 \\ &= \ln 2[G_V^2 q^2 A \tau_{1/2}(1+\lambda^*/\lambda)]^{-1},\end{aligned} \tag{23}$$

and an analogous expression for ξ^*. For mirror nuclei, it is an excellent approximation to assume that $\langle 1\rangle = 1$ and thus Eq. (23) can be used to obtain an experimental value for the reduced Gamow-Teller matrix element $\langle\sigma\rangle$; the Fermi matrix element $\langle 1\rangle^*$ vanishes for the excited state decay. The experimental matrix elements are shown in the second column of Table II; the largest uncertainty arises from the estimated experimental error in the determination of λ^*/λ. The estimated value of excited state Gamow-Teller matrix element $\langle\sigma\rangle^*$ depends strongly upon the branching ratio λ^*/λ; thus $\langle\sigma\rangle^*$ is not as well known as $\langle\sigma\rangle$.

Table II also lists several theoretical predictions of the Gamow-Teller matrix elements. The third column of Table II contains the single-particle j-j coupling values. A number of authors[27] have emphasized that the correct j-j wave functions to use for light nuclei are eigenfunctions of the total angular momentum J and the total isotopic spin T. Thus column four lists values computed with eigenfunctions of the total isotopic spin; the entries in this column are explained in more detail below. The supermultiplet theory of Wigner[28] predicts the same matrix elements as the pure single-particle values; the supermultiplet predictions are shown in column five.

The ground state of Be^7 has a spin of 3/2 and a total isotopic spin of 1/2; the ground-state configuration is assumed to be (in the notation of Mayer and Jensen[29]): $(\pi p_{3/2})^2(\nu p_{3/2})$. The ground state of Li^7 has the same total spin and isotopic spin and an analogous configuration. Mayer and Jensen have constructed the isotopic spin wave function in j-j coupling for the Be^7 and Li^7 ground states.

The first excited state of Li^7 has a total spin and isotopic spin both equal to one-half; the configuration of the excited state may be either $(\nu p_{3/2})^2(\pi p_{3/2})$ or $(p_{3/2})^2(p_{1/2})$, according to the shell model.[29] The Li^{7*} configuration involving $(p_{1/2})$ can be described by two independent wave functions which correspond to the two $p_{3/2}$ nucleons being coupled to an intermediate angular momentum of zero or one. The antisymmetrized eigenfunctions of total spin and isotopic spin can be found by the method of Mayer and Jensen[29] or by direct construc-

[24] For temperatures of the order of 2×10^7 °K, these terms will decrease the predicted lifetime by about 5%, as we have seen in Sec. II. At lower temperatures, the correction terms in the expansion of $F(Z,W)$ are completely negligible. For low densities and high temperatures, such as exist in our sun, electron screening in continuum orbits will be much less than in bound atomic orbits.

[25] R. K. Bardin, C. A. Barnes, W. A. Fowler, and P. A. Seeger, Phys. Rev. 127, 583 (1962); J. W. Butler and R. O. Bondelid, *ibid.* 121, 1770 (1961).

[26] A. Sosnovskii, P. Spivak, Yu. Prokoviev, I. Kutikov, and Yu. Dobrynin, Soviet Phys.—JETP 35, 739 (1959).

[27] See, for example, M. Mizushima and M. Umezawa, Phys. Rev. 85, 37 (1952) and B. H. Flowers, Phil. Mag. 43, 1330 (1952). The first entry in column four of Table II has also been obtained by A. Winther and O. Hofoed-Hansen, Kgl. Danske Videnskab. Selskab, Mat.-fys. Medd. 27, no. 14 (1953). The latter authors pointed out the necessity of using isotopic spin eigenfunctions for the calculation of Gamow-Teller matrix elements.

[28] E. P. Wigner, Phys. Rev. 56, 519 (1939).

[29] M. G. Mayer and J. H. D. Jensen, *Elementary Theory of Nuclear Shell Structure* (John Wiley & Sons, Inc., New York, 1953).

TABLE II. Comparison of experimental and theoretical nuclear matrix elements.

Matrix element	Experimental	Single-particle j-j	Isotopic spin[a] j-j	Supermultiplet theory
$\langle\sigma\rangle^2$	1.48±0.15	1.67	0.90	1.67
$\langle\sigma\rangle^{*2}$	1.42±0.22	1.33	$(p_{3/2})^3$: 0.44 $J_{in}=0(p_{1/2})$: 0.19 $J_{in}=1(p_{1/2})$: 0.12	1.33

[a] Entries in this column are explained in the text.

tion. The excited state Gamow-Teller matrix element was computed using each of the three possible j-j isotopic spin wave functions; the results are listed below.

$$(\nu p_{3/2})^2(\pi p_{3/2}): \quad \langle\sigma\rangle^{*2}=4/9; \tag{24a}$$

$$(p_{3/2})^2{}_{J_{in}=0}(p_{1/2}): \quad \langle\sigma\rangle^{*2}=5/27; \tag{24b}$$

$$(p_{3/2})^2{}_{J_{in}=1}(p_{1/2}): \quad \langle\sigma\rangle^{*2}=12/100. \tag{24c}$$

The configuration label indicates which of the three Li^{7*} j-j eigenfunctions was used; the quantity J_{in} gives the value of the intermediate angular momentum to which the two $p_{3/2}$ nucleons are coupled in the $(p_{3/2})^2(p_{1/2})$ configuration. A similar notation has been used in Table II.

The $p_{1/2}$ configuration yields in both cases a smaller value of $\langle\sigma\rangle^{*2}$ than the $(p_{3/2})^3$ configuration. This is because the orthogonality of the $p_{3/2}$ and $p_{1/2}$ single-particle wave functions permits only parts of the Be^7 and Li^{7*} wave functions to contribute to the matrix element if the Li^{7*} state contains, unlike the Be^7 state, a $p_{1/2}$ single-particle component.

The trend of the numbers in Table II is in agreement with results of extensive intermediate coupling calculations[6] of other nuclear parameters for light nuclei. The intermediate coupling calculations show that L-S coupling is better satisfied than j-j coupling for nuclei near the beginning of the $1p$ nuclear shell.[30] This fact is reflected in Table II by the excellent agreement between the supermultiplet predictions and the experimental values; the isotopic spin j-j eigenfunctions yield matrix elements that are too small. It would be interesting to see how well the intermediate coupling wave functions that have been used to compute other nuclear parameters for mass number seven reproduce the experimental Gamow-Teller matrix elements.

[30] C. W. Kim has confirmed this result by analyzing the ft values of light even-A nuclei (private communication).

If the imperfect overlap between initial and final atomic states caused the atomic factor A to be decreased significantly from the value adopted in Sec. III, the theoretical matrix elements listed in Table II would appear to be too small when compared with the experimental values. However, the agreement between theory and experiment is similar to the agreement found for nuclear magnetic moments,[27,29] and for magnetic moments there is, of course, no atomic overlap correction. We conclude that there is no evidence for a large decrease in the total electron capture probability due to an imperfect overlap between initial and final atomic states. An intermediate coupling calculation with the same wave functions that were used to calculate the Be^7 and Li^7 magnetic moments would allow one to place quantitative limits on the decrease in A, and thus the total electron capture probability, due to the imperfect atomic overlap.

ACKNOWLEDGMENTS

I am grateful to Professor M. H. Ross for many instructive conversations. I appreciate helpful comments from C. W. Kim, M. P. Klein, Professor E. J. Konopinski, and Professor H. J. Martin. This work was prompted by a letter from Dr. R. Davis, Jr. pointing out the importance of knowing accurately the Be^7 electron capture rate for the sun.

Note added in proof. R. W. Kavanagh [Nuclear Phys. 15, 411 (1960)] measured the $Be^7(p,\gamma)B^8$ cross section and W. A. Fowler [Mem. Soc. Roy. Liege 3, 207 (1960), Ser. 5] used Kavanagh's results to show that proton capture by Be^7 is rare in the sun.

THE REACTION $D(p, \gamma)He^3$ BELOW 50 KEV

G. M. Griffiths, M. Lal, and C. D. Scarfe
Physics Department, University of British Columbia, Vancouver, British Columbia
Received February 4, 1963

ABSTRACT

The yield and angular distribution of the γ rays from the reaction $D(p, \gamma)He^3$ have been measured with thick heavy-ice targets in the energy range from 24 kev to 48 kev. Assuming a simplified energy dependence the results have been analyzed to give cross sections for p-wave and s-wave capture. At 25 kev in the laboratory system, the cross sections are

$$\sigma_p = (2.9\pm0.3)\times10^{-32}\ \mathrm{cm}^2,$$
$$\sigma_s = (1.3\pm0.3)\times10^{-32}\ \mathrm{cm}^2.$$

The astrophysical S-factors in the center-of-mass system below 40 kev have been found to be

$$S_p = [(0.127\pm0.013)+0.0079\,E]\ \text{ev barns}$$

for E in center-of-mass kilovolts and

$$S_s = 0.12\pm0.03\ \text{ev barns}$$

independent of energy giving a total S-factor for low energies

$$S = S_s+S_p = (0.25\pm0.04)+0.0079\,E\ \text{ev barns}$$

with E in center-of-mass kilovolts.

INTRODUCTION

The low-energy cross section for the reaction $D(p, \gamma)He^3$ may be of interest in a number of astrophysical processes as was pointed out by Salpeter (1952, 1955). As the temperature rises in a condensing star this reaction will be one of the first to take place. Consequently if there is a significant amount of deuterium in the interstellar gas the energy release by this reaction will modify the early stages of condensation of the star before it reaches the main sequence. Once on the main sequence the reaction is the second step in the proton–proton chain of hydrogen burning which accounts for the main energy release in the smaller main-sequence stars. Also, since the reaction competes with the neutron-producing reaction $D(d, n)He^3$, it has an effect on the number of neutrons produced from an initial deuterium concentration, in the early stages of contraction of a star, as pointed out by Salpeter (1955). The neutron yield is of interest as it may account for some of the anomalous abundances of xenon isotopes in meteorites (Reynolds 1960*a*, *b*) as discussed by Cameron (1962). The role played by deuterium is dependent on its initial concentration in the interstellar gas, which is not accurately known at the present time. It has been common to assume that its primordial concentration is the same as the present terrestrial value of 1.4×10^{-4} times the hydrogen concentration, and Cameron (1962) has suggested that the interstellar gas is enriched in deuterium as a result of Type II supernova explosions. On the other hand

In order to proceed further it is necessary to have explicit energy dependences for σ_p, σ_s, and ϵ. The stopping cross sections, ϵ, which were used, were those of Phillips (1953) with an error of $\pm 5\%$ from 10 to 70 kev averaged with values of Reynolds *et al.* (1953) above 40 kev. Both these measurements are for water vapor and are somewhat higher than measurements reported for D_2O ice (Wenzel and Whaling 1952). Phillips (1953) has suggested that a density effect may account for the difference; however, this seems rather unlikely as there is no known reason why density effects or the difference between H_2O and D_2O could account for the experimental difference. In order to carry the integrals to zero energy the value of the stopping cross section was smoothly extrapolated to zero at zero energy from the lowest measured point of Phillips at 10 kev.

For the nuclear cross sections the following energy dependences have been used:

$$\sigma_p = a_p' \frac{(Q+\frac{2}{3}E)^3}{v} \left|\frac{F_1}{\rho}\right|^2 = a_p \frac{(Q+\frac{2}{3}E)^3}{E} \frac{|F_1|^2}{\rho} = a_p f_p, \tag{7}$$

$$\sigma_s = a_s' \frac{(Q+\frac{2}{3}E)^3}{v} \left|\frac{F_0}{\rho}\right|^2 = a_s \frac{(Q+\frac{2}{3}E)^3}{E} \frac{|F_0|^2}{\rho} = a_s f_s, \tag{8}$$

where a_p', a_s', a_p, and a_s are assumed to be energy-independent cross section factors. These are approximations to the energy dependence given by a direct capture theory, which should be adequate for low bombarding energies. The cross section is dependent on the probability that the proton and deuteron approach to within some interaction radius R given approximately by the Coulomb function $(F_l/\rho)^2$ evaluated at R for the lth partial wave. Calculations were done for two radii, the radius given by the Coulomb energy difference between the two mass-three nuclei, 2.26 fermis, and the sum of the radii of the proton and deuteron, 5.8 fermis. The energy dependences given for these two radii are not greatly different and the results given here are based on the 5.8-fermi radius. The capture cross section is inversely proportional to the flux of incident particles given by v with the normalization for the Coulomb functions used here (Block *et al.* 1951; Tubis 1958). The cross section also depends on the energy and multipolarity of the radiation through the factor $(Q+\frac{2}{3}E)^{2L+1}$, where E is the laboratory bombarding energy. For the analysis it is assumed that both contributions arise from dipole ($L = 1$) radiation, with electric dipole following p-wave capture for the main part of the cross section with a $\sin^2\theta$ angular distribution, and magnetic dipole following s-wave capture for the smaller isotropic part. The energy dependences of the two expressions given in (7) or in (8) are the same and the forms on the right-hand side were used for convenience to define the energy-independent cross section factors a_p and a_s.

After substituting (7) and (8) into (6), the energy-dependent integrals can be evaluated in terms of the known functions ϵ, f_p, and f_s. Introducing the notation

$$I_p(E_1) = \int_0^{E_1} \frac{f_p}{\epsilon} dE$$

and a similar relation for the s-wave integral, the experimental number of counts per incident proton at 90° to the incident beam and the ratio of counts at 0° to the counts at 90° can be written as

$$N(90^0, E_1) = \frac{\xi\omega}{2\pi} [(1+\tfrac{1}{2}Q_2)a_p I_p + a_s I_s], \tag{9}$$

$$\frac{N(0°, E_1)}{N(90°, E_1)} = \frac{(1-Q_2)a_p I_p + a_s I_s}{(1+\tfrac{1}{2}Q_2)a_p I_p + a_s I_s}. \tag{10}$$

It is clear that equation (10) determines the ratio a_s/a_p in terms of the measured 0° to 90° ratio. This can then be substituted into (9) to give the cross section factor a_p. The cross sections are then given by (7) and (8).

The three points at the lower right in Fig. 4 show the a_s/a_p ratios obtained from the measurements given in Table I. In addition, the values obtained from the higher-energy data (Griffiths *et al.* 1962) are also shown. Within

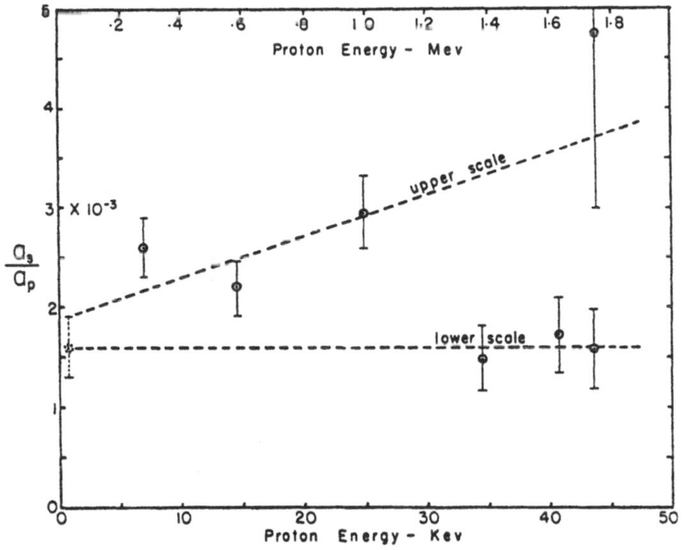

FIG. 4. The ratio of cross section factors for s-wave and p-wave capture obtained from the angular distribution data; the three points at the lower right are from the present low-energy measurements, the other points are from previous higher-energy measurements. The dotted point at the left is the average of the low-energy values.

the rather large errors the low-energy points are consistent with the assumption that a_s/a_p is independent of energy with an average value of $(1.60\pm0.30)\times10^{-3}$. However the higher-energy points are not consistent with this assumption. This indicates that the assumed energy dependences are not adequate in the higher-energy range. This will be discussed in a later paper where it is shown that, above about 100 kev, the simplified energy dependence is not adequate but that a more detailed direct capture theory does give a better fit to the experimental data. Below 70 kev the simplified functions give adequate fits to the experimental data and so should be satisfactory for the present results obtained for energies below 50 kev. This is confirmed by the fact that the a_s/a_p ratio from higher energies extrapolates to the lower-energy region, as shown in Fig. 4, in reasonable agreement with the low-energy ratio, considering the large errors.

Substituting $a_s/a_p = 1.60\times10^{-3}$ into (9) then gives the a_p values in Table I. The a_s values in Table I are just 1.60×10^{-3} times the a_p values. The errors shown on the yield data in Table I are counting statistical errors plus an error estimate for the extrapolation to the initial yield. These errors have been carried over directly to the a_p without inclusion of the additional error in a_s/a_p. The weighted mean of the a_p values, obtained using a weighting factor equal to the square of the reciprocal of the statistical error, is shown at the bottom of Table I. By substituting in equation (9) it can be shown that the 18.8% error in a_s/a_p contributes an additional 4.5% error to a_p and an additional 18.8% error to a_s. Combining the errors quadratically then gives

$$a_p = (6.00\pm0.47)\times10^{-37},$$
$$a_s = (0.96\pm0.19)\times10^{-39}.$$

Finally substituting the individual a_p and a_s values from Table I into (7) and (8) gives the cross sections shown by the experimental points in Fig. 5.

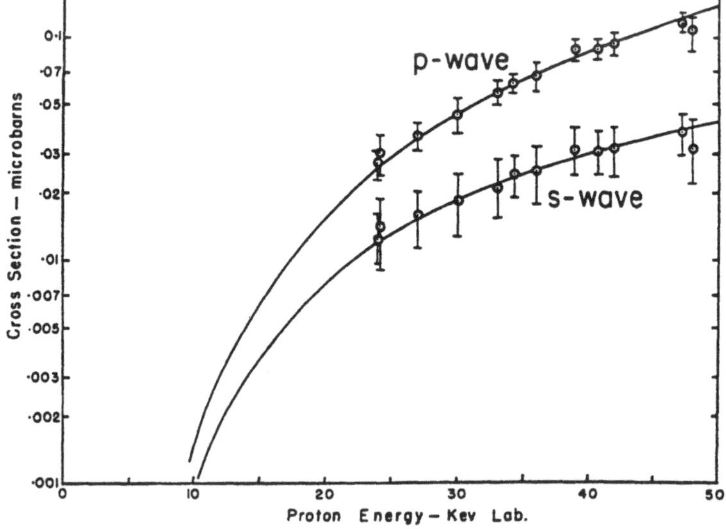

FIG. 5. The total cross sections for *p*-wave and *s*-wave capture obtained from unfolding the thick target yield curve. The solid lines show the energy dependence assumed in the analysis normalized to the experimental data.

$He^3(\alpha,\gamma)Be^7$ Reaction*

P. D. PARKER AND R. W. KAVANAGH
California Institute of Technology, Pasadena, California
(Received 8 May 1963)

The $He^3(\alpha,\gamma)Be^7$ reaction has been investigated using gaseous helium targets behind thin nickel foils and using monoenergetic alpha-particle beams with energies from 0.42 to 5.80 MeV. The prompt gamma rays were detected using NaI(Tl) scintillation spectrometers, and the resulting pulse-height spectra were analyzed to determine both the total capture cross section and the branching ratio between the cascade and the crossover transitions as functions of energy. These cross-section measurements have been used to obtain a new value for the zero-energy intercept of the cross-section factor $S(E)$ for this reaction: $S_0=0.47\pm0.05$ keV-b, and $(dS/dE)_0=-2.8\times10^{-4}$ b.

INTRODUCTION

THE direct-capture reaction $He^3(\alpha,\gamma)Be^7$ is a particularly interesting member of the class of direct-capture reactions because of the large energy range ($0.0\leq E_\alpha\leq6.8$ MeV) over which it is nonresonant and because of the extensive information that is available from elastic-scattering investigations[1-3] about the phase shifts for the various partial waves necessary to describe the initial-state wave function. A detailed discussion of the use of these phase shifts in the calculation of theoretical cross sections and angular distributions for such direct-capture reactions is presented in the paper immediately following this one.[4]

Furthermore, the $He^3(\alpha,\gamma)Be^7$ reaction is of interest in nuclear astrophysics since it serves as a He^3-burning reaction for the termination of the proton-proton chain. In the proton-proton chain, once He^3 has been produced via the reactions

$$H^1(p,\beta^+\nu)D^2(p,\gamma)He^3,$$

the chain is normally terminated by converting the He^3 into He^4 through one of the two following He^3-burning reactions:

$$He^3(He^3,2p)He^4 \qquad (1)$$

$$He^3(\alpha,\gamma)Be^7(e^-,\nu)Li^7(p,\alpha)He^4 \qquad (2a)$$

$$He^3(\alpha,\gamma)Be^7(p,\gamma)B^8(\beta^+\nu)Be^{8*}(\alpha)He^4. \qquad (2b)$$

Hence, the terminations via the $He^3(\alpha,\gamma)Be^7$ reaction produce one He^4 for every $H^1(p,\beta^+\nu)D^2$ reaction, compared to the yield of only one He^4 for every two $H^1(p,\beta^+\nu)D^2$ reactions through the $He^3(He^3,2p)He^4$ termination. Therefore, since the rate of the proton-proton chain is governed by the $H^1(p,\beta^+\nu)D^2$ reaction, completion of the chain via the $He^3(\alpha,\gamma)Be^7$ reaction instead of the $He^3(He^3,2p)He^4$ reaction doubles the rate of He^4 production and nearly doubles the rate of energy generation [a factor of 1.95 for the Li^7 branch (2a) and a factor of 1.46 for the B^8 branch (2b) due to energy losses in the form of neutrinos].[5,6] This difference in the efficiencies of these two terminations makes a knowledge of the relative importance of these reactions essential to the study and generation of models of stellar interiors for stars operating on the proton-proton chain.

Previous measurements on the $He^3(\alpha,\gamma)Be^7$ reaction and the mirror reaction $H^3(\alpha,\gamma)Li^7$ have been reported by Holmgren and Johnston.[7] However, recent work by Griffiths[8] is in marked disagreement with the total cross-section measurements of Holmgren and Johnston for the $H^3(\alpha,\gamma)Li^7$ reaction. The present experiment has been carried out to extend the cross-section measurements for the $He^3(\alpha,\gamma)Be^7$ reaction in order to check the predictions of the direct-capture theory and to redetermine the role of this reaction in astrophysics.

PROCEDURE

Figure 1 indicates the construction of the gas target system used in this experiment. The monoenergetic incident beam was obtained by magnetic analysis from the Kellogg Radiation Laboratory 3-MV electrostatic accelerator and from the 6-MV ONR-CIT tandem accelerator. Alpha-particle beams were used instead of He^3, in spite of the resultant increases in energy losses, because of the large reduction in gamma-ray background

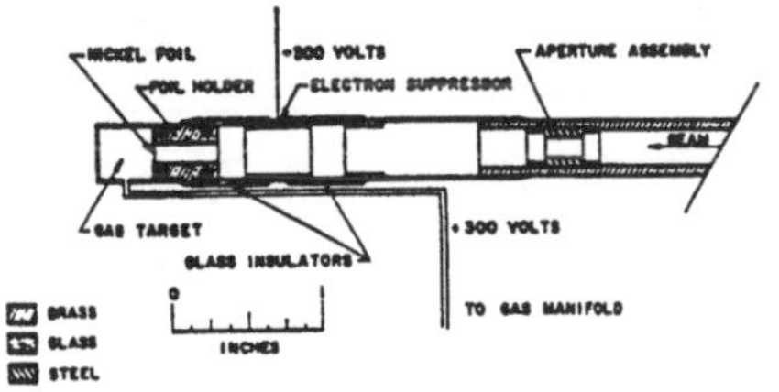

FIG. 1. Scale drawing of the gas-target assembly.

* Supported by the U. S. Office of Naval Research.

[1] P. D. Miller and G. C. Phillips, Phys. Rev. 112, 2048 (1958).
[2] C. M. Jones, A. C. L. Barnard, and G. C. Phillips, Bull. Am. Phys. Soc. 7, 119 (1962).
[3] T. A. Tombrello and P. D. Parker, Phys. Rev. 130, 1112 (1963).
[4] T. A. Tombrello and P. D. Parker, following article, Phys. Rev. 131, 2582 (1963).
[5] W. A. Fowler, Astrophys. J. 127, 551 (1958).
[6] W. A. Fowler, Mem. Soc. Roy. Sci. Liege 3, 207 (1960).
[7] H. D. Holmgren and R. L. Johnston, Phys. Rev. 113, 1556 (1959).
[8] G. M. Griffiths, R. A. Morrow, P. J. Riley, and J. B. Warren, Can. J. Phys. 39, 1397 (1961).

that it was thereby possible to achieve. The target gas was contained in the small gas cell (approximately 1 cm in length and 1 cm in diameter) and was separated from the high-vacuum beam tube by a thin nickel foil, typically 6250 Å thick.[9] At beam energies below 1.5 MeV, this was replaced by a 5000-Å nickel foil on which 1000 to 2000 Å of copper had been evaporated to improve the thermal conductivity; in this way we were able to use beams of at least 0.4 μA over the entire energy range of this experiment. The target gas could be switched rapidly between He^3 and He^4, the latter being used for background runs. This alternation was carried out from four to ten times during a series of runs at each energy in order to average out small background variations.

The target thickness was determined from a measurement of the length of the gas cell and measurements of the temperature and pressure in the gas cell. The pressure was measured using either a mercury manometer or a bourdon gauge which was calibrated with a mercury manometer; target pressures varied from 100 to 494 mm. In the determination of the target temperature, local heating by the incident beam was taken into account by measuring the change in the effective thickness of the gas target as a function of the incident beam intensity. This was accomplished by using a thick B^{10} target in the back of the target chamber and observing the shift of the position of the 1.518-MeV resonance in $B^{10}(\alpha,p)C^{13}$ as a function of the incident beam intensity. These measurements indicated an effective target temperature of 345°K at a beam current of 0.45 μA. Simple conductivity calculations suggest that most of this heating is due to energy losses in the entrance foil and the resultant heating of the gas near the foil.

Cylindrical 3-in.×3-in. NaI(Tl) scintillators were used to detect the gamma radiation. The resulting pulses were fed through conventional electronics and stored in

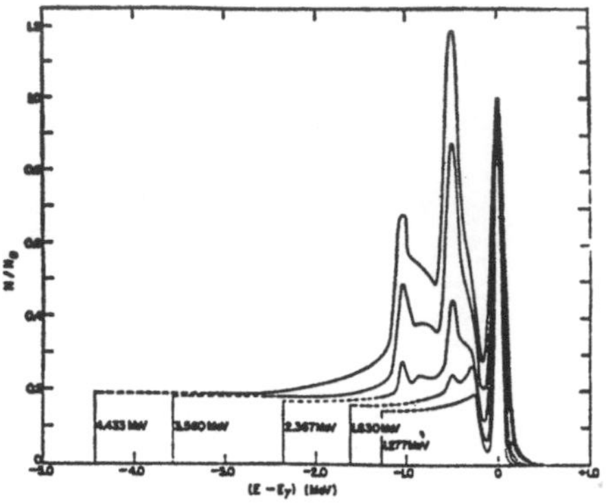

FIG. 3. The line-shape response calibration of the solid 3-in.×3-in. NaI(Tl) crystal as measured experimentally using various nuclear reactions to produce the monoenergetic gamma rays. The gamma-ray energy is indicated at the low-energy end of each function. From such a calibration, it is possible to interpolate the line-shape response at any intermediate energy.

a multichannel pulse-height analyzer. The various target-crystal geometries are shown in Fig. 2. The poor geometry (with the crystal face typically only ⅜ of an inch from the center of the target) was dictated by the small cross sections involved and unfortunately prevented the observation of any angular anisotropy of the radiation that was smaller than about 25%. No anisotropies of this magnitude were observed, and the data were analyzed under the assumption that the radiation was isotropic. The maximum predicted anisotropies[4] in the region of this experiment are only of the order of 7%, and hence, are well below what we could have observed. Furthermore, any change in the crystal-target geometry to improve our angular resolution to allow us to see this effect would drastically reduce the counting rate, while the anisotropy is already smaller than the relative uncertainties in the points of this experiment.

For each of the NaI(Tl) crystals, the total effective efficiency was calculated for these geometries using the tables of Grodstein[10] and taking into account the absorption of the materials between the crystal and the target gas. Further, the crystals were calibrated with respect to their line-shape response by using various monoenergetic gamma rays in the energy range $0.432 \leq E_\gamma \leq 4.433$ MeV. These measured response functions were then plotted as functions of $(E-E_\gamma)$, as suggested by Okano,[11] in order to simplify the interpolation used to determine these functions at intermediate energies. Such a plot is shown in Fig. 3. The photofraction (i.e., the fraction of the total gamma-ray spectrum located in the full-energy peak) of each of the crystals was then measured experimentally by observing these monoenergetic gamma rays using unshielded crystals

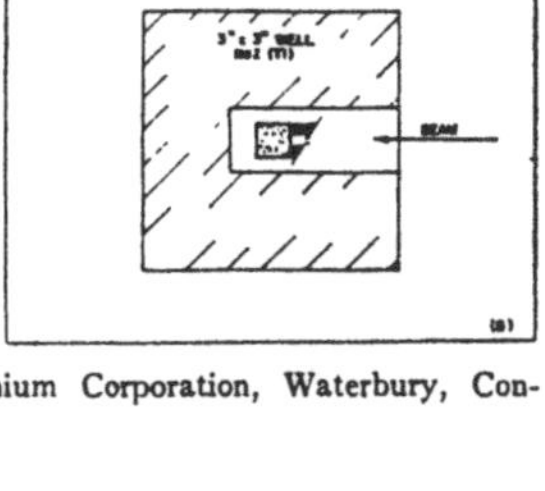

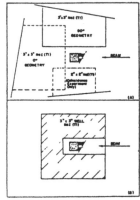

FIG. 2. The target-detector geometry for the various NaI(Tl) crystals used. (A) The configurations employed with the solid cylindrical crystals, including the coincidence experiment. (B) The arrangement used with the 3-in.×3-in. crystal with a ¾-in.×2-in. well.

[9] Obtained from Chromium Corporation, Waterbury, Connecticut.

[10] G. W. Grodstein, Natl. Bur. Std. (U. S.), Circ. 583 (1957).

[11] K. Okano, J. Phys. Soc. Japan 15, 28 (1960).

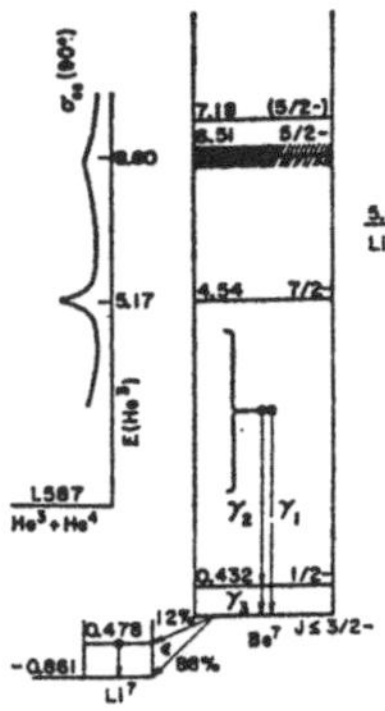

FIG. 4. Level diagram for Be^7 indicating the two possible gamma transitions in the region of this experiment as well as the various Q values involved and the locations and quantum numbers of the various excited states.

and the zero-intercept method of Zerby and Moran.[12] The photofraction was then combined with the total effective efficiency to yield the effective photoefficiency.

After applying dead-time corrections for the multichannel analyzer and total-time corrections for time-dependent background, the resulting spectra were analyzed by subtracting the He^4 spectrum from the He^3 spectrum to obtain the net $He^3(\alpha,\gamma)Be^7$ gamma-ray spectrum. The net spectrum was then analyzed on a computer to extract the various gamma-ray contributions, using the line-shape calibration of the scintillators. The Be^7 level diagram in Fig. 4 indicates that in the region investigated in this experiment there are two possible gamma-ray transitions, the crossover transition going directly to the ground state and the cascade transition proceeding through the 432-keV first excited state. Once the contributions of these transitions had been sorted out by obtaining a least-squares fit to the net experimental spectrum using the appropriate line-shape responses, an application of the effective photoefficiency yields immediately the number of each of the gamma rays produced in the target. A combination of this with the target thickness, as measured above, and with the number of incident alpha particles, as measured with a current integrator, yields the cross section for producing such gamma rays. A graphical example of the data reduction and least-squares unfolding of the experimental spectra is shown in Fig. 5 for the case where $E_{c.m.}=1248$ keV and $E_{\gamma_1}=2890$ keV with the 3-in.×3-in. crystal at 0° to the incident beam.

From such an analysis, it is then possible to obtain independent values for the cross sections of both transitions. The data can then either be presented in terms of these two cross sections or be expressed in terms of independent values for the total cross section and the branching ratio. The latter set of parameters proved more suitable because for almost half of the cases it was impractical to carry out such a least-squares analysis of the net spectrum. Such cases occur, for example, at beam energies below 1.0 MeV because of the low counting rates and resulting poor statistics, and above 3.5 MeV because of the alteration of the line-shape response of the crystals by absorbers placed between the crystals and the target to reduce the intensity of low-energy radiation from such sources as Coulomb excitation in the platinum beam stopper. For these cases it was possible to extract only one independent number, and, since the branching ratio remained essentially constant over the entire range of this experiment, the total cross section was the logical parameter to present for such data.

Since the target gas was located behind a foil that was 250 to 550 keV thick to the incident alpha particles, the center-of-mass energies at which these measurements were made were determined either by actually measuring the thicknesses of the entrance foils using the 1.518-MeV resonance in $B^{10}(\alpha,p)C^{13}$ or by using the fact that for capture reactions the energy of the prompt gamma radiation (allowing for Doppler shifting) is just the sum of the center-of-mass energy and the Q value for the reaction. In the cases where both of these methods were used, the agreement between the two was within the experimental error.

RESULTS

The experimental results are presented in Figs. 6–8. Figure 6 is a plot of the total capture cross section as a function of the center-of-mass energy. As a check on the unfolding techniques described above, a coincidence measurement was made to obtain an independent value for the cross section for the cascade transition by using

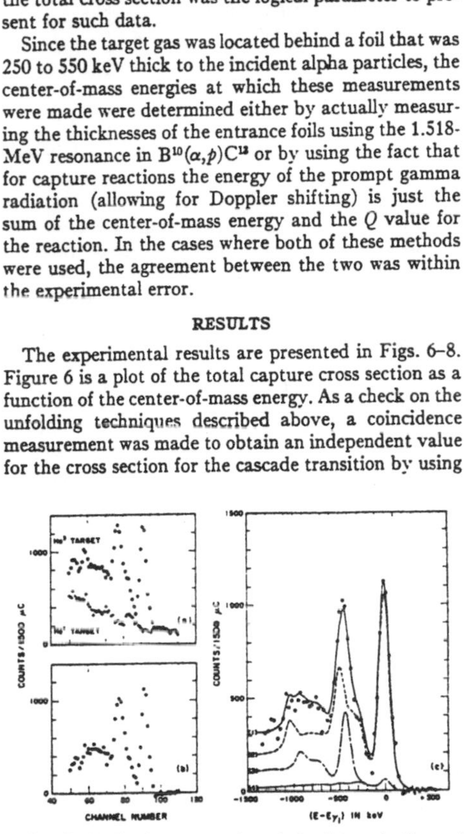

FIG. 5. Graphical representation of the data reduction and analysis for the case of $E_{c.m.}=1248$ keV. (a) The dots represent the raw NaI(Tl) spectrum obtained using He^3 as the target gas while the crosses represent the raw spectrum obtained under identical conditions using He^4 gas in the target chamber. (b) The dots represent the net $He^3(\alpha,\gamma)Be^7$ gamma-ray spectrum obtained by subtracting the He^4 spectrum from the He^3 spectrum in (a). (c) The net experimental $He^3(\alpha,\gamma)Be^7$ spectrum is represented by the dots. Curve (1) is the least-squares fit to the net spectrum obtained by varying the normalizations of curves (2) and (3), where (2) is the response function for the crossover transition and (3) is the response function for the cascade transition including the effects of coincident summing. Curve (4) is just the contribution of such summing to the cascade response function.

[12] G. D. Zerby and H. S. Moran, Nucl. Instr. Methods **14**, 115 (1961).

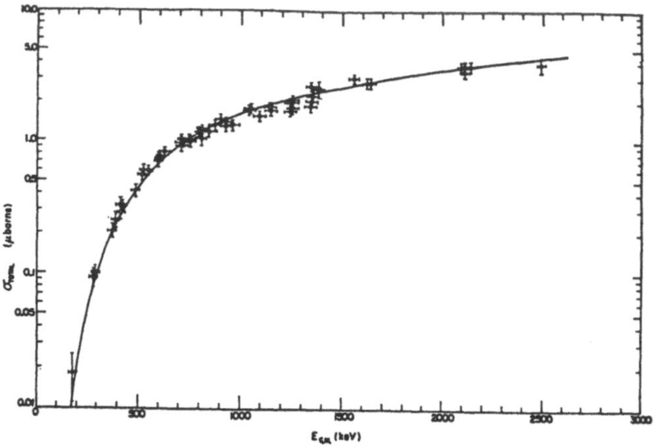

FIG. 6. The total capture cross section as a function of the center-of-mass energy. The circles are the experimental measurements, and the solid curve is the prediction of the theory in Ref. 4. The solid square at $E_{c.m.}=1378$ keV is the coincidence measurement.

two NaI(Tl) scintillators and requiring a coincidence between the two members of the cascade transition. The cross section derived in this way agreed well with the cascade cross sections obtained by the unfolding process. Converted to a total cross section by using the measured branching ratio, the coincidence point is shown in Fig. 6 as the solid square at $E_{c.m.}=1378$ keV. Figure 7 is a plot of the branching ratio (ρ), defined as the ratio of the cascade cross section to the cross section for the crossover transition. The data are consistent with a constant value for ρ over this region, and a statistical analysis indicates that $(73\pm3)\%$ of the transitions go directly to the ground state while the remaining $(27\pm3)\%$ cascade through the first excited state. The error bars in Figs. 6 and 7 (typically $\pm10\%$) are the total errors of the individual points and include such things as counting statistics, and uncertainties in the crystal efficiencies, the target thickness and the beam integration.

The solid curves in Figs. 6 and 7 are the predictions for the total capture cross section and the branching ratio based on the theory described in detail in the following paper.[4] It is sufficient to emphasize here that the agreement obtained between the theory and the experiment was achieved with the use of only two energy-independent parameters, the reduced widths of the two final states. Hence, only the normalization of these curves is subject to fitting, the curvature being completely defined by the independent elastic scattering experiments.

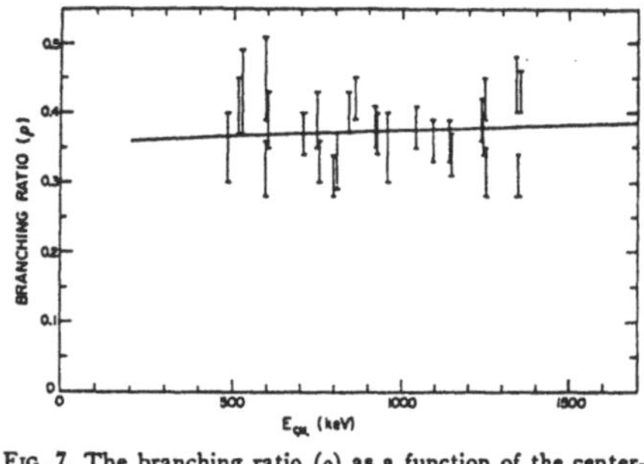

FIG. 7. The branching ratio (ρ) as a function of the center-of-mass energy, where $\rho\equiv\sigma(\text{cascade})/\sigma(\text{crossover})$. As in Fig. 6, the bars are the experimental values and the solid curve is from Ref. 4.

Finally, in order to apply our results to an analysis of the termination of the proton-proton chain we must extrapolate from our measurements to obtain cross sections at center-of-mass energies of the order of 10 to 20 keV. Such an extrapolation is most easily accomplished by first removing the penetration factor from the cross section and then working with the more nearly energy-independent "S factor,"[13]

$$S(E_{c.m.})=\sigma(E_{c.m.})E_{c.m.}\exp(31.28Z_0Z_1A^{1/2}/E_{c.m.}^{1/2})$$
$$=\sigma(E_{c.m.})E_{c.m.}\exp(163.78/E_{c.m.}^{1/2})\text{ keV-b}.$$

This function is shown in Fig. 8 where the error bars and the solid curve represent the conversion of the measured cross sections and the theoretical prediction, respectively, and consequently this figure contains exactly the same information as Fig. 6, except that the dependence on the Coulomb barrier has been removed.

The shape of the theoretical curve having been defined by the elastic scattering experiments, the normalization of the curve shown here has been determined as the best fit to all of the experimental points. As such we are able to use the direct-capture theory together with the experimental points to obtain intercept parameters for $S(E)$ of

$$S_0=0.47\pm0.05\text{ keV-b},$$
$$(dS/dE)_0=-2.8\times10^{-4}\text{ b},$$

[13] E. M. Burbidge, G. R. Burbidge, W. A. Fowler, and F. Hoyle, Rev. Mod. Phys. 29, 547 (1957).

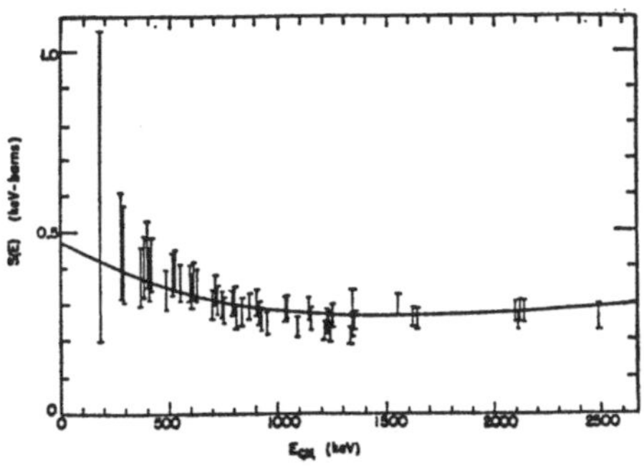

FIG. 8. The cross-section factor $S(E)$ as a function of the center-of-mass energy. The bars and solid curve are again the experimental values and the prediction of Ref. 4, respectively. At the intercept, $S_0=0.47\pm0.05$ keV-b, and $(dS/dE)_0=-2.8\times10^{-4}$ b.

which are considerably more accurate than our lowest energy points.

There may be some question as to whether the normalization for extrapolation purposes should not have been made to just the low-energy measurements. If such a normalization is made using only the data for $E_{c.m.}\leq625$ keV, one obtains an intercept of $S_0=0.51\pm0.07$ keV-b.

A combination of the S-factor intercepts of the other proton-proton-chain reactions with this new value for the $He^3(\alpha,\gamma)Be^7$ reaction allows us to determine the importance of each of the He^3-burning reactions as terminations for the proton-proton chain. A detailed discussion of the termination of this chain will be published elsewhere[14]; it is sufficient to note here that the new intercept is appreciably smaller than the value of 1.2 keV-b previously published by Holmgren and Johnston,[7] and consequently that in the temperature range below 16×10^{6}°K this new value reduces the importance ascribed in Refs. 5 and 6 to $He^3(\alpha,\gamma)Be^7$ as a He^3-burning reaction in stars like the sun.

ACKNOWLEDGMENTS

We are happy to acknowledge the assistance of the staff of the Kellogg Radiation Laboratory in carrying out this experiment. In particular, we are grateful to Professor William A. Fowler and Dr. T. A. Tombrello for their stimulus and helpful discussions.

[14] P. D. Parker, J. N. Bahcall, and W. A. Fowler, Astrophys. J. (to be published).

TERMINATION OF THE PROTON-PROTON CHAIN IN STELLAR INTERIORS*

PETER D. PARKER, JOHN N. BAHCALL, AND WILLIAM A. FOWLER
California Institute of Technology, Pasadena, California
Received September 9, 1963; revised in proof January 14, 1964

ABSTRACT

The roles of the various H^2-burning and He^3-burning reactions in the termination of the proton-proton chain in stellar interiors have been re-examined using (1) recent experimental determinations of the cross-sections for the $H^2(p, \gamma)He^3$ and $He^3(\alpha, \gamma)Be^7$ reactions, (2) theoretical studies of the rates of the relevant beta-decay reactions, and (3) a re-examination of the previous cross-section factor determinations for the other possible reactions in the proton-proton chain.

I. INTRODUCTION

A number of review articles have discussed the termination of the proton-proton chain, for example, Bethe (1939), Salpeter (1952), Burbidge, Burbidge, Fowler, and Hoyle (1957), and Fowler (1960). However, recent experimental studies of the $H^2(p, \gamma)He^3$ reaction by Griffiths, Lal, and Scarfe (1963) and recent theoretical and experimental studies of the termination of the chain via the $He^3(\alpha, \gamma)Be^7$ reaction (Kavanagh 1960; Bahcall 1962*b;* Parker and Kavanagh 1963) make it desirable to re-examine the roles of various reactions in the termination of the proton-proton chain. In the course of this re-examination it has been found that, contrary to statements in some previous review articles, further experimental investigation of the Li^4 system is necessary before one can definitely rule out the possibility that Li^4 plays an important role in the termination of the proton-proton chain.

In the calculations described below we have adopted many of the definitions and notations of Burbidge *et al.* (1957). The cross-section factor, $S(E)$, is defined by the following equation:

$$S(E) = \sigma(E)E \exp\,(31.28\, Z_1 Z_0 A^{1/2}/E^{1/2})\ \text{keV-barns}\,, \tag{1}$$

where σ is the reaction cross-section in barns, E is the center-of-mass energy in keV, Z_1 and Z_0 are the charges of the interacting nuclei in units of the proton charge, and A is the reduced mass, $A_1A_0/(A_1 + A_0)$, in atomic mass units. The mean reaction rate, P, of a non-resonant nuclear reaction can be written

$$P = 7.20 \times 10^{-19} n(1)n(0) f S \tau^2 e^{-\tau}\,(1 + \delta_{10})^{-1}(AZ_1Z_0)^{-1}\ \text{reactions cm}^{-3}\,\text{sec}^{-1}\,, \tag{2a}$$

$$P = 2.62 \times 10^{29} \rho^2 \frac{X_1 X_0}{A_1 A_0} f S \tau^2 e^{-\tau}(1 + \delta_{10})^{-1}(AZ_1Z_0)^{-1}\ \text{reactions cm}^{-3}\,\text{sec}^{-1}\,, \tag{2b}$$

where

$$\tau = 42.48\left(Z_1^2 Z_0^2 \frac{A}{T_6}\right)^{1/3}. \tag{2c}$$

In equations (2*a*), (2*b*), and (2*c*) T_6 is the temperature in millions of degrees; δ_{ij} is the Kronecker delta,

$$\delta_{ij} = 1 \quad \text{if} \quad i = j\,, \qquad \delta_{ij} = 0 \quad \text{if} \quad i \neq j\,;$$

* Supported in part by the Office of Naval Research and in part by the National Aeronautics and Space Administration.

TABLE 1

REACTION PARAMETERS (PROTON-PROTON CHAIN)

Reaction	Q-Value* (MeV)	S_0 (keV-barns)	(dS/dE) (barns)	$\tau_j(i)$† (yr.)	Reference
$H^1(p, \beta^+\nu)H^2$	$+1.442$	$(3.36\pm0.4)\times10^{-22}$	$+2.7\times10^{-24}$	8.85×10^{9}	Fowler (1960)
$H^2(p, \gamma)He^3$	$+5.493$	$(2.5\pm0.4)\times10^{-4}$	$+7.9\times10^{-6}$	4.43×10^{-8}	Griffiths *et al.* (1963)
$H^2(d, p)H^3$	$+4.032$ }	100	$+0.525$	5.32×10^{5}	Arnold *et al.* (1954)
$H^2(d, n)He^3$	$+3.268$ }				
$H^3(\tau, p)He^4$	$+18.352$	6.7×10^{3}	$+27$	1.15×10^{-4}	Arnold *et al.* (1954)
$He^3(d, p)He^4$	$+18.352$	6.7×10^{3}	$+27$	8.86×10^{8}	Arnold *et al.* (1954)
$He^3(\tau, 2p)He^4$	$+12.859$	1.1×10^{3}		5.28×10^{5}	Good *et al.* (1954)
$He^3(\alpha, \gamma)Be^7$	$+1.586$	(0.47 ± 0.05)	-2.8×10^{-4}	9.70×10^{5}	Parker *et al.* (1963)
$Be^7(e^-, \nu)Li^7$	$+0.861$ (89.7%) }			3.86×10^{-1}	{ Bahcall (1962*b*)
$Be^7(e^-, \nu)Li^{7*}$	$+0.383$ (10.3%) }				{ Taylor *et al.* (1962)
$Li^7(p, \alpha)He^4$	$+17.347$	120	0.0	1.80×10^{-6}	Sawyer and Philips (1953)
$Li^7(p, \gamma)Be^8(\alpha)He^4$	$+17.347$	<1		$>2\times10^{-3}$	Sawyer and Philips (1953)
$Be^7(p, \gamma)B^8$	$+0.135$	(0.030 ± 0.01)	0	8.81×10^{1}	Kavanagh (1960)
$B^8(\beta^+\nu)Be^{8*}(\alpha)He^4$	$+18.074$			3×10^{-8}	Bahcall (1962*a*)
$He^3(e^-, \nu)H^3$	-0.018			2×10^{10}	Bahcall and Wolf (1963)
$H^3(p, \gamma)He^4$	$+19.813$	1.5×10^{-3}	$+2.7\times10^{-6}$	1.68×10^{-8}	Perry and Bame (1955)
$He^3(p, \beta^+\nu)He^4$	$+19.795$				

* König, Mattauch and Wapstra (1962).

† τ calculated for $X_H = X_{He} = 0.5$, $\rho = 100$ gm cm^{-3} and $T_6 = 15$.

THE EFFECT OF Be^7 K-CAPTURE ON THE SOLAR NEUTRINO FLUX*

ICKO IBEN, JR., KENNETH KALATA, AND JUDAH SCHWARTZ
Massachusetts Institute of Technology
Received May 9, 1967

ABSTRACT

All other things being equal, bound-electron capture by the Be^7 nucleus near the solar center decreases the calculated rate of B^8 neutrino emission from the Sun.

I. INTRODUCTION

Under conditions that prevail near the center of the Sun, Be^7 disappears several hundred times more rapidly by the $Be^7(e^-,\nu)Li^7$ reaction than by the $Be^7(p,\gamma)B^8$ reaction. The equilibrium abundance of Be^7 is thus essentially that obtained by equating the $Be^7(e^-,\nu)Li^7$ rate with the $He^3(He^4,\gamma)Be^7$ rate. The $Be^7(p,\gamma)B^8$ rate and, consequently, the $B^8(\beta^+\nu)Be^{8*}$ rate are then proportional to this equilibrium abundance.

It has been universally assumed that, under solar conditions, the Be^7 nucleus captures electrons solely from the continuum. There is, however, a finite probability that Be^7 exists as an atom with one or two bound K-shell electrons. The nuclear electron-capture probability is larger when bound electrons are taken into account and the calculated equilibrium abundance of the Be^7 nucleus is correspondingly reduced. The $Be^7(p,\gamma)B^8$ rate and, hence, the $B^8(\beta^+\nu)Be^{8*}$ rate are reduced by the same factor. Since the current experiment to detect solar neutrinos (Davis 1964; Bahcall 1966) is primarily sensitive to neutrinos from the $B^8(\beta^+\nu)Be^{8*}$ reaction, it is of interest to examine quantitatively the influence of K-capture by Be^7.

II. FIRST APPROXIMATION—NO SCREENING

The free-electron capture probability may be written as

$$\omega_f = \frac{|\psi_f(0)|^2}{2|\psi_{\text{lab}}(0)|^2}\,\omega_{\text{lab}}\,, \tag{1}$$

where $|\psi_f(0)|^2$ is the free-electron density at the Be^7 nucleus in the star, $2|\psi_{\text{lab}}(0)|^2$ is the electron density at the Be^7 nucleus in a *neutral*, unscreened atom in the ground state, and $\omega_{\text{lab}} = \ln(2)/53.6$ days $= 1.5 \times 10^{-7}$ sec^{-1}. We have, approximately,

$$|\psi_f(0)|^2 = n_e\langle 2\pi\eta\rangle\,, \tag{2}$$

where $n_e = \rho/(\mu_e M_H)$ = mean electron number density at a given point in the star, $\langle\eta\rangle$ = average over the electron Maxwell-Boltzmann distribution of $4e^2/\hbar v_e$. In these expressions, ρ = matter density (gm/cm^3), μ_e = electron molecular weight, M_H = mass of hydrogen nucleus (gm), e = electron charge (e.s.u.), $2\pi\hbar$ = Planck's constant (ergs/sec), and v_e = electron velocity (cm sec^{-1}).

Inserting the estimate for $|\psi_{\text{lab}}(0)|^2$ given by Bahcall (1962), we have

$$\omega_f \cong 4.24 \times 10^{-9}(\rho/\mu_e)T_6^{-1/2}\text{sec}^{-1}\,, \tag{3}$$

where T_6 = temperature in 10^6 ° K.

* Supported in part by the National Science Foundation (GP-6387) and in part by the National Aeronautics and Space Administration (NsG-496).

We now examine, on the assumption that screening may be neglected, the extent to which Be^7 may be only partially ionized. For simplicity and without much loss in accuracy, we neglect all excited states. That is, we assume that only the ground state enters into the partition function for each ionization state. The probabilities f_1 and f_2 that one or two K-shell electrons are associated with any given Be^7 nucleus are

$$f_1 = \lambda/[1 + \lambda + 0.25\lambda^2 \exp(-\Delta\chi/kT)],$$
$$f_2 = 0.25\lambda\,[\exp(-\Delta\chi/kT)]f_1, \tag{4}$$

where

$$\lambda = n_e(h^2/2\pi mkT)^{3/2} \exp(\chi_1/kT). \tag{5}$$

Here k = Boltzmann's constant, m = electron mass, χ_1 = fourth ionization potential of the Be^7 atom = 216.6 eV, χ_2 = third ionization potential of the Be^7 atom = 153.1 eV, and $\Delta\chi = \chi_1 - \chi_2$ = 63.5 eV.

The probability per second that a Be^7 nucleus will capture an electron is now

$$\omega_0 = \omega_f[1 + f_1|\psi_1(0)/\psi_f(0)|^2 + f_2|\sqrt{(2)}\psi_2(0)/\psi_f(0)|^2], \tag{6}$$

where $|\psi_1(0)|^2 = (\pi a^3)^{-1}$ = electron density at the nucleus for the triply ionized Be^7 atom in the ground state. Here $a = (a_0/4) = 0.25 \times$ (first Bohr radius for hydrogen). Since

$$\lambda|\psi_1(0)/\psi_f(0)|^2 = [(8e^2/2a)/kT] \exp(\chi_1/kT) = (5.07/T_6) \exp(2.515/T_6), \tag{7}$$

we have

$$\omega_0 = \omega_f[1 + (5.07/T_6) \exp(2.515/T_6)S], \tag{8}$$

where

$$S = [1 + 0.25\lambda \exp(-0.735/T_6)|\sqrt{(2)}\psi_2(0)/\psi_1(0)|^2]$$
$$\div [1 + \lambda + 0.25\lambda^2 \exp(-0.735/T_6)], \tag{9}$$
$$\lambda = 0.246(\rho/\mu_e T_6^{3/2}) \exp(2.51/T_6).$$

Assuming that $|\psi_2(0)| \sim |\psi_{1ab}(0)|$, so that

$$2|\psi_2(0)/\psi_1(0)|^2 \sim 1.74,$$

we have

$$S \cong [1 + 0.435\lambda \exp(-0.735/T_6)]/[1 + \lambda + 0.25\lambda^2 \exp(-0.735/T_6)]. \tag{10}$$

For small λ,

$$\omega_0/\omega_f \cong 1 + (5.07/T_6) \exp(2.515/T_6), \tag{11}$$

whereas for large λ (two *or* more bound electrons),

$$\omega_0/\omega_f \cong 1 + 35.8\mu_e T_6^{1/2}/\rho. \tag{12}$$

Values of ω_0/ω_f at various points in a solar model constructed by Sears (1964) are shown in Table 1. Note that within the inner one-tenth of the Sun's mass, where most of the B^8 neutrino flux is produced, the average value of ω_0/ω_f is $\sim$4/3. This means that the B^8 neutrino flux calculated by Sears should be reduced by this same factor (provided all other reaction rates are unchanged).

III. SECOND APPROXIMATION—WITH SCREENING

In the Debeye-Hückel approximation, the average potential presented to an electron by a nucleus of charge Ze is given by

$$V = -(Ze^2/r) \exp(-r/R). \tag{13}$$

The screening radius R obeys

$$R^{-2} = (4\pi e^2/kT) \sum_j Z_j^2 n_{0j} , \tag{14}$$

where n_{0j} is the average number density of particles of charge $Z_j e$. In a medium composed primarily of hydrogen (abundance by mass X) and helium,

$$R \cong 0.63 \times 10^{-8} \left[\frac{4T_6}{\rho(3+X)}\right]^{1/2} = 1.19 \left[\frac{64T_6}{\rho(3+X)}\right]^{1/2} a . \tag{15}$$

In order to compute the effect of bound-electron capture, we must first obtain, as a function of the screening radius R, the ground-state energy and the wave function describing an electron in the field of the screened Be^7 nucleus. Before presenting results of exact solutions of the equation

$$H\psi = [-(\hbar^2/2m)\nabla^2 - (4e^2/r) \exp(-r/R)]\psi = E\psi , \tag{16}$$

it is worthwhile to illustrate the approximate nature of the solutions obtained by applying the variational principle.

TABLE 1

EFFECT OF BOUND ELECTRONS ON THE $Be^7(e^-,\nu)Li^7$ RATE

Mass Fraction	ρ	T_6	X	λ	ω_0/ω_f	R/a	σ_R	C^2_R	λ_R	ω_R/ω_f
0.0..........	158	15.7	0.36	0.498	1.294	1.64	0.203	0.642	0.438	1.172
0.1..........	83	12.8	.58	.428	1.385	1.98	.294	.735	.372	1.256
0.2..........	59	11.3	.65	.394	1.454	2.18	.334	.773	.340	1.312
0.3..........	43	10.1	.68	.355	1.531	2.41	.380	.805	.304	1.378
0.4..........	31	9.0	.69	.316	1.626	2.68	.426	.835	.270	1.458
0.5..........	22	8.1	.70	.273	1.741	3.00	.473	.860	.232	1.556
0.6..........	15	7.1	.71	.238	1.790	3.40	.521	.887	.201	1.605
0.7..........	9.4	6.2	.71	.192	2.097	4.02	.582	.912	.162	1.858
0.8..........	5.0	5.1	.71	.150	2.485	5.00	.651	.936	.126	2.190
0.9..........	1.8	3.9	0.71	0.094	3.340	7.28	0.745	0.973	0.080	2.950

As a trial function, we choose

$$\psi_a(r) = (\pi a^3)^{-1/2} \exp(-r/a) . \tag{17}$$

Minimizing the expression

$$E(a) = \langle\psi_a|H|\psi_a\rangle \tag{18}$$

with respect to variations in a, we find that the minimum value of $E(a) = E_R$ (= approximate ground-state energy) occurs when $a = a_R$, where

$$(a_R/a)[1 + (3a_R/2R)] = (1 + a_R/2R)^3 . \tag{19}$$

Then

$$E_R = -(2e^2/a_R)(1 - a_R/2R)/(1 + a_R/2R)^3 , \tag{20}$$

and the square of the trial wave function at the origin becomes

$$|\psi_R(0)|^2 = (a/a_R)^3 |\psi_1(0)|^2 . \tag{21}$$

In Figure 1 the functions $\sigma_R = -E_R/\chi_1$ and $C_R^2 = (a/a_R)^3$ are plotted against the parameter (R/a). It is interesting that, even for relatively large screening radii, the effect of screening on the ground-state energy is considerable. Note that the bound-state energy vanishes when the screening radius equals the first Bohr radius.

Results of exact solutions to equation (16) are also shown in Figure 1. Note that, as might be expected, energies obtained by the variational technique are quite accurate over a wide range in (R/a), whereas the wave function at the origin is not approximated particularly well by the variational solution except for large screening radii.

We shall not attempt to derive energies and wave functions for the ground state of the doubly ionized Be^7 atom. Instead, we content ourselves with the approximations: $\chi_2 \rightarrow \sigma_R \chi_2$, $\Delta\chi \rightarrow \sigma_R \Delta\chi$, and $|\psi_2(0)|^2 \rightarrow C_R^2 |\psi_2(0)|^2$. The total K-capture probability for the Be^7 nucleus is then approximated by

$$\omega_R = \omega_f[1 + (5.07/T_6) \exp (2.515\sigma_R/T_6)S_R] , \tag{22}$$

where

$$S_R = C_R^2[1 + 0.435\lambda_R \exp (-0.735\sigma_R/T_6)]/[1 + \lambda_R + 0.25\lambda_R^2 \times \exp (-0.735\sigma_R/T_6)] , \tag{23}$$

$$\lambda_R = 0.246(\rho/\mu_e T_6^{3/2}) \exp (2.515\sigma_R/T_6) .$$

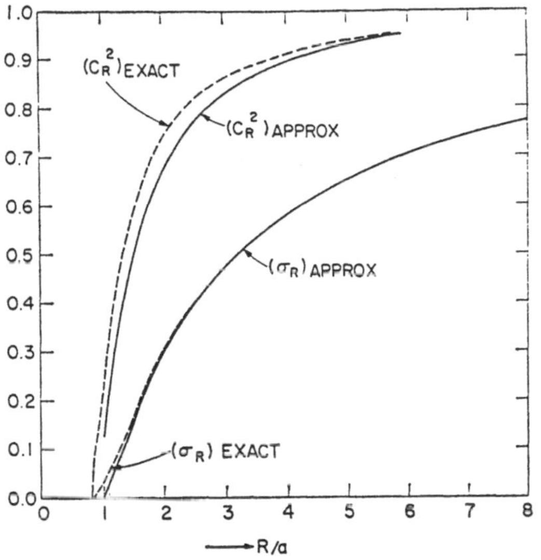

FIG. 1.—Properties of the ground state of the triply ionized Be^7 atom as a function of screening radius R. R is scaled in units of $a = 0.528 \times 10^{-8}$ cm/4 = first Bohr radius of the unscreened triply ionized atom. $C_R^2 = |\psi_R(0)/\psi_1(0)|^2$ and $\sigma_R = -E_R/\chi_1$. Approximate results are obtained with the variational technique.

Values for ω_R/ω_f at various points in the Sears solar model are presented in Table 1. In the inner one-tenth of the Sun's mass, bound-electron capture enhances the $Be^7(e^-,\nu)Li^7$ rate and decreases the calculated B^8 neutrino flux by a factor of about $\frac{5}{4}$. Thus, bound-electron capture has an effect on the solar B^8 neutrino flux which is of the same magnitude as, but which acts in the opposite sense to, the effect due to the recently reported (Parker 1966) increase in the experimentally determined estimate of the $Be^7(p,\gamma)B^8$ cross-section factor.

REFERENCES

Bahcall, J. N. 1962, *Phys. Rev.*, **128**, 1297.
———. 1966, *Phys. Rev. Letters*, **17**, 398.
Davis, R., Jr. 1964, *Phys. Rev. Letters*, **12**, 302.
Parker, P. D. 1966, *Ap. J.*, **145**, 960.
Sears, R. L. 1964, *Ap. J.*, **140**, 477.

THE RATE OF THE PROTON-PROTON REACTION*

JOHN N. BAHCALL AND ROBERT M. MAY
California Institute of Technology, Pasadena

Received January 31, 1968

ABSTRACT

The rate of the proton-proton reaction has been recalculated. Our value for the low-energy cross-section factor is $S(E = 0) = (3.78 \pm 0.15) \times 10^{-25}$ MeV barns, which is 12.5 per cent larger than the value in current usage. The effect of our larger calculated $S(E = 0)$ on the ^{8}B solar neutrino flux and on the rate of nuclear-energy generation in the proton-proton chain is briefly discussed.

The rate at which a proton beta decays in the vicinity of another proton to form a deuteron determines the conditions under which nuclear-energy generation takes place in main-sequence stars of one solar mass or less. The rate of this fundamental reaction

$$p + p \rightarrow {}^2D + e^+ + \nu \tag{1}$$

was first calculated by Bethe and Critchfield (1938). Salpeter (1952) gave a much more accurate evaluation and an authoritative discussion of the uncertainties in the calculation, using the best nuclear parameters then available. Subsequent authors (e.g., Fowler 1960; Fowler, Caughlan, and Zimmerman 1967) have kept the value of the reaction rate abreast of revisions in the value of the Gamow-Teller coupling constant. Recently, Bahcall and Shaviv (1968) have shown that the derived values of various important solar quantities, such as the neutrino production and central temperature, are sensitive to the precise rate of reaction (1). Because of its fundamental importance, we have recalculated the rate of the proton-proton reaction using the more accurate two-nucleon data now available, a variety of wave functions with which we estimated the likely errors in the calculation of the nuclear matrix element, the very recent remeasurement of the neutron half-life, improved numerical techniques for evaluating phase-space integrals, and an estimate of the radiative corrections. A complete discussion of the details of our calculations is in preparation.

Our results for the low-energy cross-section factor (~12.5 per cent larger than the number currently in general use) and its first derivative are given in equation (8); the significantly lower ^{8}B solar neutrino flux implied by our values is given in equation (9). We present in equations (10) an expression for the rate of nuclear-energy generation by the proton-proton chain.

It is both conventional and convenient to express the theoretical results in terms of the cross-section factor $S(E)$ ($S(E) = (\sigma/E)e^{-2\pi\eta}$, cf. Salpeter 1952; Fowler *et al.* 1967). We find (Bahcall and May 1968), using Salpeter's notation,

$$S(E) = 4\pi^2 m_p c\, a \ln 2\,[\Lambda^2(E)/\gamma^3]\,[f_{p\text{-}p}/(f\tau_{1/2})_n]\left[1 - \frac{(f\tau_{1/2})_n(1+\delta)}{2(f\tau_{1/2})_{0^+\rightarrow 0^+}(1+\delta')}\right], \tag{2}$$

where m_p is the proton mass, a the fine-structure constant, $\Lambda(E)$ the radial nuclear matrix element, $\gamma^{-1}(=4.316 \times 10^{-13}$ cm) the radius of the deuteron, and $f_{p\text{-}p}$ the phase-space integral for reaction (1). Here $(f\tau_{1/2})_n$ and $(f\tau_{1/2})_{0^+\rightarrow 0^+}$ are, respectively, the well-

* Supported in part by the National Science Foundation (GP-7976, formerly GP-5391) and the Office of Naval Research (Nonr-220(47)).

known "$f\tau$ values" (cf. Bahcall 1966*a*) without theoretical radiative corrections for the neutron decay and for the superallowed decays of $0^+ \rightarrow 0^+$ transitions such as the beta decay of ^{14}O. The quantities δ and δ' are, respectively, the radiative corrections (e.g., Berman and Sirlin 1962; Abers, Norton, and Dicus 1967; Johnson, Low, and Suura 1967) to the Fermi part of the neutron decay rate and the total correction to the $0^+ \rightarrow 0^+$ rate. We have written equation (2) in the above form in order to make explicit the effect of uncertainties in various quantities on the computed values of $S(E)$ and hence on the total reaction rate. We observe that, in order to evaluate $S(E)$, we need to (1) calculate $\Lambda(E)$; (2) calculate phase-space integrals such as f_{p-p}; (3) estimate the radiative corrections; and (4) use the most accurate values for the neutron and $0^+ \rightarrow 0^+$ $(f\tau)$ values.

In order to calculate $\Lambda(E)$, we used a deuteron wave function constructed by McGee (1966), which fits not only the static properties of the deuteron but also the low-energy n-p scattering data up to 300 MeV. McGee's wave function is essentially indistinguishable from other numerical wave functions such as that of Partovi (1964). We used numerical proton-proton wave functions obtained from solving the Schroedinger equation for the two protons with nuclear potentials of various assumed shapes and (Noyes 1964; Gursky and Keller 1964) a scattering length $a = -7.82\ (\pm 0.04)f$ and an effective range $\rho_p = 2.79(\pm 0.06)f$. We find

$$\Lambda^2(E) = (7.08 \pm 0.14)\ [1 + 2.2\ E(\text{MeV})]\ . \qquad (3)$$

The above result for $\Lambda^2(0)$ is about 4 per cent larger than the value due to Friedman and Motz (1953), adopted by Salpeter (1952). In making our error estimate in equation (3), we have made use of numerical calculations (Bahcall and May 1968) with different deuteron wave functions, various assumed shapes for the proton-proton potential, and have let the proton and deuteron parameters range over a conservative estimate of the domain of their likely uncertainties.

The numerical calculation of the phase-space integral f_{p-p} can be carried out to a sufficiently high accuracy (much better than 1 per cent) so that the remaining uncertainty does not make a significant contribution to the over-all error in $S(E)$. Using the computer program developed by Bahcall (1966*a*), we obtain

$$f_{p-p}(E) = 0.142[1 + 9.04\ E(\text{MeV})]\ , \qquad (4)$$

where E(MeV) is the center-of-mass energy, in MeV, of the two colliding protons. The above result includes corrections for nuclear size and relativistic effects and is based on the atomic-mass difference between two protons and a deuteron of (1.4421 ± 0.0002) MeV (Lauritsen and Ajzenberg-Selove 1966). Our value for f_{p-p} is 2 per cent smaller than the one used by Salpeter (1952).

The value of $(f\tau)_{0^+ \rightarrow 0^+}$ can be obtained from the accurate experiment (Bardin *et al.* 1962) on the decay of ^{14}O. The $f\tau$ value without radiative corrections is

$$(f\tau)_{^{14}O} = 3074(1 \pm 0.005)\ \text{sec.} \qquad (5)$$

The most recent and accurate measurement of the neutron half-life (Christensen *et al.* 1967) and the accurately calculated f value for the neutron (Bhalla 1966; Bahcall 1966*a*) give us

$$(f\tau_{1/2})_n = 1098(1 \pm 0.015)\ \text{sec.} \qquad (6)$$

The value for the neutron lifetime (10.80 ± 0.16) min given by Christensen *et al.* (1967) was noted by them to be inconsistent with the earlier result of (11.7 ± 0.3) min obtained by Sosnovsky *et al.* (1959). We adopt the lifetime measurement of Christensen *et al.* because it has a smaller quoted probable error and is more recent. The new lifetime increases the calculated $S(0)$ by about 10 per cent compared to the value obtained by Fowler *et al.* (1967).

A number of radiative processes are associated with the weak interaction vertex that describes the beta decay of reaction (1). These radiative processes may be approximately taken account of by using, in the derivation of equation (2), the value of an effective G_A extracted in a slightly unorthodox way from the neutron and ^{14}O decays. The appropriate value of G_A^2 to use in deriving equation (2) is

$$G_A^2(\text{eff}) = \frac{2\pi^3 \ln 2 (\hbar/m_e c)^7 m_e^2 c^3}{3(f\tau)_n}\left[1 - \frac{(f\tau)_n(1+\delta)}{2(f\tau)_{0^+\to 0^+}(1+\delta')}\right]. \quad (7)$$

This expression takes account of the modification of the Fermi matrix elements by the radiative corrections; the axial-vector matrix element is then $G_A^2(\text{eff})\ \langle\sigma\rangle^2$. The underlying assumption involved in the use of equation (7) is that the radiative corrections to the axial-vector part of the neutron decay are the same as for the proton beta decay in reaction (1). One can see from the work of previous authors (e.g., Berman *et al.* 1962; Abers *et al.* 1967; Johnson *et al.* 1967) that it is an excellent approximation to set $(1 + \delta)$, $(1 + \delta')$ equal to unity in equations (2) and (7). The value of $G_A^2(\text{eff})$ obtained from equation (7) with the $f\tau$ values given in equations (5) and (6) is about 2 per cent larger than the value that would have been obtained if, as is customary, radiative corrections were included for all matrix elements entering equation (7). The constant g used by Salpeter can also be calculated from equations (5)–(7) and is 6.9×10^{-7} sec^{-4} compared to the value of 7.5×10^{-4} that was the best estimate at the time Salpeter's calculation was made and the value of 6.3×10^{-4} that was used by Fowler *et al.* (1967).

Inserting the results of equations (3)–(6) in equation (2), we find

$$S(E = 0) = (3.78 \pm 0.15) \times 10^{-25}\ \text{MeV barns}\ , \quad (8a)$$

$$\left(\frac{1}{S}\frac{dS}{dE}\right)_{E=0} = (11.2 \pm 0.1)\text{MeV}^{-1}, \quad (8b)$$

and (cf. Salpeter 1952; Bahcall 1966*b*)

$$S_{\text{eff}} = (3.78 \pm 0.15) \times 10^{-25}\ \text{MeV barns}\ (1 + 0.417\tau^{-1} + 12.6\tau^{-2} + 36.6\tau^{-3})\ , \quad (8c)$$

where $\tau = 33.80\ T_6^{-1/3}$ and T_6 is the temperature in units of 10^6 °K. The result (8a) is about 12.5 per cent larger than the value quoted by Fowler *et al.* (1967). Our result for $[(dS/dE)/S]_{E=0}$ is about 30 per cent larger, due mainly to an improved estimate of the energy dependence of $\Lambda(E)$ (cf. eq. [3]).

The effect of the above changes on the solar neutrino fluxes can be estimated from the work of Bahcall and Shaviv (1968; see esp. Fig. 5). We find for the important ^{8}B flux that the most probable value (computed by correcting approximately their model MP) is

$$\phi_\nu(^8\text{B}) = 0.9 \times 10^{+7}\ \text{cm}^{-2}\ \text{sec}^{-1} \quad (9)$$

compared to the value of $1.3 \times 10^{+7}$ cm^{-2} sec^{-1} obtained by Bahcall and Shaviv with $S(E = 0) = 3.36 \times 10^{-25}$ MeV barns.

Our results (eq. [8]) also imply that the nuclear-energy-generation rate for the proton-proton chain is given by

$$E_{p\text{-}p} = (4.72 \pm 0.2)\ 10^{+6}\chi_e\rho X_{\text{H}}^2 T_6^{-2/3} f_{\text{scr}}(\exp - \tau)$$
$$\times\ (1 + 0.417\tau^{-1} + 12.6\tau^{-2} + 36.6\tau^{-3})\ \text{ergs gm}^{-1}\ \text{sec}^{-1}\ . \quad (10a)$$

The quantity χ_e is defined by equation (48) of Parker, Bahcall, and Fowler (1964) and is (cf. Fowler *et al.* 1967 for the most accurate values of $\langle i,j\rangle$)

$$\chi_e = \left[0.981 \frac{\langle {}^3\mathrm{He},{}^3\mathrm{He}\rangle X^2_{^3\mathrm{He}}}{\langle {}^1\mathrm{H},{}^1\mathrm{H}\rangle} + 1.922 \frac{\langle {}^3\mathrm{He},{}^4\mathrm{He}\rangle X_{^3\mathrm{He}} X_{^4\mathrm{He}}}{\langle {}^1\mathrm{H},{}^1\mathrm{H}\rangle} F_{e-}({}^7\mathrm{Be}) + 1.430 \frac{\langle {}^3\mathrm{He},{}^4\mathrm{He}\rangle}{\langle {}^1\mathrm{H},{}^1\mathrm{H}\rangle} X_{^3\mathrm{He}} X_{^4\mathrm{He}} F_p({}^7\mathrm{Be})\right]. \quad (10b)$$

The factor f_{scr} is the electron screening factor of Salpeter (1954) and is also defined by equations (4) of Parker *et al.* (1964).

We are grateful to E. S. Abers and R. P. Feynman for informative discussions of the radiative corrections, to C. A. Barnes, William A. Fowler, and R. McCray for stimulating conversations, and to B. A. Zimmerman for valuable assistance with the numerical calculations.

REFERENCES

Abers, E. S., Norton, R. E., and Dicus, D. A. 1967, *Phys. Rev. Letters*, **18**, 676.
Bahcall, J. N. 1966*a*, *Nucl. Phys.*, **75**, 10.
———. 1966*b*, *Ap. J.*, **143**, 259.
Bahcall, J. N., and May, R. M. 1968 (in preparation).
Bahcall, J. N., and Shaviv, G. 1968, *Ap. J.*, (in press).
Bardin, R. K., Barnes, C. A., Fowler, W. A., and Seeger, P. A. 1962, *Phys. Rev.*, **127**, 583.
Berman, S. M., and Sirlin, A. 1962, *Ann. Phys.*, **20**, 20.
Bethe, H. A., and Critchfield, C. L. 1938, *Phys. Rev.*, **54**, 248.
Bhalla, C. P. 1966, *Phys. Letters*, **19**, 691.
Christensen, C. J., Nielsen, A. J., Bahnsen, A., Brown, W. K., and Rustach, B. M. 1967, *Phys. Letters*, **26B**, 11.
Fowler, W. A. 1960, *Mém. Soc. Roy. Sci. Liège*, ser. 5, **3**, 207.
Fowler, W. A., Caughlan, G. R., and Zimmerman, B. A. 1967, *Ann. Rev. Astr. and Ap.*, **5**, 525.
Friedman, E., and Motz, L. 1953, *Phys. Rev.*, **89**, 648.
Gursky, M. L., and Keller, L. 1964, *Phys. Rev.*, **136**, B1693.
Johnson, K., Low, F. E., and Suura, H. 1967, *Phys. Rev. Letters*, **18**, 1224.
Lauritsen, T., and Ajzenberg-Selove, F. 1966, *Nucl. Phys.*, **78**, 1.
McGee, I. J. 1966, *Phys. Rev.*, **151**, 772.
Noyes, H. P. 1964, *Phys. Rev. Letters*, **12**, 171.
Parker, P. D., Bahcall, J. N., and Fowler, W. A. 1964, *Ap. J.*, **139**, 602.
Partovi, F. 1964, *Ann. Phys.* (*N.Y.*), **27**, 79.
Salpeter, E. E. 1952, *Phys. Rev.*, **88**, 547.
———. 1954, *Australian J. Phys.*, **7**, 373.
Sosnovsky, A. N., Spivak, P. E., Prokofiev, Y. A., Kutikov, I. E., and Dobrinin, Y. P. 1959, *Nucl. Phys.*, **10**, 395.

THE ^{7}Be ELECTRON-CAPTURE RATE

JOHN N. BAHCALL* AND CHARLES P. MOELLER
California Institute of Technology, Pasadena, California
Received July 3, 1968

ABSTRACT

The effect of plasma and bound-electron screening on the continuum capture rate of ^{7}Be$(e^-,\nu)^7$Li is calculated and shown to be negligible for situations in which the proton-proton chain is likely to be important. A more accurate expression for the continuum capture rate is then derived, making use of the numerical work of Bahcall and May. Convenient formulae are presented for use in stellar-model calculations. The average effect of bound-electron capture is also evaluated for several solar models.

I. INTRODUCTION

The predominant reaction by which ^{7}Be is destroyed in the proton-proton chain, under the conditions existing near the center of the Sun, is the electron-capture reaction ^{7}Be$(e^-,\nu)^7$Li. The equilibrium abundance of ^{7}Be is accurately given by equating the rates of ^{3}He$(\alpha,\gamma)^7$Be and ^{7}Be$(e^-,\gamma)^7$Li. The number of ^{7}Be$(p,\gamma)^8$Be reactions occurring per unit of time, which is expected (Bahcall 1964, 1966) to determine the counting rate in experiments designed to detect solar neutrinos (Davis 1964; Davis, Harmer, and Hoffman 1968), is therefore inversely proportional to the ^{7}Be electron-capture rate. Some time ago, Bahcall (1962) computed the ^{7}Be capture rate considering the capture of continuum electrons and neglecting the plasma screening by the ionized gas of the star. More recently, Iben, Kalata, and Schwartz (1967) computed the capture rate of bound electrons in ^{7}Be III and ^{7}Be IV using the Debye-Hückel approximation to estimate the screening effect of the ionized plasma on the rate of bound-electron capture. They found that bound-electron capture increases the total capture rate in the solar interior by 17–25 per cent in the solar interior and that plasma screening strongly affects the rate of bound-electron capture.

We show in § II that the effect of plasma and bound-electron screening is negligible on the *dominant* process of continuum electron capture. We then derive in § III convenient analytic expressions for the total capture rate, making use of the results of Bahcall and May (1968) to calculate more accurately the rate of the continuum process. We also present in § IV the results of some solar-model calculations of the average effect of bound-electron capture on the solar rate of ^{7}Be$(e^-,\nu)^7$Li.

II. SCREENING CALCULATIONS

a) Plasma Screening

We adopt, following Iben *et al.* (1967), the Debye-Hückel approximation for the plasma screening. In this approximation the *S*-wave component of the continuum electron wave function satisfies the equation (cf. Landau and Lifschitz 1958):

$$\left[-\frac{\hbar^2}{2m}\nabla^2 - \frac{4e^2}{r}\exp\left(\frac{-r}{R}\right) - E_R\right]\psi_R(r) = 0\,. \quad (1)$$

Here R is the screening radius, defined by

$$R = [(4\pi e^2/kT)\Sigma_a n_{0a} z_a{}^2]^{-1/2} \quad (2a)$$

* Alfred P. Sloan Foundation Fellow.

^{3}He(^{3}He, $2p$)^{4}He Total Cross-Section Measurements Below the Coulomb Barrier*

M. R. Dwarakanath and H. Winkler†
California Institute of Technology, Pasadena, California 91109
(Received 25 June 1971)

Measurements of total cross sections for ^{3}He(^{3}He, $2p$)^{4}He have been made for center-of-momentum energies between 80 keV and 1.1 MeV. A continuously recirculating differentially pumped gas target system was employed to minimize uncertainties in energy loss and straggling. A calorimetric device was used to integrate the beam current within the target gas. Proton angular distributions were measured at seven energies. The measured cross-section factor, $S(E)$, was fitted to a linear function of energy for $E_{c.m.} < 500$ keV:

$$S(E_{c.m.}) = S_0 + S_0' E_{c.m.},$$

where $S_0 = (5.0^{+0.6}_{-0.4})$ MeV b and $S_0' = (-1.8 \pm 0.5)$ b. The formula $S(E_{c.m.}) = S_0 + S_0' E_{c.m.} + \frac{1}{2} S_0'' E_{c.m.}^2$, with $S_0 = 5.2$ MeV b, $S_0' = -2.8$ b, and $S_0'' = 2.4$ b MeV^{-1} gives a good representation of $S(E)$ over the entire range of energies studied here.

I. INTRODUCTION

For solar-model calculations, it is necessary to have accurate total cross sections for the reaction ^{3}He(^{3}He, $2p$)^{4}He at energies well below the Coulomb barrier.[1] This reaction completes one branch of the proton-proton chain of nuclear reactions, the dominant energy-generation mechanism in certain main sequence stars such as our sun. The rate of this reaction is an important factor in determining the relative importance of the different branches of the p-p chain in the sun. In particular, the branch terminated by the sequence ^{7}Be(p, γ)^{8}B($e^+\nu$)-^{8}Be(α)^{4}He produces neutrinos with energies up to 14.1 MeV, and the detection of these neutrinos constitutes a direct test of solar models.[2] Such a test is quantitative only if the nuclear information used in the models is accurate. The cross section at 20 keV (c.m.) for the reaction ^{3}He-(^{3}He, $2p$)^{4}He is estimated to be about 3×10^{-13} b and thus cannot be measured directly in the laboratory. Hence, the cross section or rather the cross-section factor has to be extrapolated from measurements at higher energies. The cross-section factor $S(E)$ is the total cross section with the dominant energy dependence factored out

$$S(E) \equiv \sigma(E) E\, e^{2\pi\eta}; \quad \eta = \frac{Z_0 Z_1 e^2}{\hbar v},$$

with Z_0, Z_1, v, and E equal to the charge numbers, the relative velocity, and the center-of-momentum energy of the interacting particles. $S(E)$ varies only slowly with energy for nonresonant processes and so is more suitable for extrapolation than the rapidly varying $\sigma(E)$.

The total cross section for $^3He(^3He, 2p)^4He$ was first measured by Good, Kunz, and Moak[3] in 1953. The chief uncertainty in their measurements lies in the bombarding energy, as they employed a thick target of ^3He atoms trapped in an aluminum foil. The uncertainty in the energy is reflected as a large uncertainty in $S(E)$ through the exponential term. Bacher and Tombrello[4] have investigated the reaction in detail for $0.5 < E_{c.m.} < 18$ MeV. At energies above the Coulomb barrier, they interpret their results as demonstrating that the reaction proceeds through the intermediate step, $^5Li + p$. However, a different mechanism is required to explain the different shape of the proton energy spectra at bombarding energies well below the Coulomb barrier. Thus, the data at higher energies cannot be extrapolated to the energy regions of astrophysical interest. Neng-Ming *et al.*[5] have reported measurements in the energy region $0.25 < E_{c.m.} < 0.85$ MeV. Concurrently with the present work, Bacher and Tombrello[6] extended their measurements to lower energies by modifying their gas-cell target system. These results will be compared in a later section.

The present work describes measurements for $0.08 < E_{c.m.} < 1.1$ MeV, and employs a new technique for rare-gas targets. It has the advantage of eliminating the entrance window for the primary beam and at the same time using limited quantities of the rare gas. Beam integration was carried out inside the gas target by a calorimetric method. The protons and α particles were identified by a ΔE-E counter telescope with a transmission gas proportional counter as the first element of the telescope. Preliminary results of this work were given earlier.[7]

II. EXPERIMENTAL APPARATUS

Singly charged ^3He ions were obtained from the California Institute of Technology 3-MV and 500-kV Van de Graaff electrostatic generators. The energy calibrations of the magnetic analyzers for the accelerated beam were checked by observing well-known (p, γ) and $(p, \alpha\gamma)$ resonances with both protons and the molecular ions, HH^+ and HHH^+. The energy of the beam from the accelerator was thus known to within ±1 keV.

The differentially pumped gas target apparatus is shown in Figs. 1 and 2. The beam entered the target region through three canals. The main pressure drop occurred across canal A (see Figs. 1 and 2) which was 3 mm in diameter and 2 cm long. This canal connected the target chamber to a large-volume chamber A, which was pumped by a set of three Roots blowers in cascade. The output of the third Roots blower developed a sufficiently high pressure to allow recirculation of the gas. However, the gas coming out of the pumps was contaminated both by traces of oil vapor from the pumps and by air leaking into the system through small real or virtual leaks. The gas was cleaned of impurities by passing it through a zeolite adsorption trap at liquid-nitrogen temperature. The chilled gas coming out of the trap was allowed to exchange heat with the incoming gas before the gas was fed back to the target chamber, typically at a pressure of 20 Torr. The adsorber was very efficient in removing all gases and vapors except helium. The number of impurity atoms was found to be less than 1% of the number of helium atoms by observing the elastic scattering of protons from the gas target.

The pressure in chamber A was typically 0.1 Torr. This chamber was connected to another large volume, chamber B, by canal B which was 3.5 mm in diameter and 10 cm long. Chamber B was pumped by a 10-cm oil diffusion pump with its backing line connected to chamber A. This allowed the recovery of most of the gas streaming through canal B. Since the pressure in chamber B was typically 2×10^{-5} Torr, the beam pipe from the accelerator could be connected directly to this chamber through a third impedance, canal C. The gas leaking through canal C was about 1% of the total charge per hour and was not recovered. The total recirculating gas charge was approximately 500 cm³ STP for a target pressure of 20 Torr.

The beam energy loss up to the point of entering canal A was negligibly small. In order to minimize the subsequent energy loss of the beam, canal A was positioned as close as possible to the target region. The tip of canal A was 7 mm from the center of the target chamber, thus permitting a target thickness of up to 1 cm to be used. The energy loss of the beam at the target center was about 15 keV and the uncertainty in this was only 3 keV under the most adverse conditions. An aperture, slightly smaller than the bore of canal A, was placed at the entrance of the canal; the canal therefore also served as an antiscattering baffle. The fraction of degraded beam was mea-

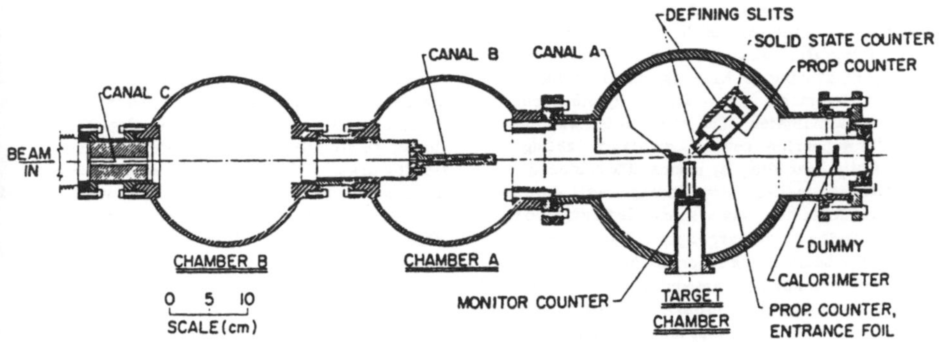

FIG. 1. A horizontal section of the differentially pumped gas target apparatus with counter telescope positioned at 45° to beam direction.

sured to be less than 0.5% for 1-MeV protons. The temperature and pressure of the gas were continuously monitored. Gas pressure could be measured accurately to ±0.1 Torr on an aneroid gauge, the accuracy of which was previously ascertained by an absolute gauge of the McLeod type. As the gas is continuously flowing out of the target region, the temperature here is estimated to be only 1 to 3°C higher than the gas temperature at the wall of the target chamber where the temperature was monitored. Reduction of gas pressure in the target region due to space-charge repulsion and hard collisions is not important at the target pressures and beam currents employed. This has been verified experimentally by employing beam currents differing by a factor of about 4. The gas pressure in the target region is essentially unmodified by the gas flow through the canal as the center of the target is situated several canal radii away from the end of the canal. There is

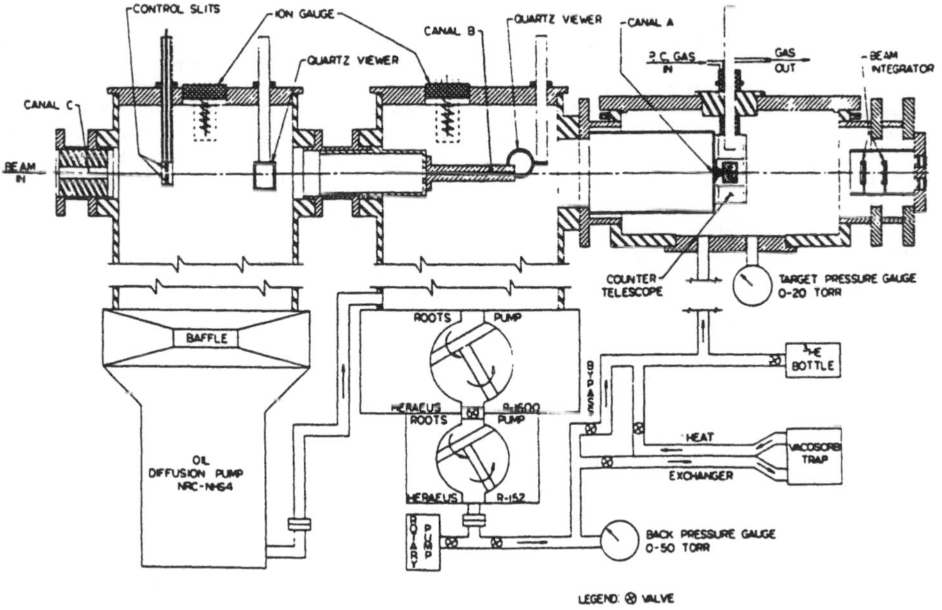

FIG. 2. A vertical section of the differentially pumped gas target apparatus with counter telescope positioned at 90° to beam direction. Only the upper part of the figure is drawn to scale.

thus an uncertainty of about 2–3% in the target density which is included in the quoted errors.

Beam integration was carried out inside the gas target by measuring the heat dissipated by the beam in a low-mass, high-conductivity metal cup called the calorimeter. The calorimeter was mechanically supported from a large flange, acting as an infinite heat sink, by means of a soldered thin stainless-steel tube constituting a well-defined heat leak. Some heat was also carried away from the calorimeter by the target gas. The heat produced by the beam in the calorimeter was measured by determining the amount of electrical energy put into a dummy of similar construction to maintain the calorimeter and the dummy at the same temperature during beam integration. The electrical energy supplied to the dummy did not exactly equal the energy put into the calorimeter by the beam because of a small departure from symmetry in the construction of the two heat leaks; however, the proportionality constant between the calorimeter and dummy was determined by supplying a measured amount of electrical energy to the calorimeter. This proportionality constant is called the calibration constant K. The number of incident particles N_B is given by

$$N_B = \frac{KW_e}{E_c},$$

where K is the calibration constant, W_e is the electric energy supplied to dummy, and E_c is the beam energy at the calorimeter. The energy loss of the beam in the gas, before hitting the calorimeter, is important in this method of beam integration. Energy losses were calculated from tables.[8,9] The errors introduced into N_B and E, the energy at the center of the target, by uncertainties in the specific energy loss contribute to errors in $S(E)$ in opposite directions. Therefore, systematic errors in the specific energy-loss data are partially compensated in $S(E)$.

The accuracy of the calorimetric beam-integration device was checked by measuring the known[10] differential cross section for $^{40}A(p,p)^{40}A$ at 135° in the laboratory with proton energies below 1.8 MeV (pure Rutherford scattering). For a beam power range from 30 mW to 3 W and varying conditions of beam stability, the uncertainty in N_B was found to be less than ±5% for runs of 4 min or longer. A more detailed description of this beam-integration device is given elsewhere.[11,12]

The reaction particle spectra were measured with a ΔE-E counter telescope in order to separate the protons from the α particles, both of which have continuous energy spectra from zero to the three-body end point. The transmission (ΔE) counter limits the observation of particle spectra to energies not stopped inside it; for this reason it is desirable to use as thin a ΔE detector as possible. A gas proportional counter with a 0.6-mg/cm² Ni foil entrance window was used in the present experiment and the window was chosen so that it was sufficiently thick to stop the strong flux of elastically scattered ^{3}He's. Continuously flowing argon +5% CO_2 at a pressure of about 70 Torr was used as the proportional-counter gas and the gas multiplication factor was between 100 and 1000. The resolution of the proportional counter was adequate to separate completely protons and α particles of the same energy. The $E' \equiv (E - \Delta E)$ counter was a 1500-μ silicon surface-barrier detector. Proton energies greater than 600 keV and α energies greater than 2 MeV could be measured with this telescope. The low-energy detection limit of the telescope was checked with 600-keV protons elastically scattered from argon. The counter telescope could be rotated about the center of the target chamber and the most backward angle accessible was 140°. At angles smaller than 45° the proportional-counter body eclipsed the beam path to the integrator. Absolute differential cross-section measurements were, therefore, restricted to the range of angles 45–140°. For measurements at more forward angles, a monitor counter fixed at 90° to the beam and covered with a foil to stop elastically scattered particles, was used to provide a measure of the beam intensity.

The solid angle subtended by the detector and the effective thickness of the target were defined by two apertures. The first aperture was circular with radius $a \approx 5.5$ mm and was placed directly in front of the surface-barrier detector at a distance $D \approx 80.5$ mm from the center of the target. The second aperture was rectangular with width $=2W$ and height much larger than its width, and was placed at a distance $d \approx 67.5$ mm from the circular aperture along the line joining the center of the circular aperture and the target center. The two apertures were placed perpendicular to this line. The rectangular slit was placed with its long axis perpendicular to the reaction plane. The effective product of solid angle and target thickness, for the telescope set at an angle ψ to the beam direction, is given by

$$\langle \Omega l \rangle_\psi = \frac{\pi a^2 2W}{Dd} \csc\psi \left[1 + O\left(\frac{a^2}{d^2}\right)\right].$$

III. EXPERIMENTAL PROCEDURE AND ANALYSIS

The reaction particle spectra were measured using the counter telescope described above and a Nuclear Data two-dimensional pulse-height analyzer operating in a 64 × 64 channel configuration. Pulses from the ΔE and E'detectors were fed to the x and y analog-to-digital converters of the analyzer after amplification. A gating signal from a low-level discriminator on the E' counter pulses was required for the two-dimensional storage of a count. Particle spectra appeared on the cathode-ray-tube display of the analyzer as two nearly rectangular hyperbolas, well separated from each other and from background. The raw particle spectra were obtained by summing all the counts in the various ΔE channels corresponding to a given E' channel about the locus of the selected hyperbola. The raw particle spectra are distorted because of the energy loss ΔE, before reaching the surface-barrier counter. The true energy spectrum $N(E)dE$ was computed from the measured spectrum $N(E')dE'$ by using the relation

$$N(E)dE = \left(N(E') \times \frac{dE'}{dE}\bigg|_{E'}\right) dE\,,$$

where $(dE'/dE)|_{E'}$ is the slope calculated from a graph of E', the energy at the surface-barrier detector, against E, the energy at the target center. Corrected proton-energy spectra at 90° for $E_{^3He} = 1.99$ and 0.19 MeV are shown in Figs. 3 and 4. The arrow at a proton energy of ~600 keV shows the lower limit of proton energies observed by the counter telescope. Counting statistics are shown for a few representative points. The dashed curve in Fig. 4 is the spectrum calculated by assuming a statistical distribution and isotropic angular distribution in the center-of-momentum system. The peak in the proton spectrum near the three-body end point, in Fig. 3, corresponds to protons from the two-body breakup: $^5Li + p$. The relatively flat part of the spectrum corresponds to the breakup of ^5Li in flight and to contributions from other mechanisms. At $E_{^3He} = 0.19$ MeV, the proton spectrum shows a much smaller effect of the final-state interaction in the $^4He + p$ system.

Total proton yields were obtained by summing all the counts in the observed spectrum. A correction for the unobserved low-energy part of the spectrum was made by assuming that the spectral shape in this narrow interval is given correctly by phase-space considerations. Though this assumption is not valid for the higher bombarding energies, it gives a rough estimate of the correction, which is itself small (~5%). The error in the total number of protons counted, introduced by this assumption is less than ±2%. The differential cross section was calculated from the total proton yield $Y(E, \psi)$ at true bombarding energy E and angle ψ, by using the relation

$$\frac{d\sigma}{d\Omega}\bigg|_{E,\psi} = \frac{Y(E, \psi)}{2N_B n_T \langle \Omega l \rangle_\psi}\,,$$

where n_T is the number density of the target nuclei

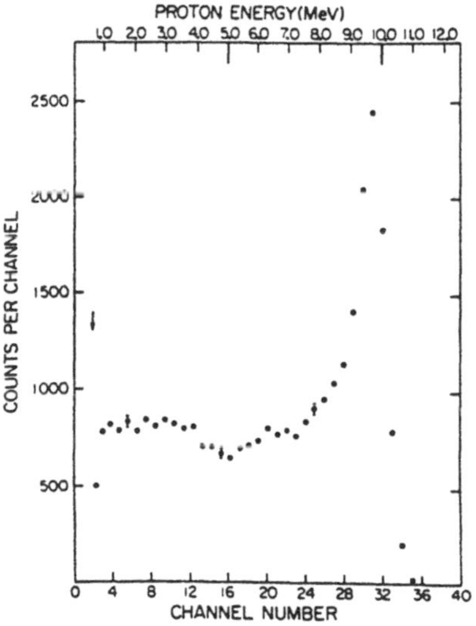

FIG. 3. Proton-energy spectrum, corrected for energy loss in the proportional-counter entrance foil, at $E_{^3He} = 1.99$ MeV and $\psi = 90°$. Arrow at $E_p = 0.6$ MeV indicates the low-energy detection limit of the telescope.

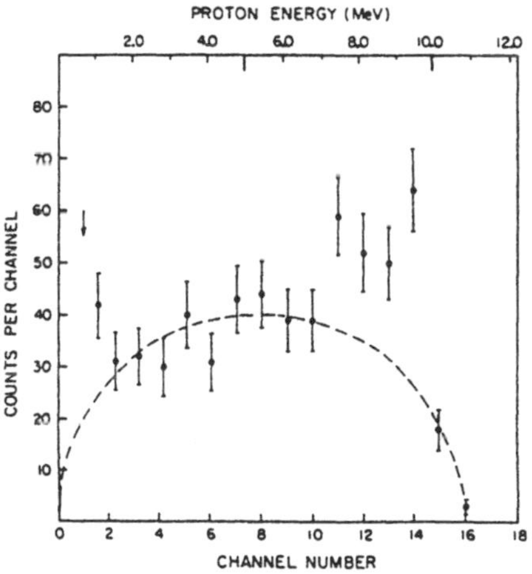

FIG. 4. Proton-energy spectrum at $E_{^3He} = 0.19$ MeV and $\psi = 90°$. The dashed curve shows the spectrum calculated by assuming a statistical spectral distribution and isotropic angular distribution in the center-of-momentum system.

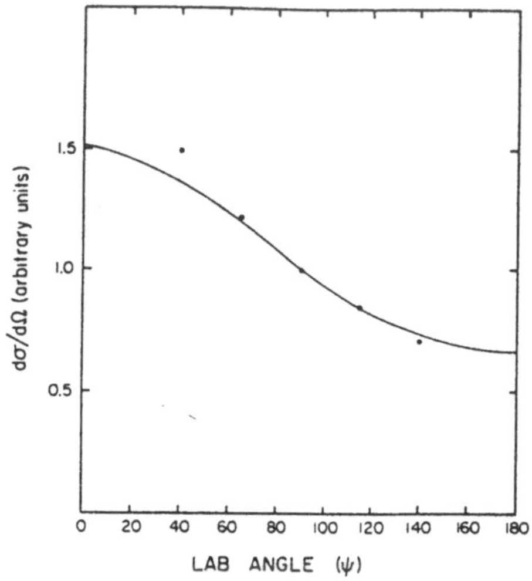

FIG. 5. Proton angular distribution in the laboratory for $E_{^3He} = 2.0$ MeV. The solid curve shows the angular distribution calculated by assuming a statiscal spectral distribution and isotropic angular distribution of protons in the center-of-momentum system and normalized to 1.0 at $\psi = 90°$.

at the target region and all other quantities are as defined earlier. The factor $\frac{1}{2}$ appears in the cross section because two protons are released in each reaction.

Proton angular distributions were obtained at seven energies over the range of angles 20–140°

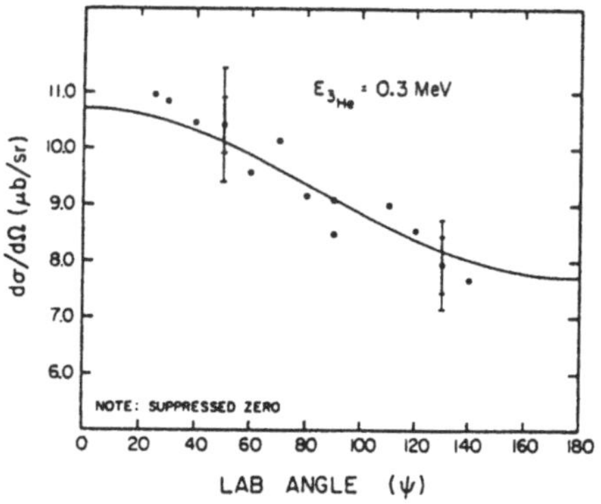

FIG. 6. Proton angular distribution in the laboratory for $E_{^3He} = 0.3$ MeV. The larger error bar indicates the absolute error in the measured differential cross sections and the smaller error bar indicates the relative error between measurements at different angles. Solid curve has the same meaning as in Fig. 5.

using the monitor counter and, where possible, the beam integrator. Figures 5 and 6 show proton angular distributions at $E_{^3He} = 2.0$ and 0.3 MeV. The solid curve shows angular distributions in the laboratory system, calculated under the assumption of statistical spectral distribution and isotropic angular distributions in the center of momentum. Figure 6 shows two sets of error bars at a few points. The larger bars denote the uncertainty in the absolute differential cross section while the smaller error bars give the relative errors.

The total reaction cross section was obtained by numerical integration of the measured angular distributions at energies where such measurements were made. The total cross section (σ) is greater than 4π times the differential cross section at 90° [$4\pi\sigma(90°)$] by 8% at $E_{^3He} = 2.0$ MeV, and by only 1% at $E_{^3He} = 0.3$ MeV. The ratio $\sigma/4\pi\sigma(90°)$ varies smoothly and slowly with energy. At energies where no angular distributions were measured, the total cross sections were obtained from $4\pi\sigma(90°)$ and the interpolated (or extrapolated) value of the ratio $\sigma/4\pi\sigma(90°)$. The variation of σ with center-of-momentum energy is shown in Fig. 7. The cross-section factor $S(E_{c.m.})$ was calcu-

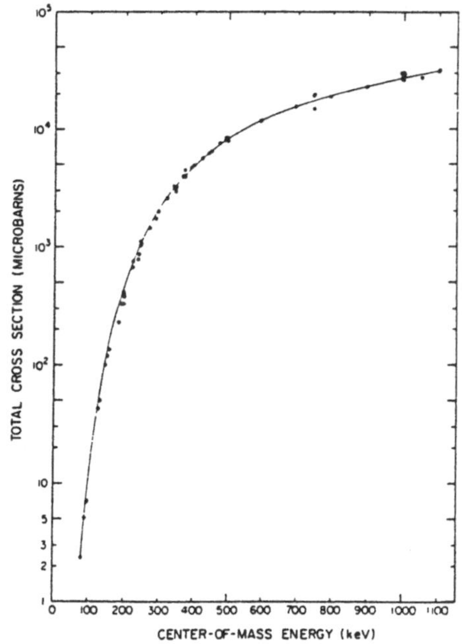

FIG. 7. $^3He(^3He, 2p)^4He$ total cross sections as a function of the center-of-momentum energy. The solid line is a guide to the eye.

lated from $\sigma(E_{c.m.})$ using the defining formula[13]

$$S(E_{c.m.}) = \sigma(E_{c.m.}) \times E_{c.m.} \times \exp\left(\frac{4.860}{E_{c.m.}^{1/2}}\right),$$

$E_{c.m.}$ in MeV.

Figure 8 shows the measured cross-section factors as a function of the c.m. energy. Total errors, including both counting statistics and estimated systematic errors, are shown at a few points. The figure also shows the cross-section factors as calculated from the measurements of Good, Kunz, and Moak,[3] of Bacher and Tombrello,[4,6] and of Neng-Ming *et al.*[5] The cross-section factors of the present experiment were fitted to the function

$$S(E_{c.m.}) = S_0 + S_0' E_{c.m.}$$

by a standard least-squares routine over different ranges of energies. For $E_{c.m.} < 500$ keV, $S_0 = (5.0^{+0.6}_{-0.4})$ MeV b and $S_0' = (-1.8 \pm 0.5)$ b. The errors indicated in S_0 and S_0' include both statistical and systematic errors in the measured total cross section, as well as uncertainties in the bombarding energy. These errors have been discussed in Ref. 11. The data of Bacher and Tombrello[6] give $S_0 = 5.4 \pm 0.5$ MeV b and $S_0' = -1.8 \pm 0.5$ b, and thus agree within the quoted errors.

All the cross-section factors measured in this work ($0.08 < E_{c.m.} < 1.1$ MeV) are well represented by the formula

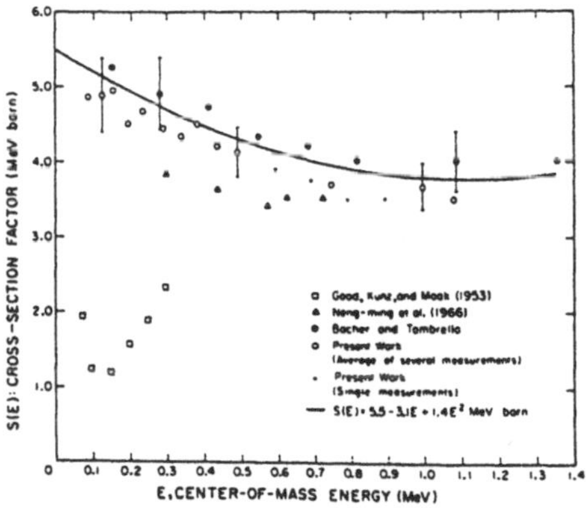

FIG. 8. ^{3}He(^{3}He,2p)^{4}He cross-section factors as a function of center-of-momentum energy. Error bars indicated are typical, include statistical and estimated systematic errors in both measured total cross sections and center-of-momentum energy. Solid line is the curve $S(E_{c.m.}) = 5.5 - 3.1E_{c.m.} + 1.4E_{c.m.}^2$ which gives the best quadratic representation of the data of the present work and the data of Bacher and Tombrello.

$$S(E_{c.m.}) = S_0 + S_0' E_{c.m.} + \tfrac{1}{2} S_0'' E_{c.m.}^2 ,$$

with $S_0 = 5.2$ MeV b, $S_0' = -2.8$ b, and $S_0'' = 2.4$ b MeV^{-1}. However, the parameters that give a good fit to the experimental data are not unique (χ^2 has a shallow minimum in parameter space) because the variation in $S(E)$ is not very large compared with the errors in the measured $S(E)$. The combined data of the present work and of Bacher and Tombrello can be represented by the parameter values $S_0 = 5.5$ MeV b, $S_0' = -3.1$ b, and $S_0'' = 2.8$ b MeV^{-1}. The $S(E)$ as given by this latter set of parameters is shown as a solid line in Fig. 8.

IV. DISCUSSION

The results of Good, Kunz, and Moak[3] are in serious disagreement with the present work. Their cross-section factor decreases rapidly from a value of 2.4 MeV b at $E_{c.m.} = 300$ keV to 1.2 MeV b at $E_{c.m.} = 150$ keV. Below this energy the cross-section factor increases sharply. This discrepancy is attributed to incomplete information on the distribution of ^{3}He in their target. The measurements of Neng-Ming *et al.*[5] are in general agreement with the measurements reported here. The values of the cross-section factors agree within the combined errors of the two measurements, their $S(E_{c.m.})$ being systematically lower. They quote much larger errors in their energy determination than in the present work. This error is reflected in the values for $S(E_{c.m.})$.

Bacher and Tombrello[6] have extended their measurements with a gas cell to $E_{c.m.} = 152$ keV. They minimized the uncertainty in the bombarding energy by experimentally measuring the energy loss at precisely the experimental energies. Their results are in very good agreement with the results given here. The absolute values of $S(E_{c.m.})$ agree within the limits of the combined errors,

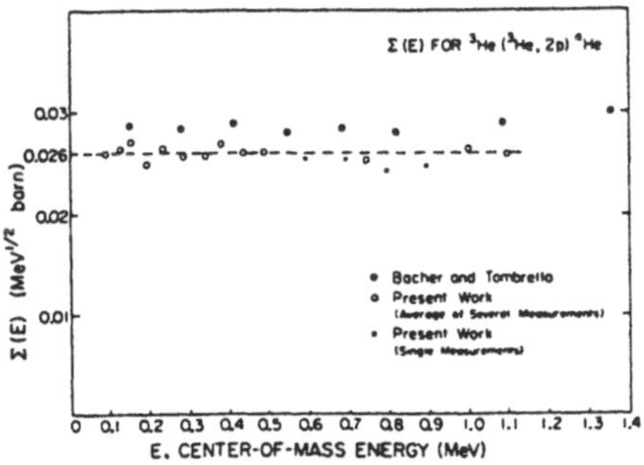

FIG. 9. $\Sigma(E)$, defined in text, as a function of the center-of-momentum energy. The dashed line denotes the average value of Σ for the present work.

their values being systematically higher by about 8%. The agreement in the energy variation of the cross-section factor is particularly impressive. The small systematic difference in the absolute values of $S(E_{c.m.})$ is not serious considering that the results have been obtained by two very different techniques.

Bacher and Tombrello[6] observed that the reaction mechanism is not entirely sequential at energies below the Coulomb barrier, in contrast to its behavior at higher energies. The proton spectra exhibit less and less of the effects of the final-state interaction in the (^{4}He + p) system at lower energies; a different reaction mechanism appears to dominate at low energies. May and Clayton[14] proposed a model in which a neutron from one of the ^{3}He nuclei "tunnels" to the other ^{3}He nucleus even when the two nuclei are outside the range of nuclear interaction. At low energies this mechanism might be expected to dominate over the other mechanisms which require a greater overlap of the ^{3}He nuclei, and may thus explain the change in shape of the proton spectra. May and Clayton calculated $S(E)$ for this reaction by considering a final-state interaction in the ($2p$) system rather than in the (^{4}He + p) system. They obtained a negative value for S_0' as observed experimentally, but the calculated absolute value of S_0' was too small to fit the data.

Since the astrophysical cross section requires an extrapolation to energies below those at which measurements have been made, it is of interest to see whether the observed negative slope of $S(E)$ at low energies could be the result of factoring out a zero-radius s-wave penetration factor ($2\pi\eta e^{-2\pi\eta}$) from the total cross section. We define a quantity, $\Sigma(E)$, which should be independent of energy and is somewhat analogous to $S(E)$, by replacing the zero-radius s-wave penetration factor by finite-radius penetration factors according to the following expression for the cross section:

$$\sigma(E) = \frac{1+\delta_{ij}}{2}\frac{\Sigma(E)}{\sqrt{E}}\left(\sum_l \omega_l \int_0^\infty P_l(r)T_l(r)r^2 dr\right),$$

where the first term on the right is chosen so as to give unity for reactions between identical particles. The $1/\sqrt{E}$ is the usual flux term and the density of final states is taken as independent of energy over the range of the present investigation, since the reaction has a large positive Q value. ω_l is the statistical weight for the relative motion of the interacting particles with angular momentum l. $P_l(r)$ is the probability per unit volume of finding the interacting particles at a separation r and $T_l(r)$ is the probability that the reaction occurs at a separation r in the lth partial wave.

A correct evaluation of the integral requires a knowledge of the wave functions and is thus fairly complicated. As a simplification, we parametrize the integral as

$$\int_0^\infty P_l(r)T_l(r)r^2 dr = P_l(R_l) = \frac{1}{F_l^2(kR_l) + G_l^2(kR_l)},$$

where F_l and G_l are the regular and irregular Coulomb wave functions and the effective reaction radii R_l are adjustable parameters. Choosing different radius parameters for the different partial waves presumably takes into account the differences in the different $T_l(r)$. If we retain only the s-wave interaction and choose $R_0 = 0$, we obtain

$$\sigma(E) = \frac{\Sigma(E)}{\sqrt{E}}\, 2\pi\eta e^{-2\pi\eta}$$

$$= \left[\sqrt{2}\,\pi Z_0 Z_1\left(\frac{e^2}{\hbar c}\right)(\mu c^2)^{1/2}\Sigma(E)\right]\frac{1}{E}e^{-2\pi\eta}.$$

The quantity in square brackets is readily recognized as $S(E)$. It may be noted that, when comparing $S(E)$ for isobaric analog reactions, the $S(E)$ should be normalized by the factor $1/Z_0Z_1$ before a comparison of the $S(E)$ values is made.

For the ^{3}He(^{3}He, $2p$)^{4}He reaction we assume that only s waves and p waves are important at the low energies of these measurements and write

$$\sigma(E) = \frac{\Sigma(E)}{\sqrt{E}}\left[P_0(R_0) + 9P_1(R_1)\right],$$

or

$$\Sigma(E) = \sigma(E)\sqrt{E}\left[P_0(R_0) + 9P_1(R_1)\right]^{-1}.$$

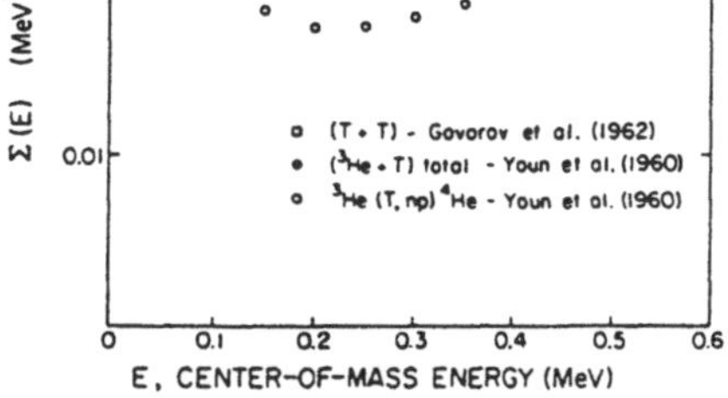

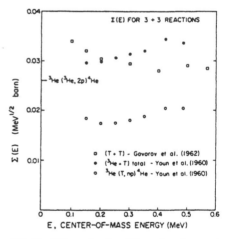

FIG. 10. $\Sigma(E)$ as a function of the center-of-momentum energy for the T + T and ^{3}He + T reactions. The nearly constant value of Σ for the ^{3}He(^{3}He, $2p$)^{4}He reaction of Fig. 9 is indicated on the ordinate.

The two free parameters R_0 and R_1 were adjusted to make $\Sigma(E)$ independent of energy. The best set of parameters to do this was $R_0 = 3.8$ fm and $R_1 = 3.0$ fm. The resulting $\Sigma(E_{c.m.})$ is shown in Fig. 9 as a function of $E_{c.m.}$. The average value of Σ is 0.026 MeV$^{1/2}$b. Figure 9 also shows $\Sigma(E_{c.m.})$ calculated from the data of Bacher and Tombrello.

The same radius parameters were used to fit the data of Govorov *et al.*,[15] on the mirror reaction $^3H(^3H, 2n)^4He$. The resulting $\Sigma(E_{c.m.})$ is approximately independent of energy (see Fig. 10) and the average value of Σ for this reaction is about 0.03 MeV$^{1/2}$b, which is very close to the value obtained for $^3He(^3He, 2p)^4He$. The same prescription was also applied to the data of Youn *et al.*[16] on the analog reaction $^3He + {}^3H$. In this reaction, the interacting particles are not identical and so we can have both singlet and triplet spin states for a given l. The statistical weights are therefore different and we have for $^3He + {}^3H$, assuming that the T_l's are independent of spin,

$$\sigma(E) = \tfrac{1}{2}\frac{\Sigma(E)}{\sqrt{E}}[4P_0(R_0) + 12P_1(R_1)],$$

or

$$\Sigma(E) = \sigma(E)\sqrt{E}\,[2P_0(R_0) + 6P_1(R_1)]^{-1}.$$

The existence of a bound state in the (np) system complicates this reaction further. For this reason the cross-section data for $^3He(^3H, np)^4He$ and the total reaction cross section for $^3He + {}^3H$ were fitted separately, with the same radius parameters as found for $^3He + {}^3He$. These are also shown in Fig. 10.

The resulting $\Sigma(E)$ is nearly independent of energy in each of these cases and the values of Σ agree to within $\pm 25\%$. By way of comparison, the experimentall; determined value of S_0 for the reaction $^3He(^3He, 2p)^4He$ is about 30 times the S_0 value for the nirror reaction $^3H(^3H, 2n)^4He$. The S factors, normalized for the change in Z_0Z_1 differ by a factor of about 7, which appears to be a result of using zero-radius penetrabilities.

The present study of the reaction $^3He(^3He, 2p)^4He$ yields a value of S_0 which is about 4.5 times larger than the value of $S_0 = 1.1$ MeV b adopted by Parker, Bahcall, and Fowler.[1] The effect of this change in S_0 is to reduce the originally predicted ^8B solar neutrino flux by a factor of about 2.

ACKNOWLEDGMENTS

We wish to thank Professor William A. Fowler for suggesting this problem and for his keen interest, help, and encouragement during the course of this work. Many illuminating discussions with Professor T. A. Tombrello, Professor C. A. Barnes, and Dr. A. D. Bacher are gratefully acknowledged.

*Work supported in part by the National Science Foundation Grant Nos. GP-28027, GP-19887.

†Present address: California State College, Los Angeles, California 90032.

[1]P. D. Parker, J. N. Bahcall, and W. A. Fowler, Astrophys. J. 139, 602 (1964).

[2]G. Shaviv, J. N. Bahcall, and W. A. Fowler, Astrophys. J. 150, 725 (1967); R. Davis, Jr., Phys. Rev. Letters 12, 303 (1964).

[3]W. M. Good, W. E. Kunz, and C. D. Moak, Phys. Rev. 83, 845 (1951).

[4]A. D. Bacher and T. A. Tombrello, Rev. Mod. Phys. 37, 433 (1965).

[5]W. Neng-Ming, V. N. Novatskii, G. M. Osetinskii, C. Nai-Kung, and I. A. Chepurchenko, Yadern. Fiz. 3, 1064 (1966) [transl.: Soviet J. Nucl. Phys. 3, 777 (1966)].

[6]A. D. Bacher and T. A. Tombrello, unpublished, as quoted in T. A. Tombrello, *Nuclear Research with Low-Energy Accelerators*, edited by J. B. Marion and D. M. Van Patter (Academic, New York, 1967).

[7]H. C. Winkler and M. R. Dwarakanath, Bull. Am. Phys. Soc. 12, 16, 1140 (1967).

[8]W. Whaling, in *Handbuch der Physik*, edited by S. Flügge (Springer, Berlin, 1958), Vol. 34, p. 193.

[9]L. C. Northcliffe, Ann. Rev. Nucl. Sci. 13, 67 (1963).

[10]J. Cohen-Ganouna, M. Lambert, and J. Schmoukev, Nucl. Phys. 40, 82 (1963).

[11]M. R. Dwarakanath, Ph.D. thesis, California Institute of Technology, 1968 (unpublished).

[12]H. Winkler and M. R. Dwarakanath, to be published.

[13]W. A. Fowler, G. R. Caughlan, and B. A. Zimmerman, Ann. Rev. Astron. Astrophys. 5, 525 (1967).

[14]R. M. May and D. D. Clayton, Astrophys. J. 153, 855 (1968).

[15]A. M. Govorov, L. Ka-Yeng, G. M. Osetinskii, V. I. Salatskii, and I. V. Sizov, Zh. Eksperim. i Teor. Fiz. 42, 383 (1962) [transl.: Soviet Phys.—JETP 15, 266 (1962)].

[16]L-G. Youn, G. M. Osetinskii, N. Sodnom, A. M. Govorov, I. V. Sizov, and V. I. Salatskii, Zh. Eksperim. i Teor. Fiz. 39, 225 (1960) [transl.: Soviet Phys.—JETP 12, 163 (1961)].

Reaction Rates in the Proton-Proton Chain*

Ralph W. Kavanagh

"The Sun is much, much bigger than even an elephant . . ."
—G. Gamow

The four decades since George Gamow took his Ph.D. at Leningrad have been, with his significant participation, a golden age for nuclear physics and astrophysics, which have overlapped to produce a remarkably detailed and persuasive picture of the mechanisms responsible for the sustained generation of energy in stars. Plausible processes have been put forward for the creation of the elements in stars, in the observed abundance ratios, out of an original cosmos of hydrogen. The advent of large-memory, high-speed computers has made it feasible to construct precise models of evolving stars that start from a given initial mass and composition, and change with time to match the present radius, mass, age, and luminosity. Besides being constrained by physical laws governing radiation transport and hydrostatic equilibrium, these models require as input a knowledge of numerous nuclear-reaction rates, or "cross sections."

* Supported in part by the National Science Foundation [GP-9114] and the Office of Naval Research [Nonr-220(47)].

(1967), in a similar experiment, spanning bombarding energies from 1 to 3.2 MeV, found $S_{17}(0) = 0.026 \pm 0.003$ keV-barn.

Both of these extrapolated values were based on a theoretical direct-capture calculation of Tombrello (1965), so that the small disagreement is experimental. Tombrello also deduced $S_{17} = 0.012 \pm 0.003$ keV-barn from a single-particle-model treatment of the ground state of ^{8}B and the measured (Imhof *et al.* 1959) cross section for the mirror reaction, ^{7}Li$(n, \gamma)^8$Li.

The spread in these values prompted a fourth similarly instrumented measurement (Kavanagh *et al.* 1969), achieving improved statistical precision and greater bombarding-energy range, $0.165 \leq E_p \leq 10.0$ MeV. The cross section *vs.* energy is shown in Figure 4 and the extracted S_{17} values in Figure 5. The data are in close agreement with those of Parker, as is the similarly extrapolated value, $S_{17}(0) = 0.0335 \pm 0.003$ keV-barn, $(1/S\, dS/dE)_0 = -0.1\%$ per keV. (For the purist, it should also be noted that another common factor is contained in the post-1966 reports, namely the cross section for ^{7}Li$(d, p)^8$Li (193 ± 15 mb at 770 keV) used for normalization of the experimental efficiency factor, target thickness times solid angle.) The shape of the theoretical curve used to perform the extrapolations is apparently not very dependent on the parameters introduced (e.g., reduced width for ^{8}B and channel-spin ratio), so that an average value for the three recent measurements of $S_{17} = 0.032 \pm 0.003$ should be reliable.

We are now in a position to summarize the status of the nuclear data regarding rates and branching among the *pp*-chains—each crucial reaction has been carefully remeasured (or recalculated) at least twice by different groups with good agreement throughout. Thus the nuclear cross sections that enter into model studies of the sun appear to be well established.

The effects of small changes in the nuclear parameters, in the neighborhood of their "best" values (prior to 1969), have been calculated by Bahcall *et al.* (1969). They find, for example, the following dependence for the neutrino-capture rate in chlorine:

$$\sum \varphi\sigma \propto S_{11}^{-2.5} S_{33}^{-0.37} S_{34}^{0.8} (1 + 3.5 S_{17}^{1.0} \tau_{e7}^{1.0}).$$

In the model calculations, there remains one important and questionable parameter, the optical opacity, which requires knowledge of the abundances in the solar core of heavy elements (which dominate the opacity) and their electronic-transition probabilities.

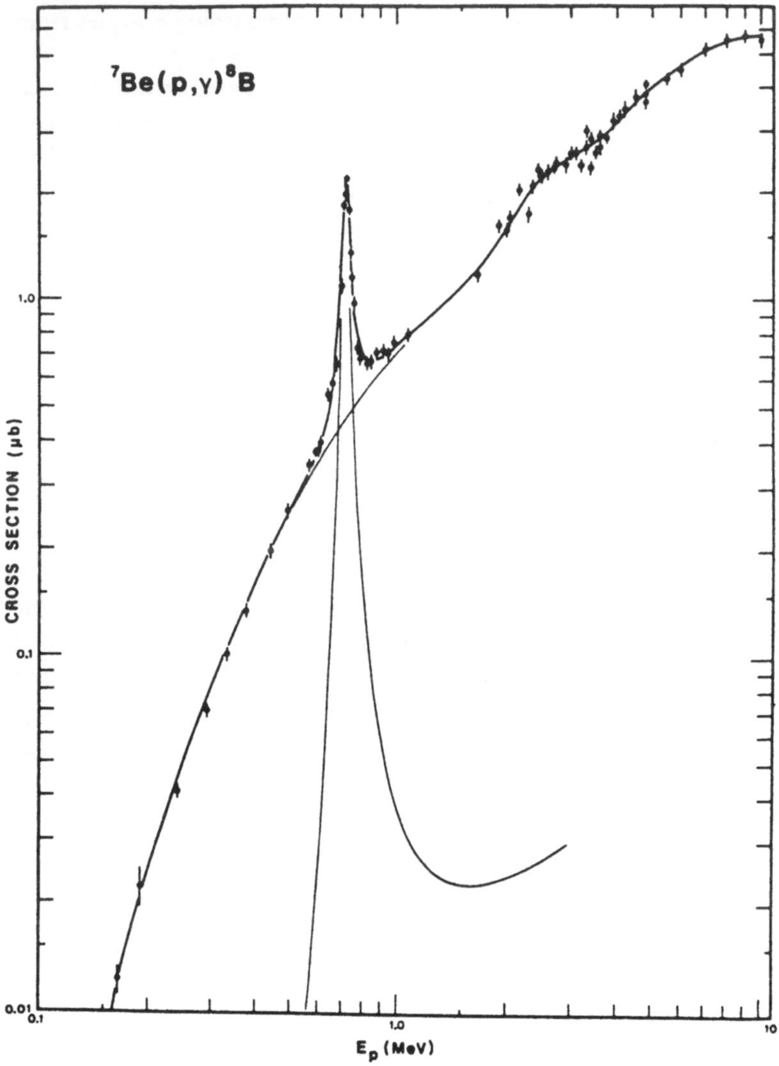

Figure 4. **The ^{7}Be(p, γ) cross section from Kavanagh, Tombrello, Mosher, and Goosman (1969). The curve through the data below 1 MeV is the sum of a nonresonant term, $(S(E)/E) \exp -(E_G/E)^{1/2}$, and a resonant term of *R*-matrix form as indicated.**

The primeval abundances are normally taken to be those found from solar-photosphere and meteorite studies, on the reasonable (but not certain!) assumption of initial homogeneity and negligible surface modification. Subsequent mixing of the evolving core with extra-core material appears to be unimportant (Bahcall *et al.* 1968, Iben 1969).

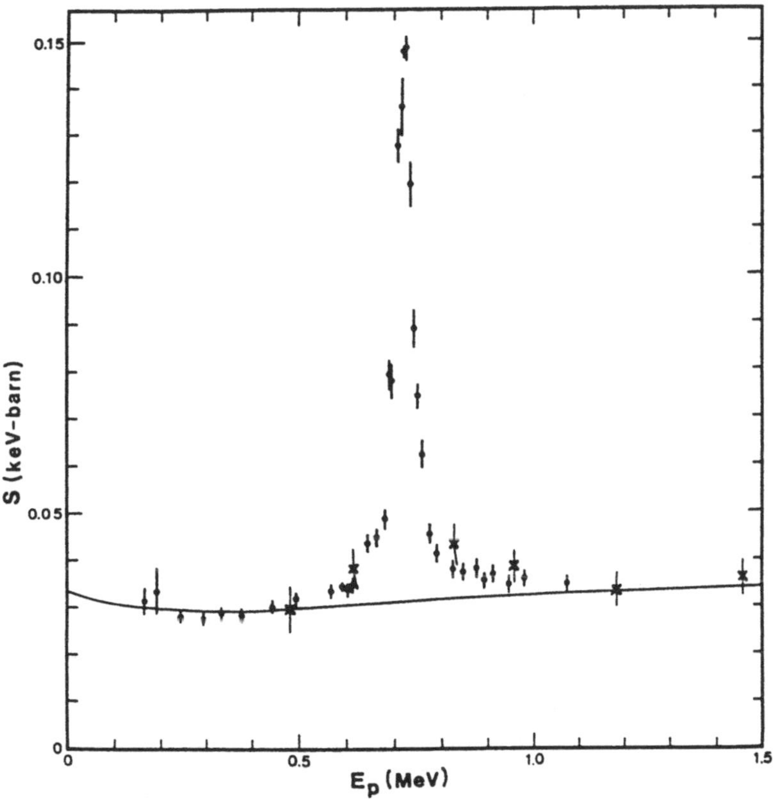

Figure 5. The data of Fig. 4 are shown here for the low-energy region, multiplied by $E \exp (E_G/E)^{1/2}$. The curve is the function $S_{17}(E)$ used in Fig. 4, and is a normalized replica of a theoretical curve given by Tombrello (1965). The crosses are the data of Parker (1966, 1968). The data of Vaughn et al. (1967), in the region $\geq$ 1 MeV (not shown here), are systematically lower by about 25%.

Regarding the transition probabilities (f-values), the currently developing technique of beam-foil spectroscopy (Bashkin 1968, Whaling *et al.* 1969) has already substantially modified the previously used f-values for iron derived from arc and spark spectroscopy. Similar modifications are imposed by recent shock-tube measurements (Huber and Tobey 1968). The net result has been to increase the iron abundance, the core opacity, and the central temperature, raising the ^8B neutrino flux by about 50 percent. At this writing, the most recent calculation incorporating these changes is that of Bahcall and Ulrich (1970), whose model B predicts the value $\sum \varphi\sigma = 7.6 \times 10^{-36}$

Low-Energy $^3He(\alpha, \gamma)^7Be$ Cross-Section Measurements

J. L. Osborne, C. A. Barnes, R. W. Kavanagh, R. M. Kremer,
G. J. Mathews,[a] and J. L. Zyskind
W. K. Kellogg Radiation Laboratory, California Institute of Technology, Pasadena, California 91125

and

P. D. Parker
A. W. Wright Nuclear Structure Laboratory, Yale University, New Haven, Connecticut 06520, and W. K. Kellogg Radiation Laboratory, California Institute of Technology, Pasadena, California 91125

and

A. J. Howard
Trinity College, Hartford, Connecticut 06106, and W. K. Kellogg Radiation Laboratory, California Institute of Technology, Pasadena, California 91125

(Received 22 March 1982)

The cross section and branching ratio for $^3He(\alpha, \gamma)^7Be$ have been measured from $E_{c.m.} = 165$ to 1170 keV by counting prompt γ rays from a windowless, recirculating, ^3He gas target. Absolute cross sections were also measured at $E_{c.m.} = 945$ and 1250 keV by measuring the ^7Be activity produced in a ^3He gas cell with a Ni entrance foil. The inferred zero-energy intercept is $S_{34}(0) = 0.52 \pm 0.03$ keV b. The effect of this extrapolated value on the solar-neutrino problem is discussed.

PACS numbers: 25.60.-t, 24.50.+g, 27.10.+h

The Brookhaven National Laboratory solar-neutrino experiment has measured[1] a neutrino flux of 2.2 ± 0.3 SNU (solar neutrino unit, 1 SNU $= 10^{-36}$ captures per ^{37}Cl atom per second), well below the 7.6 ± 3.3 SNU (3σ error) predicted by current solar models.[2] Since the detector relies on the endoergic ($Q = -0.81$ MeV) neutrino-capture reaction

$$^{37}Cl + \nu_e \rightarrow {}^{37}Ar + e^-,$$

it is most sensitive to the energetic neutrinos from the decay of ^8B. The production of ^8B proceeds through the weak (0.02%) branch of the proton-proton chain,

$$^3He(\alpha, \gamma)^7Be(p, \gamma)^8B,$$

and the theoretical calculation of the solar-neutrino flux depends on accurate experimental determinations of the low-energy cross sections for these radiative captures, extrapolated to solar temperatures. Preliminary indications of an energy-independent cross-section factor[3] for the reaction $^3He(\alpha, \gamma)^7Be$, contrary to theoretical expectations, have led to several recent theoretical investigations[4] of this reaction and, although the methods of calculation have varied, all have been in agreement with the shape calculated in 1963 by Tombrello and Parker.[5] More recent measurements by Kräwinkel *et al.*[6] agree with this energy dependence but have indicated that the cross-section factor for $^3He(\alpha,\gamma)^7Be$ may be 30–50% lower than previous values (Parker and Kavanagh[7] and Nagatani *et al.*[8]). Because the predicted solar-neutrino detection rate is nearly proportional to the $^3He(\alpha, \gamma)^7Be$ reaction rate, this lower cross section would reduce the theoretical neutrino flux to 5.0 ± 2.1 SNU (3σ error) providing a partial solution to the solar-neutrino problem.[2] The significant disagreement with

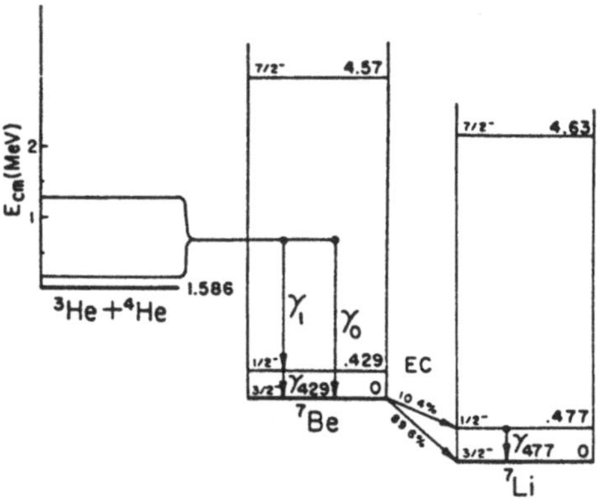

FIG. 1. Level scheme for the $^3He + {}^4He$ reaction and the electron-capture decay of ^7Be, showing the observed γ-ray transitions.

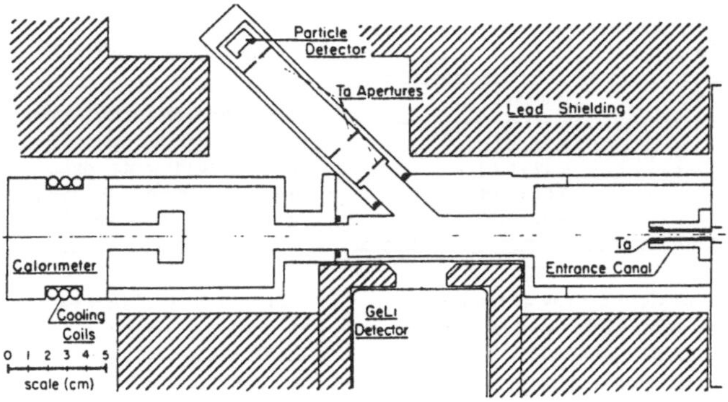

FIG. 2. ^{3}He-gas-target chamber for the windowless, differentially pumped system.

previous values has led us to reexamine this reaction.

We have used two complementary techniques for this measurement. The first method involved the observation of the prompt γ rays (γ_0, γ_1, and γ_{429} in Fig. 1) using α beams from the Office of Naval Research–California Institute of Technology (ONR-CIT) JN electrostatic accelerator and the windowless, differentially pumped, recirculating gas-target system described previously.[9] A new ^{3}He target chamber was designed to accommodate a 100-cm^3 Ge(Li) detector, a rugged silicon surface-barrier particle detector, and a single-cup calorimeter for beam-current integration (see Fig. 2). The windowless target allows measurements to be made at very low energies by reducing the γ-ray background and by eliminating the additional beam-energy spread from straggling in an entrance foil. The energy of the direct-capture γ rays is determined by the energy of the captured α, and provides an independent measurement of the beam energy. The disadvantages of this method are primarily associated with the low target pressure (typically 2.5 to 3 Torr), which requires an extended target, high beam currents (up to 40 μA), and long running times (36 h at the lowest energies). The detection efficiency in the extended target was measured as a function of γ-ray energy and source position by moving calibrated ^{56}Co and ^{152}Eu sources along the beam path. The problem of beam-current integration was solved by using a single-cup solid copper calorimeter. The beam stop was connected to a large, water-cooled, constant-temperature heat sink by a copper rod 2.5 cm long by 1 cm^2 in cross section. The difference in temperature between the beam stop and the heat sink, ΔT, yields the number of incident beam particles, n_α, by the relation

$$n_\alpha = (A/E_c)\int \Delta T\, dt,$$

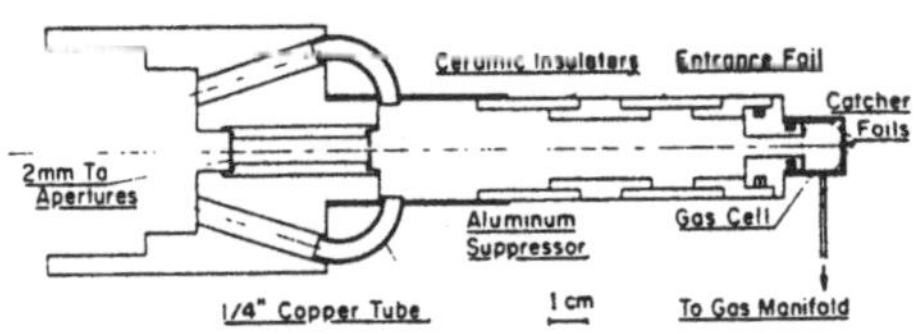

FIG. 3. Gas cell used in the activation measurement. The chamber and detector were surrounded on all sides by 10 cm of lead.

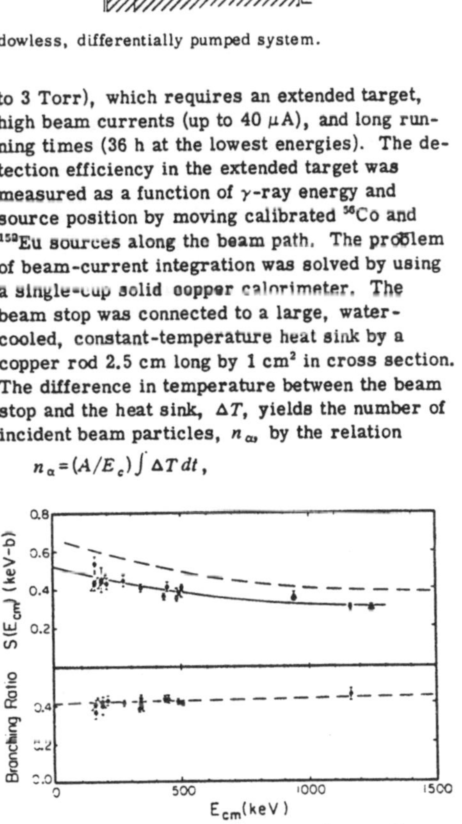

FIG. 4. Experimental results for the ^{3}He(α,γ)^{7}Be cross-section factor, $S_{34}(E_{c.m.})$, and branching ratio. Triangles indicate the activation measurements. Dashed curves are from the calculations of Liu, Kanada, and Tang (Ref. 4). The theoretical S-factor curve of Ref. 5 is shown normalized to the present data (solid curve).

TABLE I. Comparison of measured values of the cross-section factor zero-energy intercept for the reaction $^3He(\alpha,\gamma)^7Be$.

Reference	$S_{34}(0)$ (keV b)
6	0.30 ± 0.03
7	0.47 ± 0.05
8	0.61 ± 0.07
Present	0.52 ± 0.03

where E_c is the particle energy at the calorimeter and A is the calibration constant in watts per kelvin. The calorimeter was calibrated over a wide range of energies and beam currents and was found to agree with electrical integration to within 1%. The particle detector was used to monitor relative beam current, target pressure, and contaminant gases. By surrounding the detector with 10 cm of lead, γ-ray background was reduced by a factor of $\sim 10^3$ at $E_\gamma \simeq 1.5$ MeV.

In the second measurement, a gas-target cell (Fig. 3) filled with ^{3}He gas was bombarded by α's from the ONR-CIT EN tandem accelerator through a window consisting of 0.66 μm Ni plus 0.2 μm Cu. The ^{7}Be produced by the radiative capture was implanted in the 0.025-mm-Pt catcher foil at the end of the cell. The Pt catcher foil was removed from the cell at the end of a run, and the γ rays from the 10% electron-capture branch to the first excited state of ^{7}Li (γ_{477} in Fig. 1) were counted with a shielded Ge(Li) detector and compared with the yield from a calibrated ^{7}Be source in the same geometry. This method has the advantages of directly measuring the angle-integrated cross section, σ_{tot}, and of using conventional electrical current integration. Prompt γ rays were monitored using a NaI(Tl) detector during bombardment to establish effective beam energy in the target. Foil thickness and straggling were also measured by examining the known narrow resonances in $^{24}Mg(\alpha,\gamma)$ and $^{14}N(\alpha,\gamma)$ at $E_\alpha = 3.198$ and 2.348 MeV, respectively, in the same geometry. The energy loss in the foil and gas agreed with that implied by the prompt γ-ray energy to within experimental uncertainty (5 keV).

The results of the experiment are shown in Fig. 4. The cross-section factor was calculated from the expression

$$S_{34}(E_{c.m.}) = \sigma_{tot}(E_{c.m.})E_{c.m.}e^{2\pi\eta},$$

where η is the Sommerfeld parameter. The cross-section factor, extrapolated to zero energy by fitting the theoretical expression of Ref. 5 to the experimental data, is $S_{34}(0) = 0.52 \pm 0.03$ keV b (1σ error). Because of the similar shapes of the other theoretical cross-section curves, this value is also consistent with their extrapolated values.

Comparison of this experiment with other measurements is shown in Table 1. The present measurements agree with the earlier results of Refs. 7 and 8, and with the adopted value, $S_{34} = 0.52 \pm 0.15$ (3σ error), used for the most recent solar-neutrino flux calculations of Bahcall *et al.*[2] The substantial disagreement among the values in Table I is currently being studied and will be discussed in the complete report on the present experiments.[10]

We extend sincere thanks to Martin Weiss, M. D. for his expert repair of an important part of the apparatus. This work was supported in part by the National Science Foundation under Grant No. PHY 79-23638.

[a]Present address: University of California, Lawrence Livermore National Laboratory, Livermore, Cal. 94550.

[1]B. T. Cleveland, R. Davis, Jr., and J. K. Rowley, in Proceedings of the Neutrino Miniconference, University of Wisconsin Report No. 186, 1980 (unpublished), p. 38.

[2]J. N. Bahcall *et al.*, to be published.

[3]H. Kräwinkel *et al.*, Universität Münster Jahresbericht, 1978 (unpublished), p. 35.

[4]B. T. Kim, T. Izumoto, and K. Nagatani, Phys. Rev. C 23, 33 (1981); Q. K. K. Liu, H. Kanada, and Y. C. Tang, Phys. Rev. C 23, 645 (1981); R. D. Williams and S. E. Koonin, Phys. Rev. C 23, 2773 (1981).

[5]T. A. Tombrello and P. D. Parker, Phys. Rev. 131, 2578 (1963).

[6]H. Kräwinkel *et al.*, Z. Phys. A 304, 307 (1982).

[7]P. D. Parker and R. W. Kavanagh, Phys. Rev. 131, 2578 (1963).

[8]K. Nagatani, M. R. Dwarakanath, and D. Ashery, Nucl. Phys. A128, 325 (1969).

[9]M. R. Dwarakanath and H. Winkler, Phys. Rev. C 4, 1532 (1971).

[10]J. L. Osborne *et al.*, to be published.

Proton capture cross section of ^{7}Be and the flux of high energy solar neutrinos

B. W. Filippone*
Argonne National Laboratory, Argonne, Illinois 60439
and Department of Physics, The University of Chicago, Chicago, Illinois 60637

A. J. Elwyn† and C. N. Davids
Argonne National Laboratory, Argonne, Illinois 60439

D. D. Koetke
Valparaiso University, Valparaiso, Indiana 46383
(Received 25 July 1983)

The low energy cross section for the $^7Be(p,\gamma)^8B$ reaction has been measured by detecting the delayed α particles from the ^8B beta decay. Detailed discussion is presented of the analysis of the radioactive ^{7}Be target including the use of two independent methods to determine the ^{7}Be areal density. The direct capture part of the cross section is subtracted from the total cross section to deduce resonance parameters for the 1^+ first excited state in ^{8}B. The zero-energy astrophysical S factor inferred from the present experiment is compared with previous values. The effect on the ^{37}Cl solar neutrino capture rate, predicted by the standard solar model, is also discussed.

I. INTRODUCTION

For the past 20 years Davis *et al.*[1] have attempted to detect neutrinos originating from nuclear reactions in the solar interior by measuring the capture rate in a ^{37}Cl detector. These neutrinos provide the only direct probe of the conditions at the center of the sun. The agreement between the calculated and measured rates has not been good, differing by factors of 2–10. A key input to the calculations are the rates for the nuclear reactions involved in the proton-proton chain.[2] The reaction sequence

$$^7Be+p\rightarrow {}^8B+\gamma\ ,$$

$$^8B\rightarrow {}^8Be^*+e^++\nu_e\ ,$$

$$^8Be^*\rightarrow 2\alpha$$

is expected to terminate the proton-proton chain only once in $\sim$13 000 cycles through the chain. It is this sequence, however, which contributes $\sim$75% to the predicted solar neutrino capture rate of 7–8 SNU (Refs. 3 and 4) (1 SNU$\equiv 10^{-36}$ ν captures/target atom sec) in a ^{37}Cl neutrino detector because such a detector is mainly sensitive to the high energy neutrinos from ^{8}B decay. The measured capture rate is 1.8±0.3 SNU.[1] This discrepancy represents the so-called solar neutrino problem. To calculate the rate of the $^7Be(p,\gamma)^8B$ reaction in the solar interior one needs measured cross sections as low in energy as possible in order to permit accurate extrapolation to the very low energies ($\sim$20 keV) at which most of the reactions occur in the sun.

The low energy cross section for the $^7Be(p,\gamma)^8B$ reaction, and the resulting cross section factor,

$$S(E_{c.m.})=\sigma(E_{c.m.})E_{c.m.}\exp[(E_G/E_{c.m.})^{1/2}]\ , \qquad (1)$$

where

$$E_G=(2\pi\alpha Z_1Z_2)^2\mu c^2/2=13798.8\ \text{keV}\ ,$$

(where μ is the reduced mass of the incident channel and α is the fine structure constant) has been measured in five independent experiments since 1960. The pioneering experiment of Kavanagh[5] measured the cross section to ±40% at proton energies of 0.8 and 1.4 MeV by detecting the high energy positrons from the ^{8}B decay. Tombrello[6] used these data to calculate an extrapolated value of $S_{17}(0)=0.021\pm0.008$ keV b. Parker[7] achieved substantial background reduction by detecting the α particles from $^8Be^*$ breakup following ^8B β^+ decay. This technique has been used in all subsequent $^7Be(p,\gamma)^8B$ experiments. The measurements of Parker were made at eight proton energies in the range $E_p=0.483$–1.952 MeV with an overall uncertainty of ±10%. Using the calculation of Ref. 6, a value of $S_{17}(0)=0.043\pm0.004$ keV b was obtained. These data were later reanalyzed[8] to give a modified value of 0.035±0.004 keV b. An extensive series of measurements from $E_p=0.165$ to 10.0 MeV was reported by Kavanagh *et al.*[9] and by Kavanagh.[10] The extrapolated zero-energy S factor, again based on the calculation of Ref. 6, was 0.0335±0.003 keV b. In the experiment of Vaughn *et al.*[11] the cross section was determined at 20 proton energies from $E_p=0.953$ to 3.281 MeV. Two methods of analysis, which depended on the assumed resonance structure in ^{8}B, yielded $S_{17}(0)=0.0263\pm0.0027$ and 0.0226±0.0043 keV b. The most recent measurement[12] of the $^7Be(p,\gamma)^8B$ reaction was performed at only one energy ($E_p=360$ keV). If this one point is used to normalize the calculation of Ref. 6, the inferred value of $S_{17}(0)$ is

0.045 ± 0.011 keV b.

It is important to note that nearly all of the experiments discussed above[5,7–11] relied on the value of the $^7Li(d,p)^8Li$ cross section at the $E_d = 0.77$ MeV resonance to determine the absolute $^7Be(p,\gamma)^8B$ cross section (see Sec. III). The resulting S factors depend linearly on the $^7Li(d,p)^8Li$ cross section. The values for this cross section used in the above analyses ranged from 176 to 211 mb; however, two recent measurements of this cross section have obtained 146 ± 13 (Ref. 13) and 148 ± 12 mb.[14]

In light of the key role played by the $^7Be(p,\gamma)^8B$ reaction in the determination of the ^{37}Cl solar neutrino capture rate, we have measured the cross section down to $E_{c.m.} = 117$ keV ($E_p = 134$ keV) by detecting the β^+-delayed α particles. Two independent methods were used to determine the absolute value of the cross section, one of which is independent of the $^7Li(d,p)^8Li$ reaction. The zero-energy S factor from the present experiment is combined with solar model calculations to determine a new value for the predicted ^{37}Cl solar neutrino capture rate.

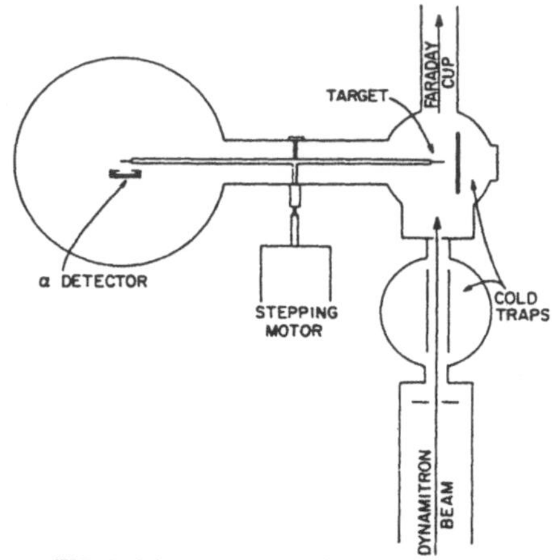

FIG. 1. Schematic diagram of the target apparatus.

II. EXPERIMENTAL PROCEDURE

Proton beams were obtained from the Argonne National Laboratory 4.5 MV dynamitron accelerator, with currents between 4 and 8 μA, limiting the beam power on target to $\lesssim 5$ W. Following magnetic analysis the beam was collimated to produce a beam spot of 3.2 mm on target. For the lower energy points a quadrupole doublet was used to focus the beam. In order to ensure a stable, uniform beam spot, a sawtooth-wave voltage was applied to two orthogonal sets of parallel plates (at incommensurate frequencies) prior to beam collimation. While most of the measurements utilized the H^+ beam, the lowest energy point ($E_{c.m.} = 117$ keV) employed an H_2^+ molecular ion beam. This beam permitted a measurement at a proton energy below the normal energy range of the machine. The energy scale of the dynamitron was determined to $\pm 0.5\%$ from thick target yield curves at resonances in the $^{19}F(p,\alpha\gamma)^{16}O$ reaction at proton energies of 224.4, 340.46, and 872.11 keV.

The experimental apparatus, shown schematically in Fig. 1, is essentially the same as that used for the $^7Li(d,p)^8Li$ cross section measurement of Ref. 14. The target was mounted on a rotating arm and could be transferred (transfer time ~ 0.3 sec) from the bombardment chamber to the counting chamber by a signal to the stepping motor. A collimated silicon surface barrier detector with 300 mm^2 active area and 23 μm depletion depth was mounted in the counting chamber to detect the β^+-delayed α particles. The detector was mounted in a "near" geometry configuration of ~ 2 mm target-collimator distance. In this tight geometry the use of a very thin detector is required because of the large background of electrons produced by the 478 keV γ rays from the ^7Be target ($\sim 3 \times 10^8$ γ/sec). To help reduce this background the detector mount and collimator were made from low Z material and the detector itself was of the ring-mount-type in which the thin silicon crystal is supported only by a ceramic ring. The system was pumped by a turbomolecular pump in the counting chamber and large area liquid nitrogen-cooled surfaces in the bombardment chamber to a pressure of $\lesssim 5 \times 10^{-7}$ Torr. The use of an oil-free pump and cold surfaces near the beam limited the proton energy loss due to carbon buildup on the target to $\lesssim 3$ keV throughout the course of the experiment.

The solid angle of the detector in the "near" geometry was measured to be $(23.1 \pm 1.5)\%$ of 4π sr. This was determined by comparing the yield of α particles from the $^7Li(d,p)^8Li \rightarrow {}^8Be^* \rightarrow 2\alpha$ reaction in the "near" geometry to that in a "far" geometry in which the target-collimator distance was 22.2 mm. An average value from the measured geometry and a calibrated ^{241}Am source then gave the solid angle in the "far" geometry. The current integration of the beam and the control of the timing cycle are described in Ref. 14. The intervals of the timing cycle were the following: $t_1 \equiv$ beam on target $= 1.50$ sec; $t_2 \equiv$ transfer to counting $= 0.52$ sec; $t_3 \equiv$ counting $= 1.50$ sec; $t_4 \equiv$ transfer to bombardment $= 0.52$ sec.

The ^7Be used in the preparation of the target was produced via the $^7Li(p,n)^7Be$ reaction at the Argonne dynamitron. A proton beam of 3.6 MeV bombarded a chemically purified and isotopically enriched 25 mg/cm^2 ^7Li metal target, produced by vacuum evaporation onto a water-cooled Cu backing. Approximately 120 mCi of ^7Be were produced from the 5000 μA h bombardment. After bombardment ^7Be was chemically separated and purified by repeated solvent extraction and ion exchange. It was then deposited onto a 0.25 mm thick Pt disk by the molecular plating method,[15] which involves high voltage electrodeposition from an organic solution. The target was then flamed red-hot in air for several minutes to convert the ^7Be to beryllium oxide and to remove any volatile contaminants. The final target consisted of ~ 80 mCi of ^7Be (~ 0.23 μg) and ~ 7 μg of solids. Details of the target

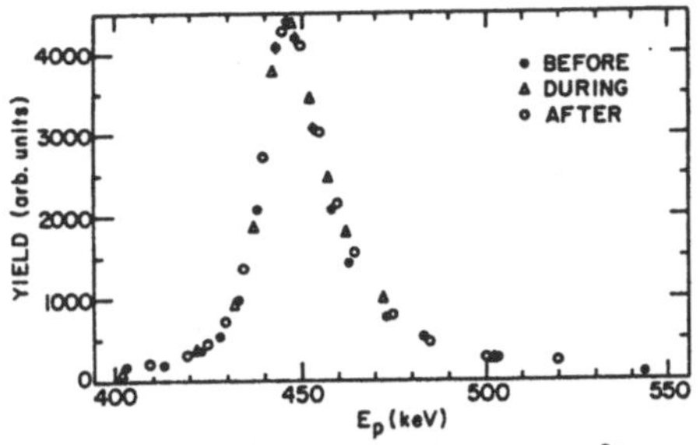

FIG. 2. $^7Li(p,\gamma)^8Be$ excitation function for the ^7Be target. These data represent three separate measurements performed before, during, and after the $^7Be(p,\gamma)^8B$ experiment. The relative uncertainties are approximately the size of the points.

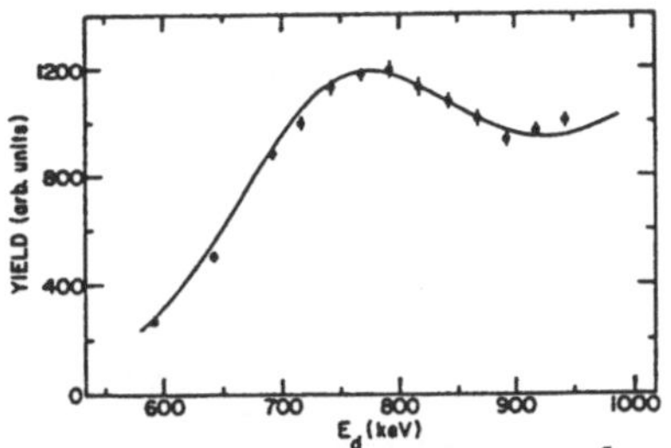

FIG. 3. $^7Li(d,p)^8Li$ excitation function for the ^7Be target. The solid curve is an excitation function for the $^7Li(d,p)^8Li$ reaction for a thin ^7LiF target normalized to the data.

preparation will be published elsewhere.[16]

The energy thickness of the target and the energy loss due to carbon buildup on the target were measured in studies of the shape of the resonance curve for the $^7Li(p,\gamma)^8Be$ reaction at the 441.4 keV resonance ($\Gamma_{lab}=12.2\pm0.5$ keV). For this analysis it was assumed that the lithium and beryllium reside at the same location in the target since essentially all of the ^7Li in the target is due to ^7Be decay (see below). This assumption was verified in a separate experiment whose aim was to investigate resonances in the $^7Be(\alpha,\gamma)^{11}C$ reaction[17] using the same target. The energy loss of the α beam agreed well (when the relative proton and α stopping powers were taken into account) with that obtained from the $^7Li(p,\gamma)^8Be$ data (within the $\pm15\%$ uncertainty in the energy loss). Resonance curves for the $^7Li(p,\gamma)^8Be$ reaction are shown in Fig. 2. These data were taken with a 25.4×25.4 cm NaI(Tl) crystal at 90° to the beam, and were corrected for a continuous buildup of carbon (to a maximum of 3 keV) as well as for the increasing amount of ^7Li from the decay of ^7Be.

The most difficult quantity to determine in the experiment is the ^7Be areal density. In the past, most workers measured the yield of the $^7Li(d,p)^8Li$ reaction at the 0.77-MeV resonance, and, with the known resonant cross section, determined the buildup of ^7Li from the ^7Be decay in the target. With this value, and by use of the relation

$$N_{^7Li}(t)=N_{^7Be}(0)(1-e^{-\lambda t})+N_{^7Li}(0)\ , \qquad (2)$$

where λ ($=1.505\times10^{-7}$ sec^{-1}) is the ^7Be decay constant, the ^7Be areal density, or, if the same detector geometry is used for both the $^7Li(d,p)$ and $^7Be(p,\gamma)$ reactions, the product of ^7Be areal density and detector solid angle, can be obtained. One previous experiment[12] made use of the total γ-ray activity of the target and the area of the target spot to determine the ^7Be areal density; however, none of the previous experiments utilized both methods. Because of inconsistencies in the measured value of the $^7Li(d,p)^8Li$ cross section[18] both methods were employed in the present experiment.

An excitation curve for the $^7Li(d,p)^8Li$ reaction from deuteron bombardment of the ^7Be target is shown in Fig. 3. This measurement utilized the same apparatus (Fig. 1) as was used for the $^7Be(p,\gamma)$ reaction. The molecular ion beam D_3^+ was used because a time- and source-dependent contamination of the mass 2 beam (D^+) with molecular hydrogen (H_2^+) was observed in backscattering experiments from a thin gold foil. The yield of α particles at the peak of the 0.77 MeV resonance as a function of $(1-e^{-\lambda t})$ is displayed in Fig. 4. The measurements were made over a period of five weeks during which time the $^7Be(p,\gamma)^8B$ experiment was performed. A linear least-squares fit to these data shows that there was essentially no ^7Li in the target on the final day of chemical separation and purification. In fact, the intercept corresponded, within several hours, to the time of the final electroplating and flaming of the target. The slope of the curve [see Eq. (2)] yields a product of ^7Be areal density and detector solid angle fraction ($\Omega/4\pi$) of $(2.24\pm0.15)\times10^{16}/cm^2$. This was determined by use of a (d,p) cross section (σ_{dp}) of 157 ± 10 mb (see Sec. IV) and the expression for the cross section described in Sec. III.

For the second method used to determine the ^7Be areal

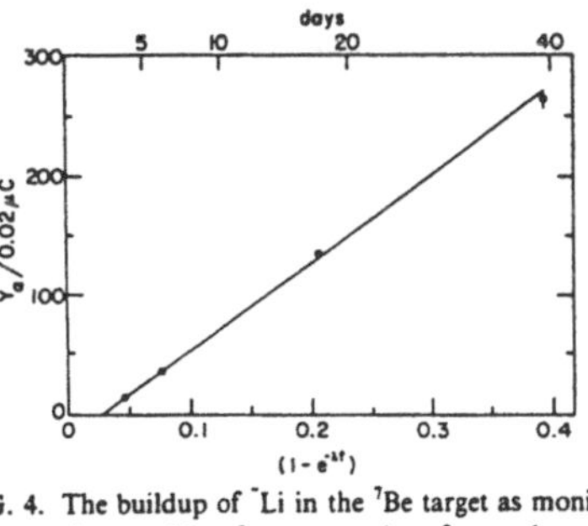

FIG. 4. The buildup of ^7Li in the ^7Be target as monitored by the increasing yield of α particles from the sequence $^7Li(d,p)^8Li\rightarrow{}^8Be^*\rightarrow2\alpha$ as a function of $(1-e^{-\lambda t})$, where λ is the ^7Be decay constant. The solid line is a linear least-squares fit to the data.

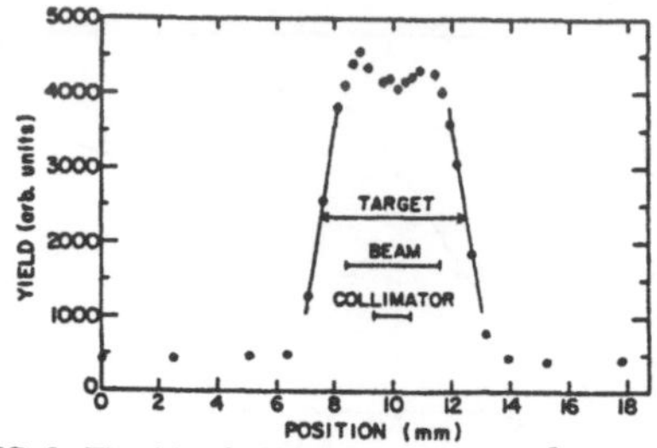

FIG. 5. The yield of 478 keV γ rays from the ^{7}Be target as a function of position for the 1.3 mm Pb collimator. Also shown are the diameters of the target, beam spot, and collimator.

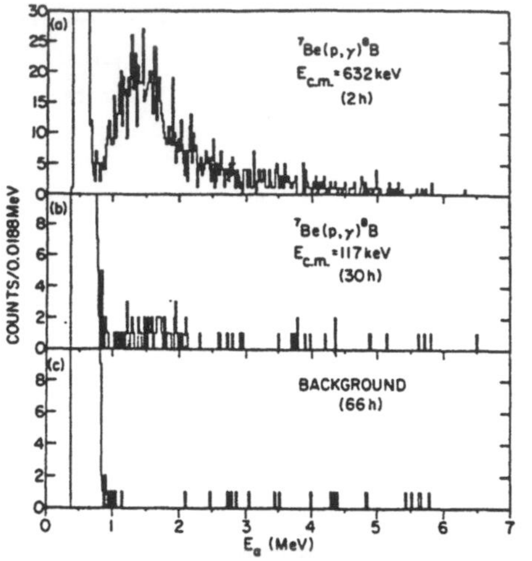

FIG. 6. (a) and (b) Delayed α-particle spectra for the reaction sequence ^{7}Be(p,γ)^{8}B$\rightarrow$^{8}Be*$\rightarrow$$2\alpha$, and (c) a background spectrum.

density, the total number of ^{7}Be nuclei in the target was found by measuring the yield of 478 keV γ rays in a Ge(Li) detector whose absolute full energy peak efficiency was determined using calibrated γ-ray sources of ^{54}Mn, ^{133}Ba, and ^{137}Cs. After correcting for decay and using a ^{7}Be branching ratio (to the 478 keV state in ^{7}Li) of (10.35 ± 0.10)%, the average of two measurements gave an initial total activity of 81.2 ± 6.0 mCi. The area of the target was determined by scanning the γ-ray activity of the target spot with a NaI(Tl) detector and two 10 cm thick Pb collimators with 0.5 and 1.3 mm diam apertures. Figure 5 shows a γ-ray scan for the 1.3 mm diam collimator. The nonuniformities in the target were on the order of 5–10%, but when averaged over the beam spot they amounted to $\lesssim 5\%$ uncertainty in the areal density. From a total of six scans with different target orientations the diameter of the target spot was found to be 4.95 ± 0.15 mm by calculating the shape of the curve expected from a circular aperture intersecting a circular target spot. The correction for the finite size of the apertures in the present experiment differed from the FWHM by less than 0.05 mm. The product of ^{7}Be areal density and solid angle fraction which results from the separate measurements of ^{7}Be activity, target diameter, and solid angle is $(2.40\pm0.27)\times10^{16}/\text{cm}^2$. This is in excellent agreement with the value extracted from the ^{7}Li(d,p)^{8}Li yield curve in Fig. 4, as discussed above.

III. RESULTS AND ANALYSIS

The spectra of β^+-delayed α particles observed from the ^{7}Be(p,γ)^{8}B reaction at $E_{\text{c.m.}}=632$ and 117 keV are shown in Fig. 6. These energies correspond to the highest and lowest measured cross sections. Also shown in this figure is a background spectrum taken with the beam off and the ^{7}Be target in front of the α detector. The equivalent running times for the three spectra are 2, 30, and 66 h, respectively. The effect of the high flux of γ rays that come from the decay of ^{7}Be can be seen to extend out to ~1.0 MeV in the background spectrum. Most of this low energy background was biased out of the spectra to reduce dead times to $\ll 1\%$. The events in the background spectrum at higher energies are primarily due to natural radioactivity in the detector, mount, and chamber. The integrated α particle yields were obtained from the spectra by accepting all α particles above an energy cut of 0.95 MeV. Detailed studies of the low energy portion of the spectrum of delayed α particles from ^{8}B and ^{8}Li decay were performed, and indicated that the data were fit very well by a Gaussian from the peak in the spectrum down to zero energy. From such a Gaussian extrapolation it was found that the energy cut of $E_\alpha>0.95$ MeV should include 92 ± 4% of the total yield of α particles at $E_{\text{c.m.}}=632$ keV. Analysis of the spectra at other energies indicated that the spectra were slightly shifted in energy because the α particles originate from different depths in the target due to the different recoil energy of the ^{8}B. Reaction kinematics and standard range and energy loss tables[20] were used to calculate this effect. These calculations suggested that at the highest and lowest energies in the experiment the $E_\alpha>0.95$ MeV cut should include 89 ± 4% and 95 ± 4% of the total yield, and the data were corrected for this effect.

The background due to natural α emitters and the ^{7}Be γ rays was measured to be 0.42 ± 0.09 counts/1000 cycles at channels above the 0.95 MeV energy cut, where a cycle is one complete revolution of the target arm. This background amounted to $\lesssim2\%$ of the integrated α particle yield for all but the two lowest energy points, where it was 10–15% of the yield. An additional source of background in the experiment is due to possible deuteron contamination of the proton beam. This can be a problem because the (d,p) reaction on the ^{7}Li in the target produces delayed α particles with a cross section 10^2–10^5 times that for the $^7Be(p,\gamma)^8B$ reaction. This background was measured by bombarding thick ^7LiF targets with the H^+ and H_2^+ beams. For the H^+ beam the deuteron contamination was found to be $<1\times10^{-8}$ at the 90% confidence level. Using the low energy $^7Li(d,p)^8Li$ cross section measurements of Ref. 21 this amounts to $<0.5\%$ of the integrated α particle yield for all proton energies, and was neglected. While deuterons are not, in fact, expected in an H^+ beam because of the difference in magnetic rigidity except via rare, complex multiple scattering trajectories, this is not true for the H_2^+ beam, which has the same rigidity as a $^2H^+$ beam. Thus, one would expect the H_2^+ beam to be contaminated by at least the natural abundance of deuterium (on the assumption of equal production of the monatomic and diatomic beams expected to be approximately valid for the ion source used in the experiment). Because of this, the H_2^+ beam could only be used at the lowest proton energy, $E_{H_2^+}=0.3$ MeV. This energy is only ~ 50 keV above the (d,p) reaction threshold and the cross section is very small. The background due to contamination of the H_2^+ beam was measured to be only 2% of the integrated α particle yield at $E_{H_2^+}=0.3$ MeV, and was consistent with that expected from the natural abundance of deuterium.

The total cross section for the $^7Be(p,\gamma)^8B$ reaction was calculated from the formula

$$\sigma(\bar{E}_p)=\frac{Y_\alpha(E_i)B(^8B)}{2N_pN_{^7Be}(t)(\Omega/4\pi)}\ , \tag{3}$$

where $\bar{E}_p$ is the average energy of the beam in the target (see discussion below), $Y_\alpha(E_i)$ is the background subtracted α particle yield at incident energy E_i, N_p is the total number of incident protons, $N_{^7Be}(t)=N_{^7Be}(0)\exp(-\lambda_{^7Be}t)$ is the number of ^7Be nuclei/cm^2 in the target at time t after chemical separation, and $\Omega/4\pi$ is the solid angle fraction of the α detector. The factor of 2 in the denominator accounts for the two α particles produced in each reaction. The remaining factor is defined as

$$B(^8B)=\frac{\lambda t_1\{1-\exp[-\lambda(t_1+t_2+t_3+t_4)]\}}{(1-e^{-\lambda t_1})[e^{-\lambda t_2}-e^{-\lambda(t_2+t_3)}]}\ , \tag{4}$$

where λ is the ^8B decay constant and t_i are the time intervals of the timing cycle described in Sec. III. This factor accounts for ^8B decay during bombardment, transfer, and counting, as well as ^8B buildup from previous cycles. A half-life[22] of 769 ± 4 msec was assumed for ^8B, which gives $B(^8B)=3.829\pm0.013$. Equations (3) and (4) were also used to determine the ^7Be areal density-solid angle product from the known $^7Li(d,p)^8Li$ cross section as discussed in Sec. II. For this determination a ^8Li half-life[22] of 842 ± 6 msec was assumed, giving $B(^8Li)=3.632\pm0.014$.

Because of the strong energy dependence in Eq. (1), the effective or average energy of the beam must be known in order to determine the S factor. This energy ($\bar{E}_p$) is defined as the energy at which the cross section, evaluated at this energy, equals the cross section averaged over the target thickness; that is,

$$\sigma(\bar{E}_p)=\frac{\int_{E_i-\Delta E}^{E_i}\sigma(E')\left[\frac{dE}{dx}(E')\right]^{-1}dE'}{\int_{E_i-\Delta E}^{E_i}\left[\frac{dE}{dx}(E')\right]^{-1}dE'}\ , \tag{5}$$

where E_i is the incident beam energy corrected for energy loss in the front carbon layer, ΔE is the energy loss in the target, and $(dE/dx)(E')$ is the energy dependent stopping power. To solve this equation the energy dependence of the cross section must be known. For most of the energy range of the experiment the energy dependence of the cross section was taken from Eq. (1) with a constant S factor. Near the $E_p=0.73$ MeV resonanoe in the $^7Be(p,\gamma)^8B$ reaction (corresponding to the first excited state in ^8B), a Breit-Wigner term was included in the energy dependence.

The energy loss of the beam in the target was measured to be 15 ± 2 keV at the 441.4 keV resonance in the $^7Li(p,\gamma)^8Be$ reaction. When this energy loss is combined with the measured mass of material and the area of the target (see Sec. II) an approximate stopping power at $E_p\approx440$ keV of 0.4 ± 0.1 keV/μg cm^2 was determined. This value implies that the average atomic number of the material in the target $\bar{Z}$ is <20. This is consistent with a composition which, aside from the ^7Be and oxygen, likely includes light metal oxides, nitrides, nitrates, etc., as well as carbon, all of which could be present as possible contaminants. It is not necessary, however, to know the exact chemical composition of the target for two reasons. First, the energy loss is measured at one energy (441 keV), and second, for $\bar{Z}<20$ the energy dependence of the stopping power (taken from the global fits of Andersen and Ziegler[23]) varies by $\lesssim15\%$ in the energy range of the present experiment. It should be noted that a change of 20% in the stopping power gives a change of only 2.4 keV in $E_{c.m.}$ even at the lowest energy. Therefore, by choosing some element or composition with $\bar{Z}<20$ (^7BeO was used) to characterize the energy dependence of the stopping power, an effective thickness τ can be derived from the known energy loss at 441 keV:

$$\tau=\int_{E_i-\Delta E}^{E_i}\left[\frac{dE}{dx}(E')\right]^{-1}dE'\ , \tag{6}$$

where in this case $E_i=441$ keV and $\Delta E=15\pm2$ keV. This value for τ can then be used to determine the energy loss ΔE at any given energy E_i by solving the above equation by successive approximation. Once the energy loss at bombarding energy E_i is known, Eq. (5) can be used to calculate the average energy. The results of such calculation indicated that in all cases the average energy deviated from the formula $\bar{E}_p=E_i-\Delta E/2$ by $<1\%$. The good agreement with this formula results because, for the small energy losses involved (the maximum was 25 keV at $E_i=146$ keV), the stopping power changes by $\sim5\%$ and the cross section is nearly linear with energy.

As mentioned earlier, the lowest energy point ($\bar{E}_p=134$ keV) was obtained with an H_2^+ molecular ion beam. As the molecule enters the target the electron is stripped off and the two protons move apart under their mutual electrostatic repulsion—the so-called Coulomb explosion. The net effect of this explosion is to give an overall energy spread to the beam which is symmetric about the beam energy. Simple kinematic considerations, in addition to experimental measurements,[24] indicate that the energy distribution is a semicircle centered at the beam energy and that the radius of this semicircle (the maximum energy deviation) is proportional to the square root of the incident beam energy. Calculations of the average energy were performed with such an energy spread by integrating Eq. (5) over all possible energies from the distribution. Even if the energy spread were 4—5 times that which would be expected on the basis of the above discussion (the expected spread is $\Delta E = \pm 2$ keV), because this spread is symmetric about the beam energy, the calculated average energies would differ by $\ll 1\%$ from those calculated without inclusion of the Coulomb explosion.

The center of mass energy has been obtained from the average energy with the expression $E_{c.m.} = 0.87444\bar{E}_p$. The uncertainty attached to $E_{c.m.}$ was ± 3 keV or 0.5%, whichever was larger. This includes uncertainties in the energy calibration, the extrapolated stopping powers, the extracted energy loss from the $^7Li(p,\gamma)^8Be$ resonance, as well as uncertainties in the energy lost in the carbon build-up on the target. The above error in $E_{c.m.}$ amounts to an additional uncertainty in the *S* factors of 11% and 5% at the two lowest energy points, and only 3% or less at all other energies.

The cross section and derived *S* factor for the $^7Be(p,\gamma)^8B$ reaction as a function of $E_{c.m.}$ are displayed in Figs. 7 and 8. The error bars correspond to relative errors due to counting statistics. Also shown is a least-squares best-fit normalization of the calculation of Tombrello[6] to the off-resonance data. In Sec. IV we describe the calculation of the zero-energy *S* factor $S_{17}(0)$ from these data and note the effect on the expected ^{37}Cl solar neutrino capture rate.

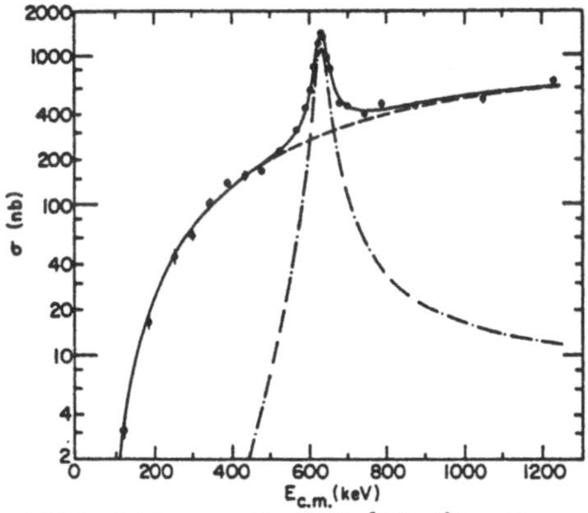

FIG. 7. Total cross section for the $^7Be(p,\gamma)^8B$ reaction as a function of the center of mass energy. The dashed curve is the nonresonant direct capture part of the cross section, the dashed-dot curve is the resonant cross section using the parameters discussed in the text, and the solid curve is the sum of the two. If not shown, the error bars (statistical only) are smaller than the data points.

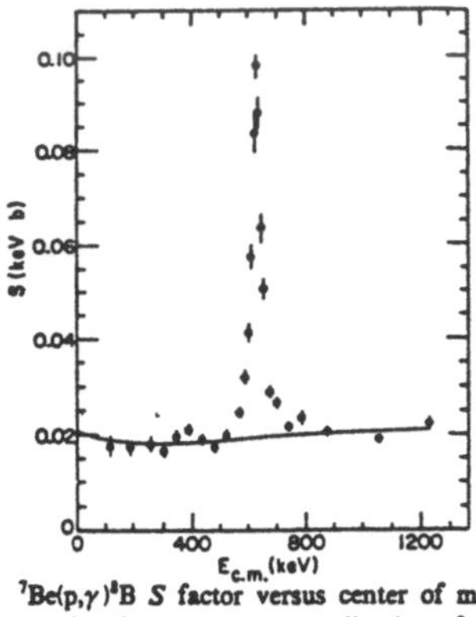

FIG. 8. $^7Be(p,\gamma)^8B$ *S* factor versus center of mass energy. The solid curve is a least-squares normalization of the calculation of Ref. 6 to the off-resonance data.

IV. DISCUSSION

The cross section data display a prominent resonance at $E_{c.m.} = 0.63$ MeV atop the direct capture part of the cross section. This resonance corresponds to the 0.77 MeV first excited state in ^8B. By subtracting the nonresonant part from the total cross section, the resonant cross section can be used to obtain the resonance parameters. A least-squares normalization of the calculation of Tombrello[6] to the data for $E_{c.m.} < 400$ keV (where the resonance should give little contribution) yields the nonresonant part of the cross section. The resonant cross section can then be fit to a resonance curve of the single-level Breit-Wigner form, including energy-dependent widths determined by the penetrability and γ ray multipolarity. Such a fit yields a center of mass resonance energy of 632 ± 10 keV, a total width (assuming $\Gamma_p \gg \Gamma_\gamma$) of $\Gamma_{c.m.} = 37 \pm 5$ keV, and a peak resonant cross section of 1180 ± 120 nb. With these values the radiative width of the resonance is found to be $\Gamma_\gamma = 0.025 \pm 0.004$ eV. This is in fair agreement with a number of shell model calculations which predict values of Γ_γ from 0.019 to 0.021 eV.[25]

As discussed in Sec. I, most previous workers have determined $S_{17}(0)$ by a least-squares normalization of the calculation of Tombrello[6] to experimental cross sections. Such a normalization to the data from this experiment gives a value of $S_{17} = 0.0217 \pm 0.0025$ keV b, where the error includes uncertainties in $E_{c.m.}$, the ^7Be areal density, the solid angle, the fraction of α particle yield above the energy cut, as well as in the normalization procedure. The data used for this normalization were in the range $E_{c.m.} < 525$ keV and $E_{c.m.} > 850$ keV. That this range is effectively outside the resonance region was checked by determining the resonant contribution to the cross section for these energies from the resonant parameters discussed above. The contributions were $\lesssim 5\%$. It should be noted that when the recent calculation of S_{17} by Barker[26] is normalized to the data, consistently lower (by 10–15%) values of $S_{17}(0)$ are obtained because of a somewhat different energy dependence in S_{17}.

Before we can compare our results with previous experiments, it is necessary to discuss our adopted value for σ_{dp} of 157 ± 10 mb. The experimental values for σ_{dp} are not in good agreement (see Fig. 6 in Ref. 14). A weighted average of the modern σ_{dp} measurements gives a chi-squared per degree of freedom (χ_ν^2) for the data set of 3.2. These experimental values include one measurement[19] with an unjustifiably small uncertainty and another[7] which lies 2.8 standard deviations above the weighted mean. If the uncertainty in Ref. 19 is doubled and the measurement of Ref. 7 is deleted from the data set, the weighted mean becomes 157 mb with $\chi_\nu^2=1.1$. We have therefore adopted a value of $\bar{\sigma}_{dp}=157\pm10$ mb where an overall systematic uncertainty has been added. This value is not very different from the average of $\bar{\sigma}_{dp}=166\pm12$ mb (1σ) adopted in Ref. 4.

A comparison of the previous ^{7}Be(p,γ)^{8}B measurements with the present experiment is shown in Fig. 9. Where appropriate, a ^{7}Li(d,p)^{8}Li cross section of 157 mb has been assumed. Aside from the measurement of Ref. 12, where the cross section was measured at a single energy with $\pm25\%$ uncertainty, the agreement between the experiments is fairly good. A weighted average of all the data gives $S_{17}(0)=0.0238\pm0.0023$ keV b. The effect of this new average value for $S_{17}(0)$ on the predicted capture rate in the ^{37}Cl solar neutrino experiment can be calculated using the solar model code of Ref. 3. Using this value for $S_{17}(0)$ the predicted capture rate is 5.6 SNU, ~20% lower than the value of 7.0 SNU determined in Ref. 3 which used a value for $S_{17}(0)$ of 0.032 keV b. In order to obtain a best estimate value for the ^{37}Cl capture rate, one should also include a new measurement[27] of S_{34}— the ^{3}He(α,γ)^{7}Be S factor—in the analysis. If this is done for the calculation of Ref. 3 the predicted rate is 5.9 SNU. A similar modification of the calculation of Ref. 4 gives 6.5 SNU using the present average value for $S_{17}(0)$, and 6.9 SNU when the new value[27] of $S_{34}(0)$ is included. Here, the dependence of capture rate on input parameters given in Ref. 28 has been used. It is important that continued studies of the nuclear reactions of the proton-proton chain be undertaken in order to eliminate nuclear physics uncertainties as a possible source of the remaining disagreement between theory and experiment.

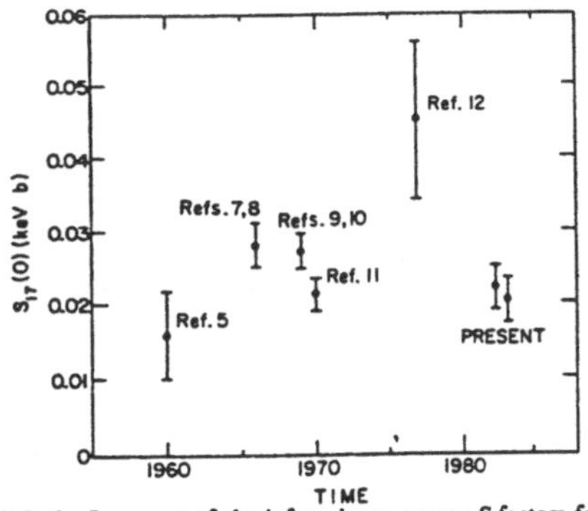

FIG. 9. Summary of the inferred zero-energy S factors from the measurements of the ^{7}Be(p,γ)^{8}B reaction. The two values for the present measurement are for the two independent methods used to determine the ^{7}Be areal density.

ACKNOWLEDGMENTS

We thank R. Evans, M. Finn, W. Ray, Jr., and the dynamitron staff for assistance with the experiment. The efforts of M. Wahlgren towards the preparation of the radioactive target are greatly appreciated. We would also like to acknowledge helpful discussions with J. Schiffer, R. Armani, and E. Kanter. This work was supported by the U. S. Department of Energy under Contract W-31-109-Eng-38.

*Present address: Kellogg Laboratory, California Institute of Technology, Pasadena, CA 91106.

[1]R. Davis, Jr., in Proceedings of the Informal Conference on Status and Future of Solar Neutrino Research, edited by G. Friedlander, BNL Report DNL 50879, 1978, Vol. 1, p. 1; R. Davis, Jr., B. T. Cleveland, and J. K. Rowley, in *Science Underground*, Proceedings of the Workshop on Science Underground, Los Alamos, 1982, edited by M. M. Nieto *et al.*, AIP Conf. Proc. No. 96 (AIP, New York, 1983).

[2]D. D. Clayton, in *Principles of Stellar Evolution and Nucleosynthesis* (McGraw-Hill, New York, 1968).

[3]B. W. Filippone and D. N. Schramm, Astrophys. J. 253, 393 (1982).

[4]J. N. Bahcall, W. F. Huebner, S. H. Lubow, P. D. Parker, and R. K. Ulrich, Rev. Mod. Phys. 54, 767 (1982).

[5]R. W. Kavanagh, Nucl. Phys. 15, 411 (1960).

[6]T. A. Tombrello, Nucl. Phys. 71, 459 (1965).

[7]P. D. Parker, Phys. Rev. 150, 851 (1966).

[8]P. D. Parker, Astrophys. J. 153, L85 (1968).

[9]R. W. Kavanagh, T. A. Tombrello, J. M. Mosher, and D. R. Goosman, Bull. Am. Phys. Soc. 14, 1209 (1969).

[10]R. W. Kavanagh, in *Cosmology, Fusion and Other Matters*, edited by F. Reines (Colorado University Press, Boulder, 1972), p. 169.

[11]F. J. Vaughn, R. A. Chalmers, D. Kohler, and L. F. Chase, Jr., Phys. Rev. C 2, 1657 (1970).

[12]C. Wiezorek, H. Kräwinkel, R. Santo, and L. Wallek, Z. Phys. A 282, 121 (1977).

[13]A. J. Elwyn, R. E. Holland, C. N. Davids, and W. Ray, Jr., Phys. Rev. C 25, 2168 (1982).

[14]B. W. Filippone, A. J. Elwyn, W. Ray, Jr., and D. D. Koetke, Phys. Rev. C 25, 2174 (1982).

[15]D. C. Aumann and G. Müllen, Nucl. Instrum. Methods 115, 75 (1974), and references therein.
[16]B. W. Filippone and M. Wahlgren (unpublished).
[17]G. Hardie, A. J. Elwyn, B. W. Filippone, R. E. Segel, and M. Wiescher (unpublished).
[18]See Ref. 14 for a complete discussion of experimental values.
[19]A. E. Schilling, N. F. Mangelson, K. L. Nielson, D. R. Dixon, M. W. Hill, G. L. Jensen, and V. C. Rogers, Nucl. Phys. A263, 289 (1976).
[20]L. C. Northcliffe and R. F. Schilling, Nucl. Data Tables 7, 233 (1970).
[21]C. R. McClenahan and R. E. Segel, Phys. Rev. C 11, 370 (1975).
[22]F. Ajzenberg-Selove, Nucl. Phys. A320, 1 (1979).
[23]H. H. Andersen and J. F. Ziegler, *The Stopping and Ranges of Ions in Matter* (Pergamon, New York, 1977).
[24]D. S. Gemmell, Chem. Rev. 80, 301 (1980).
[25]See Ref. 26 for a discussion of theoretical Γ_γ values.
[26]F. C. Barker, Aust. J. Phys. 33, 177 (1980); and private communication.
[27]R. G. H. Robertson, P. Dyer, T. J. Bowles, R. E. Brown, N. Jarmie, C. J. Maggiore, and S. M. Austin, Phys. Rev. C 27, 11 (1983).
[28]J. N. Bahcall and R. L. Sears, Annu. Rev. Astron. Astrophys. 10, 25 (1972).

ASTROPHYSICAL $S(E)$ FACTOR OF ^{3}He(^{3}He, 2p)^{4}He AT SOLAR ENERGIES*

A. KRAUSS, H.W. BECKER, H.P. TRAUTVETTER and C. ROLFS

Institut für Kernphysik, Universität Münster, Münster, W. Germany

Received 7 November 1986

Abstract: The ^{3}He(^{3}He, 2p)^{4}He reaction has been investigated in the energy range $E_{c.m.} = 17.9$ to 342.5 keV. The studies involved high-current accelerators with well-known beam characteristics and windowless gas target systems. At low energies the studies required the measurement of coincidences between the two protons in the exit channel as well as precautions for contributions from cosmic background and electronic noise. The data extend into the thermal energy region of the sun, e.g., $\sigma = 7 \pm 2$ pb at $E_{c.m.} = 24.5$ keV, and upper limits for the reaction yield have been obtained down to $E_{c.m.} \simeq 17.9$ keV. No evidence for a suggested low-energy resonance has been found. The astrophysical $S(E)$ factor at zero energy is $S(0) = 5.57 \pm 0.32$ MeV · b. The implications of the data for the solar-neutrino problem are discussed.

E | NUCLEAR REACTION ^{3}He(^{3}He, 2p), E (cm) = 17.9–342.5 keV; measured $\sigma(E, \theta)$; deduced astrophysical $S(E)$ factor. Windowless gas targets, beam calorimeter, cosmic background rejection, electronic shielding.

1. Introduction

In main-sequence stars of low mass, such as the sun, energy is produced predominantly by the hydrogen burning pp chain [1,2]). The ^{3}He(^{3}He, 2p)^{4}He reaction ($Q = 12.860$ MeV) is one of the reactions involved in this chain. It can be visualized as a direct process in the entrance channel leading to three possible decay-modes in the exit channel: (i) a direct breakup into the 3 particles of the exit channel (2p+^{4}He), (ii) a sequential breakup into p+^{5}Li with the subsequent decay ^{5}Li→p+^{4}He, and (iii) a sequential breakup into (2p)+^{4}He with the subsequent decay (2p)→p+p. The early studies [3-5]) were carried out over a wide range of beam energies and to as low as $E_{c.m.} \simeq 90$ keV (fig. 1). From the observed particle spectra it followed that the sequential processes prevail at the higher energies ($E_{c.m.} \geq 1$ MeV) but the direct breakup is dominant at lower energies. The observed energy dependence of the astrophysical $S(E)$ factor,

$$S(E) = \sigma(E)E \exp(2\pi\eta),$$

where $2\pi\eta = 4.860/E^{1/2}$ (with the c.m. energy E in MeV), has been fitted to the

* Supported in part by the Deutsche Forschungsgemeinschaft (Ro429/15-1).

polynomial function [5]) (solid curve in fig. 1)

$$S(E) = S(0) + \dot{S}(0)E + \tfrac{1}{2}\ddot{S}(0)E^2$$
$$= 5.5 - 3.1E + 1.4E^2 \text{ (MeV} \cdot \text{b)} .$$

May and Clayton [6]) suggested a mechanism for this $^3He + ^3He$ interaction at low beam energies, in which a neutron tunnels from one ^3He to the other, unimpeded by the Coulomb barrier, up to a radial distance where the nuclei overlap appreciably. In this model, a diproton remains and subsequently fissions into 2 protons. The calculated $S(E)$ factor (dotted curve in fig. 1) described the observed energy dependence of the data very well, thus providing confidence in the extrapolation via the above polynomial function.

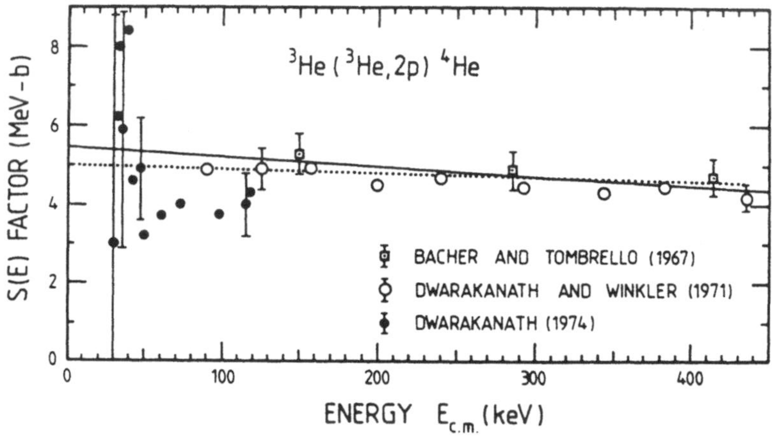

Fig. 1. Astrophysical $S(E)$ factor of $^3He(^3He, 2p)^4He$ as obtained in previous work [4,5,13]). The solid curve is a polynomial fit to the data [5]) and the dotted curve a theoretical calculation [6]) normalized to $S(0) = 5.0$ MeV · b.

The magnitude of the $S(E)$ factor is of special interest in relation to the solar-neutrino problem [7-9]). Based partially on theoretical arguments, it has been suggested [10,11]) that a low-energy resonance might exist in this reaction. If it is sufficiently low and narrow in reaction energy, it might have been unobserved in previous direct measurements. Such a low-energy resonance would significantly enhance the $^3He + ^3He$ route in the pp chain [2,7]) (chain I, $\simeq 86\%$) at the expense of the alternative $^3He + ^4He$ branch (chains II and III, $\simeq 14\%$). If so, the discrepancy between predicted and observed solar-neutrino fluxes [7-9]) might be accounted for or at least decreased. This expected resonance would correspond to an excited state in ^6Be near the $^3He + ^3He$ threshold ($E_x \simeq 11.6$ MeV). However, the search for this state, using a variety of other nuclear reactions, has not been successful: none of these reactions that populate ^6Be showed any excited state near the $^3He + ^3He$ threshold [ref. [12])

and references therein]. In 1974 Dwarakanath [13]) carried out a search for this hypothetical resonance state in a more direct way by extending the $^3He(^3He, 2p)^4He$ reaction studies down to $E_{c.m.} = 30$ keV. Although the data (fig. 1) might suggest an increase in the $S(E)$ factor at the lowest energies, the large uncertainties in the data points (about 200%) precluded any confirmation of the existence of this resonance at least down to $E_{c.m.} = 40$ keV. Below 40 keV the available data neither confirmed nor ruled out its existence.

In the absence of such a resonance the calculated solar-neutrino flux for the ^{37}Cl experiment, N_ν, depends [7-9]) on the $S(0)$ factor of the $^3He(^3He, 2p)^4He$ reaction as $N_\nu \propto S(0)^{-0.42}$. From the available data, Fowler *et al.* [14,15]) recommended $S(0) = 5.5$ MeV · b (fig. 1). Kavanagh [16]) pointed out that the absolute $S(E)$ values reported by refs. [3,13]) are systematically about 20% lower than the combined results of refs. [4,5]). Thus, the lower value of $S(0) = 5.0$ MeV · b has been recommended [16]). From similar considerations, Bahcall *et al.* [8]) used $S(0) = 4.7$ MeV · b for the calculation of the solar-neutrino flux.

In view of the importance of this reaction for the solar-neutrino problem and of the above discussions, a renewed in estigation was carried out to low energies including the thermal energy range in the sun (21.9 ± 6.2 keV for a central temperature of 15.5×10^6 K). The experimental equipment, set-up and procedures are similar to those described recently [17]) and thus only differences will be discussed below. Details of all aspects of this work beyond those reported in this paper and the recent paper [17]) can be found in ref. [18]). The experimental equipment and set-up are described in sect. 2, followed by the measurment of the energy dependence of the relative cross sections in sect. 3 and by the determination of the absolute cross section in sect. 4. The results are discussed in sect. 5.

2. Experimental equipment and set-up

2.1. ACCELERATORS

The 350 kV accelerator at the Institut für Kernphysik in Münster provided beams of $^3He^+$ and $^3He^{++}$ ions at energies $E_{lab} = 60$ to 700 keV with particle currents of up to 90 and 0.15 μA, respectively. The ^3He gas in the terminal of the accelerator was recirculated. The energy calibration of the accelerator facility [19]) has been checked during the course of the experiments and is known to better than ±0.4 keV. The measurements have been extended to significantly lower energies using the 100 kV accelerator at the Dynamitron Tandem Laboratorium in Bochum ($E_{lab} = 41$ to 100 keV; $I(^3He^+) \leq 350$ μA). Details of the accelerator, the beam transport system and the beam characteristics have been described recently [17]). The energy is known to an accuracy of $\Delta E/E = \pm 5 \times 10^{-4}$ and is stable to within ±20 eV, leading to an error in cross section of $\Delta\sigma/\sigma = \pm 4\%$ (2%) *at* $E_{c.m.} = 15$ keV (25 keV) (sect. 3.3).

at each energy below $E_{c.m.} = 80$ keV with ^{3}He as well as with ^{4}He gas in the gas target system: the resulting yields were subtracted.

With the extended gas target set-up and 3 Si detector pairs (fig. 2), an excitation function was obtained at $E_{lab}(^3\text{He}) = 60$ to 350 keV via pp-coincidences observed in each detector pair (fig. 4 and sect. 3.2). The ^{3}He gas pressure in the chamber was 0.50 Torr. The number of incident projectiles N_p was determined in these measurements via the 20 W calorimeter, where a −4.2% offset correction [17]) was taken into account. The observed beam power in connection with the beam energy arriving at the calorimeter was used to deduce N_p. Here, the energy loss of the beam in the target gas up to the place of the calorimeter has to be taken into account (e.g., 2.7 and 6.1 keV at $E_{lab}(^3\text{He}) = 60$ and 350 keV, respectively).

In a third experiment at the 100 kV accelerator ($E_{lab}(^3\text{He}) = 50$ to 100 keV; $I(^3\text{He}^+) \leq 350\ \mu$A), only two detector pairs were used for the measurement of pp-coincidences. The fifth detector monitored, via singles spectra, the deuterium contamination in the experiment (sects. 2.4 and 3.2). Here, the plastic scintillator and the Faraday cages were used for identification and suppression of background events (sect. 3.2). The ^{3}He gas pressure in the extended target chamber was again 0.50 Torr and the number of incident projectiles was determined with the 20 W calorimeter. The lowest data point at $E_{c.m.} = 24.5$ keV required a total running time of 6 days, or an accumulated charge of about 180 coulombs.

The latter set-up was finally used to determine upper limits for the $^3\text{He}(^3\text{He}, 2p)^4\text{He}$ proton yields at $E_{lab}(^3\text{He}) = 41$ to 59 keV. Here, the ^{3}He gas pressure in the extended target chamber was 2.2 Torr, whereby the two detector pairs together covered the

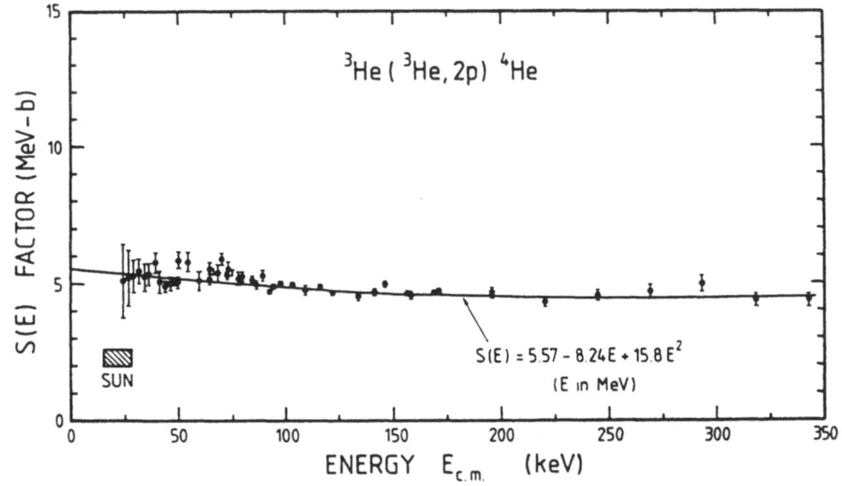

Fig. 6. The energy dependence of the $S(E)$ factor data for $^3\text{He}(^3\text{He}, 2p)^4\text{He}$ (with relative errors only, table 1) is shown. The absolute scale was obtained from fig. 7. The solid curve is the result of a polynomial fit to the data and is described by the relation given. Also indicated (dashed area) is the thermal energy region in the sun.

above. The weighted average of $S(E)=4.73\pm0.16$ MeV · b at $E_{c.m.}=170.5$ keV was taken as the standard in the present work (fig. 7 and table 1). It is in good agreement with previous work (fig. 1).

5. Discussion

Cross-section values, or upper limits thereof, have been obtained at $E_{c.m.}=17.9$ to 342.5 keV and range from $\leqslant 1.5$ pb to 3.2 mb, respectively, i.e., over more than nine orders of magnitude. The data extend into the thermal energy region of the sun (fig. 6). The $S(E)$ factor data at $E_{c.m.}\geqslant 24.5$ keV (table 1) have been fitted with the polynomial function given in sect. 1. The best fit (with a reduced χ^2 of 2.6) was obtained with the parameters $S(0)=5.57\pm0.31$ MeV · b, $\dot{S}(0)=-8.24\pm1.24$ b and $\ddot{S}(0)=31.6\pm8.2$ b/MeV. The errors correspond to one standard deviation except for $S(0)$, where the 3.4% uncertainty in the absolute cross section was included (sect. 4). The above $S(0)$ value is in excellent agreement with the value of 5.5 MeV · b [ref. [5])] recommended in the compilations [14,15]) and does not support $S(0)=$ 4.7 MeV · b used by Bahcall *et al.* [8]) in the calculation of the solar neutrino flux: $N_\nu(\text{theo})=5.8\pm2.2$ SNU, $N_\nu(\text{exp})=2.1\pm0.3$ SNU. With the relation $N_\nu\propto S(0)^{-0.42}$ and $S(0)=5.57$ MeV · b, the predicted neutrino flux reduces to 5.4 SNU.

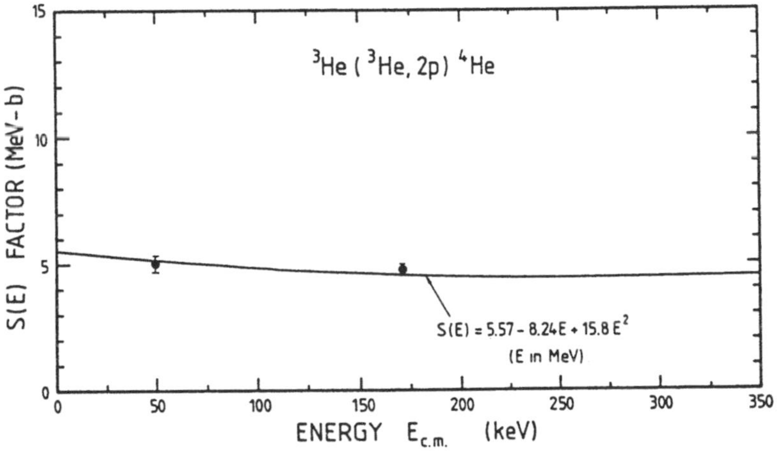

Fig. 7. The absolute cross section measurements (i.e. $S(E)$ values) obtained at $E_{c.m.}=49.0$ and 170.5 keV (sect. 4) are shown together with the polynomial form from fig. 6 normalized to these two points.

The data from previous and present work are consistent with a non-resonant reaction mechanism, at least down to $E_{c.m.}=24.5$ keV (fig. 6). Thus, the suggested resonance, if it exists, must be located at lower energies ($E_R<24$ keV) and must be narrow ($\Gamma_R\ll 24$ keV). Since the thermal energy range at the solar centre is $\Delta E_0=$ 12.4 keV (for $T=15.5\times10^6$ K),i.e., $\Delta E_0\gg\Gamma_R$, the reaction rate $\langle\sigma v\rangle$ of this resonance can be described by the formalism for narrow resonances [1,2,14]:

$$\langle\sigma v\rangle_R = (2\pi/\mu\, kT)^{3/2}\hbar^2(\omega\gamma)_R \exp(-E_R/kT)$$

where μ is the reduced mass and $(\omega\gamma)_R$ the resonance strength. In comparison, the non-resonant $S(E)$ factor leads to the reaction rate:

$$\langle\sigma v\rangle_{NR} = (2/\mu)^{1/2}\,\Delta E_0\, kT)^{-3/2} S(E_0) \exp(-E_0/kT)$$

where $E_0 = 21.9$ keV is the thermal energy in the sun. The solar-neutrino flux depends on the total reaction rate, $\langle\sigma v\rangle = \langle\sigma v\rangle_{NR} + \langle\sigma v\rangle_R$, as given before ($N_\nu \propto \langle\sigma v\rangle^{-0.42}$). Thus, the solar-neutrino problem would be solved for the condition $\lambda = \langle\sigma v\rangle_R/\langle\sigma v\rangle_{NR} \simeq 9$. This "enhancement factor" λ depends only on the location E_R and strength $(\omega\gamma)_R$ of the suggested resonance.

An experimental search for this resonance was carried out at $E_{c.m.} \leq 27.1$ keV (sect. 3.5 and table 1). The actual energy ranges ΔE and the upper limits on the associated reaction yields Y are given in table 2. Using the thick-target expression for narrow resonances, $Y_\infty = \frac{1}{2}\lambda^2\omega\gamma\varepsilon$ (λ = DeBroglie wavelength, ε = c.m. stopping power [24]), upper limits for the resonance strength were deduced (table 2). A variation of E_R within the given energy range ΔE changed the $\omega\gamma$ values by at most 25%. For the condition $\lambda = 9$ the calculated $\omega\gamma$ strengths are also given in table 2, where E_R was varied within the extrema of a given energy range ΔE. The ratios $(\omega\gamma)_{exp}/(\omega\gamma)_{cal}$ (table 2) show that the solar-neutrino problem cannot be solved by the suggested resonance, except if it is located near or below $E_{c.m.} = 16$ keV. This conclusion is consistent with the results of Dwarakanath [13], although the present work reduces the limits on such a resonance by about a factor 30. Improved measurements at $E_{c.m.} \leq 16$ keV require other experimental techniques than those used in the present work.

The authors would like to thank L. Buchmann, J. Görres, M. Hilgemeier, A. Redder, U. Schröder, S. Seuthe and K. Wolke for assistance during the course of

TABLE 2

Properties of a resonance at low energies

Energy range ΔE (keV) [a] $E_1 \to E_2$	Yield Y [b]	Resonance strength $(\omega\gamma)_R$ (MeV)			$(\omega\gamma)_{exp}/(\omega\gamma)_{cal}$	
		experiment	calculation [c]			
			$E_R = E_1$	$E_R = E_2$	$E_R = E_1$	$E_R = E_2$
16.2 → 18.9	≤1.1	$\leq 1.7\times10^{-17}$	1.5×10^{-17}	1.1×10^{-16}	≤1.1	≤0.15
18.8 → 2.15	≤2.3	$\leq 4.7\times10^{-17}$	1.1×10^{-16}	8.3×10^{-16}	≤0.43	≤0.056
21.4 → 24.4	≤5.5	$\leq 1.4\times10^{-16}$	8.3×10^{-16}	7.5×10^{-15}	≤0.17	≤0.019
24.0 → 27.1	≤17	$\leq 5.3\times10^{-16}$	6.3×10^{-15}	6.5×10^{-14}	≤0.084	≤0.008

[a]) Energies in c.m. system.
[b]) In units of 10^{-18} reactions per incident particle.
[c]) For an enhancement factor $\lambda = 9$.

the experiments. The loan of the NE102A plastic scintillator by Dr. E. Kuhlmann (Bochum) is highly appreciated. They also thank K. Brand and the technical staff at the Dynamitron Tandem Laboratorium in Bochum for technical help at the 100 kV accelerator. Finally, we would like to thank Profs. R.W. Kavanagh and T.A. Tombrello for enlightening comments on the manuscript.

References

1) W.A. Fowler, Rev. Mod. Phys. **56** (1984) 149
2) C. Rolfs, W.S. Rodney and H.P. Trautvetter, Rep. Progr. Phys. (in print, 1987)
3) N.M. Wang, V.M. Novatskii, G.M. Osetinskii, N.K. Chien and I.A. Chepurchenko, Sov. J. Nucl. Phys. **3** (1966) 777
4) A.D. Bacher and T.A. Tombrello, quoted by T.A. Tombrello, Nuclear research with low-energy accelerators, ed. J.B. Marion and D.M. van Patter (Academic Press, 1967) p. 195
5) M.R. Dwarakanath and H. Winkler, Phys. Rev. **C4** (1971) 1532
6) R. May and D.D. Clayton, Astrophys. J. **153** (1968) 855
7) J.N. Bahcall and R. Davis, Essays in nuclear astrophysics, ed. C.A. Barnes, D.D. Clayton and D.N. Schramm (Cambridge Press, 1982) p. 243
8) J.N. Bahcall, W.F. Huebner, S.H. Lubow, P.D. Parker and R.K. Ulrich, Rev. Mod. Phys. **54** (1982) 767;
J.N. Bahcall, B.T. Cleveland, R. Davis and J.K. Rowley, Astrophys. J. **292** (1985) L79
9) P.D. Parker, Physics of the sun, ed. P.A. Sturrock (Reidel, 1985) p. 15
10) V.N. Fetisov and Y.S. Kopysov, Phys. Lett. **B40** (1972) 602; Nucl. Phys. **A239** (1975) 511
11) W.A. Fowler, Nature **238** (1972) 24
12) A.B. McDonald, T.K. Alexander, J.E. Beene and H.B. Mak, Nucl. Phys. **A288** (1977) 529
13) M.R. Dwarakanath, Phys. Rev. **C9** (1974) 805
14) W.A. Fowler, G.R. Caughlan and B.A. Zimmerman, Ann. Rev. Astr. Ap. **13** (1975) 69
15) M.J. Harris, W.A. Fowler, G.R. Caughlan and B.A. Zimmerman, Ann. Rev. Astr. Ap. **21** (1983) 165
16) R.W. Kavanagh, Essays in nuclear astrophysics, ed. C.A. Barnes, D.D. Clayton and D.N. Schramm (Cambridge Press, 1982) p. 159
17) A. Krauss, H.W. Becker, H.P. Trautvetter, C. Rolfs and K. Brand, Nucl. Phys. (in print)
18) A. Krauss, thesis, Universität Münster (1986)
19) T. Freye, H. Lorenz-Wirzba, B. Cleff, H.P. Trautvetter and C. Rolfs, Z. Phys. **A281** (1977) 211
20) C. Rolfs, J. Görres, K.U. Kettner, H. Lorenz-Wirzba, P. Schmalbrock, H.P. Trautvetter and W. Verhoeven, Nucl. Instr. Meth. **157** (1978) 19
21) H.W. Becker, L. Buchmann, J. Görres, K.U. Kettner, H. Kräwinkel, C. Rolfs, P. Schmalbrock, H.P. Trautvetter and A. Vlieks, Nucl. Instr. Meth. **198** (1982) 277
22) A. Redder, H.W. Becker, H. Lorenz-Wirzba, C. Rolfs, P. Schmalbrock and H.P. Trautvetter, Z. Phys. **A305** (1982) 325
23) A. Vlieks, M. Hilgemeier and C. Rolfs, Nucl. Instr. Meth. **213** (1983) 291
24) J.F. Ziegler, Helium stopping powers and ranges in all elemental matter (Pergamon Press, 1977)
25) G.G. Ohlsen, Nucl. Instr. Meth. **37** (1965) 240
26) J. Görres, K.U. Kettner, H. Kräwinkel and C. Rolfs, Nucl. Instr. Meth. **177** (1980) 295

THE FATE OF ^{7}Be IN THE SUN

C. W. JOHNSON, E. KOLBE, S. E. KOONIN, AND K. LANGANKE
W. K. Kellogg Radiation Laboratory, California Institute of Technology, Pasadena, CA 91125
Received 1990 August 22; accepted 1991 December 16

ABSTRACT

We reexamine the electron- and proton-capture rates of ^{7}Be important to the solar neutrino "problem." Although the assumptions implied by the traditional Debye approximation for plasma screening are not valid, a careful numerical study changes the electron capture rate by less than 2%. We extrapolate experimental data on the proton capture reaction to astrophysically relevant energies using an energy dependence that includes d-wave scattering and is shown to be relatively independent of the model space and interaction used. We find that the solar proton capture rate is lowered by approximately 7% from the currently accepted value.

Subject headings: nuclear reactions, nucleosynthesis, abundances — Sun: abundances

1. INTRODUCTION

The solar neutrino "problem" (see Bahcall & Ulrich 1988 for a review) is still unresolved. Both the Homestake (Davis 1987) and Kamiokande (Hirata et al. 1989) experiments measure a flux of high-energy (above 0.814 and 9.3 MeV, respectively) neutrinos considerably smaller than predicted by the standard solar model. Most of the expected high-energy neutrinos originate from the beta-decay of ^{8}B, which in turn is produced via ^{7}Be$(p, \gamma)^8$B.

In this paper we reexamine the two processes that determine the fate of ^{7}Be in the Sun. This nuclide is consumed by either

$$^7\text{Be} + e^- \rightarrow {}^7\text{Li} + \nu \tag{1}$$

or

$$^7\text{Be} + p \rightarrow {}^8\text{B} + \gamma \,. \tag{2}$$

The former has a lifetime in the core of the Sun of $\tau_e \approx 80$ days (depending on temperature and density [Bahcall & Moeller 1969]; this core lifetime is by coincidence approximately the same as the laboratory value) and the latter $\tau_p \approx 200$ yr. Since $\tau_p \gg \tau_e$, the concentration of ^{7}Be is proportional to τ_e, implying that the high-energy neutrino flux is proportional to the product of τ_e and the (p, γ) rate. This latter is of particular interest as it is believed to be the most uncertain nuclear physics input to the solar neutrino problem (Bahcall & Ulrich 1988).

In § 2 we calculate the effect of plasma screening on the bound-electron contribution to the electron-capture rate $\Lambda_e = 1/\tau_e$. Traditionally the Debye-Hückel approximation (DH) was used for the plasma-screened potential for the bound electron. However, there are three assumptions for DH that are weakly violated in the core of the Sun. When the same assumptions are strongly violated in laboratory plasma, experiments show that DH fails dramatically. We have therefore pursued a careful numerical study. We find that, despite these concerns, DH describes the electron capture rate to within 2%.

In § 3 we present the results of a microscopic 3-cluster calculation of the astrophysical S-factor for the (p, γ) reaction that takes into account both d- and s-wave entrance channels; the d-waves are unimportant at solar energies but are non-negligible at the energies at which experiments are performed, and thus must be accounted for when extrapolating downward. The energy dependence of the S-factor is calculated using two different interactions, and found to be in accord with each other and with previous calculations; the overall scale is then fit to six experimental data sets. Our final conclusion is that the p-capture cross-section is approximately 7% lower than the value used in the standard Solar model. Unfortunately, this is still far from resolving the solar-neutrino problem.

2. ^{7}Be$(e, \nu)^7$Li

The ^{8}B solar-neutrino flux is inversely proportional to the ^{7}Be electron-capture rate, which in turn is proportional to the density of electrons at the nucleus., $\Lambda_e \propto |\psi_e(0)|^2$. In the laboratory there are only bound electrons, but in the solar plasma continuum electrons contribute as well ($\Lambda_e = \Lambda_{\text{cont}} + \Lambda_{\text{bound}}$) and in fact dominate. In the Sun, $\Lambda_{\text{bound}}/\Lambda_{\text{cont}} \approx 20\%$ (Iben, Kalata, & Schwartz 1967).

The density of electrons at the nucleus is determined by the screened Coulomb potential. The continuum density (and contribution to the rate) is insensitive to screening (Bahcall & Moeller 1969). However, screening is much more important for bound-electron capture, as it reduces the density by about 64% (Iben et al. 1967).

To find the bound electron wave-function ψ_e, one solves the Schrödinger equation

$$\hat{H}\psi_e = \left(-\frac{\hbar^2}{2m_e}\nabla^2 - V\right)\psi_e = E\psi_e \,. \tag{3}$$

Previously, Iben et al. (1967, hereafter IKS) and others used the DH screened potential (also often referred to as the static screened Coulomb potential), which is found by solving

$$(\nabla^2 - q_D^2)V = -4\pi\rho \,, \tag{4}$$

where $R_D = 1/q_D$ is the Debye screening length and ρ is the charge density that gives rise to the unscreened potential. If one takes $\rho = \rho_N = Z\delta^3(r)$, one obtains the standard form

$$V_{\text{DH}} = Z \exp(-q_D r)/r \,. \tag{5}$$

(Here and throughout we take the unit charge $e = 1$.) From this IKS calculated both variational and numerical forms of ψ_e in equation (3). Bahcall & Moeller (1969), like IKS, used V_{DH} in the Schrödinger equation (eq. [3]) to calculate the bound-state electron wave function; they then use $\rho = \rho_N + \rho_e$ with $\rho_e = -|\psi_e|^2$ in (4) to obtain V_{DH} and use this screened potential in equation (3) to calculate the wave function of *continuum* electrons.

Experiments suggest that the DH potential can fail for bound electrons in plasmas. In particular, Goldsmith, Griem, & Cohen (1984) looked for evidence of line shifts in a laboratory plasma; from the work of Rogers, Graboske, & Harwood (1970), who used the DH potential, one estimates a shift in the Lyman-α line of oxygen of about 0.06 Å, but Goldsmith et al. (1984) found no line shift to within 2 σ uncertainty of 0.02 Å.

This experimental result is not surprising when one considers the assumptions that go into using the DH potential:

1. The mean interparticle spacing $\lambda_m = N_e^{-1/3}$ is much smaller than the Debye length R_D (Landau & Lifshiftz 1980),

$$\lambda_m \ll R_D$$

2. In addition, for applications of a screened potential to bound electrons, the number of plasma electrons within a sphere with the radius of the Bohr orbit must be much larger than 1 (Hummer & Mihalas 1988),

$$N_e(4\pi/3)(a_0/Z)^3 \gg 1 ,$$

or equivalently

$$\lambda_m \ll a_0/Z .$$

3. Finally, using the DH potential assumes that the plasma electrons and ions move on a time scale $1/\omega_{plasma}$ much shorter than the bound electron, so that the plasma "sees" two static point charges (Hummer & Mihalas 1988).

$$\omega_{plasma} \gg \omega_e$$

All three assumptions are strongly violated in the laboratory plasma of Goldsmith et al. (1984), which had an electron density of $N_e = 5 \times 10^{-3}$ Å^{-3} and an electron temperature of $T_e = 0.070$ keV. In the core of the Sun, by way of comparison, $N_e = 60$ Å^{-3} and $T = 1.3$ keV. The assorted scale lengths can be easily computed: for the laboratory experiment, using oxygen ($Z = 8$), one obtains $R_D = 3.1$ Å, $\lambda_m = 6$ Å, and $a_0/Z = 0.07$ Å. For the plasma in the solar core, $R_D = 0.218$ Å, $\lambda_m = 0.255$ Å, and $a_0/Z = 0.133$ Å. Clearly neither assumption (1) or (2) are satisfied in either plasma, although they are more strongly violated in the case of the laboratory experiment.

The time scales are estimated as follows. As the electronic component of the plasma moves the fastest, one estimates $\omega_{plasma} = (4\pi n_0 e^2/m_e)^{1/2}$ (Jackson 1975); and $\omega_e = Z^2e^2/\hbar a_0$. Note that the plasma ions will have a much smaller frequency and longer time scale. In atomic units ($e^2/\hbar a_0 = 1$) one finds that for the laboratory experiment, $\omega_{plasma} = 9.7 \times 10^{-2}$ and $\omega_e = 64$, while for the solar core $\omega_{plasma} = 10$ and $\omega_e = 16$. Thus assumption (3) is also violated, again more strongly for the laboratory experiment.

The failure of the DH potential for laboratory plasmas has been previously addressed by theory. Theimer & Kepple (1970) and Skupsky (1980) accounted for assumption (3) by calculating the bound electron to interact self-consistently with the (still classical) plasma, that is, V_{DH} calculated in equation (2) is calculated using $\rho = Z\delta(r) - |\psi_e(r)|^2$, and then ψ_e calculated in equation (1) using that V_{DH}. Davis & Blaha (1982) then corrected, at least in part, for assumptions (1) and (2) by treating both the free and bound electrons in a self-consistent Hartree calculation with occupation numbers given by a Fermi-Dirac thermal distribution. Exchange and correlation effects for the free electrons were included in an approximate way, but fully antisymmetrized wavefunctions were not used because degeneracy effects were expected to be small for the plasmas in which they were interested (conditions similar to those of the experiments of Goldsmith et al. 1984). All three papers found significant deviations from DH results for laboratory plasmas.

The question arises whether the DH potential fails also for conditions in the solar core, where assumptions (1), (2), and (3) are not quite satisfied. Our detailed numerical calculations, described below, show that in fact DH gives $|\psi_e(0)|^2$ to within a few percent.

Our self-consistent thermal Hartree calculation is similar to that of Davis & Blaha (1982). First, consider the continuum (plasma) electrons. Given a screened electrostatic potential ϕ surrounding a nucleus of charge Z, Schrödinger's equation is integrated to give the continuum electron wavefunctions. The charge density due to the continuum electrons is calculated from these wavefunctions, weighted by the usual thermal Fermi-Dirac distribution, with a chemical potential set to match the average charge density, that is, the electron charge density $-\rho_\infty$ at a large distance from the nucleus Z. In addition, the "hole" in the background (ionized nuclei of charge Z_i) charge distribution is calculated using the distribution $\rho_\infty[1 - \exp(-Z_i\phi/kT)]$. Beyond a certain radius, about $\frac{1}{3}$ to $\frac{1}{2}$ Å, it becomes computationally taxing to sum a sufficient number of partial waves, and we use instead the Thomas-Fermi approximation to arrive at the continuum electron density, an approximation whose validity at large radii we confirmed numerically. We then enforce charge conservation on this charge distribution, so that the integral of the charge density in excess of the average density ρ_∞ is exactly $-Z$, by simply setting the charge density to ρ_∞ beyond an appropriate cutoff radius. This cutoff radius was typically about 1 Å, much larger than both the Debye screening length and the characteristic size of a bound electron orbit, and the discontinuity in density was small. From the excess charge distribution we solved Poisson's equation to obtain the potential screening the nucleus. This process was iterated until convergence.

For a pure plasma (no bound electron), the self-consistent potential was indistinguishable from V_{DH}, independent of the potential used to initiate the iterations. We found that the density of continuum electron calculated in the self-consistent potential (or V_{DH}), is about 2.4% less than that for the density in a pure Coulomb potential. (This is slightly larger than reported by Bahcall & Moeller 1969). This result was not affected by the approximate introduction of exchange forces via the local Slater approximation, $V_{ex}(r) = -3^{1/3}\pi^{-1/3}\rho(r)^{1/3}$.

Next, consider the bound electron. We take the opposite of assumption (3), that the bound electron moves much faster than the plasma electrons. This is not quite true for our conditions, but acts as a limit; and as we shall see, the effect is negligible.

We begin with an analytic treatment in the spirit of Debye-Hückel, which we will call "self-consistent Debye-Hückel" (SCDH). The self-consistent potential V to be used in equation (3) is

$$V = V_N + V_{pN} + V_{pe} , \tag{6}$$

where the first two terms are those of IKS, with $V_N = Z/r$ and V_{pN} solving

$$(\nabla^2 - q_D^2)(V_N + V_{pN}) = -4\pi\rho_N ; \tag{7}$$

that is, $V_N + V_{pN}$ gives V_{DH} (5). The third term is new: V_{pe} is the solution to

$$(\nabla^2 - q_D^2)(V_e + V_{pe}) = -4\pi\rho_e , \tag{8}$$

where V_e is the potential generated by the bare electron,

$$\nabla^2 V_e = -4\pi\rho_e . \tag{9}$$

The total energy of the system can then be written as

$$E = \langle \hat{H} \rangle = T + U_{IKS} + \tfrac{1}{2}U_{pN} + \tfrac{1}{2}U_{pe} \tag{10}$$

where

$$T = \frac{\hbar^2}{2m_e}\int d^3r|\nabla\psi_e|^2 , \tag{11}$$

$$U_{IKS} = \int d^3r\rho_e(V_N + V_{pN}) . \tag{12}$$

$$U_{pN} = \int d^3r\rho_N V_{pN} = -Z^2 q_D ; , \tag{13}$$

$$U_{pe} = \int d^3r\rho_e V_{pe} , \tag{14}$$

where again $\rho_e = -|\psi_e|^2$. The factors of $\frac{1}{2}$ in equation (10) are corrections for self-energy terms of the plasma; one treats the plasma as a dielectric medium and calculates the work necessary to insert the nucleus-electron system into the medium (Jackson 1975). The term $-\frac{1}{2}Z^2 q_D$ is well-known as the interaction energy between the bare nucleus and the plasma; because it is independent of the electron wave function it can be ignored (but will be important later for confirming our results).

An approximate solution to equations (3), (4), and (6) can be obtained with a variational wave function,

$$\psi_e(r) = (\pi a^3)^{-1/2} \exp(-r/a) , \tag{14}$$

where a is chosen to minimize $E = \langle \hat{H} \rangle$. Note for a $^7Be^{+++}$ ion in free space, $a = a_0/4$, where $a_0 = 0.531$ Å is the Bohr radius. Solving the Debye equation (eq. [4]), one obtains

$$V_{pe} = \frac{1}{r}\left[1 - \left(\frac{q_D a}{2}\right)^2\right]^{-2} \times \left[\exp\left(\frac{-2r}{a}\right)\left\{1 + \left[1 - \left(\frac{q_D a}{2}\right)^2\right]\frac{r}{a}\right\} - \exp(-q_D r)\right] + \frac{1}{r}\left[1 - \exp\left(\frac{-2r}{a}\right)\right] - \frac{1}{a}\exp\left(\frac{-2r}{a}\right) . \tag{15}$$

Integrating, one finds the new term in the potential is

$$U_{pe} = -\frac{1}{a}\left[\frac{5}{16} + \frac{1}{2}\left[1 - \left(\frac{q_D a}{2}\right)^2\right]^{-2} \times \left\{\frac{1}{4} - \left(1 + \frac{q_D a}{2}\right)^{-2} + \frac{1}{8}\left[1 - \left(\frac{q_D a}{2}\right)^2\right]\right\}\right] . \tag{16}$$

We can check this result by taking the limit as $q_D a \to 0$, which corresponds to the spatial extent of the bound state becoming small compared to the screening length; i.e., the plasma "sees" a point charge of $Z - 1$. In this limit, equation (16) is $-\frac{1}{2}q_D$. One also has (Iben et al. 1967) $U_{IKS} = -(z/a)(1 + q_D a/2)^{-2}$ so that

$$\lim_{q_D a \to 0} U_{IKS} = -\frac{Z}{a} + Zq_D . \tag{17}$$

The first term of this limit is a self-energy that we discard. Adding the nucleus-plasma interaction energy $-Z^2 q_D/2$, the total potential energy is

$$U = Zq_D - \tfrac{1}{2}Z^2 q_D - \tfrac{1}{2}q_D = -\tfrac{1}{2}(Z - 1)^2 q_D , \tag{18}$$

which is exactly what it should be.

In the core of the Sun, $q_D = 4.46$ Å^{-1}, and for our variational wavefunction (14) we find $a_{IKS} = 0.162$ Å, and $a_{sc} = 0.156$ Å. We also solved the system of equations (7)–(10) numerically, and obtained the same "self-consistent Debye-Hückel potential which is V_{DH} plus V_{pe} given in equation (15), although of course the numerical bound-state wavefunction and binding energy differed from that in the variational calculation. One must then fold in the population of bound states in the solar plasma, which is governed by the Boltzmann factor. We follow the recipe of IKS to finally obtain in both the variational and numerical cases

$$\Lambda_{IKS}/\Lambda_{cont} = 1.15 \tag{19}$$

while

$$\Lambda_{sc}/\Lambda_{cont} = 1.17 , \tag{20}$$

a 1.7% difference.

The final step is to treat both continuum and bound electron quantum mechanically (our Hartree calculation, as opposed to DH and SCDH), and it is here that we find a small but definite deviation from a classical (DH) treatment of the plasma.

Our calculation is exactly that of our full self-consistent calculation of the continuum electrons is described above, but with a bound-state electron included. Once again charge conservation is enforced. We find that the self-consistent correction to the DH screening potential in which the bound electron moves is better approximated (though not exactly) by $-1/ZV_{pN}$ than by the V_{pe} given in equation (15). Again, the self-consistent solution is independent of the initial potential used to start the iterations. The density of continuum electrons at the origin is virtually unaffected by the presence of the bound state.

We give our results for a bound electron in Table 1. The units are those of a $^7Be^{+++}$ ion in free space. We compare the numerical results using the DH potential, the self-consistent DH potential (eq. [15]), and the full quantum-mechanical Hartree calculation. Note that one needs not just the binding energy E_b of the electron as calculated in the Schrödinger equation, but also, as in (10), account for the change in the plasma

TABLE 1

NUMERICAL RESULTS FOR BOUND ELECTRON IN $^7Be^{+++}$ AT THE SUN'S CORE

Potential	$\lvert\psi(0)\rvert^2/\lvert\psi_{free}(0)\rvert^2$	E_{bind}/E_{free}	E_{sys}/E_{free}	Rate
DH	0.67	0.22	0.22	18.2%
SCDH	0.72	0.38	0.30	19.9%
Hartree	0.79	0.42	0.34	21.9%

NOTES.—"Free" denotes $^7Be^{+++}$ in free space. The rate is relative to the continuum contribution. See text for details.

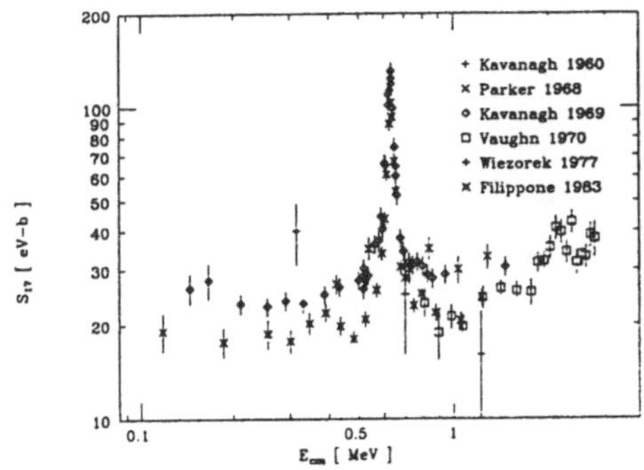

FIG. 1.—World's data on $^7Be(p, \gamma)^8B$, given as astrophysical *S*-factor

self energy due to the introduction of the bound electron; the relevant quantity is the "binding energy of the system," E_{sys}. The rate is given in units of the continuum electron rate, and is calculated using the formula of IKS. The full Hartree calculation gives a total electron capture rate 3.7% larger than that using DH; however, subtracting off the 2.4% lowering of the continuum rate by use of DH continuum wavefunctions, the total change in the electron capture rate is only 1.3%—negligible in the context of the solar neutrino problem.

3. $^7Be(p, \gamma)^8B$

We now turn to the production of ^8B via the (p, γ) reaction. There have been six measurements of the $^7Be(p, \gamma)^8B$ cross section, shown in Figure 1: Kavanagh (1960), Parker (1966, 1968), Kavanagh et al. (1969), Vaughn et al. (1970), Wiezorek et al. (1977), and Filippone et al. (1983). The lowest experimental data point is at center-of-mass $E = 117$ keV; to determine the reaction rate in the solar core, the data must be extrapolated to lower energies ($E \approx 0$–20 keV) using an energy dependence calculated in a direct-capture model.

There are three sources of uncertainty to be considered: uncertainty in the theoretical energy dependence (which in turn depends on the model space and interaction used), uncertainty in the one-parameter (overall normalization) fit to the experimental data, and uncertainty in the normalization of the experimental data. We consider only the first two in this paper. Barker & Spear (1986) have raised some questions about the third, noting that many experimental normalizations can be traced to a stopping power of protons in lithium that Barker & Spear (1986) consider suspect. However, we note that Filippone et al. (1983) found no change within their quoted uncertainties if their data were normalized in a manner independent of this stopping power.

The standard solar model (Bahcall & Ulrich 1988) currently assigns to the $^7Be(p, \gamma)^8B$ reaction a zero-energy *S*-factor of $S_{17}(E = 0) = 0.0243 \pm 0.018$ keV barn^{-1} (the uncertainty is that of Parker 1986; Bahcall & Ulrich 1988 quote a 3 σ uncertainty, which is 22%) and an energy derivative of $S'_{17}(0) = -3 \times 10^{-5}$ barns. These values have been obtained from extrapolations that use the energy dependence of a potential model of structureless $^7Be + p$ fragments and consider only *s*-wave scattering in the entrance channel (Tombrello 1965).

Robertson (1973), and later Barker (1980) and others, criticized this extrapolation, pointing out that *d*-wave capture cannot be neglected at the energies where the data are taken ($E \leq 1$ MeV). These authors showed that *d*-wave capture modifies the energy dependence of the low-energy cross section to the extent that the extrapolated value of $S_{17}(0)$ is reduced by about 15%. One weakness in this analysis is that there are essentially no experimental constraints on the *d*-wave potential. Barker found only a slight dependence on the *d*-wave depth and so equated the *d*-wave and *s*-wave interactions, the parameters of the latter being determined by the experimental $^7Li + n$ scattering and capture cross sections. While Filippone et al. (1983) adopted an extrapolated $S_{17}(0)$ using only *s*-wave capture (Tombrello 1965), they noted that use of Barker's (1980) energy dependence gave an extrapolated value 10%–15% lower.

Two recent microscopic calculations of the $^7Be(p, \gamma)^8B$ reaction have treated the various *s*- and *d*-wave capture contributions consistently within the same 8-nucleon model space (Descouvemont & Baye 1988; Kolbe, Langanke, & Assenbaum 1988). While these approaches, which use an effective nucleon-nucleon interaction, are not accurate enough to predict absolute cross sections, the calculated energy dependences agree qualitatively with that of Barker (1980) and so confirm the importance of *d*-wave capture contributions. However, neither work extrapolated the experimental data to low energies using the calculated energy dependence.

3.1. *Microscopic Calculation of the Energy Dependence*

In this section, we present microscopic calculations that extend the approach of Kolbe et al. (1988) to a full dynamical 3-cluster treatment of the reaction. In detail, our model space is spanned by fully antisymmetric $p + {}^3He + {}^4He$ cluster wave functions

$$|\Psi^{J^\pi}\rangle = \mathcal{A}\left\{\sum_I [(\Phi^{I_1}_{Be} \otimes \Phi^{I_2}_p)^I \otimes Y_l(\hat{r})]^{J^\pi} g^{J^\pi}_{l,I}(r)\right\} \tag{21}$$

with

$$|\Phi^{I_1}_{Be}\rangle = \mathcal{A}\{\Phi_\alpha[\Phi^{I'}_{He} \otimes Y_L(\hat{r}')]^{I_1} \tilde{g}^{J^\pi}_{L,I}(r')\} \,. \tag{22}$$

Here, we have formally written the 8-nucleon wave function as a $^7Be + p$ cluster function, while the internal degrees of freedom of the ^7Be fragment are described by a $^3He + \alpha$ cluster function. In equation (21), $\Phi^{I_2}_p$ is the spin-isospin function of the proton (spin $I_2 = \frac{1}{2}$), while in equation (22), Φ_α and $\Phi^{I'}_{He}$ describe the internal degrees of freedom of the α particle (assumed to be $T = S = 0$) and of the ^{3}He-nucleus ($I' = \frac{1}{2}$). The separations of the $p + {}^7Be$ and $^3He + {}^4He$ clusters are r and r', respectively. Correspondingly, $g(r)$ and $\tilde{g}(r)$ are the relative wave functions between these cluster fragmentations; they might be different for different values of the channel spin I. Thus, the relative wave function $\tilde{g}(r')$ in equation (22) carries an index I. Further, as we will in the following consider only the ground-state of ^7Be, $L = 1$ and $I_1 = 3/2$, so that the channel spin I in equation (21) can thus take the values $I = 1, 2$. To reduce the computational effort, we have neglected coupling of channels with different orbital angular momenta. Thus, $l = 1$ for the ^8B ground state, while $l = 0, 2$ for the important $^7Be + p$ scattering states, allowing *E*1 capture into the ^8B ground state.

Our calculations use either the Minnesota force (Chwieroth et al. 1973) or the Hasegawa-Nagata force (Furitani et al. 1980) as the effective nucleon-nucleon interaction. We have adjusted one of the parameters in each of these interactions (the exchange mixture parameter u in the Minnesota force and the Majorana exchange parameter m of the medium-ranged Gaussian in the force of Furitani et al. 1980) to reproduce the binding energy of the ^{8}B ground state relative to the ^{7}Be + p threshold. The properties of the individual clusters (p, ^{3}He, ^{4}He) do not depend on these interaction parameters.

We determine the dynamical degrees of freedom in our approach—the relative wave functions g and $\tilde{g}$ in equations (21), 22)—by solving the many-body Schrödinger equation assuming fixed internal cluster structures. Different procedures for doing so have been used for the bound and scattering states. In both cases we first calculate the ^{7}Be ground state by a standard two-cluster RGM treatment of the seven-nucleon problem, equation (22). To reduce the numerical effort necessary to calculate the ^{8}B wave functions, we have then expanded the radial wave function $\tilde{g}$ in a minimal number of basis wave functions. As in Kolbe et al. (1988), we succeeded in representing $\tilde{g}$ by a sum of only two radial harmonic oscillator states $u_N^L(r, \beta)$ with different width parameters β and quantum numbers ($L = 1, N = 2n + L = 1$)

$$\tilde{g}_{L=1}^{J^\pi,I}(r) = \alpha_1 u_{N=1}^{L=1}(r, \beta_1) + \alpha_2 u_{N=1}^{L=1}(r, \beta_2) . \tag{23}$$

When inserted in equation (22), this expression well-reproduces the properties of the ^{7}Be ground state obtained in the full RGM approach.

In solving the 3-cluster problem, we have expanded the relative wave function $g_{lI}^{J^\pi}$ in a basis of 24 harmonic oscillator states $u_N^l(r, \gamma)$:

$$g_{l,I}^{J^\pi}(r) = \sum_{m=1}^{6} \sum_{n=0}^{3} \beta_{m,n}^{J^\pi I} u_{N=l+2n}^{l}(r, \gamma_m) . \tag{24}$$

At a radius beyond the range of the nuclear forces and the influence of the Pauli principle, $g_{lI}^{J^\pi}$ is matched to the appropriate asymptotic boundary condition for bound and scattering states, respectively.

In our bound state calculation, we have allowed the parameters α_i and $\beta_{m,n}^{J^\pi I}$ in equations (23) and (24) to vary, particularly allowing α_1, α_2 to be different for any triplet of indices (m, n, I). The values of these parameters were determined by minimizing the ^{8}B ground state energy. If we adopt $u = 1.1315$ in the Minnesota force and $m = 0.3714$ in the Hasegawa-Nagata force, our approach can reproduce the experimental binding energy of the ^{8}B ground state relative to the p + ^{7}Be threshold ($E_B = 138$ keV). These values of (u, m) are typical for studies of light nuclear reactions.

For our two different effective interactions, the parameters in the ^{7}Be ground state (22) are

$$\alpha_1 = 0.01390 \quad \beta_1 = 2.319 \text{ fm}$$

$$\alpha_2 = -0.03334 \quad \beta_2 = 1.026 \text{ fm}$$

for the Minnesota force and

$$\alpha_1 = 0.01229 \quad \beta_1 = 2.331 \text{ fm}$$

$$\alpha_2 = -0.06986 \quad \beta_2 = 1.046 \text{ fm}$$

for the Hasegawa-Nagata force.

For the scattering states, we solved the 3-cluster problem by introducing a set of ^{7}Be pseudostates. These are obtained by diagonalizing the microscopic seven-nucleon Hamiltonian in the two-dimensional Hilbert space spanned by the basis functions (eq. [22]) considering the coefficients α_1, α_2 as variables and using the same width parameters as above. The resulting lower eigenstate is identical to the ^{7}Be ground state defined above. The upper pseudostate generally does not correspond to a physical level, but rather is just a tractable way of accounting for distortion in the scattering states (Shen et al. 1985). Upon inserting the two ^{7}Be configurations into equation (21), we define p + ^{7}Be channels, which, after the appropriate asymptotic boundary condition is imposed, can be interpreted as the (physical) p + ^{7}Be system and an inelastic p + ^{7}Be* channel. Note that the inelastic channel is closed at the low p + ^{7}Be energies of interest in this paper. The corresponding coupled-channel problem can be solved with standard techniques (Wildermuth & Tang 1977).

At low energies, the ^{7}Be(p, γ)^{8}B cross section is dominated by $E1$ capture into the ^{8}B ground state. We have calculated the respective many-body matrix elements of the electric dipole operator in the long-wavelength approximation, truncating the integral over the radial relative coordinate at 200 fm.

It is convenient to present the cross section in terms of the astrophysical S-factor:

$$S(E) = \sigma(E)E \exp\{2\pi\eta(E)\} , \tag{25}$$

where in the present case the Sommerfeld parameter is given by $2\pi\eta(E) = 117.47/\sqrt{E}$ with the energy expressed in keV. Furthermore, for applications to the solar neutrino problem, the relevant quantities are $S(0)$ and its first two derivatives at $E = 0$. For these latter two we use the normalization-independent parameterization given by Williams & Koonin (1981), namely

$$\frac{1}{S}\frac{dS}{dE} = a + bE .$$

These three quantities are given in Table 2. We give the theoretical $S_{17}(0)$ only for completeness; in the next subsection, we use experiment to set the overall normalization. While the Minnesota force predicts an S-factor 5% smaller than that of the Hasegawa-Nagata force, the two calculations predict nearly identical energy dependences for the low-energy cross section. Therefore we adopt values of $a = -1.00$ MeV^{-1} and $b = 8.96$ MeV^{-2} with "theoretical uncertainties" of 1% or less.

Our calculations with the Hasegawa-Nagata force can be compared with those of Kolbe et al. (1988), who adopted the same effective interaction but a less flexible model space. Our cross sections are about 10% lower than those of Kolbe et al. (1988) because of a smaller amplitude in the asymptotic ^{8}B ground state. The calculated energy dependences are similar, as are the relative contributions of $E1$ capture from the

TABLE 2

LOW-ENERGY PARAMETERS OF COUPLED-CLUSTER CALCULATION OF S_{17}

Parameter	Minnesota	Hasegawa-Nagata
$S(0)$ (keV) barn^{-1}	0.02514	0.02384
a (MeV^{-1})	−0.99	−1.01
b (MeV^{-2})	9.01	8.90

NOTES.—a and b are from the parameterization of Williams & Koonin 1981; see text.

$p + {}^7\text{Be}$ d-wave. Distortion effects are important in the ^{8}B ground state, but negligible in the low-energy $^7\text{Be} + p$ scattering states.

Barker adjusted his parameters for his model using $^7\text{Li}(n, \gamma)^8\text{Li}$; therefore, for comparison, we have calculated the $^7\text{Li}(n, \gamma)^8\text{Li}$ capture cross sections into the ^{8}Li ground state using the Hasegawa-Nagata force. The calculation is identical to that for $^7\text{Be}(p, \gamma)^8\text{B}$, except for the appropriate changes in the isospin quantum numbers and in the internal cluster parameters for the ^{3}H and ^{7}Li nuclei. To reproduce the binding energy of ^{8}Li, we have adjusted the Majorana exchange parameter in the effective interaction to $m = 0.3684$. With a flux of neutrons with a (center-of-mass energy) Maxwell-Boltzmann distribution with $kT = 21.3$ keV (which corresponds to a distribution of laboratory energies with $kT = 25$ keV), we find a capture cross section of 30.6 mbarns. This should be compared to the experimental values of 40.2 ± 2 mbarns (Imhof et al. 1959) and 45.4 ± 3.0 mbarns (Lynn, Jurney, & Raman 1991) which have been derived by scaling by $1/v$ and by adopting the experimental branching ratio for capture into the excited 1^+ state ($10.6 \pm 10\%$) of ^{8}Li as we calculate only capture to the ground state. Note, however, that our present results should not be overinterpreted as our calculation does not reproduce the scattering length a_2 in the $I = 2$ $^7\text{Li} + n$ channel, which, in turn, yields the dominant contribution (91%) to the low-energy capture cross section. We find $a_2 = 0.26$ fm, while the experimental value is -3.59 ± 0.06 fm. On the other hand, we find a good agreement in the $I = 1$ channel: $a_1 = 1.25$ fm, to be compared with the experimental value 1.09 ± 0.2 fm.

We argue, however, that our results for the $^7\text{Li}(n, \gamma)^8\text{Li}$ reaction do not really bear on the quality of our $^7\text{Be}(p, \gamma)^8\text{B}$ results. The $^7\text{Li}(n, \gamma)^8\text{Li}$ reaction is sensitive to the complete (i.e., interior) wave functions in the initial and final channels, as penetration of the neutron is not inhibited by the Coulomb barrier. In contrast, the $^7\text{Be}(p, \gamma)^8\text{B}$ reaction at low energies is a direct capture process sensitive mainly to the asymptotic forms of the wave functions, especially to the amplitudes of the Whittaker functions of the ^{8}B ground state and to the s- and d-wave phase shifts. The latter dominate the energy dependence of the low-energy cross section, while the spectroscopic amplitude determines its absolute magnitude.

3.2. *Extrapolation of Experimental Data*

Our most important results are the energy dependences and the subsequent extrapolations to $E = 0$. Not only do our calculated energy dependences with two different effective interactions agree well with each other, but they are also in good agreement with the microscopic GCM calculation of Descouvemont & Baye (1988) (who used a somewhat less flexible microscopic 3-cluster approach and yet a third effective interaction), and the phenomenological potential model results of Barker (1980). We can therefore conclude that the energy dependence of the low-energy $^7\text{Be}(p, \gamma)^8\text{B}$ cross section as calculated in all of these approaches are reasonable and that they differ from the experimental data only by a normalization factor (as can be embodied in the proton spectroscopic factor of ^{8}B).

It is important to note that these calculations can only be trusted for $E < 430$ keV, above which the inelastic excitation of the $1/2^-$ state in ^{7}Be is possible, and that no calculation to date includes this inelastic channel. Because flux is lost to this channel, optical-potential models (Tombrello 1965 or Barker 1980) should use a complex, not real, potential above 430 keV. As noted above, our 3-cluster model only includes $3/2^-$ states and so also does not describe this inelastic channel; an extension of the present microscopic calculation is in principle straightforward, but computationally very taxing. The loss of flux could be nonnegligible: the isospin-conjugate reaction, $^7\text{Li}(n, n')^7\text{Li}$, shows an abrupt rise from threshold in the excitation function for 480 keV photons (Ajzenberg-Selove 1984). For this reason we only use our calculation of the $^7\text{Be}(p, \gamma)^8\text{B}$ reaction at energies where the inelastic channel is not open.

In view of the foregoing, we are justified in using our energy dependence, consistently including the s- and d-wave capture contributions, to extrapolate the experimental low-energy data below 430 keV to $E = 0$, with the normalization constant determined by a least-squares fit. Only the data of Kavanagh et al. (1969) (hereafter Kav69), Wiezorek et al. (1977), and Filippone et al. (1983, hereafter Fil83) allow this procedure. [In fact, Parker 1968 has one point at $E_{\text{c.m.}} = 422$ keV; however, one obtains the same result by either (a) normalizing our theoretical curves by the single point at 422 keV or (b) normalizing Kav69 and Fil83 as described below.] Figure 2 shows the fits to Kav69 and Fil83.

However the remaining three experiments (Kavanagh 1960; Parker 1966, 1968; Vaughn et al. 1970) provide valuable information and should not be excluded. The experiments of Kav69 and Fil83 include data both below and above 430 keV. Therefore we normalized the high-energy data of Kav69 and Fil83 to match the experiments of Kavanagh (1960), Parker (1966, 1968), and Vaughn et al. (1970). (We only used the Kav69 data, which is plotted in Kavanagh 1972, up to 1460 keV.) This normalization is then automatically extrapolated to $E = 0$ by the low-energy fit to Kav69 and Fil83. This procedure not only avoids the use of suspect theoretical calculations above 430 keV, but also obviates the need to carefully subtract the resonance at 630 keV (below 430 keV the resonance contributes no more than 3% to the cross section). The extrapolated values of S_{17} of Kavanagh (1960), Parker (1968), and Vaughn et al.

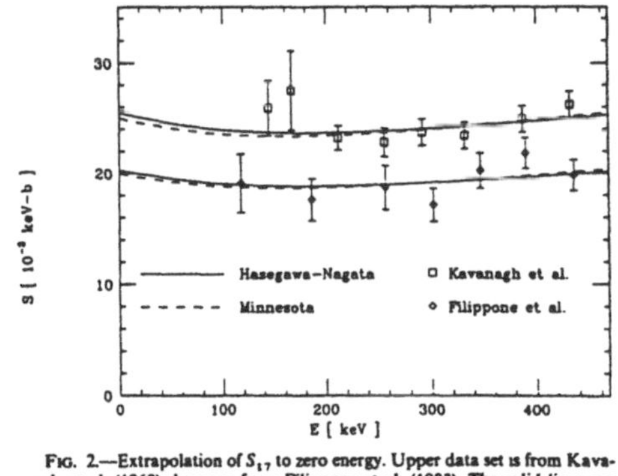

FIG. 2.—Extrapolation of S_{17} to zero energy. Upper data set is from Kavanagh et al. (1969); lower is from Filippone et al. (1983). The solid lines are calculated using the Hasegawa-Nagata interaction; the dashed lines are calculated using the Minnesota interaction. The error bars do not include the systematic error from uncertainty in σ_{dp}.

TABLE 3

EXTRAPOLATION OF EXPERIMENTAL *S*-FACTOR TO $E = 0$

EXPERIMENT	$S_{17}(0)$ (eV bar^{-1}) Previous[a]	Present
Kav 60	16 ± 6	15 ± 6
Par 68	28 ± 3	27 ± 4
Kav 69	27.3 ± 2.4	25.2 ± 2.4
Vau 70	21.4 ± 2.2	19.4 ± 2.8
Wie 77	45 ± 11	41.5 ± 9.3
Fil 183	22.2 ± 2.8	20.2 ± 2.3

[a] From Filippone 1986; Parker 1986.

(1970) using either Kav69 or Fil83 for the three "high-energy" experiments were consistent within 4%, 3%, and 7%, respectively.

Before presenting our results, we comment on the normalization of the data. Most of the experimental cross-sections were normalized using the broad 0.77 MeV resonance in ^{7}Li(d, p)^{8}Li. The measured value of σ_{dp} range from 138 to 211 mbarns (see Filippone 1986). We use the currently adopted value of 157 ± 10 mbarns.

Another method of normalization is via direct measurement of ^{7}Be activity. Wiezorek et al. (1977) used only this method. Note that while the extrapolated value from their experiment, which measured the (p, γ) cross-section at only energy, is nearly twice that of all other experiments, because of the large error bars this experiment has a nearly negligible effect on our results. Filippone et al. (1983) used both σ_{dp} and ^{7}Be activity to normalize their cross-sections and found consistent results within their uncertainties. Following Parker (1986) we normalize Fil83, and the remaining four experiments, with $\sigma_{dp} = 157 \pm 10$ mbars.

In Table 3 we give our extrapolated values of $S_{17}(0)$ for each of the six experiments, as well as the previously adopted extrapolations (Filippone 1986; Parker 1986). The "theoretical uncertainty," that is, the difference between using the energy-dependence calculated using the Minnesota or Hasegawa-Nagata force, was less than 2%. Our new values were consistently lower than those of previous extrapolations by up to 10%. (The extrapolated value of Kavanagh [1960] did not change, but this is likely due to rounding off to two digits.)

The six experimental values of $S_{17}(0)$ in Table 3 were averaged in two different ways, both of which gave the same result of 0.0224 keV barn^{-1}. The first was a simple weighted average. The combined uncertainty is then calculated as 0.0013 keV barn^{-1}. However, $\chi^2/(N-1)$, where the number of experiments, N, is 2.1, implying that the error bars on some or all of the experiments were underestimated. Following accepted practice (see, e.g., section IV.C.2 of Particle Data Group 1990), we multiply σ by $\sqrt{\chi^2/(N-1)}$ and obtain an uncertainty of $\sigma = 0.0019$ keV barn^{-1}.

This analysis ignores the fact that all but one experiment was normalized by $\sigma_{dp} = 157 \pm 10$ mbarns; this contribution to the uncertainty should be treated as an overall systematic uncertainty and should not be used while averaging the experiments. For our second analysis we ignored the experiment of Wiezorek et al. (1977), which was normalized by ^{7}Be activity, and performed a weighted average of the remaining five experiments, using, however, uncertainties calculated without $\Delta\sigma_{dp}$. The result was 0.0224 ± 0.0010 keV bar with $\chi^2 = 8.5$; multiplying the uncertainty by $\sqrt{\chi^2/N-1}$ yields 0.0016 which *then* is combined, in quadrature, with the 6.4% "systematic" uncertainty in σ_{dp} to obtain 0.0021. We adopt this latter, more conservative uncertainty.

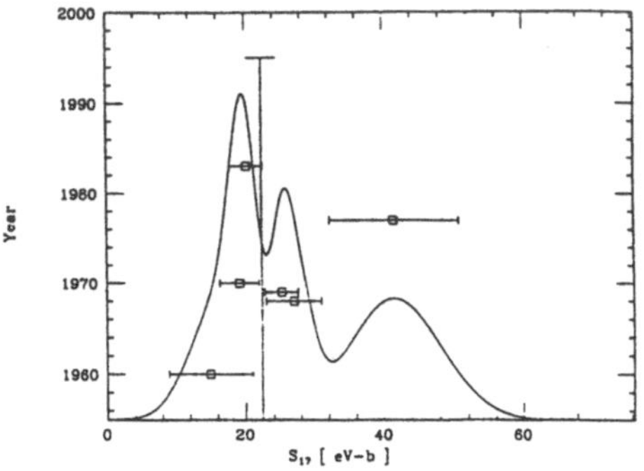

FIG. 3.—Ideogram of the extrapolated measurements of $S_{17}(0)$. Each experiment is represented by a Gaussian with width σ (shown also as the error bars) and with area proportional to $1/\sigma$. Vertical line is adopted value of $S_{17} = 22.4$ eV barn^{-1}.

To illustrate the uncertainties, we present in Figure 3 an ideogram of the values of $S_{17}(0)$ extrapolated from each experiment, expressed as a Gaussian with width σ and total area proportional to $1/\sigma$ (see Particle Data Group 1990).

In summary, then, our final result is $S_{17}(0) = 0.0224 \pm 0.0021$ barn^{-1}. This corresponds to a 3 σ uncertainty of 28%. The "theoretical uncertainty" we have argued is quite small, and we believe that our quoted uncertainty accurately reflects uncertainties in the experiments themselves as well as uncertainty in σ_{dp}.

Some authors have advocated even lower values of S_{17}. Turck-Chieze et al. (1988) adopt a value of 0.0209 keV barn^{-1}, by reducing the previously adopted value of S_{17} by 15% to adjust for d-wave capture, following a statement of Filippone et al. (1983) that inclusion of d-wave capture reduces S_{17} by 10%–15%. We find inclusion of d-wave capture only reduces extrapolated S_{17} by 10% or less. Barker & Spear (1986) propose a greater reduction in $S_{17}(0)$ to 0.017 keV barn^{-1}. They discard Kavanagh et al. (1969) and Parker (1968), include d-wave capture, and finally advocate a significantly lower value of σ_{dp}. While one might exclude the experiment of Wiezorek et al. (1977) as being spurious—and we in fact found it made no difference in the final result—clearly Parker (1968) and Kavanagh et al. (1969) are consistent with each other and should not be excluded. For reasons discussed in the second paragraph of this section, we do not use a lower σ_{dp}.

4. CONCLUSION

We have reexamined the two nuclear processes determining the fate of ^{7}Be (and hence the production of ^{8}B) in the Sun. For the dominant electron capture reaction that makes ^{7}Be unavailable for transmutation into ^{8}B, we have carefully reconsidered the approximations that go into computing the electron density and performed a self-consistent, thermal Hartree calculation. The net increase in the solar rate for this process relative to previous calculations is small. For the proton-capture reaction, which produces the ^8B and thus the high-energy neutrinos seen (or rather not seen!) in terrestrial detectors, we have performed a microscopic cluster calculation of the energy dependence of the astrophysical *S*-factor that includes capture from *d*-waves. Our extrapolations of the experimental data using all six experiments decreases the proton capture rate, and hence the ^8B neutrino flux, is from the currently accepted value by about 7%. While not insignificant, this reduction is still far from explaining the solar neutrino "problem."

We thank C. Carraro for a helpful discussion of the material in § 2, and J. N. Bahcall, F. C. Barker, B. W. Filippone, and R. W. Kavanagh for comments and helpful suggestions on § 3; in particular we thank B. W. F. and R. W. K. for providing the statistical and systematic uncertainties in their measurements. This work was supported in part by the National Science Foundation grants PHY86-04197 and PHY88-17296.

REFERENCES

Ajzenberg-Selove, F. 1984, Nucl. Phys. A, 413, 1
Bahcall, J. N., & Moeller, C. P. 1969, ApJ, 155, 511
Bahcall, J. N., & Ulrich, R. K. 1988, Rev. Mod. Phys., 60, 297
Barker, F. C. 1980, Australian J. Phys., 33, 177
Barker, F. C., & Spear, R. H. 1986, ApJ, 307, 847
Chwieroth, F. S., Brown, R. E., Tang, Y. C., & Thompson, D. R. 1973, Phys. Rev. C, 8, 938
Davis, J., & Blaha, M. 1982, J. Quant. Spectros. Rad. Trans., 27, 307
Davis, R. 1987, in Proceedings of the Seventh Workshop On Ground Unification, Toyama, Japan 1986, ed. J. Arafune (Singapore: World Scientific), 237; Rowley, J. K., Cleveland, B. T., & Davis, R. 1984, in AIP Conf. Proc. No. 126, Solar Neutrinos and Neutrino Astronomy, ed. M. L. Cherry, K. Lande & W. A. Fowler (New York: AIP), 1
Descouvemont, P., & Baye, D. 1988, Nucl. Phys. A, 487, 420
Filippone, B. W. 1986, Annu. Rev. Nucl. Sci., 36, 717
Filippone, B. W., Elwyn, A. J., Davids, C. N., & Koetke, D. D. 1983, Phys. Rev. Lett., 50, 412; Phys. Rev. C, 28, 2222 (Fil83)
Furutani, H., Kanada, H., Kaneko, T., & Nagata, S. 1980, Prog. Theor. Phys. Suppl., 68, 215
Goldsmith, S., Griem, H. R., & Cohen, L. 1984, Phys. Rev. A, 30, 2775
Hirata, K. S., et al. 1989, Phys. Rev. Lett., 63, 16
Hummer, D. G., & Mihalas, D. 1988, ApJ, 331, 794
Iben, I., Kalata, K., & Schwartz, J. 1967, ApJ, 150, 1001 (IKS)
Imhof, W. L., et al. 1959, Phys. Rev., 114, 1037
Jackson, J. D. 1975, Classical Electrodynamics, 2nd Ed. (New York: Wiley).
Kavanagh, R. W. 1960, Nucl. Phys., 15, 411
——— 1972, in Cosmology, Fusion, and Other Matters, ed. F. Reines (Boulder: Colorado Associated Univ. Press), 169
Kavanagh, R. W., Tombrello, T. A., Mosher, J. M., & Goosman, D. R. 1969, Bull. Am. Phys. Soc., 14, 1209 (Kav69)
Kolbe, E., Langanke, K., & Assenbaum, H. J. 1988, Phys. Lett., 214, 169
Landau, L. D., & Lifshitz, E. M. 1980, Statistical Physics Part 1 (Oxford: Pergamon)
Lynn, J. E., Jurney, E. T., & Raman, S. 1991, Phys. Rev., C 44, 764
Parker, P. D. 1966, Phys. Rev., 150, 851
———. 1968, ApJ, 153, L85
———. 1986, in Physics of the Sun, Vol. I, ed. P. A. Sturrock, T. E. Holzer, D. M. Mihalas, & R. K. Ulrich (Dordrecht: Reidel), 15
Particle Data Group 1990, Phys. Lett. A, 239, 1
Robertson, R. G. H. 1973, Phys. Rev. C, 7, 543
Rogers, F. J., Graboske, H.C., Jr., & Harwood, D. J. 1970, Phys. Rev. A, 1, 1577
Shen, P. N., Tang, Y. N., Fujiwara, Y., & Kanada, H. 1985, Phys. Rev. C, 31, 2001
Skupsky, S. 1980, Phys. Rev. A, 21, 1316
Theimer, O, & Kepple, P. 1970, Phys. Rev. A, 1, 957
Tombrello, T. A. 1965, Nucl. Phys., 71, 459
Turck-Chièze, S., Cahen, S., Casse, M., & Doom, C. 1988, ApJ, 335, 415
Wildermuth, K., & Tang, Y. C. 1977, A Unified Theory of the Nucleus (Braunschweig: Vieweg)
Wiezorek, C., Kräwinkel, H., Santo, R., & Wallek, K. 1977, Z. Phys., A, 282, 121
Williams, R. D., & Koonin, S. E. 1981, Phys. Rev. C, 23, 2773
Vaughn, F. J., Chalmers, R. A., Kohler, D., & Chase, L. F., Jr. 1970, Phys. Rev. C, 2, 1657

IV. Physics Beyond the Standard Model

Introduction: Alexei Smirnov

Solar neutrino experiments provide unique information on neutrino propagation in vacuum and in matter. On the way from the central parts of the sun to the terrestrial detectors, the neutrinos can decay, oscillate, undergo resonant flavor conversion, or have their spins precess or flip resonantly. These processes may be manifested in different ways, including an energy-dependent suppression of the electron neutrino flux, the appearance of muon or tau (anti)neutrinos or even electron antineutrinos from the sun, and time variations of the fluxes.

The indicated processes imply the existence of physics beyond the standard model: lepton number violation, non zero neutrino masses and mixing angles, the existence of large magnetic moments of neutrinos, and interactions of neutrinos with new particles.

Oscillations, conversion, and spin-precession all permit the reconciliation of the data from solar neutrino experiments with predictions of standard solar models, i.e. they provide (different) solutions to the solar neutrino problem. Observations of the indicated effects will allow us to measure the neutrino parameters. Negative results of the searches will exclude interesting regions of neutrino parameter space and establish bounds on new physics. In this section, we have collected papers which contain the key theoretical ideas concerning how physics beyond the standard electroweak model can be investigated in solar neutrino studies. To a great extent, the reprinted papers have determined the subsequent development of the field.

As was first shown by Pontecorvo, leptonic charge nonconservation leads to neutrino mixing, and consequently, oscillations of neutrinos. "If the oscillation length is large ... from the point of view of detection possibilities an ideal object is the sun." This statement from Pontecorvo's 1967 paper[1] can be considered as the birth certificate for the subject of solar studies of new neutrino physics. Two short paragraphs from this paper contain several important remarks. If the characteristic length of the oscillations is smaller than the neutrino production region, then one predicts an averaged effect: the suppression of the original electron neutrino flux by a factor of two. Pontecorvo quotes the unpublished Pomeranchuk remark that, if the oscillation length is much larger than the production region, one can expect time variations of signals related to the annual changes of the distance between the sun and the Earth (also Paper 1.A.IV). Oscillations can explain the discrepancy between observations and theoretical predictions (Paper 1.A.IV, also review [2]). In practice, averaging over the neutrino energy is usually more important, if the oscillation length is much smaller than the distance between the sun and the Earth (Paper 2.A.IV). If the oscillation length is comparable with the Earth-sun separation, i.e. the difference of the squares of the masses has certain "just-so" values (Paper 4.A.IV), the observable suppression in some energy regions can be more than a factor of two (Papers 1-4.A.IV).

Neutrinos might decay, e.g. into a neutrino of another type and a massless or light scalar boson, either inside the sun or on the way from the sun to the Earth (Paper 1.B.IV). (Such a decay would signify spontaneous violation of lepton number). The decay of solar neutrinos can be reconciled with observations of neutrinos from SN1987A if the neutrinos have rather large vacuum mixing angles (Paper 3.B.IV). Matter effects modify properties of the decays (Paper 2.B.IV), in particular, the decay of an electron neutrino into an electron antineutrino and a scalar becomes possible (Paper 2.B.IV).

As was discovered by Wolfenstein (1978), neutrino oscillations are modified in matter: the effect is related to elastic forward scattering of neutrinos. The evolution equation for mixed neutrinos in matter was derived in (Paper 1.C.IV). The expected phenomena depend on the sign of the scattering amplitude; the correct sign was found in (Paper 3.C.IV). Matter effects change the flavor mixing (as well as the effective neutrino masses), and consequently, can enhance or suppress the amplitudes of oscillations (Papers 1,2.C.IV). Moreover, there is always some energy at which mixing becomes maximal (Papers 2,4.C.IV). The enhancement of oscillations in a medium with constant density has a resonant character: for small vacuum mixing the dependence of the amplitude of oscillations on energy is strongly peaked (Papers 4,5.C.IV). Resonant enhancement of oscillations can occur for solar neutrinos when they cross the Earth [3]. The enhancement regenerates the electron neutrino flux and the rotation of the Earth leads to time variations of the regeneration (day-night and seasonal effects [3]).

In 1985, Mikheyev and Smirnov showed that in matter with a varying density neutrinos may undergo resonant flavor conversion (the MSW-effect) – the effect is related to the change in the amount of mixing (Papers 4,5.C.IV). If the *adiabaticity* condition is satisfied (Papers 4,5.C.IV), the flavor of the neutrino state will follow the density change. Strong transformation of one neutrino species into another type may take place even for very small vacuum mixing angles if the *resonance* condition is fulfilled, i.e. neutrinos cross the resonant layer (Papers 4,5.C.IV). In the first papers, one can find a physical description of the effect (Paper 6.C.IV), calculations of the suppression factors (Paper 5.C.IV, also 7.C.IV), and estimations of the neutrino mass squared and mixing angles for which one expects strong effects inside the sun (Papers 4,5.C.IV). A detailed analysis of the adiabaticity criteria and adiabatic propagation were first given by Messiah (Paper 8.C.IV).

In Papers 5–7.C.IV, the MSW-effect was described in terms of the *eigenstates* of neutrinos. An equivalent and complimentary description was given independently by Cabibbo [4] and by Bethe (Paper 9.C.IV) in terms of the *eigenvalues* (energy levels). In this approach, the resonance is identified with *level-crossing phenomena.* For non-adiabatic level crossing, the effect can be described by the Landau-Stueckelberg-Zener formula (Papers 10,11.C.IV). Important improvements of this formula for solar applications have been obtained by deriving exact solutions of the evolution equations for an exponentially varying density [5]. An especially simple solution has been given by Pizzochero [6].

Resonant conversion has a number of analogies. For example, it is similar to oscillation transfer between two coupled pendulums when their eigenfrequencies change in time [7],(12C.VI). At the Berkeley high-energy physics conference (1986), Weinberg demonstrated a real device which reproduces the effect (Paper 12.C.IV).

If the neutrino has a nonzero magnetic moment, its spin will precess in the transverse magnetic field of the sun, and left handed neutrinos will be converted into right handed neutrinos that are not detectable in radiochemical experiments (Paper 1.D.IV). Spin-flip in the convection zone of the sun, where the magnetic field changes with time, can result in time variations of the neutrino signal, e.g. in an anti-correlation with solar activity (Paper 2.D.IV). Another signature of spin precession is a semi-annual variation related to the inclination of the place of the elliptic and the existence of a zero in the solar magnetic field near the equatorial plane (Paper 2.D.IV).

Resonance phenomena similar to those occurring in the MSW-effect can take place in neutrino spin precession in matter due to (transition) magnetic moments (Papers 3,4.D.IV). If the neutrino components have different flavors, the helicity and the flavor of the neutrino state will be changed simultaneously [8]. This resonant spin-flavor conversion can result in both energy dependence of the suppression and in time variation of the signal. Moreover if flavor mixing also exists, one predicts the appearance of an antineutrino signal from the sun (Paper 3.D.IV).

Neutrino mixing and the effects of oscillations or resonant conversion can be induced by new interactions beyond those that occur in the standard electroweak model. In supersymmetric models with R-parity violation, an effective mixing appears as the result of exchange of squarks between neutrinos and quarks (Papers 2,3.E.IV). Even for massless neutrinos (Paper 3.E.IV) an appreciable conversion of solar neutrinos may take place. Mixing of neutrinos can be induced also by interactions with the gravitational potential if the weak equivalence principle is violated (Paper 1.E.IV).

The reader can find additional references in some reviews [9].

References

[1] B. Pontecorvo, ZETF, 53, 1717 (1967), Sov. Phys. JETP, 26, 984 (1968).

[2] S. M. Bilenkii, and B.M. Pontecorvo, Phys. Rep., 41, 225 (1978).

[3] S. P. Mikheyev, and A. Yu. Smirnov, Proc. of the 6th Moriond Workshop on Massive Neutrinos in Astrophysics and Particle Physics, (Tignes, France) eds. O. Fackler and J. Tran Than Van, p. 355.

J. Bouches et al, Z. Phys., C32, 499, (1986).

[4] N. Cabibbo, Summary talk given at 10th Int. Workshop on Weak Interactions and Neutrinos, (Savonlinna, Finland, June 1985) unpublished.

[5] T. Kaneko, Prog. Theor.Phys., 78 532 (1987).

S. Toshev, Phys. Lett., B196, 170 (1987).

S. T. Petcov, Phys. Lett., B200, 373, (1988).

[6] P. Pizzochero, Phys. Rev. D36, 2993 (1987).

[7] S. P. Mikheyev, and A. Yu. Smirnov, in Proceedings of Twelfth International Conference on Neutrino Physics and Astrophysics, eds. T. Kitagaki and H.Yuta (Singapore: World Scientific) p.177 (1986).

[8] J. Schechter, and J. W. F. Valle, Phys.Rev. D24, 1883 (1981).

[9] S. P. Mikheyev, and A. Yu. Smirnov, Usp. Fiz. Nauk, 153, 3 (1987); Sov. Phys. Usp., 30, 759, (1987); Prog. Part. Nucl. Phys., 23, 41 (1989).

T. K. Kuo, and J. Pantaleone, Rev. Mod. Phys., 61, 937, (1989).

S. M. Bilenky, and S.T. Petcov, Rev. Mod. Phys., 59, 671, (1987).

J. N. Bahcall, Neutrino Astrophysics (Cambridge Univ. Press: Cambridge), Chapter 9 (1989).

IV. PHYSICS BEYOND THE STANDARD MODEL - Smirnov

A. Vacuum Oscillations

1. "Neutrino Astronomy and Lepton Charge," V. N. Gribov and B. M. Pontecorvo, *Phys. Lett.* **B28**, 493-496 (1969).

2. "Lepton Non-Conservation and Solar Neutrinos," J. N. Bahcall and S. C. Frautschi, *Phys. Lett.* **B29**, 623-625 (1969).

3. "Realistic Calculations of Solar Neutrino Oscillations," V. Barger, R. J. N. Phillips, and F. Whisnant, *Phys. Rev.* **D24**, 538-541 (1981).

4. " 'Just So' Neutrino Oscillations," S. L. Glashow and L. M. Krauss, *Phys. Lett.* **B190**, 199 (1987).

B. Neutrino Decay

1. "Are Neutrinos Stable Particles?" J. N. Bahcall, N. Cabibbo and A. Yahil, *Phys. Rev. Lett.* **28**, 316-318 (1972).

2. "Neutrino Decay in Matter," Z. G. Berezhiani and M. I. Vysotsky, *Phys. Lett.* **B199**, 281-285 (1987).

3. "Neutrino Mixing, Decays and Supernova 1987A," J. A. Frieman, H. E. Haber and K. Freese, *Phys. Lett.* **B200**, 115 (1988).

C. Neutrino Oscillations in Matter. Resonant Flavor Conversion.

1. "Neutrino Oscillations in Matter," L. Wolfenstein *Phys. Rev.* **D17**, 2369-2374 (1978).

2. "Matter Effects on Three Neutrino Oscillations," V. Barger, K. Whisnant, S. Pakvasa and R. J. N. Phillips, *Phys. Rev.* **D22**, 2718-2720 (1980).

3. "On the Detection of Cosmological Neutrinos by Coherent Scattering," P. Langacker, J. P. Leville and J. Sheiman *Phys. Rev.* **D27**, 1228, 1231 (1983).

4. "Resonance Enhancement of Oscillations in Matter and Solar Neutrino Spectroscopy," S. P. Mikheyev and A. Y. Smirnov, *Sov. Jour. Nucl. Phys.* **42**, 913-917 (1985).

5. "Resonant Amplifications of ν-Oscillations in Matter and Solar-Neutrino Spectroscopy," S. P. Mikheyev and A.Y. Smirnov, *Nuovo Cimento* **C9**, 24 (1986).

6. "Neutrino Oscillations in Variable-Density Medium and ν-Bursts Due to Gravitational Collapse of Stars," S.P. Mikheyev and A. Y. Smirnov, *Sov. Phys. JETP* **64**, 4-7 (1986).

7. "Treatment of $\nu_{\odot}$-Oscillations in Solar Matter: The MSW Effect," A. Messiah, *Proceedings of the 6th Moriond Workshop on Massive Neutrinos in Particle Physics and Astrophysics*, ed. O. Fackler and J. Tran Thanh Van, Tignes, 373, 376 and 378 (1986).

8. "Mikheyev-Smirnov-Wolfenstein Enhancement of Ocillations as Possible Solution to the Solar Neutrino Problem," S.P. Rosen and J. M. Gelb, *Phys. Rev.* **D34**, 969-974 (1986).

9. "Possible Explanation of the Solar Neutrino Puzzle," H. Bethe, *Phys. Rev. Lett.* **56**, 1305-1308 (1986).

10. "Nonadiabatic Level Crossing in Resonant Neutrino Oscillations," S. J. Parke, *Phys. Rev. Lett.* **57**, 1275-1278 (1986).

11. "Adiabatic Conversion of Solar Neutrinos," W. C. Haxton, *Phys. Rev. Lett.* **57**, 1271-1274 (1986).

12. "Summary and Outlook," S. Weinberg, *Int. Journal of Modern Physics* **A2**, 304-305 (1987).

D. **Neutrino Spin-Flip Phenomena**

1. "Effect of Neutrino Magnetic Moment on Solar Neutrino Observations," A. Cisneros, *Astro. & Space Sci.* **10**, 87-92 (1971).

2. "Neutrino Electrodynamics and Possible Consequences for Solar Neutrinos," M.B. Volshin, M.I. Vysotsky and L.B. Okun, *Sov. Phys. JETP* **64**, 446-452 (1986).

3. "Resonant Spin-Flavor Precession of Solar and Supernova Neutrinos," C.S. Lim and W.J. Marciano, *Phys. Rev.* **D37**, 1368-1373 (1988).

4. "Resonant Amplifications of Neutrino Spin Rotation in Matter and the Solar Neutrino Problem," E.K. Akhmedov, *Phys. Lett.* **B213**, 64-68 (1988).

E. **Flavor Mixing by New Interactions**

1. "Testing the Principle of Equivalence with Neutrino Oscillations," M. Gasperini, *Phys. Rev.* **D38**, 2635-2637, (1988).

2. "Mikheyev-Smirnov-Wolfenstein Effect with Flavor-Changing Neutral Interaction," E. Roulet, *Phys. Rev.* **D44**, R935-938 (1991).

3. "On the MSW Effect with Massless Neutrinos and No Mixing in the Vacuum," M.M. Guzzo, A. Masiero, and S.T. Petcov, *Phys. Lett.* **B260**, 154-156 (1991).

NEUTRINO ASTRONOMY AND LEPTON CHARGE

V. GRIBOV* and B. PONTECORVO
Joint Institute for Nuclear Research, Dubna, USSR

Received 20 December 1968

It is shown that lepton nonconservation might lead to a decrease in the number of detectable solar neutrinos at the earth surface, because of $\nu_e \rightleftarrows \nu_\mu$ oscillations, similar to $K^0 \rightleftarrows \tilde{K}^0$ oscillations. Equations are presented describing such oscillations for the case when there exist only four neutrino states.

Recently there became known the results of the beautiful experiment of Davis et al. [1], in which deep underground a search was made of sun neutrinos.

Using a spectrometer proportional counter [2,3] to detect ^{37}A produced in the reaction $\nu + {}^{37}Cl \rightarrow {}^{37}A + e^-$ [3,4], (which is expected to take place in 390 000 litres of C_2Cl_4), Davis et al. so far were not able to detect solar neutrinos. It was shown by them that the neutrino flux at the earth from ^8B decay in the sun [5] is smaller than 2×10^6 cm^{-2} sec^{-1}. This limit is definitely smaller than the theoretical predictions [6,7]. However, various astrophysics and nuclear physics uncertainties do not allow to draw the conclusion that we are faced with a catastrophic discrepancy [7]. The purpose of this note is to emphasize again that the result of sun neutrino experiments are related not only to the above mentioned uncertainties but also, and in a marked way, to properties which are so far unknown [8] of the neutrino as an elementary particle. The question at issue is: are (is) lepton charges (charge) conserved exactly ?. The question which, as we shall see, is relevant to neutrino astronomy, is certainly not far-fetched from an elementary particle physics point of view. As a matter of fact the most significant and recent experiments on lepton conservation give upper limits for the constants of hypothetical interactions nonconverving lepton charge which are surprisingly large.

The most accurate information can be obtained from the experiments, in which a search was made for the processes $^{48}Ca \rightarrow {}^{48}Ti + e^- + e^-$ [9], $\nu_\mu + p \rightarrow \mu^+ + n$ [10], $\mu^+ \rightarrow e^+ + \gamma$ [11]. At a confidence level of about 70% one finds respectively the following upper limits for the corresponding interaction constants f_1, f_2, f_3 phenomenologically responsible for such processes [e.g.12].

$$f_1/G < 0.02; \qquad f_2/G < 0.15; \qquad f_3/G < 0.005,$$

where $G = 10^{-5}/M_p^2$ is the Fermi weak interaction constant.

In a period of development of physics in which such quantum numbers as P, C, PC were found to be not good, it is natural to question the exact validity of any symmetry [e.g. 13]. The relatively high upper limits for f_1, f_2, f_3 show that there is once more plenty of room for a violated conservation law and suggest the lepton charge(s) as the first candidate(s) for the nonconserved quantum number(s).

In previous publications [8,14] there was shown that lepton nonconservation leads to the possibility of oscillations in vacuum between various neutrino states, and, generally speaking, acts in the sense of decreasing the number of detectable solar neutrinos with respect to the number expected theoretically under the assumption that lepton charges are strictly conserved.

This effect, which incidentally would be in the right direction if the necessity should definitely arise of accounting for unexpectedly small values of detected solar neutrinos is due to the fact that in the presence of oscillations, part of the neutrinos are sterile, that is practically unobservable. It turns out that the study of solar neutrino oscillations is the most sensitive way of investigating the question of lepton charge conservation.

In ref. 8 possible oscillations $\nu_e \rightleftarrows \tilde{\nu}_e$, $\nu_\mu \rightleftarrows \tilde{\nu}_\mu$, $\nu_e \rightleftarrows \nu_\mu$ have been discussed. In view of applications to neutrino astronomy we would like to point out here that the first two types of

* Leningrad Physical-Technical Institute Leningrad, USSR.

oscillations should not be considered if it is required that in nature there are only four neutrino states.

In order to study the oscillations for this case, we shall consider in approximation zero (V-A theory) four neutrino states with mass zero, which are described by two two-component spinors ν_e and ν_μ. In such approximation it is convenient to think of two exactly conserved lepton charges (muon and electron charges).

Lepton nonconservation leads to virtual or real transitions between the above mentioned neutrino states. All the possible transitions may be described with the help of an interaction Lagrangian

$$L_{int.} = m_{e\bar{e}}\bar{\nu}_e'\nu_e + m_{\mu\bar{\mu}}\bar{\nu}_\mu'\nu_\mu + m_{e\bar{\mu}}\overline{\nu_e'} + \qquad (1)$$

$$+ \text{Herm. conjug.}$$

where $\nu' = \bar{\nu}C$ is the charge conjugated spinor. For the charge conjugated spinors there was adopted the notation ν' instead of $\tilde{\nu}$ to avoid confusion with $\bar{\nu}$.

Below for simplicity it will be assumed that $m_{e\bar{e}}, m_{\mu\bar{\mu}}, m_{e\bar{\mu}}$ are real values, i.e. *CP*-invariance is assumed. Otherwise, the formulae become somewhat more complicated and in the present note we shall not give them for the general case. Interaction (1) can be easily diagonalized. The diagonal states are:

$$\varphi_1 = \cos\xi\,(\nu_e + \nu_e') + \sin\xi\,(\nu_\mu + \nu_\mu')$$
$$\varphi_2 = \sin\xi\,(\nu_e + \nu_e') - \cos\xi\,(\nu_\mu + \nu_\mu') \qquad (2)$$

where

$$\text{tg}\,2\xi = \frac{2m_{e\bar{\mu}}}{m_{e\bar{e}} - m_{\mu\bar{\mu}}}\ .$$

These states correspond to two Majorana neutrinos (i.e. four states when the spin orientation is taken into account) with the masses m_1 and m_2

$$m_{1,2} = \tfrac{1}{2}\left[m_{e\bar{e}} + m_{\mu\bar{\mu}} \pm \sqrt{(m_{e\bar{e}} - m_{\mu\bar{\mu}})^2 + 4m_{e\bar{\mu}}^2}\right] \qquad (3)$$

(if $m_2 < 0$, the real state with the positive mass $-m_2$ is $\varphi_2' = \gamma_5\varphi_2$).

The two component spinors ν_e and ν_μ now are not describing anymore particles with zero mass but must be expressed in terms of four-component Majorana spinors φ_1 and φ_2

$$\nu_e = \tfrac{1}{2}(1+\gamma_5)[\varphi_1\cos\xi + \varphi_2\sin\xi]$$
$$\nu_\mu = \tfrac{1}{2}(1+\gamma_5)[\varphi_1\sin\xi - \varphi_2\cos\xi]\ . \qquad (4)$$

In this case the (V-A) lepton current, to which weak processes are due, can be written as usual

$$j_\alpha = \bar{e}\gamma_\alpha\nu_e + \bar{\mu}\gamma_\alpha\nu_\mu\ . \qquad (5)$$

The mass difference between Majorana neutrinos described by φ_1 and φ_2 leads to the oscillations $\nu_e \rightleftarrows \nu_\mu$, $\nu_e' \rightleftarrows \nu_\mu'$ (in the usual notations $\tilde{\nu}_e \rightleftarrows \tilde{\nu}_\mu$). If at the time $t = 0$, one electron neutrino is generated, the probability of observing it at the time t is

$$|\nu_e(t)|^2 = |\nu_e(0)|^2\left\{\frac{m_-^2 + 2m_{e\bar{\mu}}^2}{m_-^2 + 4m_{e\bar{\mu}}^2} + \frac{2m_{e\bar{\mu}}^2}{m_-^2 + 4m_{e\bar{\mu}}^2}\cos 2\Delta t\right\}$$

where

$$m_- = m_{e\bar{e}} - m_{\mu\bar{\mu}};$$

$$\Delta = \frac{1}{2p}(m_1^2 - m_2^2) = \frac{(m_{e\bar{e}} + m_{\mu\bar{\mu}})}{2p}\sqrt{m_-^2 + 4m_{e\bar{\mu}}^2}$$

and P is the neutrino momentum.

It should be emphasized that the oscillations take place only if $m_{e\bar{\mu}}$ and at least one of the values $m_{e\bar{e}}$ and $m_{\mu\bar{\mu}}$ are different from zero. In the absence of oscillations there are two possibilities. If $m_{e\bar{\mu}} = 0$, then $\xi = 0$ and there exist two Majorana neutrinos (without oscillations). If $m_{e\bar{e}} = m_{\mu\bar{\mu}} = 0$, but $m_{e\mu} \neq 0$, it is natural to attribute an opposite sign of the lepton charge (only one!) to charged leptons of equal electrical charge (say, e^- and μ^-) [15] and to consider, (instead of the degenerated states φ_1 and $\varphi_2' = \gamma_5\varphi_2$ with the mass $m = m_{e\bar{\mu}}$), the states with a definite lepton charge $\psi = \nu_e + \nu_\mu'$, $\psi' = \nu_e' + \nu_\mu$ (this is the four-component neutrino theory with parity nonconservation [16]).

If $m_{e\bar{\mu}}$ and one of the values $m_{e\bar{e}}$, $m_{\mu\bar{\mu}}$ are different from zero, i.e. if oscillations take place, a very attractive case arises when $m_{e\bar{e}}, m_{\mu\bar{\mu}} \ll \ll m_{e\bar{\mu}}$. In such a case

$$\xi \approx \frac{\pi}{4}, \qquad \varphi_1 \approx \frac{1}{\sqrt{2}}(\psi + \psi'), \qquad \varphi_2 \approx \frac{1}{\sqrt{2}}(\psi - \psi') \qquad (7)$$

and the oscillations are entirely similar to the $K^0 \rightleftarrows \tilde{K}^0$ oscillations, φ_1 and φ_2 being analogous to K_1^0 and K_2^0. According to (6) the oscillation amplitude in this case is the largest possible one. The two ψ spin states, ν_{left} and ν_{right} are approximately the same as the observable "phenomenological" particles ν_e and ν_μ' (or $\tilde{\nu}_\mu$), similarly $\tilde{\nu}_{left} \approx \nu_\mu$ and $\tilde{\nu}_{right} \approx \nu_e' \equiv \tilde{\nu}_e$. A very simple picture of neutrino oscillations, similar to the $K^0 \rightleftarrows \tilde{K}^0$ oscillations arises also if m_{ee}-

and $m_{\mu\bar{\mu}}$ are no longer small in comparison with $m_{e\bar{\mu}}$, but are equal ($m_{e\bar{e}} = m_{\mu\bar{\mu}}$) in other words if there is a μ-e symmetry. In such a case $\xi = \frac{1}{4}\pi$ and relations (7) are exact.

In ref. 8 and also in unpublished work of Kobzarev and Okun there was discussed mainly the possibility that the neutrino oscillations are due to the so called milliweak interaction [17] which, in addition to *PC*, would violate lepton charge conservation as well.

The oscillations might be also induced by a (first order) superweak interaction changing the lepton charge by two units [18]. This interaction reminds of the Wolfenstein [19] superweak interactions, changing the strangeness by two units and might be closely related to it.

Let us remark here that the "sterility" of the neutral leptons generated, say, as a result of $\nu_e \rightleftarrows \nu_\mu$ oscillations is not absolute (as it would be in the case of the oscillations [8] $\nu_e \rightleftarrows \bar{\nu}_e$, $\nu_\mu \rightleftarrows \bar{\nu}_\mu$ corresponding to the existence of more than four neutrino states) and in the case of solar neutrinos arises simply from the fact that low energy ν_e transform mainly into ν_μ which cannot interact with matter (their energy is smaller than the muon mass).

Unfortunately, nothing can be said about the mass values $m_{e\bar{e}}$, $m_{\mu\bar{\mu}}$, $m_{e\bar{\mu}}$ and about the oscillation length $1/\Delta$, (see 8), even if they were connected to a definite "etiquette" (milliweak, superweak), as the cut off energy is unknown.

Returning now to neutrino astrophysics, we are going to consider only the simple cases $m_{e\bar{e}}, m_{\mu\bar{\mu}} \ll m_{e\bar{\mu}}$ or $m_{e\bar{e}} = m_{\mu\bar{\mu}}$, when the oscillations are similar to the oscillations in the K^0 meson beams.

What can be said about the oscillation length? In reactor experiments the existence of an oscillation smaller than the lengths involved in the problem (the diameter of the reactor and the distance detector-reactor, of the order of a few metres) would lead to a decrease by a factor two in the number of active particles reaching the detector since the number of sterile particles is equal to the number of active particles ($\bar{\nu}_e$) at large distances. Such a circumstance would have the effect that the cross section for the reaction $\bar{\nu}_e + P \rightarrow n + e^+$, measured in the experiment of Nezrik and Reines [20] would have been two times smaller than the cross section expected for exact two-component neutrinos. Since there is no such a discrepancy, it can be stated that reactor experiments exclude oscillations with a length smaller than a few meters.

From the point of view of observing neutrino oscillations, the ideal object is the sun. If the oscillation length is much smaller than the radius of the sun region effectively generating neutrino (say, one tenth of the sun radius R_0, or about 0.1×10^6 km for ν_e from ^{8}B) it will be impossible to observe oscillations because of the smearing effect. At the earth surface the only effect consists in the following: the flux of observable neutrino must be two times smaller than the total sun neutrino flux.

As it was pointed out by I. Pomeranchuk, if the oscillation length of solar neutrinos is comparable with or larger than the radius of the sun region which generates neutrinos but smaller than the earth-sun distance, there might arise time variations of the intensity of detectable neutrinos of the earth surface. These variations are connected with the fact that the sun-earth distance varies with time. In order to observe the oscillations mentioned above it is necessary to make measurements over relative distances (times), comparable with the oscillation length (period). If the oscillation length is of the order of $0.1\ R_0 = 0.1 \times 10^6$ km, there will arise time variations) in the intensity of detectable neutrinos with the period of the order of days. If the oscillation length is of the order of 5×10^6 km (the difference in the semi-axis of the earth orbit), variations will arise with periods of the order of a hundred days.

Summarizing, the presence of neutrino oscillations leads to a decrease of the detectable neutrino intensity by a factor δ. This factor will be 2, if the oscillation length is smaller than the radius of the sun region effectively generating neutrinos.

Otherwise, equal to 2 will be the value $\delta_{aver.}$ averaged over time variations. If the oscillation length is comparable with the earth-sun distance, the variations are not observable, but δ could turn out to be much larger than 2 (of course, the statement is true if a detector of more or less monoenergetic neutrinos is assumed).

In conclusion we wish to express our warm gratitude to I. Kobzarev, L. Okun, B. Ioffe, A. Mukhin for critical and illuminating discussions.

References

1. R. Davis, D. Harmer and K. Hoffman, Phys. Rev. Letters 20 (1968) 1205.
2. B. Pontecorvo, Helv. Phys. Acta 23 Suppl. III (1950) 97;
   B. Pontecorvo, D. Kirkwood and G. Hanna, Phys. Rev. 75 (1949) 982.
3. R. Davis, Phys. Rev. Letters 12 (1964) 303.
4. B. Pontecorvo, Chalk. River Report P. D. (1946) 205.
5. W. Fowler, Astrophys. J. 127 (1958) 551;
   J. Bahcall, W. Fowler, I. Iben and R. Sears, Astrophys. J. 137 (1963) 344.

6. J. Bahcall, N. Bahcall, W. Fowler and G. Sharir, Phys. Letters 26B (1968) 1.
7. J. Bahcall, N. Bahcall and G. Sharir, Phys. Rev. Letters 20 (1968) 1209.
8. B. Pontecorvo, J. Expl. Theoret. Phys. 53 (1967) 1717.
9. R. Bardin, P. Gollon, J. Ullman and C. Wu, Phys. Letters 26B (1967) 112.
10. J. Bienlein et al., Phys. Letters 13 (1964) 80; G. Bernardini, Int. Conf. on High energy phys., Dubna 2 (1964) 48.
11. S. Parker, H. Anderson and C. Reg, Phys. Rev. 133B (1964) 768.
12. B. Pontecorvo, Uspekhi Fiz. Nauk 95 (1968) 503.
13. L. Okun, School of Theoretical and experimental physics, Nor Amberd, 1966. Ed. of Acad. of Armenian SSR.
14. B. Pontecorvo, Zh. Exp. i Teor. Fiz. 34 (1958) 247.
15. Ya. Zeldovich, Doklad Akad. Nauk 86 (1952) 505; E. Konopinski and H. Mahmoud, Phys. Rev. 92 (1953) 1045.
16. I. Kawakami, Prog. Theor. Phys. 19 (1958) 459; E. Lipmanov, J. Exp. i Teor. Fiz. 37 (1959) 1054; A. Sokolov, Phys. Letters 3 (1963) 21.
17. L. B. Okun and C. Rubbia, Talk given at the Heidelberg Int. Conf. on Elementary Particles, 1967.
18. B. Pontecorvo, Phys. Letters 26B (1968) 630.
19. L. Wolfenstein, Phys. Rev. Letters 13 (1964) 562.
20. F. Nezrik and F. Reines, Phys. Rev. 142 (1966) 852.

LEPTON NON-CONSERVATION AND SOLAR NEUTRINOS *

J. N. BAHCALL ** and S. C. FRAUTSCHI
California Institute of Technology, Pasadena, California, USA

Received 2 July 1969

The consequences for experiments involving solar neutrinos of a small amount of lepton non-conservation are examined following the discussion of Gribov and Pontecorvo.

Gribov and Pontecorvo [1, 2] have recently suggested that a tiny amount of lepton non-conservation might lead to a decrease in the detectable number of solar neutrinos at the earth's surface and have linked their suggestion with the unexpectedly low counting rate in the ^{37}Cl solar-neutrino experiment of Davis et al. [3]. Oscillations between electron neutrinos and muon neutrinos, similar to those observed between K^0 and $\bar{K}^0$, were proposed by Gribov and Pontecorvo as a possible consequence of lepton non-conservation. The oscillation distances are supposed to be so large that the oscillations are observable only when distances much greater than laboratory dimensions are involved. Hence the relevance and necessity of experiments involving solar neutrinos. In the scheme of Gribov and Pontecorvo, electron neutrinos produced in the solar interior oscillate between electron and muon neutrino states as they travel from the sun to the earth. Some of the original neutrinos are in the undetectable muon-neutrino state when they arrive on earth.

We examine the experimental consequences of neutrino oscillations for solar-neutrino experiments. Specifically, we calculate the dependence of the capture rate of the most important neutrino sources on the parameters describing lepton non-conservation [1]. Our results show that even with simplifying assumptions and neglecting all astrophysical uncertainties the magnitude of lepton non-conservation cannot be uniquely inferred from the ^{37}Cl experiment. A useful test is provided by experiments more sensitive to the low-energy pep neutrinos (from the reaction $p + e + p \rightarrow {}^2D + \nu$). Since the estimated production rate of pep neutrinos in the sun depends only on the solar luminosity [4, 5], their capture rate was believed previously to constitute a lower limit on the possible result of solar-neutrino experiments, given the usual hypothesis that nuclear fusion reactions are responsible for the solar luminosity. Our most important result is that the capture rate at earth for pep neutrinos depends sensitively on the precise amount of lepton non-conservation. Experiments involving mainly neutrinos from either the standard proton-proton reaction or the pep reaction would provide the most valuable information about lepton conservation.

Estimates are provided for the expected magnitude of lepton non-conservation on the basis of several hypotheses, and the various experimental consequences are outlined. An interesting result of the analysis is that milliweak and superweak theories of lepton non-conservation, based on an analogy to *PC* violation, can be distinguished experimentally in some cases.

The neutrino flux $\phi_\nu(R)$ observed at a distance R from the place where it was produced is given by [1]

$$\phi_\nu^{\text{eff}} = \tfrac{1}{2}\phi_\nu^{0}[1 + \cos D(R)/E_\nu] , \qquad (1)$$

where ϕ_ν^0 is the flux that would be observed if lepton number were conserved exactly and E_ν is the neutrino energy (in units of the electron mass). In writing eq. (1), we have made the simplifying assumption that the masses of the electron and muon neutrinos are the same. With this assumption

$$D \approx \left(\frac{m_{ee}\, m_{e\mu}}{m_e^2}\right)\left(\frac{R}{1\text{AU}}\right) \times 10^{24} . \qquad (2)$$

* Work supported in part by the U.S. Atomic Energy Commission under Contract AT(11-1)-68 of the San Francisco Operations Office, U.S. Atomic Energy Commission; by the National Science Foundation [GP-9433, GP-9114]; and by the Office of Naval Research [Nonr-220(47)].

** Alfred P. Sloan Foundation Fellow.

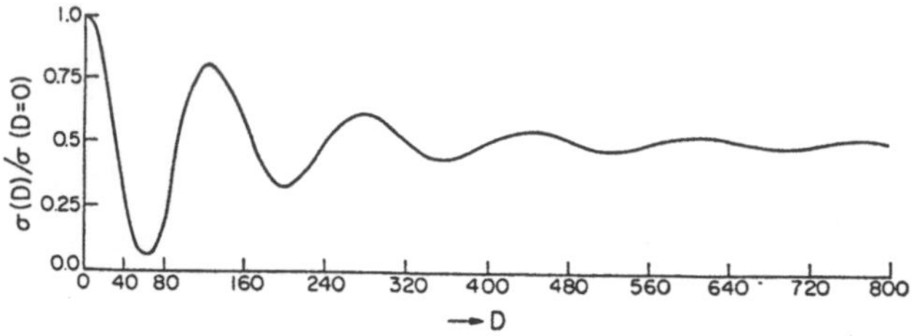

Fig. 1.

The quantities m_{ee} and $m_{e\mu}$ are, respectively, the diagonal and off-diagonal components of the neutrino mass matrix defined by Gribov and Pontecorvo [1], m_e is the electron's mass and 1 AU is the mean earth-sun distance. The assumption of equal masses for electron and muon neutrinos maximizes the fraction of the ν_e flux that oscillates; thus it corresponds to the most favorable conditions for detecting lepton non-conservation. Note that for $m_{ee}, m_{e\mu}$ of the order of the K_1, K_2 mass difference, $D \sim 500$.

Gribov and Pontecorvo discussed two kinds of average that should be considered for neutrino fluxes obeying eq. (1): a) an average over the emitting region of the sun; and b) an average (pointed out by I. Pomeranchuk [1]) over the time of reception. The average over the emitting region involves changes in $R \sim \pm 0.03\, R_\odot$ (where $R_\odot$ is the radius of the sun) and the average over the time of reception involves changes $\sim \pm 0.02$ AU caused by seasonal variations of the earth-sun distance.

There is another average, involving the energy E_ν of the emitted neutrinos, that was not considered by Gribov and Pontecorvo. It is actually the most important average in the ^{37}Cl experiment because the fractional range in energies, $\Delta E_\nu / E_\nu$, is greater than the fractional range in distance, $\Delta R/R$.

About 80% of the expected counting rate in the ^{37}Cl experiment arises from ^{8}B neutrinos produced in the solar interior [5]. In fig. 1 we show the dependence on D of the cross section for absorption of ^{8}B neutrinos by ^{37}Cl after eq. (1) has been averaged in the following way:

$$\frac{\bar{\sigma}(D)}{\sigma(D=0)} = \frac{\int_0^{E_\nu^{\max}} \frac{1}{2} dE_\nu\, \phi(E_\nu)\, \sigma_{\text{abs}}(E_\nu)(1+\cos D/E_\nu)}{\int_0^{E_\nu^{\max}} dE_\nu\, \phi(E_\nu)\, \sigma_{\text{abs}}(E_\nu)} \tag{3}$$

The cross section for absorption, $\sigma_{\text{abs}}(E_\nu)$, has contributions from each of the states of ^{37}Ar that are populated from neutrino capture by ^{37}Cl. Fig. 1 was calculated for the most important transition [6], the superallowed transition from ^{37}Cl to ^{37}Ar. The mean energy, $\bar{E}$, of ^{8}B neutrinos captured via this superallowed transition is $\bar{E} \sim 18\, m_e c^2$. The cross sections for the population of other states in ^{37}Ar have a similar dependence on D to that shown in fig. 1 with only small ($\lesssim$25%) shifts in $\bar{E}$. Note that D does not cause any appreciable effect on the expected counting rate until $D/\bar{E} \sim \frac{1}{2}\pi$ or $D \sim 30$. For $D >> 300$, the expected reduction in ^{8}B flux at the earth is a factor of two independent of the precise values of the neutrino masses.

The average over the emitting region of the sun is required only when $D(0.06\, R_\odot/R)/\bar{E}_\nu \sim \frac{1}{2}\pi$ or $D \sim 10^5$; the variation in the earth-sun distance causes changes in the observed flux when $(D/\bar{E}_\nu)(0.03) \sim \frac{1}{2}\pi$ or $D \sim 10^3$. For values of D this large, however, the cosine factor in eq. (1) changes sign so many times in the average over energy, eq. (3), that always (cf. fig. 1): $\bar{\sigma}(D)/\bar{\sigma}(D=0) = 0.5$.

Thus the value of D could not be determined from the ^{37}Cl experiment alone even if there were no uncertainties in the parameters used to predict the neutrino fluxes. Fig. 1 shows that many different values of D can lead to a significant reduction in $\bar{\sigma}_{^8\text{B}}(D)$ with respect to $\bar{\sigma}_{^8\text{B}}(D=0)$.

Detailed numerical calculations show that the cross sections for ^{13}N and ^{15}O, also continuum sources of neutrinos [5, 6], exhibit a similar behavior to that for ^{8}B (fig. 1) if D is scaled inversely as the ratio of the average neutrino-absorption energies: $(\bar{E}_\nu/m_e c^2) \sim 2$, 2.5 and 18 for ^{13}N, ^{15}O and ^{8}B neutrinos, respectively.

The counting rate from the pep neutrinos ($p + e + p \rightarrow {}^2D + \nu$; $E_\nu = 2.82\, m_{e\mu}{}^2$) constitutes a minimum rate compatible with current ideas regarding thermonuclear reactions in main sequence stars because this reaction is closely linked to the observed solar luminosity [4]. The main broadening of the neutrino line emitted in the pep reaction is caused by the thermal motion of the two colliding protons. For $D < 10^2$, one finds

$$\phi_\nu^{\text{eff}}(\text{pep}) \approx 0.5\, \phi_\nu^{\text{o}}(\text{pep})\, [1 + \cos(D/2.82)]\,; \qquad (4)$$

for $D \gg 10^2$, $\phi_\nu^{\text{eff}}(\text{pep}) \approx 0.5\phi_\nu^{\text{o}}(\text{pep})$. For the other important line source of neutrinos, ^{7}Be, we find

$$\phi_\nu^{\text{eff}}(^7\text{Be}) = 0.5\, \phi_\nu^{\text{o}}(^7\text{Be})\, [1 + \cos(D/1.69)]\,, \qquad (5)$$

for $D \lesssim 10^3$.

Eq. (4) has the important consequence that the flux of neutrinos from the pep reaction is not uniquely determined by the solar luminosity for values of $D \neq 0$. The most suitable experiment to investigate the value of D is one, such as the ^{7}Li experiment discussed by Bahcall [7] or the ^{87}Rb experiment proposed by Sunyar and Goldhaber [8], in which the main counting rate is from p-p or pep neutrinos for which the astrophysical uncertainties are minimal.

It is of interest to make a crude estimate (ignoring the well-known divergence difficulties) of the magnitude of D in order to obtain some hint as to how large this number might be in nature. Pontecorvo [2], and Gribov and Pontecorvo [1], have pointed out the possible analogy between PC non-conservation and the non-conservation of lepton number. If we adopt this analogy, there are at least three possibilities:

1) In the 'milliweak' theory [2] of PC non-conservation ($H_{PC\,\text{violating}} \sim 10^{-3} H_{\text{weak}} \sim 10^{-9} H_{\text{strong}}$), both m_{ee} and $m_{\text{e}\mu}$ are of order 10^{-1} eV (i.e., 10^{-9} times a typical strong interaction mass such as m_π). D is then of order 10^{10}, and all observed solar ν_{e} fluxes will be reduced by $\sim 50\%$ from the values expected if D were zero.

2) In the superweak theory [9] of PC non-conservation, the observed amount of PC violation suggests $H_{\text{superweak}} \sim 10^{-8}$ eV. If both m_{ee} and m_{e} are of this order, we estimate $D \sim 10^{-4}$, and the effect is unobservable.

3) One might also consider, in closer analogy with K^0 particles, a superweak model in which the lepton non-conserving term $m_{\text{e}\mu}$ is $\sim 10^{-8}$ eV, but the neutrino masses arise mainly from virtual transitions such as $\nu_{\text{e}} \to \pi^+ + \pi^0 + \cdot + \text{e}^- \to \nu_{\text{e}}$. In this case one expects $m_\nu \sim \sim 10^{-13} m_\pi \sim 10^{-5}$ eV, and $D \sim 1$. $D \sim 1$ is large enough to allow $\cos(D/E_\nu)$ to deviate from 1 for the relatively low-energy pep neutrinos detected in a ^{7}Li experiment. But one also expects $|m(\nu_\mu) - m(\nu_{\text{e}})| \sim 10^{-5}$ eV $\gg m_{\text{e}\mu}$ in this case (due to the difference between μ and e masses in intermediate states), so the neutrino eigenstates would be nearly pure ν_{e} and ν_μ, only a small fraction ($\sim 10^{-3}$) of the ν_{e} flux would oscillate, and the effect would again be unobservable.

The possibility of discriminating between milliweak and superweak models is very interesting. Only further experiments involving p-p or pep neutrinos from the sun can clarify the situation.

This work was initiated while one of us (JNB) was a member of the Institute for Advanced Studies, Princeton. It is a pleasure to acknowledge stimulating discussions with S. L. Adler.

References

1. V. Gribov and B. Pontecorvo, Phys. Letters 28B (1969) 493.
2. B. Pontecorvo, Soviet Phys. JETP 26 (1968) 984.
3. R. Davis Jr., D. S. Harmer and K. C. Hoffman, Phys. Rev. Letters 20 (1968) 1205.
4. J. N. Bahcall, N. A. Bahcall and G. Shaviv, Phys. Rev. Letters 20 (1968) 1209.
5. J. N. Bahcall, N. A. Bahcall and R. K. Ulrich, Astrophys. J. 156 (1969) 559.
6. J. N. Bahcall, Phys. Rev. Letters 13 (1964) 332; J. N. Bahcall, Phys. Rev. 135 (1964) B137.
7. J. N. Bahcall, submitted to Phys. Rev. Letters.
8. A. W. Sunyar and M. Goldhaber, Phys. Rev. 120 (1960) 871.
9. L. Wolfenstein, Phys. Rev. Letters 13 (1964) 562.

Realistic calculations of solar-neutrino oscillations

V. Barger and K. Whisnant
Physics Department, University of Wisconsin—Madison, Madison, Wisconsin 53706

R. J. N. Phillips
Rutherford Laboratory, Chilton, Didcot, Oxon, England

(Received 1 December 1980)

We reexamine the possible effects of oscillations on the apparent solar-neutrino flux, integrated over the theoretical solar spectrum weighted by the ^{37}Cl or ^{71}Ga neutrino capture cross section. Spectral, thermal, and distance averaging do not reduce all oscillations to their mean values. The averaged $\nu_e \to \nu_e$ transition probability can fall as low as 0.1. Annual variations are evaluated.

The discrepancy between the observed[1] solar-neutrino capture rate 2.1 ± 0.3 SNU (solar-neutrino units) and the calculated[2] rate 7.5 ± 1.5 SNU has been widely attributed to neutrino oscillations.[3,4] If all oscillation wavelengths are much shorter than the mean radius $\bar{R} = 1.5 \times 10^{11}$ m of the Earth's orbit, the formulas are simple and mixing of three neutrinos can suppress the averaged transition probability $\langle P(\nu_e \to \nu_e)\rangle$ by factors down to $\frac{1}{3}$. If any oscillation wavelengths are comparable to $\bar{R}$, however, the picture becomes both more interesting and more complicated. The averaged transition probability can fall below $\frac{1}{3}$ (still with three neutrinos)[4] and there can be annual variations related to the eccentricity of the Earth's orbit.[3-5] In this circumstance calculations must include[4] integration over the solar emission spectrum weighted by the detector capture cross section.

We present updated and rather complete calculations of this kind for existing ^{37}Cl and future ^{71}Ga detectors. The spectrum-averaged value $\langle P(\nu_e \to \nu_e)\rangle_s$ for the ^{37}Cl experiment can be as low as 0.1 for two neutrinos[4,6] or 0.05 for three neutrinos, for certain mass differences and symmetrical mixing. Annual variations related to orbital eccentricity come mainly from a narrow ^7Be line; we evaluate the maximal possible variations.

We suppose that the electron neutrino ν_e is a linear superposition of nondegenerate mass eigenstates ν_i with masses m_i $(i = 1, 2, \ldots, n)$:

$$|\nu_e\rangle = \sum_i U_{ei} |\nu_i\rangle , \quad (1)$$

where U is a unitary mixing matrix. Given an initial relativistic ν_e of energy E, the probability of finding ν_e after flight path L is

$$P(\nu_e \to \nu_e) = 1 - \sum_{i<j} 4|U_{ei}|^2 |U_{ej}|^2 \sin^2(\tfrac{1}{2}\Delta_{ij}) , \quad (2)$$

where $\Delta_{ij} = \frac{1}{2}\delta m_{ij}^2 L/E$ and $\delta m_{ij}^2 = m_i^2 - m_j^2$. If all wavelengths $4\pi E/\delta m_{ij}^2$ are much less than the size of the emitting solar core $\Delta L \sim 10^8$ m or the annual variation of Earth-Sun distance $\Delta L \sim 5 \times 10^9$ m (assuming a long time average), the oscillations in Eq. (2) are not resolved and the averaged transition probability is then

$$\langle P(\nu_e \to \nu_e)\rangle = 1 - \sum_{i<j} 2|U_{ei}|^2 |U_{ej}|^2 . \quad (3)$$

This simple formula has lower bound $\langle P(\nu_e \to \nu_e)\rangle \geq 1/n$ for n-neutrino mixing. If the oscillation wavelengths are not all $\ll \Delta L$, we must use Eq. (2) to make an energy average over the source and detector.

The solar emission spectrum for the standard model[2] is shown in Fig. 1, with the thresholds and rates for the $\nu_e + {}^{37}Cl \to e^- + {}^{37}Ar$ and $\nu_e + {}^{71}Ga \to e^- + {}^{71}Ge$ neutrino capture reactions. We examine what happens to the factors $\sin^2(\frac{1}{2}\Delta_{ij})$ in Eq. (2) for $L = \bar{R} = 1.5 \times 10^{11}$ m when averaged over the spectrum weighted by the capture cross section. Figure 2 shows the averaged[7] value of $\sin^2(\frac{1}{2}\Delta) = \sin^2(\frac{1}{4}\delta m^2 L/E)$ at $L = \bar{R}$ vs δm^2, for ^{37}Cl and ^{71}Ga detectors. Changing L would simply rescale δm^2. For an ideal narrow line spectrum the result would be a sinusoid oscillating between 0 and 1; the figure shows results for the real solar spectrum. Consider for example the ^{37}Cl case.

(i) For small enough δm^2 there is no effect: $\sin^2(\frac{1}{2}\Delta)$ is small throughout the range of E.

(ii) As δm^2 increases above 10^{-13} eV^2, the first contributions to $\sin^2(\frac{1}{2}\Delta)$ arise from the region near 1 MeV (dominated by the ^7Be line at $E = 0.862$ MeV); they are approximately sinusoidal at first, bounded by 0.2.

(iii) As δm^2 increases above 10^{-11} eV^2, the ^8B continuum around 8 MeV gives an additional sinusoid that rises initially to about 0.8.

(iv) As δm^2 increases beyond 10^{-10} eV^2, the broad spectral contributions damp toward constants, and the narrow lines from ^7Be electron capture and pep fusion give superimposed oscillations that are finally damped by thermal broad-

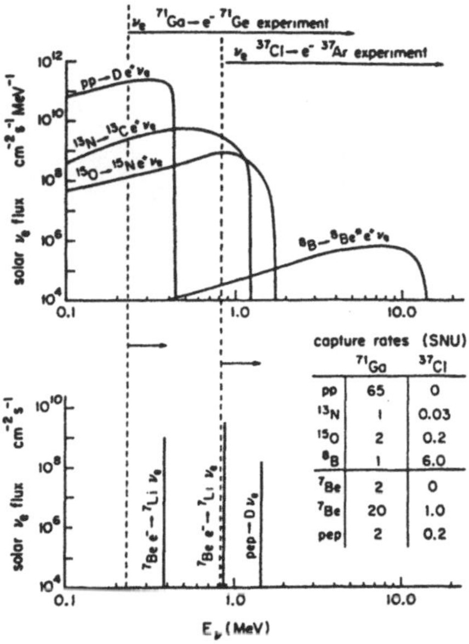

	^{71}Ga	^{37}Cl
pp	65	0
^{13}N	1	0.03
^{15}O	2	0.2
^{8}B	1	6.0
^{7}Be	2	0
^{7}Be	20	1.0
pep	2	0.2

FIG. 1. Continuous bands and discrete lines of the solar emission spectrum in the standard model, based on Ref. 2. Thresholds for ^{37}Cl and ^{71}Ga detectors are indicated.

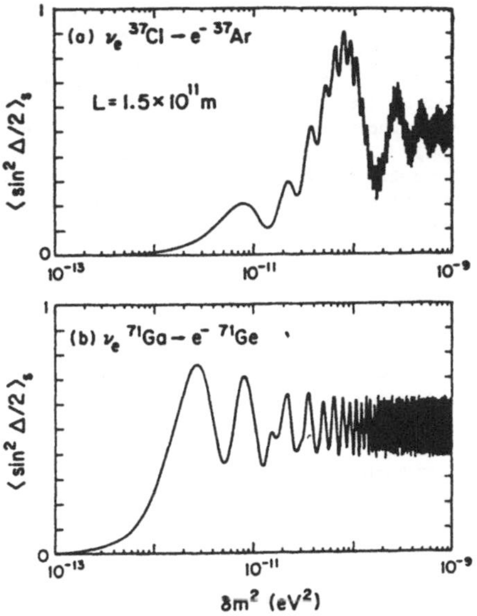

FIG. 2. Spectrum-averaged values of $\sin^2(\frac{1}{2}\Delta)$ vs δm^2 at $L = 1.5\times10^{11}$ m for (a) ^{37}Cl detectors and (b) ^{71}Ga detectors.

ening ($\Delta E \sim 1$ keV) for $\delta m^2 \gtrsim 10^{-8}$ eV^2. In the ^{71}Ga results of Fig. 2(b), the dominant *pp* flux enters first and the persistent oscillation comes mainly from a narrow ^{7}Be line.

The spectrum-averaged transition probability $\langle P(\nu_e \to \nu_e)\rangle_s$ can be read directly from Eq. (2) and Fig. 2 for any neutrino mixing, at any fixed L. We use the subscript s to emphasize spectrum averaging as opposed to complete oscillation averaging for which $\langle \sin^2(\frac{1}{2}\Delta)\rangle = \frac{1}{2}$ and Eq. (3) applies. It is instructive to consider some particular examples.

Two-neutrino mixing. Assuming maximal mixing $|U_{ei}|^2 = \frac{1}{2}$, the complete oscillation average of Eq. (3) achieves its lower bound $P = \frac{1}{2}$. With spectrum averaging alone, however, we have

$$\langle P(\nu_e \to \nu_e)\rangle_s = 1 - \langle \sin^2(\tfrac{1}{2}\Delta_{12})\rangle_s \tag{4}$$

that can fall to 0.1 for δm_{12}^2 near 0.8×10^{-10} eV^2, for a ^{37}Cl detector.

Three-neutrino mixing. With symmetrical mixing $|U_{ei}|^2 = \frac{1}{3}$ we have

$$P(\nu_e \to \nu_e) = 1 - \tfrac{4}{9}[\sin^2(\tfrac{1}{2}\Delta_{12}) + \sin^2(\tfrac{1}{2}\Delta_{23}) + \sin^2(\tfrac{1}{2}\Delta_{31})]\,. \tag{5}$$

The complete oscillation average then reaches its lower bound $\langle P\rangle = \frac{1}{3}$. The spectrum average, however, depends on two independent δm_{ij} parameters, and we distinguish several cases (with the convention $m_1 < m_2 < m_3$).

(i) All $|\delta m_{ij}^2| \gg 4\pi E/L$; L averaging sets all $\sin^2(\frac{1}{2}\Delta_{ij})$ to $\frac{1}{2}$ as discussed earlier, the complete oscillation average.

(ii) Only $\delta m_{32}^2, \delta m_{31}^2 \gg 4\pi E/L$; the Δ_{23} and Δ_{31} oscillations average out completely, while Δ_{12} is subject only to spectral averaging, giving

$$\langle P(\nu_e \to \nu_e)\rangle_s = \tfrac{5}{9} - \tfrac{4}{9}\langle \sin^2(\tfrac{1}{2}\Delta_{12})\rangle_s\,. \tag{6}$$

Hence $\langle P\rangle_s$ can fall to 0.15 for a ^{37}Cl detector for $\delta m_{12}^2 = 0.8\times10^{-10}$ eV^2. In this mass regime symmetrical three-neutrino mixing does not allow $\langle P\rangle_s$ to fall quite as low as two-neutrino mixing does.

(iii) $\delta m_{31}^2 \approx \delta m_{32}^2 \approx 4\pi E/L \gg \delta m_{21}^2$. Here $\sin^2\frac{1}{2}\Delta_{21}$ is essentially zero and $\langle P\rangle_s = 1 - \frac{8}{9}\langle \sin^2(\frac{1}{2}\Delta_{31})\rangle_s$ can fall to 0.12 for a ^{37}Cl detector. A similar result holds if $\delta m_{31}^2 \approx \delta m_{21}^2 \gg \delta m_{32}^2$.

(iv) All oscillation lengths $4\pi E/\delta m_{ij}^2 \gtrsim L$. The three oscillatory terms on the right-hand side of Eq. (5) are subject to spectrum averaging only, and we now have two independent mass scales δm_{21}^2 and δm_{31}^2 to consider. The results for $\langle P\rangle_s$ can in principle be deduced from Fig. 2, but it is more transparent to make a contour plot in the δm_{21}^2, δm_{31}^2 plane, as we show in Fig. 3.

In these contour plots $\delta m_{31}^2 > \delta m_{21}^2 > 0$ by convention (i.e., $m_3 > m_2 > m_1$). There is a symmetry about the diagonal $\delta m_{21}^2 = \frac{1}{2}\delta m_{31}^2$, in the sense that $\langle P\rangle_s$ is unchanged by the substitution $\delta m_{21}^2 \to \delta m_{31}^2 - \delta m_{21}^2$ in the symmetric mixing model. Figure 3 shows that long-wavelength oscillation effects can take $\langle P\rangle_s$ below 0.1 for suitable δm_{ij}^2 for the ^{37}Cl detector (in fact values as low as 0.05 are possible) and close to 0.1 for the ^{71}Ga detector.

We see that $\langle P\rangle_s$ values can differ quite widely between ^{37}Cl and ^{71}Ga detectors for long-wavelength regimes. If such a difference were established experimentally (modulo uncertainties in the solar model) it would be evidence for a long-wavelength oscillation, and contour plots such as in Fig. 3 could be used to restrict the allowed ranges of δm_{ij}^2.

The Earth-Sun distance L varies annually between perihelion, $L = R_P$, and aphelion, $L = R_A$, $\Delta L = R_A - R_P = \bar{R}/30$. To use Fig. 2 that was calculated for the mean $L = \bar{R} = 1.5\times 10^{11}$ m, we must multiply the scale of δm^2 by the varying factor $\bar{R}/L$. In principle we can then read the annual variation in $\langle\sin^2(\frac{1}{2}\Delta)\rangle_s$ directly from Fig. 2. The steeper the local slope $d\langle\sin^2(\frac{1}{2}\Delta)\rangle_s/d(\ln\delta m^2)$, the faster the time variation of the spectrum average of $\sin^2(\frac{1}{2}\Delta)$. The time variation is dominated by the ^7Be spectral line at $E = 0.862$ MeV, through its contribution of $0.14\sin^2(\frac{1}{4}\delta m^2 L/E)$.

The most dramatic annual variation results for $(\Delta L\delta m^2)/(2E) \gtrsim \pi$ corresponding to a half-cycle or more of the ^7Be oscillation. The lowest δm^2 that achieves this is about 2×10^{-10} eV^2, in which case the annual variation could be 1 SNU for the ^{37}Cl detector and 20 SNU for the ^{71}Ga detector. Only a few such cycles could be discriminated experimentally. A case for an annual variation in the solar-neutrino data has been presented by Ehrlich.[5]

The contribution of a narrow line to annual variations is simply expressed through Eq. (2), taking E as the line energy and multiplying by its overall weight factor. The contribution of the continuum is usually ignored, because the line widths $\Delta E/E$ here are much larger than the eccentricity $\Delta L/L$. In calculations of annual changes, we find the continuum variations are less than 5% of their total contribution, and can be neglected in a first approximation.

For the symmetric solar minimizing solution with three neutrinos, the capture rate s (in SNU) varies approximately with L as

$$s = \alpha/3 + \beta - \beta V, \tag{7}$$

$$V = \tfrac{4}{9}[\sin^2(\tfrac{1}{2}\Delta_{12}) + \sin^2(\tfrac{1}{2}\Delta_{23}) + \sin^2(\tfrac{1}{2}\Delta_{31})]\,.$$

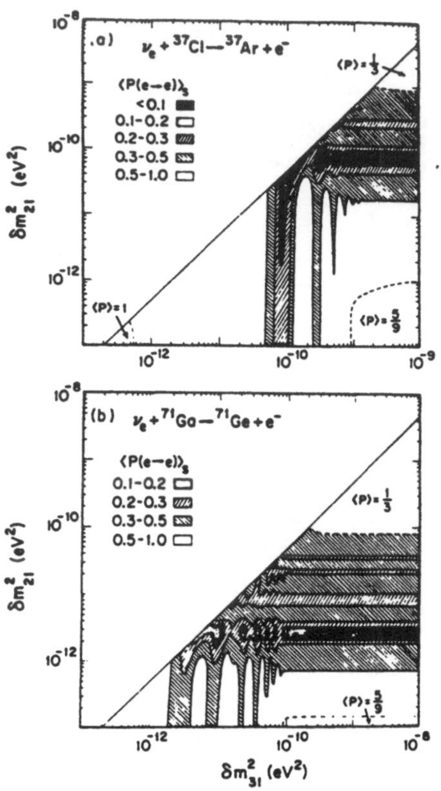

FIG. 3. Contour plot of the spectrum average $\langle P(\nu_e \to \nu_e)\rangle_s$ vs δm_{21}^2 and δm_{31}^2 for symmetrical three-neutrino mixing at $L = 1.5\times 10^{11}$ m for the (a) ^{37}Cl detector and (b) ^{71}Ga detector. The sector $\delta m_{31}^2 > 2\delta m_{21}^2$ is shown; other sectors are defined by symmetry. In the ^{37}Cl results, the rapid oscillations due to the ^7Be and pep lines are included as oscillation averages.

Here α sums the continuum, ^7Be (0.384 MeV) and pep contributions, while β is due to the ^7Be (0.862 MeV) line. For the standard solar model,[2] $\alpha = 6.5$, $\beta = 1$ SNU for ^{37}Cl and $\alpha = 73$, $\beta = 20$ SNU for ^{71}Ga. The maximum variation of V is 1 (for an optimum choice of the δm_{ij}^2), and the corresponding variation is $\Delta s = \beta$. The effect $\Delta s/\bar{s} = 3\beta/(\alpha + \beta)$ could be as large as 40% for ^{37}Cl and 65% for ^{71}Ga. The maximum variation $\Delta s = 1$ SNU for ^{37}Cl is difficult to test with present experimental uncertainties ± 0.5 SNU per quarter year.

The following δm^2 categories give large variations in V and hence in s:

$$\text{(i)}\ \delta m_{21}^2 \simeq \delta m_{31}^2 \gtrsim 2\times 10^{-10}\ \text{eV}^2, \quad 0 \leqslant V \leqslant 1$$

$$\text{(ii)}\ \delta m_{21}^2 \ll \delta m_{31}^2 \gtrsim 2\times 10^{-10}\ \text{eV}^2, \quad 0 \leqslant V \leqslant \tfrac{8}{9}$$

$$\text{(iii)}\ \delta m_{31}^2 \gg \delta m_{21}^2 \gtrsim 2\times 10^{-10}\ \text{eV}^2, \quad \tfrac{4}{9} \leqslant V \leqslant \tfrac{8}{9}.$$

If all δm^2 are much larger or much smaller than 2×10^{-10} eV^2, the variation of s with orbital distance is negligible.

We are indebted to John Bahcall for kindly providing details of solar spectra and neutrino capture cross-section calculations. We thank G. J. Stephenson for a stimulating discussion and R. Ehrlich for communication regarding his analysis of the solar-neutrino data. This research was supported in part by the University of Wisconsin Research Committee with funds granted by the Wisconsin Alumni Research Foundation, and in part by the Department of Energy under Contract No. DE-AC02 76ER00881-179.

---

[1]R. Davis, Jr., in *Proceedings of the Telemark Neutrino Mass Miniconference*, edited by V. Barger and D. Cline (University of Wisconsin, Madison, 1980).

[2]J. N. Bahcall *et al.*, Phys. Rev. Lett. 45, 945 (1980); J. N. Bahcall, Ref. 1.

[3]V. N. Gribov and B. Pontecorvo, Phys. Lett. 28B, 493 (1969); S. M. Bilenky and B. Pontecorvo, Phys. Rep. 41C, 225 (1978).

[4]J. N. Bahcall and S. C. Frautschi, Phys. Lett. 29B, 623 (1969).

[5]R. Ehrlich, Phys. Rev. D 18, 2323 (1978); R. Ehrlich, private communication; R. Barbieri, J. Ellis, and M. K. Gaillard, Phys. Lett. 90B, 249 (1980).

[6]G. J. Stephenson (private communication) drew our attention to this result.

[7]For these averages, we used standard solar model spectra and capture cross sections given in Ref. 2.

"JUST SO" NEUTRINO OSCILLATIONS

Sheldon L. GLASHOW [1,2]
Lyman Laboratory of Physics, Harvard University, Cambridge, MA 02138, USA

and

Lawrence M. KRAUSS [2,3,4]
Department of Physics and Astronomy, J.W. Gibbs Laboratory, Yale University, New Haven, CT 06511, USA

Received 23 February 1987

We examine in detail the possibility that the low rates observed in the chlorine solar neutrino experiment are due to vacuum oscillations between just two neutrino species – a possibility which remains viable over a finite mass range. We calculate the expected signals, seasonal variations, and the effects of time-averaging for both chlorine and gallium based solar neutrino detectors. These provide unique signals for such oscillations, and thus also allow them to be distinguished from the recently proposed resonant neutrino conversion process, as well as from other possible causes of the observed suppression. Finally we display several striking signatures of oscillations relevant to future experiments which may be sensitive to the incident neutrino energy spectrum.

The problem of the anomalously small flux of neutrinos detected from the sun by Davis and his collaborators persists [1]. The observed rate of 2.1 ± 0.3 SNU seems to be at least a factor of three less than that predicted by standard solar models [2]. Moreover, recent seismic observations of the sun seem to favour models which would produce larger fluxes. To date no compelling astrophysical solution of this problem has been forthcoming. Recent theoretical work [3] has revitalized interest in the possibility that the solution to the problem lies in the properties of neutrinos themselves. The recognition that it may be possible for electron neutrinos to almost completely resonantly convert to another neutrino type inside the sun, even for small mixing angles, has spurred a new interest in neutrino masses and oscillations.

[1] Research supported in part by NSF Grant #PHY-82-15249.

[2] Visiting Scientist, Physics Department, Boston University, Boston, MA 02215, USA.

[3] Research supported in part by DOE Contract #DE-AC02-76ER0 3075, and also in part by a Presidential Young Investigator Award.

[4] Visiting Scientist, Smithsonian Astrophysical Observatory, Cambridge, MA 02138, USA.

Resonant conversion can take place efficiently in the sun for neutrino masses greater than about 10^{-3} eV and less than about 10^{-1} eV [4] for a wide range of mixing angles. While this possibility is extremely exciting, it unfortunately makes any results from the forthcoming gallium solar neutrino experiments less definitive, since virtually any observation in gallium would be consistent with some mass-mixing angle combination. Hence, if a suppression is observed in the solar neutrino signal in gallium, it will now be more difficult to disentangle whether the origin of this suppression in astrophysical or due to neutrino conversion, at least until detectors which can observe the spectrum of solar neutrinos [5] are built.

In this context we re-investigate here the original suggestion that the observed suppression in chlorine is due to neutrino oscillations on the journey between the sun and earth [6,7]. We first explicitly confirm the result [7] that if mixing angles are large, oscillations between just two generations can reduce the high-energy electron neutrino flux at the earth by up to a factor of 10. Moreover, in this case the range of possible suppression in gallium is both limited and predictable. If yearly averaging of the signal is per-

Are Neutrinos Stable Particles?*

John N. Bahcall, Nicola Cabibbo,† and Amos Yahil‡
Institute for Advanced Study, Princeton, New Jersey 08540
(Received 17 December 1971)

It is pointed out that neutrinos with a finite mass could be unstable. We discuss the consequences of this possibility for solar-neutrino experiments.

It is generally assumed in textbooks on nuclear physics or elementary particles that neutrinos are stable particles. This assumption is necessary for the validity of the astrophysical conclusions that have been drawn[1] from the unexpectedly low counting rate in the currently operating Brookhaven solar-neutrino experiment.[2] In the context of solar-neutrino problems, it is convenient to adopt a limited definition of "stability"; we shall call a neutrino stable if the lifetime of a 1-MeV particle (typical energy expected for solar neutrinos) exceeds $\sim 5\times10^2$ sec (the Sun-to-Earth light travel time). We show by examples that our present knowledge of neutrinos is insufficient to establish whether or not neutrinos are stable even in the very limited sense defined above. Our results provide a new possibility[3] for the interpretation of the ^{37}Cl experiment of Davis and his associates[2]; future experiments with solar neutrinos (see Table I) can determine whether the low counting rate in the ^{37}Cl experiment is due to neutrino decay or astrophysical uncertainties. We note in passing that our examples showing that ν_e need not be stable apply equally well to ν_μ, but we concentrate on ν_e since it is of most direct interest for solar-neutrino experiments.

Possible decay modes.—In order for a neutrino to be unstable, it must have a finite mass which may, however, be tiny. We assume for simplicity that the masses for all the decay products of ν_e are identically zero. The simplest example of a possible decay mode for ν_e is

$$\nu_e \to \nu' + \varphi, \tag{1}$$

where φ is a massless scalar or pseudoscalar boson[4] and ν' is some other neutrino that could be related to ν_μ or $\bar\nu_\mu$. If the coupling is of the form

$$h = g(\bar\psi_{\nu'}\psi_{\nu_e})\varphi \text{ or } g(\bar\psi_{\nu'}\gamma_5\psi_{\nu_e})\varphi, \tag{2a}$$

the mean decay lifetime for a neutrino of energy E and rest mass m is

$$\tau(E) = (16\pi\hbar/g^2mc^2)E/mc^2. \tag{2b}$$

The dimensionless coupling constant g can be written conveniently in terms of the relevant physical variables[5] as

$$g^2 = 1.7\times10^{-14}\frac{E}{1\text{ MeV}}\cdot\left(\frac{60\text{ eV}}{mc^2}\right)^2\frac{500\text{ sec}}{\tau(1\text{ MeV})}.$$

Although the above value of g^2 suggest a rather weak interaction, an interaction of the same strength that allowed $\mu\to e+\varphi$ would lead to a partial lifetime shorter than the total observed muon lifetime and must be excluded. If ν' is identical to $\bar\nu_\mu$ or is unrelated to muon neutrinos, this exclusion can be achieved by known selection rules. If ν' is identified with ν_μ, then the exclusion of the $\mu\to e+\varphi$ coupling requires an *ad hoc* assumption. We note also that values of g^2 orders of magnitude larger than 10^{-14} are possible without

TABLE I. Implications for future experiments (see Ref 8). All counting rates are expressed in units of 10^{-36} captures per target atom per second.

Assumed ^{37}Cl capture rate	Assumed explanations	^{7}Li capture rate	^{87}Rb capture rate	ν–e scattering
1.5	Standard solar models incorrect	15 ± 1 (Ref. 1)	5×10^2	Mostly low-energy neutrinos are scattered
	Most neutrinos decay [τ(1 MeV) = 33 sec]	5	1	Mostly high-energy ^{8}B neutrinos are scattered
≤ 0.4	Standard solar models very wrong	9	4×10^2	Only low-energy neutrinos are scattered
	Almost all neutrinos decay [τ(1 MeV) = 18 sec]	1	3×10^{-1}	Only high-energy ^{8}B neutrinos are scattered: intensity very low

violating any known experimental results, permitting much smaller values of τ or m than 500 sec or 60 eV.

More complicated decay schemes than Reaction (1) are in principle possible,[6] e.g., (i) $\nu_e \to \nu' + \varphi_1 + \varphi_2$; (ii) $\nu_e \to \nu' + V$; and (iii) $\nu_e \to \nu_1 + \nu_2 + \nu_3$, where V is a massless vector particle and φ_i and ν_i are other massless scalars and neutrinos, respectively. Process (iii) can be ruled out since it would require (for a ν_e unstable in our sense) a muon partial lifetime (via $\mu^+ \to e^+ + \bar{\nu}_\mu + \nu_1 + \nu_2 + \nu_3$) that is much shorter than the observed total lifetime. From a cursory examination of muon decay data,[7] we estimate that if ν_e is unstable through processes (i) or (ii) then $\tau(1\text{ MeV}) \gtrsim 10^{-1}$ sec, so that we cannot exclude the possible role of (i) or (ii) in solar-neutrino experiments.

Astrophysical consequences.— The major astrophysical consequences of our considerations are illustrated in Table I. We consider two general classes of explanation for the unexpectedly low counting rate in the ^{37}Cl experiment: (1) Something is wrong with the theory of solar models (see Ref. 1 for specific possibilities); and (2) most of the neutrinos produced in the sun decay before they reach the Earth. In order to make our discussions more definite, we have calculated the expected counting rates for three kinds of targets (^{7}Li, ^{87}Rb, and ν-e scattering) that have been proposed[8] as possible solar-neutrino detectors; our calculations are given for two assumed levels of counting rates in the ^{37}Cl experiment, namely, the 1.5-SNU (1 solar-neutrino unit = 10^{36} captures per target particle per second) level tentatively reported by Davis, Rogers, and Rodeka[2] and $\lesssim 0.4$ SNU (which represents approximately the ultimate sensitivity of the experiment). We calculated rows 2 and 4 of Table I assuming that the standard solar models are correct and that the failure to observe the predicted 9 SNU is due entirely to neutrino decay. For rows 1 and 3, we have made the opposite assumption, namely, that neutrinos are stable. We adopted for row 1 the suggestions for modifying the standard solar models given in Ref. 1. To obtain row 3, we assumed that only p-p and p-e-p neutrinos are produced in the Sun. The predicted rate for such low-energy neutrinos is practically model independent[1,9] and should contribute 0.3 SNU to the ^{37}Cl experiment.

Note that the expected capture rates in the different experiments considered depend strongly on whether the discrepancy between theory and observation for the ^{37}Cl experiment is ascribed to the solar models or neutrino decay. This difference is easily understood. The low-energy neutrinos from the basic p-p and p-e-p reactions[9] are most strongly affected by decay; on the other hand, the flux of high-energy ^{8}B neutrinos is most sensitive to changes in astrophysical parameters[1] but is least affected by decay. The ratio of neutrino flux at Earth to neutrino flux near the production region in the Sun is a very sensitive function of energy if decay is important since

$$\varphi_{\text{Earth}}(E) = \varphi_{\text{solar}}(E) \exp[-(500\text{ sec})/\tau(E)]. \quad (3)$$

The exponential dependence upon neutrino lifetime in Eq. (3) suggests that a ratio of $\varphi_{\text{solar}}/\varphi_{\text{Earth}} \sim$ a few is unlikely and that the most likely outcome of the ^{37}Cl experiment, if neutrino decay is important, is an unmeasurably low capture rate.

Finally, we note that there is even a possibility that one of the decay products of ν_e, i.e., ν', could be detected in ν'-e scattering[10] experiments. This possibility adds to the physical as well as astrophysical interest of electron-neutrino scattering experiments.

We are grateful to Dr. R. Davis, Jr., and the participants in the Princeton University Astronomy Department student luncheons for stimulating conversations. One of us (N.C.) would like to thank the Director of the Institute for Advanced Study, Carl Kaysen, and the members of the school of Natural Sciences for their kind hospitalty.

---

*Research sponsored in part by the National Science Foundation under Grant No. GP-16147 A No. 1.

†Permanent address: Istituto di Fisica, Università di Roma, Rome, Italy.

‡On leave of absence from Tel-Aviv University, Ramat-Aviv, Tel Aviv, Israel.

[1]See, for example, J. N. Bahcall, Phys. Rev. Lett. 17, 398 (1966); J. N. Bahcall and R. K. Ulrich, Astrophys. J. 170, 479 (1971). Additional references are given in these papers and in Z. Abraham and I. Iben, Jr., Astrophys. J. 170, 157 (1971).

[2]R. Davis, Jr., Phys. Rev. Lett. 12, 303 (1964); R. Davis, Jr., D. S. Harmer, and K. C. Hoffman, Phys. Rev. Lett. 20, 1205 (1968); R. Davis, Jr., L. C. Rogers, and V. Rodeka, Bull. Amer. Phys. Soc. 16, 631 (1971).

[3]The decay processes discussed in the present paper should not be confused with the oscillatory processes ($\nu_e \rightleftarrows \nu_\mu$) discussed earlier by B. Pontecorvo, Zh. Eksp. Teor. Fiz. 53, 1717 (1967) [Sov. Phys. JETP 26, 984 (1968)]; V. Gribov and B. Pontecorvo, Phys. Lett. 28B, 493 (1969). The oscillatory processes typically lead to only a factor-of-2 reduction in the terrestrially detected flux of electron neutrinos when properly averaged over the spectrum of energies of solar neutrinos

[see J. N. Bahcall and S. C. Frautschi, Phys. Lett. 29B, 623 (1969)].

[4]A possible role for a massless pseudoscalar field in *CP*-violating interactions has been proposed by F. Gürsey and A. Pais, unpublished.

[5]The most severe limit that we know for m is $\lesssim 60$ eV, given by K. Bergkvist, in *Topical Conference on Weak Interactions, CERN, Geneva, Switzerland, 14–17 January 1969* (CERN Scientific Information Service, Geneva, Switzerland, 1969), p. 91.

[6]The lifetime for processes (i) and (ii) would be proportional to $m_\nu^{-4}E$. Specifically, for process (i), with $h = k(\bar{\psi}_\nu \psi_\nu)\varphi_1\varphi_2$, $\tau = 3(2^8\pi^3)m_\nu^{-4}k^{-2}E$. In order to have instability in our sense one would need a fairly strong interaction (~ electromagnetic interactions) which, however, cannot be excluded by present data on muon decay. The lifetime for process (iii) is proportional to $m_\nu^{-6}E$, which would require, for instability in our sense, a very strong interaction that can be excluded by present data.

[7]We refer to the accurate e^+ spectra used for a determination of the ρ value. For a list of references, see M. Roos *et al.*, Phys. Lett. 33B, 32 (1970).

[8]The ^{87}Rb experiment was first proposed by A. W. Sunyar and M. Goldhaber, Phys. Rev. 120, 871 (1970); the ^{7}Li experiment by J. N. Bahcall, Phys. Lett. 13, 332 (1964); and the ν-e scattering experiment by F. Reines and W. R. Kropp, Phys. Rev. Lett. 12, 457 (1964). The counting rates given in Table I have been calculated in the usual way by averaging the relevant neutrino absorption cross section over the neutrino spectrum of each β-decay source [see J. N. Bahcall, Phys. Rev. 135, B137 (1964)]; the cross sections used were taken from J. N. Bahcall, Phys. Lett. 13, 332 (1964), and Phys. Rev. Lett. 23, 251 (1969); and G. V. Domogatsky, Lebedev Institute Report No. 153, 1971 (unpublished).

[9]Most of the expected low-energy neutrinos are from the basic reactions $p + p \to {}^2\text{H} + e^+ + \nu_e$ and $p + e^- + p \to {}^2\text{H} + \nu_e$. The *astrophysical* basis for expecting the calculated number of these low-energy neutrinos is very secure; the calculation depends mainly on the measured solar luminosity and the fact that four protons are ~ 25 MeV heavier than an α particle (see Ref. 1 and other references cited therein).

[10]This could happen, for example, if $\nu' = \nu_\mu$ (or $\nu' = \bar{\nu}_\mu$) and muon neutrinos are scattered by electrons, as predicted by some versions of weak-interaction theory; see S. Weinberg, Phys. Rev. Lett. 19, 1264 (1967); S. L. Glashow, J. Iliopoulos, and L. Maiani, Phys. Rev. D 2, 1285 (1970); G. 't Hooft, Phys. Lett. 37B, 195 (1971).

NEUTRINO DECAY IN MATTER

Z.G. BEREZHIANI [1] and M.I. VYSOTSKY
Institute of Theoretical and Experimental Physics, 117259 Moscow, USSR

Received 3 September 1987

We demonstrate that matter can induce decay of the neutrino into the antineutrino and a light scalar particle (majoron): $\nu \to \bar{\nu} + \alpha$, or vice versa, $\bar{\nu} \to \nu + \alpha$. Diagonal as well as non-diagonal transitions on neutrino types are possible. The decay probabilities depend on the neutrino energies in an unusual way. Applications for (1) accelerator neutrinos passing through the earth, (2) solar neutrinos, (3) neutrino emission accompanied by the gravitational collapse of the stellar core, are discussed.

It is well known that the oscillations of neutrinos in matter could be different from vacuum oscillations [1,2]. We shall demonstrate that the coherent interaction of neutrinos with matter could be important for their decay also. In particular, a transition of the left-handed neutrino into a right-handed one (right neutrino in the case of the Dirac neutrino or antineutrino in the case of the Majorana neutrino) with the emission of massless (or sufficiently light) scalar particle can become energetically allowed, even in the case of strictly massless neutrinos.

Coherent scattering could be taken into account directly in the "current×current" lagrangian which describes elastic neutrino scattering in terms of neutral currents [#1]. In the rest frame of matter only the time components of the electron, proton and neutron vector currents remain nonzero after averaging over the matter, e.g. $\langle \bar{e}\gamma_\mu(1+\gamma_5)e\rangle = \delta_{0\mu} n_e$. Consequently neutrino propagation in the medium is described by the Dirac equation which is analogous to that describing electron motion in an external field $eA_\mu = (\rho, 0)$. Therefore, for the left-handed neutrino and CP conjugated right-handed antineutrino, $\bar{\nu}_R = C\nu_L$, energy levels we have correspondingly: $E = \sqrt{\bar{p}^2 + m^2} \pm \rho$. Here $\bar{p}$ and m are neutrino momentum and mass and $\rho = \rho_e - \frac{1}{2}\rho_n$ for ν_e and $\rho = -\frac{1}{2}\rho_n$ for ν_μ and ν_τ, where $\rho_{e,n} = \sqrt{2} G_F n_{e,n}$ and n_e and n_n are electron and neutron densities, correspondingly. So if there exists an interaction which changes neutrino chirality – for example, radiation of a massless or very light scalar particle φ – the decay $\nu_L \to \bar{\nu}_R + \varphi$ is induced in matter (or vice versa, depending on the neutrino type and the relative neutron density of matter). We will also demonstrate that even if this scalar particle interacts diagonally with a neutrino in vacuum flavour-nondiagonal transitions would be induced in matter.

Let us illustrate the above statements within the triplet majoron model [3]. We briefly remind this model. A triplet of scalar fields $\phi = (\phi^{++}, \phi^{+}, \phi^{0})$ is introduced in the standard $SU(2)_L \otimes U(1)$ model of electroweak interactions without right-handed neutrino field components. The field ϕ has Yukawa couplings with the doublets $\ell_L: \binom{\nu_e}{e}_L, \binom{\nu_\mu}{\mu}_L$ (for the sake of simplicity we shall discuss the case of two generations). If the lagrangian does not contain couplings of the type $H\tau\varepsilon H\phi$ (where H is the Higgs doublet) then the global U(1) symmetry could be spontaneously broken by a non-zero vacuum average u of the neutral component ϕ^0 of the triplet field. In this way the neutrino acquires the Majorana masses. Simultaneously a massless Goldstone boson (majoron) $\alpha = (1/\sqrt{2})\,\mathrm{Im}\,\phi^0$ appears in the particle spectrum. A light Higgs boson $\varphi = (1/\sqrt{2})\,\mathrm{Re}\,\phi^0 - u$ with a mass $\sqrt{\lambda} u$ appears also (λ is an interaction constant of the field ϕ) [#2]. The lagrangian of the ϕ^0 interactions has the following form:

[1] Institute of Physics, Academy of Sciences GSSR, 380077 Tbilisi, USSR.

[#1] The charged (ν_e, e) current interactions must also be Fierz-transformed into the form $(G_F/\sqrt{2})\bar{\nu}_e\gamma_\mu(1+\gamma_5)\nu_e \bar{e}\gamma_\mu(1+\gamma_5)e$.

[#2] The energy loss of red giants due to majoron emission puts the constraint: $u < 100$ keV [4]. The fields ϕ^{++} and ϕ^{+} must be heavy, $M \sim M_W$.

$$\mathcal{L}=\overline{(\nu_e\nu_\mu)_L}\hat{h}(u+\mathrm{i}\alpha+\varphi)\begin{pmatrix}\tilde{\nu}_e\\ \tilde{\nu}_\mu\end{pmatrix}_R+\text{h.c.}\,, \tag{1}$$

where $\tilde{\psi}_R\equiv C\psi_L$ is the antineutrino field and $\hat{h}=\left(\begin{smallmatrix}h_e & h_{e\mu}\\ h_{e\mu} & h_\mu\end{smallmatrix}\right)$ is the matrix of the Yukawa coupling constants.

If one transforms the neutrino mass matrix into diagonal form and arrives at the states with definite masses m_1 and m_2 ($\nu_1=c\nu_e+s\nu_\mu$, $\nu_2=-s\nu_e+c\nu_\mu$, where $c=\cos\theta$, $s=\sin\theta$, $\sin^2 2\theta=4h_{e\mu}^2/[4h_{e\mu}^2+(h_\mu-h_e)^2]$) the Yukawa couplings (1) are diagonalized simultaneously:

$$h_{1,2}=m_{1,2}/u=\tfrac{1}{2}[h_e+h_\mu\mp\sqrt{4h_{e\mu}^2+(h_\mu-h_e)^2}\,]\,,\quad \hat{h}=\begin{pmatrix}h_1 & 0\\ 0 & h_2\end{pmatrix}. \tag{2}$$

We demonstrate that α and φ couple diagonally with ν_1 and ν_2. So the heaviest neutrino can not decay in vacuum.

The next step is to take into account effects of matter. For the ultrarelativistic neutrino ($p\equiv|\bar{p}|\gg m$) the evolution of states in matter after subtraction of an unessential common phase is described by the following Schrödinger equation [1]:

$$\mathrm{i}\frac{\mathrm{d}}{\mathrm{d}t}\begin{pmatrix}\nu_1\\ \nu_2\end{pmatrix}=H\begin{pmatrix}\nu_1\\ \nu_2\end{pmatrix}, \tag{3}$$

with the hamiltonian

$$\mathcal{H}=\begin{pmatrix}m_1^2/2p+c^2\rho_e-\tfrac{1}{2}\rho_n & cs\rho_e\\ cs\rho_e & m_2^2/2p+s^2\rho_e-\tfrac{1}{2}\rho_n\end{pmatrix}. \tag{4}$$

Eigenstates of this hamiltonian are

$$\nu_1^m=c_m\nu_1+s_m\nu_2\,,\quad \nu_2^m=-s_m\nu_1+c_m\nu_2\quad (c_m\equiv\cos\theta_m,\, s_m=\sin\theta_m)\,,$$

where

$$\sin^2 2\theta_m=\sin^2 2\theta/[\sin^2 2\theta+(\cos 2\theta-\xi)^2]\,, \tag{5}$$

and $\xi=(m_2^2-m_1^2)/2p\rho_e$. Eigenvalues of the hamiltonian (4) look like

$$\lambda_{1,2}=\tfrac{1}{2}\{(m_1^2+m_2^2)/2p+\rho_e-\rho_n\mp\sqrt{[(m_1^2-m_2^2)/2p+\cos 2\theta\,\rho_e]^2+\sin^2 2\theta\rho_e^2}\}\,. \tag{6}$$

Antineutrino evolution in the same matter is described by an analogous hamiltonian with the change $\rho_{e,n}\to-\rho_{e,n}$ (when we interchange ν and $\tilde{\nu}$ the zero angle scattering amplitude $f(0)$ changes sign). Eigenstates and eigenvalues for $\tilde{\nu}_R$ are given by formulae (5), (6) with the change $\rho_{e,n}\to-\rho_{e,n}$ ($\tilde{c}_m=\cos\tilde{\theta}_m$, $\tilde{s}_m=\sin\tilde{\theta}_m$).

For eigenstates in matter we obtain from (1):

$$\mathcal{L}=\overline{(\nu_1^m\nu_2^m)_L}\begin{pmatrix}c_m\tilde{c}_m h_1+s_m\tilde{s}_m h_2 & s_m\tilde{c}_m h_2-c_m\tilde{s}_m h_1\\ c_m\tilde{s}_m h_2-s_m\tilde{c}_m h_1 & s_m\tilde{s}_m h_1+c_m\tilde{c}_m h_2\end{pmatrix}\begin{pmatrix}\tilde{\nu}_1^m\\ \tilde{\nu}_2^m\end{pmatrix}(\mathrm{i}\alpha+\varphi)+\text{c.c.} \tag{7}$$

Therefore, nondiagonal $\tilde{\nu}_{1,2}^m\to\nu_{2,1}^m$ as well as diagonal $\tilde{\nu}_{1,2}^m\to\nu_{1,2}^m$ transitions are possible.

For the forthcoming statements we need an expression for the neutrino decay probabilities. We shall give a general expression from which all needed decay probabilities could easily be obtained. We shall consider the decay of the Majorana neutrino ν_1 with momentum $\bar{p}$ and mass m_1 and an interaction potential with matter ρ_1 into a massless scalar α and the Majorana neutrino ν_2 with a mass m_2 and a potential ρ_2, which is described by the amplitude $h\tilde{\nu}_{1L}\tilde{\nu}_{2R}\alpha$. For the width in a reference system where matter is at rest we obtain:

$$\Gamma=\frac{h^2}{16\pi}\frac{[m_1^2+(\rho_1-\rho_2)\sqrt{\bar{p}^2+m_1^2}]}{\sqrt{\bar{p}^2+m_1^2}}\left(1-\frac{m_2^4}{[m_1^2+(\rho_1-\rho_2)^2+2(\rho_1-\rho_2)\sqrt{\bar{p}^2+m_1^2}]^2}\right). \quad (8)$$

We start our analysis from the case of dense matter (or light neutrinos) $\rho\gg(m_1^2,m_2^2)/2E$ ($\xi\ll1$). In this case the eigenstates of the hamiltonian (4) are directly the current ones: $\nu_1^m=\nu_e$, $\nu_2^m=\nu_\mu$. For the neutrino decay probabilities we obtain:

$$\Gamma_{\bar{\mu}\mu}=(h_\mu^2/16\pi)\rho_n\,,\quad \Gamma_{\bar{e}e}=(h_e^2/16\pi)(2\rho_e-\rho_n)\,,\quad \Gamma_{\mu\bar{e}}=\Gamma_{e\bar{\mu}}=(h_{e\mu}^2/16\pi)(\rho_e-\rho_n)\,. \quad (9)$$

where $\Gamma_{\bar{\mu}\mu}\equiv\Gamma(\bar{\nu}_\mu\to\nu_\mu)$ etc. Negative probability means that ν and $\bar{\nu}$ must be interchanged (if $\rho_n>2\rho_e$, then $\bar{\nu}$ decays into ν).

Now, let us turn to the case of a heavier neutrino (or low densities of matter) $\xi\gg1$. For the diagonal transitions $\nu_{1L}^m\to\bar{\nu}_{1R}^m$, $\nu_{2L}^m\to\bar{\nu}_{2R}^m$ we obtain:

$$\Gamma_{1\bar{1}}=(h_1^2/2\pi)(\rho_e\cos^2\theta-\tfrac{1}{2}\rho_n)\,,\quad \Gamma_{2\bar{2}}=(h_2^2/2\pi)(\rho_e\sin^2\theta-\tfrac{1}{2}\rho_n)\,. \quad (10a)$$

Nondiagonal transitions have the following width:

$$\Gamma_{1\bar{2}}=\Gamma_{\bar{1}2}=(h_{12}^2/16\pi|\bar{p}_1|)(m_1^4-m_2^4)/m_1^2\,,$$

$$h_{12}=\tfrac{1}{2}\sin2\theta_m(h_1-h_2)\Big|_{\xi\gg1}=[(\sin2\theta)/2\xi](h_1-h_2)=h_{e\mu}/\xi\,. \quad (10b)$$

which are suppressed compared with diagonal ones (10a).

The probabilities of matter-induced transitions depend on the initial neutrino energy in an unusual way. They do not depend on energy when $\xi\ll1$ (see (9)). When $\xi\gg1$ diagonal transitions are also energy independent. As for nondiagonal probabilities (10b), they are proportional to E/m. For ordinary vacuum decays $\Gamma=\Gamma_0 m/E$, where Γ_0 is the decay probability in the rest frame of the decaying particle.

Before we discuss some applications let us give the experimental limits on the majoron coupling constants:

$$h_e<2\times10^{-3} \quad (11)$$

($2\beta_0$ decay with majoron emission) [5]

$$h_e^2+h_{e\mu}^2<4.5\times10^{-5},\quad h_\mu^2+h_{e\mu}^2<2.4\times10^{-4} \quad (12,13)$$

(lepton decays of π- and K-mesons) [6].

(1) The passing of the accelerator neutrinos through the earth. The $\bar{\nu}_\mu$ are heavier in matter than the ν_μ. So, when a bunch of accelerator produced $\bar{\nu}_\mu$ pass through the earth some of them can decay into ν_μ. The charged current gives a distinct signature of the effect: together with μ^+ from the reaction $\bar{\nu}_\mu p\to n\mu^+$, μ^- from the reaction $\nu_\mu n\to\mu^- p$ appears. Let us estimate the value of the effect. We take 10 g/cm³ for the earth density, the relative number of neutrons is given by the electron number $Y_e\equiv n_e/(n_p+n_n)\approx0.5$. The region $\xi\gg1$ is more profitable. With the maximal value of h_μ from (10a) and (13) we obtain:

$$\tau_{\bar{\nu}_\mu}=4\pi/h_{\bar{\mu}\mu}^2\rho_n\approx(3\times10^{12}\text{ cm})/(3\times10^{10}\text{ cm/s})\approx100\text{ s}\,. \quad (14)$$

As the diameter of the earth is of the order of 10^9 cm not more than 0.03 % of $\bar{\nu}_\mu$ could transform into ν_μ.

(2) Solar neutrinos. Electron neutrinos produced in the core of the sun fall on the earth and, in particular, are detected in the Davis experiment. The deficit of ν_e in the Davis experiment in comparison with the standard solar model prediction can be understood if one supposes that ν_e decays inside the sun: $\nu_e\to(\bar{\nu}_e,\bar{\nu}_\mu)\alpha$. Let us make an estimate for the matter-induced decay. As the maximal allowed value for h_e^2 is an order of magnitude lower than h_μ^2 then in the domain $\xi\ll1$ neglecting ρ_n in comparison with ρ_e we obtain:

$$\Gamma_{\nu_e\to\bar{\nu}_\mu}=(h_{e\mu}^2/16\pi)\rho_e\,. \quad (15a)$$

When $\xi \gg 1$, maximum neutrino mixing is profitable, $\theta \approx 45°$. Then from (10a) we obtain:

$$\Gamma_{\nu_e \to \bar{\nu}_e, \bar{\nu}_\mu} = (h^2_{e\mu}/4\pi)\rho_e . \qquad (15b)$$

Taking the maximally allowed value for $h^2_{e\mu}$ from (12) and using a density of the sun of 100 g/cm³ at a distance of 2×10^{10} cm we obtain in the case (15b):

$$N_{\bar{\nu}}/N_{\nu_e} \approx 5\% , \qquad (16)$$

which is too small to explain the three-times deficit of ν_e in the Davis data but probably rather large enough to be detected in the future with new solar neutrino detectors.

(3) *Neutrinos from the gravitational collapse of a stellar core.* In supernovae the densities and distances have such values that in spite of the limits (10)–(13) neutrino decays could occur. The neutrinos emitted during the collapse have a typical energy of about 10 MeV. Taking a typical density of 10^{12} g/cm³ we obtain $\rho_e \approx 10^{-1}$ eV and $\xi \ll 1$. Typical distances in supernovae are $R \sim 10^7$ cm and decays take place if $h^2 > 10^{-9}$. The relative number of electrons in supernovae, Y_e, varies between 0.3 at $R \sim 10^6$ cm and 0.5 at $R \sim 10^7$ cm [7]. Without taking decays into account, approximately equal numbers of ν's and $\bar{\nu}$'s of different types are expected; the decays change this prediction drastically. In particular, at $Y = 0.3$, $\bar{\nu}_\mu$ and $\bar{\nu}_e$ decay into ν_μ and ν_e, respectively, and no antineutrinos fall on the earth (let us remind that the detectors which are under operation now are sensitive to $\bar{\nu}_e$ only [8]; however, the detectors under construction will be sensitive to ν_e as well).

It is noteworthy that the effects of neutrino decay in matter can be relevant also to the case of Dirac neutrinos. Indeed, in the standard SU(2)⊗U(1) model with right-handed neutrino fields ν_R one can introduce together with the standard H an additional Higgs doublet $H' = \binom{H^{0'}}{H^{-'}}$ with Yukawa couplings of the type $h\bar{e}_L \nu_R H'$ + h.c. and impose the global U(1) symmetry $H' \to \exp(i\alpha)H'$. The breaking of this symmetry by the vacuum average of the scalar $H'(\langle H^{0'} \rangle \ll \langle H^0 \rangle = 250$ GeV) leads to the appearance of neutrino Dirac masses and the corresponding Goldstone boson β (the diron – by analogy with the majoron) simultaneously appears in the particle spectrum. Then the transitions $\nu_L \to \nu_R + \beta$ are possible in matter (or vice versa, $\nu_R \to \nu_L + \beta$, depending on the neutrino type and the neutron concentration). It is obvious that due to the sterility of ν_R the experimental consequences of the diron scheme substantially differ from that for the case "Majorana neutrinos + majoron".

Let us discuss now the case of neutrino decays in vacuum. They are possible if we incorporate in the majoron scheme also the small Dirac mass terms. Another way is the introduction of two (or more) majoron-type global U(1) symmetries acting distinctly on the different neutrino flavours. In vacuum the neutrino-type nondiagonal transitions due to couplings of the type $h\bar{\nu}_1 \nu_2 \alpha$ + h.c. ($\nu_1 = c\nu_e + s\nu_\mu$,...) are possible only, e.g. $\nu_1 \to \bar{\nu}_2 + \alpha$, $\bar{\nu}_1 \to \nu_2 + \alpha$ if $m_1 > m_2$. The corresponding width in the rest frame of the neutrino source is $\Gamma = \Gamma_0 m/E$, where $\Gamma_0 = (h^2/16\pi)(m_1^4 - m_2^4)/m_1^3$.

The vacuum decays can explain the deficit of the solar ν_e flux in the Davis experiment in two different ways.

(1) Neutrino mixing is negligible and $m_{\nu_e} > m_{\nu_\mu}$. Supposing that $\tau_{\nu_e} = \tau^0_{\nu_e} E_{\nu_e}/m_{\nu_e} \approx 500$ s for $E_{\nu_e} \approx 5$ MeV we conclude that one third of the boron neutrinos emitted in the core of the sun decay in flight from the sun to the earth. In this scenario the flux of pp neutrinos (with energies lower than 420 keV) is diminished in e^{10} times. This is a strong (but negative) prediction for the Ga–Ge detector experiments in preparation.

(2) Neutrino mixing is substantial, $\nu_e = c\nu_1 + s\nu_2$, and $m_1 > m_2$. If the heavier neutrino ν_1 decays into $\bar{\nu}_2$ with a lifetime $\tau_{\nu_1} \ll 500$ s, the component $\nu_2 = -s\nu_e + c\nu_\mu$ reaches the earth only. The magnitude of the ν_e flux directly measures the neutrino mixing angle: $N_{\text{Davis}}/N_0 = \sin^4\theta$.

In conclusion, we would like to emphasize that matter-induced neutrino decays are drastically distinct from vacuum decays and ν oscillations. They can take place even for strictly massless neutrinos. The experimental signatures of the decays are also very distinct: the neutrino changes chirality, which can not take place as a result of ν oscillations.

Numerical estimates show that in an experiment with an accelerator $\bar{\nu}_\mu$ beam a little part could transform into ν_μ; a flux of ν_e from the sun could contain a few percent of $\bar{\nu}_e$ and $\bar{\nu}_\mu$ and the content of neutrinos and antineutrinos of different flavours coming into the earth from stellar collapse could differ widely in number because of ν decays in dense regions of the star. Finally, note that the existence of the majoron leads to the absence of relic neutrinos: they annihilate into majorons in the early stages of thè evolution of the universe if the Yukawa constant h is larger than 10^{-5}.

We are grateful to S.I. Blinnikov, A.A. Gerasimov, A.Yu. Smirnov, M.Yu. Khlopov and V.A. Tsarev for useful discussions.

References

[1] L. Wolfenstein, Phys. Rev. D 17 (1978) 2369; D 20 (1979) 2634.
[2] S.P. Mikheev and A.Yu. Smirnov, Yad. Fiz. 42 (1985) 1441.
[3] G.B. Gelmini and M. Roncadelli, Phys. Lett. B 99 (1981) 411.
[4] H. Georgi, S.L. Glashow and S. Nussinov, Nucl. Phys. B 193 (1980) 297.
[5] D.O. Caldwell et al., Phys. Rev. Lett. 54 (1985) 281.
[6] V. Barger, W. Keung and S. Pakvasa, Phys. Rev. D 25 (1982) 907.
[7] D.K. Nadyozhin, Astrophys. Space Sci. 49 (1977) 399; 51 (1977) 283; 53 (1978) 131;
R.L. Bowers and J.R. Wilson, Astrophys. J. 263 (1982) 366;
S.W. Bruenn, Astrophys. J. Suppl. 58 (1985) 771.
[8] L.N. Alekseeva, in: Chastizi i Kosmologiya, I.Ya.I.AN SSSR (Moscow, 1984) p. 58.

NEUTRINO MIXING, DECAYS AND SUPERNOVA 1987A ☆

Joshua A. FRIEMAN
Stanford Linear Accelerator Center, Stanford University, Stanford, CA 94305, USA

Howard E. HABER
Santa Cruz Institute for Particle Physics, University of California, Santa Cruz, CA 95064, USA

and

Katherine FREESE
Institute for Theoretical Physics, University of California, Santa Barbara, CA 93106, USA

Received 4 May 1987; revised manuscript received 1 October 1987

We discuss the evolution of neutrino beams in the presence of both flavor mixing and decays, and give a careful discussion of the implications of the supernova 1987A observations for neutrino mixing angles and lifetimes. Although the observation of electron antineutrinos from SN 1987A naively implies a lower bound on the electron neutrino lifetime $\tau_{\nu_e} > 5.7\times10^5(m_{\nu_e}/\mathrm{eV})$ s, neutrino mixing must be taken into account when using this to place constraints on particle physics models with neutrino decay. In models with large mixing angles, a short lifetime for the neutrino with large electron neutrino component cannot be ruled out.

The idea of fast neutrino decays has been proposed in the context of extended majoron models [1–3], and it has been suggested by Bahcall and collaborators [4–6] that electron neutrino decay in flight from the sun may explain the solar neutrino problem. Our aim here is to investigate the constraints placed on particle physics models with fast neutrino decay by the SN 1987A observations [7,8]. To do so, we generalize the standard neutrino mixing formalism to include the non-radiative decay of the heavy mass eigenstate, and study the combined effects of mixing and decay on the electron antineutrino signal of supernovae.

Naively, the observation of electron antineutrinos of energy as low as $\simeq 8$ MeV from SN 1987A, with a count rate within a factor 3 of that expected gives a lower bound on the electron neutrino lifetime, $\tau_{\nu_e} > 5.7\times10^5(m_{\nu_e}/\mathrm{eV})$ s. For electron neutrino decay to a massless neutrino ν_x and massless pseudoscalar o at tree level, with decay rate $\Gamma = g^2_{\nu_e\nu_x o} m^2_{\nu_e}/16\pi E_{\nu_e}$ in the lab frame, the lower bound on the electron neutrino lifetime from the supernova corresponds to an upper limit on the coupling constant, $g_{\nu_e\nu_x o} < 2.4\times10^{-10}\ (m_{\nu_e}/\mathrm{eV})^{-1}$. However, this argument neglects the importance of neutrino flavor mixing, which is generally present in models in which neutrino decay occurs. It turns out that, as one might expect, in the case of relatively large mixing angles, neutrino decay does not drastically reduce the electron antineutrino detection rate from SN 1987A, and is consistent with expectations from supernova theory.

We first consider the electron antineutrino signal of supernovae in the absence of neutrino mixing and decay. The energetics of the collapse, and, therefore, the gross features of the neutrino emission from type II supernovae are thought to be insensitive to the details of theoretical collapse models. Essentially, the binding energy of the resultant neutron star or black hole, of order a few times 10^{53} erg, must be released

☆ Work supported by the US Department of Energy and the National Science Foundation, supplemented by funds from the National Aeronautics and Space Administration.

Neutrino oscillations in matter

L. Wolfenstein
Carnegie-Mellon University, Pittsburgh, Pennsylvania 15213
(Received 6 October 1977; revised manuscript received 5 December 1977)

The effect of coherent forward scattering must be taken into account when considering the oscillations of neutrinos traveling through matter. In particular, for the case of massless neutrinos for which vacuum oscillations cannot occur, oscillations can occur in matter if the neutral current has an off-diagonal piece connecting different neutrino types. Applications discussed are solar neutrinos and a proposed experiment involving transmission of neutrinos through 1000 km of rock.

I. INTRODUCTION

There exists considerable interest in the possibility that one type of neutrino may transform into another type while propagating through the vacuum.[1] A number of experiments have been proposed to search for such oscillations.[2,3] In order for such *vacuum oscillations* to occur, it is necessary that at least one neutrino have a nonzero mass and that the neutrino masses be not all degenerate. In addition, there must be a nonconservation of the separate lepton numbers (like electron number and muon number) so that the different neutrino types as defined by the weak charged current are mixtures of the mass eigenstates.

In this paper we show that even if all neutrinos are massless it is possible to have oscillations occur when neutrinos pass through matter. This can happen as a result of coherent forward scattering provided that this scattering is partially off diagonal in neutrino type. The phenomenon is analogous to the regeneration of K_S from a K_L beam passing through matter, A simple model is given in Sec. II from which it is seen that the oscillation length in matter of normal density is of the order 10^9 cm or larger. One of the proposed experiments[2] to test the hypothesis of vacuum oscillations involves the detection of neutrinos 1000 km distant from their source at Fermilab. Since the neutrinos would pass through the earth, we show that this experiment could possibly also test the hypothesis that the neutral current changes neutrinos from one type to another.

If neutrinos have a mass and vacuum oscillations do occur, these oscillations may be modified when neutrinos pass through matter. For the case of electron-type neutrinos ν_e such a modification takes place even if the neutrino scattering is described by a standard theory. The effect of the medium in this case, discussed in Sec. III, arises from the coherent forward scattering of ν_e as a result of its charged-current interaction with electrons. It is shown that the modification is large only if the vacuum oscillation length is larger than 10^9 cm.

Our results are summarized in Sec. IV and their significance with respect to gauge models of current interest is discussed.

II. OSCILLATIONS FOR MASSLESS NEUTRINOS

We consider here a simple model involving two types of massless neutrinos for which the effective neutral-current Hamiltonian is

$$H_w = \frac{G}{\sqrt{2}} L_\lambda J_\lambda , \tag{1}$$

$$L_\lambda = \cos^2\alpha[\bar{\nu}_a\gamma_\lambda(1+\gamma_5)\nu_a + \bar{\nu}_b\gamma_\lambda(1+\gamma_5)\nu_b] + \sin^2\alpha[\bar{\nu}_a\gamma_\lambda(1+\gamma_5)\nu_b + \bar{\nu}_b\gamma_\lambda(1+\gamma_5)\nu_a] , \tag{2a}$$

$$J_\lambda = g_p\bar{p}\gamma_\lambda p + g_n\bar{n}\gamma_\lambda n + \bar{g}_e\bar{e}\gamma_\lambda e + \cdots , \tag{2b}$$

where the ellipses represent terms in J_λ of no interest for our present considerations. The essential term in H_w is the off-diagonal term proportional to $\sin^2\alpha$, where ν_a and ν_b are neutrino types defined by the charged-current interaction. We have also assumed $\nu_a - \nu_b$ symmetry, which simplifies the discussion and maximizes the effect of interest.

The neutrino current may be rewritten

$$L_\lambda = \cos^2\alpha(\bar{\nu}_1\nu_1 + \bar{\nu}_2\nu_2) + \sin^2\alpha(\bar{\nu}_1\nu_1 - \bar{\nu}_2\nu_2) , \tag{3}$$

$$|\nu_1\rangle = (|\nu_a\rangle + |\nu_b\rangle)/\sqrt{2} ,$$
$$|\nu_2\rangle = (|\nu_a\rangle - |\nu_b\rangle)/\sqrt{2} , \tag{4}$$

where we have omitted the Lorentz indices. A beam originally ν_a propagating through matter is described by

$$|\nu_a(x)\rangle = (|\nu_1\rangle e^{ikn_1x} + |\nu_2\rangle e^{ikn_2x})/\sqrt{2} , \tag{5}$$

where the indices of refraction,

$$n_i = 1 + (2\pi N/k^2) f_i(0) , \tag{6}$$

are different because of the $\sin^2\alpha$ term. Because the interaction is weak, for practical purposes we can consider $f_i(0)$ as real; even though the absorption is negligible, the real part of the index of refraction may yield interesting effects. The probability that ν_a emerges as ν_a (that is, not transformed to ν_b) is

$$|\langle\nu_a|\nu_a(x)\rangle|^2 = \tfrac{1}{2}[1+\cos k(n_1-n_2)x]. \qquad (7)$$

A direct calculation gives the simple result

$$k(n_1-n_2) = 2G\sin^2\alpha(g_pN_p+g_nN_n+g_eN_e), \qquad (8)$$

where N_p is the number of protons per unit volume, etc. Writing $N_i=\rho_iN_0$, where $N_0=6\times10^{23}$ cm^{-3},

$$k(n_1-n_2)\approx 5\times10^{-9}\text{ cm}^{-1}[\sin^2\alpha(g_p\rho_p+g_n\rho_n+g_e\rho_e)]. \qquad (9)$$

The effective oscillation length l, defined by $|k(n_1-n_2)l|=2\pi$, has a minimum value of the order 10^9 cm, provided ρ_ig_i is of the order unity.

In contrast to the case of vacuum oscillations this length is independent of neutrino energy. This can be seen from the relation

$$|k(n_1-n_2)|\propto\left|\frac{f(0)}{k}\right| = \left|\frac{d\sigma/d(\cos\theta)(0)}{2\pi k^2}\right|^{1/2} = \left|\frac{d\sigma/dq^2(0)}{2\pi}\right|^{1/2}.$$

For neutrino elastic scattering it is well known that $d\sigma/dq^2$ is independent of energy for $q^2=0$.

As a specific example, consider $\nu_a=\nu_\mu$ and ν_b as some new neutrino such as the postulated[4] ν_τ. The extreme case of $\sin^2\alpha=1$ corresponds to the idea that in neutral-current scattering the outgoing neutrino is never the same as the incoming one. In our previous discussion[5] it seemed impossible to check this idea unless the hadronic neutral current had a special non-Hermitian character. Here we see that a proposed experiment[2] looking for ν_μ 1000 km distant from their origin could be considered a test of this idea. Assuming a density of 4 g/cm^3 ($\rho_n\approx\rho_p=\rho_e=2$) for the rock through which the neutrinos pass, we have from Eqs. (7) and (9), with $\sin^2\alpha=1$ and $x=10^8$ cm,

$$|\langle\nu_\mu|\nu_\mu(x)\rangle|^2=\tfrac{1}{2}[1+\cos(g_p+g_n+g_e)],$$
$$|\langle\nu_\tau|\nu_\mu(x)\rangle|^2=\tfrac{1}{2}[1-\cos(g_p+g_n+g_e)]. \qquad (10)$$

If $(g_p+g_n+g_e)$ is less than unity we have approximately

$$|\langle\nu_\tau|\nu_\mu(x)\rangle|^2\approx\tfrac{1}{4}(g_p+g_n+g_e)^2. \qquad (11)$$

TABLE I. Transformation probabilities $P=|\langle\nu_b|\nu_\mu(x)\rangle|^2$ for ν_μ passing through 1000 km of rock for various choices of neutral currents with $\sin^2\alpha=1$. Within accuracy shown results hold for $\nu_b=\nu_\tau$ [Eq. (10)] or $\nu_b=\nu_e$ [Eqs. (18), (15), (16)]. A, B stand for solutions in Ref. 6 for g_p and g_n; WS stands for Weinberg-Salam values. For the same cases the transmission probability f for neutrinos from the sun is shown.

g_p, g_n	g_e	$g_p+g_n+g_e$	P	f
A	0.45	−0.6	0.09	0.98
A	−0.45	−1.5	0.46	0.61
B	0.45	0.8	0.15	0.60
B	−0.45	−0.1	0.002	0.97
WS	WS	−0.5	0.06	0.97

The quantities g_p, g_n, and g_e must be determined from neutral-current scattering experiments. In Table I we give results for the transformation probability given by Eq. (10) using values of g_p and g_n derived[6] from ν_μ scattering experiments and setting $|g_e|=0.45$, a value in the middle of the allowed range of values.[7] We also show results when $g_p+g_n+g_e$ has the value it has in the Weinberg-Salam theory, a value which is independent of $\sin^2\theta_W$. Transformation probabilities anywhere between 0 and 50% are seen to be possible. If such transformations were found, the model discussed here could be distinguished from the original idea of vacuum oscillations by a study of the dependence on neutrino energy as noted above.

We now discuss the modification required when one of the neutrinos (ν_a) is ν_e. In this case there is a contribution to the coherent scattering from the charged-current term[8]

$$\frac{G}{\sqrt{2}}\bar{\nu}_e\gamma_\lambda(1+\gamma_5)e\bar{e}\gamma_\lambda(1+\gamma_5)\nu_e = \frac{G}{\sqrt{2}}\bar{\nu}_e\gamma_\lambda(1+\gamma_5)\nu_e(\bar{e}\gamma_\lambda e+\cdots), \qquad (12)$$

where the second line is obtained via the Fierz transformation, and we have left out the axial-vector electron current which is of no interest here. The Hamiltonian contributing to coherent scattering can now be written

$$H=G/\sqrt{2}\{\tfrac{1}{2}[\bar{\nu}_e\gamma_\lambda(1+\gamma_5)\nu_e - \bar{\nu}_b\gamma_\lambda(1+\gamma_5)\nu_b]\bar{e}\gamma_\lambda e + \sin^2\alpha[\bar{\nu}_e\gamma_\lambda(1+\gamma_5)\nu_b+\bar{\nu}_b\gamma_\lambda(1+\gamma_5)\nu_e](g_p\bar{p}\gamma_\lambda p+g_n\bar{n}\gamma_\lambda n+g_e\bar{e}\gamma_\lambda e)+\cdots\}, \qquad (13)$$

where the ellipses include diagonal terms symmetric under the interchange of ν_e and ν_b which do not contribute to the difference $n_e - n_b$. The eigenstates for propagation through matter are no longer given by Eq. (3) but are

$$|\nu_1\rangle = \cos\theta|\nu_e\rangle + \sin\theta|\nu_b\rangle ,$$
$$|\nu_2\rangle = -\sin\theta|\nu_e\rangle + \cos\theta|\nu_b\rangle , \qquad (14)$$

$$\tan 2\theta = 2\sin^2\alpha(g_p\rho_p + g_n\rho_n + g_e\rho_e)/\rho_e . \qquad (15)$$

Equation (9) which determines the oscillation length l is also modified,

$$2\pi/l = k(n_1 - n_2)$$
$$= 5\times 10^{-9}\ \text{cm}^{-1}\rho_e\left(\frac{1+\tan^2 2\theta}{4}\right)^{1/2} . \qquad (16)$$

Thus for matter of normal density ($\rho_e = 1$) there is a maximum oscillation length of about 2.5×10^9 cm. In place of Eq. (7) the oscillation probability is determined by

$$|\langle\nu_e|\nu_e(x)\rangle|^2 = \cos^4\theta + \sin^4\theta$$
$$+ 2\cos^2\theta\sin^2\theta\cos[k(n_1 - n_2)x] . \qquad (17)$$

For the application discussed above involving ν_μ going through 1000 km of earth with now ν_e taking the role of ν_τ, the quantitative results are changed very little. Starting with an equation for ν_μ such as Eq. (17) we have for the transformation probability

$$|\langle\nu_e|\nu_\mu(x)\rangle|^2 = \tfrac{1}{2}\sin^2 2\theta\{1 - \cos[k(n_1 - n_2)x]\} . \qquad (18)$$

If the argument of the cosine is less than unity, we find, using Eqs. (15) and (16) and setting $\sin^2\alpha = 1$, $x = 10^8$ cm, $\rho_e = \rho_p = \rho_n = 2$,

$$|\langle\nu_e|\nu_\mu(x)\rangle|^2 \approx \tfrac{1}{4}(g_p + g_n + g_e)^2 ,$$

which is the same result as Eq. (11). The effects of the decrease in the oscillation length and the decrease in the amplitude of oscillation approximately cancel each other as long as the transformation probability is not too large. For the cases discussed before shown in Table I, the results from Eq. (18) agree with those of Eq. (10) within 2%. On the other hand, if we consider the minimum in the oscillation, where $[k(n_1 - n_2)x] = \pi$, we find from Eq. (17) that this minimum is $\cos^2 2\theta$ in contrast to Eq. (7) which has a minimum of zero. From Eq. (15), setting $\rho_p = \rho_n = \rho_e$ and $\sin^2\alpha = 1$, this minimum value is $[1 + 4(g_p + g_n + g_e)^2]^{-1}$.

A very interesting use of the vacuum oscillation theory is for the problem of the apparent deficiency of solar neutrinos in terrestrial experiments.[9] According to that theory, ν_e may oscillate into one or more other neutrino types during the trip from sun to earth. Here we consider the possibility that ν_e may transform into ν_b while passing from the interior of the sun to the surface as a result of the coherent off-diagonal scattering. Since the oscillation length is less than $2.5\rho_e^{-1}\times 10^9$ cm from Eq. (16), the oscillating term in Eq. (17) will average to zero as a result of averaging over the radial position of the neutrino source. Thus the fraction of ν_e emerging from the solar surfaces is

$$f = \cos^4\theta + \sin^4\theta = \tfrac{1}{2}[1 + (1 + \tan^2 2\theta)^{-1}] . \qquad (19a)$$

From Eq. (15)

$$\tan 2\theta = +2\sin^2\alpha(g_p + g_e + y g_n) , \qquad (19b)$$

where y is the ratio of neutrons to protons. The value of y changes from a value of about 0.41 at the solar center to about 0.14 at the solar surface as X, the hydrogen mass fraction, changes from 0.42 to 0.75. Using an appropriate average value of y and assuming the extreme case discussed above with $\sin^2\alpha = 1$, we have calculated the values of f shown in Table I for various possible neutral-current couplings. The most extreme case gives a reduction of about 40% in ν_e flux. This is in contrast to the vacuum oscillation theory which gives a reduction of 50% for a large range of parameters and an even larger reduction for a very special choice of parameter.[10] In the case of the vacuum oscillation theory a reduction well below 50% is also possible by the introduction of additional types of neutrinos[11]; in our model this does not give much further reduction since the minimum is determined by the ratio of the effective neutral off-diagonal current to the charged current. A further reduction could be obtained if the diagonal neutral current were not symmetric with respect to ν_e and ν_b, but this alternative requires particularly artificial assumptions.

So far we have assumed the neutrinos defined by the charged current are orthogonal to each other. However, in many models involving violation of lepton number there also occurs a nonorthogonality among the neutrinos associated with different charged leptons.[12] To be specific, we label as ν_e the neutrino associated with the electron and let

$$|\nu_\mu'\rangle = \cos\phi|\nu_b\rangle + \sin\phi|\nu_e\rangle$$

be the neutrino associated with the muon. Here ν_b and ν_e are the basic orthogonal neutrinos so that

$$\langle\nu_e|\nu_\mu'\rangle = \sin\phi .$$

Assuming[13] the Hamiltonian (13) we can express ν_μ' in terms of the eigenstates (14) for propagation in matter as

$$|\nu'_\mu\rangle = \cos(\theta+\phi)|\nu_2\rangle + \sin(\theta+\phi)|\nu_1\rangle .$$

The oscillations of ν'_μ are determined by

$$|\langle\nu'_\mu|\nu'_\mu(x)\rangle|^2 = \cos^4(\theta+\phi)+\sin^4(\theta+\phi)+2\cos^2(\theta+\phi)\sin^2(\theta+\phi)\cos[k(n_1-n_2)x] . \quad (20)$$

It is interesting to note that oscillations occur even in the absence of any nondiagonal neutral current, that is, for the case $\sin^2\alpha = 0$ which gives, from Eq. (15), $\theta = 0$. The minimum in the transmission in this case is given by $(1-\sin^2 2\phi)$; since ϕ^2 is limited to about 0.01 by the limits on the nonorthogonality of ν_e and ν'_μ, the oscillation amplitude must be very small. From Eq. (16) the minimum occurs at a distance of about 5×10^8 cm of terrestrial matter. Another way to look at this oscillation (the case with $\theta = 0$) is to consider it as resulting from nondiagonal charged-current scattering. To see this we choose $|\nu'_\mu\rangle$ as one of the basic vectors, in which case $|\nu_e\rangle$ must be written as

$$|\nu_e\rangle = \sin\phi|\nu'_\mu\rangle + \cos\phi|\nu_c\rangle ,$$

where $|\nu_c\rangle$ is the other basic vector. The charged-current coupling of ν_e to electrons now includes a term which transforms ν'_μ to ν_c, which is proportional to $\sin 2\phi$.

III. MODIFICATION OF VACUUM OSCILLATIONS IN MATTER

In this section we shall adopt the assumptions of the vacuum-oscillation theory and consider how these oscillations appear when neutrinos pass through matter. We further assume that the neutral current has the same general form as in the standard model; this means it is diagonal and symmetric with respect to neutrino types. Considering just two types of neutrinos, it is then given by Eqs. (1) and (2) with $\sin^2\alpha = 0$.

With these assumptions the only interesting case is that in which one of the neutrinos is ν_e. Considering the case of ν_e and ν_μ, vacuum oscillations require that the eigenstates in vacuum are mixtures

$$|\nu_1\rangle = |\nu_e\rangle\cos\theta_v - |\nu_\mu\rangle\sin\theta_v ,$$
$$|\nu_2\rangle = |\nu_e\rangle\sin\theta_v + |\nu_\mu\rangle\cos\theta_v , \quad (21)$$

with distinct masses m_1 and m_2 $(m_1 > m_2)$. Neutrino oscillations result from the difference in the phase factors governing the time dependence of ν_1 and ν_2,

$$|\nu_i t\rangle \sim \exp(-itm_i^2/2k) .$$

The characteristic oscillation length in the vacuum is $l_v(k) = 4\pi k/(m_1^2 - m_2^2)$. For propagation through matter we must also consider the phase factors arising from coherent scattering. Since the neutral-current scattering is diagonal and symmetric, it just causes an overall phase shift of no physical importance. However, we must include the charged-current scattering which singles out ν_e. As a result we have in the ν_1-ν_2 representation, omitting the neutral current,

$$i\frac{d}{dt}\begin{bmatrix}\nu_1\\ \nu_2\end{bmatrix} = \begin{bmatrix}\frac{m_1^2}{2k} - GN_e\cos^2\theta_v & -GN_e\sin\theta_v\cos\theta_v \\ -GN_e\sin\theta_v\cos\theta_v & \frac{m_2^2}{2k} - GN_e\sin^2\theta_v\end{bmatrix} \times \begin{bmatrix}\nu_1\\ \nu_2\end{bmatrix}, \quad (22)$$

where $GN_e = k(n_e - 1)$ and n_e is the index of refraction associated with the charged-current scattering. The eigenstates for propagation in matter are

$$|\nu_{1m}\rangle = |\nu_e\rangle\cos\theta_m - |\nu_\mu\rangle\sin\theta_m ,$$
$$|\nu_{2m}\rangle = |\nu_e\rangle\sin\theta_m + |\nu_\mu\rangle\cos\theta_m ,$$
$$\tan 2\theta_m = \tan 2\theta_v\left(1 - \frac{l_v}{l_0}\sec 2\theta_v\right)^{-1} , \quad (23a)$$
$$l_0 \equiv 2\pi/GN_e = 2.5\times10^9\ \mathrm{cm}/\rho_e . \quad (23b)$$

The oscillation length in matter is

$$l_m(k) = l_v(k)\left[1 + \left(\frac{l_v(k)}{l_0}\right)^2 - 2\cos 2\theta_v\left(\frac{l_v(k)}{l_0}\right)\right]^{-1/2} , \quad (24a)$$

and the transformation probability is given by

$$|\langle\nu_e|\nu_\mu(x)\rangle|^2 = \tfrac{1}{2}\sin^2(2\theta_v)(l_m/l_v)^2 \times [1 - \cos(2\pi x/l_m)] . \quad (24b)$$

As long as $l_v \ll l_0$, it is seen from Eqs. (23) and (24) that $l_m \approx l_v$, $\theta_m \approx \theta_v$, and therefore the oscillations in the medium will be very much the same as in the vacuum. For $l_v \gg l_0$ it is seen that $l_m \approx l_0$ independent of θ_v and therefore from Eq. (24b) the amplitude of the oscillation is very small. Some examples of the effect of the medium for the intermediate case $l_v = l_0$ are illustrated in Table II. Independent of the value of l_v/l_0, it follows from Eq. (24b) that as long as $(2\pi x/l_m) < 1$, the oscilla-

TABLE II. Transformation probabilities $|\langle\nu_e|\nu_\mu(x)\rangle|^2$ for a vacuum oscillation length $l_v = l_0 = (2.5\times10^9\ \mathrm{cm})\rho_e^{-1}$ for three values of the vacuum mixing angle θ_v. Results are shown for vacuum oscillations (vac) and oscillations for neutrinos passing through matter (mat) with an electron number density of $6\times10^{23}\rho_e\ \mathrm{cm}^{-3}$.

	$\theta_v=45°$		$\theta_v=60°$		$\theta_v=15°$	
x/l_0	vac	mat	vac	mat	vac	mat
0.1	0.095	0.093	0.072	0.067	0.024	0.025
0.2	0.345	0.301	0.259	0.196	0.086	0.095
0.3	0.655	0.472	0.491	0.249	0.164	0.205
0.4	0.905	0.479	0.679	0.169	0.226	0.342
0.5	1.000	0.317	0.750	0.041	0.250	0.492

tion probability in the medium is approximately the same as in vacuum.

IV. DISCUSSION AND SUMMARY

This paper has demonstrated the possible importance of coherent forward scattering when neutrinos pass through matter. It follows from our discussion that for matter of normal density the coherent scattering produces a phase change of the order of π after neutrinos have traversed a distance of the order of 10^9 cm of normal matter. This distance, which is independent of energy, is much less than the mean free path, which is of the order 10^{14} cm for a 1-GeV neutrino and varies inversely as the energy. The reason for this large difference, of course, is that the coherent scattering effect depends on the scattering amplitude proportional to G whereas the mean free path depends on the cross section proportional to G^2.

In Sec. II we considered a test for the extreme hypothesis that the neutral current always changed ν_μ to another type of neutrino. In this case a significant fraction of ν_μ may be transformed to the other type of neutrino after passing through 1000 km or more of terrestrial rock. The quantitative results depend on the detailed form of the neutral current and in general the fraction is somewhat smaller when the other neutrino is ν_e. When this extreme hypothesis is applied to the passage of neutrinos from the center of the sun, it is found that up to 40% of these ν_e may have transformed to another type of neutrino on arrival at the surface of the sun. At best this extreme hypothesis would provide only a partial answer to the deficiency of solar neutrinos.

Nondiagonal neutral currents of the form assumed in Eq. (2) occur in many gauge models that extend the standard Weinberg-Salam model to heavy leptons and right-handed currents. Indeed, it has been shown on general grounds[14] that only a very special class of gauge theories can avoid having flavor-changing neutral currents; the nondiagonal neutral current in Eq. (2) is an example of a current which changes "lepton flavor." However, in any realistic model we can think of, the value of the parameter $\sin^2\alpha$ in Eq. (2) would be much less than unity (of the order 0.1 or less). This follows from the empirical constraints of hadron-lepton universality and limits on lepton-number-nonconserving processes. The lower value of $\sin^2\alpha$ has two important phenomenological consequences relative to the extreme hypothesis discussed above:

(1) The characteristic oscillation length in normal matter (when ν_e is not involved) determined from Eq. (9) now has a minimum value of 10^{10} cm. Thus, the effect of these oscillations on an experiment using a path length of 1000 km of terrestrial rock will be negligible.

(2) The amplitude of oscillations when ν_e is involved as determined from Eqs. (15) and (17) will be small. Thus the effect of the oscillations on terrestrial experiments with a path length of 1000 km or on solar neutrinos will be negligible.

An example of a gauge model which has such a term is the model with right-handed charged leptons coupled to heavy neutral leptons.[15] In this model the value of $\sin^2\alpha$ for ν_i coupling to ν_j is $m_i m_j/m^2$ where m_i is the mass of the charged lepton associated with ν_i and m is some mean mass of the neutrals. Thus for ν_μ coupled to ν_τ, $\sin^2\alpha$ would be of the order $m_\mu m_\tau/m^2$, which is less than or of the order of m_μ/m_τ. In this model the different diagonal terms have different coefficients; thus even for the case of ν_μ and ν_τ the mixing is not complete and the eigenstates in matter are given by Eq. (12) with a small value of θ instead of by Eq. (3). Thus, in this model, as in other realistic gauge models, the phenomenological consequences of the nondiagonal coherent scattering appear to be very hard to detect.

In Sec. III we assumed that vacuum oscillations occurred. This means that the eigenstates of the neutrino mass matrix are mixtures of different neutrino types as defined by the charged current and that the eigenvalues are nondegenerate. On the other hand, we assumed that the neutral current was diagonal or that the nondiagonal pieces were negligible. With these assumptions the oscillation phenomenon in matter differs from that in vacuum when ν_e is involved because the charged-current ν_e-electron scattering gives ν_e a different index of refraction from other neutrinos. The qualitative conclusions are:

(1) For characteristic vacuum oscillation lengths much smaller than 10^9 cm, oscillations in normal matter are essentially the same as in vacuum.

(2) For characteristic vacuum oscillation lengths

much larger than 10^9 cm. oscillations involving ν_e in normal matter have a much smaller amplitude than in vacuum.

(3) For characteristic vacuum oscillation lengths of the order of 10^9 cm. the quantitative results in matter are quite different from in vacuum as illustrated in Table II.

In general, if one is considering the possibility of large vacuum oscillation lengths, as in the discussion of solar neutrinos, the oscillations should be calculated for the actual vacuum path[16] ignoring the passage through matter. Thus, in the detailed solar neutrino calculations[10] the effective distance over which neutrino oscillations take place is from the solar surface to the earth's surface; there are no significant oscillations inside the sun or in traversals through the earth.

ACKNOWLEDGMENTS

I wish to thank E. Zavattini for asking the right question, and J. Ashkin, J. Russ, J. F. Donoghue, L. F. Li, S. Adler, and D. Wyler for discussions. This research was supported in part by the U. S Energy Research and Development Administration.

---

[1]B. Pontecorvo, Zh. Eksp. Teor. Fiz. 53, 1771 (1967) [Sov. Phys.—JETP 26, 984 (1968)]; V. Gribov and B. Pontecorvo, Phys. Lett. 28B, 493 (1969).

[2]A. K. Mann and H. Primakoff, Phys. Rev. D 15, 655 (1977).

[3]H. W. Sobel *et al.*, in *Proceedings of the International Neutrino Conference, Aachen, 1976*, edited by H. Faissner, H. Reithler, and P. Zerwas (Vieweg, Braunschweig, West Germany, 1977), p. 678; L. Sulak (private communication); F. Boehm (private communication).

[4]For a discussion of the existence of ν_τ within the context of gauge models, see J. F. Donoghue and L. Wolfenstein, Phys. Rev. D 17, 224 (1978) and references therein.

[5]L. Wolfenstein, Nucl. Phys. B91, 95 (1975).

[6]P. Q. Hung and J. J. Sakurai, Phys. Lett. 72B, 208 (1977). In their notation $g_p = \frac{1}{2}(3\gamma + \alpha)$ and $g_n = \frac{1}{2}(3\gamma - \alpha)$. For illustration we have used the central values of their solutions A and B.

[7]See, for example, Fig. 1 in J. Blietschau *et al.*, Phys. Lett. 73B, 232 (1978), which shows that $|g_e|$ is restricted to lie below 0.7 from the limited data on $\nu_\mu e$ and $\bar{\nu}_\mu e$ scattering.

[8]I am indebted to Dr. Daniel Wyler for pointing out the importance of the charged-current terms.

[9]J. N. Bahcall and R. Davis, Science 191, 264 (1976) and references therein.

[10]J. N. Bahcall and S. C. Frautschi, Phys. Lett. 29B, 623 (1969).

[11]B. Pontecorvo, Zh. Eksp. Teor. Fiz. Pis'ma Red. 13, 281 (1971) [JETP Lett. 13, 199 (1971)].

[12]See, for example, B. W. Lee *et al.*, Phys. Rev. Lett. 38, 937 (1977); H. Fritzsch, Phys. Lett. 67B, 451 (1977).

[13]The exact symmetry between ν_e and ν_b may no longer seem natural in this case, but a small asymmetry is not important for our results.

[14]S. L. Glashow and S. Weinberg, Phys. Rev. D 15, 1958 (1977).

[15]T. P. Cheng and L.-F. Li, Phys. Rev. D 16, 1425 (1977); J. D. Bjorken, K. Lane, and S. Weinberg, *ibid.* 16, 1474 (1977); S. Treiman, F. Wilczek, and A. Zee, *ibid.* 16, 152 (1977).

[16]Clearly the vacuum is defined by $l_0 \gg l_v$ or $\rho_e \ll 2.5 \times 10^8$ cm/l_v from Eq. (24b).

Matter effects on three-neutrino oscillations

V. Barger and K. Whisnant
Physics Department, University of Wisconsin, Madison, Wisconsin 53706

S. Pakvasa
Physics Department, University of Hawaii at Manoa, Honolulu, Hawaii 96822

R. J. N. Phillips
Rutherford Laboratory, Chilton, Didcot, Oxon, England
(Received 4 August 1980)

We evaluate the influence of coherent forward scattering in matter upon neutrino oscillations in the three-neutrino picture. We write down the exact solution and also approximate first-order solutions that exhibit general features more transparently. Oscillation characteristics in matter that could be observed in deep-mine experiments are discussed and illustrated using an oscillation solution suggested by solar and reactor data.

I. INTRODUCTION

Interest in neutrino oscillations[1] has been heightened recently by indications from beam-dump experiments,[2-4] from reanalysis[5] of old reactor data,[6] and from a new reactor experiment with reduced sensitivity to spectrum uncertainties.[7] If such oscillations can be clearly resolved, they will not only show that neutrinos are massive but also provide information about their mass differences and mixing matrix.

Deep-mine experiments that measure high-energy events from atmospheric neutrinos offer unique opportunities to probe oscillations in the range $L/E \sim 1-10^5$ m/MeV (where L is the path length and E the energy) that is sensitive to mass-squared differences $\delta m^2 \gtrsim 10^{-5}$ eV^2. Wolfenstein[8] has pointed out, however, that the standard vacuum oscillations can be significantly modified by coherent forward scattering from electrons in matter (that selectively affect ν_e and $\bar{\nu}_e$ components) when the path integral of electron number density N_e is of order $\int N_e dL \sim 10^9 N_A$ cm^{-2} where $N_A = 6\times 10^{23}$. Thus with electron densities in typical terrestrial matter of order $N_e \sim 2N_A$ cm^{-3}, matter effects can occur over the distance of the earth's radius, i.e., in deep-mine events produced by upward neutrinos.

Wolfenstein has given a complete analytic solution for matter corrections to oscillations of two neutrinos.[8] We examine the properties of this solution in detail in Sec. II. We then derive a general solution for matter oscillations with any number of neutrinos, but its implications are not immediately transparent. We therefore also write down first-order approximate solutions that exhibit general properties rather simply. For the case of three neutrinos, we give an exact solution in closed form. In Sec. III we illustrate the properties of the three-neutrino oscillations in matter based on a vacuum-oscillation solution suggested by solar and reactor data.

II. OSCILLATIONS IN MATTER

A. General equations

Consider a set of neutrino charged-current eigenstates ν_α ($\alpha = e, \mu, \tau, \ldots$) and mass eigenstates ν_i ($i = 1, 2, 3, \ldots$) at time $t=0$, distinguished by their suffixes and related by a unitary transformation,

$$|\nu_\alpha\rangle = \sum_i U_{\alpha i} |\nu_i\rangle . \tag{1}$$

Then, for a relativistic neutrino beam energy E, we recall the standard amplitude A and probability P for $\nu_\alpha - \nu_\beta$ transitions after a time t *in vacuo*,

$$\begin{aligned} A(\nu_\alpha - \nu_\beta) &= \sum_i U_{\alpha i} \exp(-\tfrac{1}{2} i m_i^2 t/E) U^\dagger_{i\beta} , \\ P(\nu_\alpha - \nu_\beta) &= |A(\nu_\alpha - \nu_\beta)|^2 , \end{aligned} \tag{2}$$

where m_i are the mass eigenvalues and the interference terms in $|A|^2$ are oscillatory. We usually write L/E in place of t/E in Eq. (2), where L is the length of the flight path (units $\hbar = c = 1$).

To treat neutrino evolution in matter, we consider an arbitrary state vector in neutrino-flavor space,

$$|\psi(t)\rangle = \sum_i \psi_i(t) |\nu_i\rangle . \tag{3}$$

For an initial state ν_α at time $t=0$, $\psi_i(0) = U_{\alpha i}$, and the $\nu_\alpha - \nu_\beta$ transition amplitude is

$$A(\nu_\alpha - \nu_\beta) = \sum_i U^\dagger_{i\beta} \psi_i(t) . \tag{4}$$

The time evolution is controlled by the equation

$$i d\psi_j(t)/dt = m_j^2/(2E)\psi_j(t) - \sum_k \sqrt{2}\, G N_e U_{ej} U^\dagger_{ke} \psi_k(t) \equiv H_{kj}\psi_k(t), \tag{5}$$

where N_e is the number of electrons per unit volume and G is the weak coupling constant. In Eq. (5) we have dropped a common overall phase in the ψ_i which produces no observable effects; this includes the phase shift from neutral-current scattering which is the same for all neutrino flavors. The coefficient $\sqrt{2}GN_e$ in Eq. (5) differs from the value GN_e given in Ref. 8; this correction is confirmed by Ref. 9. This equation applies strictly for *neutrinos*; for *antineutrinos*, change the sign of the $\sqrt{2}GN_e$ term and substitute U^* for U.

We assume that the vacuum oscillations are already prescribed so that U and the m_i are known. The problem of propagation is therefore to diagonalize the matrix H defined in Eq. (5). Let us suppose that it can be diagonalized by proceeding to a new set of basis state $|\nu_i'\rangle = V^\dagger_{i\alpha}|\nu_\alpha\rangle = V^\dagger_{i\alpha}U_{\alpha j}|\nu_j\rangle$, where V is a unitary matrix. Then if the eigenvalues of H corresponding to these states $|\nu_i'\rangle$ are written as $M_i^2/2E$, the solution for a uniform medium is given in principle by Eq. (2) with m_i^2 and U replaced by M_i^2 and V, respectively.

There is another way to solve the propagation problem without explicitly constructing the matrix V. For n neutrinos, Eq. (5) has n independent solutions for the row vectors $\psi_j(t)$. We choose the set of solutions $\psi_j^{(i)}$ $(i=1,\ldots,n)$ that are pure mass eigenstates at time $t=0$:

$$\psi_j^{(i)}(t=0) = \delta_{ij}\,. \tag{6}$$

If these row vectors are assembled into an $n\times n$ matrix X according to

$$X_{ij}(t) = \psi_j^{(i)}(t)\,, \tag{7}$$

then X satisfies the matrix equation

$$i dX/dt = XH \tag{8}$$

with the boundary condition $X(t=0)=1$. An analytical solution to Eq. (8) is possible for constant N_e,

$$X(t) = \exp(-iHt)\,. \tag{9}$$

Row i of X describes the state that started as the mass eigenstate $|\nu_i\rangle$ at $t=0$; column j describes the amplitude that it has evolved at time t into mass eigenstate $|\nu_j\rangle$. In the presence of matter the transition amplitude of Eq. (2) is modified to

$$A(\nu_\alpha - \nu_\beta) = \sum_{ij} U_{\alpha i} X_{ij} U^\dagger_{j\beta}\,. \tag{10}$$

The resulting transition probability requires only knowledge of the original mixing matrix U and the mass-squared differences $\delta M^2_{ij} = M_i^2 - M_j^2$, $\delta m^2_{ij} = m_i^2 - m_j^2$ and $M_i^2 - m_j^2$. For computation, the matrix X can be rewritten via Lagrange's formula as

$$X = \sum_k \left[\prod_{j\neq k} \frac{(2EH - M_j^2 1)}{\delta M_{kj}^2}\right] \exp\left(-i\frac{M_k^2 L}{2E}\right). \tag{11}$$

B. Two-neutrino case

We take the charged-current (CC) eigenstates to be ν_e, ν_μ and parametrize the 2×2 matrices U, V that relate them to the mass eigenstates ν_i and matter eigenstates ν_i' by

$$U_{e1} = U_{\mu 2} = \cos\alpha, \quad U_{e2} = -U_{\mu 1} = \sin\alpha,$$
$$V_{e1} = V_{\mu 2} = \cos\alpha', \quad V_{e2} = -V_{\mu 1} = \sin\alpha'. \tag{12}$$

There is just one mass-squared difference $\delta m^2 = m_1^2 - m_2^2$. Vacuum oscillations of two neutrinos determine only $|\delta m^2|$ and $\sin^2 2\alpha$, leaving the sign of δm^2 and the quadrant of 2α unresolved. By convention we take $\alpha < 45°$. The ambiguity in the sign of δm^2 is resolved by the effects of matter.

It is convenient to define the oscillation length in vacuum l_V and a characteristic length l_M for matter effects by

$$l_V = 4\pi E/\delta m^2, \quad l_M = 2\pi/(\sqrt{2}GN_e)\,. \tag{13}$$

For *antineutrinos*, the sign of l_M is reversed. We note that l_V or l_M can have either sign. With E in MeV, δm^2 in eV^2, and N_e in cm^{-3}, the oscillation lengths in meters are given by

$$l_V = 2.48E/\delta m^2, \quad l_M = 1.77\times 10^7 N_A/N_e\,. \tag{14}$$

For terrestrial matter[10] $N_e \sim 2N_A$ cm^{-3} in the mantle ($3.5\times 10^6 < r < 6.4\times 10^7$ m) and $N_e \sim 5N_A$ cm^{-3} in the core ($r < 3.5\times 10^6$ m).

Following Ref. 8, the solution in the *neutrino case* is given by

$$\tan 2\alpha' = \sin 2\alpha/(\cos 2\alpha - l_V/l_M) \tag{15}$$

for the oscillation angle in matter, and

$$\delta M^2 = \delta m^2[1 - 2(l_V/l_M)\cos 2\alpha + (l_V/l_M)^2]^{1/2} \tag{16}$$

for the matter eigenmass-squared difference. The associated oscillation length in matter is

$$l = l_V[1 - 2(l_V/l_M)\cos 2\alpha + (l_V/l_M)^2]^{-1/2}\,. \tag{17}$$

The physical consequences are clear.

(i) If $|l_V| \ll |l_M|$, i.e., the vacuum-oscillation length is very short on the matter scale, then $\alpha' \simeq \alpha$, $\delta M^2 \simeq \delta m^2$, and matter corrections are negligible.

(ii) If $|l_V| \gg |l_M|$, then $\alpha' \simeq 0$ and the matter corrections damp out all oscillation effects.

(iii) At intermediate values $|l_V| \sim |l_M|$ the matter corrections are very significant and differ between neutrinos and antineutrinos. Moreover, matter effects resolve the vacuum-oscillation ambiguity in the sign of δm^2.

(iv) For matter corrections to be observable, the distance traversed in matter must also be an appreciable fraction of l_M. Hence matter corrections are very small in all terrestrial contexts, except when neutrinos traverse a substantial fraction of the earth's diameter and have energies

$$E\ (\text{MeV}) \gtrsim 10^5\,|\delta m^2\ (\text{eV}^2)|\,. \tag{18}$$

(v) For given δm^2 the vacuum-oscillation length l_V depends on E, whereas l_M does not; hence, there is always some energy range where matter effects are important.

(vi) There is always some energy where $l_V/l_M = \cos 2\alpha$ and hence $\alpha' = 45°$ for either ν or $\bar{\nu}$ depending on the sign of δm^2. Hence, there is always some energy where ν or $\bar{\nu}$ matter mixing is maximal. At this energy, the diagonal transition probability vanishes at a distance

$$L = \frac{l_M}{2}\cot 2\alpha\,. \tag{19}$$

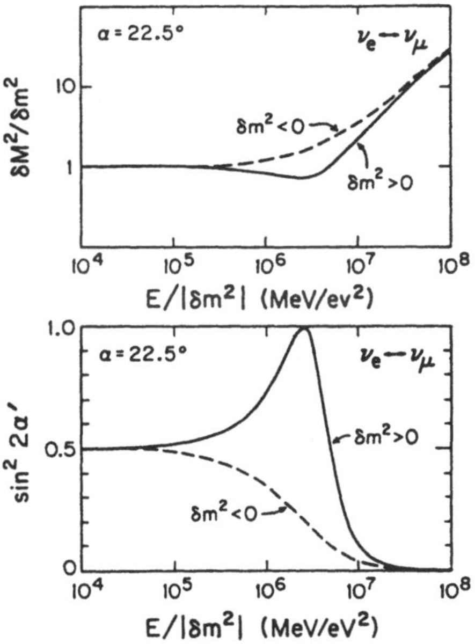

FIG. 1. Matter-to-vacuum eigenmass-squared difference ratio and matter amplitude $\sin^2 2\alpha'$ for oscillations of two neutrinos with vacuum amplitude $\sin^2 2\alpha = 0.5$ ($\alpha = 22.5°$).

With $\alpha = 22.5°$ and $N_e = 2N_A$ cm^{-3}, this distance is $L \simeq 5\times10^6$ m, which would correspond to deep-mine events about 10° below the horizontal direction.

Some of these results are illustrated in Fig. 1, showing the ratio $\delta M^2/\delta m^2$ (describing the correction to the oscillation wavelength) and $\sin^2 2\alpha'$ (describing the oscillation amplitude) versus $E/|\delta m^2|$. This illustration is based on $\alpha = 22.5°$ and $N_e = 2N_A$. As expected, there is little matter correction for E (MeV) $< 10^5\,|\delta m^2$ (eV^2)$|$. The mixing becomes maximal in one channel (ν or $\bar{\nu}$) at one energy. At sufficiently large energy where $l_V \gg l_M$ the mixing is damped out. When δm^2 changes sign, ν and $\bar{\nu}$ exchange roles. The transition probabilities are given simply by

$$\begin{aligned} P(\nu_e \to \nu_\mu) &= P(\nu_\mu \to \nu_e) = 1 - P(\nu_e \to \nu_e) \\ &= 1 - P(\nu_\mu \to \nu_\mu) = \sin^2(2\alpha')\sin^2(\tfrac{1}{4}\delta M^2 L/E)\,. \end{aligned} \tag{20}$$

Figure 2 compares vacuum- and matter-oscillation results for $P(e \to e)$ in the two-neutrino case at a fixed distance $L = 5\times10^6$ m.

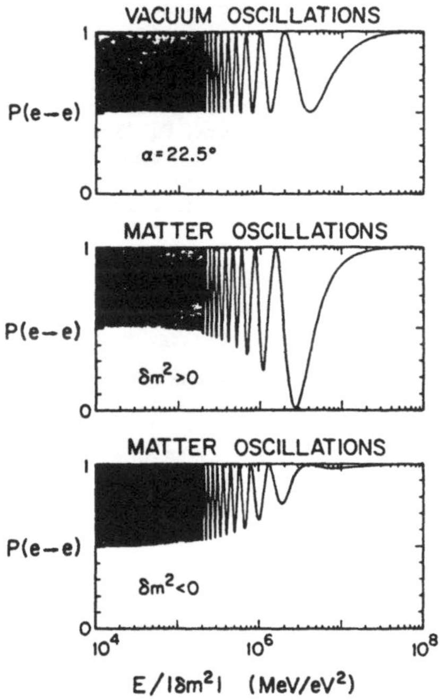

FIG. 2. Comparison of vacuum and matter transition probability $P(e \to e)$ for two neutrinos at fixed $L = 5\times10^6$ m for density $N_e = 2N_A$.

On the detection of cosmological neutrinos by coherent scattering

Paul Langacker
Department of Physics, University of Pennsylvania, Philadelphia, Pennsylvania 19174

Jacques P. Leveille*
Randall Laboratory, University of Michigan, Ann Arbor, Michigan 48109

Jon Sheiman†
Department of Physics, University of Pennsylvania, Philadelphia, Pennsylvania 19174
(Received 1 September 1982)

There have been several proposals that the sea of cosmological relic neutrinos could be detected by the energy, momentum, or angular momentum transferred during their coherent interaction with matter. We show that all but one of these proposals is incorrect and that the one exception leads to an effect that is probably immeasurably small. We first review the existing limits on the cosmological neutrinos and describe the expectations from the standard hot big-bang model. We then prove a general theorem that if the time average of the neutrino flux is spatially homogeneous (which is expected for cosmological neutrinos), then to first order in the weak coupling the energy and momentum transfer to any microscopic or macroscopic target is zero. Similarly, the angular momentum transfer is zero unless the target has a nonzero polarization or current density (but in this case the effect is probably immeasurably small). No assumption is made concerning the isotropy of the neutrino flux. Finally, we reexamine the individual proposals using the language of geometrical optics and show where all but one of them was incorrect. In particular, we prove that the momentum and angular momentum transferred by refraction from a homogeneous incoming beam to a (microscopically) isotropic target vanish to order $n-1$, with n the index of refraction.

I. INTRODUCTION

A. Cosmological neutrinos

The standard hot big-bang cosmological model[1] is now widely accepted theoretically. In addition to explaining the observed expansion of the universe, it successfully predicts the abundances of primordial helium and deuterium,[2] explains the observed background microwave radiation (which is characterized by a blackbody temperature $T_\gamma \simeq 2.7\,^\circ\mathrm{K}$, and a corresponding number density $n_\gamma \simeq 400/\mathrm{cm}^3$), and—when combined with grand unified theories[3]—may give a dynamical explanation of the small baryon asymmetry $(N_B - N_{\bar{B}})/N_\gamma \simeq N_B/N_\gamma \simeq 10^{-10\pm 1}$ observed in the present universe.

The big-bang model also predicts the existence of a sea of cosmological relic neutrinos, analogous to the sea of microwave photons, left over from the very early universe. The idea is that soon after the big bang the neutrinos would have been kept in thermodynamic equilibrium by charged- and neutral-current scattering processes. When the temperature fell to the decoupling temperature $T_D \simeq 1$ MeV (at a time $t_\nu \simeq 1$ sec after the big bang), however, the weak-reaction rates would have become slow compared to the expansion rate of the universe and the neutrinos would have essentially decoupled from all other particles (and from each other). Except for a red-shifting of their momenta due to the expansion of the universe, these neutrinos would have remained undisturbed until the present time. If these neutrinos could be detected they would provide a direct experimental probe of the period 1 sec after the big bang (by way of contrast, the microwave photons provide information on the much later period $T \sim 4000\,^\circ\mathrm{K}$, $t \sim 4\times 10^5$ yr, when the photons decoupled after electrons and protons recombined into neutral hydrogen). Unfortunately, our conclusions in this paper are that the relic neutrinos predicted by the standard model are essentially impossible to detect directly.

Let us now consider the expected properties of the relic neutrinos in more detail.[4] At the decoupling temperature T_D the neutrinos would still have been in equilibrium, so that the phase-space density of the ith type of neutrino ν_i and antineutrino $\bar{\nu}_i$ would be $F_{\nu_i}(p)d\tau$ and $F_{\bar{\nu}_i}(p)d\tau$, respectively, where diffraction can be ignored, one can describe the propagation of a neutrino "ray" through matter by geometrical optics.

The index of refraction for $\nu(\bar{\nu})$ is given (for small $n-1$) by

$$n_{\nu,\bar{\nu}}-1=\frac{2\pi}{p^2}\sum_a N_a f^a_{\nu,\bar{\nu}}(0) , \quad (1.11)$$

where N_a is the number density of scatterers of type a and $f^a_\nu(0)$ $[f^a_{\bar{\nu}}(0)]$ is the forward-scattering amplitude for νa $[\bar{\nu}a]$ elastic scattering. f^a is easily computed to give (for a target at rest)

$$f^a_{\nu,\bar{\nu}}(0)=\mp\frac{1}{\pi}\frac{G_F E}{\sqrt{2}}K(p,m_\nu)(g^a_V+g^a_A\vec{\sigma}_a\cdot\hat{p}) , \quad (1.12)$$

where the upper (lower) sign refers to neutrinos (antineutrinos),[24] $E=(p^2+m_\nu{}^2)^{1/2}$ is the neutrino energy,

$$K(p,m_\nu)\equiv\frac{1}{4E}(E+m_\nu)\left[1+\frac{p}{E+m_\nu}\right]^2$$

$$\rightarrow\begin{cases}1, & m_\nu=0\\ \frac{1}{2}, & p\ll m_\nu ,\end{cases} \quad (1.13)$$

and g^a_V and g^a_A are the vector and axial-vector couplings in the effective Lagrangian

$$-L=\frac{G_F}{\sqrt{2}}\bar{\nu}\gamma_\mu(1+\gamma_5)\nu\bar{\psi}_a\gamma^\mu(g^a_V+g^a_A\gamma^5)\psi_a \quad (1.14)$$

for $\nu a\rightarrow\nu a$. For massive neutrinos we have assumed helicity -1 $(+1)$ for ν $(\bar{\nu})$, as would be the case for neutrinos that decoupled while relativistic.[25] (The sign of the $\vec{\sigma}\cdot\hat{p}$ term changes for the opposite helicities.)

For a target with Z protons and electrons and $A-Z$ neutrons (assuming $\langle\vec{\sigma}_p\rangle=\langle\vec{\sigma}_n\rangle=0$, but allowing $\langle\vec{\sigma}_e\rangle\neq0$, as in a ferromagnet) and the neutral-current parameters of the $SU_2\times U_1$ model one obtains[24]

$$n_{\nu_e,\bar{\nu}_e}-1=\mp\frac{G_F}{\sqrt{2}}\frac{EKN}{p^2}(3Z-A+Z\langle\vec{\sigma}_e\rangle\cdot\hat{p}) , \quad (1.15)$$

where N is the number density of target atoms, and

$$n_{\nu_l,\bar{\nu}_l}-1=\mp\frac{G_F}{\sqrt{2}}\frac{EKN}{p^2}(Z-A-Z\langle\vec{\sigma}_e\rangle\cdot\hat{p}) \quad (1.16)$$

for $\nu_l=\nu_\mu$ or ν_τ. The difference between (1.15a) and (1.15b) is due to the charged-current contribution to $g^e_{V,A}$.

For iron ($Z=26, A=56, N\sim0.85\times10^{23}/\text{cm}^3$), for example,

$$n_{\nu_e,\bar{\nu}_e}-1=\mp2.3\times10^{-10}(1+0.85\langle\vec{\sigma}_e\rangle\cdot\hat{p}) ,$$
$$n_{\nu_l,\bar{\nu}_l}-1=\pm3.1\times10^{-10}(1+1.2\langle\vec{\sigma}_e\rangle\cdot\hat{p}) \quad (1.17)$$
$$(i=\mu,\tau)$$

for massless neutrinos with $E=p=5.2\times10^{-4}$ eV, while for massive neutrinos the coefficients are increased by $m_i/2p$. For $m=E=30$ eV $p=5.2\times10^{-4}$ eV the coefficients in (1.16) become 6.6×10^{-6} and 8.9×10^{-6}, respectively. We therefore see that $n-1$ is very tiny for all reasonable parameters for the relic neutrinos.

Several experiments have been proposed to measure the energy momentum, or angular momentum transferred from the neutrino sea to a macroscopic target due to the refractive bending of a neutrino ray as it passes through the target or the total external reflection that might be expected for neutrinos incident on a surface with angle (with respect to the plane) $\theta<\theta_c\equiv\sqrt{2(1-n)}$ (for $n<1$). However, we will show that all but one of these proposals are incorrect. The one exception[18] is technically correct but, as noted by the author, is probably impractically difficult.

D. Outline

Rather than initially discussing each of the proposals separately, we will begin (in Sec. II) by giving a general analysis in the Born approximation of the energy, momentum, and angular momentum transferred from a quantum-mechanical wave (e.g., a neutrino wave) to a target. We assume that the unperturbed target (which may be either macroscopic, as in the coherent scattering proposals, or microscopic) is static. Furthermore, we assume that the time average of the incident wave is spatially homogeneous over the size of the detector (which is certainly expected to be true for cosmological neutrinos). We do not need to make any assumption concerning the isotropy[26] of the incident wave, however; our results would apply even to a unidirectional neutrino beam. We show that under these assumptions the total momentum and energy transfer to the target vanishes in the Born approximation (i.e., to first order in the interaction strength). Furthermore, the net angular momentum transfer vanishes to the same order unless the target is either polarized or carried an electric current.

We expect the Born approximation to be valid for sufficiently weak scattering. For the scattering of

Resonance enhancement of oscillations in matter and solar neutrino spectroscopy

S. P. Mikheev and A. Yu. Smirnov

Institute for Nuclear Research, USSR Academy of Sciences

(Submitted 4 December 1984)
Yad. Fiz. 42, 1441–1448 (December 1985)

Matter can enhance neutrino oscillations (increase the mixing parameter $\sin^2 2\theta_m$). For small mixing angles in vacuum the enhancement displays a resonance behavior in the neutrino energies or the density of the medium. This resonance effect is important for solar neutrinos in a wide range of oscillation parameters $\Delta m^2 = 10^{-4}$–10^{-8} eV^2 and $\sin^2 2\theta > 10^{-4}$. It leads to a strong suppression of the neutrino flux even for small $\sin^2 2\theta$.

1. INTRODUCTION

Interactions of neutrinos with a medium change the picture of neutrino oscillations.[1,2] The effect of the medium has a coherent nature and is to a significant degree analogous to the well known transmission (coherent) regeneration of K mesons.

In a medium with constant density the general form of the evolution of the neutrino states is the same as in the case of vacuum oscillations and the effect of the interactions reduces to a change of the strength of the oscillations (the mixing angle θ_m) and the oscillation length l_m (Refs. 1–4). Relations between the oscillation parameters in matter θ_m, l_m and in a vacuum θ, l_v have been obtained as functions of the matter density ρ. It has been pointed out[1–3] that the interaction can lead to both suppression ($\theta_m < \theta$) and enhancement ($\theta_m > \theta$) of oscillations.

This effect has been studied for the case of propagation of an oscillating neutrino beam in the sun,[1,5] in supernovae,[5,6] and in the Earth.[7] In particular, it has been pointed out that oscillations are suppressed inside the sun for oscillation lengths l_v larger than the solar radius and that only the distance from the surface of the sun needs to be taken into account when the effect is calculated for the case of the Earth.

In the present study we show that under certain conditions the effect of the medium has a resonance behavior. In relation to this we study oscillations in a medium with varying density. We discuss the resonance enhancement of oscillations in supernovae and mainly in the sun.

As is well known, the solar neutrino flux measured in the chlorine-argon experiment[8] is about 4 times smaller than that predicted by the standard solar model.[9] Neutrino oscillations have been considered as one of the possible reasons for this discrepancy.[10] If the vacuum oscillation length l_v is comparable to the distance from the sun to the Earth the averaged effect would not be observed at the Earth and in the case of maximal mixing the electron neutrino flux could be decreased by a factor of 10.[11] For values of l_v much smaller than the distance from the sun to the Earth the effect of the oscillations would be averaged both over the neutrino energies and over the creation point in the sun. In this case for maximal mixing the neutrino-flux suppression factor is equal to the number of neutrino types. In any case, the attribution of the solar neutrino defect to oscillations requires maximal or nearly maximal mixing. In the present article we shall show that a significant decrease of the neutrino flux can occur even for small mixings, as small as $\sin^2 2\theta \simeq 10^{-4}$–$10^{-3}$, if the effect of matter on "ordinary" vacuum oscillations is taken into account. Here the dependence of the flux-suppression factor on the neutrino energy can be complicated and may involve both the neutrino-oscillation parameters and the matter distribution in the sun.

The detailed analysis of the factors affecting the solar neutrino spectrum and, in particular, oscillations in matter is very important right now in view of new experiments which are being planned[12] to determine the neutrino fluxes in various energy ranges.

In Section 2 we study the resonance enhancement of oscillations and in Section 3 we consider oscillations in a medium with varying density. The effect occurring in the sun and in supernove is discussed in Sections 4 and 5, respectively. The results are summarized in the Conclusion.

2. RESONANCE ENHANCEMENT OF OSCILLATIONS IN MATTER

Let us consider oscillations of two types of neutrino $\nu_e = c\nu_1 + s\nu_2$ and $\nu_a = -s\nu_1 + c\nu_2$, where $\nu_a = \nu_\mu$ or ν_τ (flavor oscillations) or $\nu_a = \bar{\nu}_e$ (ν–$\bar{\nu}$ oscillations), ν_1 and ν_2 are states of definite mass m_1 and m_2, $s \equiv \sin\theta$, and $c \equiv \cos\theta$. Matter modifies the vacuum oscillations if the amplitudes for the forward elastic scattering of ν_e and ν_a on the electrons or nucleons of the medium are different, that is, if $\Delta f(0) = f_e(0) - f_a(0) \neq 0$. In the case of $\nu_e \leftrightarrow \nu_\mu$ (ν_τ) oscillations a difference arises because both neutral and charged curents participate in interactions of electron neutrinos with the electrons of the medium, while only neutral currents participate in muon and tau neutrino interactions. Here $\Delta f(0) = \sqrt{2}Gk$ (Ref. 3). If the transition $\nu_e \leftrightarrow \bar{\nu}_e$ occurs, then, since $\bar{\nu}_{eL}$ do not participate in weak interactions in the standard model, charged currents on electrons and neutral currents on electrons and nucleons contribute to $\Delta f(0)$. Then $\Delta f(0) = (1/\sqrt{2})Gk$ for a medium with equal numbers of electrons, protons, and neutrons and $\Delta f(0) = \sqrt{2}Gk$ for a medium without neutrons. Below as an example we shall consider $\nu_e \leftrightarrow \nu_\mu$ (ν_τ) transitions.

In a medium the neutrino oscillation parameters are different from the vacuum parameters θ and l_v. First of all, the difference between the interactions of electron and muon neutrinos with the medium implies that $\nu_1 \leftrightarrow \nu_2$ transitions occur, that is, that the eigenstates in the matter ν_1^m and ν_2^m are different from the vacuum eigenstates ν_1 and ν_2. Secondly, in a medium the refractive index is not equal to 1 and is different for ν_1^m and ν_2^m waves. This leads to both a change of the oscillation length and a change of the mixing angle θ_m, which is now defined in terms of ν_1^m and ν_2^m, implying that it is different from the vacuum mixing angle θ. The corresponding relations for a medium with constant density are[1,2]

$$\sin^2 2\theta_m = \frac{\sin^2 2\theta}{1-2(l_\nu/l_0)\cos 2\theta + (l_\nu^2/l_0^2)}, \quad (1)$$

$$l_m = \frac{l_\nu}{\sqrt{1-2(l_\nu/l_0)\cos 2\theta + (l_\nu^2/l_0^2)}}. \quad (2)$$

Here $l_\nu = 4\pi E/\Delta m^2$ is the vacuum oscillation length,

$$l_0 = 2\pi\left[\frac{\Delta f(0)}{k}\rho\frac{Y_e}{m_N}\right]^{-1}$$

is a quantity characterizing the medium, which can be viewed as the "proper" oscillation length in matter, ρ is the matter density, m_N is the nucleon mass, and Y_e is the number of electrons per nucleon.

It follows from expressions (1) and (2) that the effect of the matter can vary depending on the sign of Δm^2, $\Delta f(0)$, and $\cos 2\theta$. For definiteness we shall assume that $\theta < 45°$; then if Δm^2 and $\Delta f(0)$ have opposite signs ($l_\nu/l_0 < 0$), we will have $\sin^2 2\theta_m < \sin^2 2\theta$ for any neutrino energy and any matter density, that is, vacuum oscillations are suppressed by the medium. If Δm^2 and $\Delta f(0)$ have the same sign ($l_\nu/l_0 > 0$), then for $l_\nu/l_0 < 2\cos 2\theta$ we obtain $\sin^2 2\theta_m > \sin^2 2\theta$, which implies that oscillations are enhanced. We note that enhancement of the oscillations for certain values of Δm^2 and $\Delta f(0)$ can occur either for neutrinos or for antineutrinos, since $\Delta f(0)$ changes sign in the replacement $\nu \leftrightarrow \bar{\nu}$.

Our "observation" is that the dependence of $\sin^2 2\theta_m$ on l_ν/l_0 (see Fig. 1) has a resonance behavior for small values of $\sin^2 2\theta$. For l_ν/l_0 close to zero the matter has practically no effect: $\sin^2 2\theta_m \simeq \sin^2 2\theta$. As l_ν/l_0 is increased the amplitude of the oscillations grows and at the point

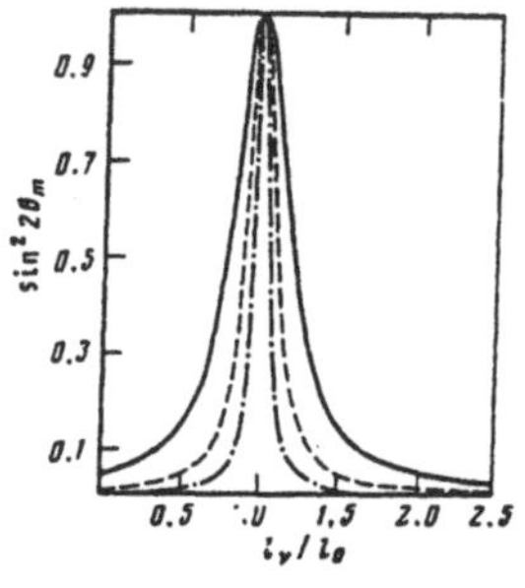

FIG. 1. Dependence of the mixing parameter $\sin^2 2\theta_m$ on the ratio l_ν/l_0. The solid line is for $\sin^2 2\theta = 4\times10^{-2}$, the dashed line is for $\sin^2 2\theta = 10^{-2}$, and the dot-dash line is for $\sin^2 2\theta = 2.5\times10^{-3}$.

$$l_\nu/l_0 = \cos 2\theta \quad (3)$$

it reaches the maximum value $\sin^2 2\theta_m = 1$. For larger l_ν/l_0 the amplitude ($\sin^2 2\theta_m$) falls off so that in the limit $l_\nu/l_0 \gg 1$ vacuum oscillations are suppressed by the factor $(l_\nu/l_0)^2$: $\sin^2 2\theta_m \simeq \sin^2 2\theta/(l_\nu/l_0)^2$. In other words, when the eigenfrequency of the system (in our case l_ν) coincides with the eigenfrequency of the surrounding medium ($l_0\cos 2\theta \simeq l_0$ for small θ) there is a resonance increase of the oscillation amplitude.

The width of the resonance curve at half-max is

$$\Delta(l_\nu/l_0) = (l_\nu/l_0)_{res}\,\mathrm{tg}\,2\theta = \sin 2\theta. \quad (4)$$

Therefore, the smaller the vacuum neutrino mixing, the narrower the resonance peak. As θ decreases, the position of the peak approaches $l_\nu/l_0 = 1$.

For a fixed value of l_ν the resonance corresponds to the maximum value of the oscillation length in the matter:

$$l_m = l_\nu/|\sin 2\theta|. \quad (5)$$

Using the definitions of l_ν and l_0, the ratio l_ν/l_0 can be expressed in terms of the neutrino energy and the density of the medium as

$$\frac{l_\nu}{l_0} = \frac{E\rho}{\Delta m^2 a}, \quad (6)$$

where $a = (m_N/Y_e)[k/2\Delta f(0)]$. (If l_ν and l_0 are expressed in m, E in MeV, and Δm^2 in eV^2, then $a = 7.14\times10^6$ for $Y_e = 1$.) Using (6), the resonance condition can be written as

$$E\rho = a\Delta m^2\cos 2\theta. \quad (7)$$

Two different manifestations of the resonance enhancement of the oscillations can be distinguished. The first occurs when a beam of neutrinos with a continuous energy spectrum passes through a medium of constant density. In this case the enhancement of the neutrino oscillations will occur in the portion of the spectrum near

$$E_{res} = a\frac{\Delta m^2\cos 2\theta}{\rho}, \quad (8)$$

and of width

$$\Delta E = E_{res}\,\mathrm{tg}\,2\theta = a\frac{\Delta m^2\sin 2\theta}{\rho}. \quad (9)$$

The second occurs when a monoenergetic beam of neutrinos passes through a medium of varying density. Then a significant enhancement of the oscillations occurs in the layer (which we shall henceforth refer to as the resonance layer) with

$$\rho \sim \rho_{res} = a\frac{\Delta m^2\cos 2\theta}{E} \quad (10)$$

and width

$$\Delta\rho_{res} = \rho_{res}\,\mathrm{tg}\,2\theta = a\frac{\Delta m^2\sin 2\theta}{E}. \quad (11)$$

Of course, the resonance enhancement will be sizeable if the resonance layer is sufficiently thick and oscillations of maximum amplitude can be developed:

$$r_{res} \gg l_m - l_v / |\sin 2\theta|. \qquad (12)$$

As a rule, in real objects both cases occur: the spectrum is continuous and the density varies.

We stress the fact that the relations (1) and (2) obtained for a medium with constant density are, in general, inapplicable in the second case. Nevertheless, the concepts of the resonance layer, its density, and its width introduced using (1) and (2) are useful for the qualitative analysis and interpretation of the exact solutions of the problem with varying density.

3. OSCILLATIONS IN A MEDIUM WITH VARYING DENSITY

We shall generalize the analysis of Wolfenstein[1,2] to the case of a spatially varying density. The equations for the evolution of the ν_e and ν_μ wave functions[1] can be used to obtain a system of differential equations directly involving the probability:

$$\dot{P} = -2MI, \quad \dot{I} = -mR + M(2P-1), \quad \dot{R} = mI, \qquad (13)$$

where $P \equiv \langle \nu_e | \nu_e \rangle$ is the desired probability of finding a ν_e, R and I are the real and imaginary parts of the matrix element $\langle \nu_\mu | \nu_e \rangle = R + iI$, and $2M$ and m are parameters:

$$2M = \frac{2\pi}{l_v} \sin 2\theta, \quad m = \frac{2\pi}{l_v}\left(\cos 2\theta - \frac{l_v}{l_0}\right). \qquad (14)$$

The initial conditions (in the case of ν_e creation) for the system are

$$P(0) = 1, \quad I(0) = R(0) = 0. \qquad (15)$$

Assuming that the neutrino velocity is $v_\nu \simeq c = 1$, in (13) we have transformed to the time scale with the time dependence contained in the parameter $m[l_0(\rho(t))]$.

Let us for certain special conditions [distributions $\rho(t)$] discuss the solutions of the system (13)–(15) which are of interest for specific applications.

Let us consider a finite object (a layer of matter) in which the density varies monotonically in a range from ρ_{max} to $\rho_{min} = 0$. Let a beam of oscillating neutrinos ν_e of fixed energy E be incident on this object. What is the state of the beam (P) after passing through the layer?

Case A. Let the resonance conditions be satisfied in this object, that is, for the given beam ρ_{res} lies in the range $\rho_{min} < \rho_{res} < \rho_{max}$. There are no edge effects—the resonance layer is far from the edge of the object $\rho_{max} - \rho_{res} \gg \Delta\rho_{res}$ and $\rho_{res} - \rho_{min} \gg \Delta\rho_{res}$ and the density varies slowly (the adiabatic regime)

$$\left(\frac{\partial \rho}{\partial r}\right)^{-1} \rho \gg l_0 / \mathrm{tg}^2\, 2\theta. \qquad (16)$$

For the resonance layer (16) reduces to $r_{res} > l_m$, where $r_{res} = (\partial\rho/\partial r)^{-1} \Delta\rho_{res}$ [see (12)]. Under these conditions the initial beam of electron neutrinos (for small mixing angle θ) passing through the object is almost completely transformed into a beam of muon neutrinos—after leaving the object $P \simeq \sin^2\theta$.

This result is related to the fact that in the process of travelling through the object the mixing angle in the matter changes by $\pi/2$. In fact, the slowness of the density variation (16) (the adiabatic regime) ensures that the system manages to adjust itself to the variation of the density ρ, so that the evolution of the system can be traced using the solutions (1) and (2) for constant density. It follows from (1) that at the beginning ($\rho = \rho_{max}$)$\theta_m \simeq \pi/2$ (θ is small), that is, the ν_e coincides with the eigenstate in the matter $\nu_2^m(t)$; just after leaving the object ($\rho = 0$)$\theta_m \simeq \theta \simeq 0$ and the ν_e practically coincides with the other eigenstate $\nu_1^m(t)$. Accordingly, $\nu_2^m(t) \simeq \nu_\mu$, that is, $\nu_i \simeq \nu_e \simeq \nu_2^m(0), \nu_2^m(t)$ is the eigenstate in the matter, so $\nu_f \simeq \nu_2^m(t)$, but ν_e is rotated relative to $\nu_2^m(t)$ by $\pi/2$ and therefore $\nu_f \simeq \nu_\mu$.

We note that if the initial and final densities are the same (for example, $\rho_i = \rho_f = \rho_{min}$), because of the adiabatic nature of the situation the system passing through the object emerges in the initial state $\nu_i = \nu_f = \nu_e$, in spite of the fact that oscillations of maximum intensity occurred inside the object.

If the resonance layer is located near the edge of the object the transition $\nu_e \to \nu_\mu$ will not be complete. For example, for $\rho_{max} = \rho_{res}$, $P \simeq 1/2$.

Case B. As in case A, let $\rho_{min} < \rho_{res} < \rho_{max}$, but the adiabaticity condition is no longer satisfied: $(\partial\rho/\partial r)^{-1}\rho < l_0 / \tan^2 2\theta$, so the density varies rapidly. In this case the transition $\nu_e \to \nu_\mu$ is incomplete: $P > \sin^2\theta$. For the resonance layer this inequality implies that $r_{res} < l_m$, that is, the layer is thin and large-amplitude oscillations cannot develop.

Case C. Let $\rho_{res} > \rho_{max}$, so the resonance conditions are not satisfied in the object. In the limit $\rho_{res} - \rho_{max} \gg \Delta\rho_{res}$ the effect of the matter can be neglected: $\langle P \rangle \simeq 1/2 \sin^2 2\theta$. It becomes noticeable for $\rho_{res} - \rho_{max} \simeq \Delta\rho_{res}$. If the density varies sufficiently slowly and θ is small, then in the range from $\rho_{max} + \Delta\rho_{res}$ to ρ_{max} the probability P decreases by about a factor of 1/2 and the shape of the curve $P(\rho_{res})$ is similar to that of the resonance curve $\sin^2 2\theta(\rho)$.

Combinations of these cases arise for the specific objects considered below.

4. RESONANCE ENHANCEMENT OF ν OSCILLATIONS IN THE SUN

Let us first determine the ranges of the oscillation parameters Δm^2 and $\sin^2 2\theta$ for which resonance enhancement will occur. Neutrinos of energy in the range 0–14 MeV are created in the sun and the matter density falls off monotonically from the maximum value 156 g/cm^3 at the center to zero according to the standard solar model.[9] The maximum value of $\xi \equiv E/\Delta m^2$ at which the resonance condition (7) holds corresponds to the maximum solar density and is equal to 4×10^4 (if the energy is expressed in MeV and Δm^2 in eV^2). The maximum energy of the neutrinos created in the sun determines the maximum value $\Delta m^2 \simeq 3 \times 10^{-4}$ eV^2 for which resonance enhancement of the oscillations occurs. A shift of the resonance layer from the center of the sun to the periphery corresponds to an increase of ξ. Then the relation between r_{res} and l_m changes. As $\xi \equiv E/\Delta m^2$ increases the oscillation length grows ($l_m \sim l_v \sim E/\Delta m^2$) and the width of the resonance layer decreases ($\Delta\rho_{res} \sim \rho$). Therefore, if $r_{res} > l_m$ in the central regions (small $E/\Delta m^2$) and the adiabatic regime occurs, the ratio r_{res}/l_m falls off with increasing ξ, the equation $r_{res} = l_m$ is satisfied at some ξ_{ad}, and for $\xi > \xi_{ad}$ the adiabaticity condition is violated. The lowest neutrino energy (we take E_{min} to be equal to the threshold energy in the gallium-germanium experiment: $E_{min} = 0.234$ MeV) and ξ_{ad}, that is, satisfaction of the conditions (7) and (12), determine the lower limit on the range of Δm^2 at 10^{-8}–10^{-9} eV^2 (see Fig. 2).

Let us find the attenuation factor for the flux of solar neutrinos reaching the earth $\langle P_{ee}\rangle$ as a function of $\sin^2 2\theta$ and $\xi \equiv E/\Delta m^2$. It is determined by the transition probability $\langle P\rangle$ averaged over the period of the oscillations. (The averaging arises because of the finite energy resolution of the apparatus.)

The system of equations (13)–(15) was solved for the density distribution given by the standard model[9] and the results are shown in Fig. 3. The dashed lines correspond to a point source at the center of the sun and the solid lines to the distribution of sources of *pp* neutrinos[9] ($R_\nu \sim 0.2R_\odot$).

The shape of the curves is easily understood on the basis of the analysis of Section 3. Let us first discuss the case of a point source. The point $\xi_m = 5\times10^4$ corresponds to a resonance at the center of the sun. Situation *C* occurs for $\xi < \xi_m$—there is no resonance and as ξ decreases the probability $\langle P\rangle$ approaches the value $1/2\sin^2 2\theta$ corresponding to vacuum oscillations. The transition to resonance enhancement occurs in the range of ξ from $\xi_m - \tan 2\theta\xi_m$ to ξ_m. The conditions of case A are satisfied for ξ in the range from $\xi_m + \xi_m \tan 2\theta$ to ξ_{ad} and for the values of ξ for which the resonance layer lies far from the "edge" of the object $\langle P\rangle$ reaches its minimum value: $\langle P\rangle = \sin^2\theta$.[11]

The case of a thin layer (case B) occurs for $\xi > \xi_{ad}$. As ξ increases, the inequality $r_{res} < l_m$ gets stronger, so that the suppression of the neutrino flux weakens and $\langle P\rangle \to 1$.

The curves for the case of an extended source differ most from those considered above in the range of small

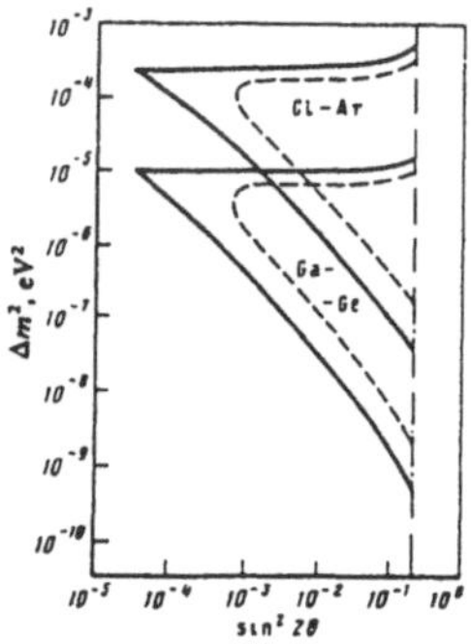

FIG. 2. Ranges of values of the parameters Δm^2 and $\sin^2 2\theta$ for which the observed solar neutrino flux in the chlorine-argon and gallium-germanium experiments is decreased by (a) more than 10% (solid lines) and (b) by a factor of 3 (dashed lines).

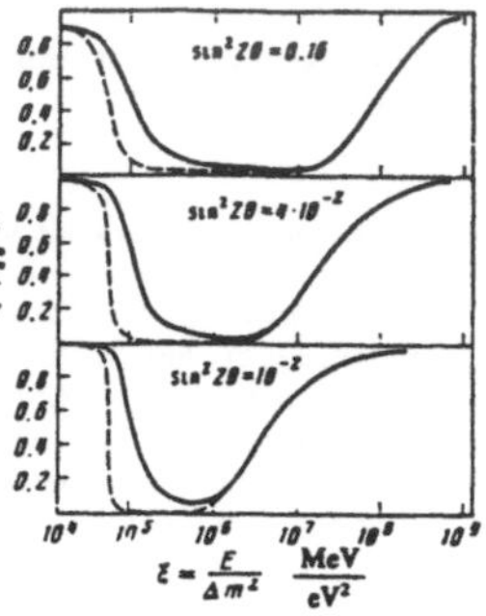

FIG. 3. Suppression factors for the solar neutrino flux for various values of the mixing angle $P_{ee}(E/\Delta m^2, \sin^2 2\theta)$. The dashed lines correspond to the case of a point neutrino source at the center of the sun and the solid lines correspond to the case of *pp*-neutrino sources distributed according to the standard model of the sun.

$\xi \equiv E/\Delta m^2$. This is because $\rho_{max} < \rho_{center}$ for neutrinos created at the periphery of the generation region, so that ξ_m is shifted to larger values; on the other hand, for $\xi = 5\times10^4$ the resonance condition is not satisfied for such neutrinos.

As θ decreases, the curve $P(\xi)$ in the region near the transition to resonance enhancement ($\xi \sim \xi_m$) becomes steeper, ξ_{ad} is shifted to smaller values, and the value of the maximum suppression falls off as $\sin^2\theta$.

Let us determine how this affects the number of events detected at the earth in the chlorine-argon and gallium-germanium experiments. The number of events is proportional to the interaction cross section $\sigma(E)$ and the neutrino flux $F_\nu(E)$. Let us calculate the ratio

$$R(\Delta m^2, \sin^2 2\theta) = \frac{\int F_\nu(E)\sigma(E)dE}{\int F_\nu^0(E)\sigma(E)dE}, \qquad (17)$$

where $F_\nu^0(E)$ is the neutrino flux in the absence of oscillations and $F_\nu(E) = P(E/\Delta m^2, \sin^2 2\theta)F_\nu^0$. Different regions of the calculated curves $P(E/\Delta m^2)$ will affect the neutrino flux depending on the value of Δm^2. For 10^{-5} eV$^2 < \Delta m^2 < 3\times10^{-4}$ eV^2 only the oscillations of boron neutrinos will undergo resonance enhancement. In the range $(10^{-9}$ eV$^2)/\sin^2 2\theta < \Delta m^2 < 10^{-5}$ eV^2 all solar neutrinos lie in the resonance enhancement region. Finally, for $\Delta m^2 < (10^{-9}$ eV$^2)/\sin^2 2\theta$ a resonance layer exists only for *pp* neutrinos. In Fig. 2 the dashed lines show the limits of the region of parameters where the suppression factor of the expected number of events in the chlorine-argon and gallium-germanium experiments is greater than 3.

5. RESONANCE ENHANCEMENT OF OSCILLATIONS IN COLLAPSING STARS

Let us briefly consider supernovae, where resonance conditions can also occur. Only suppression of the oscillations in supernova matter has been discussed before,[4,5] but a range of values of the oscillation parameters can be found for which the resonance condition (7) is satisfied. Actually, the density of the matter in the center of a collapsing star is $\sim 10^{15}$ g/cm^3, the fraction of electrons is 10^{-3}, and the neutrino energy is ~ 10 MeV. Then for small mixing angles ($\sin^2 2\theta \lesssim 10^{-2}$) relation (7) is satisfied for $\Delta m^2 \sim 10^6$ eV^2. Similarly, in a supernova shell where the density is 10^9 g/cm^3 the resonance condition is satisfied for $\Delta m^2 \sim 10^3$ eV^2. These values lie outside the existing experimental limits on the oscillation parameters because of the small mixing angles.

6. CONCLUSION

The effect of the medium on neutrino oscillations can lead not only to suppression, but also to a significant enhancement. When the vacuum mixing is small this enhencement has the form of a resonance in the neutrino energies and the density of the medium. The conditions for resonance enhancement can be met in the sun and also in the centers and outer shells of collapsing stars. In the sun, resonance enhancement of the oscillations leads to significant changes in the neutrino flux in a wide range of values of the oscillation parameters Δm^2 and $\sin^2 2\theta$. Detailed spectroscopy of the solar neutrinos would give information on the oscillation parameters in the ranges $\Delta m^2 = 3\times 10^{-4}$—$10^{-8}$ eV^2 and $\sin^2 2\theta > 10^{-4}$.

In conclusion, the authors would like to thank V. S. Berezinskiĭ, E. A. Gavryuseva, G. T. Zatsepin, V. A. Kuz'min, N. V. Sosnin, A. E. Chudakov, and M. E. Shaposhnikov for many useful discussions. The authors are grateful to L. Wolfenstein for a remark concerning the adiabatic regime of neutrino propagation.

[1)] $\langle P \rangle_{\min} = \sin^2\theta$ and not $1/2 \sin^2 2\theta$, as might be expected for vacuum oscillations. The reason for this lies in the "initial conditions." If the ν_e, ν_μ mixing angle is θ and the initial condition is characterized by the angle θ_0, $\nu_i = \cos\theta_0 \nu_e + \sin\theta_0 \nu_\mu$, then the probability for finding ν_e averaged over the period is $\langle P \rangle = \cos^2\theta_0 - 1/2 \sin 2\theta \text{-} \sin 2(\theta + \theta_0)$. The maximum value $\langle P \rangle = \cos^2\theta_0$ is attained for $\theta_0 = -\theta$ and the minimum value $\sin^2\theta$ is attained at $\theta_0 = -\theta + \pi/2$.

[1] L. Wolfenstein, Phys. Rev. D 17, 2369 (1978).
[2] L. Wolfenstein, Neutrino-78, 1978, p. C3.
[3] V. Barger *et al.*, Phys. Rev. D 22, 2718 (1980).
[4] L. Wolfenstein, Phys. Rev. D 20, 2634 (1979).
[5] S. Pakvasa, in: DUMAND-80, 1981, Vol. 2, p. 457.
[6] H. J. Haubold, Astrophys. Spac Sci. 82, 457 (1982).
[7] P. V. Ramana Murthy, in *Proc. of the Eighteenth Int. Cosmic Ray Conf.*, Vol. 7, 125 (1983).
[8] R. Davis, Jr. *et al.*, in *Proc. of the Neutrino Miniconference*, Telemark, 1982, p. 23.
[9] J. N. Bahcall *et al.*, Rev. Mod. Phys. 54, 767 (1982).
[10] V. N. Gribov and B. Pontecorvo, Phys. Lett. 28B, 493 (1969); S. M. Bilenky and B. Pontecorvo, Phys. Rep. 41C, 225 (1978).
[11] V. Barger *et al.*, Phys. Rev. D 24, 538 (1981).
[12] I. R. Barabanov *et al.*, Neutrino-77, Vol. 1, 1977, p. 20.

Translated by Patricia Millard

from *Resonant Amplifications of ν-Oscillations in Matter and Solar-Neutrino Spectroscopy*, by Mikheyev and Smirnov

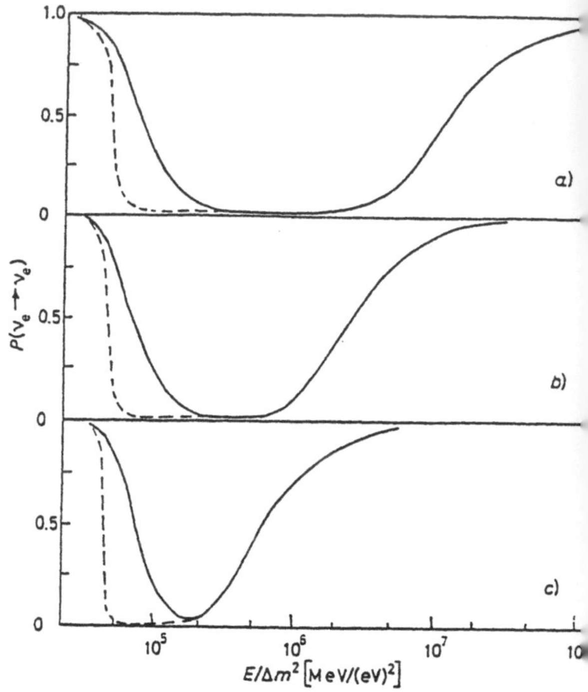

Fig. 3. – The suppression factor $P(E/m^2, \sin^2 2\theta)$ for neutrino production region wi $R_{pr} = 0.2 \cdot R_S$ (full line) and for pointlike neutrino source in the centre of the Su (dashed line). *a*) $\sin^2 2\theta = 4 \cdot 10^{-2}$, *b*) $\sin^2 2\theta = 10^{-2}$, *c*) $\sin^2 2\theta = 2.5 \cdot 10^{-3}$.

the neutrino spectrum is shifted in the region of large values $E/\Delta m^2$. A $10^{-5}\,(\text{eV})^2 < \Delta m^2 < 3 \cdot 10^{-4}\,(\text{eV})^2$ only ^{8}B-neutrinos fall in the resonant regio (see fig. 2). In the interval $10^{-9}\,(\text{eV})^2/\sin^2 2\theta < \Delta m^2 < 10^{-5}\,(\text{eV})^2$ all solar neu trinos undergo resonant amplification of oscillation. At $\Delta m^2 < 10^{-9}\,(\text{eV})^2/\sin^2 2$ the resonance layer exists only for pp neutrinos.

4. – Discussion and results.

Now some remarks on the matter effect for 3 neutrino mixing. In the mos natural case with the mass hierarchy ($m_1 \ll m_2 \ll m_3$) the results coincide wit the considered ones or can be obtained from them in a simple way. Mor interesting situation is when at least two masses fall inside the resonant interval $\Delta m^2_{max} > m_2^2, m_3^2 > \Delta m^2_{min}$. In this case the curve $P(E)$ will be the superpositio of two considered ones with $\Delta m_1^2 = m_2^2$ and $\Delta m_2^2 = m_3^2$ and with the correspondin mixing angles.

One should remark that the resonance conditions can be fulfilled in th cores as well as in envelopes of collapsing stars. In contrast with the existin

Neutrino oscillations in a variable-density medium and ν-bursts due to the gravitational collapse of stars

S. P. Mikheev and A. Yu. Smirnov

Institute of Nuclear Research, Academy of Sciences of the USSR
(Submitted 24 December 1985)
Zh. Eksp. Teor. Fiz. **91**, 7–13 (July 1986)

Under certain conditions, the propagation of a beam of oscillating neutrinos in a variable-density medium takes the form of an almost complete transformation of the initial type of neutrino into another type. The depth of the oscillations is then negligible. The transformation can occur in the cores and envelopes of collapsing stars.

1. INTRODUCTION

The interaction between neutrinos and matter modifies the picture of ν-oscillations.[1-3] The effect of the medium is analogous in a number of significant respects to the coherent regeneration of K mesons, and also to the appearance of a refractive index.

In a medium of constant density, the overall character of the oscillations is the same as in vacuum, i.e., there is a change in only the oscillation length and depth. A number of astrophysical[1,4,5] and geophysical[1,3,6] applications of these oscillations have been considered in the constant-density approximation.

In the present paper, we examine qualitatively new effects that appear in a variable-density medium.

2. EQUATION FOR THE TRANSFORMATION PROBABILITY; RESONANCE CONDITION

We shall consider the mixing of two types of neutrino. To be specific, we shall suppose that they are ν_e and ν_μ. Suppose the ν_e are created in the source. We shall seek the probability $P(t)$ of an oscillatory $\nu_e \to \nu_e$ transformation in a time t (or distance $r \simeq ct$ from the source). The equations describing the evolution of the wave functions of the ν_e and ν_μ in the medium[1] (essentially, the Schroedinger equations) can be used to show that the probability $P(t)$ is the solution of the following equation:

$$M\dddot{P} - \dot{M}\ddot{P} + M(M^2 + 4\bar{M}^2)\dot{P} - 2\bar{M}^2\dot{M}(2P-1) = 0\,, \tag{1}$$

where

$$M = (2\pi/l_\nu)(\cos 2\theta - l_\nu/l_0), \quad 2\bar{M} = (2\pi/l_\nu)\sin 2\theta, \tag{2}$$

$l_\nu = 4\pi E/\Delta m^2$ is the oscillation length in vacuum, E is the neutrino energy, $\Delta m^2 = m_1^2 - m_2^2$ is the difference between the squares of the masses, θ is the mixing angle, l_0 is a characteristic length of the medium, given by

$$l_0^{-1} = \frac{\rho}{m_N}\sum_i X_i \frac{\Delta f_i(0)}{2\pi E}, \tag{3}$$

$\Delta f_i(0) = f_{ie}(0) - f_{i\mu}(0)$ is the difference between the ν_e and ν_μ forward-scattering amplitudes for the ith component of the medium ($i = e,p,n$), X_i is the abundance of the ith component per nucleon, ρ is the density of the medium, and m_N is the nucleon mass. In accordance with (3), we have $l_0^{-1} \sim G_F\rho/m_N$, where G_F is the Fermi constant.

The initial conditions for (1) are:

$$P(0) = 1, \quad \dot{P}(0) = 0, \quad \ddot{P} = -2\bar{M}^2. \tag{4}$$

If the density is constant ($M = 0$), we find from (1) that

$$\ddot{P} + (M^2 + 4\bar{M}^2)\dot{P} = 0. \tag{5}$$

This equation has a periodic solution of the form

$$P = 1 - A\sin^2(\pi r/l_m)$$

with oscillation length

$$l_m = 2\pi(M^2 + 4\bar{M}^2)^{-1/2} \tag{6}$$

and oscillation depth determined by the mixing angle in the medium:[1)]

$$A = \sin^2 2\theta_m = \bar{M}^2 l_m^2/\pi^2. \tag{7}$$

We recall that the mixing angle θ_m relates the ν_e and ν_μ states with the neutrino eigenstates ν_1^m and ν_2^m in the medium. In the medium states with the specific masses of ν_1 and ν_2 are not eigenstates of the Hamiltonian and themselves oscillate).

Using (2) and (6), we can rewrite (7) in the form

$$\sin^2 2\theta_m = \sin^2 2\theta[(\cos 2\theta - l_\nu/l_0)^2 + \sin^2 2\theta]^{-1}. \tag{8}$$

from which it follows that, for low values of $\sin^2 2\theta$, the dependence of $\sin^2 2\theta_m$ on l_ν/l_0 ($l_\nu/l_0 \sim \rho E$) exhibits a resonance. When

$$l_\nu/l_0 = \cos 2\theta\,, \tag{9}$$

$\sin^2 2\theta_m$ is a maximum, i.e., $\sin^2 2\theta_m = 1$. The quantity $\sin^2 2\theta_m$ falls rapidly as l_ν/l_0 departs from $\cos 2\theta \simeq 1$. Condition (9), for which the mixing angle in the medium is equal to 45° ($\sin^2 2\theta_m = 1$), will be referred to as the resonance condition. Correspondingly, the values of ρ and E for which (9) is satisfied will be referred to as the resonance values. From (9) and (3), we have

$$\rho_R = m_N \Delta m^2 \cos 2\theta / 2^{1/2} G_F E. \tag{10}$$

The width $\Delta\rho_R$ of the resonance layer will be defined as the density interval around ρ_R in which $\sin^2 2\theta_m > 1/2$. From (8) we have

$$\Delta\rho_R = \rho_R \tan 2\theta. \tag{11}$$

The resonance becomes narrower as the mixing angle decreases. Similarly, we may introduce a resonance energy E_R and a resonance width $\Delta E_R = E_R \tan 2\theta$.

The physical meaning of the resonant behavior of $\sin^2 2\theta_m$ is as follows. Suppose that a constant-density layer intercepts a neutrino flux with a continuous energy spectrum. Neutrinos with energy $E = E_R(\rho)$ will then oscillate with maximum oscillation depth (despite the small mixing in vacuum). The energy dependence of the oscillation amplitude is determined by the resonance curve (8): $\sin^2 2\theta_m = f[l_\nu(E)]$.

In the case of a variable-density medium, the oscillation depth does not depend on θ_m alone, and even in the resonance layer ($\rho = \rho_R$), it may turn out to be low (see below).

3. MEDIUM WITH SLOWLY-VARYING DENSITY (ADIABATIC REGIME); NONOSCILLATORY TRANSFORMATION IN THE ν-BEAM

Consider a layer of a medium satisfying the following conditions:

(1) The density distribution $\rho(r)$ has no singularities and the derivative $d\rho/dr$ is a smooth function.

(2) The variation of density with r is relatively slow and such that, for a given mixing angle, the width Δr_R of the resonance layer in the r direction is less than the oscillation length

$$\Delta r_R = (d\rho/dr)^{-1}\Delta\rho_R \geqslant l_m^R/2. \qquad (12)$$

This will be referred to as the adiabatic condition.

(3) The density distribution $\rho(r)$ is a monotonic function of ρ_0 up to $\rho_{min} \simeq 0$. We shall also assume that the resonance density for neutrinos of given energy E falls into the interval $\rho_0 - \rho_{min}$, so that the neutrinos generated in the region with $\rho = \rho_0$ will traverse the resonance layer.

Under this conditions, the solution $P(r)$ is universal with respect to the density distribution. The universality can conveniently be expressed in terms of the following dimensionless parameters. Let

$$n = [\rho(r) - \rho_R]/\Delta\rho_R, \qquad (13)$$

where ρ_R and $\Delta\rho_R$ are determined by the values of E, θ, and Δm^2. We shall measure distance in units of n rather than r. We note that, at resonance, $n = 0$, and, as $\rho \to \infty$, $n \to \infty$, while for $\rho \to 0$, $n \to -(\tan 2\theta)^{-1}$. The initial conditions are set at

$$n_0 = [\rho(r_0) - \rho_R]/\Delta\rho_R, \qquad (14)$$

where n_0 is the number of resonance layers that can be fitted between the point at which the neutrino is created and the resonance layer. When $\rho(r_0) \gg \rho_R$, we have

$$n_0 \approx \rho(r_0)/\rho_R \sin 2\theta \approx 1/\sin 2\theta_m^0$$

[see also (8)], i.e., n_0 is equal to the reciprocal of the mixing parameter at the point at which the ν is created, and increases with distance from the resonance layer and/or with decreasing $\sin^2 2\theta$.

We also introduce the further variable

$$m = (d\rho/dr)^{-1}\Delta\rho_R/l_m^R, \qquad (15)$$

i.e., the number of oscillation lengths in the resonance layer. This number increases as the distribution $\rho(r)$ becomes shallower or $\sin^2 2\theta$ decreases.

When conditions (1)–(3) are satisfied, the solution $P(r)$ is a function of n, n_0, and m, i.e., $P \simeq P(n, n_0, m)$, and is not very dependent on the density distribution $\rho(r)$.

Let us examine some of the properties of the function $P(n, n_0, m)$:

(a) P is an oscillating function of n with period T; by definition, $T \simeq 1/m$ in the resonance layer and, as we depart from resonance, $T \sim 1/[m(n^2+1)^{1/2}]$.

(b) P oscillates around its mean value

$$\bar{P}(n, n_0) = [1 + n_0(n_0^2+1)^{-1/2} n(n^2+1)^{-1/2}]/2; \qquad (16)$$

(c) The amplitude of the P oscillations is $A_P = |P_{max} - P_{min}|$ and is a maximum in the resonance layer

$$A_P^R \approx (n_0^2+1)^{-1/2} \qquad (17)$$

and decreases with distance from the layer. A_P is practically independent of m.

(d) At exit from the layer, for $\rho \to 0$,

$$\bar{P}_0 \to [1 - n_0(n_0^2+1)^{-1/2}\cos 2\theta]/2. \qquad (18)$$

Consider the limit of large n_0. Increasing n_0 means that the point at which the ν is created becomes more distant (in ρ) from the resonance layer. The oscillation amplitude is then shown by (17) to decrease ($A_P \sim 1/n_0$) and the average of $P(n, n_0)$ tends to the asymptotic value given by

$$\bar{P}_a(n) = [1 + n(n^2+1)^{-1/2}]/2. \qquad (19)$$

At exit, $\bar{P}_a \to \sin^2\theta$ (see Fig. 1).

Thus, for small mixing angles θ in vacuum and large n_0 (so that the ν-flux is generated quite far from the resonance layer), the neutrino propagation process takes the form of a virtually nonoscillatory transformation of one type of neutrino into the other.

The interpretation of these results is as follows. The neutrino oscillations in the medium take place around the eigenstates ν_1^m and ν_2^m, i.e., the neutrino oscillates, as in the vacuum, around states of particular mass of ν_1 and ν_2. If the density of the medium varies, there is also a variation in the

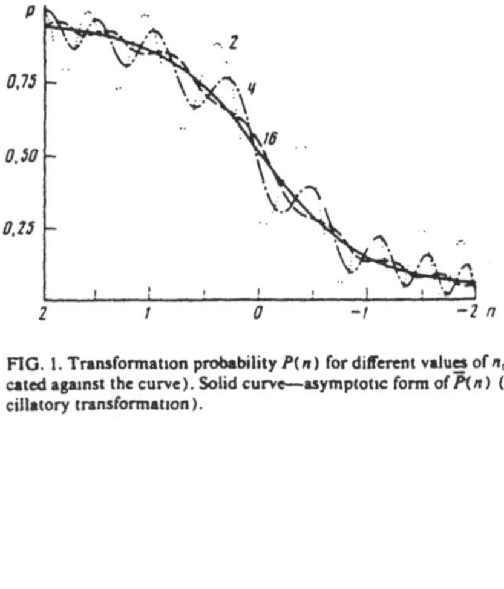

FIG. 1. Transformation probability $P(n)$ for different values of n_0 (indicated against the curve). Solid curve—asymptotic form of $\bar{P}(n)$ (nonoscillatory transformation).

eigenstates ν_1^m and ν_2^m or, more precisely, there is a change in the mixture of ν_e and ν_μ. When $\rho \gg \rho_R$, it can be shown from (8) that $\theta_m \simeq \pi/2$, i.e., ν_2^m becomes practically identical with ν_e. When $\rho = \rho_R$, we have $\theta_m = \pi/4$ and ν_2^m contains equal admixtures of ν_e and ν_μ. When $\rho \ll \rho_R$, we have $\theta_m = \theta$ and, when θ is small, ν_2^m consists mostly of ν_μ. When the density changes from $\rho \gg \rho_R$ to $\rho \ll \rho_R$, the basis ν_1^m, ν_2^m rotates through 90° relative to the basis ν_e, ν_μ. The adiabatic condition then shows that the neutrino state $\nu(t)$ is altered as a result of the variation in density: $\nu(t)$ follows ν_1^m and ν_2^m, while the impurities ν_1^m and ν_2^m in $\nu(t)$ undergo only a small change.

Let us suppose that the neutrinos ν_e are generated in a region with $\rho_0 \gg \rho_R$ and then cross layers with continuously decreasing density ($\rho \to 0$). Initially, $\theta_m^0 \simeq 90°$, $\sin^2 2\theta_m^0 \simeq \sin^2 2\theta(\rho_R/\rho_0)^2 \ll 1$, and $\nu(0) = \nu_e \simeq \nu_2^m$. As the density tends to zero, the state ν_2^m rotates through 90°, as noted above, and because of the adiabatic property, ν_2^m rotates together with $\nu(t)$. When θ is small, $\nu(t) \simeq \nu_2^m \simeq \nu_\mu$ in the final state. Thus, ν_e becomes transformed into ν_μ. As the distance of the point of creation of the neutrinos from the resonance layer increases (in ρ), $\sin^2 2\theta_m^0$ decreases, and the difference between $\nu(t)$ and ν_2^m also decreases. There is an attendant reduction in the oscillation depth. In the limit of large ρ_0, the function $\nu(t)$ becomes practically identical with ν_2^m, i.e., the eigenstate in the medium, and, consequently, there are no oscillations. This limiting case corresponds to a nonoscillatory transformation.

4. APPLICATIONS OF OSCILLATION EFFECTS

We shall now formulate the general conditions for the above effects to produce observable consequences.

(1) Both amplification of oscillations and significant changes in the properties of the ν-beam in the adiabatic regime are due to the crossing of the resonance layer by the neutrinos. The resonance condition is satisfied for a particular sign of l_ν/l_0 or $\Delta f(0)\cos 2\theta/\Delta m^2$ [see (8)]. Since replacing ν with $\bar{\nu}$ produces a change in the sign of $\Delta f(0)$, the resonance condition is satisfied in a given medium only for a neutrino or an antineutrino. If the oscillation effects are amplified in the ν-channel, they are suppressed in the $\bar{\nu}$-channel, and vice versa.

(2) the matter effect occurs in a charge-asymmetric medium. If the particle and antiparticle densities are equal, we have $\Sigma_i n_i \Delta f_i(0) = 0$.

(3) The typical scale over which the influence of matter is significant is $l \gtrsim l_0 = A/G_F\rho$, where $A = 3.5\times10^4$ km. Hence, it follows that the thickness of the medium must be greater than $d = l\rho > 3.5\times10^9$ g/cm^2.

These conditions and the adiabatic condition are satisfied in the sun as well as in the envelopes and cores of collapsing stars.

5. NEUTRINO FLUXES FROM COLLAPSING STARS

The effects examined above can occur in a wide range of values of Δm^2 and $\sin^2 2\theta$ in the outer layers of the cores (above the neutrino sphere) and in the envelopes of collapsing stars.

For the purposes of estimates, we shall use the density distribution in the envelope at the beginning of the collapse, by analogy with the situation prevailing in white dwarfs:

$$\rho \approx \rho_0[(R_s/r)-1]^3, \tag{20}$$

where $\rho_0 \simeq (8\text{–}10)\times10^5$ g/cm^3 and $R_s = 5\times10^8$ cm. The maximum density in the interior is $\rho_{max} \simeq 10^9$ g/cm^3. For massive stars, there is, in addition to (20), an extended hydrogen envelope with $\rho \simeq 10^{-9}$ g/cm^3. The envelope may begin to expand during the collapse process with velocity $v \simeq 5000$ km/s.

For collapsing cores, we shall use the model density distribution and the neutron and electron densities given in Ref. 7. The density in the neutrino sphere will be taken to be $\rho \simeq 10^{12}$–10^{13} g/cm^3.

We have used these density distributions to calculate the ν-parameter ranges (see Fig. 2) for which the adiabatic transformation conditions (1)–(3) of Section 3 are satisfied. The upper limits for Δm^2 are determined by the maximum density. The lower limits for $\sin^2 2\theta$ follow from the adiabatic conditions. The values of Δm^2_{max} for $\nu_e \leftrightarrow \nu_\mu$ and $\nu_e \leftrightarrow \nu_s$ (ν_s is the sterile state) are different:

$$\Delta m^2_{max}(\nu_e \leftrightarrow \nu_s) = 3\cdot10^{-2}\Delta m^2_{max}(\nu_e \leftrightarrow \nu_\mu).$$

This difference is largely due to the strong neutronization of the core ($n_n \gg n_e$). We note that in the case of the ($\nu_e \leftrightarrow \nu_s$)-oscillations, the ratio l_ν/l_0 changes sign for $\rho \simeq 10^9$ g/cm^3 in the core. Here, we have a cancellation of the effects on neutrons, on the one hand, and on electrons and protons, on the other. This means that, if most of the conversion in the core is to the neutrino, most of the conversion in the envelop will be to the antineutrino.

Since, under the above conditions, the neutrino state is practically the same as one of the eigenstates in the medium, there is no loss of coherence due to the spreading of the wave packets corresponding to the different ν_i^m.

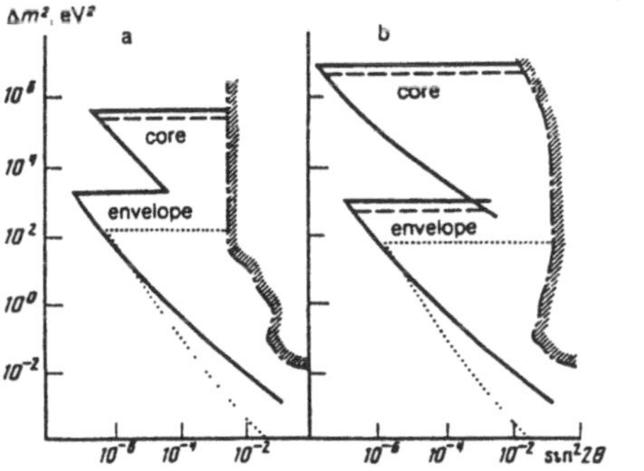

FIG. 2. Range of ν-parameters in which the cores and envelopes of collapsing stars exhibit strong adiabatic transformation: a—$\nu_e \leftrightarrow \nu_\mu$, b—$\nu_e \leftrightarrow \nu_s$ oscillations. Below the shaded line—effect at exit from the envelope: $\bar{P}_\nu < (\sin^2 2\theta)/2$, dotted line—effect of an expanding envelope at the end of the ν-burst ($t = 10$ s); dot-dash lines—experimental limits.

For the ranges of values of the parameters Δm^2, $\sin^2 2\theta$ given in Section 3 (Fig. 2), the factor representing the suppression of the flux of neutrinos of the original type approaches something between $(1/4)\sin^2 2\theta$ and $(1/2)\sin^2 2\theta$.

We must now consider the consequences of the oscillation effect in matter from the point of view of detection of ν-bursts due to collapses.

(a) Suppose that conditions (1)–(3) are satisfied in the $\bar{\nu}_e \leftrightarrow \bar{\nu}_\alpha$ channel, where $\bar{\nu}_\alpha = \bar{\nu}_\mu$ or $\bar{\nu}_\tau$. In that case, the $\bar{\nu}_e$ flux will be almost completely transformed into the $\bar{\nu}_\alpha$ flux, and vice versa; $\bar{\nu}_e$ and $\bar{\nu}_\alpha$ will exchange their spectra. It is assumed that the $\bar{\nu}_\mu$ and $\bar{\nu}_\tau$ have harder spectra than the $\bar{\nu}_e$ (the total fluxes are roughly equal for the entire burst). The exchange of the spectra between the $\bar{\nu}_e$ and $\bar{\nu}_\alpha$ means that there is a substantial increase in the number of events in scintillation counters recording $\bar{\nu}_e$ $(\sigma_\nu \sim E_\nu^2)$ (Ref. 8).

(b) If the resonance condition is satisfied for the neutrinos, $\nu_e \to \nu_\alpha$ $(\nu_\alpha = \nu_\mu, \nu_\tau)$, an effect analogous to that just described may be expected in systems based on radiochemical methods (ν_e detection).

The $\nu_\alpha \leftrightarrow \nu_e$ or $\bar{\nu}_\alpha \leftrightarrow \bar{\nu}_e$ spectrum exchange in the stellar core leads to a substantial release of energy in the envelope because of $\nu_e e$- or $\bar{\nu}_e e$-scattering. This may play a definite part in the mechanism responsible for the shedding of the envelope.

(c) Resonance in the $\nu_e \to \nu_s$ or $\bar{\nu}_e \to \bar{\nu}_s$ channels. The sterile neutrino fluxes appear to be very low. The effect due to the precession of spin in the magnetic field is negligible for $H \lesssim 10^{12}$ G and $\Delta m^2 \lesssim 10^3$ eV^2. The $\nu_e \to \nu_s$ or $\bar{\nu}_e \to \bar{\nu}_s$ transformations will therefore give rise to strong (by several orders of magnitude) suppression of the ν_e or $\bar{\nu}_e$ fluxes.

We emphasize the importance of simultaneous experiments on the detection of ν_e and $\bar{\nu}_e$.

We note that, since l_ν/l_0 has opposite signs in cores and envelopes, a strong effect may be present for both $\bar{\nu}_e$ and ν_e with different energies.

In the case of mixing of three or more neutrino types, it is possible that two or more differences Δm^2 will fall into the range of strong adiabatic conversion. A combination of the above effects may then be observed.

We note that matter effects may not remain constant in the course of a ν-burst ($\Delta t \simeq 5$–20 s). This time dependence will reflect the variation in the structure of the core and envelope.

The authors are indebted to L. Wolfenstein, G. T. Zatsepin, A. Yu. Ignat'ev, D. K. Nadeizhin, V. A. Rubakov, V. G. Ryasnoĭ, and M. E. Shaposhnikov for useful discussions.

[1)]Apart from θ_m, the oscillation depth is also found to depend on the initial conditions. The formula (7) then corresponds to the initial condition (4).

[1]L. Wolfenstein, Phys. Rev. D 17, 2369 (1978).
[2]L. Wolfenstein, Phys. Rev. D 20, 2634 (1979).
[3]V. Barger, K. Whisnant, S. Pakvasa, and R. K. N. Phillips, Phys. Rev. D 22, 2718 (1980).
[4]S. Pakvasa, DUMAND-80 2, 45 (1981).
[5]H. J. Haubold, Astrophys. Space Sci. 82, 457 (1982).
[6]P. V. Ramana Murphy, Proc. Eighteenth Intern. Conf. on Cosmic Rays, Bangalore, 1983, Vol. 7, p. 125.
[7]V. S. Imshennik and D. K. Nadezhin, Preprint ITEF-98, 1980.
[8]A. E. Chudakov, O. G. Ryazhskaya, and G. T. Zatsepin, Proc. Thirteenth Intern. Conf. on Cosmic Rays, Denver, 1973, Vol. 3, p. 2007.

Translated by S. Chomet

TREATMENT OF $\nu_{\odot}$-OSCILLATIONS IN SOLAR MATTER THE MSW EFFECT

Albert Messiah
Institut de Recherche Fondamentale,
CEN-Saclay, 91191 Gif-sur-Yvette Cedex, France

ABSTRACT

Mikheyev and Smirnov, following Wolfenstein's theory of neutrino oscillations in the presence of matter, have found that the change of flavour of solar neutrinos may be spectacularly enhanced in the presence of solar matter, when the parameters of the neutrino mass operator fall in a suitable range *(MSW effect)*. It is shown that this effect can be readily deduced from the adiabatic solution of the equation of flavour evolution. A complete study of the two-flavour case is given, permitting to calculate, for any set of values of the mass operator parameters, the ν_e suppression factor at the site of detection on earth. The adiabatic approximation holds over a wide range of the parameters, leading to especially simple expressions. Our calculations cover the whole range, including domains where the adiabatic approximation is no longer valid. Some of the results, presented in a form most suited for an analysis of solar neutrino experiments, are displayed for illustration and discussed.

If $q/\delta m^2 > 10^9$, the vacuum oscillation length exceeds $R_\odot$: $B(x) \gg A$ during most of the travelling within the sun, keeping flavour ν_e up to the solar surface, from where the usual theory of ν-oscillations applies.

2. Outline of the solution

The Schrödinger equation amounts to a system of two (or n, if the number of flavours is n) coupled differential equations, with a Hamiltonian, M(x), which *does not stay constant in "time"*, due to the presence of the matter term:

$$M'(x) = B(x)\ \vec{\sigma}.\vec{e} \qquad B(x) \propto n_e(x)$$

It, at any rate, can be solved numerically.

The solution takes a very simple form, though, if the eigenvectors of M(x) *rotate slowly enough (adiabatic approximation*[4]*)*: the system "follows adiabatically" the eigenvectors of M(x).

To be precise, let

$$\varpi(x) = \frac{|\text{rotation velocity of eigenvector}|}{|\text{level spacing}|} \tag{2.8}$$

If, in a given range $[x_1, x_2]$, $\varpi(x)$ stays small, i.e. $\varpi < \varpi_{max} \ll 1$, the components of $|\psi(x)\rangle$ along the rotating eigenvectors deviate no more than $\approx \varpi_{max}$ from the strict adiabatic result, according to which each one merely picks up a well-defined phase factor.

Hence the method to be followed:

. Solve the eigenvalue problem of M(x) and calculate $\varpi(x)$

. If $\varpi(x) \ll 1$ over the whole range [0,1], the adiabatic solution gives $\langle P_{\nu_e \mapsto \nu_e} \rangle$ to within $\approx \varpi_{max}^2$

. If not, one needs a better solution, namely apply the adiabatic approximation in all ranges of x where $\varpi(x) < \epsilon$, say, and resort to numerical integration in the remainder. One thus obtains $\langle P_{\nu_e \mapsto \nu_e} \rangle$ to within $\approx \epsilon$ at worst.

3. Set up of the two-flavour formalism

Our first task is to solve the eigenvalue problem of M(x), as given by Eq.(1.6-7), and to build the unitary operator $\hat{U}(x'',x')$ which rotates the eigenvectors into one another as "time" x develops.

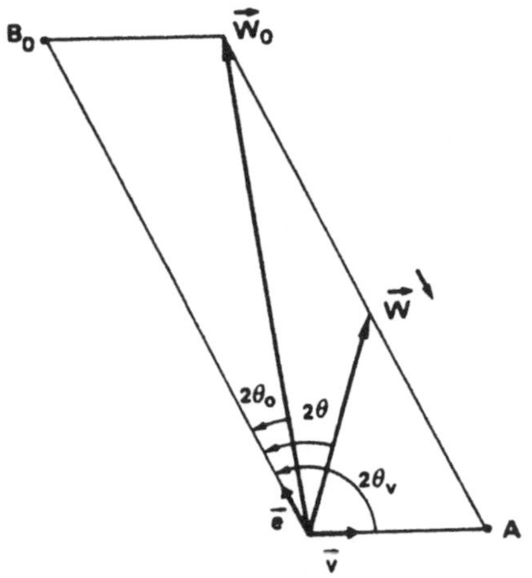

Fig.1b : $\theta_v > \pi/4$
$|e\rangle$ closer to lower mass

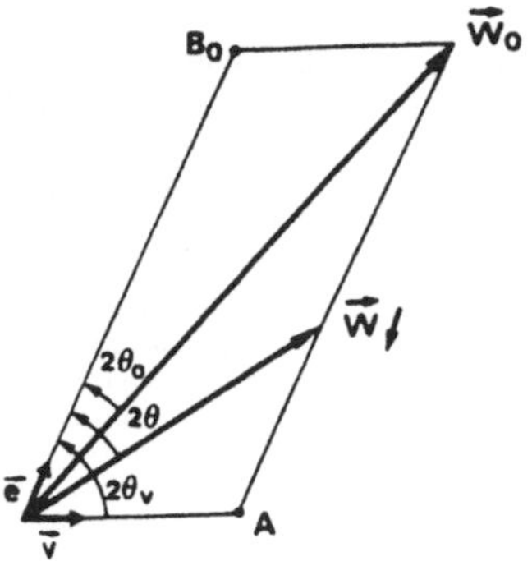

Fig.1a : $\theta_v < \pi/4$
$|e\rangle$ closer to higher mass

and the explicit form of the "rotating axis" operator

$$\hat{U}(x'',x') = \exp\left[i(\theta'' - \theta')\vec{\sigma}.\vec{k}\right] \qquad (3.3)$$

$$\hat{U}(x'',x')|w_\pm\rangle_{x'} = |w_\pm\rangle_{x''} \qquad \hat{U}(x'',x')\vec{\sigma}.\vec{w}' = \vec{\sigma}.\vec{w}''\,\hat{U}(x'',x') \qquad (3.4)$$

When x goes from 0 to 1 - cf. Fig.1a or 1b - B(x) goes from B_0 to 0, and $\vec{W}$ from $\vec{W}_0$ to its value in vacuum A $\vec{v}$. The eigenvalue W varies accordingly, and the "mixing angle in solar matter" $\theta(x)$ goes from its initial value θ_0 to θ_v, mixing angle in vacuum.

$$W(x) = A\sqrt{\kappa^2+2\kappa\cos 2\theta_v+1} \qquad \kappa(x) \equiv B(x)/A$$

$$\cos 2\theta(x) = (\kappa+\cos 2\theta_v)\Big/\sqrt{\kappa^2+2\kappa\cos 2\theta_v+1} \qquad (3.5)$$

The Schrödinger Eq. for U(x,x') will be solved by considering the evolution in the "rotating picture".

Mikheyev-Smirnov-Wolfenstein enhancement of oscillations as a possible solution to the solar-neutrino problem

S. P. Rosen and J. M. Gelb
Theoretical Division, Los Alamos National Laboratory, Los Alamos, New Mexico 87545
(Received 7 April 1986)

Mikheyev and Smirnov have observed that neutrino oscillations in the Sun can be greatly enhanced through the mechanism of Wolfenstein matter oscillations. We develop a qualitative understanding of this phenomenon in the small-mixing-angle limit and carry out extensive calculations in order to apply it to the solar-neutrino problem. Our simple theoretical model agrees remarkably well with the calculations. After determining those values of Δm^2 and $\sin^2 2\theta$ in the small-mixing-angle limit for which the ^{8}B plus ^{7}Be neutrino capture rate in ^{37}Cl is suppressed by a factor 2–4, we predict the corresponding capture rate for *pp* plus ^{7}Be neutrinos in ^{71}Ga. The gallium capture rate can range from no reduction to a factor of 10 reduction. We also determine the modified spectrum of ^{8}B neutrinos arriving at Earth and discuss the importance of this spectrum as a means of choosing between oscillations and the solar model as the cause of the solar-neutrino problem, and also as a means of distinguishing between different sets of oscillation parameters.

I. INTRODUCTION

Mikheyev and Smirnov[1] have recently observed that, given the high densities of matter encountered in the interior regions of the Sun, the oscillations between neutrino flavors caused by small mass differences can be greatly enhanced by Wolfenstein matter oscillations.[2] In particular, intrinsically small mixing angles can be converted by this enhancement mechanism into large effective mixing angles, and a neutrino born as an electron neutrino ν_e in the core of the Sun can emerge from it as an almost pure muon neutrino ν_μ. This means that instead of regarding the solar-neutrino problem as a probe of oscillations with very small mass differences ($\Delta m^2 \simeq 10^{-11}$ eV^2), and relatively large mixing angles ($\sin^2 2\theta \simeq 0.3$), we can also use it to explore very small mixing angles ($\sin^2 2\theta \simeq 0.001-0.1$) and somewhat larger, but still terrestrially inaccessible, mass differences in the range of $10^{-8}-10^{-4}$ eV^2.

In this paper we describe a simplified model which explains the principal features of the Mikheyev-Smirnov-Wolfenstein (MSW) enhancement mechanism, and we report on a series of calculations in which it is assumed to be the cause of the reduction in the capture rate of ^{8}B plus ^{7}Be (upper energy line) neutrinos in ^{37}Cl by a factor 2–4, as observed by Davis.[3] We locate that region in the $(\Delta m^2, \sin^2 2\theta)$ diagram which yields the requisite reduction in the Davis experiment, and we then predict the capture rate for *pp* plus ^{7}Be (both energy lines) neutrinos in ^{71}Ga for each point in this region. We also calculate the corresponding modified spectra of ^{8}B neutrinos arriving at Earth.

II. OSCILLATIONS AND THE SOLAR MODEL

On the basis of our calculations we can develop the following scenario for the solar-neutrino problem. There are two lines in the small-$\sin^2 2\theta$ region of the $(\Delta m^2, \sin^2 2\theta)$ plane for which the ^{8}B plus ^{7}Be capture rate in ^{37}Cl is reduced by a factor $\simeq 3$ [see Fig. 5(a)]: (i) $\Delta m^2 \simeq 10^{-4}$ for all $\sin^2 2\theta$ and (ii) $\log_{10}(\sin^2 2\theta) + \log_{10}(\Delta m^2) \simeq -7.5$ [Fig. 5(a) is a log-log plot.]

From calculations of the modified spectra of ^{8}B neutrinos arriving at Earth [Figs. 6(a) through 6(f) and Table II], we see that the first line corresponds to the case in which the higher-energy ^{8}B neutrinos are suppressed, while the lower-energy ones are unaffected. (The separation between "high" and "low" occurs in the neighborhood of 5–7 MeV, depending on the value of $\sin^2 2\theta$.) By contrast, the second line corresponds to a suppression of all ^{8}B neutrinos, but most especially in the low- and middle-energy range. It follows that for oscillation parameters on the first line, the spectra of *pp* plus ^{7}Be neutrinos will be unaltered, and the capture rate in ^{71}Ga will be exactly as in the standard model, namely, about 105 solar neutrino units[4] (SNU). Parameters on the second line, however, modify the *pp* and ^{7}Be spectra to varying degrees, and they can give reductions in the ^{71}Ga capture rate by as much as a factor of 10. The calculated capture rates as functions of Δm^2 and $\sin^2 2\theta$ are shown in Table I.

The impact of the first-line solution on ^{71}Ga can also be understood on the basis of the spatial distribution of neutrino production. In Sec. IV we demonstrate that the first line corresponds to the occurrence of the MSW effect in the core of the Sun, and that the second line corresponds to the effect occurring in the main body of the Sun. A much higher percentage of ^{8}B neutrinos is created in the core than is the case for *pp* or ^{7}Be neutrinos.[4]

The MSW mechanism is a beautiful device for enhancing oscillations with intrinsically small mixing angles, and it presents us with a most attractive solution to the solar-neutrino puzzle. From this point of view, the ^{71}Ga experiment will enable us to determine whether the first or the second line is the correct solution, and a measurement of

TABLE I. Various expected capture rates for $pp-{}^{7}\mathrm{Be}$ in $^{71}\mathrm{Ga}$ in SNU for parameters that yield a factor of 3 reduction in $^{8}\mathrm{B}+{}^{7}\mathrm{Be}$ in $^{37}\mathrm{Cl}$ (underlined values). The rest were filled in for completeness. 100 SNU was taken to be 100%.

Δm^2 eV^2 \ $\sin^2 2\theta$	$10^{-3.0}$	$10^{-2.5}$	$10^{-2.0}$	$10^{-1.5}$	$10^{-1.0}$	$10^{-0.5}$
1.1×10^{-4}	100	100	100	100	100	100
1.0×10^{-4}	100	100	100	100	100	100
9.5×10^{-5}	100	100	100	100	100	100
5.8×10^{-5}	100	100	100	100	100	100
5.0×10^{-5}	100	100	100	100	100	100
1.7×10^{-5}	100	100	100	100	100	95
3.6×10^{-6}	65	60	60	60	50	45
1.1×10^{-6}	45	15	10	10	20	25
3.5×10^{-7}	70	40	10	5	5	20
1.0×10^{-7}	85	70	40	10	5	20

the spectrum will provide the final confirmation for it. The spectrum can also be used to remove ambiguities between the MSW mechanism and older attempts to solve the puzzle.

For example, should the outcome of the ^{71}Ga experiment be exactly as predicted by the standard solar model, then the first-line MSW solution will most likely be correct. However, unless we measure the spectrum of ν_e from ^{8}B decay arriving at Earth, we will not be able to exclude some modification of the solar model as an explanation of the Davis experiment. The first-line solution leads to a suppression of the higher-energy half of the spectrum, whereas a modification of the solar model will reduce the overall normalization, but not the shape of the spectrum.

Should the outcome of the ^{71}Ga experiment be a significantly reduced capture rate, as compared with the predictions of the standard solar model, then we can certainly conclude that neutrino oscillations are taking place. In the event that the reduction factor for ^{71}Ga turns out to be smaller than that for ^{37}Cl (say, $\frac{3}{2}$ instead of 3) or much larger (say, 5–10, see Table I), it would be most likely that the second-line MSW solution is correct. If however, the reduction factors for the two experiments are approximately equal to one another, then we will have to choose between MSW-enhanced oscillations with parameters in the range $\sin^2 2\theta \simeq 10^{-2}$–$10^{-1.5}$ and $\Delta m^2 \simeq 4$–1×10^{-6} eV^2 (see Table I), and the possibility of large mixing angles ($\sin^2 2\theta \simeq 0.3$–$0.4$) with small $\Delta m^2 (\simeq 10^{-10}$–$10^{-12}$ eV^2). Again we can use the spectrum to make the choice.

If we plot the probability that a ν_e emitted in an ^{8}B decay in the Sun remains a ν_e when it arrives at Earth, as a function of energy, i.e., $P(\nu_e\to\nu_e;\ E_\nu)$, then we find that for oscillation parameters in the MSW range, the lower-energy neutrinos are heavily suppressed, but the probability function climbs steadily to a value of 0.4–0.5 at the high-energy end [see Figs. 6(c) and 6(d)]. In the case of large mixing angles and $\Delta m^2 \simeq 10^{-10}$ eV^2, the oscillation lengths for both ^{8}B and pp neutrinos are much smaller than the astronomical unit, and so the suppression is independent of energy ($P\simeq 1-\frac{1}{2}\sin^2 2\theta$). As Δm^2 becomes smaller, some oscillatory structure will develop in $P(\nu_e\to\nu_e;\ E_\nu)$ for ^{8}B neutrinos, but at the high-energy end the probability will generally tend to a value close to unity.

III. SLAB MODEL

We now describe a simple model for the MSW effect which accounts for most of the qualitative features, and even the quantitative features, of our computations in the case of small mixing angles.

According to Wolfenstein[2] the time development of vacuum and matter oscillations is described by the differential equation

$$i\frac{d}{dt}\begin{pmatrix}\nu_e\\ \nu_\mu\end{pmatrix}=\begin{pmatrix}A & B\\ B & D\end{pmatrix}\begin{pmatrix}\nu_e\\ \nu_\mu\end{pmatrix} \tag{3.1}$$

in the basis of eigenstates $|\nu_e\rangle$ and $|\nu_\mu\rangle$. The elements of the "Hamiltonian" matrix are

$$A\equiv\frac{m_1{}^2c^2+m_2{}^2s^2}{2E}+\sqrt{2}G_F N_e\ ,$$
$$B\equiv\frac{\Delta m^2}{2E}cs\ , \tag{3.2}$$
$$D\equiv\frac{m_1{}^2s^2+m_2{}^2c^2}{2E}\ ,$$

where m_1 and m_2 are the masses of the neutrino mass eigenstates $|\nu_1\rangle$ and $|\nu_2\rangle$, respectively, and $\Delta m^2 \equiv m_2{}^2-m_1{}^2$ is measured in units of eV^2. The mixing angle θ is defined by $c\equiv\cos\theta$, $s\equiv\sin\theta$, and

$$|\nu_e\rangle=c\,|\nu_1\rangle+s\,|\nu_2\rangle\ ,$$
$$|\nu_\mu\rangle=-s\,|\nu_1\rangle+c\,|\nu_2\rangle\ . \tag{3.3}$$

G_F is the Fermi constant, N_e is the density of electrons in the Sun, and E is the neutrino energy (momentum) measured in units of MeV. The location of the neutrino in the Sun at any time t is R, where $R\simeq ct=t$ ($\hbar=c\equiv1$).

It should be noted that the matter oscillation contribution to A, namely, $\sqrt{2}G_F N_e$ in (3.2), differs from that originally given by Wolfenstein in the factor of $\sqrt{2}$ and in sign. The factor of $\sqrt{2}$ has been noted by several authors[5] and it comes about because, when one evaluates the coherent (i.e., spin-nonflip) part of the W^+-exchange diagram for $\nu_e - e$ scattering in the electron rest frame, one must keep in mind that the incident neutrino is polarized, essentially 100%. As for the sign, there are now three independent arguments which agree with the one above: one is a direct calculation by Langacker;[6] another, given by Wolfenstein,[7] is based on the observation that the forward amplitude for $\bar{\nu}_e - e$ scattering below the W pole in the s channel must have a definite sign; and the third is based on the facts that, in the Glashow-Weinberg-Salam model, the W-exchange diagram has the opposite sign to the Z^0-exchange diagram, and that Z^0 exchange gives rise to an attractive force between ν_e and e (Ref. 8). As emphasized by Langacker, the positive sign means that MSW enhancement will now occur when the dominant component of ν_e is the lighter of the two neutrinos.

As the neutrino travels through the Sun, the density of electrons, N_e, changes and hence the matrix element A in (3.1) is a function of time. If we eliminate ν_μ from the coupled equations of (3.1) and take out a time-dependent phase factor from ν_e,

$$\nu_e(t)=p(t)\exp\left[-i\int_0^t A(t')dt'\right], \tag{3.4}$$

we find that $p(t)$ obeys the second-order equation

$$\ddot{p}-i(A-D)\dot{p}+B^2p=0 . \tag{3.5}$$

The probability that a neutrino born as a ν_e at $t=0$ remains a ν_e at time t is given by

$$P(\nu_e \xrightarrow{t} \nu_e;t)=|p(t)|^2 . \tag{3.6}$$

The relationship between the effective mixing angle for oscillations in matter, $\sin^2 2\theta_m$, and the intrinsic mixing angle $\sin^2 2\theta$ can be written as

$$\sin^2 2\theta_m=\frac{\sin^2 2\theta}{[\sin^2 2\theta+(L/L_0-\cos 2\theta)^2]}, \quad L\equiv 4\pi E/\Delta m^2, \; L_0\equiv 2\pi/(\sqrt{2}G_F N_e) . \tag{3.7}$$

Obviously the MSW enhancement occurs at $L/L_0=\cos 2\theta$ and its full width at half maximum is given by

$$2\Delta(L/L_0)=2\sin 2\theta . \tag{3.8}$$

For very small intrinsic mixing angles ($\sin^2 2\theta \simeq 10^{-3}-10^{-1}$) this is a very narrow region, and outside it $\sin^2 2\theta_m$ is very small. Thus we can adopt a simple model for small mixing angles in which a neutrino travels from the core of the Sun to the narrow region around the enhancement without oscillating, undergoes maximal oscillations inside this region, and emerges from it as some admixture of ν_e and ν_μ which remains essentially unchanged all the way to Earth.

In terms of the electron density ρ_e, measured in units of Avogadro's number, the ratio L/L_0 is given by

$$L/L_0=10^{-7}(E/\Delta m^2)\sqrt{2}\rho_e \tag{3.9}$$

and the density at enhancement is

$$(\rho_e)_0=\frac{10^7}{\sqrt{2}}(\Delta m^2/E)\cos 2\theta . \tag{3.10}$$

For fixed $E/\Delta m^2$, the change in density across the MSW-enhancement region is from (3.8):

$$2\Delta\rho_e=\sqrt{2}\times 10^7(\Delta m^2/E)\sin 2\theta = 2(\rho_e)_0\tan 2\theta . \tag{3.11}$$

The corresponding spatial extent of the region is therefore

$$2\Delta R_0\equiv 2\Delta t_0=2[\Delta\rho_e/(d\rho_e/dR)]_0 . \tag{3.12}$$

We now examine the model in which the neutrino oscillates in the region

$$(R_0-\Delta R_0,R_0+\Delta R_0)\equiv(t_0-\Delta t_0,t_0+\Delta t_0)$$

and nowhere else; $R_0=t_0$ is the location of the MSW enhancement.

An equivalent way of writing the enhancement condition ($L/L_0=\cos 2\theta$) is $A-D=0$ [see (3.1) and (3.2)]: therefore the probability amplitude $p(t)$ of (3.5) obeys a simple oscillatory equation $\ddot{p}+B^2p=0$ in the neighborhood of the enhancement. If we assume that just before it hits the enhancement region $p=e^{i\alpha}$, then from continuity across the enhancement region boundary we find that

$$p(t)=e^{i\alpha}\cos B(t-t_0+\Delta t_0) \quad (t_0-\Delta t_0\le t\le t_0+\Delta t_0) . \tag{3.13}$$

Thus $p(t)$ emerges from the enhancement region with the value

$$p(t_0+\Delta t_0)=e^{i\alpha}\cos(2B\Delta t_0) \tag{3.14}$$

and the probability for its being an electron neutrino at Earth is

$$P(\nu_e\rightarrow\nu_e)_{\text{Earth}}\equiv\cos^2\left[\frac{\sin^2 2\theta}{h_0}\frac{\Delta m^2}{2E}\right], \tag{3.15}$$

where h_0 is the scale height of the solar density distribution at enhancement:

$$h_0\equiv-\left[\frac{1}{\rho_e}\frac{d\rho_e}{dR}\right]_0 . \tag{3.16}$$

The essential feature of this model, and the one that determines its domain of validity, is that the distance $2\Delta R_0$ over which maximal mixing occurs ($\sin^2 2\theta_m \simeq 1$) is assumed to be small compared with the oscillation length at the point of enhancement:

$$2\Delta R_0 \ll L_m\equiv 4\pi E/(\Delta m^2\sin 2\theta)=\pi/B . \tag{3.17}$$

Thus the argument of the cosine function in (3.14) and (3.15) is always less than π, and the change in the nature of the incident neutrino takes place in some fractional part of the maximally oscillating wave. In other words, the region sensitive to MSW enhancement is so small that it locks in only a part of the wave.

This is essentially the reverse of the adiabatic approximation,[9] in which the region of large mixing angles $\sin^2 2\theta_m$ must be large compared with the oscillation length L_m at the point of enhancement. This condition makes it possible for the eigenvectors of the "Hamiltonian" (or effective mass) matrix (3.1) and (3.2) to rotate

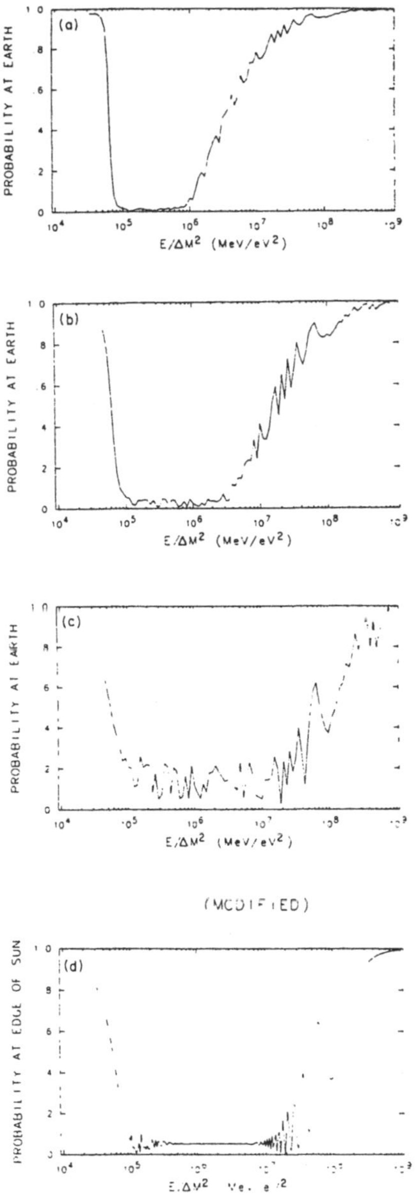

FIG. 1. Probability that a single-electron neutrino created at the center of the Sun arrives at Earth versus $E/\Delta m^2$ for a given $\sin^2 2\theta$. (a) $\sin^2 2\theta = 0.01$; (b) 0.04; (c) and (d) 0.20. (d) is modified: probability at edge of the Sun and more $E/\Delta m^2$ points help smooth out the plot.

slowly as the neutrino moves through the enhancement region; consequently an eigenvector associated with the larger (smaller) of the eigenvalues at one point in the path of the neutrino maintains its association with the larger (smaller) eigenvalue at another point.

In terms of $\sin^2 2\theta$ and $E/\Delta m^2$, the criterion for validity of our slab model is that

$$\sin^2 2\theta/(E/\Delta m^2) < 2\pi h_0 \ . \tag{3.18}$$

For a fixed, small value of $\sin^2 2\theta$, the probability for $\nu_e \Longrightarrow \nu_e$ at Earth (3.15) starts out at 1, goes down to zero, and climbs steadily back to 1 as $E/\Delta m^2$ increases through the range of values consistent with (3.18). The scale for this behavior is set by $\sin^2 2\theta$: the smaller $\sin^2 2\theta$, the smaller the region in $E/\Delta m^2$ for which the probability remains small. These qualitative features are well illustrated by the results of computer calculations shown in Fig. 1, especially for the regime in which the probability climbs from small values back to large ones. Earlier parts of the curve are better described by the adiabatic approximation.

IV. APPLICATIONS OF THE SLAB MODEL

Another aspect of the slab model concerns the relation between $\sin^2 2\theta$ and Δm^2 for a fixed suppression. If the scale height h_0 is a constant, as happens for the exponential density distribution found in a large part of the Sun beyond the core (see Fig. 7), then for a *fixed* value of $P(\nu_e \to \nu_e)$, we must have, from (3.15), that

$$(\sin^2 2\theta)(\Delta m^2) = \text{const}$$

or (4.1)

$$\log_{10}(\sin^2 2\theta) + \log_{10}(\Delta m^2) = \text{const} \ .$$

Furthermore this property survives when we integrate the product $P(\nu_e \to \nu_e) \times \text{flux} \times \text{capture cross section}$ over the energy range of ^{8}B neutrinos. By a simple change of variables we can show in this case that the quantity fixed by a given total suppression of the ^{8}B capture rate [in other words, the analog of the argument of the cosine function in (3.15)] is

$$Q = \left| \frac{\sin^2 2\theta}{h_0} \frac{\Delta m^2}{E_{\max}} \right| , \tag{4.2}$$

where $E_{\max}$ is the maximum energy of the ^{8}B neutrinos. Furthermore, Q has the general property that as the integral of $P(\nu_e \to \nu_e; E)$ increases, so the value of Q must decrease.

Given the scale height of the density distribution in the body of the Sun, we can use (4.2) to determine the constant appearing on the right-hand side of (4.1). If we express the density at R as an exponential function

$$\rho_e(R) = \rho_e(0)\exp(-R/\beta) \ , \tag{4.3}$$

where R and β are measured in meters, then we find from Fig. 7 that

$$\beta \simeq R_{\text{Sun}}/10 \simeq 7 \times 10^7 \text{ m} \ . \tag{4.4}$$

The scale height is simply the inverse of β, and so

$$h_0=(7\times 10^7)^{-1}m^{-1}\simeq 3\times 10^{-15}\text{ eV}\ . \qquad (4.5)$$

It then follows from (4.2) that

$$\log_{10}(\sin^2 2\theta)+\log_{10}(\Delta m^2\ (\text{eV})^2)=\log_{10}Q+\log_{10}h_0 +\log_{10}(E_{max}(\text{eV})) =\log_{10}Q-7.4\ . \qquad (4.6)$$

Since Q is a number of order 1 for suppression factors in the range 2—4, we obtain

$$\log_{10}(\sin^2 2\theta)+\log_{10}(\Delta m^2)\simeq -7.4\ . \qquad (4.7)$$

Considering the crudeness of our model, the agreement between (4.7) and the second solution in Fig. 5(a) is extremely good. Because we neglect energy spreading, spatial spreading, and multineutrino types (all of which are included in Fig. 5), we can expect some discrepancies.

We also note that as $P(\nu_e\to\nu_e)$ increases (i.e., the suppression factor decreases), Q decreases and so the constant on the right-hand side of (4.6) decreases. Thus the second line in Fig. 5 should move down the page as the suppression factor decreases, and up the page as it increases. This is exactly what happens in the computer calculations.

From (3.18), the criterion for the validity of the slab model for the exponential density distribution is

$$(\sin^2 2\theta)(\Delta m^2)<2\pi h_0 E\ (\text{eV})\simeq 10^{-6.7}\ . \qquad (4.8)$$

The range of parameters on the second line are well within this limit.

As has been emphasized by Bethe,[9] the first line solution of Fig. 5 corresponds to the adiabatic approximation, which is valid when the inequality in Eq. (3.18) is reversed. This approximation accounts for the earlier (in $E/\Delta m^2$) parts of the curves of Fig. 1 in which the probability for an original ν_e to remain a ν_e at Earth starts out close to unity, falls rapidly, and then remains close to zero. Since the region of rapid fall occurs close to $E/\Delta m^2\approx 10^5$ for a large range of values for $\sin^2 2\theta$, the solution is approximately constant as a function of mixing angle. The value of $\Delta m^2(\approx 10^{-4}\text{ eV}^2)$ indicates that the enhancement occurs in the core region of the Sun.

V. COMPUTER CALCULATIONS

Equation (3.5) plus the initial conditions $\langle\nu_e(0)|\nu_e(0)\rangle=1$ and $\langle\nu_\mu(0)|\nu_\mu(0)\rangle=0$ form the basis of our computations. It is convenient to represent p as $w+is$. Equation (3.5) then becomes two coupled second-order differential equations in w and s. The above initial conditions can be shown to be equivalent to $w(0)=1$ and $s(0)=\dot w(0)=\dot s(0)=0$.

We use Runge-Kutta techniques for two coupled second-order differential equations to solve for P_{edge}. We use the following parameters:

$$\begin{aligned} &x\equiv R/R_{Sun}(R_{Sun}\cong 6.96\times 10^5\text{ km})\ ,\\ &L\equiv 4\pi E/\Delta m^2\\ &\quad =2.5\times 10^{-3}E/\Delta m^2[\text{MeV}/(\text{eV})^2]\text{ km}\ ,\\ &L_0\equiv 2\pi/(\sqrt{2}G_F N_e)\cong 1.77\times 10^4/\rho_e(x)\text{ km}\ ,\\ &\Delta x=0.00001\ , \end{aligned} \qquad (5.1)$$

where $\rho_e(x)$ is the hydrogen/helium corrected electronic density[10] in units of Avogadro's number (see Fig. 7).

Once we know w, $\dot w$, s, and $\dot s$ at the edge of the Sun ($P_{edge}\equiv w^*w+s^*s$), the probability of Earth, P_{Earth}, can be computed making use of the equation for vacuum-only oscillations;[11] we obtain

$$\begin{aligned} P_{Earth}=&\sin^2 2\theta\sin^2(\pi R_{ee}/L)(1-2P_{edge})+P_{edge}+\sin 2\theta\cos 2\theta\sin^2(\pi R_{ee}/L)(2/B)(w\dot s-s\dot w)\\ &+\sin 2\theta\sin(\pi R_{ee}/L)\cos(\pi R_{ee}/L)(2/B)(w\dot w+s\dot s)\ , \end{aligned} \qquad (5.2)$$

where R_{ee} is the distance from the edge of the Sun to Earth ($\cong 1.5\times 10^8$ km) and L is as defined in (5.1). The last two terms in (5.2) are interference terms which can be constructive or destructive and help explain why the vacuumlike oscillations in Fig. 2 are not bounded by $\sin^2 2\theta$.

The energy spectrum for ^{8}B neutrinos was taken from Bahcall,[12] that for ^{7}Be from Bahcall *et al.*,[4] and the latest *pp* spectrum was personally given to us by Bahcall. The most recent normalizations were used (4.0×10^6 no./cm^2/sec for ^{8}B, 3.2×10^9 for both ^{7}Be lines combined, and 6.1×10^{10} for *pp*) (Ref. 13). Chlorine cross sections were taken from Bahcall[14] and gallium cross sections were taken from Bahcall.[12]

We wrote a series of programs of increasing complexity using the data described above. The simplest program computes $P(\nu_e\to\nu_e)$ at Earth as a function of $E/\Delta m^2$ for a given $\sin^2 2\theta$ assuming that a single electron neutrino is produced at the center of the Sun; typical results are shown in Figs. 1(a), 1(b), and 1(c). The roughness of the oscillations about the envelope are an artifact of the plot, but the oscillations themselves are real and can be smoothed out by taking finer steps in $E/\Delta m^2$, as illustrated in Fig. 1(d).

The next program follows the probability $P(\nu_e\to\nu_e)$ for an electron neutrino as it travels through the Sun with a given mixing angle and oscillation length. Figure 2(a) beautifully displays the regions of suppression, amplification, and vacuum-only oscillations as described by Mikheyev and Smirnov; Fig. 2(b) shows a similar graph for a smaller oscillation length, demonstrating a larger suppression of ν_e. Figures 2(c) and 2(d) show examples of "double enhancement." Neutrinos produced behind the center of the Sun can sometimes undergo enhancement twice due to the symmetric nature of the Sun. However, the derivative of the density switches sign traversing the center. This can lead to further suppression, Fig. 2(c), or raise the probability back up near one, Fig. 2(d), depending on the phase of the wave arriving at the second region of enhancement (thereby dependent on $E/\Delta m^2$).

Subsequent programs take into account the dependence

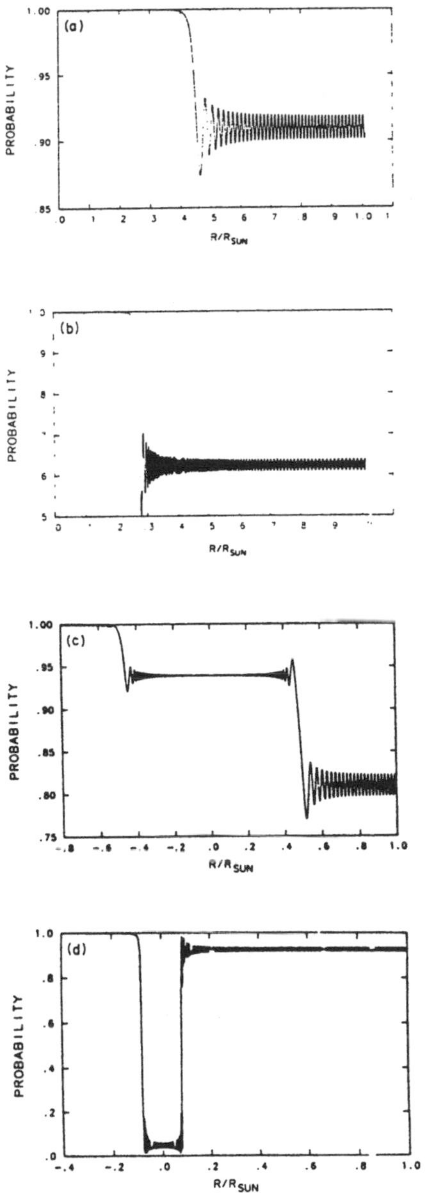

FIG. 2. Probability of a single-electron neutrino created at the center of the Sun at each point in the Sun for a given $\sin^2 2\theta$ and $E/\Delta m^2$. All plots are for $\sin^2 2\theta = 0.001$. (a) $E/\Delta m^2 = 3\times10^6$ MeV/eV^2; (b) 6×10^5; (c) "double enhancement" $E/\Delta m^2 = 5\times10^6$ MeV/eV^2; (d) "double," 1×10^5.

of P_{Earth} on where the neutrinos are produced. ^{8}B neutrinos are produced predominantly at the center of the Sun, but *pp* neutrinos can be produced as far out as one quarter of the solar radius. For comparison, Fig. 3(a) shows the modified results for P_{Earth} with the same parameters as were used to generate Fig. 1(a); note that the region of suppression is now slightly smaller than before. A greater shrinking of this region occurs for *pp* neutrinos as can be seen from a comparison of Figs. 3(b) and 1(a). The method by which spatial dependence is taken into account is described in Appendix A.

The most comprehensive programs allow for the capture cross sections (see Appendix B) and the energy spectra of the ^{8}B and *pp* neutrinos, and they yield capture rates as functions of Δm^2 for specific values of $\sin^2 2\theta$. Typical results for $\sin^2 2\theta = 0.01$ are shown in Fig. 4 in which an averaging process (see Appendix B) has been used to smooth out the curves. Our most important results are contained in Fig. 5, which determines the values of $\sin^2 2\theta$ and Δm^2 for a given reduction of the ^{8}B capture rate; Table I, which gives the predicted capture rates for the ^{71}Ga experiment for the parameters shown in Fig. 5(a); and Fig. 6, which shows the modified spectra of ^{8}B neutrinos at Earth corresponding to these parameters. (Figure 6 is presented in numerical form in Table II.) We

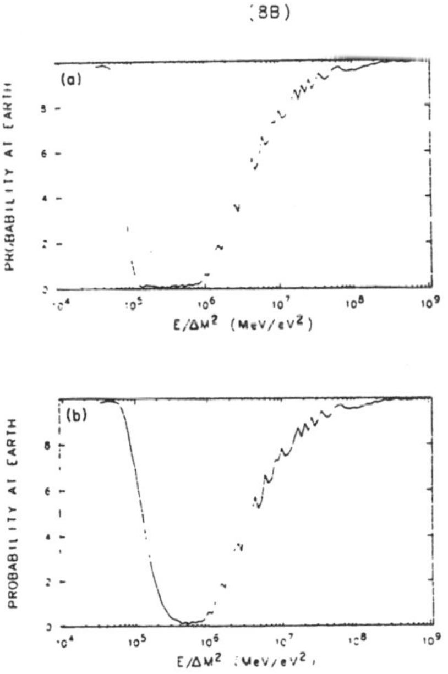

FIG. 3. Similar to Fig. 1 but with spatial spreading (see Appendix A) taken into account. Both plots are for $\sin^2 2\theta = 0.01$. (a) ^{8}B neutrinos; (b) *pp* neutrinos. Not depicted: ^{7}Be neutrinos whose effect lies somewhere in between. All subsequent programs take spatial spreading into account.

Possible Explanation of the Solar-Neutrino Puzzle

H. A. Bethe

Newman Laboratory of Nuclear Studies, Cornell University, Ithaca, New York 14853,[a] *and Institute for Theoretical Physics, University of California, Santa Barbara, Santa Barbara, California 93106*

(Received 27 December 1985; revised manuscript received 27 January 1986)

Mikheyev and Smirnov have shown that electron neutrinos above a certain minimum energy E_m may all be converted into μ neutrinos on their way out through the sun. We assume here that this is the reason why Davis and collaborators, in their experiments, find many fewer solar neutrinos than predicted. The minimum energy E_m is found to be about 6 MeV, the mass m of the μ neutrino must be greater than that of the electron neutrino, $m_2^2 - m_1^2 = 6\times 10^{-5}$ eV^2, and there is a very minor restriction on the neutrino mixing angle.

PACS numbers: 96.60.Kx, 12.15.Ff, 14.60.Gh

Mikheyev and Smirnov (MS), in a very important paper,[1] have discovered a mechanism by which a large fraction of the neutrinos ν_e emitted in the sun may be converted into ν_μ when traversing the sun, and thereby be rendered unobservable. We shall first derive the MS results in a different but equivalent way, and then draw conclusions from it. Like MS, we shall consider only two neutrino flavors, ν_e and ν_μ; the third flavor, ν_τ, will be discussed at the end.

It is generally believed that the neutrinos ν_e and ν_μ are not the ones that propagate in free space, ν_1 and ν_2. Conventionally, we write

$$|\nu_e\rangle = |\nu_1\rangle\cos\theta + |\nu_2\rangle\sin\theta, \quad |\nu_\mu\rangle = -|\nu_1\rangle\sin\theta + |\nu_2\rangle\cos\theta, \tag{1}$$

and assume $\theta < 45°$ The masses are m_1 and m_2. With a transformation to the ν_e, ν_μ representation, the square of the mass is then given by the matrix

$$\mathcal{M} = \tfrac{1}{2}(m_1^2 + m_2^2)\begin{pmatrix}1 & 0\\ 0 & 1\end{pmatrix} + \tfrac{1}{2}(m_2^2 - m_1^2)\begin{pmatrix}-\cos 2\theta & \sin 2\theta\\ \sin 2\theta & \cos 2\theta\end{pmatrix}. \tag{2}$$

Wolfenstein[2] has pointed out that in matter, the masses are changed as a result of the weak-current interaction. The neutral weak current acts equally on ν_e and ν_μ and is therefore unimportant for our purpose. But the charged current will exchange ν_e with the electrons in the matter, while it has no effect on ν_μ. This exchange gives an extra term in the Hamiltonian,[2]

$$H_{\text{int}} = (G/\sqrt{2})\bar{\nu}_e\gamma_\lambda(1+\gamma_5)\nu_e[\bar{e}\gamma_\lambda(1+\gamma_5)e], \tag{3}$$

where G is the Fermi constant of weak interactions. For electrons at rest, only γ_4 ($=1$) contributes, and for neutrinos $1+\gamma_5 = 2$; so (3) is equivalent to a potential energy for ν_e

$$V = G\sqrt{2}N_e, \tag{4}$$

where N_e is the number of electrons per unit volume.

The momentum k of an electron neutrino is then related to its energy E by

$$k^2 + m^2 = (E - V)^2 = E^2 - 2EV, \tag{5}$$

with neglect of the small V^2. Thus, V is equivalent to an addition to m^2 of

$$m_i^2 = 2EV = 2\sqrt{2}(GY_e/m_n)\rho E = A, \tag{6}$$

where m_n is the mass of the nucleon and Y_e the number of electrons per nucleon in the matter, generally $Y_e = \frac{1}{2}$.

When m_i^2 is added to $\mathcal{M}$ the mass matrix (2) becomes

$$\mathcal{M} = \tfrac{1}{2}(m_1^2 + m_2^2 + A)\begin{pmatrix}1 & 0\\ 0 & 1\end{pmatrix} + \frac{1}{2}\begin{pmatrix}A - \Delta\cos 2\theta & \Delta\sin 2\theta\\ \Delta\sin 2\theta & -A + \Delta\cos 2\theta\end{pmatrix}, \tag{7}$$

where

$$\Delta = m_2^2 - m_1^2. \tag{7a}$$

The eigenvalues are

$$m_\nu^2 = \tfrac{1}{2}(m_1^2 + m_2^2 + A) \pm \tfrac{1}{2}[(\Delta\cos 2\theta - A)^2 + \Delta^2\sin^2 2\theta]^{1/2}. \tag{8}$$

The splitting between the two mass eigenvalues is given by the square root.

The splitting has a minimum as a function of A provided that

$$\Delta = m_2^2 - m_1^2 > 0 \tag{9}$$

[$A > 0$ according to (6), and $\cos 2\theta > 0$ by assumption]. The minimum occurs when

$$A = \Delta \cos 2\theta. \tag{10}$$

Figure 1 shows the two eigenvalues (8) of m_ν^2 as a function of A. At low A, i.e., small matter density ρ, the electron neutrino has the smaller mass, but when A reaches the value (10), the two curves would cross if it were not for the coupling term $\Delta \sin 2\theta$. The near-crossing point (10) is the resonance of MS. At larger A, beyond the crossing point, the electron neutrino has larger mass than the μ neutrino.

For these statements to be true, it is essential that the ν_e-e interaction has the positive sign, as indicated in (3) and (4). This sign was given by Wolfenstein in his first paper,[2] but then unfortunately was reversed in his second paper,[3] and the incorrect sign was taken over by MS.[1] They therefore claimed that the crossing of the curves in Fig. 1 (resonance) would occur only if $\Delta < 0$, i.e., ν_e heavier than ν_μ. I am grateful to Paul Langacker for pointing out the correct sign to me. The main features of the theory, however, are independent of the sign, and it is the great contribution of MS to have discovered the significance of the curve crossing (resonance).

Assume now that an electron neutrino is produced in the sun at sufficiently high density that $A > \Delta \cos 2\theta$. Then its m^2 will clearly be given by the + sign in (8). As the neutrino moves outward, A will decrease, Eq. (6), and it will finally hit the resonance (10). At that point, its mass will continue to follow the upper curve in Fig. 1, and it will therefore *emerge from the sun as a μ neutrino* (which cannot be detected). This is the result obtained by MS.

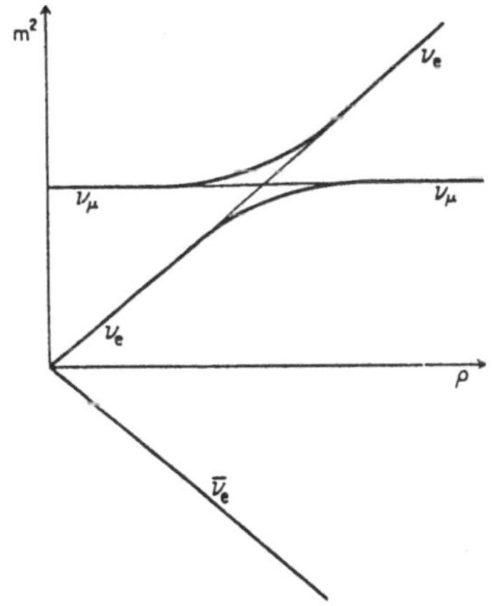

FIG. 1. The masses of two flavors of neutrinos as a function of density. The curves nearly cross at one point. The electron-antineutrino mass $\bar{\nu}_e$ is also shown.

What happens in the resonance is that, for the upper curve in Fig. 1, the state vector which was originally almost in the direction of $|\nu_e\rangle$ turns slowly to the direction $|\nu_\mu\rangle$. Evaluating (6) for $Y_e = \frac{1}{2}$ gives

$$A = 0.76 \times 10^{-7} \rho E \tag{11}$$

if ρ is in grams per cubic centimeter, E in megaelectronvolts, and A in electronvolts squared.

So far the MS theory. Now I propose to take this theory seriously; i.e., we assume that this conversion of ν_e into ν_μ is indeed the cause of the depletion of observable neutrinos from the sun. For any given neutrino energy E, the resonance (10) occurs at a definite density ρ_E. From (10),

$$E\rho_E = 1.3 \times 10^7 \Delta m^2 \cos 2\theta = \Lambda. \tag{12}$$

There is a critical energy $E_c = \Lambda/\rho_c$, where ρ_c is the density at the center of the sun (a definition to be modified later). All neutrinos of energy $E > E_c$ have to go through the resonance; they will emerge as ν_μ and hence be undetectable. The less energetic neutrinos, $E < E_c$, will not go through the resonance and will emerge unscathed as ν_e.

This means that Davis and his collaborators will observe only the solar neutrinos of energy below E_c, but will observe these at full strength. On this assumption, we shall now determine E_c from experiment, using the data from Bahcall *et al.*[4] Table I of that paper gives the composition of the solar neutrino units (SNU) of neutrinos detectable by ^{37}Cl as follows: (i) from ^8B, 4.3 SNU; (ii) from all other nuclear species (*pep*, ^7Be, ^{13}N, and ^{15}O), 1.6 ± 0.2 SNU. All these other neutrinos have maximum energies of 2.8 MeV or less, which will be found to lie below E_c, so that they will be fully detectable. The ^8B neutrinos have a continuous spectrum extending to 14.0 MeV.

The number observed by Davis *et al.* is[4] 2.1 ± 0.3 SNU. Subtracting the expected number from other species, we get for the "observed" neutrinos from ^8B

$$S(^8B, \text{obs}) = 0.5 \pm 0.5. \tag{13}$$

Therefore only a fraction of the ^8B neutrinos are observed, viz.,

$$F(^8B) = (12 \pm 12)\%. \tag{13a}$$

According to our theory then, 12% of the neutrinos emitted by ^8B should be below the critical neutrino energy E_c for the sun. This permits a determination of E_c.

We assume that all the ^{8}B decays lead to the same state of ^{8}Be, in an allowed transition. Then the spectrum of neutrinos should be $x^2(1-x)^2\,dx$, where $x=\epsilon_\nu/Q$, and $Q=14.0$ MeV is the energy release in the β decay. We assume that the neutrino detection probability is simply proportional to x^2 provided that $x>x_1=\epsilon_{th}/Q$, where ϵ_{th} is the detection threshold which is 0.82 MeV for the ^{37}Cl detector. Then, if we set $E_c/Q=x_2$,

$$\int_{x_1}^{x_2} x^4(1-x)^2\,dx = F(^8\mathrm{B})\int_{x_1}^{1} x^4(1-x)^2\,dx$$

$$= 9.4\times 10^{-3} F(^8\mathrm{B}). \qquad (14)$$

This determines $x_2=0.42$, or

$$E_c = 5.9^{+1.1}_{-3}\ \mathrm{MeV}. \qquad (15)$$

This is safely above the energy of all neutrinos other than those from ^{8}B.

Again taking the theory seriously, (12) permits the determination of $\Delta=m_2^2-m_1^2$. For ρ_c, I take the solar density at a radius which includes half the ^{8}B reactions,

$$\rho_c = 130\ \mathrm{g/cm^3}$$

according to the standard solar model of Bahcall *et al.* Then

$$\Delta\cos 2\theta = +5.9\times 10^{-5}\ \mathrm{eV^2}. \qquad (16)$$

This is small, and in the range predicted by MS. If we assume $m_1=0$ and $\cos 2\theta=1$, then

$$m_2 = 0.008\ \mathrm{eV}. \qquad (16a)$$

With Δ as small as (16), it is likely to be extremely difficult to confirm this result by laboratory experiments. Thus the astrophysical evidence, from the observed deficiency of solar neutrinos, seems to be the best, perhaps the only, way to determine[5] the elusive Δ.

With use of the conclusion (13a), it is now possible to predict the result of future experiments using detectors other than ^{37}Cl. We merely need to take the prediction by Bahcall *et al.* for the SNU from other species, and add 12% of the SNU from ^{8}B. For instance, for the ^{71}Ga detector,[6]

SNU from other species = 106

12% of SNU from ^{8}B = 1.6

Total = 108 SNU. (17)

For this detector, the conversion of ν_e into ν_μ decreases the expected SNU by only 10%. Observation of solar neutrinos by the Ga detector, therefore, would be the best confirmation (or disproof) of theory presented here.[7]

We have tactily assumed that the mass of the neutrino follows the upper curve in Fig. 1. As MS point out, this is true only if the change of density near the crossing point is adiabatic. According to (8), the width of the resonance is $\Gamma=2\Delta\sin 2\theta$, and so in the resonance A goes from $\Delta\cos 2\theta-\frac{1}{2}\Gamma$ to $\Delta\cos 2\theta+\frac{1}{2}\Gamma$. Since A is proportional to ρ, this corresponds to a relative change of density of

$$\delta\rho/\rho = \Gamma/A\cos 2\theta = 2\tan 2\theta, \qquad (18)$$

and this happens in a distance

$$\delta r = 2\left[-\frac{1}{\rho}\frac{d\rho}{dr}\right]^{-1}\tan 2\theta, \qquad (18a)$$

where the bracketed quantity is taken from the density distribution in the sun. To make the transition adiabatic, δr must be large compared to the neutrino oscillation distance at the resonance,

$$L=\frac{2\pi}{\delta k}=\frac{2\pi\times 2E}{\Delta\sin 2\theta}=\frac{4\pi\times 1.3\times 10^7}{\rho\tan 2\theta}\ \mathrm{cm}, \qquad (19)$$

where (5), (10), and (11) have been used. The adiabatic condition $\delta r \gg L$ then requires

$$\tan^2 2\theta > 0.8\times 10^8\left(-\frac{1}{\rho^2}\frac{d\rho}{dr}\right)$$

$$=0.8\times 10^8\frac{d}{dr}\left(\frac{1}{\rho}\right). \qquad (20)$$

In the relevant region of the sun, i.e., where neutrinos between 5 and 14 MeV go through the resonance, the density is between 130 and 50. In the standard model of the sun,[8] in this density region

$$\frac{d}{dr}\left(\frac{1}{\rho}\right)=(0.8\text{–}2)\times 10^{-12}\ \frac{\mathrm{cm^2}}{\mathrm{g}}, \qquad (20a)$$

the larger number being the relevant one. Then (20a) gives $\tan^2 2\theta > 1.6\times 10^{-4}$,

$$\theta > 0.0065\ \mathrm{rad} = 0.4^\circ. \qquad (21)$$

Thus even quite small mixing angles are compatible with the present theory.

MS speculate about the effect of the neutrino resonance in supernova stars. If our theory of the solar-neutrino puzzle is correct, we can make definite predictions about the neutrinos in a supernova. These neutrinos range in energy from 5 MeV up; hence, according to (12), their resonance density in matter will range downward from the sun's central density, 130. Therefore, there is no chance for resonance conversion of ν_e into ν_μ in the dense core of a supernova where $\rho > 10^{10}$ g/cm^2, just as originally predicted by Wolfenstein.[3]

[If the mass of ν_τ is 2 eV, as we shall estimate

below, the crossing between ν_τ and ν_e is at about 10^7 g/cm^3, still much below the supernova core density.]

However, the enormous number of neutrinos escaping from the collapsing core will all pass through a density region around 100 on their way out. Here the ν_e will be converted into ν_μ. Conversely, ν_μ emitted from the core will be converted into ν_e. So electron neutrinos will be observable (if we ever catch the neutrinos from a supernova), but they will not have started out as ν_e.

Antineutrinos $\bar{\nu}_e$ have an interaction of the opposite sign with matter. Thus they will *not* have a resonance with $\bar{\nu}_\mu$ but will escape from the star unchanged.

Gell-Mann, Ramond, and Slansky,[9] and also Yanagida[10] have proposed a "seesaw model" in which the mass of a neutrino of flavor i is related to that of the quark of the same flavor by

$$m(\nu_i) = m^2(q_i)/M, \tag{22}$$

where M is a superheavy mass. For definiteness, I replace q_i by the charged lepton of the same flavor, l_i; then

$$\frac{m(\nu_\tau)}{m(\nu_\mu)} = \frac{m^2(\tau)}{m^2(\mu)} \simeq 300 \tag{23}$$

and

$$m(\nu_\tau) \simeq 2.5 \text{ eV}. \tag{24}$$

Such a neutrino could have useful cosmological consequences. If I chose, instead of the τ lepton, some geometric mean between bottom and top quark, let us say $m(q_\tau) = 12$ GeV, then I would get $m'(\nu_\tau) = 100$ eV, giving cosmologists a choice between 2 and 100 eV. The mass of the electron neutrino, using (22), would be

$$m(\nu_e) = 2\times 10^{-7} \text{ eV}, \tag{25}$$

clearly unmeasurable. The superheavy mass would be

$$M \simeq 10^{18} \text{ eV} = 10^9 \text{ GeV}, \tag{26}$$

less than the usually assumed $M \simeq 10^{14}$ GeV.

The oscillation distance between ν_μ and ν_e in vacuum is

$$L_{e\mu} = \frac{\pi}{\delta k} = \frac{2\pi E}{m_2^2 - m_1^2} \simeq 4000 \text{ km} \tag{27}$$

(for $E = 200$ MeV), clearly impractical for a laboratory experiment. But if (24) is the correct mass of the τ neutrino, and if we take $E = 3$ GeV, then

$$L_{\mu\tau} \simeq L_{e\tau} \simeq 600 \text{ m} \tag{27a}$$

which may be a feasible distance for a laboratory experiment. However, the fraction of ν_μ which would convert into ν_τ at this (most favorable) distance is only $\sin^2 2\theta_{\mu\tau}$ which may perhaps be 1%. If ν_τ is to be detectable, its energy must be greater than the mass of the τ lepton (1.8 GeV). But if we use energies as large as this, there will be many τ and ν_τ produced in the first place, and it will be exceedingly difficult to find out if 1% of the ν_μ have been converted into ν_τ.

Coming back to the neutrinos escaping from a supernova and thus going from very high to low ρ, there is first a crossing of the masses of ν_e and ν_τ at a density around 10^7 g/cm^3, and then a crossing of ν_e and ν_μ at around $\rho = 10^2$. Thus the history of escaping neutrinos will be as follows:

$$\nu_e \to \nu_\tau; \quad \nu_\tau \to \nu_e \to \nu_\mu; \quad \nu_\mu \to \nu_e, \tag{28}$$
$$\bar{\nu}_e \text{ remains } \bar{\nu}_e.$$

If supernova neutrinos are measured on Earth, ν_e and $\bar{\nu}_e$ can be measured which indicate the ν_μ and $\bar{\nu}_e$ originally produced in the supernova core.

I am much indebted to Ray Davis for drawing my attention to the paper by Mikyeyev and Smirnov. I also thank Paul Langacker for pointing out a mistake in sign in the first version of my paper (as well as that of MS), and drawing my attention to Refs. 9 and 10, and Michael Turner for checking my arithmetic.

This research was supported in part by the National Science Foundation under Grants No. PHY82-09011 and No. PHY82-17853, supplemented by funds from the National Aeronautics and Space Administration.

---

[a] Permanent address.

[1] S. P. Mikheyev and A. Yu. Smirnov, in Proceedings of the Tenth International Workshop on Weak Interactions, Savonlinna, Finland, 16–25 June 1985 (unpublished).

[2] L. Wolfenstein, Phys. Rev. D 17, 2369 (1978).

[3] L. Wolfenstein, Phys. Rev. D 20, 2634 (1979).

[4] J. N. Bahcall *et al.*, Astrophys. J. 292, L79 (1985).

[5] S. P. Rosen and J. M. Gelb, unpublished, have pointed out a second solution in which the conversion of ν_e to ν_μ takes place near the surface of the sun. Δ is smaller, 10^{-7} to 10^{-4} eV^2, and θ is directly related to Δ. The Ga detector is predicted to give less SNU than in (17), so that the Ga experiment can decide between the two solutions.

[6] The predicted SNU from ^{8}B has been corrected according to the experiments of D. Krofcheck *et al.*, Phys. Rev. Lett. 55, 1051 (1985).

[7] H. H. Chen, private communication, points out that in the planned "Sudbury experiment" it may be possible to measure the elastic scattering of neutrinos by neutrons (in the deuteron), due to the neutral weak current. This would measure the sum of ν_e and ν_μ from the sun.

[8] J. N. Bahcall *et al.*, Rev. Mod. Phys. 54, 767 (1982).

[9] M. Gell-Mann, P. Ramond, and R. Slansky, in *Supergravity*, edited by P. van Nieuwenhuizen and D. Z. Freedman (North-Holland, Amsterdam, 1979).

[10] T. Yanagida, in Proceedings of the Workshop on Unified Theory and Baryon Number of the Universe, Tsukuba, Ibaraki, Japan, 1979 (unpublished).

Nonadiabatic Level Crossing in Resonant Neutrino Oscillations

Stephen J. Parke
Fermi National Accelerator Laboratory, Batavia, Illinois 60510
(Received 27 May 1986)

Analytic results are presented for the probability of detecting an electron neutrino after passage through a resonant oscillation region. If the electron neutrino is produced far above the resonance density, this probability is simply given by $\langle P_{\nu_e}\rangle \approx \sin^2\theta_0 + P_x\cos2\theta_0$, where θ_0 is the vacuum mixing angle. The probability is averaged over the production as well as the detection positions of the neutrino and P_x is the Landau-Zener transition probability between adiabatic states. Finally, this result is applied to resonance oscillations within the solar interior.

PACS numbers: 96.60.Kx, 12.15.Ff, 14.60.Gh

Recently Mikheyev and Smirnov[1] and Bethe[2] have revived interest in the solar-neutrino deficit by demonstrating that electron neutrinos produced in the sun can be efficiently rotated into muon neutrinos by passage through a resonant oscillation region. This mechanism may solve the solar-neutrino puzzle. In this paper, I present an analytic result for the probability of detecting an electron neutrino after passage through one or more resonant oscillation regions. This result is then used to show the regions of parameter space, the difference of the squared masses versus the vacuum mixing angle, for which the solar-neutrino puzzle is solved.

A neutrino state is assumed to be a linear combination of the two flavor states $|\nu_e\rangle$ and $|\nu_\mu\rangle$:

$$|\nu,t\rangle = C_e(t)|\nu_e\rangle + C_\mu(t)|\nu_\mu\rangle. \quad (1)$$

If the neutrinos are massive, then the mass eigenstates need not be identical to the flavor eigenstates, so that the Dirac equation which governs the evolution of the neutrino state is not necessarily diagonal in the flavor basis. This leads to the well known phenomena of vacuum neutrino oscillations. In the presence of matter, the nondiagonal nature of this evolution is further enhanced by coherent forward scattering which can lead to resonant neutrino oscillations. Wolfenstein[3] has derived the Dirac equation for this process, in the ultrarelativistic limit, in terms of the vacuum mass eigenstates. Here, I use his result, in the flavor basis, after discarding a term proportional to the identity matrix, as this term only contributes an overall phase factor to the state $|\nu,t\rangle$. The resulting Schrödinger-type wave equation is

$$i\frac{d}{dt}\begin{pmatrix} C_e \\ C_\mu \end{pmatrix} = \frac{1}{2}\begin{pmatrix} -\Delta_0\cos2\theta_0+\sqrt{2}G_FN & \Delta_0\sin2\theta_0 \\ \Delta_0\sin2\theta_0 & \Delta_0\cos2\theta_0-\sqrt{2}G_FN \end{pmatrix}\begin{pmatrix} C_e \\ C_\mu \end{pmatrix}, \quad (2)$$

where $\Delta_0 = (m_2^2 - m_1^2)/2k$, $m_{1,2}$ are the neutrino masses, k is the neutrino momentum, θ_0 is the vacuum mixing angle, N is number density of electrons, and G_F is the Fermi constant. The constraints $\Delta_0 > 0$ and $\theta_0 < \pi/4$ are assumed. At an electron density, N, the matter mass eigenstates are

$$|\nu_1,N\rangle = \cos\theta_N|\nu_e\rangle - \sin\theta_N|\nu_\mu\rangle,$$
$$|\nu_2,N\rangle = \sin\theta_N|\nu_e\rangle + \cos\theta_N|\nu_\mu\rangle, \quad (3)$$

which have eigenvalues $\pm\Delta_N/2$, where

$$\Delta_N = [(\Delta_0\cos2\theta_0 - \sqrt{2}G_FN)^2 + \Delta_0^2\sin^2 2\theta_0]^{1/2}, \quad (4)$$

and θ_N satisfies

$$\Delta_N\sin2\theta_N = \Delta_0\sin2\theta_0. \quad (5)$$

These states evolve in time by the multiplication of a phase factor, if the electron density is a constant. For such a constant density there are three regions of interest: (i) Well below resonance, $\sqrt{2}G_FN \ll \Delta_0\cos2\theta_0$, where the matter mixing angle is $\theta_N \sim \theta_0$ and the oscillation length is $L_0 = 2\pi/\Delta_0$. Typically, this is the region that the electron neutrinos are detected in. (ii) At resonance, $\sqrt{2}G_FN = \Delta_0\cos2\theta_0$, where the matter mixing angle is $\theta_N = \pi/4$ and the resonant oscillation length is $L_R = L_0/\sin2\theta_0$, which for small vacuum mixing angle can be many times the vacuum oscillation length. (iii) Far above resonance, $\sqrt{2}G_FN \gg \Delta_0\cos2\theta_0$, where the matter mixing angle $\theta_N \sim \pi/2$, and the oscillation length $L_N = 2\pi/\Delta_N$ is much smaller than the vacuum oscillation length L_0. For the situation of current interest the electron neutrinos are produced above resonance, pass through resonance, and are detected in the vacuum.

If the electron density varies slowly, the states which evolve independently in time (the adiabatic states) are

$$\exp(-i\tfrac{1}{2}\int^{t}\Delta_N\,dt)|\nu_1,N(t)\rangle$$

and

$$\exp(+i\tfrac{1}{2}\int^{t}\Delta_N\,dt)|\nu_2,N(t)\rangle.$$

Therefore, it is convenient to use these states, as the basis states, in the region for which there are no transitions (away from the resonance region). As a neutrino goes through resonance these adiabatic states may be mixed, but on the other side of resonance, the neutrino state can still be written as a linear combination of these states. That is, a basis state produced at time t, going through resonance at time t_r, and detected at time t' is described by

$$\exp(-i\tfrac{1}{2}\int_{t_r}^{t}\Delta_N\,dt)|\nu_1,N(t)\rangle \rightarrow a_1\exp(-i\tfrac{1}{2}\int_{t_r}^{t'}\Delta_N\,dt)|\nu_1,N(t')\rangle + a_2\exp(+i\tfrac{1}{2}\int_{t_r}^{t'}\Delta_N\,dt)|\nu_2,N(t')\rangle,$$

$$\exp(+i\tfrac{1}{2}\int_{t_r}^{t}\Delta_N\,dt)|\nu_2,N(t)\rangle \rightarrow -a_2^*\exp(-i\tfrac{1}{2}\int_{t_r}^{t'}\Delta_N\,dt)|\nu_1,N(t')\rangle + a_1^*\exp(+i\tfrac{1}{2}\int_{t_r}^{t'}\Delta_N\,dt)|\nu_2,N(t')\rangle,$$

where a_1 and a_2 are complex numbers such that $|a_1|^2+|a_2|^2=1$. The relationship between the coefficients, for these two basis states, is due to the special nature of the wave equation, Eq. (2). The phase factors have been chosen so that coefficients a_1 and a_2 are characteristics of the transitions at resonance and are not related to the production and detection of the neutrino state.

Hence, the amplitude for producing, at time t, and detecting, at time t', an electron neutrino after passage through resonance is

$$A_1(t)\exp(-i\tfrac{1}{2}\int_{t_r}^{t'}\Delta_N\,dt) + A_2(t)\exp(+i\tfrac{1}{2}\int_{t_r}^{t'}\Delta_N\,dt),$$

where

$$A_1(t)=\cos\theta_0[a_1\cos\theta_N\exp(+i\tfrac{1}{2}\int_{t_r}^{t}\Delta_N\,dt) - a_2^*\sin\theta_N\exp(-i\tfrac{1}{2}\int_{t_r}^{t}\Delta_N\,dt)],$$

$$A_2(t)=\sin\theta_0[a_2\cos\theta_N\exp(+i\tfrac{1}{2}\int_{t_r}^{t}\Delta_N\,dt) + a_1^*\sin\theta_N\exp(-i\tfrac{1}{2}\int_{t_r}^{t}\Delta_N\,dt)].$$

Thus the probability of detecting this neutrino as an electron neutrino is given by

$$P_{\nu_e}(t,t')=|A_1(t)|^2+|A_2(t)|^2+2|A_1(t)A_2(t)|\cos(\int_{t_r}^{t'}\Delta_N\,dt+\Omega)$$

with $\Omega=\arg(A_1^*A_2)$. After averaging over the detection position, the detection averaged probability is given by

$$P_{\nu_e}(t)=\tfrac{1}{2}+\tfrac{1}{2}(|a_1|^2-|a_2|^2)\cos2\theta_N\cos2\theta_0-|a_1a_2|\sin2\theta_N\cos2\theta_0\cos(\int_{t_r}^{t}\Delta_N\,dt+\omega)$$

with $\omega=\arg(a_1a_2)$. The last term shows that the phase of the neutrino oscillation at the point the neutrino enters resonance can substantially affect this probability. Therefore, we must also average over the production position to obtain the fully averaged probability of detecting an electron neutrino as

$$\langle P_{\nu_e}\rangle=\tfrac{1}{2}+(\tfrac{1}{2}-P_x)\cos2\theta_N\cos2\theta_0, \tag{6}$$

where $P_x=|a_2|^2$, the probability of transition from $|\nu_1,N\rangle$ to $|\nu_2,N\rangle$ (or vice versa) during resonance crossing. The adiabatic case[4] is trivially obtained by setting $P_x=0$. Also, if the electron neutrinos are produced at a density much greater than the resonance density, so that $\cos2\theta_N\sim-1$, then

$$\langle P_{\nu_e}\rangle\approx\sin^2\theta_0+P_x\cos2\theta_0. \tag{7}$$

Thus for small θ_0 the probability is just equal to the probability of level crossing during resonance passage.

Similar calculations can also be performed for the case of double resonance crossing (neutrinos from the far side of the sun). Here we must average not only over the production and detection positions of the neutrino but also over the separation between resonances. This sensitivity to the separation of the resonances can be understood as the effect of the phase of the oscillation as the neutrino enters the second resonance region. The fully averaged probability of detecting an electron neutrino is the same as Eq. (6) with P_x replaced by $P_{1x}(1-P_{2x})+(1-P_{1x})P_{2x}$ (the classical probability result). Therefore, the generalization to any number of resonance regions, suitably averaged, is obvious.

To calculate the probability P_x, I make the approximation that the density of electrons varies linearly in the transition region. That is, a Taylor-series expansion is made about the resonance position and the second- and higher-derivative terms are discarded;

$$N(t) \approx N(t_r) + (t - t_r)\, dN/dt|_{t_r}. \tag{8}$$

In this approximation the probability of transition between adiabatic states was calculated by Landau and Zenner.[5] This is achieved by solving the Schrödinger equation, Eq. (2), exactly in this limit. The solution is in terms of Weber (parabolic cylinder) functions. Application of the Landau-Zenner result to the current situation gives

$$P_x = \exp\left[-\frac{\pi}{2}\frac{\sin^2 2\theta_0}{\cos 2\theta_0}\frac{\Delta_0}{|(1/N)\, dN/dt|_{t_r}}\right]. \tag{9}$$

This expression, together with Eq. (6), are the main analytical results of this paper and demonstrate that only the electron number density, at production, and the logarithmic derivative of this density, at resonance, determine the probability of detecting an electron neutrino in the vacuum. It should be emphasized here that this result assumes that the neutrino state is produced before significant transitions take place and thus Eq. (9) is not valid for neutrinos produced in the transition region.

From Eq. (9) the size of the transition region can be determined. There are significant transitions ($P_x > 0.01$) if $\theta_0 < \theta_{\rm crit}$, where $\theta_{\rm crit}$ satisfies

$$\frac{\sin^2 2\theta_{\rm crit}}{\cos 2\theta_{\rm crit}} = 3\frac{1}{\Delta_0}\left|\frac{1}{N}\frac{dN}{dt}\right|_{t_r}. \tag{10}$$

Hence, the maximum separation between the eigenstates for which transitions take place is $\Delta_0 \sin 2\theta_{\rm crit}$. Therefore, the transition region is defined by

$$\Delta_N < \Delta_0 \sin 2\theta_{\rm crit}. \tag{11}$$

This can only happen if $\theta_0 < \theta_{\rm crit}$. In this transition region, the maximum variation of the electron number density from the resonant value is $\pm\delta N$, where

$$\delta N / N(t_r) = \sin 2\theta_{\rm crit}.$$

Thus, the size of the transition region is

$$|t - t_r| = \frac{\sin 2\theta_{\rm crit}}{|(1/N)\, dN/dt|_{tr}}.$$

This is the maximum $|t - t_r|$ for which the linear approximation must be good, so that Eq. (9) gives a reasonable estimate of the probability of crossing. For an exponential density profile, the Taylor-series expansion is an expansion in $\sin 2\theta_{\rm crit}$, so that for small $\theta_{\rm crit}$ this is an excellent approximation.

For the sun, the density profile is exponential except for the region near the center. In Fig. 1, I have plotted the probability contours for detection of an electron neutrino at the Earth in the $\Delta_0/\sqrt{2}G_F N_c$ vs $\sin 2\theta_0$ plane for an exponential density profile. N_c is the electron number density at the point at which the electron neutrinos are produced. This plot depends only on the properties of the sun and this dependency is only through the combination $R_s N_c$, where R_s is the scale height. For Fig. 1, I have used an N_c corresponding to a density of 140 g/cm^3 and $Y_e = 0.7$. The scale height R_s is 0.092 times the radius of the sun.

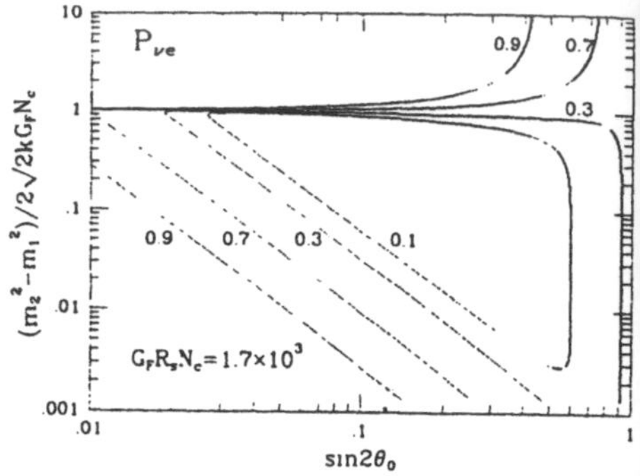

FIG. 1. Probability-contour plot for detecting an electron neutrino at the Earth which was produced in the solar interior.

Above the line $\Delta_0/\sqrt{2}G_F N_c = 1/\cos 2\theta_0$, the neutrinos never cross the resonance density on their way out of the sun. Here, the probability of detecting an electron neutrino is close to the standard neutrino-oscillation result. Below this line, the effects of passing through resonance come into play. Inside the 0.1 contour, there is only a small probability of transitions between the adiabatic states as the neutrino passes through resonance. To the right-hand side of this contour, the probability of detecting a neutrino grows, not because of transitions, but because both adiabatic states have a substantial mixture of electron neutrino at zero density. To the left-hand side and below this contour, the probability grows because here there are significant transitions between the adiabatic states as the neutrino crosses resonance. The diagonal lines of these contours have slope of −2 because of the form of P_x. It is only the intercept of these lines which depends on the product $R_s N_c$. Therefore, if one wishes to change the production density, which is held fixed in this plot, only these lines need to be shifted. In fact, a line labeled with P_e "crosses" $\Delta_0/\sqrt{2}G_F N_c = 1$ when a small θ_0 satisfies

$$\frac{\sin^2 2\theta_0}{\cos 2\theta_0} = \frac{-\sqrt{2}\ln P_e}{\pi G_F R_s N_c}. \tag{12}$$

Note that I find the probability of detecting an electron

neutrino, which crosses resonance, to be greater than 0.25 when $\theta_0 < 0.01$.

This isoprobability plot can easily be converted into an approximate iso-SNU (solar neutrino units) plot for the Davis *et al.* experiment.[6] The predicted result for this experiment[7] is 6 SNU, with 4.3 SNU coming from the ^{8}B neutrinos and 1.6 SNU from the lower-energy neutrinos (*pep*, ^{7}Be, ^{13}N, and ^{15}O), whereas Davis *et al.* observe 2.1 ± 0.3 SNU. Roughly speaking, the 2 SNU contour, in the $m_2^2 - m_1^2$ vs $\sin 2\theta_0$ log-log plot, will be a triangle, similar to the 0.3 contour of Fig. 1, with rounded corners. The three straight sections of this triangle are approximately given below. The horizontal line is given by choosing the parameters so that all the low-energy neutrinos and only 12% of the ^{8}B neutrinos are observed. This gives the constraints obtained by Bethe,[2]

$$m_2^2 - m_1^2 \approx 8 \times 10^{-5}\ \text{eV}^2, \quad 0.03 < \sin 2\theta_0 < 0.6. \tag{13}$$

For the vertical line, the probability of detecting an electron neutrino is nearly independent of energy, if $1 > \Delta_0/\sqrt{2}G_F N_e > 10^{-1}$. Therefore, we need to reduce all neutrinos by 30%.[4] This is achieved when

$$8 \times 10^{-8}\ \text{eV}^2 < m_2^2 - m_1^2 < 1 \times 10^{-5}\ \text{eV}^2, \quad \sin 2\theta_0 \approx 0.9. \tag{14}$$

For the diagonal line, we need to arrange that the Davis experiment only observed 50% of the ^{8}B neutrinos and none of the lower-energy neutrinos.[8,9] This is achieved when the probability for the mean ^{8}B neutrino, weighted by the detector cross section (energy ~ 10 MeV), is 0.5. This gives the following constraint:

$$(m_2^2 - m_1^2)\sin^2 2\theta_0 \approx 3 \times 10^{-8}\ \text{eV}^2, \quad 0.03 < \sin 2\theta_0 < 0.6. \tag{15}$$

To summarize, Eqs. (13)–(15) give regions of parameter space for which the expected result from the Davis experiment is ~ 2 SNU.

Since the proposed gallium experiment observes lower-energy neutrinos, from the *pp* process, these three regions will be distinguishable by use of the results of this experiment. More precise iso-SNU plots, for both experiments, are being generated taking into account the production energy and production position distributions of the neutrinos from the various processes within the solar interior.

I would like to acknowledge discussions with T. Walker and R. Kolb. Fermilab is operated by the Universities Research Association Inc., under contract with the United States Department of Energy.

[1] S. P. Mikheyev and A. Yu. Smirnov, in Proceedings of the Tenth International Workshop on Weak Interactions, Savonlina, Finland, June, 1985 (to be published).

[2] H. A. Bethe, Phys. Rev. Lett. 56, 1305 (1986).

[3] L. Wolfenstein, Phys. Rev. D 17, 2369 (1978) and 20, 2634 (1979).

[4] V. Barger, R. J. N. Phillips, and K. Whisnant, Phys. Rev. D 34, 980 (1986).

[5] L. D. Landau, Phys. Z. U.S.S.R. 1, 426 (1932); C. Zenner, Proc. Roy. Soc. A137, 696 (1932).

[6] J. N. Bahcall *et al.*, Astrophys. J. 292, L79 (1985).

[7] J. N. Bahcall *et al.*, Rev. Mod. Phys. 54, 767 (1982).

[8] S. P. Rosen and J. M. Gelb, Phys. Rev. D 34, 969 (1986).

[9] E. W. Kolb, M. S. Turner, and T. P. Walker, Fermilab Report No. FERMILAB-Pub-86/69-A, May 1986 (to be published).

Adiabatic Conversion of Solar Neutrinos

W. C. Haxton
Institute for Nuclear Theory, Department of Physics, University of Washington, Seattle, Washington 98195
(Received 31 March 1986)

The adiabatic conversion of solar electron neutrinos to muon neutrinos via the Mikheyev-Smirnov-Wolfenstein mechanism is discussed as a level-crossing problem. Solutions to the ^{37}Cl puzzle obtained numerically by Mikheyev and Smirnov, Rosen and Gelb, and others are shown to form the boundary of the region in the δm^2-$\sin^2 2\theta$ plane where adiabatic conversion occurs. Most of the region within this boundary is ruled out by the nonzero signal in Davis's experiment. The Landau-Zener description of the adiabatic boundary is employed.

PACS numbers: 96.60.Kx, 12.15.Ff, 14.60.Gh

Mikheyev and Smirnov[1] (MS) have uncovered a mechanism by which the electron neutrinos produced in the solar core could be efficiently converted into neutrinos of a different flavor. The mechanism depends on an effective density-dependent electron-neutrino mass arising from the weak charged-current interactions with solar electrons, a phenomenon first discussed by Wolfenstein.[2] This offers a plausible particle-physics solution to the solar-neutrino puzzle: Neutrino oscillations governed by small vacuum mixing angles can produce the needed factor-of-3 suppression in the ^{37}Cl counting rate.[3,4]

Recently Bethe[5] rederived the MS result from an elegant formulation based on the adiabatic approximation. In this approximation one describes the evolution of the neutrino wave function in terms of the stationary eigenstates of a time-independent Hamiltonian evaluated for the appropriate instantaneous electron density. For the case of two-state mixing, it is attractive to assume that, in vacuum, the ν_μ is composed primarily of the heavier-mass eigenstate. In this case adiabatic propagation of the neutrino through the Sun will lead to a large $\nu_e \rightarrow \nu_\mu$ amplitude provided that the ν_e produced in the solar core is predominantly the heavy eigenstate of the instantaneous Hamiltonian appropriate for the core density. Such level crossing[5] occurs if

$$\beta\rho(t=0) > \cos 2\theta \sim 1, \tag{1}$$

where β and ρ are the dimensionless quantities $\beta = 2E/\delta m^2 R_s$ and $\rho(t) = \sqrt{2} G_F n_e(t) R_s$ with E the neutrino energy, $\delta m^2 = m_2^2 - m_1^2$, R_s the Sun's radius, and $n_e(t)$ the instantaneous electron density at time t; t is measured in units of R_s and thus runs from 0 to 1 for a neutrino propagating from the Sun's center to its surface. An additional constraint comes from the requirement of adiabatic propagation. For small mixing angles θ the oscillation frequency moves through a pronounced minimum at a critical density that determines the level-crossing point,

$$\beta\rho(t_c) \equiv \beta\rho_c = \cos 2\theta . \tag{2}$$

The adiabatic condition relates the allowed mixing angles to the rate of change of the density:

$$\tan^2 2\theta \gtrsim \frac{\sin^3 2\theta}{[(\beta\rho - \cos 2\theta)^2 + \sin^2 2\theta]^{3/2}} \times |\rho_c^{-2} d\rho/dt|, \tag{3}$$

a constraint that becomes particularly stringent at t_c:

$$\gamma = \tan^2 2\theta \left| \frac{1}{\rho^2}\frac{d\rho}{dt}\right|_{t_c} \Bigg|^{-1} \gtrsim 1. \tag{4}$$

The term adiabatic conversion will be used to describe neutrino oscillations when both the level-crossing condition [Eq. (1)] and the adiabatic condition [Eq. (4)] are satisfied.

Bethe noted that as Davis has measured[3] a *nonzero* signal of 2.0 ± 0.3 SNU (solar neutrino units), some of the solar neutrinos that contribute to the ^{37}Cl counting rate are not converted to ν_μ. Equation (1) permits one to exclude low-energy neutrinos from adiabatic conversion for an appropriate choice of δm^2. Bethe found that $\delta m^2 \sim (0.008\ \text{eV})^2$, corresponding to a critical neutrino energy $E_c^B \sim 6$ MeV, provides the correct suppression of the experimental signal. Equation (4) then requires $\sin^2 2\theta \gtrsim 8\times 10^{-4}$. This result is quite similar to a solution obtained by Mikheyev and Smirnov,[1] who determined $\nu_e \rightarrow \nu_\mu$ amplitudes by numerically integrating the Schrödinger equation. Bethe's analysis identifies this solution with a boundary in the δm^2-$\sin^2 2\theta$ plane where the level-crossing requirement for adiabatic conversion fails for low-energy neutrinos.

The correspondence between large-amplitude $\nu_e \rightarrow \nu_\mu$ oscillations and adiabatic conversion is intuitively very attractive. However, in the numerical studies of Mikheyev and Smirnov,[1] Rosen and Gelb[6] (RG), and others,[6] a second solution to the solar-neutrino puzzle was obtained. In this solution δm^2 runs from

10^{-4} to 10^{-7} eV^2 as $\sin^2 2\theta$ runs from 10^{-3} to 0.3. I now argue that the second solution also corresponds to a boundary of the region of adiabatic conversion.

Solutions consistent with the Davis experiment are not permitted for $\delta m^2 > (0.008\ eV)^2$ because of Eq. (1). Values of $\delta m^2 < (0.008\ eV)^2$ lead to level crossing but, from Eq. (1) alone, would appear to reduce the ^{37}Cl counting rate by too great a factor. However, consider the additional constraint of Eq. (4). The denominator appearing in Eq. (4) is a monotonically increasing function of t, running from 0 at $t=0$ to 1 at $t=0.9$ for a neutrino produced at the Sun's center. Suppose that we select some mixing angle in the allowed range $8\times10^{-4} \lesssim \sin^2 2\theta \lesssim 1$. Equation (4) then defines a minimum permissible ρ_c (maximum t_c) that we denote by $\rho_c^{\gamma=1}(\theta)$. For fixed $\sin^2 2\theta$, Eq. (4) is satisfied by any δm^2 and E that lead to $\rho_c > \rho_c^{\gamma=1}$. That is,

$$\delta m^2 R_s/2E > \rho_c^{\gamma=1}(\theta)/\cos 2\theta. \tag{5}$$

For initial "back-of-the-envelope" purposes I will assume that the condition $\gamma=1$ defines a sharp boundary between the adiabatic and nonadiabatic regions. (This approximation will be improved below.) Then, for fixed δm^2, neutrinos of energy E *above* a critical value will not undergo adiabatic conversion because the adiabatic approximation is no longer valid [Eq. (4)]. Davis's result is consistent with an absence of neutrinos with energies below $E_c \sim 10.4$ MeV. Thus a second solution to the solar-neutrino problem is obtained:

$$\delta m^2 = (5.9\times10^{-9}\ eV^2)\frac{\rho_c^{\gamma=1}(\theta)}{\cos 2\theta}, \tag{6}$$

where the adiabatic condition determines δm^2 as a function of θ. For the spectrum of neutrinos sampled by the ^{37}Cl detector, this solution connects smoothly to Bethe's solution at $\delta m^2 \sim (0.008\ eV)^2$ and $\sin^2 2\theta \sim 8\times10^{-4}$, but runs to much smaller δm^2 and large mixing angles. Values satisfying Eq. (6) for a range of mixing angles are given in the third column of Table I and are very similar to the corresponding exact results of Refs. 1 and 6. Thus the solutions found by Mikheyev and Smirnov can be interpreted as those δm^2 and $\sin^2 2\theta$ that define the boundaries of the region in the δm^2-$\sin^2 2\theta$ plane where adiabatic conversion occurs, an appealing result.

The level-crossing solution discussed by Bethe leads to a reduced flux of higher-energy neutrinos and thus, in various radiochemical experiments, mimics the signal expected of a nonstandard solar model. The critical densities for this solution are achieved in the solar core. For the adiabatic-boundary solution the strongest suppression occurs (except for the small-angle hybrid solution discussed below) for the low-energy neutrinos. Thus a distinctively low counting rate will result in the gallium experiment. The critical densities become progressively lower (i.e., the crossing point occurring nearer the surface) as one moves away from the level-crossing solution to smaller δm^2.

TABLE I. Values of δm^2, $\rho_c^{\gamma=1}(\theta)$, and $\sin^2 2\theta$ of the adiabatic-boundary solution for Eq. (6) (δm^2) and for the calculation that includes the Landau-Zener profile of the adiabatic boundary and presence of hybrid solutions for small $\sin^2 2\theta$ (δm^2_{LZ}). The right-hand side of Eq. (3) was taken from a smooth fit to the tabulated density profile of Ref. 4. $\rho_c^{\gamma=1}$ can be compared to the Sun's central density, $\rho(0)=2.8\times10^4$.

$\sin^2 2\theta$	$\rho_c^{\gamma=1}$	δm^2 (eV^2)	δm^2_{LZ} (eV^2)
8×10^{-4}	1.1×10^4	6.5×10^{-5}	7.3×10^{-5}
1×10^{-3}	9.8×10^3	5.5×10^{-5}	6.1×10^{-5}
3×10^{-3}	3.6×10^3	2.2×10^{-5}	1.1×10^{-5}
6×10^{-3}	1.9×10^3	1.1×10^{-5}	5.4×10^{-6}
1×10^{-2}	1.1×10^3	6.5×10^{-6}	3.1×10^{-6}
0.03	0.033	1.9×10^{-6}	9.9×10^{-7}
0.06	0.016	9.9×10^{-7}	4.7×10^{-7}
0.1	0.084	5.7×10^{-7}	2.5×10^{-7}
0.3	0.02	1.5×10^{-7}	9.0×10^{-8}

To help the reader visualize these results, the general constraints imposed by Eqs. (1) and (4) are plotted in Fig. 1. The boundaries of the region of adiabatic conversion depend on two parameters, $\delta m^2/E$ and $\sin^2 2\theta$. Above the horizontal line no level crossing occurs: Even in the solar core the electron neutrino is dominantly the lighter of the two instantaneous mass eigenstates. Below the diagonal line the propagation of the neutrino in the vicinity of the crossing point is highly nonadiabatic. Of course, the sharp boundaries actually represent transition regions. It is easy to improve the simple arguments above to account for the widths and shapes of these boundaries. The behavior near the level-crossing boundary is governed by the overlap of ν_e with the heavy-mass eigenstate $\nu_H(t)$,

$$\langle \nu_e \nu_H(t)\rangle = \tfrac{1}{2}\sqrt{2}[1 \pm (1+\alpha^2)^{-1/2}]^{1/2}, \tag{7}$$

where $\alpha(t) = \sin 2\theta/[\beta\rho(t) - \cos 2\theta]$ and the + (−) sign is taken for α positive (negative). At the level-crossing boundary $\alpha(t=0)$ is infinite. However, for small $\sin 2\theta$, modest variations in $\delta m^2/E \propto \beta^{-1}$ drive $\alpha(t=0)$ to small values and the overlap to ~ 0 (above the boundary) or to ~ 1 (below boundary). The representation of the transition region by a sharp boundary in $\delta m^2/E$ is thus a good approximation for small $\sin 2\theta$, and Bethe's E_c^B is well defined.

The behavior near the diagonal (adiabatic approximation) boundary is determined by the probability of remaining on the heavy-mass trajectory while crossing

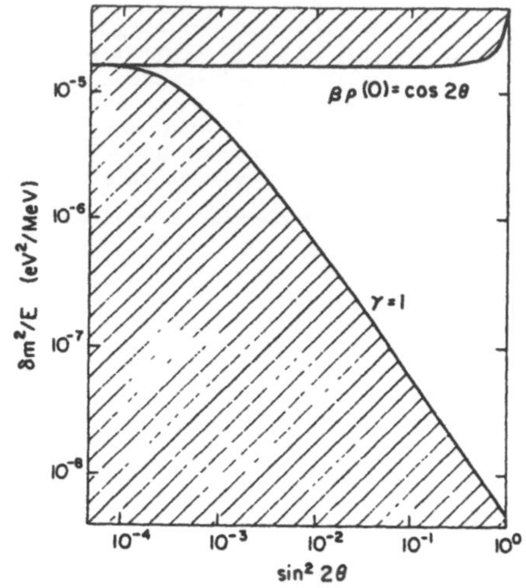

FIG. 1. The shaded areas define those values of $\delta m^2/E$ and $\sin^2 2\theta$ that fail to satisfy the constraints imposed by Eq. (1) (upper region) and by the condition of adiabatic propagation [Eq. (3), lower region]. The conversion of solar $\nu_e \to \nu_\mu$ is highly efficient in the unshaded region. Solutions consistent with Davis's experiment lie along the boundaries for an appropriate choice of an average E.

the critical point. This is given approximately by the Landau-Zener[7] factor

$$P_{LZ} = 1 - e^{-\pi\gamma/2}.$$

This result, which is familiar from level-crossing problems in atomic physics, is derived by the approximation of the off-diagonal mixing element [and therefore $n_e(t)$] by a linear function of t having the correct derivative at the crossing point t_c. The Landau-Zener factor is an excellent approximation to the numerical boundary profiles given, effectively, as a function of $\delta m^2/E$ in Figs. 1 of Ref. 6. A single parameter γ governs the physics of the adiabatic region, the boundary region, and the highly nonadiabatic region below the diagonal line in Fig. 1.

One sees that the definition $\gamma = 1$ of the adiabatic boundary in Eq. (4) is somewhat arbitrary: The condition for equal probabilities of remaining on the electron trajectory or jumping to the muon trajectory, $e^{-\pi\gamma/2} = 0.5$, yields $\gamma = 0.44$. Furthermore, there is an explicit quadratic dependence of γ on $\delta m^2/E$ and an implicit dependence through $d\rho/dt_c$. This dependence is much more gentle than that of Eq. (7), so that the notion of a sharp cutoff energy $E_c \sim 10.4$ MeV is an oversimplification. However, if we fold the neutri spectrum with the appropriate Landau-Zener facto the arguments leading to Eq. (6) can be generalized incorporate these corrections. The results are given Table I as δm^2_{LZ}. Near the intersection with the lev crossing solution (i.e., for small $\sin^2 2\theta$) low-ener neutrinos (^{7}Be, *pep*, CNO) fail to satisfy the lev crossing condition and thus contribute substantially the ^{37}Cl capture rate. Excluding these small $-\sin^2$ hybrid solutions, one must reduce the naive coef cient in Eq. (6) from 5.9 to approximately 2.9 to repr duce on average the last column of Table I. Altern tively, if, in Eq. (6), $\rho_c^{\gamma-1}$ is replaced by $\rho_c^{\gamma-0.44}$, th simple formula becomes quite accurate, yieldi values of δm^2 approximately 13% smaller than t δm^2_{LZ}.

Efficient adiabatic conversion of solar ν_e's to ν_μ will occur throughout much of the unshaded region Fig. (1). One quickly realizes the implications of t Mikheyev-Smirnov-Wolfenstein mechanism in view the *nonzero capture rate* for the ^{37}Cl experimen Davis's result excludes, in the case of two-state mi ing, a substantial portion of the $\sin^2 2\theta$-δm^2 plane co responding, roughly, to the region bounded by t Bethe and RG solutions to the ^{37}Cl puzzle and by th line $\sin^2 2\theta \sim 0.4$ (a requirement imposed by Eq. (7) $\rho \to 0$). The excluded masses and mixing angles a totally unexplored in terrestrial experiments.

To reach this conclusion one must attribute the pr duction of ^{37}Ar in the Homestake tank to solar neutr nos. The capture rate of 2.0 ± 0.3 SNU is determine by subtracting from the ^{37}Ar production rate of 0.4 atom/day an estimate of the rate from known bac grounds, 0.08 ± 0.03 atom/day.[8] Although this est mate is likely reliable,[8,9] the possibility of unidentifie backgrounds cannot be excluded. In principle, Dav could, at the limit of his sensitivity, be detecting n solar neutrinos. Although a null result in the Dav experiment would be consistent with adiabatic conve sion, no definite conclusions about neutrino oscilla tions could then be drawn: A nearly zero capture ra is also consistent with the minimum astronomic model of Bahcall,[10] where the Sun is presumed to pr duce all of its energy through the *pp* I cycle. (The *p* neutrino flux yields a ^{37}Cl capture rate of only 0.2 SNU.)

For this and other reasons the Ga experiment be comes a crucial test of neutrino physics. The G detector will be sensitive to a flux that is distinctivel solar, the low-energy *pp* neutrinos (0.42-MeV en point) produced copiously in the *pp* I cycle. Th predicted standard-solar-model capture rate is 12 SNU. The minimum astronomical rate consistent wit the assumption of steady-state hydrogen burning is 7 SNU.[10] For $\sin^2 2\theta \gtrsim 10^{-2}$ the adiabatic-boundar solution yields a very suppressed *pp* neutrino flu

Counting rates far below the minimum astronomical value would result, providing a compelling argument for neutrino masses and mixing. As one continues along this solution to somewhat smaller angles so that the *pp* neutrinos begin to cross the level-crossing boundary in Fig. 1, the Ga counting rate increases. These hybrid solutions are special in that both $\sin^2 2\theta$ and δm^2 could be determined from the results of the Ga and Cl experiments.[11] For still smaller values of $\sin^2 2\theta$ the entire *pp* flux lies above the region of adiabatic conversion, guaranteeing that the Ga counting rate exceeds the minimum astronomical value. As in the case of the level-crossing solution discussed by Bethe, an experiment capable of measuring the spectrum of ^8B neutrinos would be needed to distinguish oscillations from nonstandard solar physics.

Any substantial counting rate in the Ga experiment will unambiguously rule out a large region in δm^2 and $\sin^2 2\theta$ where adiabatic conversion of the *pp* neutrinos must take place. The region is similar in size and shape to that tested by the ^{37}Cl experiment but is shifted, as the effective E in $\delta m^2/E$ is about an order of magnitude smaller, to lower δm^2. Adiabatic conversion for δm^2 as small as 10^{-8} eV^2 can take place. For those larger δm^2 that are also tested in the ^{37}Cl experiment, the adiabatic region for the Ga experiment will extend to somewhat smaller mixing angles.

I thank H. Bethe, E. Fischbach, S. Koonin, S. P. Rosen, and L. Wilets for helpful discussions. This work was supported in part by the U. S. Department of Energy.

[1]S. P. Mikheyev and A. Yu. Smirnov, Nuovo Cimento Soc. Ital. Fis. **9C**, 17 (1986); A. Smirnov, in Proceedings of the Workshop on Massive Neutrinos in Physics and Astrophysics, 21st Recontre de Moriond, Tignes, France, 1986 (to be published).

[2]L. Wolfenstein, Phys. Rev. D **16**, 2369 (1978).

[3]J. N. Bahcall, B. T. Cleveland, R. Davis, Jr., and J. K. Rowley, Astrophys. J. **292**, 279 (1985).

[4]J. N. Bahcall, W. F. Huebner, S. H. Lubow, P. D. Parker, and R. K. Ulrich, Rev. Mod. Phys. **54**, 767 (1982). I thank B. Filippone for providing the solar density profile for large radii.

[5]H. A. Bethe, Phys. Rev. Lett. **56**, 1305 (1986). Also see A. Messiah, in Proceedings of the Workshop on Massive Neutrinos in Physics and Astrophysics, 21st Recontre de Moriond, Tignes, France, 1986 (to be published).

[6]S. P. Rosen and J. M. Gelb, Phys. Rev. D **34**, 969 (1986); W. Hampel, to be published; J. Bouchez *et al.*, Saclay Report No. DPhPe 86-10, 1986 (to be published).

[7]L. D. Landau, Phys. Z. Sowjetunion **2**, 46 (1932); C. Zener, Proc. Roy. Soc. London, Ser. A **137**, 696 (1932). I thank S. Koonin and L. Wilets for mentioning these papers.

[8]J. K. Rowley, B. T. Cleveland, and R. Davis, Jr., in *Solar Neutrinos and Neutrino Astronomy*, edited by M. L. Cherry, W. A. Fowler, and K. Lande, AIP Conference Proceedings No. 126 (American Institute of Physics, New York, 1985), p. 1.

[9]E. L. Fireman, B. T. Cleveland, R. Davis, Jr., and J. K. Rowley, in Ref. 8, p. 22.

[10]J. N. Bahcall, in Ref. 8, p. 60.

[11]H. Bethe, private communication.

GeV (or whatever it is), or speculations about new long range feeble forces. But solar neutrinos and superstrings are two topics about which I wanted to express some opinions.

First, the solar neutrino problem. (This by the way was reviewed by Altarelli and Goldhaber, whose printed reports will doubtless be more detailed than my remarks.) The problem is that in experiments looking for the capture of electron neutrinos from the sun in the reaction $\nu_e + {}^{37}\mathrm{Cl} \rightarrow e^- + {}^{37}\mathrm{A}$, Davis has found a rate of 2.0 ± 0.3 solar neutrino units, whereas the theoretical prediction by Bahcall and his colleagues is 5.8 ± 2.2 solar neutrino units. For a long time it has been realized that one possible explanation for this is neutrino flavor oscillations: the neutrinos may start in the sun as electron type neutrinos, but turn into an incoherent mixture of electron, muon, and tau type by the time they reach the earth. The muon and tau type are not detected in this kind of experiment, and therefore one sees a reduced rate. Now this is a perfectly plausible explanation as far as the neutrino mass difference that is required. For a solar neutrino of energy 5 MeV, there will be many flavor oscillations between the earth and the sun provided the neutrino mass difference is much greater than $(5\ \mathrm{MeV}/1\ \mathrm{A.U.})^{1/2}$, or 10^{-6} eV. As I'll indicate a little later, this is a reasonable mass difference even if the masses arise from lepton nonconservation at a grand unification scale. The problem is that with N species of neutrinos, the maximum reduction you can possibly get from this kind of oscillation (without fine tuning the distance between the earth and the sun) is $1/N$, and that means that the mixing angles have to just so arrange themselves to frustrate Davis. A more natural explanation, it had seemed to many of us, was that either the solar theory was wrong or the experiment was wrong, or both.

The plausibility of solar neutrino oscillations as an explanation for the reduced neutrino rate has recently been greatly increased as a result of a remarkable suggestion made in June 1985, at the conference at Savonlinna in Finland, by Mikheyev and Smirnov, based on earlier work by Wolfenstein. (Hence the name, MSW effect.) The MSW effect describes how neutrinos starting in the sun as electron neutrinos can, even if the mixing angle is quite small, and for quite a broad range of masses and mixing angles, turn into a different kind of neutrino, μ or τ, with nearly 100% efficiency, through a resonance induced by interaction with electrons in the sun.

To see how this works, all you need is a pair of weakly coupled mechanical oscillators, one whose frequency is variable, corresponding to ν_e, and the other with fixed frequency, corresponding, say, to ν_τ. (An apparatus of this sort was demonstrated in the lecture at Berkeley.) My reason for guessing that it is the ν_τ rather than the ν_μ that is relevant will be explained shortly; however, it would make no difference in this discussion if it were ν_μ rather than ν_τ.

First, start with the frequencies unequal, and with the ν_e mode excited. This corresponds to the situation at the center of the sun; the effective ν_e and ν_τ masses are different here both because we assume that the vacuum masses are unequal, and also because the electron neutrinos have a W-exchange interaction with electrons that is absent for τ (or μ) neutrinos.

Next, slowly change the frequency of the ν_e mode. This corresponds to what happens as ν_e rises through the sun—its effective mass changes because it encounters lower electron densities.

Eventually, at some distance from the sun's center, the frequencies of the two modes may become equal. In this resonant case, the ν_τ mode becomes strongly excited. Continue slowly changing the frequency of the ν_e mode, and stop when the frequencies are again quite different, corresponding to the neutrino leaving the sun, where the effective masses reach their vacuum values. If the passage through the resonance has been sufficiently slow, most of the oscillation energy will be found in the ν_τ mode. If not, it stays in the ν_e mode. (This actually works.)

Now as you can see from this demonstration there are a number of conditions that have to be met to have a sizable reduction in the ν_e counting rate. First, the mass difference has to be small enough (and of the right sign: $m(\nu_\tau) > m(\nu_e)$) so that the resonance actually occurs in the sun. If the mass difference is too great the sun's density will not be large enough to compensate for it. So there's an upper bound on the neutrino mass difference, roughly 10^{-2} eV, for the resonance to occur in the sun. Also the transition has to be slow: the neutrino must go through the resonance sufficiently slowly, so as to follow one eigenmode continuously. This requires that the mixing angle, which determines the width of the resonance, can't be too small, because if it's too small the resonance is too narrow, and the neutrino zips right through it without feeling the effect of the resonance. Finally, the mixing angle can't be too big, because if it's too large then even if the MSW effect occurs with 100% efficiency the neutrinos wind up at the earth with still a strong mixture of the electron mode.

These three conditions define a roughly triangular-shaped region in the mass-difference/mixing angle plane within which there is a large (say, a factor 3) supression in the ν_e counting rate in the $^{37}Cl \rightarrow {}^{37}Ar$ reaction. The boundaries of this region depend upon the particular reaction used to detect the neutrinos, because different reactions are sensitive to neutrinos of different energy, and the efficiency of the MSW effect depends on the energy of the neutrino because there is a large relativistic time dilation. Thus for instance there is a large reduction in the counting rate in the gallium as well as the chlorine experiments for parameters on the low-mixing-angle side of the triangle for the chlorine experiments.

Now one would of course like to know whether the MSW effect is really responsible for the low ν_e rate in chlorine. Assuming that someone hasn't made a terrible error, there are two leading hypotheses that we would like to test against each other. One, hypothesis, A, is the one that I've just described—that the electron neutrino has changed into some other neutrino species with high efficiency, perhaps 60% to 70% efficiency, through resonant oscillations in the sun. The other hypothesis, B, is that most of the electron neutrinos with enough energy to show up in the chlorine reaction are simply missing: they never were produced in the sun. The usual way of understanding this is that, since the electron neutrinos to which the Davis experiment is sensitive are primarilly from the beta decay of ^8B, all we have to explain is why the ^8B is not produced, which one can arrange by supposing that the temperature at the center of the sun is a little lower than we had thought.

EFFECT OF NEUTRINO MAGNETIC MOMENT ON SOLAR NEUTRINO OBSERVATIONS

ARTURO CISNEROS*

Lauritsen Laboratory, California Institute of Technology, Pasadena, Calif., U.S.A.

(Received 9 June 1970; in final form 28 July, 1970)

Abstract. Neutrino spin precession effects in the magnetic field of the Sun are considered as an explanation of the outcome of Davis' solar neutrino experiments. Theoretically, it is possible to account for a neutrino magnetic moment only as the result of the interaction of the electromagnetic field with charged particles into which the neutrino can transform virtually. The currently accepted theory of weak interactions (the two component neutrino and *V–A* interactions) forbids a resulting magnetic moment interaction with the electromagnetic field for all such virtual processes. Modifications of this theory are considered to find out whether an appreciable precession effect is permitted within the experimentally established limits. It is found that the value for the neutrino magnetic moment evaluated under these theoretically anomalous circumstances is still so small that only the largest possible estimate for the magnetic field strength in the Sun's interior would cause the required effect.

The results of the solar neutrino experiments of Davis, *et al.* (1968) are in conflict with the predictions of the accepted models of stellar evolution, the upper limit on neutrinos detected being approximately one half of the number expected on the basis of these models. It is most likely that the explanation involves conventional physics, such as uncertainties in the opacity of the Sun's interior. But it is at least conceivable that a fundamentally new physical phenomenon is involved. One such suggestion has been made by Pontecorvo (1968). The idea is to introduce muon and electron neutrino mixing similar to that of the neutral kaons. For an appropriate oscillation length of the transformation $\nu_e \leftrightarrow \nu_\mu$, the solar neutrinos arriving at the Earth would be an equal mixture of the two types of neutrinos. Only the electron neutrinos would be detected; resolving the discrepancy since the detection rate would be one half of that expected without mixing.

It was noted by the author and by Werntz (1970) that if the neutrino were to have a magnetic moment, the magnetic field inside the Sun would cause precession of the neutrino spin. If the product of the magnetic moment of the neutrino and the average value of the magnetic field in the Sun were large enough to cause appreciable precession; the solar neutrinos arriving at the Earth would be an equal amount mixture of positive and negative helicity states. This would give another possible unconventional explanation of the outcome of Davis' experiment, since only left handed neutrinos would be detected. One further effect is the deflection of a particle with a magnetic moment traveling through a magnetic field gradient. It is found that this is completely negligible in comparison with the precession effect.

* The author has received scholarship support from the Latin American Scholarship Program of American Universities during the preparation of this work.

To get an idea of the numbers involved, suppose the average magnetic field strength inside the Sun were 10^6 G, which is one of the largest possible estimates. Then one finds it necessary to have a magnetic moment of at least 10^{-13} Bohr magnetons to cause a one half turn precession of the spin. The present experimental limit (cf. Bernstein *et al.*, 1963) on the value of the magnetic moment of the neutrino is $\mu < 10^{-10}\ \mu_B$, where μ_B is the Bohr magneton. We see that phenomenologically there is room for precession effects in solar neutrinos*, though admittedly the average field inside the Sun may be less than 10 G, in which case even the present limits on μ would rule out a precession.

A number of years ago Salam (1957) showed that if chiral invariance is required for all neutrino interactions, the neutrino cannot have a magnetic moment. Chiral invariant interactions always produce neutrinos of the same helicity, hence the helicity of the neutrino cannot be changed; eliminating magnetic moment interactions with photons. The currently accepted theory of weak interactions, namely the $V-A$ theory of Feyman and Gell-Mann (1958), is chiral invariant; so it will be necessary to consider modifications of this theory to allow the neutrino to have a magnetic moment. Though this is ugly, the $V-A$ theory has only been tested to an accuracy of about two percent, so phenomenologically there is room for some modification. Apart from considerations regarding the actual form of the interaction, the two component theory of the neutrino which is the currently accepted representation is also chiral invariant. Chiral invariance may be relaxed in our neutrino representation by assuming that the neutrino is a four component Dirac spinor, which is a linear combination of the two opposite helicity positive energy solutions of the zero mass Dirac equation, the antineutrino being represented by a linear combination of the negative energy solutions. We shall consider below various modifications of the $V-A$ interaction which lead to a magnetic moment for the neutrino.

The magnetic moment is evaluated by letting an initial neutrino v with four momentum p transform virtually into particles which can interact with an electromagnetic potential a which gives a momentum transfer q. The virtual intermediate particles then combine to give a final neutrino v' with momentum $p+q$. We write the matrix element $\langle v'|M|v\rangle$ for the process and calculate the coefficient of the magnetic moment interaction term $(a\cdot\gamma q\cdot\gamma - q\cdot\gamma a\cdot\gamma)$ in M. This coefficient gives the corresponding contribution to the magnetic moment.** Since all higher than first order processes involving weak interactions are divergent, the quantity we are trying to evaluate is divergent as the integral is taken over the intermediate momenta of the virtual particles. To get a

* Precession effects are allowed for a value of the neutrino magnetic moment equal to the experimental limit and assuming a magnetic field in the interior of the Sun equal to that measured at sunspots.

** The effect of the magnetic field on the external particle wave functions and on the propagators is negligible for the particle energies and magnetic fields under consideration. This effect becomes important for electron propagators when the magnetic field is of the order of $B = m^2c^3/eh = 4.4 \times 10^{13}$ G, see, for example, T. Erber, *Rev. Mod. Phys.* **38**, 626 (1966). The effect of the magnetic field on the neutrino wave function becomes important when $\mu B/E$ is of the order of 1, where μ is the neutrino magnetic moment calculated in the weak field limit and E is the energy of the neutrino. This means that B must be of the order of $E\,(\text{MeV}) \times 2 \times 10^{24}$ G, if we use the experimental limit for μ.

number we must cut off the integral at some finite momentum. Mohapatra *et al.* (1968) estimated the weak cut off to be of the order of four proton masses in considering effects due to second order weak interactions such as the $K_s^0 - K_L^0$ mass difference. We shall use this cut off in our calculations.

The first process we consider is indicated in Figure 1 by its Feynman diagram; it is

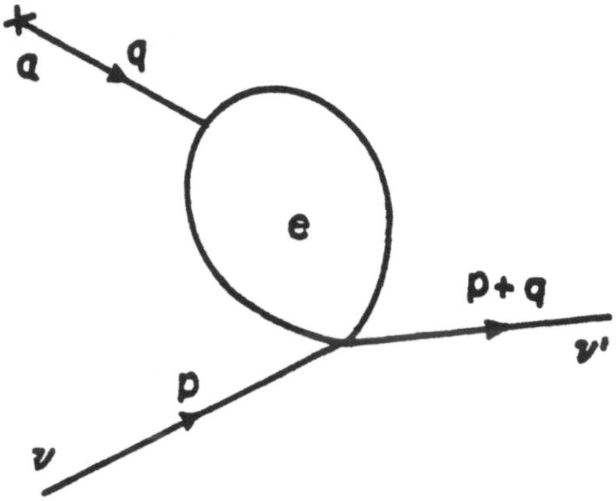

Fig. 1. First order weak process contributing to the electromagnetic interactions of the neutrino.

of first order in the weak coupling constant. The incoming neutrino emits and re-absorbs an electron which effects the interaction with the electromagnetic field. Let $\psi_e O \psi_\nu$, where ψ_e is the electron field, O is an operator which defines the form of the interaction and ψ_ν is the neutrino field, be the weak current transforming a neutrino into an electron. The matrix element for the diagram is then given by

$$\langle \nu' |M| \nu \rangle = \left\langle \nu' \left| \int \frac{Ge}{\sqrt{2}} O \frac{s\cdot\gamma + \frac{1}{2}q\cdot\gamma + m}{(s+\frac{1}{2}q)^2 - m^2} a\cdot\gamma \frac{s\cdot\gamma - \frac{1}{2}q\cdot\gamma + m}{(s-\frac{1}{2}q)^2 - m^2} O \frac{d^4 s}{(2\pi)^4} \right| \nu \right\rangle.$$

It is important to point out that for the process we are now considering there is essentially electron-neutrino elastic scattering at the weak interaction vertex. Gell-Mann *et al.* (1969) have stressed that the diagonal terms in the weak interaction are on a somewhat separate theoretical basis from the non-diagonal terms; such as those responsible for muon decay, pion decay and nuclear beta decay. Considering that the $V-A$ theory has been tested only very weakly (cf. Steiner, 1970) for the diagonal terms, we may introduce wider modifications of this theory than those permitted by the two percent accuracy tests which apply to the non-diagonal terms.

Returning to our calculation, if we have a vector or axial vector interaction; O will be γ_α or $\gamma_5\gamma_\alpha$ respectively, where the γ's are the Dirac matrices. In both of these cases the resulting magnetic moment interaction with the electromagnetic field is zero. Any combination of vector and axial vector will similarly give zero magnetic moment contribution for this virtual process, including $V-A$ interaction, as expected. We note further that since the neutrino appears only on external lines in the process, permitting it to have a non-zero mass will not change the answer.

Other forms of the weak interaction do give a magnetic moment. We now assume we have either a scalar or pseudoscalar interaction where O will be given by 1 or γ_5 respectively; and we shall take the interaction strength to be that given by the Fermi constant $G=10^{-5}/M_p^2$ where M_p is the proton mass. After doing the calculation we find the magnetic moment to be $\mu=\mu_B(Gm^2/8\sqrt{2}\pi^2)\ln\lambda/m$, where m is the mass of the electron, μ_B is the Bohr magneton and λ is the cut-off energy. For $\lambda=4M_p$ we get $\mu=2.4\times10^{-13}\ \mu_B$. We may also consider a tensor interaction with $O=\frac{1}{2}(\gamma_\mu\gamma_\nu-\gamma_\nu\gamma_\mu)$; and all other assumptions being unchanged, the magnetic moment is found to be $\mu=9.6\times10^{-13}\ \mu_B$.

We note that we have obtained these numbers giving full strength to the interaction. If we were to take into consideration the experimental limits placed on the strength of the scalar or pseudoscalar interaction, the number evaluated above would be reduced by a factor of 10^{-3}. For example, the ratio of the lifetime for pion decay into muon or electron which is accurately predicted by the $V-A$ theory is very sensitive to a small admixture of scalar or pseudoscalar interaction. The above argument applies if we assume the $V-A$ interaction still to be the main effect in diagonal weak interactions and allow other interactions only to the extent that they do not violate the two percent limit deviation. On the other hand, as was mentioned earlier, we are not completely sure that $V-A$ is still the main effect in diagonal processes or that its strength is given by the Fermi constant. At present, the experimental bound on the neutrino-electron cross section is still 40 times* the standard theoretical value; leaving the possibility that the strength assumed above may be increased by $\sqrt{40}$, and the magnetic moment by the same factor.

We consider one other process which is of second order in the weak interaction; its Feynman diagram is given in Figure 2. In order to have an interaction with the elec-

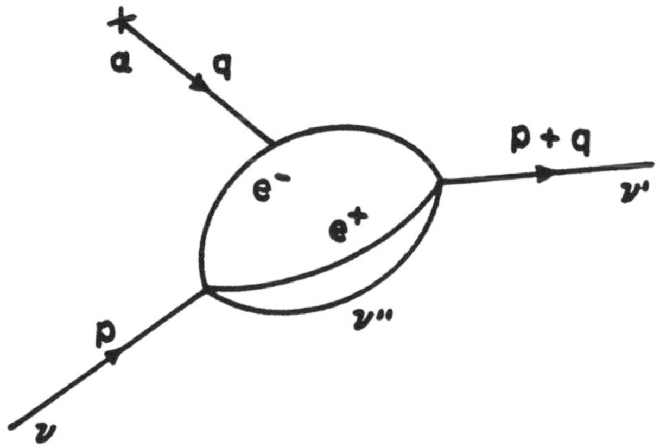

Fig. 2. Second order weak process contributing to the electromagnetic interactions of the neutrino.

* Spin flip scattering from real or virtual photons in the Sun's interior is negligible for neutrinos with this magnetic moment.

tromagnetic field we let an incoming neutrino transform virtually into an electron, a positron and a neutrino. One of the charged virtual particles can effect the interaction with the electromagnetic field.

After calculating the matrix M we find the magnetic moment is zero if the $V-A$ theory is assumed, but only a slight modification is needed in order to get a non-vanishing contribution. We replace the operator $\gamma_\mu(1-\gamma_5)$ appearing in the weak current by $\gamma_\mu(1-\omega\gamma_5)$ where ω may be different from 1. As in the first process we considered, the integrals over the intermediate momenta are divergent. In this case, the term with the highest divergence is quadratic in the cut-off energy*. Nevertheless, the magnetic moment is smaller than that obtained in the first process because we have a second order weak effect. The magnetic moment we get is $\mu=(1-\omega^4)\times 3\times 10^{-19}\ \mu_B$. If the experimental limits on the deviation from exact $V-A$ weak interactions are taken into account, $1-\omega^4$ is of the order of a few percent. On the other hand, the different status of the diagonal weak interaction terms may be taken into account by allowing a larger strength by a factor of $\sqrt{40}$ and permitting other interaction forms. This will allow a magnetic moment larger by a factor of 40 since we have a second order effect. Considering scalar, pseudoscalar or tensor interactions does not change the moment obtained above for pure vector interaction, that is $\omega=0$, by more than one order of magnitude. The calculation was also made letting the intermediate neutrino have a mass. It was found that the introduced term had a weaker divergence in the cut-off energy than the term independent of the neutrino mass. This additional contribution to the magnetic moment would be several orders of magnitude below the number given above.

The conclusion is that only under extreme assumptions we get a magnetic moment for the neutrino, large enough for spin precession effects to explain the results of Davis' experiment. These assumptions violate the usual features of the weak interactions such as the $V-A$ hypothesis and the two component theory of the neutrino. In addition a large value for the magnetic field strength in the interior of the Sun is needed. Thus the effect is possible but quite unlikely.

Acknowledgements

The author is indebted to Professor W. A. Fowler and C. W. Werntz for valuable discussions, and in particular to Professor S. C. Frautschi who suggested considering the problem.

References

Bernstein, J., Ruderman, M., and Feinberg, G.: 1963, *Phys. Rev.* **132**, 1227.
Davis, R., Jr., Harmer, D. S., and Hoffman, K. C.: 1968, *Phys. Letters* **20**, 1205.

* A quadratic divergence in the cut-off energy is of course very sensitive to the cut-off value; but this value would have to be increased by a factor of 10^3 to get a moment of the order or $10^{-13}\ \mu_B$. The magnetic moment calculated from the first diagram where we did get $10^{-13}\ \mu_B$, is only logarithmically divergent in the cut-off value.

Feinberg, G.: 1965, *Proc. Roy. Astron. Soc.* **A285**, 257.
Feynman, R. P. and Gell-Mann, M.: 1958, *Phys. Rev.* **109**, 193.
Gell-Mann, M., Goldberger, M. L., Kroll, N. M., and Low, K. E.: 1969, *Phys. Rev.* **179**, 1518.
Mohapatra, R. N., Subba Rao, J., and Marshak, R. E.: 1968, *Phys. Rev. Letters* **20**, 1081.
Pontecorvo, B.: 1968, *Sov. Phys. JETP* **26**, 984.
Salam, A.: 1957, *Nuovo Cim.* **5**, 299.
Steiner, H. J.: 1970, *Phys. Rev. Letters* **24**, 746.
Werntz, C. W.: 1970, Private communication.

Neutrino electrodynamics and possible consequences for solar neutrinos

M. B. Voloshin, M. I. Vysotskiĭ, and L. B. Okun'

Institute of Theoretical and Experimental Physics

(Submitted 25 April 1986)

Zh. Eksp. Teor. Fiz. **91**, 754–765 (September 1986)

The neutrino electromagnetic moment matrix and the possibility that some of the elements of this matrix are of the order of 10^{-10} of the Bohr magneton are discussed. Flavor oscillations and spin precession are examined for a neutrino in a magnetic field in the presence of matter. The interaction between solar neutrinos and the magnetic field in the interior of the convective zone of the Sun can lead, in this case, to the 11-year and semiannual variations in the neutrino flux, shown experimentally to be correlated with the magnetic activity of the Sun.

1. INTRODUCTION

There has been continuing interest in the mass matrix that determines the masses of neutrinos, in neutrino oscillations, and, when Majorana terms are present, in the probability of double β-decay without the emission of a neutrino. In this paper, we discuss a further class of static characteristics of neutrinos, namely, the electromagnetic moment (EMM) matrix μ that appears in the Lagrangian for the interaction between a neutrino and the electromagnetic field $F_{\mu\nu}$:

$$L_{int} = {}^1/_2\mu_{ij}(\bar{\nu}_R)_i\sigma_{\mu\nu}F_{\mu\nu}(\nu_L)_j + \text{h.c.} = {}^1/_2\mu_{ab}(\bar{\nu}_R)_a\sigma_{\mu\nu}F_{\mu\nu}(\nu_L)_b + \text{h.c.} \quad (1)$$

The subscripts L and R indicate left- and right-handed fields, the subscripts i, j label the current states of the neutrinos, $i, j = e, \mu, \tau, \ldots$ (this will be referred to as the flavor basis), and a, b refer to the eigenstate basis of the neutrinos mass matrix, $a, b = 1, 2, 3, \ldots$. In the latter basis, the real (imaginary) part of the diagonal elements of the matrix μ_{ab} are the magnetic (electric) dipole moments of the mass state $\nu_1, \nu_2, \nu_3, \ldots$ (with masses $m_1 < m_2 < m_3$), and the off-diagonal elements of μ_{ab} describe the decay of neutrinos into lighter neutrinos and γ-rays, e.g., $\nu_2 \to \nu_1\gamma$.

We shall examine the possibility that some of the elements of the matrix μ may be of the order of 10^{-10} μ_B ($\mu_B = e\hbar/(2m_ec)$ is the Bohr magneton), which may already have been seen experimentally in the form of the very specific variations in the solar neutrino flux recorded by the Davis group.[1] This question has been examined in our previous brief communications.[2–4] Here, we would like to give a more complete and closed presentation of the possible manifestations of the EMM in solar-neutrino experiments.

Briefly, the effect is that the EMM ensures that the helicity of the neutrinos is partially modified during their passage through the toroidal magnetic field **H** in the convective zone of the Sun, and the observed solar activity is ascribed to processes in this zone. This results in a reduction in the flux of left-handed neutrinos, as detected by the Cl–Ar method. The quantity |**H**| and, consequently, the reduction in the flux reach their respective maxima at the maximum of the 11-year cycle of solar activity. Moreover, **H** changes sign at the solar equator, so that, because of the inclination of the Earth's orbit to this equator, there should also be semiannual variation in the recorded neutrino flux.

In Section 2, we shall discuss electroweak interaction models, in which EMM values of the order of 10^{-11}–10^{-10} of the Bohr magneton can be obtained, and we summarize experimental and astrophysical limits on the matrix μ. In Section 3, we consider the behavior of the neutrino helicity when the EMM interact with the magnetic field in the presence of matter, which may be very significant for neutrino propagation in the solar interior. In Section 4, we give a brief review of aspects of the theory of the Sun that are relevant to this question, and estimate the time variation in the recorded neutrino flux. Finally, in Section 5, we discuss possible further studies of this effect, using the new solar-neutrino detectors that are being built at present.

2. THE NEUTRINO EMM MATRIX

1. In standard $SU(2)_L \times U(1)$ theory, if the right-handed neutrino ν_R is an $SU(2)$ singlet, the matrix μ can be found from the diagrams shown in Fig. 1, and turns out to be proportional to the neutrinos mass matrix m (Refs. 5, 6):

$$\mu = \frac{3eG}{8\sqrt{2}\pi^2} m = \frac{3m_eG}{4\sqrt{2}\pi^2}\mu_B m \simeq 2.7\cdot 10^{-10}\mu_B \frac{m}{m_N}. \quad (2)$$

where G is the Fermi constant and m_N the nucleon mass. The EMM described by this formula are exceedingly small, e.g., $\mu_{ee} \sim 10^{-17}\mu_B$ for $m_{ee} = 30$ eV. The reason for the fact that μ and m are mutually proportional (and, hence, μ is small) in this scheme is that the W boson in the diagrams of Fig. 1 interacts only with left-handed currents. This means that the change in helicity demanded by an interaction such as (1) must occur on the external neutrino line, $\hat{p}\nu_L = m\nu_R$. Hence, it is clear that the neutrino EMM can be substantially increased if the theory contains right-handed currents. The change in helicity can then occur on the charged fermion line, so that the matrix μ will contain the charged lepton masses instead of the neutrino masses.

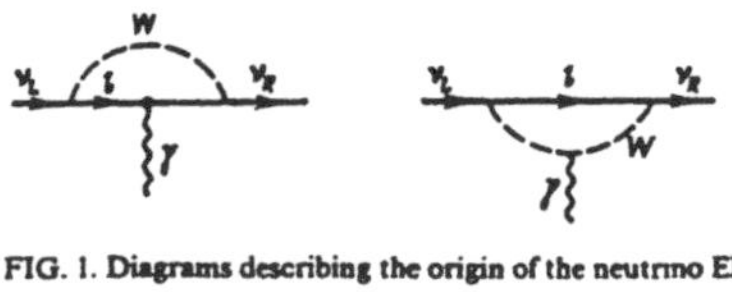

FIG. 1. Diagrams describing the origin of the neutrino EMM.

2. As an example of a model with right-handed currents, let us examine the widely discussed scheme with left-right symmetry $SU(2)_L \times SU(2)_R \times U(1)$ (see, for example, Ref. 7). The mediator of the usual weak interaction in this scheme is the charged boson W_1 containing a small admixture of the right-handed boson W_R:

$$W_1 = W_L \cos\varphi + W_R \sin\varphi.$$

The neutrino EMM matrix is determined in this model by the masses of the charged leptons and the matrices U and V that mix the neutrinos in the left-handed (U) and right-handed (V) currents[6,8]:

$$\mu_{ij} = \frac{Gm_e}{\sqrt{2}\pi^2}\mu_B \sin 2\varphi \sum_l U_{il} m_l V_{lj}, \qquad (3)$$

where the sum is evaluated over the types of charged leptons. In the minimum model[7] of this kind with the flavor basis we have $U = V = 1$, so that μ_{ij} is diagonal:

$$\mu_{\nu_l} \approx 3.6\cdot 10^{-10}\mu_B \sin 2\varphi (m_l/m_N).\ l = e,\ \mu,\ \tau, \ldots . \qquad (4)$$

In this scheme, the maximum EMM has a neutrino current state that appears in the doublet with the heaviest charged lepton. In the case of three generations, this is the ν_τ, for which the magnitude of μ corresponding to the maximum value $\sin 2\varphi = 0.1$ allowed by the experimental limit[9] ($\sin\varphi \lesssim 0.05$) is

$$\mu_{\nu_\tau} \approx 0.6\cdot 10^{-10}\mu_B. \qquad (5)$$

For the mass states of the neutrinos, the EMM values in this case are obviously proportional to the fraction of the ν_τ present in them. We note that the matrix μ_{ah} is nondiagonal in the mass basis, i.e., decays of the form $\nu_2 \to \nu_1\gamma$ should occur. It is also clear that, if there is a fourth generation, the neutrino appearing in this generation will have the highest EMM.

It is clear from the above example that matrix elements of the order of $10^{-10}\mu_B$ can be obtained in schemes involving an admixture of right-handed currents allowed by the corresponding experimental limits.[9] Of course, the mechanism for the appearance of the neutrino EMM may be quite different, and the structure of the matrix μ may be significantly different from that described by (3) and (4). In particular, in supersymmetric models, the appearance of fairly large EMM (including $\mu_{\nu_e} \sim 10^{-10}\mu_B$) seems quite possible because these schemes include heavy fermions (winos and higgsinos), whose masses effectively turn over the helicity in graphs similar to those shown in Fig. 1.[1]

We emphasize that, strictly speaking, the EMM are not necessarily related to the masses of the neutrinos themselves. In principle, it is possible to consider the EMM of massless neutrinos without introducing contradictions although, from the point of view of the natural condition $\mu \neq 0$ for $m = 0$, this seems strange and requires special compensation by counterterms in the neutrino mass diagrams.

3. Let us now consider the limits on the matrix μ that follow from reactor experiments and astrophysical estimates. The interaction (1) produces an additional contribution to the neutrino-electron scattering cross section de-

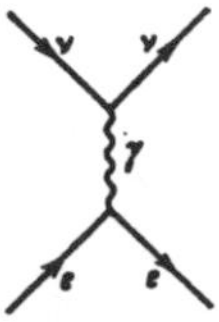

FIG. 2. Diagram showing the scattering due to the neutrino EMM.

scribed by the graph of Fig. 2. The differential cross section for νe scattering due to this mechanism is the same for ν_e and $\tilde{\nu}_e$, and is given by[10]

$$\frac{d\sigma_{em}}{dT} = \Big(\sum_j |\mu_{ej}|^2/\mu_B^2\Big)\frac{\pi\alpha^2}{m_e^2}\Big(\frac{1}{T} - \frac{1}{E_\nu}\Big), \qquad (6)$$

where E_ν is the energy of the incident neutrino and T the kinetic energy of the recoil electron, which is actually the quantity recorded experimentally.

The sum over the neutrino current states in (6) corresponds to summation over the different final states of the neutrinos in the reaction $\tilde{\nu}_e e \to \tilde{\nu}_j e$.

Let us compare (6) with the well-known expression for the $\tilde{\nu}_e e$ cross section due to the weak interaction (see, for example, Ref. 11):

$$\frac{d\sigma_W}{dT} = \frac{G^2 m_e}{2\pi}\Big[\Big(1 - \frac{T}{E_\nu}\Big)^2 (1+2\sin^2\theta_W)^2 + 4\sin^4\theta_W - (1+2\sin^2\theta_W)\sin^2\theta_W \frac{m_e T}{E_\nu}\Big], \qquad (7)$$

where θ_W is the Weinberg angle. The electromagnetic and weak amplitudes do not interfere, so that the total cross section is obtained by combining the expressions given by (6) and (7). When $T \ll E_\nu$, the cross section given by (7) becomes constant, while that in (6) behaves as T^{-1}, and when $\sin^2\theta_W = 0.22$, it becomes comparable with the weak cross section for $T \simeq 0.3$ MeV if $\mu_{eff} \equiv (\Sigma|\mu_{ej}|^2)^{1/2} = 10^{-10}\mu_B$. Thus, the problem of reducing the experimental limit for μ_{eff} involves, above all, a reduction of the threshold for the detection of recoil electrons, which is complicated by the higher background at low electron energies. The experimental data reported in Ref. 12 have been used to show[13,14] that $\mu_{eff} \lesssim 2\times 10^{-10}\mu_B$.

The best limitation on the matrix μ is obtained by considering the cooling of young white dwarfs due to the decay of plasmons γ^* to $\nu\tilde{\nu}$ pairs. The decay width due to the EMM is

$$\Gamma_\mu(\gamma^* \to \nu\tilde{\nu}) = \Big(\sum_{ij} |\mu_{ij}|^2/\mu_B^2\Big)\frac{\alpha\omega_p^3}{24 m_e^2}, \qquad (8)$$

where ω_p is the plasma frequency in the star. Analysis of astrophysical data gives the following limit[15]:

$$\Big(\sum_{ij}|\mu_{ij}|^2\Big)^{1/2} \leqslant 0.7\cdot 10^{-10}\mu_B, \qquad (9)$$

which is actually the upper limit for the norm of the matrix

μ. The fact that (8) contains precisely this quantity is due to the summation over all the types of neutrino pairs. This summation, and the neglect of the neutrino masses, is justifiable if the masses are substantially smaller than $\omega_p/2$. For young white dwarfs, for which (9) was obtained, $\omega_p \simeq 30$–40 keV, so that the last assumption seems fully acceptable. It would be interesting to repeat the analysis given in Ref. 15 with more recent astrophysical data, and deduce the upper limit for $|\mu|$, which may well turn out to be more stringent than (9).

4. To conclude this section, let us consider a further aspect of the neutrino magnetic moment. The introduction of right-handed neutrinos ν_R increases the number of neutrino degrees of freedom by a factor of two, which may give rise to difficulties in explaining the primeval abundance of ^{4}He (see, for example, the review in Ref. 16 and the book by Okun'[11]). Leaving to one side the question of how serious these difficulties really are, we note that, if we allow the nonconservation of lepton number, we can avoid the introduction of the right-handed neutrino field and, instead, use the antineutrino field. The analog of (1) then assumes the "Majorana" form:

$$L_{int} = {}^1/_2 \tilde{\mu}_{ij} (\overline{\nu^c})_{Ri} \sigma_{\mu\nu} F_{\mu\nu} \nu_{Lj} + \text{h.c.} \tag{10}$$

The matrix $\tilde{\mu}_{ij}$ should then be antisymmetric because the part of the lepton operator in (10) that is symmetric in the neutrino types will then vanish. In other words, only the off-diagonal matrix moments are possible in this case. We shall not consider specific models leading to interactions such as (10), but it is not difficult to foresee that, in such models, the condition for the experimental absence of a double β-decay without the emission of neutrinos will be a very stringent limitation.

3. SPIN PRECESSION AND NEUTRINO OSCILLATIONS IN A MAGNETIC FIELD, INCLUDING THE EFFECT OF MATTER

1. We begin by considering the motion of a neutrino with energy $E \gg m$ in a vacuum, taking oscillations into account. The standard description of these oscillations follows from the fact that the momentum p in the neutrino wave function $\psi_\nu = \exp(-iEt + ipz)\psi_0$ is a matrix in the space of the neutrino types, and is related to the mass matrix by

$$p = (E^2 - m^2)^{1/2} \approx E - m^2/2E \tag{11}$$

(we shall choose the phases of the ν_L and ν_R so that the matrix m is Hermitian). Let us extract the unimportant phase factor $\exp[iE(z-t)]$:

$$\psi_\nu = \exp[iE(z-t)]\nu(z).$$

According to (11), we then have $\nu = \exp[-im^2 z/(2E)]\nu_0$, so that $\nu(z)$ satisfies the evolution equation.

$$i\frac{d\nu}{dz} = \frac{m^2}{2E}\nu. \tag{12}$$

This implies that ν is a vector (column vector) in the space of the neutrino types. To include the interaction with the magnetic field and with matter, we recall that each of the elements of the column is a spinor:

$$\begin{pmatrix} \nu_R \\ \nu_L \end{pmatrix} \tag{13}$$

(the direction of motion of the neutrino is taken to be the spin quantization axis).

2. When the electromagnetic field $F_{\mu\nu}$ is present, the interaction (1) leads to the following modification of (12):

$$i\frac{d\nu}{dz} = \mathcal{H}\nu, \tag{14}$$

where the "Hamiltonian' $\mathcal{H}$ is

$$\mathcal{H} = \frac{m^2}{2E} - \mu P_+ \sigma \cdot (\mathbf{H} - i\mathbf{E})P_- - \mu^+ P_- \sigma \cdot (\mathbf{H} + i\mathbf{E})P_+, \tag{15}$$

and $P_\pm = (1 \pm \sigma_3)/2$ are projectors onto the ν_R and ν_L states. We note that, since $P_\pm \sigma_3 P_\mp = 0$, the longitudinal components of **H** and **E** do not appear in $\mathcal{H}$. This is readily understood because the longitudinal field components remain unaltered under the transformation to the neutrino rest system, whereas the lateral components contain the factor $\gamma = E/m \gg 1$. The derivation of the Hamiltonian (15) assumes that the field $F_{\mu\nu}$ changes little over the length E^{-1}. It is also assumed that the neutrino is not deflected by field gradients, which is valid with high precision in all realistic situations.

We note that we deliberately retained the difference between μ and μ^+ in (15) because the phases of the mass matrix were fixed. If it were possible to observe the phase difference between neutrinos oscillations in the presence and absence of the field, it would be possible to find the relative phases of the elements of the matrices m and μ. Nonzero values of these phases would correspond to *CP* violation in the neutrino sector.

It is immediately clear from (14) and (15) that, for the mass states ν_a in a transverse magnetic field $\mathbf{H}_\perp$, there are oscillations between left- and right-handed components (spin precession) with frequency $\omega(z) = |\mu_{aa}||\mathbf{H}_\perp(z)|$. If the region in which $\mathbf{H}_\perp$ is present (its direction is independent of z) intercepts a beam of the ν_L, the numbers of the ν_L and ν_R are given by

$$N_L(z) = N_0 \cos^2\left(\int^z \omega(z)\,dz\right), \quad N_R = N_0 \sin^2\left(\int^z \omega(z)\,dz\right). \tag{16}$$

In the case of Dirac masses that we are considering, precession will occur in an arbitrarily weak transverse field because the left- and right-handed neutrinos are degenerate in energy. Oscillations between the mass states $\nu_{aL} \leftrightarrow \nu_{bR}$ are possible because of the diagonal terms in μ_{ab}. However, these oscillations have appreciable amplitudes only when the mixing element $|\mu_{ab}||\mathbf{H}_\perp|$ is comparable with the difference between the diagonal elements of the Hamiltonian (19):

$$|\mu_{ab}||\mathbf{H}_\perp| \gtrsim \left|\frac{m_a^2 - m_b^2}{2E}\right|. \tag{17}$$

To estimate the order of magnitude of the various quantities in this expression, we note that $\Delta m_{ab}^2 \lesssim 10^{-7}\,\text{eV}^2$ for $|\mathbf{H}_\perp| = 10^3$ G, $|\mu_{ab}| = 10^{-10}\mu_B$, and $E = 10$ MeV.

3. Finally, let us consider the general case where the

neutrino propagates in matter (it is precisely this case that is interesting from the point of view of solar neutrinos). Only the coherent interaction between the neutrinos and matter is significant because the noncoherent neutrino interaction cross section is exceedingly small. Coherent effects are conveniently examined directly at the level of the Lagrangian. The weak interaction neutrino Lagrangian is

$$L_W = -\frac{G}{\sqrt{2}}(\bar{\nu}_e\gamma_\mu(1+\gamma_5)\nu_e)(\bar{e}\gamma_\mu(1+\gamma_5)e)$$
$$-\frac{2G}{\sqrt{2}}\sum_l \bar{\nu}_l\gamma_\mu(1+\gamma_5)\nu_l J_\mu^n$$
$$-\frac{4G}{\sqrt{2}}\sum_l T_3^R\bar{\nu}_l\gamma_\mu(1-\gamma_5)\nu_l J_\mu^n. \qquad (18)$$

The first term is due to the interaction between the charged currents of the electron-neutrino and the electron, and the second is due to the interaction with the neutral current

$$J_\mu^n = J_\mu^3 - \sin^2\theta_W J_\mu^{em} \qquad (19)$$

(J^{em} is the electromagnetic current and J^3 is the current of the third component of weak isospin). Finally, the third term in (18) occurs when the third component of weak isospin of right-handed neutrinos, T_3^R, is not zero. The next step is to average the Lagrangian (18) over matter. For matter at rest, which does not have a macrosocopic spin magnetic moment, only the average of the time component of the vector part of the current is nonzero. We thus have

$$\langle\bar{e}\gamma_\mu(1+\gamma_5)e\rangle = \delta_{\mu 0}n_e, \qquad (20)$$

where n_e is the electron density. For electrically neutral matter, (19) shows that neutral currents cancel out for protons and electrons, and only neutrons provide a contribution to $\langle J_\mu^n\rangle$:

$$\langle J_\mu^n\rangle = -{}^1/_4\delta_{\mu 0}n_n. \qquad (21)$$

As a result, the neutrino equation of motion, obtained by averaging the Lagrangian over matter, corresponds to the following Hamiltonian in (14):

$$\mathcal{H} = m^2/2E - \mu P_+\sigma(\mathbf{H}-i\mathbf{E})P_-$$
$$-\mu^+P_-\sigma(\mathbf{H}+i\mathbf{E})P_+ + C_LP_- + C_RP_+, \qquad (22)$$

where the matrices C_L and C_R for the interactions between left- and right-handed neutrinos have a diagonal form in the flavor basis:

$$(C_L)_{ij} = \delta_{ei}\delta_{ej}\sqrt{2}Gn_e - \delta_{ij}Gn_n/\sqrt{2},$$
$$(C_R)_{ij} = -\delta_{ij}\sqrt{2}Gn_nT_3^R. \qquad (23)$$

4. Let us begin by considering the case where there is no electromagnetic field, so that the helicities are not reversed. The matrix C_L then describes the effect of matter on the oscillations of left-handed neutrinos, first considered by Wolfenstein.[17] The contribution of neutrinos to C_L is then the same for all neutrino types, and is unimportant for oscillations, whereas the contribution of electrons leads to a striking and potentially exceedingly important effect, discovered by Mikheev and Smirnov,[18] which can be summarized as follows. Let us suppose that $(m^2)_{ee}$ is smaller than $(m^2)_{\mu\mu}$ or $(m^2)_{\tau\tau}$ (to be specific, we shall consider ν_e and ν_τ). If the density n_e is such that

$$\sqrt{2}Gn_e = [(m^2)_{\tau\tau} - (m^2)_{ee}]/2E, \qquad (24)$$

The difference $\mathcal{H}_{\tau\tau} - \mathcal{H}_{ee}$ between the diagonal matrix elements is then zero. This corresponds to "level crossing," well-known in quantum mechanics. If $n_e(z)$ varies sufficiently slowly and passes through the value given by (24), and if there is slight mixing ($\mathcal{H}_{e\tau}$), neutrinos of a given type will adiabatically transform into neutrinos of the other type. In our example, $\nu_1 \simeq \nu_e$ will transform to $\nu_2 \simeq \nu_\tau$. The adiabatic condition then requires that the characteristic length l for a change in n_e must satisfy the condition

$$l \gg 2E/(m^2)_{e\tau}. \qquad (25)$$

Mikheev and Smirnov have analyzed numerically the solution of the neutrino evolution equation and found[18] that the mechanism they have discovered is effective for solar neutrinos in a very wide range of values of Δm^2 and mixing angles. (Typical intervals for $E \simeq 10$ MeV are 10^{-6} eV$^2 \lesssim \Delta m^2 \lesssim 10^{-4}$ eV^2, $\sin^2 2\theta \gtrsim 10^{-3}$).

5. When spin precession in a magnetic field is considered, the difference between C_L and C_R becomes significant because it lifts the degeneracy of the left- and right-handed components. Let us consider the simplest case of one neutrino type in a uniform magnetic field $|\mathbf{H}_\perp| = H =$ const in constant-density matter, so that $\Delta C = C_L - C_R =$ const. Discarding the unimportant phase factor, the solution of (14) can then be written in the form

$$\nu(z) = \exp\{i(\mu H\sigma_1 - \Delta C\sigma_3/2)\}\nu_0$$
$$= [\cos\omega z + i(\mu H\sigma_1 - \Delta C\sigma_3/2)\omega^{-1}\sin\omega z]\nu_0, \qquad (26)$$

where $\omega = [(\mu H)^2 + (\Delta C/2)^2]^{1/2}$. If the initial state ν_0 is a pure left-handed state, the number of left- and right-handed neutrinos at the point z is then, respectively, given by

$$N_L = \cos^2\omega z + \cos^2 2\theta_H \sin^2\omega z, \quad N_R = \sin^2 2\theta_H \sin^2\omega z, \qquad (27)$$

where $\tan 2\theta_H = 2\mu H/\Delta C$. These two expressions correspond to the usual picture of the oscillations with mixing angle θ_H. It is clear that the mixing angle is a maximum ($|\theta_H| = \pi/4$) for $\Delta C = 0$, and that this angle is always less than $\pi/4$ for $\Delta C \neq 0$. The necessary condition for effective modulation of the left-handed neutrino flux is therefore

$$2\mu H > \Delta C. \qquad (28)$$

When $\mu = 10^{-10}\mu_B$ and $H = 10^3$ G, this condition demands that the electron density must be less than 10^{22} cm^{-3} in the case of the electron-neutrino, and ΔC is due to the interaction between charged currents. For the other neutrino types, the limit on the neutrino density for $T_3^R = 0$ is $n_n \lesssim 2\times 10^{22}$ cm^{-3}. If, on the other hand, $T_3^R = 1/2$, the left- and right-handed nonelectron neutrinos are degenerate even in the presence of matter of any density.

In the opposite limit from (28), i.e.,

$$\Delta C \gg 2\mu H \qquad (29)$$

spin precession is strongly suppressed, and the beam consists mostly of the ν_L.

4. THE SUN AND POSSIBLE CONSEQUENCES FOR SOLAR NEUTRINOS

1. Neutrino spin precession cannot be observed under terrestrial laboratory conditions because, when $\mu = 10^{-10}\mu_B$, the field H necessary to turn the spin through an angle of the order of unity over a path length L is given by

$$HL \sim \mu^{-1} \approx 3 \cdot 10^{11}\ \text{G} \cdot \text{cm}. \qquad (30)$$

Searches for this phenomenon must therefore be concentrated on natural neutrino sources in which there are large-scale natural magnetic fields. The nearest source of this type is the Sun, and we therefore begin with a brief account of the relevant parts of the theory of the Sun that are significant for the spin precession of solar neutrinos.

The solar radius $R_\odot$ is about 7×10^{10} cm. The core, in which the nuclear reactions providing the Sun with its energy take place, accounts for about a quarter of the radius. We note, however, that neutrino generation occurs in this region mostly as a result of the reaction $pp \to d + e^+ + \nu_e$, in which the maximum energy of the ν_e is 0.42 MeV, and is below the detection threshold of the Cl–Ar method.[1] High-energy neutrinos from ^{7}Be and ^{8}B, recorded in Ref. 1, originate from the central, hottest part of the core, whose radius amounts to only $\sim 3\cdot10^9$ cm (see Ref. 19 for the flux calculations). The radiative transfer zone (so called because of the way in which heat is removed from the solar core) extends up to $R \simeq 0.7R_\odot$. Finally, the last 2×10^{10} cm of the solar radius correspond to the convective zone in which heat is transferred by turbulent convection. This zone contains the currents responsible for global magnetic fields with the 22-year (quasi) periodicity, and the modulus of the magnetic field has an 11-year cycle. During solar-active years, the magnetic field in the convective zone has a toroidal structure (it points in the azimuthal direction). The strength of this field can be judged from the sunspot field[20] ($H \simeq 2\times10^3$–4×10^3 G). The sunspots are regions in which the lines of force of the field that has "floated up" to the surface either leave or enter the Sun. It is therefore very likely that, during the years of solar activity maximum, the magnetic field in the convective zone is of the order of a few kG. The field H may increase to some extent between the surface and the bottom of the convective zone. The 22-year component of the field cannot penetrate the radiative heat-transfer zone because of the jump in the magnetic permeability (~ 1 in the radiative transfer zone and $\sim 10^{-5}$ in the convective zone).[21] While the Sun remains quiet at the minimum of the 11-year cycle, the field in the convective zone decreases by at least an order of magnitude.[21] (A more detailed account of the structure of the Sun and its magnetic field can be found in the literature.[20,21])

2. Thus, the product of the average (within the convective zone) field H and the depth L of the zone may reach $HL \simeq 3\times10^{13}$–10^{14} G·cm. The estimate given by (30), therefore suggests that, for $\mu = 10^{-10}\mu_B$, the flux of left-handed neutrinos measured in Ref. 1 was effectively modulated during the 11-year cycle. If the product μHL does not reach $\pi/2$, the flux recorded for the active Sun should be a minimum; the maximum flux should, at any rate, be reached at the activity minimum. The possible presence of this type of anticorrelation first emerged as a result of the analysis of experimental data in Ref. 22 (see Ref. 23 for a discussion of possible time variations in the data of the Davis group with other periods). Figure 3 shows a graph taken from Ref. 1, in which the neutrino flux data are compared with the number of sunspots characterizing the solar activity. There is a clear reduction in the neutrino flux during the solar activity maximum in 1979–1980. However, the experimental uncertainties are very large and the statistical significance of the anticorrelation effect is still not clear.

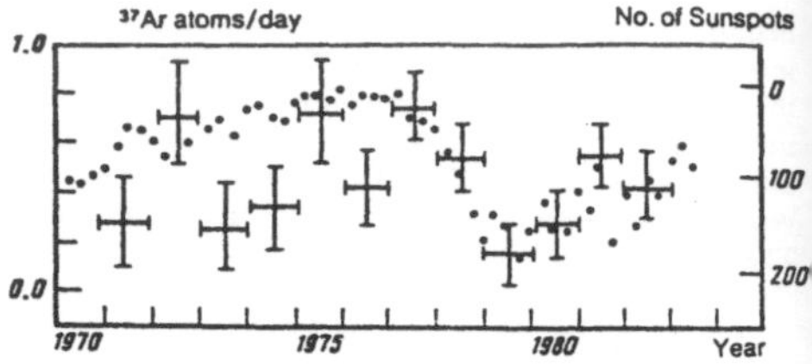

FIG. 3. Graph taken from Ref. 1 and showing the rate of production of ^{37}Ar as a function of time (annual averages are reproduced) and the number of sunspots (right-hand scale).

Another possible effect[4] requiring a shorter time of observation is the seasonal (semiannual) variation in the recorded flux of high-energy neutrinos. This effect is due to the fact that the sign of the magnetic field in the convective zone is different in the Northern and Southern Hemispheres, and is therefore zero at the equator. The transition region covers the latitude interval $\pm(5$–$7^\circ)$, which corresponds to distances of $\pm 6\cdot10^9$–8×10^9 cm from the equator. (The reduction in the field near the equator is reflected in the fact that there are no sunspots at these latitudes.) On the other hand, the plane of the Earth's orbit (the plane of the ecliptic) makes an angle of 7°15' with the plane of the solar equator. This means that, when the Earth lies in the plane of the solar equator (at the beginning of June and the beginning of December), the central part of the core, whose diameter is 3×10^9 cm and in which the boron and beryllium neutrinos originate, is seen from the Earth through the equatorial "slit" in the magnetic field, and the flux of left-handed neutrinos should be a maximum. Conversely, at the beginning of March and the beginning of September (maximum distance from the plane of the solar equator), the strong-field region will lie in the field of view of the terrestrial neutrino detector during the years of the active Sun, and the recorded flux should change because the neutrino helicity is reversed. (When $\mu HL \lesssim \pi/2$, the minimum current should occur at the beginning of March and of September.) Seasonal variations should become weaker as the solar activity falls, and may disappear altogether during the years of the quiet Sun.

We emphasize that the last effect is not well-defined for pp-neutrinos because the size of the region in which these

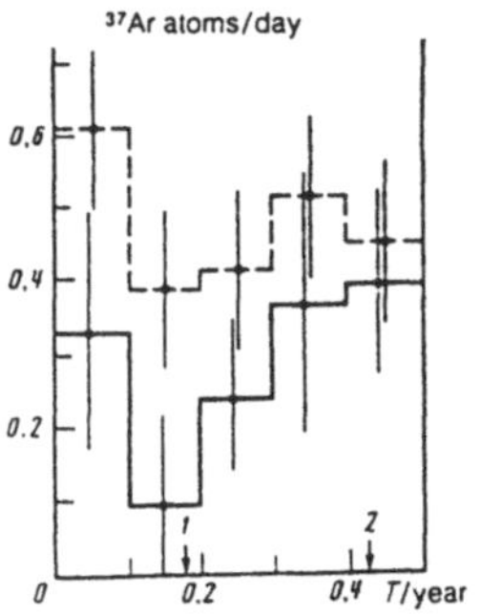

FIG. 4. Semiannual phase diagrams for the solar active years[1] 1979–1980 (solid line) and for 1975–1978 and 1983–1984 (dashed line). Arrow 1 shows the time of maximum distance of the Earth from the plane of the solar equator and arrow 2 shows the time when the Earth crosses this plane.

neutrinos originate ($R_\odot/4 \simeq 1.7\times10^{10}$ cm) is greater than the size of the equatorial gap in the magnetic field.

Figure 4 shows the semiannual phase diagrams for the 1975–1984 data.[1] It is clear that the flux was lower during the years of the quiet Sun near the beginning of March and of September. At other times, the reduction in the flux at these times is less well-defined (if is exists at all). We note, however, that the uncertainties in these data are large, and the existence of the seasonal variations cannot be deduced with a high degree of statistical confidence. Nevertheless, it is interesting that, if we remove from the 1979–1981 data, and from the data for the first half of 1982, all the series for which the dates of maximum sensitivity are closest to March 5 and September 5 (a total of 7 series), the average rate at which the ^{37}Ar atoms were created amounts to only 0.11 ± 0.08 atoms/day, whereas the average over all the counting-rate data was 0.45 ± 0.04 atoms/day. The difference between these two numbers is more than 3.5 standard deviations. (We note that the background estimated in Ref. 1 was 0.08 ± 0.03 atoms/day.)

3. At least three conditions must be met for the above effects to be present.

(1) The neutrinos entering the convective zone must have high enough magnetic moments so that $\mu HL \gtrsim 1$.

(2) The interaction with the matter present in the convective zone must not appreciably suppress the neutrino spin precession.

(3) The neutrinos must not be depolarized by the constant magnetic field in the solar interior (if it is present).

Let us consider condition (2) first. The density ρ of matter in the convective zone varies, according to the model proposed in Ref. 20, from about 0.2 g/cm^3 at the bottom of the zone to about 10^{-7} g/cm^3 at the surface. At half depth (10^{10} cm), it amounts to about 0.05 g/cm^3. About 75% of the mass in this region[31] is accounted for by hydrogen and 25% by ^{4}He. Hence, near the bottom of the zone, the electron density n_e is about 10^{23} cm^{-3} and the neutron density is $n_n \simeq 10^{22}$ cm^{-3}. In view of (28), it is clear from the above estimates that the necessary condition for (2) to be satisfied in the case of the electron-neutrinos is that the magnetic field at the bottom of the zone must be of the order of 10 kG, whereas, for the ν_μ and ν_τ, the field can be weaker by an order of magnitude (all the estimates were obtained for $\mu = 10^{-10}\mu_B$).

If it turns out that variable fields of the order of 10 kG are not present near the bottom of the convective zone, or that the $\mu_{\nu e}$ is much smaller than $10^{-10}\mu_B$ (the latter may become clear as a result of a further increase in the precision of reactor experiments), the above effects may occur provided there is considerable mixing between ν_e and, for example, ν_τ ($\theta \simeq \pi/4$), and the oscillation length is substantially less than $R_\odot$. In that case, the convective zone will receive a noncoherent mixture of the mass states

$$\nu_1 = \nu_e\cos\theta + \nu_\tau\sin\theta, \qquad \nu_2 = -\nu_e\sin\theta + \nu_\tau\cos\theta,$$

for which (a) the interaction with electrons is suppressed (by the factors $\cos^2\theta$ and $\sin^2\theta \simeq 1/2$, respectively) and (b) their EMM may be of the necessary order of magnitude because of the presence of a large ν_τ admixture (see Section 2 for a discussion of the model with left-right symmetry), which ensures that condition (1) is satisfied.

Finally, it may turn out that the Mikheev-Smirnov mechanism[18] is working, and only the mass state ν_2 enters the convective zone. The probability of detecting such neutrinos in the experiment reported in Ref. 1 is then $\sin^2\theta$ and, when $\sin^2\theta < 1/3$, the estimated initial flux of solar neutrinos must be increased as compared with Ref. 19. On the other hand, conditions (1) and (2) with small θ are obviously more readily satisfied for ν_2 than for ν_e. We note that, since the effectiveness of the Mikheev-Smirnov mechanism[18] depends on the neutrino energy [see (24)], different relationships may, in general, be possible between the field modulation depth of the fluxes of high-energy and *pp*-neutrinos.

Finally, let us consider condition (3). The fact that the flux observed by the Davis group is so low as compared with the theoretical calculations[19] is sometimes explained[24] by supposing that the solar core contains a frozen-in (primeval) magnetic field of the order of 10^7 G, and that ν_e has an EMM of about $10^{-13}\mu_B$, so that the neutrinos are completely depolarized by the magnetic field (and the flux of left-handed neutrinos is reduced by a factor of two). The estimates given in Section 3 [inequalities (28) and (29)] show that the interaction between ν_e and matter then excludes depolarization even for $\rho \gtrsim 0.1$ g/cm^3, which is much lower than the density in the solar core (up to about 150 g/cm^3 at the center). The conclusions given in Ref. 24 are therefore invalid. If we adopt our value $\mu \simeq 10^{-10}\mu_B$, the condition for the absence of depolarization by the frozen-in field is more stringent (but can be satisfied). Actually, the random primeval field in the solar interior should be $H_{pr} \sim \rho^{2/3}$ (the condition for the frozen-in flux in matter). Hence, to satisfy condition (3), it is sufficient to satisfy the inequality given by (33) for the lowest density, i.e., on the surface of the radiative transfer zone (where, we recall, $\rho \simeq 0.2$ g/cm^3). If we take into account the estimates given above, this leads us to an upper limit for the frozen-in primeval field immediately under the surface of the radiative transfer zone: $H_{pr} \lesssim 10^4$ G for ν_e and $H_{pr} \lesssim 10^3$ G for ν_μ and ν_τ. If $H_{pr} \sim \rho^{2/3}$, the corresponding limits on the frozen-in field near the solar center are $H_{pr} \lesssim 10^6$ G and $H_{pr} \lesssim 10^5$ G. As far as we know, there is no evidence against these limits for H_{pr}.

5. MAGNETIC MOMENT AND DETECTORS OF SOLAR NEUTRINOS

It is clear from the foregoing discussion that, because the neutrino mass and electromagnetic matrices are not adequately known, there is a range of possibilities, and a choice must be made between them by using different detectors with different energy thresholds. For example, if the ν_e has an EMM $< \sim 10^{-10}\,\mu_B$, and oscillations are unimportant, the boron and beryllium neutrinos detected with the Cl–Ar detector should exhibit 11-year and semiannual flux variations, whereas for *pp*-neutrinos (Ga–Ge detectors[25] are being built for these neutrinos and superconducting indium detectors are being developed[26]), there are only the 11-year variations and the semiannual variations should be weak. If oscillations and the Mikheev-Smirnov effect are significant, and other types of neutrinos (e.g., ν_τ) are the only ones to have a magnetic moment, then, because of the energy dependence of the Mikheev-Smirnov effect,[18] the 11 year variation in the observed *pp*-neutrino flux can be either reduced or amplified in comparison with variations in the high-energy neutrino flux.

The liquid argon detector ICARUS,[27] which is sensitive to *ve* scattering at neutrino energies above ~5 MeV, is very promising for studies of solar neutrinos. This detector will give information on the flavor composition of the high-energy component of solar neutrinos, and will thus settle the question of the role of oscillations and the Mikheev-Smirnov effect.

From the point of view of the time varations in the solar neutrino flux, which we have examined in this paper, the new detectors must begin to acquire the necessary data by the end of the 1980s because the next solar activity maximum is expected about 1990. As noted above, semiannual variations in the flux during this period should be particularly well-defined.

Because the solar activity amplitude is irregular, it is at present difficult to forecast the modulation depth of the neutrino flux during the forthcoming solar activity maximum (the 1979–1980 peak was strong, judging by the number of sunspots). In this respect, it is very interesting to consider the possible solution of the converse problem, i.e., the problem of monitoring the magnetic field in the interior of the Sun by measuring the neutrino flux. If it were possible to establish independently the electromagnetic parameters of the neutino (e.g., by measuring the EMM of the ν_e in reactor experiments), this would enable us to perform quantitative studies of magnetic fields in the solar interior. At any rate, the time variation in the flux of solar neutrinos, regarded as a manifestation of the electromagnetic properties of these neutrinos, appears to us to deserve further theoretical and experimental study.

We are indebted to Z. G. Bereshani, E. A. Gavryuseva, A. D. Dolgov, B. L. Ioffe, A. V. Fedotov, and M. Yu. Khlopov for useful discussions.

[1)]We are grateful to B. Gavele, who drew our attention to this point.

[2)]We are grateful to A. A. Ruzmaikin and P. V. Sasorov for a discussion of the structure of the solar magnetic field.

[3)]We are indebted to S. I. Blinnikov, A. A. Ruzmaikin, A. V. Tutukov, and L. R. Yungel'son for a discussion of the density distribution and isotopic composition in the Sun.

[1]J. J. Rowley, B. T. Cleveland, and R. Davis Jr., "Solar neutrinos and neutrino astronomy," in: AIP Conf. Proc. No. 126, Homestake, 1984, ed. by M. L. Cherry *et al.*, New York, 1985, p. 1.

[2]M. V. Voloshin and M. I. Vysotsky, Preprint ITEP-1, 1986.

[3]L. B. Okun', Preprint, ITEP-14, 1986.

[4]L. B. Okun', M. B. Voloshin, and M. I. Vysotsky, Preprint ITEP-20, 1986.

[5]B. W. Lee and R. E. Shrock, Phys. Rev. D 16, 1444 (1977).

[6]K. Fujikawa and R. Shrock, Phys. Rev. Lett. 55, 963 (1980).

[7]R. N. Mohapatra and R. E. Marshak, Phys. Lett. B 91, 222 (1980).

[8]J. E. Kim, Phys. Rev. D 14, 3000 (1976).

[9]D. P. Stoker *et al.*, Phys. Rev. Lett. 54, 1887 (1985).

[10]G. V. Domogatskii and D. K. Nadezhin, Yad. Fiz. 12, 1233 (1970) [Sov. J. Nucl. Phys. 12, 678 (1971)].

[11]L. B. Okun', Leptony i kvarki (Leptons and Quarks), Nauka, Moscow, 1981 [English Transl., North-Holland, 1982].

[12]F. Reines, H. Gurr, and H. Sobel, Phys. Rev. Lett. 37, 315 (1976).

[13]S. V. Tolokonnikov and S. A. Fayans, Izv. Akad. Nauk SSSR Ser. Fiz. 27, 2667 (1973).

[14]A. V. Kyuldjiev, Nucl. Phys. B 243, 387 (1984).

[15]P. Sutherland *et al.*, Phys. Rev. D 13, 2700 (1976).

[16]A. D. Dolgov and Ya. B. Zeldovich, Rev. Mod. Phys. 53, 1 (1981).

[17]L. Wolfenstein, Phys. Rev. D 17, 2369 (1978).

[18]S. P. Mikheev and A. Yu. Smirnov, Yad. Fiz. 42, 1441 (1986) [Sov. J. Nucl. Phys. 42, 913 (1985)].

[19]J. N. Bahcall, Rev. Mod. Phys. 50, 881 (1978); 54, 767 (1982).

[20]E. G. Gibson, The Quiet Sun, NASA, Washington, 1973.

[21]S. I. Vaĭnshteĭn, Ya. B. Zel'dovich, and A. A. Ruzmaikin, Turbulentnoe dinamov astrofizike (Turbulent Dynamics in Astrophysics), Nauka, Moscow, 1980.

[22]G. A. Bazilevskskaya, Yu. I. Stozhkov, and T. N. Charakhch'yan, Pis'ma Zh. Eksp. Teor. Fiz. 35, 273 (1982) [JETP Lett. 35, 341 (1982)].

[23]V. N. Gavrilin, Yu. S. Kopysov, and N. T. Makeev, Pis'ma Zh. Eksp. Teor. Fiz. 35, 491 (1982) [JETP Lett. 35, 608 (1982)].

[24]A. Cisneros, Astrophys. Space Sci. 10, 87 (1971).

[25]G. T. Zatsepin, in: Proc. Eighth Intern. Workshop on Weak Interactions and Neutrinos, Javea, Spain, 1982, World Scientific, 1983, p. 754.

[26]A. De Bellefon and P. Espigat, Preprint College de France LPC-85-09, 1985; L. Gonzales-Mestres and D. Perret-Gallex, Preprint LAPP-EXP 84-05TH-112, 1985.

[27]Cern-Harvard-Milano-Roma-Tokyo-Wisconsin Collaboration, ICARUS-85-01, A proposal for the Grand Sasso Laboratory, Preprint INFN/AE-85/7, September 1985.

Translated by S. Chomet

Resonant spin-flavor precession of solar and supernova neutrinos

Chong-Sa Lim*
Institute for Nuclear Study, University of Tokyo, Tanashi, Tokyo 188, Japan

William J. Marciano
Brookhaven National Laboratory, Upton, New York 11973
(Received 5 October 1987)

The combined effect of matter and magnetic fields on neutrino spin and flavor precession is examined. We find a potential new kind of resonant solar-neutrino conversion $\nu_{e_L} \to \nu_{\mu_R}$ or ν_{τ_R} (for Dirac neutrinos) or $\nu_e \to \bar{\nu}_\mu$ or $\bar{\nu}_\tau$ (for Majorana neutrinos). Such a resonance could help account for the lower than expected solar-neutrino ν_e flux and/or indications of an anticorrelation between fluctuations in the ν_e flux and sunspot activity. Consequences of spin-flavor precession for supernova neutrinos are also briefly discussed.

There has been a long-standing disagreement between the solar-neutrino ν_e flux monitored by Davis[1] and collaborators,

$$\text{average flux} = 2.1 \pm 0.3 \text{ solar-neutrino unit (SNU)} \quad (1 \text{ SNU} = 10^{-36} \text{ captures}/s \text{ atom}) , \tag{1}$$

via the reaction $\nu_e + {}^{37}\mathrm{Cl} \to e^- + {}^{37}\mathrm{Ar}$ and Bahcall's standard-solar-model prediction[2]

$$\text{predicted flux} = 7.9 \pm 2.5 \text{ SNU} \quad (3\sigma \text{ errors}) . \tag{2}$$

That discrepancy has come to be known as the solar-neutrino puzzle. Attempts to resolve it have given rise to many speculative ideas about unusual properties of neutrinos and/or the solar interior.

One rather recently proposed solution, the Mikheyev-Smirnov-Wolfenstein[3] (MSW) effect, is particularly elegant. It employs the changing solar density and the difference between ν_e and other neutrinos' interactions with matter to bring about an energy-level-crossing resonance. In that way, ν_e neutrinos propagating from the core of the Sun to its surface can encounter a resonance region (generally in the radiation zone which extends from $0.04R_\odot$ to $0.7R_\odot$, where $R_\odot \simeq 7\times 10^{10}$ cm) and be converted to ν_μ, ν_τ, or some as yet unknown flavor. Part of the MSW solution's appeal is that it naturally provides the required flux depletion [cf. Eqs. (1) and (2)] for a large range of neutrino mass differences $\Delta m^2_{21} = m_2^2 - m_1^2 \simeq 10^{-7}-10^{-4}$ eV^2 and mixing angles, $\sin^2 2\theta \gtrsim 0.001$; so, it is not contrived.[4] In fact, the required parameters are very much in keeping with theoretical prejudices and expectations.

A more speculative solution to the solar-neutrino puzzle, originally advocated by Cisneros,[5] involves endowing neutrinos with magnetic moments such that spin precession in the strong interior solar magnetic fields can lead to $\nu_{e_L} \to \nu_{e_R}$. The sterility of ν_{e_R} would then lead to an effective depletion in the measured flux. That scenario has been recently revived and improved by Okun, Voloshin, and Vysotsky[6] (OVV). Their motivation came from the observation that there appears to be an anticorrelation between sunspot activity and variations in the detected solar ν_e flux.[7,8] During times of high sunspot activity (i.e., large-magnetic-field disturbances in the convection zone $> 0.7R_\odot$), the measured flux is smallest. It is, therefore, quite natural to correlate the flux variation with spin precession, which would of course be greatest when the magnetic fields are most intense.

The precession scenario has been studied in some detail by OVV (Ref. 6). They noted that either a magnetic- or electric-dipole moment could give $\nu_{e_L} \to \nu_{e_R}$ precession, while flavor transition moments between different species could result in the combined spin-flavor precession $\nu_{e_L} \to \nu_{\mu_R}$ or ν_{τ_R} (for four component Dirac neutrinos) and $\nu_e \to \bar{\nu}_\mu$ or $\bar{\nu}_\tau$ (Ref. 9) (for Majorana neutrinos). [Precession of Majorana neutrinos gives rise, for example, to $\nu_{e_L} \to (\nu_{\mu_L})^C$ with C being the charge-conjugation operator. Since $(\nu_{\mu_L})^C$ is right handed and generally called $\bar{\nu}_\mu$, we refer to it that way.] In all cases, however, they concluded that μB, where μ is a generic dipole or transition moment and B is the transverse solar magnetic field, must at least be of order $10^{-16}-10^{-15}$ eV to make such a scenario viable. Since they argued that B could be of order 10^3 G in the Sun's convection zone where precession was envisioned, they required[6]

$$|\mu| \simeq 0.3-1\times 10^{-10} e/2m_e . \tag{3}$$

A moment that large is consistent with direct experimental bounds,[10]

$$|\mu_{\nu_e}| \leq 4\times 10^{-10} e/2m_e \tag{4}$$

(from $\overset{(-)}{\nu}{}_e e$ data[11]), but the upper range is slightly in conflict with astrophysical arguments which imply[12]

$$|\mu_{\nu_e}| < 8.5\times 10^{-11} e/2m_e \quad \text{(astrophysics bound)} . \tag{5}$$

In addition, it is generally difficult to generate such a large moment while keeping the neutrino mass very small. For example, the standard $SU(2)_L \times U(1)$ model with a singlet right-handed neutrino gives rise to[13]

$$|\mu_{\nu_e}| = \frac{3eG_\mu m_{\nu_e}}{8\sqrt{2}\pi^2} \simeq 3\times 10^{-19}(m_{\nu_e}/1\text{ eV})e/2m_e\ , \quad (6)$$

which is much too small for a viable solar-neutrino precession scenario. Recently, however, a model has been proposed by Fukugita and Yanagida[14] in which an $SU(2)_L$ charged scalar singlet can induce at the one-loop level a neutrino magnetic moment in the range of Eq. (3) without conflicting with low-energy phenomenology. Therefore, in this paper we keep an open mind regarding the magnitude of μ.

It has also been noted[6] that neutrino interactions with matter can quench spin precession. In the case of $\nu_{e_L}\to\nu_{e_R}$ precession, the ν_{e_L} and ν_{e_R} interact differently with matter. The difference in their interactions effectively splits their degeneracy and suppresses precession. To illustrate that point, we consider the evolution equation connecting the chiral components ν_{e_L} and ν_{e_R}:

$$i\frac{d}{dt}\begin{bmatrix}\nu_{e_R}\\ \nu_{e_L}\end{bmatrix} = \begin{bmatrix}0 & \mu B\\ \mu B & a_{\nu_e}(t)\end{bmatrix}\begin{bmatrix}\nu_{e_R}\\ \nu_{e_L}\end{bmatrix}, \quad (7)$$

where B is the transverse magnetic field and $a_{\nu_e}(t)$ represents the "matter" potential experienced by ν_{e_L} as it propagates through the Sun. In the standard model (for an unpolarized medium),

$$a_{\nu_e}(t) = \frac{G_\mu}{\sqrt{2}}[(1+4\sin^2\theta_W)N_e + (1-4\sin^2\theta_W)N_p - N_n]\ , \quad (8)$$

where $G_\mu = 1.166\,36\times 10^{-5}$ GeV^{-2}, $\sin^2\theta_W \simeq 0.23$ and N_f is the fermion number density. [For other neutrino species $(1+4\sin^2\theta_W)N_e \to (-1+4\sin^2\theta_W)N_e$, while for antineutrinos, the sign of $a(t)$ is reversed.] For a neutral medium $N_e = N_p$ and one finds, from Eq. (8),

$$a_{\nu_e}(t) = \frac{G_\mu}{\sqrt{2}}(2N_e - N_n)\ . \quad (9)$$

The t (or spatial) dependence comes about because the densities in Eq. (9) vary as the neutrino propagates outward through the solar interior. In addition, B will also vary with t; but it is likely to exhibit a complicated dependence.

To obtain a feel for the matter effect, we can solve Eq. (7) for constant densities and constant B. In that case, the spin-precession probability of starting with ν_{e_L} at $t=0$ and finding ν_{e_R} at time t is given by

$$P(t)_{\nu_{e_L}\to\nu_{e_R}} = \frac{(2\mu B)^2}{a_{\nu_e}^2 + (2\mu B)^2} \times \sin^2\{[a_{\nu_e}^2 + (2\mu B)^2]^{1/2} t/2\}\ . \quad (10)$$

In a vacuum where $a_{\nu_e}=0$, this expression reduces to a standard spin-precession formula with frequency μB; but if $a_{\nu_e}^2 \gg (2\mu B)^2$, precession is suppressed. Of course, to carry out a realistic calculation, one needs a density and magnetic-field profile for the solar interior. The densities of electrons and neutrons in the convection zone and upper radiation zone[15] are well approximated by

$$N_e \simeq 6N_n \simeq 2.4\times 10^{26}\exp(-r/0.09R_\odot)/\text{cm}^3\ , \quad 0.2 < r/R_\odot \le 1\ , \quad (11a)$$

while in the lower radiation zone the linear approximation

$$N_e \simeq 6\times 10^{25}\left[1 - \frac{10}{3}\frac{r}{R_\odot}\right], \quad N_n \simeq 2\times 10^{25}\left[1 - \frac{21}{5}\frac{r}{R_\odot}\right]\Big/\text{cm}^3\ , \quad 0.1 < r/R_\odot \le 0.2 \quad (11b)$$

works well. (The relative neutron density increases.) Unfortunately, little is known about magnetic fields in the core, radiation zone or convection zone, except that they may be quite large. We have examined the evolution in Eq. (7) for an average B of $\sim 10^3$ G in the convection zone (i.e., for a distance $\simeq 2\times 10^{10}$ cm) and find for $\mu \simeq 10^{-10}e/2m_e$, one can obtain a ν_e flux depletion consistent with Eqs. (1) and (2). That finding is in keeping with the results of OVV (Ref. 6) and a more recent analysis by Barbieri and Fiorentini.[16] Of course, if that scenario is realized, the ν_e flux depletion would be strongly dependent on the magnitude of the magnetic field. Hence, one could expect a strong correlation between ν_e solar flux and sunspot activity, which is a measure of the convection-zone magnetic field.

We come now to the main focus of our work, the effect of matter on spin-flavor precession. To begin, we note that even if an electromagnetic transition moment between mass eigenstates ν_1 and ν_2 exists, one expects spin-flavor precession $\nu_{1L}\to\nu_{2R}$ in magnetic fields to be suppressed by the mass difference between ν_2 and ν_1, unless[6,9]

$$\mu_{21}B > \Delta m_{21}^2/2E_\nu \quad (12)$$

with μ_{21} the transition moment and E_ν the neutrino energy. For $\mu_{21}\simeq 10^{-10}e/2m_e$, $B\simeq 10^3$ G, and $E_\nu \simeq 10$ MeV, that condition requires $\Delta m_{21}^2 \lesssim 10^{-7}$ eV^2, which is below the MSW solutions but not prohibitively small. (Of course, the condition depends on energy.) Partly because of that mass difference suppression, neutrino spin-flavor precession seems not to have been thoroughly studied. However, here we will show that matter interactions of the distinct neutrino flavors can compensate for the mass difference. In fact, for a medium of changing density, such as the Sun, a resonance region can exist where neutrino spin-flavor precession may occur unimpeded. The physics of that resonance is quite similar to the MSW resonance[3] as we shall see.

To illustrate the resonant spin-flavor precession phenomenon, we first consider the case of two generations with four-component Dirac neutrinos. (Extensions to higher generations are straightforward but cumbersome.) For definiteness, we examine the ν_e-ν_μ system; but the results hold also for ν_e-ν_τ.

Using the chiral bases $\nu_{e_L}, \nu_{\mu_L}, \nu_{e_R}, \nu_{\mu_R}$, the evolution equation for neutrino propagation through matter and a transverse magnetic field B is

$$i\frac{d}{dt}\begin{pmatrix}\nu_{e_L}\\ \nu_{\mu_L}\\ \nu_{e_R}\\ \nu_{\mu_R}\end{pmatrix}=\begin{pmatrix}H_L & BM^\dagger\\ BM & H_R\end{pmatrix}\begin{pmatrix}\nu_{e_L}\\ \nu_{\mu_L}\\ \nu_{e_R}\\ \nu_{\mu_R}\end{pmatrix}, \tag{13}$$

where the 2×2 submatrices are

$$H_L=\begin{pmatrix}(\Delta m_{21}^2/2E_\nu)\sin^2\theta+a_{\nu_e} & (\Delta m_{21}^2/4E_\nu)\sin 2\theta\\ (\Delta m_{21}^2/4E_\nu)\sin 2\theta & (\Delta m_{21}^2/2E_\nu)\cos^2\theta+a_{\nu_\mu}\end{pmatrix},\quad H_R=\begin{pmatrix}0 & 0\\ 0 & \Delta m_{21}^2/2E_\nu\end{pmatrix},\quad M=\begin{pmatrix}\mu_{ee} & \mu_{e\mu}\\ \mu_{\mu e} & \mu_{\mu\mu}\end{pmatrix}, \tag{14}$$

and θ is the neutrino mixing angle

$$\begin{pmatrix}\nu_e\\ \nu_\mu\end{pmatrix}=\begin{pmatrix}\cos\theta & \sin\theta\\ -\sin\theta & \cos\theta\end{pmatrix}\begin{pmatrix}\nu_1\\ \nu_2\end{pmatrix}. \tag{15}$$

The different matter potentials are given by (for a neutral unpolarized medium)

$$a_{\nu_e}=\frac{G_\mu}{\sqrt{2}}(2N_e-N_n),\quad a_{\nu_\mu}=\frac{G_\mu}{\sqrt{2}}(-N_n), \tag{16}$$

and M represents the electromagnetic moments in the chiral-flavor bases. Note that because ν_{e_R} and ν_{μ_R} are sterile, i.e., do not interact electroweakly with matter, they can be considered vacuum mass eigenstates.

From Eq. (13) we can easily determine the energy-level crossings of the four neutrino chiral states by equating terms along the diagonal. Those crossings are illustrated in Fig. 1, where we have used the approximation $N_n\simeq\frac{1}{6}N_e$ [cf. Eq. (11a)] and assume $\Delta m_{21}^2>0$ is the more plausible scenario. There are three potential crossing resonances. The usual MSW $\nu_{e_L}\to\nu_{\mu_L}$ oscillation resonance occurs at density (for small θ)

$$N_e\simeq\frac{\Delta m_{21}^2}{2\sqrt{2}G_\mu E_\nu}\quad\text{(MSW resonance)}, \tag{17}$$

while the $\nu_{e_L}\to\nu_{\mu_R}$ spin-flavor resonance occurs at a somewhat higher density:

$$N_e\simeq\frac{12}{11}\frac{\Delta m_{21}^2}{2\sqrt{2}G_\mu E_\nu}\quad(\nu_{e_L}\to\nu_{\mu_R}\ \text{resonance}). \tag{18}$$

The $\nu_{\mu_L}\to\nu_{e_R}$ spin-flavor resonance occurs at the much higher density:

$$N_e\simeq 12\frac{\Delta m_{21}^2}{2\sqrt{2}G_\mu E_\nu}\quad(\nu_{\mu_L}\to\nu_{e_R}\ \text{resonance}). \tag{19}$$

Note that the MSW and $\nu_{e_L}\to\nu_{\mu_R}$ resonance regions are not so far from one another and may in some cases have to be studied together.

To illustrate the above scenario, we consider an adiabatic MSW solution $\Delta m_{21}^2\simeq 10^{-4}$ eV^2 and $\sin^2 2\theta$ small. In that case, the MSW resonance occurs for $E_\nu\simeq 10$ MeV at $r\simeq 0.10R_\odot$ while the $\nu_{e_L}\to\nu_{\mu_R}$ resonance is at $r\simeq 0.065R_\odot$. [These values are directly obtained from the bare density profiles N_e and N_n (Ref. 15), without approximation (11a) or (11b).] Both are in the inner radiation zone. The $\nu_{\mu_L}\to\nu_{e_R}$ resonance requires too large a N_n and therefore does not exist in the Sun. In this example, the MSW and $\nu_{e_L}\to\nu_{\mu_R}$ resonances are far enough spaced to be treated separately. The $\nu_{e_L}\to\nu_{\mu_R}$

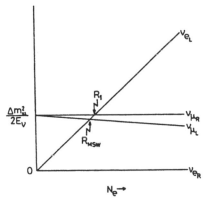

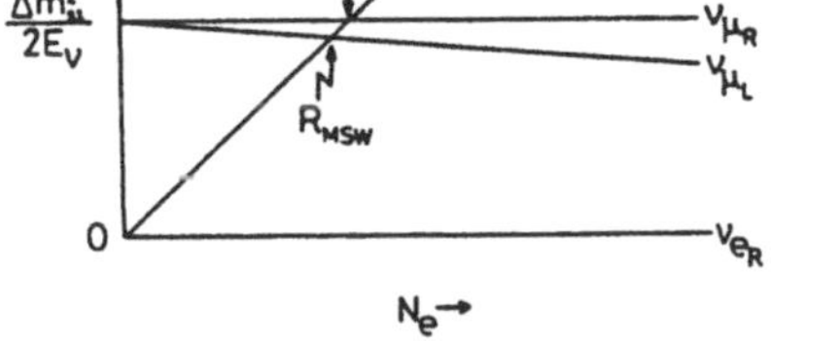

FIG. 1. The energy-level crossings of four neutrino chiral states in the two-generation model with Dirac neutrinos. The medium is assumed to be proton rich, say $N_e\simeq 6N_n$, for simplicity. The crossings denoted by R_{MSW} and R_1 correspond to the MSW resonance and the resonant $\nu_{e_L}\to\nu_{\mu_R}$ spin-flavor precession. R_2, the crossing of ν_{μ_L} and ν_{e_R}, does not appear for solar neutrinos because the required density is too high.

resonant conversion is therefore governed by a 2×2 submatrix

$$\begin{bmatrix} (\Delta m_{21}^2/2E_\nu)\sin^2\theta + a_{\nu_e} & \mu_{\mu e}^{*}B \\ \mu_{\mu e}B & \Delta m_{21}^2/2E_\nu \end{bmatrix} . \tag{20}$$

That problem is the same as the MSW resonant conversion one if we recall that a_{ν_e} gets contributions not only from the charged current but also from the neutral current in the present case [cf. Eq. (16)]. We can therefore utilize the analysis of Parke[17] to determine the average $\nu_{e_L} \rightarrow \nu_{\mu_R}$ transition probability

$$P(\nu_{e_L} \rightarrow \nu_{\mu_R}) = \tfrac{1}{2} - (\tfrac{1}{2} - P_{LZ})\cos 2\tilde{\theta}_N \cos 2\tilde{\theta} , \tag{21}$$

where

$$\tan 2\tilde{\theta} = 4E_\nu |\mu_{\mu e}| B/\Delta m_{21}^2 \cos^2\theta , \tag{22a}$$

$$\tan 2\tilde{\theta}_N = 4E_\nu |\mu_{\mu e}| B/[\Delta m_{21}^2 \cos^2\theta - 2E_\nu a_{\nu_e}(0)] , \tag{22b}$$

$$P_{LZ} = \exp\left[- \frac{4\pi E_\nu |\mu_{\mu e}|^2 B^2}{\Delta m_{21}^2 \cos^2\theta \, |d(\ln a_{\nu_e})/dr|} \right] , \tag{22c}$$

and $d(\ln a_{\nu_e})/dr$ in Eq. (22c) should be the value at the resonance point.

To be more quantitative, we have given the numerical solar-neutrino depletion results in Table I for several cases. For illustrative purpose, we have assumed $\Delta m_{21}^2 = 10^{-4}$ eV^2, $E_\nu = 10$ MeV and only $\mu_{\mu e} = 10^{-10} e/2m_e$ has been taken into account as the electromagnetic moment. Note that for relatively small B, as in the cases (a) and (b), the MSW resonance shown in the case (d) essentially provides the ν_{e_L} depletion. This is because the magnetic fields are too small for the adiabatic $\nu_{e_L} \rightarrow \nu_{\mu_R}$ transition to take place. For larger values of B, however, the $\nu_{e_L} \rightarrow \nu_{\mu_R}$ conversion only can account for the depletion, as we learn from the case (c). In the radiation zone, such large magnetic fields $\simeq 10^4 - 10^5$ G may in fact be feasible. It is also interesting to note that very large magnetic fields are entirely reasonable in the supernova and the neutron stars. The value of $P_{\nu_{\mu_R}}$ given in (c) is in good agreement with the analytic formula Eq. (22). The final example, case (e), combines both types of resonances (c) and (d), to result in enough fractions of both $P_{\nu_{\mu_R}}$ and $P_{\nu_{\mu_L}}$.

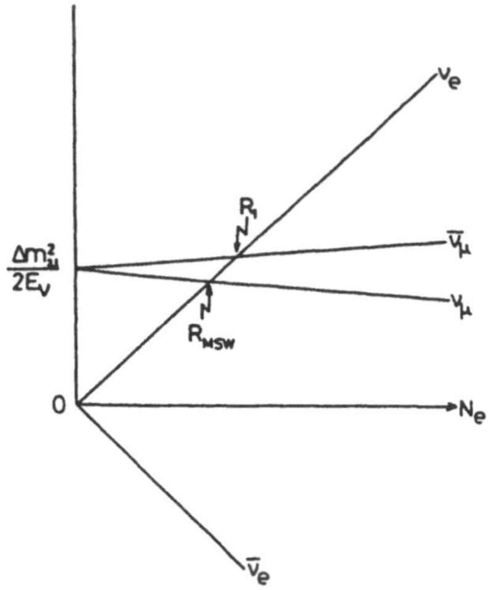

FIG. 2. The energy-level crossings in the two-generation Majorana neutrino model. The proton-rich medium has been assumed as in Fig. 1. For Majorana neutrinos the third crossing R_2 between ν_μ and $\bar{\nu}_e$ does not exist (for $N_e \geq N_n$). For $N_e < N_n$, the R_1 disappears while the R_2 crossing takes place.

In the above example, the $\nu_{e_L} \rightarrow \nu_{\mu_R}$ precession occurs deep in the Sun's radiation zone. Hence, it is unlikely to be correlated with sunspot activity. However, one could speculate that in a three-generation scenario $\nu_{e_L} \rightarrow \nu_{\tau_L}$ or ν_{τ_R} occurs in the radiation zone, while $\nu_{e_L} \rightarrow \nu_{\mu_R}$ occurs in the convection zone with the latter process correlated with sunspot activity. [The resonance in the convection

TABLE I. The probabilities $P_{\nu_{e_L}}$, $P_{\nu_{\mu_L}}$, and $P_{\nu_{\mu_R}}$ of a solar neutrino produced at the solar core as ν_{e_L} to be observed as ν_{e_L}, ν_{μ_L}, and ν_{μ_R}, respectively, at the solar surface in the two-generation model with Dirac neutrinos. For illustrative purpose, we have assumed $\Delta m_{21}^2 = 10^{-4}$ eV^2, $E_\nu = 10$ MeV, and only $\mu_{\mu e} = 10^{-10} e/2m_e$ has been switched on among electromagnetic moments. Several cases (a),(b), . . . , (e) with different sets of parameters B (G) and $\sin\theta$ are shown.

Case	$\sin\theta$	B	$P_{\nu_{e_L}}$	$P_{\nu_{\mu_L}}$	$P_{\nu_{\mu_R}}$
(a)	7.18×10^{-3}	10^3	0.33	0.67	0
(b)	7.18×10^{-3}	10^4	0.32	0.62	0.06
(c)	0	5.25×10^4	0.33	0	0.67
(d)	7.18×10^{-3}	0	0.33	0.67	0
(e)	7.18×10^{-3}	5.25×10^4	0.11	0.37	0.52

zone would necessitate, through Eq. (18), $\Delta m_{21}^2 \lesssim 2\times10^{-7}$ eV^2 (for $E_\nu \simeq 10$ MeV and small θ). This condition happens to be similar to the one derived by OVV (Ref. 6) from Eq. (12). They are, however, based on very different physical processes. For instance, in the spin-flavor precession without resonance[6] the average probability of $\nu_{e_L} \to \nu_{\mu_R}$ does not become as large as $\frac{2}{3}$ even if Δm_{21}^2 is small enough.] Careful monitoring of the ν_e flux in future experiments could shed light on such a scenario.

In the case of Majorana neutrinos, only transition moments which change lepton number by $|\Delta L|=2$ can exist. They lead to $\nu_e \to \bar{\nu}_\mu$ or $\bar{\nu}_\tau$ spin-flavor precession and if mixing is large, $\nu_e \to \bar{\nu}_e$. Since antineutrinos can interact with matter, the evolution equation for ν_e, ν_μ, $\bar{\nu}_e$, and $\bar{\nu}_\mu$ is governed by the Hamiltonian 4×4 matrix

$$H_{\text{Maj}} = \begin{bmatrix} a_{\nu_e} & \frac{\Delta m_{21}^2}{4E_\nu}\sin2\theta & 0 & \mu_\nu^* B \\ \frac{\Delta m_{21}^2}{4E_\nu}\sin2\theta & \frac{\Delta m_{21}^2}{2E_\nu}\cos2\theta + a_{\nu_\mu} & -\mu_\nu^* B & 0 \\ 0 & -\mu_\nu B & -a_{\nu_e} & \frac{\Delta m_{21}^2}{4E_\nu}\sin2\theta \\ \mu_\nu B & 0 & \frac{\Delta m_{21}^2}{4E_\nu}\sin2\theta & \frac{\Delta m_{21}^2}{2E_\nu}\cos2\theta - a_{\nu_\mu} \end{bmatrix} \tag{23}$$

Therefore, for small mixing θ, the $\nu_e \to \bar{\nu}_\mu$ level crossing occurs at somewhat higher densities [using $N_n \simeq (\frac{1}{6})N_e$]

$$N_e \simeq \frac{3\sqrt{2}}{10}\frac{\Delta m_{21}^2}{G_\mu E_\nu} \quad (\nu_e \to \bar{\nu}_\mu \text{ resonance}), \tag{24}$$

and there is no $\nu_\mu \to \bar{\nu}_e$ resonance (see Fig. 2). Most of what was said for $\nu_{e_L} \to \nu_{\mu_R}$ above carries over to $\nu_e \to \bar{\nu}_\mu$ precession. However, $\bar{\nu}_\mu$ or $\bar{\nu}_\tau$ are, in principle, detectable by measuring $\bar{\nu}$-e scattering cross section. Although that is very difficult, we are optimistic that it will one day be possible. That optimism is based on the tremendous progress in measuring $\nu_e - e$ scattering recently reported.[18] Also, if $\bar{\nu}_e$ (from mixing) is appreciable, one might try to detect $\bar{\nu}_e + p \to e^+ + n$, although it will be hard to disentangle from backgrounds.

In closing, we should emphasize that spin or spin-flavor precession of the solar neutrino is still a long shot. It requires large magnetic fields and a very large electromagnetic dipole or the transition moment. However, we feel that the existence of spin-flavor precession resonant regions makes such a scenario at least plausible. One should, therefore, keep an open mind and search for antineutrinos from the Sun as well as monitor the energy dependence of ν_e and its correlation with solar activity. Finally, we note that even if spin-flavor precession is too weak to affect solar neutrinos (say if μ is too small), it could influence other phenomena. In particular, spin-flavor precession of Majorana neutrinos could have remarkable consequences for supernova bursts[19] where very intense magnetic field ($\sim10^{12}$ G) are quite likely. For example, it could lead to large asymmetries in the ν_μ and $\bar{\nu}_\mu$ or ν_τ and $\bar{\nu}_\tau$ fluxes, which are otherwise expected to be equal. An even more dramatic possibility stems from the fact that in the core region of a supernova $N_n > N_e$ is realized. In such a neutron-rich region $\nu_\mu \leftrightarrow \bar{\nu}_e$ or $\nu_\tau \leftrightarrow \bar{\nu}_e$ become the more plausible spin-flavor precessions (rather than $\nu_e \leftrightarrow \bar{\nu}_\mu, \bar{\nu}_\tau$). That could lead to a higher than expected $\bar{\nu}_e$ flux and/or effectively higher average energy $\bar{\nu}_e$. Also, potential sources of very high-energy neutrinos (such as Cygnus X-3) could lead to $\nu_\mu \to \bar{\nu}_\tau$ or $\bar{\nu}_\mu \to \nu_\tau$ (depending on the sign of the mass difference). We, of course, strongly advocate experimental searches for all such phenomena.

We are very grateful to A. Baltz, S. P. Rosen, J. Weneser, and M. Yoshimura for useful and informative discussions. One of us (C.S.L.) would like to thank K. Nomoto and T. Sakurai for very helpful and introductory arguments on the density profiles and the magnetic fields of the Sun and the supernova. He also would like to thank N. Ohnishi, and members of Theory Group at the Institute for Nuclear Studies, University of Tokyo, for their useful discussions. W.J.M. was supported by the Department of Energy under Contract No. DE-AC02-76-CH-0016.

*Present address: Theory Group, National Laboratory for High Energy Physics (KEK), Oho-machi, Tsukuba, Ibaraki 305, Japan.

[1] J. K. Rowley, B. T. Cleveland, and R. Davis, in *Solar Neutrinos and Neutrino Astronomy, Lead High School, Lead, South Dakota,* proceedings of the Conference, edited by M. L. Cherry, K. Lande, and W. A. Fowler (AIP Conf. Proc. No. 126) (AIP, New York, 1984), p. 1.

[2] J. N. Bahcall (unpublished); S. P. Rosen (private communication).

[3]S. P. Mikheyev and A. Yu. Smirnov, Nuovo Cimento **C9**, 17 (1986); L. Wolfenstein, Phys. Rev. D **17**, 2369 (1978).

[4]For a review of the MSW effect, see S. P. Rosen and J. M. Gelb, in *Proceedings of the XXIII International Conference on High Energy Physics*, Berkeley, California, 1986, edited by S. Loken (World Scientific, Singapore, 1986), p. 909.

[5]A. Cisneros, Astrophys. Space Sci. **10**, 87 (1981)

[6]M. B. Voloshin and M. I. Vysotsky, ITEP Report No. 1, 1986 (unpublished); L. B. Okun, Yad. Fiz. **44**, 847 (1986) [Sov. J. Nucl. Phys. **44**, 546 (1986)]; L. B. Okun, M. B. Voloshin, and M. I. Vysotsky, *ibid.* **91**, 754 (1986) [**44**, 440 (1986)]; Zh. Eksp. Teor. Fiz. **91**, 754 (1986) [Sov. Phys. JETP **64**, 446 (1986)].

[7]G. A. Bazilevskaya, Yu. I. Stozhkov, and T. N. Charahchyan, Pis'ma Zh. Eksp. Teor. Fiz. **35**, 273 (1982) [JETP Lett. **35**, 341 (1982)]; G. A. Bazilevskaya *et al.*, Yad. Fiz. **39**, 856 (1984) [Sov. J. Nucl. Phys. **39**, 543 (1984)].

[8]R. Davis, in *Proceedings of the Seventh Workshop on Grand Unification*, Toyama, Japan 1986, edited by J. Arafune (World Scientific, Singapore, 1986).

[9]An early detailed discussion of Majorana neutrino spin-flavor precession (without matter effects) was given by J. Schechter and J. Valle, Phys. Rev. D **24**, 1883 (1981).

[10]A. V. Kyuldjev, Nucl. Phys. **B243**, 387 (1984); W. J. Marciano and Z. Parsa, Annu. Rev. Nucl. Part. Sci. **36**, 171 (1986). We have increased the bound quoted in the first reference because it employed an outdated reactor neutrino energy spectrum. Using more recent spectra estimates leads to the bound in Eq. (4).

[11]F. Reines, H. S. Gurr, and H. W. Sobel, Phys. Rev. Lett. **37**, 315 (1976); R. C. Allen *et al.*, *ibid.* **55**, 2401 (1985).

[12]P. Sutherland *et al.*, Phys. Rev. D **13**, 2700 (1976); M. A. B. Bég, W. J. Marciano, and M. Ruderman, *ibid.* **17**, 1395 (1978).

[13]W. J. Marciano and A. I. Sanda, Phys. Lett. **67B**, 303 (1977); B. W. Lee and R. E. Shrock, Phys. Rev. D **16**, 1444 (1977); S. T. Petkov, Yad Fiz. **25**, 641 (1977) [Sov. J. Nucl. Phys. **25**, 340 (1977)]; **25**, 698(E) (1977) [**25**, 641 (1977)]; K. Fujikawa and R. Shrock, Phys. Rev. Lett. **55**, 963 (1980).

[14]M. Fukugita and T. Yanagida, Phys. Rev. Lett. **58**, 1807 (1987).

[15]J. N. Bahcall and R. K. Ulrich, Institute for Advanced Study report, 1987 (unpublished).

[16]R. Barbieri and G. Fiorentini, University of Pisa report, 1987 (unpublished).

[17]S. Parke, Phys. Rev. Lett. **57**, 1275 (1986).

[18]Kamiokande Collaboration, K. Hirata *et al.*, University of Tokyo Report No. UT-ICEPP-87-04, 1987 (unpublished).

[19]Kamiokande Collaboration, K. Hirata *et al.*, Phys. Rev. Lett. **58**, 1490 (1987); IMB Collaboration, R. M. Bionta *et al.*, *ibid.* **58**, 1494 (1987).

RESONANT AMPLIFICATION OF NEUTRINO SPIN ROTATION IN MATTER AND THE SOLAR-NEUTRINO PROBLEM

E.Kh. AKHMEDOV
I.V. Kurchatov Institute of Atomic Energy, Moscow 123 182, USSR

Received 28 June 1988

It is shown that in the presence of matter there can occur resonant amplification of he flavor-changing neutrino spin rotation in transverse magnetic fields, which is roughly analogous to the Mikheyev–Smirnov–Wolfenstein effect in neutrino oscillations. Possible consequences for solar neutrinos are briefly discussed.

1. If the magnetic moment of a neutrino is not zero, its spin must precess under transverse magnetic fields. Therefore a sufficiently strong and extended magnetic field can convert, e.g., the left-handed neutrino ν_{eL} into the right-handed component ν_{eR} thereby making it unobservable [1–7]. In particular, the neutrino spin rotation (νSR) in the magnetic field of the sum [1,4–7] may account for the solar neutrino puzzle, the essence of which is that the experimental neutrino detection rate [8] is about a factor of three smaller than that predicted by the standard solar model [9].

Voloshin, Vysotsky and Okun [4–7] considered νSR in the toroidal magnetic field of the convective zone of the sun. They showed that the precession may result in a significant suppression of the left-handed neutrino flux from the sun. For this phenomenon to take place, the neutrino magnetic moment must be of the order of $(10^{-10}–10^{-11})$ times the Bohr magneton μ_B, assuming the transverse magnetic field strength $B_\perp$ to be $(10^3–10^4)$G. The magnetic moments of such a magnitude are compatible with the available experimental bounds and can be reached, e.g. in models with an $SU(2)_L$-singlet charged scalar particle [10,11]. Since in the periods of maximum solar activity the toroidal magnetic field strength in the convective zone of the sun is at least an order of magnitude greater than that in the quiet time, the hypothesis of νSR [4–7] gives a plausible explanation of the probable anticorrelation between the observed neutrino flux and the solar activity (11-year variations) [12]. In ref. [7] the effect of matter on νSR was also considered. It was shown that matter suppresses the precession, this suppression being rather important.

In the present paper the flavor-changing neutrino spin rotation (FCνSR) due to transition magnetic moments of neutrinos is considered. It is shown that, unlike the flavor-conserving precession, matter would not only lead to the suppression of the FCνSR but can also enhance it. Possible consequences for solar neutrinos are briefly discussed.

2. If neutrinos are Majorana particles, they cannot possess diagonal magnetic moments due to *CPT* invariance but can have off-diagonal (transition) moments μ_{12}, which are responsible for the decay $\nu_2 \to \nu_1\gamma$ (for $m_2 > m_1$). Dirac neutrinos can also have off-diagonal magnetic moments in addition to the diagonal ones. Under the transverse magnetic field the off-diagonal moments will cause the rotation of left-handed neutrinos of a given flavor into right-handed neutrinos (or antineutrinos) of another flavor. In ref. [7] it was pointed out that the FCνSR is suppressed as compared to flavor-conserving precession by a factor of

$$(2\mu_{12}B_\perp)^2/[(2\mu_{12}B_\perp)^2+(\Delta m^2/2E)^2]\ .$$

The reason for this suppression is that ν_{1L} and ν_{2R} are not degenerate, their energy difference being $\Delta E \approx \Delta m^2/2E$ (we assume neutrinos to be ultra-relativistic). However in matter the potential energies of interaction with the electrons and nucleons add to the kinetic energies of the neutrinos. The difference of the potential energies of ν_{1L} and ν_{2R} may cancel the difference of their kinetic energies resulting in the resonant amplification of the FCνSR. This phenomenon is roughly analogous to the resonant neutrino oscillations – the Mikheyev–Smirnov–Wolfenstein (MSW) effect [13,14].

Let us assume for definiteness that the neutrinos are Majorana particles, i.e. the FCνSR convert ν_{eL} into a right-handed antineutrino (e.g., $\bar{\nu}_{\mu R}$), which is not sterile. For simplicity we shall disregard neutrino oscillations assuming the neutrino mixing angle to be vanishing or very small. The results of the joint consideration of the neutrino precession and oscillations will be presented elsewhere. We shall also assume that there are only two neutrino flavors.

In the case of the uniform matter and magnetic field, one can readily obtain the probability of finding $\bar{\nu}_{2R}$ at point r, provided there were only ν_{1L}'s at the origin;

$$P(\nu_{1L}\to\bar{\nu}_{2R};r)=\sin^2 2\tilde{\theta}\sin^2(\pi r/l)\,, \tag{1}$$

where the mixing angle $\tilde{\theta}$ and the precession length l are given by

$$\tan 2\tilde{\theta}=\frac{2\mu_{12}B_\perp}{\sqrt{2}G_F(N_e-N_n)-(m_2^2-m_1^2)/2E}, \tag{2}$$

$$l=2\pi\{(2\mu_{12}B_\perp)^2+[\sqrt{2}G_F(N_e-N_n)-(m_2^2-m_1^2)/2E]^2\}^{-1/2}. \tag{3}$$

Here N_e and N_n are the electron and neutron number densities respectively, and G_F is the Fermi constant. It follows from eqs. (1), (2) that for $m_2>m_1$ the resonant amplification of the precession is possible, the resonant density being defined by the condition

$$\sqrt{2}G_F(N_e-N_n)_r=(m_2^2-m_1^2)/2E \equiv\Delta m^2/2E\,. \tag{4}$$

In the case of non-uniform matter and magnetic field there will be sizeable $\nu_{1L}\to\nu_{2R}$ conversion provided the adiabaticity condition is satisfied, i.e. the resonant precession length l is much smaller than the size of the resonant layer (which is defined as a region where $\sin^2 2\tilde{\theta}\geq\frac{1}{2}$);

$$\chi\equiv\frac{1}{\pi}\frac{l_r}{\Delta r}\approx\frac{2.5\times10^{-12}\,\text{cm}}{L_\rho}\left[1-\left(\alpha\frac{L_\rho}{L_B}\right)^2\right]\times\frac{\Delta m^2}{E}\left(\frac{\text{eV}^2}{\text{MeV}}\right)\frac{1}{(\mu_{12}B_{\perp 0})^2(\text{eV}^2)}\ll 1\,. \tag{$$

Here L_ρ and L_B are characteristic sizes over whi matter density $\rho(r)$ and $B_\perp(r)$ vary significantly the resonant region;

$$L_\rho\equiv\left(-\frac{1}{\rho}\frac{d\rho}{dr}\right)_r^{-1},\quad L_B\equiv\left(-\frac{1}{B_\perp}\frac{dB_\perp}{dr}\right)_r^{-1},$$

$$\alpha\equiv2\mu_{12}B_{\perp 0}/(\Delta m^2/2E)$$

and $B_{\perp 0}$ is the magnetic field strength at the res nance. Hereafter we shall assume $\alpha L_\rho\ll L_B$ and d regard the $(\alpha L_\rho/L_B)^2$ term in eq. (5).

For solar neutrinos, the conversion $\nu_{eL}\to\bar{\nu}_{\mu R}$ ($\bar{\nu}_{\tau R}$ will be almost complete provided ν_{eL}'s are born in t core of the sun outside the resonant region and th pass through the resonance on their way to the su face of the sun. The adiabatic conversion $\nu_{1L}\to\nu_{2R}$ d to the resonant FCνSR is roughly analogous to t $\nu_{1L}\to\nu_{2L}$ conversion in the case of resonant neutrir oscillations [13]. It should, however, be stressed th the parameter χ in eq. (5) and the corresponding ad abaticity parameter of the MSW effect have differe dependence on the neutrino energy E: the greater $\Delta m^2/E$ (and, consequently, the resonant density), th better the adiabaticity condition for the resonant o cillations is satisfied [13], whereas in the case of re onant precession the situation is opposite. Note th the parameter χ in eq. (5) depends crucially on th magnetic field strength at the resonance $B_{\perp 0}$. Unfo tunately, there is no direct experimental informatio concerning the radial dependence of $B_\perp$ inside th sun. Moreover, our theoretical knowledge of this is sue is very poor too. In this situation one is forced t choose various more or less probable magnetic fiel configurations and calculate the probabilities of th $\nu_{eL}\to\bar{\nu}_{\mu R}$ ($\bar{\nu}_{\tau R}$) conversion to be compared with th experimental data. The results of such calculations fo a number of functions $B_\perp(r)$ will be published else where; in the present paper we shall confine our selves to simple estimates.

3. In the case of the resonant FCνSR, as well as ir the case of the MSW effect, there are two major so

lutions to the solar-neutrino problem – adiabatic and moderately non-adiabatic ones.

The adiabatic solution. All the ν_{eL}'s with the energies $E > E_c \approx 6$ MeV hit the resonance and are completely converted into $\nu_{\mu R}$'s (or $\nu_{\tau R}$'s), whereas those with the energies $E < E_c$ leave the sun unscathed since their resonant densities are greater than the matter density in the center of the sun. This solution requires [#1]

$$\Delta m^2 \approx 4 \times 10^{-5} \mathrm{eV}^2 ,$$

$$Z \equiv \left(\frac{\mu_{12}}{10^{-10}\mu_B}\right) B_{\perp 0}(G) \gtrsim 10^5 . \qquad (6)$$

The predictions for the spectra of ν_{eL}'s leaving the sun are practically the same as those in the adiabatic case of the MSW effect [13,17,18]. This, in particular, means that one should expect only an insignificant ($\lesssim 10\%$) reduction of the neutrino detection rate in the ^{71}Ga experiment. Since in this case all the neutrinos hit the resonance near the core of the sun, where the magnetic field is expected to be frozen in, no observable time variations of the neutrino flux would be there.

Moderately non adiabatic solution. In this case to obtain the reduction factor of about $\frac{1}{3}$ for the detection rate in ^{37}Cl experiment the following conditions must be satisfied:

$$10^{-8} \lesssim \Delta m^2 \lesssim 10^{-5}, \quad Z \approx 10^7 \sqrt{\Delta m^2 (\mathrm{eV}^2)} . \qquad (7)$$

Note that the slab model developed for the moderately non-adiabatic case of the MSW effect in ref. [15], can be used in our case too. For the probability for ν_{eL} to leave the sun unscathed it gives

$$P(\nu_{eL} \to \nu_{eL}) \approx \cos^2[8(E/\Delta m^2)(\mu_{12} B_\perp)^2 L_\rho] ,$$

which should be compared with the corresponding expression for the MSW effect [15]

$$P(\nu_e \to \nu_e)$$
$$\approx \cos^2[(\sin^2 2\theta_0 / \cos 2\theta_0)(\Delta m^2/2E) L_\rho] .$$

From these equations it is easily seen that the high-energy part of the solar neutrino spectrum should be more suppressed that the lower-energy part provided $B_\perp(r)$ is nearly constant. This is just opposite to what is expected for the MSW effect [13,15,16]. Since the magnetic field inside the sun is indeed not uniform, the spectra of solar ν_{eL}'s seen on the earth should depend substantially on the radial dependence of $B_\perp$. There is a class of the magnetic field configurations $B_\perp(r)$ for which the distortions of the solar neutrino spectra due to the MSW effect and due to the matter-enhanced FCνSR would be quite similar. For such functions $B_\perp(r)$ it will be extremely difficult (if possible at all) to distinguish between the resonant oscillations and the resonant precession by comparing the results of the ^{37}Cl and ^{71}Ga experiments or even by measuring the solar neutrino spectrum. The distortions in the solar neutrino spectrum due to these two effects will be qualitatively different only if $B_\perp^2(r)$ decreases with increasing r slower than $\rho(r)$.

However, there is a particular case for which the matter-enhanced FCνSR can be unambiguously distinguished from the MSW effect, as well as from the ordinary (flavor-conserving) neutrino spin precession. If Δm^2 lies in the range $(10^{-8}\text{–}10^{-7})$ eV^2 and $Z \gtrsim 10^3$ or 10^4 (the latter value being relevant for the upper bound on Δm^2), almost all the ^{8}B and ^{7}Be neutrinos will hit the resonance in the convective zone of the sun, where the magnetic field is subjected to the 11-year variations. For this reason the 11 year and semi-annual variations of the chlorine detection rate should be observable in this case, the latter being due to the peculiarities of the solar magnetic field configuration [7]. At the same time all the pp neutrinos which are expected to give the main contribution to the ^{71}Ga experiment have much lower energies and, hence, will hit the resonance in the inner regions of the sun outside the convective zone. Therefore the gallium detection rate should not exhibit any sizeable time variations. This distinguishes the effect under consideration from the MSW effect for which there is no reason to expect the 11-year variations as well as from the flavor-conserving νSR for which *both* the chlorine and gallium detection rates should exhibit the 11-year variations (the former undergoing the semi-annual variations as well).

There are also some other possibilities to distinguish the phenomenon under consideration from the MSW effect. If Δm^2 lies in the range $(1\text{–}5) \times 10^{-6}$ eV^2 and the vacuum mixing angle is not too small ($\sin^2 2\theta_0 \geqslant 0.1$), the MSW effect should cause the regeneration of a part of ν_{eL}'s due to the resonant oscil-

[#1] Here and below we take $L_\rho \approx \frac{1}{10} R$, where R is the radius of the sun [15,16].

lations of neutrinos in the matter of the earth [19]. This should give rise to large time-of-night and seasonal variations of the detection rate of solar neutrinos, whereas no such variations are expected if the solar neutrino deficiency is due to the νSR.

Another possible way of distinguishing the FCνSR from the MSW effect is the use of neutral-current reactions such as νe→νe and νd→npν. The difference will be observable if one of these effects transforms ν_{eL}'s into active neutrinos, whereas the other, into sterile neutrinos. This would take place in the following cases:

(i) there are only Dirac masses of neutrinos;

(ii) there are both Dirac and Majorana masses, and the oscillations being enhanced via the MSW mechanism are the $\nu_{eL}\rightleftarrows\bar{\nu}_{eL}$ $(\bar{\nu}_{\mu L}, \bar{\nu}_{\tau L})$ oscillations.

However, here one needs the information on the nature of neutrino masses (Dirac or Majorana), which is not available presently.

Some final comments are in order. Since the magnetic field strength in the central region of the sun may reach as large values as 10^7G, the matter-enhanced FCνSR may account for the solar neutrino deficiency even if the transition magnetic moment is as small as $10^{-12}\ \mu_B$.

Our assumption of vanishing neutrino mixing angle θ_0 does not contradict, strictly speaking, to having relatively large transition magnetic moments since there is no direct connection between the off-diagonal moments and off-diagonal masses. The latter can always be cancelled by appropriate mass counter terms. We shall not discuss here the problem of naturalness that arises in this case and just note that the assumption of vanishing θ_0 was made for simplicity and is by no means necessary for the matter-enhanced FCνSR to occur. Our present treatment is valid if $\sin 2\theta_0 \ll \alpha$.

The effect we have considered may also take place in collapsing stars and in the early universe. If neutrinos are Majorana particles, it would not lead to any problems with the primordial ^{4}He abundance, since there is no neutrino species doubling.

In refs. [4–7] was shown that for the flavor-conserving νSR in the sun to take place with a sizeable probability it is necessary that $\mu_{11}B_\perp L_{\rm conv} \gtrsim 1$, where $L_{\rm conv} \approx 0.3\ R$ is the size of the convective layer. This condition imposes the lower limit on $\mu_{11}B_\perp$. We shall show now that for this precession to be effective there is, in general, and additional condition which imposes the upper bound on $\mu_{11}B_\perp$. In the case of flavor-conserving νSR the mixing angle θ is defined by

$$\tan 2\theta = 2\mu_{11}B_\perp/\sqrt{2}G_F(N_e - N_n/2)\ .$$

In the core of the sun θ is small unless $B_\perp$ is extremely large. In the convective zone θ may be close to $\frac{1}{4}\pi$ (but is always below this value). Near the surface of the sun both $B_\perp(r)$ and $\rho(r)$ fall off rapidly. If $B_\perp(r)$ decreases more rapidly than $\rho(r)$, $\theta \to 0$ in this region. Then, if the adiabaticity condition is satisfied, the precession is suppressed in the core of the sun, about a half of ν_{eL}'s will be converted into ν_{eR}'s in the convective zone, but they well be reconverted into ν_{eL}'s at the exit of this zone. Thus the neutrinos will leave the sun as almost pure ν_{eL}'s. This means that the flavor-conserving νSR will be effective in the sun only if the adiabaticity condition is violated. This amounts to the requirement

$$(\mu_{11}B_\perp)^{-1} \gtrsim L_\theta \simeq L_B L_\rho / |L_B - L_\rho|$$
$$\approx \min\{L_B, L_\rho\}\ .$$

It is reasonable to assume $L_\theta \sim L_\rho \sim \frac{1}{10}R$ which gives $\mu_{11}B_\perp L_{\rm conv} \lesssim 3$. Thus there is only rather narrow range of $\mu_{11}B_\perp$ for which the flavor-conserving νSR can account for the solar-neutrino problem.

It is interesting to note that matter can enhance resonantly this precession as well if $N_e \approx \frac{1}{2}N_n$. This relation cannot take place for stable nuclei but can be satisfied at early stage of the collapse of the stars due to the neutronization of matter. Note that the resonant condition does not depend on the neutrino energy in this case, so that the neutrino spectra will remain unchanged. Thus the matter-enhanced flavor-conserving νSR may play a role in the dynamic of supernovae and neutron stars.

The preliminary results of this work were published in ref. [20].

The author is grateful to S.T. Belyaev, M.Yu. Khlopov, L.B. Okun, A.A. Ruzmaikin, A.Yu. Smirnov, M.B. Voloshin and M.I. Vysotsky for useful discussions.

Note added. After this paper was completed, a paper by Lim and Marciano appeared [21], in which the same problem was considered. The resonant

FCνSR with the allowance for neutrino oscillations was studied numerically. However, in ref. [21] the moderately non-adiabatic regime was not considered, the ranges of parameters for which the FCνSR can account for the solar-neutrino problem were not found, and possible ways of distinguishing the resonant FCνSR from the MSW effect and from the flavor-conserving precession were not discussed.

References

[1] A. Cisneros, Astrophys. Space Sci. 10 (1979) 2634.
[2] K. Fujikawa and R.E. Shrock, Phys. Rev. Lett. 45 (1980) 963.
[3] J. Schechter and J.W.F. Valle, Phys. Rev. D 24 (1981) 1883.
[4] M.B. Voloshin and M.I. Vysotsky, Yad. Fiz. 44 (1986) 845.
[5] L.B. Okun, Yad. Fiz. 44 (1986) 847.
[6] M.B. Voloshin, M.I. Vysotsky and L.B. Okun, Yad. Fiz. 44 (1986) 677.
[7] M.B. Voloshin, M.I. Vysotsky and L.B. Okun, Sov. Phys. JETP 64 (1986) 446.
[8] J.K. Rowley, B.T. Cleveland and R. Davis Jr., AIP Conf. Proc. N 126, Solar neutrinos and neutrino astronomy (Homestake, 1984), eds. L.M. Cherry et al. (New York, 1985) p. 1.
[9] J.N. Bahcall et al., Rev. Mod. Phys. 54 (1982) 767.
[10] S.T. Petcov, Phys. Lett. B 115 (1982) 401.
[11] M. Fukugita and T. Yanagida, Phys. Rev. Lett. 58 (1987) 1807;
K.S. Babu and V.S. Mathur, Phys. Lett. B 196 (1987) 218.
[12] G.A. Bazilevskaya, Yu.I. Stozhkov and T.N. Charakhchyan, Sov. Phys. JETP Lett. 35 (1982) 341.
[13] S.P. Mikheyev and A.Yu. Smirnov, Sov. J. Nucl. Phys. 42 (1985) 913; Usp. Fiz. Nauk 153 (1983) 3.
[14] L. Wolfenstein, Phys. Rev. D 17 (1978) 2369.
[15] S.P. Rosen and J.M. Gelb, Phys. Rev. D 34 (1986) 969.
[16] E.W. Kolb, M.S. Turner and T.P. Walker, Phys. Lett. B 175 (1986) 478.
[17] H. Bethe, Phys. Rev. Lett. 56 (1986) 1305.
[18] V. Barger, R.J.N. Phillips and K. Whisnant, Phys. Rev. D 34 (1986) 980.
[19] S.P. Mikheyev and A.Yu. Smirnov, Proc. 6th Moriond Workshop on Massive neutrinos in the astrophysics and in particle physics (Tignes, Savoie, France, 1986), eds. O. Fackler and J. Tran Thanh Van, p. 355;
J. Bouchez et al., Z. Phys. C 32 (1986) 499;
A.J. Baltz and J. Weneser, Phys. Rev. D 35 (1987) 528.
[20] E.Kh. Akhmedov, Institute of Atomic Energy preprint IAE-4568/1 (1988).
[21] C.-S. Lim, W.J. Marciano, Phys. Rev. D 37 (1988) 168.

Testing the principle of equivalence with neutrino oscillations

M. Gasperini
Dipartimento di Fisica Teorica dell'Università, Corso M.D'Azeglio 46, 10125 Torino, Italy and Istituto Nazionale di Fisica Nucleare, Sezione di Torino, Torino, Italy
(Received 6 June 1988)

If the equivalence principle is violated, and gravity is not universally coupled to all leptonic flavors, a gravitational field may contribute to neutrino oscillations. The laboratory limits on the oscillation process can thus be interpreted as tests of the equivalence principle in the quantum-relativistic regime, and put severe constraints on a maximal violation of this principle in the case of massless neutrinos coupled to the Earth's gravitational field.

I. INTRODUCTION

It is well known that neutrino oscillations[1] can occur, in a vacuum, if the eigenvalues of the mass matrix are not all degenerate, and the corresponding mass eigenstates are different from the weak-interaction eigenstates $\nu_e, \nu_\mu, \ldots$ Inside matter oscillations can be induced also by weak interactions, if the leptonic current has an off-diagonal part connecting different neutrino types;[2] but even if the current is diagonal the oscillations in matter are affected by weak interactions, because the different neutrinos are differently scattered by the electrons,[2] and, in particular, the oscillations can be resonantly enhanced at a given critical density.[3]

The only other interaction a neutral lepton can feel, besides the weak, is the gravitational interaction. A gravitational field, however, cannot induce nor affect neutrino oscillations if gravity couples universally to all types of matter, according to the principle of equivalence. A gravitational field could contribute to neutrino oscillations only if the different neutrino types would be differently affected by gravity, that is only if the equivalence principle would be violated in neutrino interactions.

The validity of the principle of equivalence is very well tested for macroscopic bodies, but this does not necessarily imply that such a principle continues to hold at a microscopic scale and in the quantum regime. It has been shown in fact that the equality of inertial mass and gravitational mass is no longer valid in the context of quantum field theory at finite temperature,[4] and the possibility that this equality is violated also in the case of the gravitational interactions of antimatter[5] has been recently suggested, and shown not to be in contrast with *CPT* invariance.[6]

The propagation of a neutrino beam through a gravitational field probes the validity of the Einstein principle of equivalence for quantum and relativistic test particles. As pointed out recently, the observations of neutrinos from the supernova 1987A has provided a test of the universality of the gravitational time delay for photons and neutrinos,[7,8] and for neutrinos with different energies.[7] The aim of this paper is to point out that experiments on neutrino oscillations test the universality of the gravitational red-shift for different neutrino flavors.

The interesting result is that, from the present laboratory limits on neutrino oscillations, in the hypothesis of massless neutrinos one can exclude a maximal violation (i.e., with maximum mixing angle) of the equivalence principle up to one part in 10^{11}, that is with a precision comparable to that of the Dicke-Braginskii experiments[9,10] which test the universality of the accelerations of different macroscopic bodies in the solar gravitational field. For comparison, in the time delays of the supernova neutrinos and photons the principle of equivalence is tested[7,8] only up to 1 part in 10^3, and 1 part in 10^6 for neutrinos of different energies in the hypothesis they are massless.[7]

II. GRAVITY-INDUCED OSCILLATIONS OF MASSLESS NEUTRINOS

If neutrinos are massless, in vacuum the eigenstates of the energy E coincide with the weak-interaction eigenstates $|\nu_W\rangle$: considering for simplicity two flavors only, we can set

$$|\nu_W\rangle = \begin{bmatrix} \nu_e \\ \nu_\mu \end{bmatrix}, \quad E = \begin{bmatrix} p & 0 \\ 0 & p \end{bmatrix}, \tag{1}$$

where p is the neutrino momentum, and the time evolution is then

$$|\nu_W(t)\rangle = e^{-iEt}|\nu_W(0)\rangle \ . \tag{2}$$

There is no mixing, and no oscillation occurs.

Suppose now that neutrinos propagate through a given gravitational field. The neutrino energies are red-shifted with respect to the vacuum, $E \to E' = \sqrt{g_{44}}E$, but according to the equivalence principle the energy shift should be the same for all the neutrino flavors, because of the universality of the gravitational coupling: therefore the energy should be still diagonal in the $|\nu_W\rangle$ basis, the only effect being an overall shift of the momenta, given by $\Delta p/p = -GM/r$ (if we are working in the weak field of a static source), which does not contribute to the oscillations.

If the principle of equivalence is violated, however, and gravity is not minimally coupled to all kinds of energy, we can have different red-shifts for different neutrino types: in this case there is no reason to believe that the weak-interaction eigenstates and the gravitational eigenstates are identical or, in other words, that the shifted kinetic energy E' is diagonal in the $|\nu_W\rangle$ basis.

Therefore, let $|\nu_G\rangle$ be the eigenstates of the total energy in the presence of gravity, related to the $|\nu_W\rangle$ basis by a rotation angle θ_G, that is $|\nu_W\rangle = R(\theta_G)|\nu_G\rangle$, where

$$|\nu_G\rangle = \begin{bmatrix} \nu_{1G} \\ \nu_{2G} \end{bmatrix}, \quad R(\theta) = \begin{bmatrix} \cos\theta & \sin\theta \\ -\sin\theta & \cos\theta \end{bmatrix} \tag{3}$$

and let $p'_{1,2}$ be the corresponding eigenvalues,

$$p'_{1,2} = p\left[1 - \frac{GM}{r}\epsilon_{1,2}\right] \tag{4}$$

where the dimensionless coefficients ϵ_1 and ϵ_2 parametrize the strength of a possible violation of the equivalence principle. If this principle is violated, and $\Delta\epsilon = \epsilon_1 - \epsilon_2 \neq 0$, flavor oscillations can occur, induced by gravity, even if neutrinos are massless. Suppose in fact we have a state that, at $t=0$, is a pure $|\nu_e\rangle$, and propagates through a gravitational potential GM/r; at a time t later we have the mixture (modulo an overall phase factor)

$$|\nu(t)\rangle = \cos\theta_G|\nu_{1G}\rangle + e^{-i\Delta E_G t}\sin\theta_G|\nu_{2G}\rangle\,, \tag{5}$$

where

$$\Delta E_G = p'_2 - p'_1 = \frac{GM}{r}p\Delta\epsilon \tag{6}$$

corresponding to oscillations, with characteristic length $L_G = 2\pi/\Delta E_G$, and oscillation probability

$$P_{\nu_e\to\nu_\mu}(t) = |\langle\nu_\mu|\nu(t)\rangle|^2 = \sin^2(2\theta_G)\sin^2(\tfrac{1}{2}\Delta E_G t)\,. \tag{7}$$

Assuming that neutrinos are massless, the laboratory limits on the oscillations can be directly interpreted therefore as limits on a possible violation of the equivalence principle in the Earth's gravitational field. Among the experimental results presently available (for a recent review see Ref. 11), the best limits on $\Delta\epsilon$ can be obtained by considering the limits on $\nu_\mu \leftrightarrow \nu_e$ oscillations recently obtained[12] with neutrinos of average energy $p = 1.5$ GeV. Using Eq. (7) to interpret the data of Ref. 12, and setting $GM/R = 0.69\times10^{-9}$ for the Earth average potential, one obtains, in the case of maximal mixing ($\theta_G = \pi/4$), the upper limit

$$\Delta\epsilon \lesssim 3\times10^{-11}\,. \tag{8}$$

A maximal violation of the equivalence principle for massless neutrinos is thus excluded, at the GeV scale, with the same accuracy achieved in the case of macroscopical test bodies.[9,10] It should be pointed out, however, that the neutrino oscillations probe the universality of the gravitational coupling to different neutrino flavors, but give no informations on the possibility that neutrinos have an anomalous gravitational red-shift with respect to other fields, that is $\epsilon_1 = \epsilon_2 \neq 1$.

III. RESONANT OSCILLATIONS OF MASSLESS NEUTRINOS IN MATTER

The result (8) is valid if there is no mass contribution to the oscillations. If this is the case, one may wonder what happens in the case of the solar-neutrino problem, which can find explanations based just on the oscillations of massive neutrinos, in a vacuum[1] or inside the solar matter[13,14] according to the so-called Mikheyev-Smirnov-Wolfenstein (MSW) resonance mechanism.[4,5]

Of course the solution of the problem might be due to different mechanisms (for example, as recently suggested,[15] to the existence of exotic, weakly interacting, massive particles called in general cosmions[16]); but if one insists in looking for a solution based on neutrino oscillations, it must be noted first of all that the condition (8), in the case of solar neutrinos ($p \sim 10$ MeV) traveling to the Earth in the solar gravitational potential, is still compatible with an oscillation length smaller than the Sun-Earth distance; moreover, even without mass contributions, neutrinos can have resonant oscillations inside the Sun if the principle of equivalence is violated.

Consider in fact the most general situation in which neutrinos are massive, and the equivalence principle is violated in such a way that $|\nu_W\rangle \neq |\nu_G\rangle \neq |\nu_M\rangle$, where $|\nu_M\rangle$ are the mass eigenstates. Given a medium characterized by a gravitational potential ϕ, and electron density N_e, the propagation eigenstates are obtained by diagonalizing the matrix which contains the contributions of mass, weak and gravitational interactions to the total energy. In the $|\nu_W\rangle$ basis, and for mass eigenvalues $m_{1,2} \ll p$, this matrix can be written (modulo a multiple of the identity, which only contributes to an overall phase and does not affect oscillations)

$$\begin{bmatrix} \sqrt{2}G_F N_e - \frac{\Delta m^2}{2p}\cos^2\theta_M - \Delta E_G\cos^2\theta_G & \frac{\Delta m^2}{4p}\sin2\theta_M + \frac{\Delta E_G}{2}\sin2\theta_G \\ \frac{\Delta m^2}{4p}\sin2\theta_M + \frac{\Delta E_G}{2}\sin2\theta_G & \frac{-\Delta m^2}{2p}\sin^2\theta_M - \Delta E_G\sin^2\theta_G \end{bmatrix}, \tag{9}$$

where G_F is the Fermi coupling constant, and

$$\Delta m^2 = m_2^2 - m_1^2 ,$$
$$\Delta E_G = \phi p \Delta\epsilon = \phi p(\epsilon_1 - \epsilon_2) . \qquad (10)$$

Here $\sqrt{2}G_F N_e$ represents the charged current contribution of the weak interactions, which affects ν_e but not ν_μ, and θ_M, θ_G, are the rotation angles which diagonalize, respectively, the mass and gravitational part of the energy, such that

$$|\nu_W\rangle = R(\theta_M)|\nu_M\rangle , \quad |\nu_W\rangle = R(\theta_G)|\nu_G\rangle . \qquad (11)$$

Taking into account also a possible violation of the equivalence principle (i.e., $\Delta E_G \neq 0$), the MSW condition of maximal mixing,[2,3] for neutrino propagation in the presence of a background gravitational field, is then generalized as follows:

$$\sqrt{2}G_F N_e = \frac{\Delta m^2}{2p}\cos 2\theta_M + \phi p \Delta\epsilon \cos 2\theta_G . \qquad (12)$$

Starting from this equation with $\Delta m^2 = 0$ (corresponding to massless neutrinos, or massive but degenerate, $m_1 = m_2$), and applying the same arguments as in Ref. 13, one can find that

$$\Delta\epsilon \cos 2\theta_G \lesssim 2\times 10^{-12} \left[\frac{\text{MeV}}{p}\right] \qquad (13)$$

is a sufficient condition in order that the gravity-induced oscillations of neutrinos, created with momentum p in the solar interior, be resonantly enhanced by the solar matter, and that the passage through the resonance region is adiabatic[3,13] provided that

$$\Delta\epsilon \frac{\sin^2 2\theta_G}{\cos 2\theta_G} \gg 2\times 10^{-14} \left[\frac{\text{MeV}}{p}\right] . \qquad (14)$$

The two conditions are compatible for $\tan(2\theta_G) \gg 10^{-1}$.

A suitable deviation from universality in the gravitational coupling of massless neutrinos may therefore induce resonant oscillations in the Sun interior, and explain the neutrino puzzle. Note that the violation of the equivalence principle is to be fine-tuned according to Eqs. (13) and (14), which must be satisfied in order to apply the MSW mechanism in its original version,[3,13] but a resonant solution to the puzzle may exist even without imposing the adiabatic condition.[14]

IV. CONCLUSION

The laboratory experiments on neutrino oscillations provide interesting tests of the equivalence principle, and put severe restrictions on a maximal violation of this principle for massless neutrinos in the terrestrial gravitational field, $\Delta\epsilon \lesssim 10^{-11}$.

This limit does not preclude however the possibility that the solar-neutrino problem may find a solution based on gravity-induced oscillations of massless neutrinos.

The experimental constraints on the oscillations could be reconciled with a higher value of $\Delta\epsilon$ in two ways: either assuming that neutrinos are massive, so that the propagation eigenstates in a vacuum are not $|\nu_G\rangle$ but correspond instead to the eigenstates of the matrix (9) (with $N_e = 0$), or supposing that neutrinos are massless but the gravitational mixing is not maximal, i.e., $\sin 2\theta_G < 1$.

In the last case, however, a sufficiently high value of $\Delta\epsilon$ would imply $\cos 2\theta_G \simeq 1$, so that an oscillatory solution of the solar-ν puzzle might be no longer possible, neither in vacuum nor through a resonance [see Eq. (13)].

Therefore, in the hypothesis that neutrino oscillations are the true explanation of the solar puzzle, an experimental evidence that gravity is not universally coupled to all lepton numbers, with $\Delta\epsilon \gg 10^{-11}$ (obtained independently from the oscillations experiments), would also indirectly suggest that neutrinos have a nonzero rest mass.

ACKNOWLEDGMENT

It is a pleasure to thank A. Bottino for useful information on neutrino oscillations.

[1]S. M. Bilenky and B. Pontecorvo, Phys. Rep. **41**, 225 (1978).
[2]L. Wolfenstein, Phys. Rev. D **17**, 2369 (1978).
[3]S. P. Mikheyev and A. Yu. Smirnov, Nuovo Cimento **9C**, 17 (1986).
[4]J. F. Donoghue, B. Holstein, and R. W. Robinett, Phys. Rev. D **30**, 2561 (1984); Gen. Relativ. Gravit. **17**, 207 (1985); Phys. Rev. D **34**, 1208 (1986).
[5]T. Goldman and M. M. Nieto, Phys. Lett. **112B**, 437 (1982); T. Goldman, M. V. Hynes, and M. M. Nieto, Gen. Relativ. Gravit. **18**, 67 (1986); T. Goldman, R. J. Hughes, and M. M. Nieto, Phys. Lett. B **171**, 217 (1986).
[6]M. M. Nieto, T. Goldman, and R. J. Hughes, in *Hadrons, Quarks and Gluons*, proceedings of the XXII Rencontre de Moriond, Les Arcs, France, 1987, edited by J. Tran Thanh Van (Editions Frontières, Gif-sur-Yvette, France, 1987).
[7]M. J. Longo, Phys. Rev. Lett. **60**, 173 (1988).
[8]L. M. Krauss and S. Tremaine, Phys. Rev. Lett. **60**, 176 (1988).
[9]P. G. Roll, R. Krotkov, and R. H. Dicke, Ann. Phys. (N.Y.) **26**, 442 (1964).
[10]V. B. Braginskii and V. I. Panov, Zh. Eksp. Teor. Fiz. **61**, 873 (1971) [Sov. Phys. JETP **34**, 463 (1972)].
[11]V. Flaminio and B. Saitta, Riv. Nuovo Cimento **10**, 1 (1987).
[12]C. Angelini *et al.*, Phys. Lett. B **179**, 307 (1986).
[13]H. A. Bethe, Phys. Rev. Lett. **56**, 1305 (1986).
[14]E. W. Kolb, M. S. Turner, and T. P. Walker, Phys. Lett. B **175**, 478 (1986); S. P. Rosen and J. M. Gelb, Phys. Rev. D **34**, 969 (1986).
[15]W. H. Press and D. N. Spergel, Astrophys. J. **296**, 679 (1985).
[16]J. Ellis, CERN Report No. TH 4811/87 (unpublished).

Mikheyev-Smirnov-Wolfenstein effect with flavor-changing neutrino interactions

Esteban Roulet
National Aeronautics and Space Administration Fermilab Astrophysics Center, Fermi National Accelerator Laboratory, P.O. Box 500, Batavia, Illinois 60510-0500
(Received 18 January 1991)

We consider the effect that flavor-nondiagonal neutrino interactions with matter have on the resonant ν oscillations. It is shown that, even in the absence of ν mixing in a vacuum, an efficient conversion of the electron neutrinos from the Sun to another ν flavor can result if the strength of this interaction is $\sim 10^{-2}G_F$. We show how this can be implemented in the minimal supersymmetric standard model with R-parity breaking. Here, the L-violating couplings induce neutrino masses, mixings, and the flavor-nondiagonal neutrino interactions that can provide a Mikheyev-Smirnov-Wolfenstein-like solution to the solar-neutrino problem even for negligible vacuum mixings.

The most elegant solution to the solar-neutrino problem is based on the so-called Mikheyev-Smirnov-Wolfenstein (MSW) effect [1–4], i.e., the resonant oscillation of ν_e into ν_μ or ν_τ induced by the neutrino interactions with the medium in the Sun. The effect of the matter in neutrino oscillations is related here to the fact that electron neutrinos, unlike ν_μ or ν_τ, have charged-current interactions due to W exchange with the electrons in the medium. These interactions produce a potential energy for the ν_e that leads, when the electron density corresponds to the resonance value, to a maximum neutrino mixing angle in matter, even if the mixing angle in vacuum θ is small. Hence, neutrinos crossing a resonance may be subject to a significant flavor conversion.

It was already noted by Wolfenstein [1] that even in the absence of neutrino masses (and hence of mixing angles in vacuum) there could be matter-induced neutrino oscillations in the presence of flavor-nondiagonal neutrino interactions with the medium. Our purpose here is to study the effects that these interactions can have in the resonant neutrino conversion, and the requirements they should satisfy in order to provide a solution to the solar-neutrino problem. Furthermore, we will describe how the necessary ingredients can be obtained in the minimal supersymmetric standard model with R-parity-violating interactions, which can provide the required neutrino masses and interactions. Since the effects of the flavor-nondiagonal ν interactions mimic those of a nonvanishing vacuum mixing angle, an MSW-like solution to the solar-neutrino problem can be obtained even for vanishing small θ.

Let us first quickly review the MSW effect with vacuum neutrino mixing and ordinary neutrino interactions [5]. For definiteness we will concentrate on the two-generation case. The evolution for the two-flavor neutrinos is given by

$$i\frac{d}{dx}\nu_c = \frac{1}{2E}\mathbf{M}^2\nu_c , \qquad (1)$$

where $\nu_c^T = (\nu_e, \nu_\alpha)$, with $\alpha = \mu$ or τ, and E is the neutrino energy. The matrix $\mathbf{M}^2$ is, neglecting an overall irrelevant (as far as oscillations are concerned) phase,

$$\mathbf{M}^2 = \frac{1}{2}\left[R_\theta \begin{pmatrix} -\Delta & 0 \\ 0 & \Delta \end{pmatrix} R_\theta^\dagger + 2E\begin{pmatrix} A & 0 \\ 0 & -A \end{pmatrix}\right], \qquad (2)$$

where $A = \sqrt{2}G_F N_e$ is due to the electron-neutrino coherent forward scattering from electrons (of number density N_e) and $\Delta \equiv m_2^2 - m_1^2$ is the squared-mass difference of the vacuum mass eigenstates (ν), that are related to the current eigenstates (ν_c) by

$$\nu_c = R_\theta \nu, \text{ with } R_\theta = \begin{pmatrix} c\theta & s\theta \\ -s\theta & c\theta \end{pmatrix}. \quad (3)$$

We denote $s\theta \equiv \sin\theta$, $c\theta \equiv \cos\theta$, etc.

The matter mass eigenstates are defined by $\nu_m \equiv R^\dagger_{\theta_m}\nu_c$, with

$$R^\dagger_{\theta_m} \mathbf{M}^2 R_{\theta_m} = \frac{1}{2}\begin{pmatrix} -\Delta_m & 0 \\ 0 & \Delta_m \end{pmatrix}, \quad (4)$$

where $\Delta_m = \Delta[(a-c2\theta)^2 + s^2 2\theta]^{1/2}$ with $a = 2EA/\Delta$. The matter mixing angle θ_m satisfies

$$s^2 2\theta_m = \frac{s^2 2\theta}{(c2\theta - a)^2 + s^2 2\theta}. \quad (5)$$

Hence, there is maximum mixing in matter in the "resonance region" corresponding to $a = c2\theta$, and the width of this resonance corresponds to the electron density for which $|a - c2\theta| = |s2\theta|$.

The ν_m evolution is then determined by

$$i\frac{d}{dx}\begin{pmatrix} \nu_m^1 \\ \nu_m^2 \end{pmatrix} = \begin{pmatrix} -\Delta_m/4E & -i\frac{d}{dx}\theta_m \\ i\frac{d}{dx}\theta_m & \Delta m/4E \end{pmatrix}\begin{pmatrix} \nu_m^1 \\ \nu_m^2 \end{pmatrix}, \quad (6)$$

with

$$\frac{d\theta_m}{dx} = \frac{1}{2}\frac{s2\theta}{(a-c2\theta)^2 + s^2 2\theta}\frac{da}{dx}. \quad (7)$$

Hence, in a medium with varying density, $N_e = N_e(x)$, transitions between the matter mass eigenstates are induced by a nonzero $d\theta_m/dx$. These transitions are usually negligible unless the neutrinos are near the resonance layer, for which the diagonal elements in Eq. (6) are minimum and $d\theta_m/dx$ is enhanced. If $P = |\langle \nu_m^1 | \nu_m^2 \rangle|^2$ is the probability of $\nu_m^1 \to \nu_m^2$ conversion in the resonance crossing, the averaged probability to detect an electron neutrino that has crossed a resonance is

$$P_{\nu_e\nu_e} = \tfrac{1}{2} + (\tfrac{1}{2} - P)c2\theta c2\theta_m, \quad (8)$$

where θ_m is the matter mixing angle corresponding to the point where the ν_e was produced, while θ is the vacuum angle. Under the assumption that the electron density varies linearly in the resonance layer, the probability of level crossing at resonance is found to be [4]

$$P = e^{-\pi\gamma/2}, \quad (9)$$

where the adiabaticity parameter γ is

$$\gamma \equiv \frac{\Delta_m}{4E|d\theta_m/dx|} = \frac{\Delta}{2E}\frac{s^2 2\theta}{c2\theta}\frac{1}{d\ln N_e/dx|_r} \quad (10)$$

(the subindex r stands for the resonance value). In the adiabatic case, i.e., $\gamma \gg 1$, the off-diagonal terms in Eq. (6) can be neglected even at resonance, so that $P = 0$ and the probability of having a ν_e after adiabatic flavor conversion is $P_{\nu_e\nu_e} = (1 + c2\theta c2\theta_m)/2$.

A solution to the solar-neutrino deficit observed at the Davis experiment [6] with adiabatic neutrino evolution is obtained for $\Delta \simeq 10^{-4}$ eV^2 (if the resonant layer is not too narrow, i.e., $s2\theta \gtrsim 10^{-2}$) and also for the so-called large-angle solution ($s2\theta \sim 0.8$, 10^{-4} eV$^2 > \Delta > 10^{-8}$ eV^2). A nonadiabatic solution exists for $\Delta s^2 2\theta \simeq 10^{-7.5}$ eV^2, closing a "triangle" in the Δ-$s^2 2\theta$ plane. The reduction factor in the neutrino flux depends on the ν energy in different ways in the three regimes. Hence, experiments with other energy thresholds or capable of measuring the neutrino spectra can distinguish among the three solutions. The results of the Kamiokande detector [7] in fact disfavor the $\Delta \simeq 10^{-4}$ eV^2 adiabatic solution, while the preliminary observation of a very low neutrino rate at the gallium SAGE experiment, if confirmed, will support the nonadiabatic solution.

Let us now assume the existence of additional ν interactions leading to an effective Lagrangian including flavor-nondiagonal ν interactions

$$-\mathcal{L}_{\rm eff} = \frac{1}{\sqrt{2}}\bar{\nu}_j\gamma^\mu(1-\gamma_5)\nu_i(G^e_{ij}\bar{e}\gamma_\mu e + G^q_{ij}\bar{q}\gamma_\mu q). \quad (11)$$

We have not included other possible couplings, such as to axial-vector currents, since for an unpolarized medium only the time component of the vector current leads to a significant scattering cross section in the nonrelativistic limit. This leads to an additional contribution to the matrix $\mathbf{M}^2$ describing the ν evolution in matter:

$$\frac{M^2_{ij}}{2E} \to \frac{M^2_{ij}}{2E} + \sqrt{2}(G^e_{ij}N_e + G^p_{ij}N_p + G^n_{ij}N_n), \quad (12)$$

where $N_{n,p}$ are the neutron and proton densities and $G^n = 2G^d + G^u$, while $G^p = 2G^u + G^d$.

Clearly we expect $|G_{ij}| \ll G_F$ (see below for specific bounds), so that the effects of the diagonal elements G_{ii} in the neutrino oscillations can be neglected with respect to those of the charged-current ν_e interaction [8]. Hence, the evolution equation for the two-flavor neutrinos now becomes

$$i\frac{d}{dx}\begin{pmatrix} \nu_e \\ \nu_a \end{pmatrix} \simeq \frac{\Delta}{4E}\begin{pmatrix} -c2\theta + a & s2\theta + b \\ s2\theta + b & c2\theta - a \end{pmatrix}\begin{pmatrix} \nu_e \\ \nu_a \end{pmatrix}, \quad (13)$$

where $b = \sqrt{2}(G^e_{ea}N_e + G^p_{ea}N_p + G^n_{ea}N_n)4E/\Delta$.

Thus, the mixing angle in matter is given by

$$s^2 2\theta_m = \frac{(b + s2\theta)^2}{(c2\theta - a)^2 + (b + s2\theta)^2}, \quad (14)$$

and since we expect $|db/dx| \ll |da/dx|$, the resonance width is obtained from

$$|a - c2\theta| \simeq |b_r + s2\theta|, \quad (15)$$

where b_r is the resonance value of b, i.e., corresponding to $a = c2\theta$.

Similarly, the remaining expressions can be obtained with the substitution $s2\theta \to b + s2\theta$ in expressions (4)–(10) for the ordinary MSW effect. Since at resonance $N_e = (c2\theta/\sqrt{2})\Delta/2EG_F$, it follows that

$$b_r = \frac{2c2\theta}{G_F}(G^e + G^p + Y_n G^n)_{ea}, \quad (16)$$

with $Y_n \equiv N_n/N_e$. Note that b_r is independent of the neutrino energy, since the explicit dependence on E of b is compensated by the fact that at resonance $N_e \sim E^{-1}$

To compute the ν survival probability $P_{\nu_e\nu_e}$, Eq. (8), the knowledge of P and of $c2\theta_m$ at the neutrino production point is required. P obviously only depends on the resonance value of b, while

$$c2\theta_m = \frac{c2\theta - a}{[(c2\theta - a)^2 + (b + s2\theta)^2]^{1/2}} \tag{17}$$

is sensitive to the value of b only if the production point is near the resonance. From the previous discussion it is clear then that in the case $s2\theta < |b_r|$, the role of $s2\theta$ in the neutrino oscillations is played by b_r. Hence, to have interesting oscillation effects as in the ordinary MSW effect for which $s2\theta \gtrsim 10^{-2}$ was required, we need

$$|G^e + G^p + Y_n G^n|_{ea} \gtrsim 10^{-2}\frac{G_F}{2}. \tag{18}$$

Since b_r is independent of the ν energy, we remarkably expect the same suppression on the ν_e fluxes as a function of the neutrino energy as in the ordinary MSW effect, even if the physics responsible for the neutrino conversion is quite different. We also expect a similar general behavior of the solar ν_e fluxes as a function of the neutrino squared-mass difference Δ. For $\Delta \simeq 10^{-4}$ eV^2 there should be an important adiabatic conversion of ν flavors for $b_r \gtrsim 10^{-2}$, while for $\Delta b_r \simeq 10^{-7.5}$ eV^2 there should be a nonadiabatic regime.

One interesting model leading to nonstandard neutrino properties such as masses or flavor-changing neutrino interactions is the well-known supersymmetric extension of the standard model [9] including some R-parity-violating interactions [10]. In particular, we will concentrate on the following lepton-number-violating contributions to the superpotential:

$$\lambda_{ijk} L_i L_j E_k^c , \tag{19a}$$

$$\lambda'_{ijk} L_i Q_j D_k^c , \tag{19b}$$

where L, Q, E^c, D^c are the usual lepton and quark SU(2) doublets and singlets, respectively, and i, j, k are generation indices. Because of the contraction of the SU(2) indices in Eq. (19a), the λ_{ijk} should be antisymmetric under the exchange of i and j. In the following, we will take these Yukawa couplings to be real so that they do not violate CP.

These couplings induce a finite neutrino Majorana mass matrix through one-loop diagrams involving a lepton (quark) and a slepton (squark) line. For instance, the leptonic loop leads in the usual supersymmetric model with soft breaking terms as arising from low-energy supergravity, to [11,12]

$$\delta m_{ij} \simeq \frac{\lambda_{ikk'}\lambda_{jk'k}}{8\pi^2} m_k m_{k'} \frac{\tilde{m}}{m_0^2} , \tag{20}$$

where $\tilde{m}$ is a typical supersymmetric mass ($\sim$100 GeV if supersymmetry is to solve the naturalness problem), m_i is the lepton mass, and all sleptons were assumed almost degenerate at a mass m_0.

For the MSW solution to the solar-neutrino problem, with small vacuum mixing angles and discarding fine-tuned cancellations between the ν masses, we need the diagonal entry δm_{aa} to be $\sim 10^{-2}$–10^{-4} eV with the remaining entries in the ν mass matrix much smaller. This can be easily satisfied by many possible choices of reasonably small λ (or λ') constants. Several phenomenological constraints typically impose [13] $\lambda, \lambda' \lesssim 10^{-1}$, although some stronger bounds apply to the products of constants that are involved in rare processes such as $\mu \to e\gamma$, $\mu \to ee\bar{e}$, etc. [11]. Clearly, it is also possible to generate in this way the mixing angles required in the ordinary MSW effect, but we want here to concentrate on the resonant conversion induced by the flavor-changing neutrino interactions previously discussed.

Turning now to the neutrino interactions with the medium, we will analyze first the interactions with the electrons. The superpotential (19a) leads to the Lagrangian

$$\mathcal{L} = \lambda_{ijk}[\bar{\nu}_L^i \bar{e}_R^k e_L^j + \bar{e}_L^j \bar{e}_R^k \nu_L^i + \bar{e}_R^{k*} (\bar{\nu}_L^i)^c e_L^j - i \leftrightarrow j] + \text{H.c.} \tag{21}$$

In the case of ν_e scattering off electrons, this leads to an effective low-energy interaction of the form

$$\mathcal{L}_{\text{eff}} = -\frac{\lambda_{1j1}\lambda_{lj1}}{m_{\tilde{e}_L^j}^2} \bar{e}_R \gamma_\mu e_R \bar{\nu}_L^l \gamma^\mu \nu_L^e . \tag{22}$$

The scattering through $\tilde{e}_R$ exchange is not present because it vanishes in the s channel ($\lambda_{11k} = 0$), while for $\tilde{e}_R$ exchange in the t channel ν_e only scatters off e^+. Hence,

$$\sqrt{2} G_{1l}^e = \frac{\lambda_{1j1}\lambda_{lj1}}{2m_{\tilde{e}_L^j}^2} \tag{23}$$

and consequently the condition for resonant oscillations to take place in the Sun, Eq. (18), now reads

$$\lambda_{1j1}\lambda_{lj1} \gtrsim 1.5 \times 10^{-3} \left(\frac{m_0}{100 \text{ GeV}} \right)^2 . \tag{24}$$

If $l = 2$ ($\nu_e \leftrightarrow \nu_\mu$ conversion), the interaction is mediated by $\tilde{\tau}_L$ exchange ($j = 3$). However, the same product of couplings $\lambda_{131}\lambda_{231}$ appears in the radiative decay $\mu \to e\gamma$ and hence [11] it is severely constrained to be less than $10^{-5}(m_0/100 \text{ GeV})^2$. The allowed values of $G_{e\mu}^e$ are then too small. The remaining possibility is to have $l = 3$ ($\nu_e \leftrightarrow \nu_\tau$ conversion), mediated by $\tilde{\mu}_L$ exchange. The product $\lambda_{121}\lambda_{321}$ is bounded by the products of the individual bounds obtained in Ref. [13], i.e., $\lambda_{121}\lambda_{321} \lesssim 0.04 \times 0.09(m_0/100 \text{ GeV})^2$ (the bound from the process $\tau \to e\gamma$ is weaker). This just marginally allows parameter values for which there can be interesting oscillation effects [14]. Note also that these products of couplings are not involved in the elements of the neutrino mass matrix [see Eq. (20)] and so they are not further constrained.

For the ν-q interactions, the superpotential (19b) leads to

$$\mathcal{L} = \lambda'_{ijk}[\bar{d}_L^j \bar{d}_R^k \nu_L^i + (\bar{d}_R^k)^* (\bar{\nu}_L^i)^c d_L^j] + \text{H.c.} \tag{25}$$

(omitting terms without neutrinos). This results in a low-energy effective Lagrangian

$$\mathcal{L}_{\text{eff}} = \frac{\lambda'_{ijk}\lambda'_{lmk}}{2m_{\tilde{d}_R^k}^2} (\bar{\nu}_L^l \gamma_\mu \nu_L^i \bar{d}_L^m \gamma^\mu d_L^j) - \frac{\lambda'_{ijk}\lambda'_{ljn}}{2m_{\tilde{d}_L^j}^2} (\bar{\nu}_L^l \gamma_\mu \nu_L^i \bar{d}_R^k \gamma^\mu d_R^n) . \tag{26}$$

Since the neutrinos only interact with the down quarks in the nucleons through the exchange of down-type squarks, constraint (18) translates into

$$(1+2Y_n)G^d_{e\alpha} \gtrsim 10^{-2}\frac{G_F}{2}\,, \tag{27}$$

so that the couplings involved should satisfy $|\lambda'\lambda'| \gtrsim 10^{-3}(m_{\tilde{q}}/100\ \mathrm{GeV})^2$. In the case of $\nu_e\nu_\mu$ conversion, scattering through $\tilde{s}_L$ exchange involves the product of couplings $\lambda'_{121}\lambda'_{221}$, while the scattering through $\tilde{b}_L$ exchange involves $\lambda'_{131}\lambda'_{231}$. However, since these couplings also induce the process $\mu \to e\gamma$, they are very suppressed and $\nu_e\nu_\mu$ conversion is not allowed. Instead, it is possible to generate $\nu_e\nu_\tau$ conversion by exchange of either left or right down-type squarks, since there are no strong bounds on λ_{3jk} alone, while the bound from $\tau \to e\gamma$ is [15] $\lambda'_{1jk}\lambda'_{3jk} \lesssim 5\times 10^{-2}(m_{\tilde{q}}/100\ \mathrm{GeV})^2$. It is interesting to note that for this model the required couplings could be probed at a τ factory [15].

In conclusion, in the same way as small neutrino mixing in a vacuum can be amplified producing significant oscillations of the neutrinos that cross a resonance layer while propagating in a medium, we have shown that similar effects can be obtained in the presence of flavor-changing neutrino interactions. This has important applications to solar neutrinos, since allowed strengths for those new interactions can lead to a solution to the solar-neutrino deficit even for negligibly small vacuum mixings.

We have shown how lepton-number-violating couplings that can be present in the minimal supersymmetric extension of the standard model are able to generate the required flavor-nondiagonal neutrino interactions with quarks and leptons, as well as the necessary neutrino masses, taking into account the experimental bounds on the new couplings.

I want to thank D. Tommasini, J. Frieman, G. F. Giudice, M. Guzzo, and S. Parke for very helpful discussions. This work was supported in part by the U.S. Department of Energy and by NASA (Grant No. NAGW-1340) at Fermilab.

[1] L. Wolfenstein, Phys. Rev. D **17**, 2369 (1978).

[2] S. P. Mikheyev and A. Yu. Smirnov, Yad. Fiz. **42**, 1441 (1985) [Sov. J. Nucl. Phys. **42**, 913 (1985)].

[3] H. A. Bethe, Phys. Rev. Lett. **56**, 1305 (1986); S. P. Rosen and J. M. Gelb, Phys. Rev. D **34**, 969 (1986).

[4] S. J. Parke, Phys. Rev. Lett. **57**, 1275 (1986).

[5] For reviews, see S. M. Bilenky and S. T. Petcov, Rev. Mod. Phys. **59**, 671 (1987); T. K. Kuo and J. Pantaleone, *ibid.* **61**, 937 (1989).

[6] R. Davis, in *Neutrino '88*, Proceedings of the 13th International Conference on Neutrino Physics and Astrophysics, Boston, Massachusetts, 1988, edited by J. Schneps, T. Kafka, W. A. Mann, and P. Nath (World Scientific, Singapore, 1989), p. 518.

[7] K. S. Hirata *et al.*, Phys. Rev. Lett. **65**, 1297 (1990); **65**, 1301 (1990).

[8] Effects of a nonuniversal diagonal strength of the neutral currents have been discussed by J. Valle, Phys. Lett. B **199**, 432 (1987).

[9] For reviews, see, e.g., H. P. Nilles, Phys. Rep. **110**, 1 (1984); H. E. Haber and G. L. Kane, *ibid.* **117**, 75 (1985).

[10] C. Aulak and R. Mohapatra, Phys. Lett. **119B**, 136 (1983); F. Zwirner, *ibid.* **132B**, 103 (1983); L. J. Hall and M. Suzuki, Nucl. Phys. **B231**, 419 (1984); I. H. Lee, *ibid.* **B248**, 120 (1984); J. Ellis *et al.*, Phys. Lett. **150B**, 142 (1985); S. Dawson, Nucl. Phys. **B261**, 297 (1985); R. Barbieri and A. Masiero, *ibid.* **B267**, 679 (1986); S. Dimopoulos and L. J. Hall, Phys. Lett. B **207**, 210 (1987).

[11] R. Barbieri *et al.*, Phys. Lett. B **252**, 251 (1990).

[12] E. Roulet and D. Tommasini, Phys. Lett. B **256**, 218 (1991).

[13] V. Barger, G. F. Giudice, and T. Y. Han, Phys. Rev. D **40**, 2987 (1989).

[14] Some caution should be taken with the bounds quoted because $\lambda_{121} < 0.04$ is a 1σ bound, while $\lambda_{231} < 0.09$ is a 2σ bound. At 1σ there is no allowed value for λ_{231}, while at 2σ the bound on λ_{121} is weaker.

[15] A. Masiero, Report No. DFPD/90/TH732 (unpublished).

On the MSW effect with massless neutrinos and no mixing in the vacuum

M.M. Guzzo [a,1], A. Masiero [b] and S.T. Petcov [a,c]

[a] Scuola Internazionale Superiore di Studi Avanzati, I-34100 Trieste, Italy

[b] Istituto Nazionale di Fisica Nucleare, Sezione di Padova, I-35131 Padua, Italy

[c] Institute of Nuclear Research and Nuclear Energy, Bulgarian Academy of Sciences, 1784 Sofia, Bulgaria

Received 21 February 1991

The general properties of the resonantly enhanced neutrino transitions in matter induced by neutrino flavour changing neutral current (FCNC) interactions (rather than by nontrivial neutrino mixing in vacuum) are analysed. Such transitions can take place even if neutrinos have zero mass. It is shown, in particular, that matter-enhanced transitions of the solar neutrinos (into ν_τ, for instance) generated by neutrino FCNC interactions are not excluded by the existing data and provide an alternative solution of the solar neutrino problem.

Matter-enhanced neutrino oscillations [1–3] (see also the review articles [4,5]) and the possible solutions of the solar neutrino problem [6] that they imply [1,7,8] continue to be extensively studied [9–11]. One of the conventional approaches exploited in the relevant analyses is to assume [12] that the states of the flavour neutrinos ν_l, $l=e, \mu, \tau$, produced with definite momentum p in vacuum in weak interaction processes are coherent superpositions of the states of neutrinos ν_i, $i=1, 2, 3$, having the same momentum p and definite masses m_i, $i=1, 2, 3$, some of which are nonzero. It is also implicitly assumed that the weak interaction of the neutrinos is described by the lagrangian of the standard electroweak theory [13], in other words, that the relevant mechanism of neutrino mass generation does not lead to new nonnegligible interactions of neutrinos with the particles forming the matter. Two conditions have to be fulfilled in this case in order for the transitions (oscillations) between neutrinos possessing different flavours $\nu_l \Leftrightarrow \nu_{l'}$, $l\neq l'$, $l, l' = e, \mu, \tau$, to be possible: at least two of the neutrinos ν_i with definite mass in vacuum must be mass-nondegenerate, and nontrivial neutrino (lepton) mixing in vacuum must exist. In particular, the two neutrino oscillations, say $\nu_e \Leftrightarrow \nu_\mu$, which are characterized in vacuum by two parameters $\Delta m^2 = m_2^2 - m_1^2$ and θ, where m_1 and m_2 are the masses of the corresponding two vacuum mass eigenstate neutrinos (ν_1 and ν_2) and θ is the neutrino mixing angle in vacuum, can take place only provided $\Delta m^2 \neq 0$ and $\sin 2\theta \neq 0$.

The neutrino oscillation hypothesis [12] provides [1,7–11] a very attractive solution of the solar neutrino problem [6]. Recent analyses [9–11] have shown that when interpreted in terms of transitions (oscillations) of the solar neutrinos (ν_e) into neutrinos of one different type (ν_μ, for example), the solar neutrino data can be explained either by existence of solar matter enhanced transitions of the solar neutrinos (nonadiabatic solution [5,7,9,10]: $\sin^2 2\theta \geqslant 4\times 10^{-3}$, $\Delta m^2 \sin^2 2\theta = (3.2\pm 1.0)\times 10^{-8}$ eV^2) or by existence of "long" wavelength two-neutrino oscillations of the solar ν_e not affected by the solar matter (vacuum solution [10,11]: $\sin^2 2\theta \gtrsim 0.7$, $\Delta m^2 \simeq (0.5$–$2.5)\times 10^{-10}$ eV^2).

In the present note we discuss an alternative possibility [2] which can lead to transitions in matter between neutrinos possessing different flavours even if neutrinos have zero mass and there is no neutrino mixing in vacuum: the existence of flavour changing neutral current (FCNC) interactions of the neutrinos. We analyse the general features of such transi-

[1] Supported by CNPq.

tions using the specific context provided by the supersymmetric theories with broken R parity [14]. Taking into account the relevant phenomenological constraints we show that the neutrino FCNC interactions can cause a resonant transition of the solar electron neutrinos [#1] into, e.g., τ neutrinos with a probability compatible with the qualitative features of the solar neutrino observations. The transition in question can take place in the core of the sun and can be of adiabatic type. The corresponding transition probability does not depend on the solar neutrino momentum. We analyse also the possibility that neutrinos do have a mass, but the effects of the existence of vacuum neutrino mixing on the solar neutrino transitions are negligible, the transitions of interest being induced by neutrino FCNC interactions with the nucleons in the sun. The solution of the solar neutrino problem that one obtains in this case is practically equivalent to the nonadiabatic solution found in refs. [9,10]. Some phenomenological implications of the particular scheme with neutrino FCNC interactions considered are also discussed. The most interesting prediction for lepton physics is that decays in which the τ lepton charge is not conserved, in particular $\tau^{\pm}\to\rho^0+l^{\pm}$ and $\tau^{\pm}\to\pi^0+l^{\pm}$, should occur with branching ratios close to the existing experimental upper limits.

As is quite well known, FCNC interactions of neutrinos which are the basic ingredients of the mechanism of neutrino transitions that we will analyse, do not appear in the standard theory of electroweak interactions [13]. Such FCNC interactions arise at higher order of perturbation theory (at one or higher loop level) in most of its extensions with massive neutrinos (see, e.g., ref. [4]). The supersymmetric theories with R-parity breaking represent a remarkable exception since in their framework the neutrino FCNC interactions can occur at tree level [14]. The minimal version of these theories contains two types of terms in the superpotential which break the R-parity conservation and do not conserve, in general, the additive lepton numbers L_l, $l=e, \mu, \tau$:

$$f^{\rm lep}(x)=\tfrac{1}{2}\lambda_{ijk}[L_i, L_j]l^c_k\,, \qquad (1)$$

$$f^{l-q}(x)=\lambda'_{ijk}L_iQ_jD^c_k\,. \qquad (2)$$

[#1] This possibility is not discussed in ref. [2].

In (1) and (2) $i, j, k=1, 2, 3$, are family indices. L_i and Q_i are the lepton and quark left-handed (LH) doublet superfields, l^c_k and D^c_k are the ("charge conjugated" of the) right-handed (RH) charged lepton and charge $(-\frac{1}{3})$ quark singlet superfields and $\lambda_{ijk}=-\lambda_{ijk}$ and λ'_{ijk} are coupling constants. The phenomenological implications of the presence of each of the terms in (1) and (2) in the electroweak interaction lagrangian have been studied in detail [14]. Constraints on the coupling constants λ_{ijk} and λ'_{ijk} have also been derived. Typically one has [15]: $|\lambda_{12k}|<0.04$, $|\lambda_{13k}|<0.10$, $k=1, 2, 3$, $|\lambda_{231}|<0.09$, etc., while the constraints on the coupling constants λ'_{ijk} are, in general, less stringent. For instance, there are practically no constraints from the data on the constants λ'_{331} since the terms in (2), which it is associated with, include the fields of quarks, squarks and leptons (or sleptons) belonging to the third family. Most of the limits depend on the assumed values of the masses of the relevant supersymmetric partners of the charged leptons and quarks for which experimental lower bounds lying typically in the interval (50–120) GeV exist. The limits quoted above correspond to values of the slepton and squark masses equal to 100 GeV.

More recently, it was also pointed out [16] that the simultaneous presence of two (or more) of the terms in (1) or (2) in the electroweak interaction lagrangian [together with the existence of mixing between the LH and RH sleptons (squarks)] can lead to radiative generation (at one loop level) of nonzero neutrino masses and nontrivial neutrino mixing. In this case the theory provides a solution of the solar neutrino problem in terms of the neutrino oscillation hypothesis.

As is not difficult to convince oneself, the R-parity nonconserving terms in (1) and (2) can generate neutrino FCNC interactions with the electron and the u and d quarks. Such interactions can arise effectively at tree level due to (1) or (2) as a result of the exchange of virtual sleptons or virtual squarks, respectively. Obviously, at least two of the coupling constants λ_{ijk} (e.g., λ_{121} and λ_{231}) or λ'_{ijk} (e.g., λ'_{131} and λ'_{331}) must be nonzero in order for the interaction terms of interest to appear in the effective lagrangian of the theory. As can be shown, the constraints on the relevant couplings λ_{ijk} following from the data on the flavour changing lepton decays

($\mu^\pm \to e^\pm + \gamma$, $\mu^\pm \to e^\pm e^+ e^-$, $\tau^\pm \to e^\pm e^+ e^-$, etc.) make, however, even the coherent effects of the neutrino FCNC interactions, generated by $f^{lep}(x)$, on the propagation of the neutrinos in matter negligible. This is not the case when the neutrino FCNC interactions are due to the R-parity nonconserving couplings in $f^{l-q}(x)$ [eq. (2)] and they involve the neutrinos $\overset{(-)}{\nu}_e$ and $\overset{(-)}{\nu}_\tau$ or $\overset{(-)}{\nu}_\mu$ and $\overset{(-)}{\nu}_\tau$.

To be more specific and to simplify the discussion, let us assume that only the coupling constants λ'_{131} and λ'_{331} in eqs. (1) and (2) are different from zero and the fields entering in the corresponding terms in eq. (2) have definite mass [#2]. One has in this case for the supersymmetric R-parity nonconserving interaction lagrangian

$$\begin{aligned}
\mathcal{L}_R(x) &\\
=\lambda'_{131}[&\tilde{\nu}_{eL}(x)\bar{d}_R(x)b_L(x)+\bar{\tilde{b}}_L(x)\bar{d}_R(x)\nu_{eL}(x)\\
&+\bar{d}_R(x)\overline{\nu^c}_{eR}(x)b_L(x)-\tilde{e}_L\bar{d}_R(x)t_L(x)\\
&-\tilde{t}_L(x)\bar{d}_R(x)e_L(x)-\bar{d}_R(x)\overline{e^c}_R(x)t_L(x)]\\
+\lambda'_{331}[&\tilde{\nu}_{\tau L}(x)\bar{d}_R(x)b_L(x)+\bar{\tilde{b}}_L(x)\bar{d}_R(x)\nu_{\tau L}(x)\\
&+\bar{d}_R(x)\overline{\nu^c}_{\tau R}(x)b_L(x)-\tilde{\tau}_L(x)\bar{d}_R(x)t_L(x)\\
&-\tilde{t}_L(x)\bar{d}_R(x)\tau_L(x)-\bar{d}_R(x)\overline{\tau^c}_R(x)t_L(x)]+\text{h.c.}\,,
\end{aligned} \tag{3}$$

where $\overset{(\sim)}{\nu}_{eL}(x), \ldots, \overset{(\sim)}{b}_L(x)$ are the fields of the (scalar) electron neutrino, ..., LH bottom (scalar) quark, $\nu^c_{eR}(x)=C(\bar{\nu}_{eL})^T(x)$, $e^c_R(x)=C\bar{e}^T_L(x)$, etc., C being the charge conjugation matrix. We shall assume also in what follows that the coupling constants λ'_{131} and λ'_{331} are real and that apart from $\mathcal{L}_R(x)$ [eq. (3)] the lagrangian of the theory coincides with the lagrangian of the minimal supersymmetric extension of the standard theory [17].

The interactions described by the lagrangian (3) can lead to resonant $\overset{(-)}{\nu}_e \to \overset{(-)}{\nu}_\tau$ transitions in matter. Indeed, if $|\lambda'_{131}| \neq |\lambda'_{331}|$, as it follows from (3), the $\overset{(-)}{\nu}_e$ and $\overset{(-)}{\nu}_\tau$ neutrinos will scatter coherently with different amplitudes not only on the electrons present in matter [2], but also on the d quarks. Moreover, the process $\overset{(-)}{\nu}_e + d \to \overset{(-)}{\nu}_\tau + d$ is possible at tree level: it is mediated by the exchange of the virtual $\tilde{b}_L$ squark. Correspondingly, the system of evolution equations which describes the $\overset{(-)}{\nu}_e \to \overset{(-)}{\nu}_\tau$ transitions in matter has the form:

$$\mathrm{i}\frac{\mathrm{d}}{\mathrm{d}t}\begin{pmatrix}\overset{(-)}{A}_e(t,t_0)\\ \overset{(-)}{A}_\tau(t,t_0)\end{pmatrix}=\left(\begin{smallmatrix}+\\-\end{smallmatrix}\right)\sqrt{2}G_F$$

$$\times\begin{pmatrix}0 & \epsilon(2N_n+N_p)\\ \epsilon(2N_n+N_p) & \epsilon'(2N_n+N_p)-N_e\end{pmatrix}\begin{pmatrix}\overset{(-)}{A}_e(t,t_0)\\ \overset{(-)}{A}_\tau(t,t_0)\end{pmatrix}. \tag{4}$$

Here $\overset{(-)}{A}_e(t,t_0)$ and $\overset{(-)}{A}_e(t,t_0)$ are the amplitudes of the probabilities to find the neutrinos $\overset{(-)}{\nu}_e$ and $\overset{(-)}{\nu}_\tau$ at time t if some coherent mixture of $\overset{(-)}{\nu}_e$ and $\overset{(-)}{\nu}_\tau$ was produced in matter at time t_0. n_e, N_p and N_n are the electron, proton and neutron number densities in the point of the neutrino trajectory reached by the neutrinos at time t, $(\overset{+}{-})\sqrt{2}G_F\epsilon'(2N_n+N_p)$ is the difference between the $\overset{(-)}{\nu}_\tau - d$ and $\overset{(-)}{\nu}_e - d$ forward elastic scattering amplitudes, $(\overset{+}{-})\sqrt{2}G_F\epsilon(2N_n+N_p)$ is the $\overset{(-)}{\nu}_e + d \to \overset{(-)}{\nu}_\tau + d$ forward scattering amplitude where

$$\epsilon' = \frac{|\lambda'_{331}|^2 - |\lambda'_{131}|^2}{4m^2(\tilde{b}_L)\sqrt{2}G_F}, \tag{5}$$

$$\epsilon = \frac{\lambda'_{331}\cdot\lambda'_{131}}{4m^2(\tilde{b}_L)\sqrt{2}G_F}, \tag{6}$$

and $m(\tilde{b}_L)$ is the mass of $\tilde{b}_L$. In all cases of practical interest (propagation of neutrinos in the sun, in the earth and in supernovae) the matter is electrically neutral, so $N_p = N_e$. In (4) we did not take into account the possible effects of scattering of $\overset{(-)}{\nu}_e$ and $\overset{(-)}{\nu}_\tau$ off neutrinos which can be important for the transitions of the neutrinos emitted by the supernovae (see, e.g., ref. [5]).

Eq. (4) implies that the properties of the neutrino (antineutrino) transitions induced solely by the interactions of the neutrinos with the particles forming the matter are very different from the properties of the transitions due to nonzero neutrino masses and nontrivial neutrino mixing [1,4,5] (for brevity, we

[#2] If the quark fields in eq. (2) are current (flavour) eigenstates, nonnegligible neutrino masses and mixing would, in general, be generated radiatively. However, we are interested in the present analysis in the case when the neutrino masses and/or mixing in vacuum are not relevant for the solar neutrino problem.

V. Helioseismology

Introduction: Roger Ulrich

Helioseismology is the analysis of the sun's internal structure and dynamics through the study of surface wave motion. This analysis is closely related to the seismic study of the earth's interior through the measurement of oscillations generated by large earthquakes. Many of the detailed techniques have been carried over to the study of the sun. The greatest differences between terrestrial seismology and helioseismology come from the sun's lack of resistance to shear stresses, so that shear-mode oscillations do not occur on the sun, and from the different sources of the energy for the oscillations. The convective motion energizes the oscillations in the sun, and sudden stress release via earthquakes energizes terrestrial seismological waves. The solar oscillations are continuously excited, while terrestrial oscillations are (fortunately) only occasionally excited.

The opportunity of probing solar structure with helioseismology has become widely recognized only within the past 15 years or so. Prior to the mid-1970's neutrinos were thought to be the only sources of information from the solar interior. The first key step toward helioseismology was taken by Robert Leighton, who was experimenting with photographic subtractions as a way of isolating the solar velocities. When the temporal dependence was displayed by scanning the photographic plate slowly, the resulting images showed a clear signal of quasi-periodic oscillatory motions with a preferred period near five minutes. This discovery was announced in Paper 1.V by Leighton, Noyes and Simon.

The nature of the oscillatory motions was the subject of much speculation, with most discussions centering on the propagation of sound waves in the solar atmosphere. The five-minute period is near the cutoff frequency for vertical wave propagation in a stratified atmosphere as described by Lamb[1]. Other possible explanations included the refocusing of acoustic waves due to the chromospheric temperature rise and the transient response of the atmosphere to the arrival of convection cells. None of these explanations suggested that the oscillations would be of interest in the study of the solar interior. The apparently sharp surface of the sun caused by the abrupt rise in the opacity when the hydrogen ionizes led to the incorrect assumption that the observed motions are confined to the visible solar layers.

An early hint that the oscillations could be more than a direct effect of convective cells was provided by Frazier[2], who was studying high resolution solar images in an effort to improve our understanding of the convection zone. He and Ulrich were fellow graduate students under the direction of Louis Henyey at UC Berkeley. In an informal presentation of his thesis observations, Frazier commented that the oscillations appeared to be independent of the arrival of the granulation and suggested these waves could involve layers below the solar surface.

Following Frazier's suggestion, Ulrich extended the theory of acoustic wave propagation to include the solar interior (Paper 2.V). This work faced the problem of including the effects of the temperature and sound speed gradients in the well-known dispersion relation governing the propagation of acoustic waves. In addition, the large vertical wavelength of the observed oscillations implied by the simplified dispersion relation indicated that an assumption of vertical homogeniety is invalid.

A proper treatment of the wave structure to take the gradients into account required a normal mode analysis that was implemented numerically. The change of perspective from viewing the solar atmosphere as a uniform and infinite medium to viewing it as a specific structure to be modeled led naturally to the discovery that the 5-minute oscillations could be explained as normal modes. The analysis showed that the oscillatory power should be confined to diagonal dispersion lines on the frequency-wave number diagnostic diagram (paper V.2). This diagonal nature of the modal dispersion relation immediately explained why the existence of preferred frequencies had never been observed before–no-one had isolated a fixed horizontal wavelength so that variations in the distribution of power along the diagonal lines showed up as apparently randomly varying oscillation frequencies.

An independent study by Leibacher and Stein (Paper 3.V) also proposed that the oscillations are a result of a resonance in a cavity below the solar surface. These latter authors did not include the effect of a finite horizontal wave number so that they needed to impose an arbitrary inner boundary condition in order to achieve trapping in the cavity. The prediction by Ulrich of diagonal dispersion lines was subsequently confirmed observationally by Deubner in Paper 4.V.

The identification of the oscillations as the surface manifestation of internal modes generated a number of theoretical studies that derived constraints on the sun's interior structure. Important examples of this kind of work are references [3] and [4]. Eventually this line of investigation led to the summary paper by Christensen-Dalsgaard, Gough and Thompson, which is included (first page only) here as Paper 9.V.

The above studies have utilized spatially-resolved observations of the solar surface to measure the oscillatory behavior. An alternate approach using the resonance cell technique originally proposed by Issak[5] investigates the sun as a spatially integrated source. The oscillations detected this way are globally coherent and must involve internal eigenfunctions which penetrate to the solar center. These integrated-sunlight observations are therefore valuable in deriving constraints which are relevant to the study of solar neutrinos.

The first detection of the 5-minute oscillations in integrated-sunlight was by Claverie et. al[6] and is included (first page only) here as Paper 5.V. This paper gives measures for the first time of the frequency difference between the $\ell = 0, n$ modes and the $\ell = 2, n-1$ modes. This difference is sensitive to the gradient in the hydrogen/helium abundance ratio at the solar center and consequently provides a measurement of the sun's nuclear age. Observations of the global oscillations can in principle determine the rotation rate of the solar interior. Unfortunately, the diurnal gaps in the data which are necessarily part of the observing sequence from a single low latitude station produce side-bands which interfere with the rotationally split components from solar rotation. Grec, Fossat and Pomerantz in the paper included (first page only) as 6.V showed that observations without diurnal gaps could be obtained from the geographic south pole and that such data is extremely valuable in providing unambiguous power spectra. The paper by Duvall et. al[7] included (first page only) as 7.V obtained the first extensive measurement of the sun's internal rotation profile. The observations used for their analysis were obtained from the south pole and utilized a spatially resolved system sensitive to the intensity in the Ca II K line.

The field of helioseismology is currently advancing at a rapid pace and the interested reader will need to refer to one of the extensive recent review articles to begin a study of the subject. Good introductory reviews are included (first page only) as Papers 8.V, 9.V, and 10.V.

References

[1] H. Lamb, Proc. London Math. Soc., 7, 122 (1909).

[2] E. N. Frazier, Ap. J., 152, 557 (1968).

[3] D. O. Gough, in The Energy Balance and Hydrodynamics of the Solar Chromosphere and Corona, eds. R.M. Bonnet and P. Delache, (Clairmont-Ferrard: G. de Bussac), p. 3 (1976).

[4] R. K. Ulrich, and E. J. Rhodes, Jr., Ap.J., 218, 521 (1977).

[5] G. Issak, Nature, 189, 373, (1961).

[6] A. Claverie, G. R. Isaak, C. P. McLeod, H. B. Van der Raay, and T. Roca Cortes, Nature, 282, 519 (1979).

[7] T. L. Duvall, Jr., W. Dziembowski, P. R. Goode, D. O. Gough, J. W. Harvey, and J. W. Leibacher, Nature, 310, 22 (1984).

V. HELIOSEISMOLOGY- Ulrich

1. "Velocity Fields in the Solar Atmosphere I. Preliminary Report," R.B. Leighton, R.W. Noyes, and G.W. Simon, *Ap. J.* **135**, 474-477, 488-490 (1962).

2. "The Five-Minute Oscillations on the Solar Surface," R.K. Ulrich, *Ap. J.* **162**, 993-1002 (1970).

3. "A New Description of the Solar Five-Minute Oscillation," J. Leibacher and R.F. Stein, *Astroph. Letters* **7**, 191-192 (1971).

4. "Observations of Low Wavenumber Nonradial Eigenmodes of the Sun," F-L Deubner, *Astro. Ap.* **44**, 371-375 (1975).

5. "Solar Structure from Global Studies of the Five-minute Oscillation," A. Claverie, G.R. Isaak, C.P. McLeod, H.B. Van der Raay, and T. Roca Cortes, *Nature* **282**, 591-593 (1979).

6. "Solar oscillations: Full disk observations from the geographic south pole," G. Grec, E. Fossat, and M. Pomerantz, *Nature* **288**, 541-544 (1980).

7. "Internal rotation of the Sun," T.L. Duvall, Jr., W. Dziembowski, P.R. Goode, D.O. Gough, J.W. Harvey, and J.W. Leibacher, *Nature* **310**, 22-25 (1984).

8. "Seismology of the Sun," J. Christensen-Dalsgaard, D.O. Gough, and J. Toomre, *Science* **229**, 923-931 (1985).

9. "The Depth of the Solar Convective Zone," J. Christensen-Dalsgaard, D.O. Gough, and M.J. Thompson, *Ap. J.***378**, 413 (1991).

10. "Advances in Helioseismology," K.G. Libbrecht and M.S. Woodard, *Science* **253**, 152-157 (1991).

VELOCITY FIELDS IN THE SOLAR ATMOSPHERE
I. PRELIMINARY REPORT*

ROBERT B. LEIGHTON, ROBERT W. NOYES, AND GEORGE W. SIMON
California Institute of Technology, Pasadena, California
Received October 16, 1961

ABSTRACT

Velocity fields in the solar atmosphere have been detected and measured by an adaptation of a technique previously used for measuring magnetic fields. Data obtained during the summers of 1960 and 1961 have been partially analyzed and yield the following principal results:

1. Large "cells" of horizontally moving material are distributed roughly uniformly over the entire solar surface. The motions within each cell suggest a (horizontal) outward flow from a source inside the cell. Typical diameters are 1.6×10^4 km; spacings between centers, 3×10^4 km ($\sim 5 \times 10^3$ cells over the solar surface); r.m.s. velocities of outflow, 0.5 km sec^{-1}; lifetimes, 10^4–10^5 sec. There is a similarity in appearance to the Ca^+ network. The appearance and properties of these cells suggest that they are a surface manifestation of a "supergranulation" pattern of convective currents which come from relatively great depths inside the sun.

2. A distinct correlation is observed between local brightness fluctuations and vertical velocities: bright elements tend to move upward, at the levels at which the lines Fe λ 6102 and Ca λ 6103 are formed. In the line Ca λ 6103, the correlation coefficient is $\sim$0.5. This correlation appears to reverse in sign in the height range spanned by the Doppler wings of the Na D_1 line and remains reversed at levels up to that of Ca^+ λ 8542. At the level of Ca λ 6103, an estimate of the mechanical energy transport yields the rather large value 2 W cm^{-2}.

3. The characteristic "cell size" of the vertical velocities appears to increase with height from $\sim$1700 km at the level of Fe λ 6102 to $\sim$3500 km at that of Na λ 5896. The r.m.s. vertical velocity of $\sim$0.4 km sec^{-1} appears nearly constant over this height range.

4. The vertical velocities exhibit a striking repetitive time correlation, with a period $T = 296 \pm 3$ sec. This quasi-sinusoidal motion has been followed for three full periods in the line Ca λ 6103, and is also clearly present in Fe λ 6102, Na λ 5896, and other lines. The energy contained in this oscillatory motion is about 160 J cm^{-2}; the "losses" can apparently be compensated for by the energy transport (2).

5. A similar repetitive time correlation, with nearly the same period, seems to be present in the *brightness fluctuations* observed on ordinary spectroheliograms taken at the center of the Na D_1 line. We believe that we are observing the transformation of potential energy into wave energy through the brightness-velocity correlation in the photosphere, the upward propagation of this energy by waves of rather well-defined frequency, and its dissipation into heat in the lower chromosphere.

6. Doppler velocities have been observed at various heights in the upper chromosphere by means of the Hα line. At great heights one finds a granular structure with a mean size of about 3600 km, but at lower levels one finds predominantly *downward* motions, which are concentrated in "tunnels" which presumably follow magnetic lines of force and are geometrically related to the Ca^+ network. The Doppler field changes its appearance *very rapidly* at higher levels, typical lifetimes being about 30 *seconds*.

In an earlier paper, Leighton (1959) described a method for obtaining "spectroheliograms" whose density variation indicated the presence of Zeeman splitting due to longitudinal magnetic fields on the sun rather than actual light-intensity variations. This method is readily applicable to measurements of Doppler shifts as well as Zeeman splitting, giving a "spectroheliogram" of the line-of-sight velocity field in the region of line formation. The present paper describes some of the velocity observations made in this way during the summers of 1960 and 1961. Some of the results have been reported by Leighton (1960, 1961).

I. OBSERVATIONAL TECHNIQUES

The specific details of the method were described in the above-mentioned paper, so we shall give here only a brief description of how it has been adapted to the measurement of Doppler shifts.

* Supported in part by the joint program of the Office of Naval Research and the U.S. Atomic Energy Commission.

To observe velocity fields, one omits the quarter-wave plate and polaroids which are used in observing magnetic fields and utilizes instead a "line-shifter," comprising two glass blocks situated just below the exit slit, one for each of the two images formed by the beam splitter (Fig. 1, *a*). These blocks may be tilted equal amounts in opposite directions about an axis parallel to the slit, thus shifting the spectrum at the exit slit equal amounts in opposite directions. With the blocks initially having no tilt, the slit is precisely centered on a spectral line having a symmetrical profile. The blocks are then tilted a prescribed amount, so that the wavelength of light passed by the slit corresponds to a position on the red wing of the line for one image and an equidistant position on the violet wing for the other image (Fig. 1, *b*). A Doppler shift of the line then moves the line core *toward* the slit for one image and *away* from the slit for the other, thereby producing opposite changes in the amount of light passed by the slit to the photographic plate.

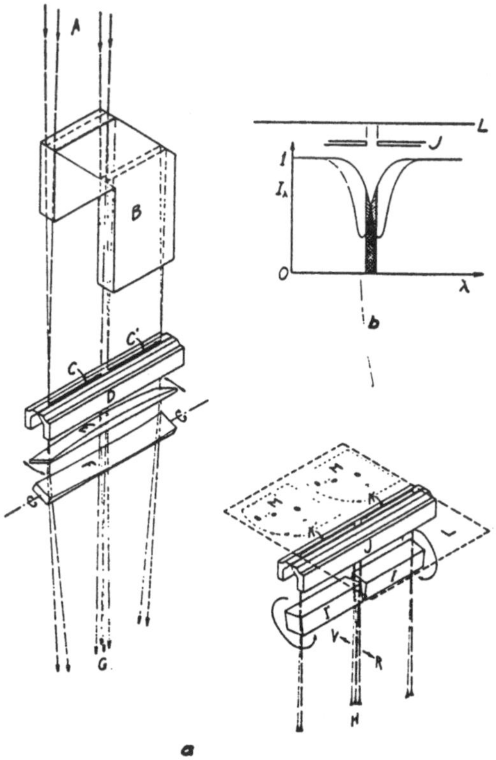

FIG. 1.—(*a*) Schematic diagram of the optical system. *A:* vertical light-beam of the 65-foot tower telescope. *B:* optical glass beam splitter (moves with spectroheliograph). *C,C'*: duplicate images. *D:* entrance slit (19.3 cm long). *E:* double-convex lens used to adjust the "tilt" of the spectral lines. *F:* double-concave lens used to adjust the "curvature" of the spectral lines. (*E* and *F* also act as a field lens for the spectrograph.) *G:* "split" light-beam proceeding toward the collimator lens. *H:* dispersed light-beam proceeding from the camera lens (*V* and *R* designate the violet and red directions along the spectrum). *I,I'*: plane-parallel glass blocks, whose equal and opposite tilt serves to "shift" the spectra of the two images *C,C'* equally in opposite directions. *J:* exit slit. *K,K'*: monochromatic image "strips" passed by exit slit. *L:* photographic plate. *M,M'*: latent images on photographic plate, built up of successive "strips" by the motion of the spectroheliograph. (*b*) Diagram of the action of the line shifter, showing the displaced line profiles. The crosshatched sections represent the portion of the line profiles admitted to the photographic plate *L* by the exit slit *J*.

The resulting two images are "canceled" with each other photographically by the same method as was previously used: a unity-gamma contact transparency is made of one of the images such that, when placed on its own negative, a uniform gray field results. When placed on the other negative, however, the Doppler shift contribution to the image is enhanced, while density variations due to variations of intrinsic brightness on the sun "cancel" to a uniform gray. Thus a "spectroheliogram" is produced in which darker-than-average areas represent velocities of approach, and lighter-than-average areas, velocities of recession or vice versa. The variations of transmission of this "Doppler plate" are, over a significant range, proportional to the velocities on the sun; the proportionality constant is determined by the calibration method described in the previous paper or, preferably, by introducing an accurately known wavelength shift by moving the exit slit by a small, precisely defined amount at some time during the exposure (Fig. 4).

In practice, several pairs of images are usually taken in succession. By photographically subtracting, in the same way as above, two *singly canceled* plates (each representing the Doppler field from a single pair of images), a "double cancellation," or *Doppler difference* plate, is obtained. If the two original image pairs were exposed under identical conditions and *if the velocity field on the sun were the same for both pairs*, the two singly canceled Doppler plates would be identical, and their Doppler difference would be an uninteresting uniform gray. However, a time interval of a few minutes elapses between exposures, due to the finite scanning speed of the spectroheliograph, during which appreciable changes in the velocity field occur. A second cancellation results in signals which sufficiently exceed the "noise" due to imperfect registration, guiding errors, and seeing fluctuations to provide a measure of *accelerations* in the solar atmosphere. Some of the most interesting results so far obtained involve changes in the photospheric velocity field which occur during the traversal time of the spectroheliograph.

If, between taking two successive pairs of images, the glass blocks are reversed in their tilt, the sign of the Doppler contribution is reversed on each image, and the "polarity" of the Doppler signal is reversed on the singly canceled plate. Then, when a second subtraction is made between two such Doppler plates of opposite polarity, the Doppler signals *add*. This *Doppler sum* would produce a fourfold enhancement of the Doppler contribution to each original image, while other signals would cancel to zero, if the velocity field did not change during the time interval between exposures.

A typical observational procedure might be as follows: the glass blocks of the line shifter are initially tilted so that the two halves of the plate receive light from opposite wings of a spectral line. At $t = 0$, the spectroheliograph is started moving south across the 17-cm solar image. (The scan direction of the instrument is roughly parallel to the axis of rotation of the sun.) It finishes an 11-cm scan 3 or 4 minutes later. Immediately the spectroheliograph is stopped, the plate changed, the line shifter either reversed or left unchanged, depending on the purpose of the observation, and the spectroheliograph again started, moving north. Several seconds are required to make these changes. The spectroheliograph finishes its traverse 3 or 4 minutes later, 6 or 8 minutes after beginning the sequence.

Each of the two pairs of images is canceled to form a Doppler plate, and the two resulting plates are again canceled to form either a Doppler difference or a Doppler sum plate, depending, respectively, on whether the shifter was untouched or reversed between exposures.

Such a Doppler difference (or sum) plate has the characteristic that the time elapsed between recording the two velocities which are subtracted (or added) varies linearly from nearly zero at one end to 6 or 8 minutes at the other end. The effects of still longer time delays may be observed by waiting a certain time between taking the two exposures. Doppler difference or sum plates with *constant* time differences over the entire image may be obtained by subtracting two pairs of images scanned in the *same direction* at the same traverse speed. The application of all these techniques will be described later with the presentation of data obtained by them.

Many sequences of such observations were obtained at the 65-foot solar tower telescope of the Mount Wilson Observatory during the summers of 1960 and 1961. For most of these observations a 17-cm diameter image, formed by the 30-cm objective, was used. In a few instances, however, the entire solar disk was recorded, using a 5-cm image. The field of view covered by a single image is about 9 by 11 cm (16′ by 20′ on the 17-cm image). The spectral lines used were Fe 6102.4, Ca 6102.8, Na 5896, Ba^+ 4554, Mg 5173, Ca^+ 8542, Mg 5528, Hα, and sometimes others. Both slits were ordinarily set at 0.05 or 0.07 mm width, corresponding to angular resolution of about 0″.6–0″.8 on the 17-cm image,[1] and a wavelength window of 0.07–0.10 A. The line shifter was usually set so that this window was centered about 0.10 A either side of the line center, except that, for Hα, considerably greater offsets were used. Precise setting of the second slit on the unshifted line is required to insure that the two images are taken at wavelengths equidistant from the line center and thus represent the same physical height in the atmosphere. The spectral line is rendered straight and parallel to the slit by means of certain optical elements which serve to introduce adjustable curvature and tilt into the lines (Fig. 1) and is photometrically centered on the slit to within 0.01 A along the entire slit length.

II. MEASUREMENT PROCEDURE

In order to extract reliable statistical information from a Doppler plate without undue amounts of calculation, a device (Fig. 2) was built which carries out the operations involved in evaluating auto-correlation (A-C) and cross-correlation (C-C) functions over a two-dimensional field. The two-dimensional A-C function $C(s, t)$ is defined by

$$C(s,t) \equiv \frac{K}{A}\int\int T(x,y)T(x+s,y+t)\,dA \equiv K\langle T(x,y)T(x+s,y+t)\rangle,$$

where A is the area, $T(x, y)$ is the transmission of the plate at the point (x, y), K is a normalization constant, and the integration area is made sufficiently great that fluctuations due to the boundaries are negligible. To obtain the A-C function over a given area A, two copies of the plate are made, in a right- and left-handed pair, so that they may be placed in register with their emulsions in contact, On one plate, the entire area except for the area of interest is masked off. The plates are fixed to separate frames, which are placed in a holder in such a way that their emulsions are in contact and in register. A motor drive slides one plate slowly past the other. Collimated light is passed through the two plates and is brought to a focus on the photocathode of a photomultiplier tube. When the plates are displaced an amount s in the x-direction and t in the y-direction, relative to each other, the photomultiplier records $C(s, t)$. In practice, t is usually zero. The signal $C(s, 0)$ from the photomultiplier is amplified and fed into a 0–10-mv chart recorder. Since the fluctuation of $C(s, 0)$ seldom exceeds 10 per cent, a 10× multiplier resistor is inserted in the circuit, along with a 90-mv bucking voltage. Thus the chart recorder's range is between 90 and 100 mv, and the full scale of the chart represents a 10 per cent change in the signal. Multiplication by other factors with appropriate bucking voltage is also sometimes used.

Two easily measured parameters of an A-C function yield relatively direct information about the field $T(x, y)$ from which the function was obtained. The full width at half-maximum (FWHM), when converted into units of km on the sun, provides a measure of the *linear size* of the elements in the field; the normalized height of the central

[1] Because of smearing effects due to the atmosphere and the spectrograph, the actual resolution attained on the 17-cm image was about 2″ under good conditions in 1960 and about 1″ in 1961. Only plates taken under very good seeing conditions were utilized for the measurements reported here.

Farther from a sunspot but still within the associated active area, the pattern of velocities has a somewhat filamentary appearance but is relatively quiescent and diffuse and changes significantly only over many minutes of time. At places well removed from active areas, on the contrary, the velocity field is of large amplitude, shows quite fine, non-filamentary structure, *and changes significantly within 20–40 sec* (Fig. 13).

We have photographed the Hα Doppler field at various offsets $\Delta\lambda$ from the line center, in order to study the changes in the properties of the velocity field with height. The most interesting changes are found within the range $0.35 \text{ A} < \Delta\lambda < 0.8 \text{ A}$ (Fig. 10, *c*, *d*). At high elevations ($\Delta\lambda \sim 0.3$ A), one sees a rather uniform, fine-grained pattern of both upward and downward velocities, with a characteristic grain size of $\sim 3.6 \times 10^3$ km, r.m.s. velocity amplitude ~ 2 km sec^{-1}, and lifetime ~ 30 sec. As one goes deeper into the atmosphere, this pattern gradually changes into a totally different one at $\Delta\lambda \sim 0.8$ A. Here, most of the disk is *quiescent*, and the only sizable velocities present are confined to a network of narrow "tunnels," through which material streams predominantly *downward* into the sun (Fig. 10, *d*). This network appears to correspond closely with the network seen on K_2 spectroheliograms.

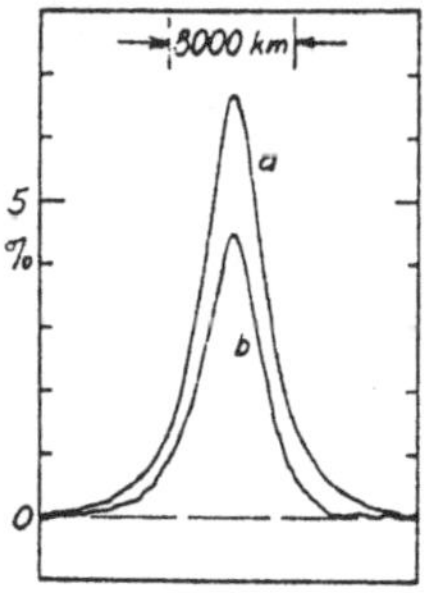

FIG. 12.—A-C curves of Na 5896 Doppler field inside and outside plage areas. *a*, Outside plage; *b* inside plage. Plate Na 5896, July 3, 1960.

When they are analyzed in greater detail, the Hα Doppler plates should yield many interesting and important results concerning mass motions in the Hα chromosphere. Such an analysis will require somewhat more precise velocity calibration than we have so far achieved, however. This matter is currently under study, and further work on the Balmer series lines is planned for the near future.

d) Oscillatory Motions in the Solar Atmosphere

We have used the "Doppler difference" procedure described in Section I to detect *changes* in the velocity field which take place over time intervals of a few minutes. On the basis of the few-minute "lifetime" of the solar granulation and because of the supposed "turbulent" nature of the velocities, it was expected that the appearance of such a velocity-difference plate would resemble that for Hα shown in Figure 13, changing gradually from a uniform gray at one end of the plate (where $\Delta t = 0$) to a granular field characteristic of the superposition of the two (now uncorrelated) velocity fields at the other end, if the total elapsed time were many minutes.

Actually, a quite different and rather striking behavior is observed, which leads us to conclude that *the local vertical velocities in the solar atmosphere are not random, but quasi-oscillatory, in their time variation:* the "signal" on a Doppler difference plate grows from a uniform gray at the end where $\Delta t = 0$, passes through a ***maximum*** of "contrast" and "grain-size" at a time $\Delta t \sim 150$ sec, diminishes to a ***minimum*** contrast and grain size at $\Delta t \sim 300$ sec, and so on. This behavior is a consistent feature of Doppler difference

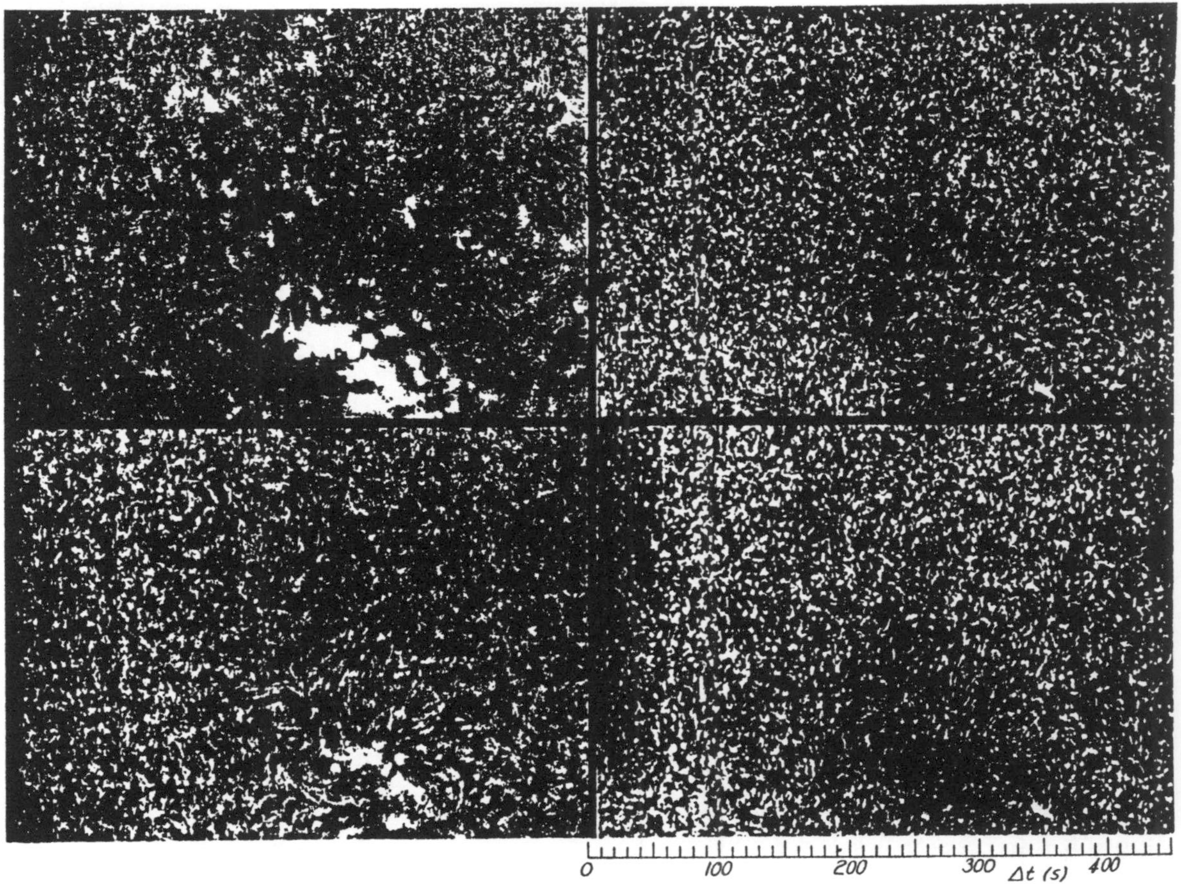

FIG. 13.—*Upper left:* original Hα plate, at line core. *Lower left:* original Hα plate of same area, $\Delta\lambda$ - 0.35 Å. *Upper right:* Hα Doppler field of same area, $\Delta\lambda$ = 0.35 Å. *Lower right:* Hα Doppler difference field, showing the rapidity with which the Hα velocity field changes: the velocity loses its correlation in time in 20-40 seconds. June 21, 1961, 11^h16^m U.T.

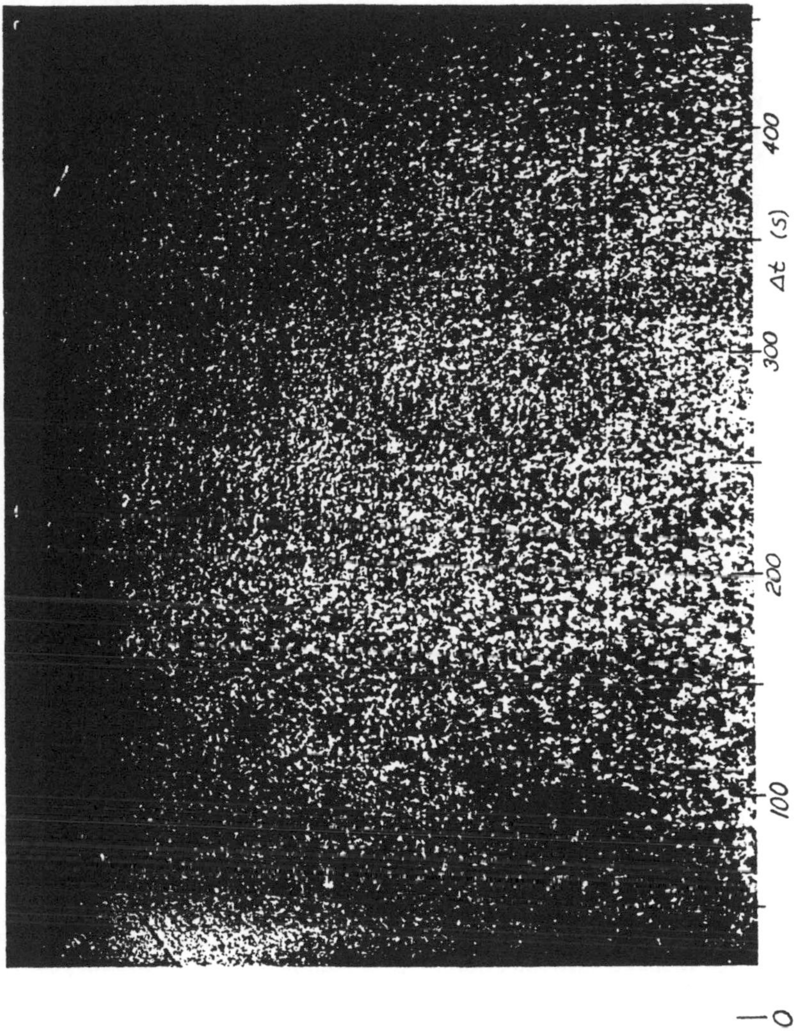

FIG. 14. Doppler difference plate, showing the oscillatory time correlation of the small scale velocity field. Ca 6103. June 10, 1960, 13^h40^m U.T.

plates taken in any of our lines except Hα and Ca^+ 8542 but is especially evident in Ca 6103, Ba^+ 4554, and Na 5896 plates. Visual estimates of the period of this oscillation on twenty different Ca 6103 plates taken in 1960 give $T = 296 \pm 3$ sec. The stated uncertainty is indicative of the internal consistency of our eye-estimates and does not include possible systematic effects involved in judging where the "minimum" of the contrast pattern falls.

Figure 14 shows a Doppler difference plate for the Ca 6103 line; Figure 15, curve *D*, shows a plot of the central-peak heights of A-C functions corresponding to various areas on this plate.[6] Each A-C function was evaluated for a long, narrow area whose long axis was parallel to the slit, so that Δt was nearly constant over each area of integration. The mean-square fluctuation is seen to be a minimum at $\Delta t = 0$ and to have a secondary minimum near $\Delta t = 300$ sec.

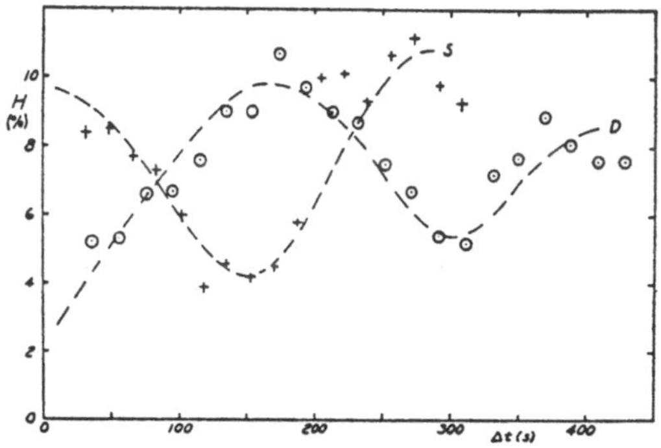

Fig. 15.—Plot of the peak heights of A-C curves (evaluated for long, narrow strips parallel to the slit) versus Δt for the Doppler difference plate of Fig. 14 (curve *D*) and the Doppler sum plate of Fig. 16 (curve *S*).

The features just described are sometimes difficult to see in Doppler difference plates such as Figure 14. Some of the reasons for this are as follows: In spite of the efforts of the photoelectric guider to keep the image stationary to within about 1″, some motions do occur which are too fast to be followed. Over-all *distortions* of the image are common as well, so that even if the south and west limbs were held *stationary* by the guider, other points on the disk would move about. Thus, when two singly canceled Doppler plates are themselves canceled to make a Doppler difference or sum plate, corresponding points on the two *plates* are not necessarily the same point on the *sun*. In the actual process of placing one plate upon its partner, it is often possible to improve the registration in one area, but at the expense of losing register in another. Thus the observed oscillatory time

[6] The meaning of the central-peak height of an A-C function for a Doppler difference plate may be seen as follows: At each point (x, y) of such a plate, the transmission fluctuation is proportional to $v(x, y, t) - v(x, y, t + \Delta t)$, where t and $(t + \Delta t)$ are the two times at which the slit scanned past that point on the two corresponding exposures. Thus the height of an A-C function evaluated for a long, narrow strip parallel to the slit will be

$$H(\Delta t) \propto \langle [v(x,y,t) - v(x,y,t+\Delta t)]^2 \rangle$$

$$= \langle [v(x,y,t)]^2 + [v(x,y,t+\Delta t)]^2 - 2\,v(x,y,t)\,v(x,y,t+\Delta t) \rangle$$

$$= 2\langle v^2 \rangle - 2\langle v(t)\,v(t+\Delta t) \rangle .$$

Thus $H(\Delta t)$ provides a measure of the *time correlation* of the local velocity, averaged over the corresponding spatial area, since $v(x, y, t)$ is, statistically, spatially homogeneous and stationary in time.

correlation is much more evident under laboratory conditions, wherein the effects of small relative motions of the plates may be seen at first hand. It is invariably true that, at this stage, suitable positions of the two plates can be found which ***minimize*** the difference signal at the places where $\Delta t = 0$ and $\Delta t \sim 300$ sec., but that ***no position exists*** which correspondingly minimizes the difference signal at $\Delta t \sim 150$ sec. A Doppler difference plate such as that of Figure 14 necessarily suffers from the fact that we must choose some *one* relative position in which the two plates match "best."

Interestingly enough, the oscillatory time correlation may also be seen on a Doppler *sum* plate, if the center of the solar disk lies near the strip where $\Delta t = 150$ sec. Figure 16 shows a Ca 6103 Doppler sum plate, and curve S of Figure 15 shows the corresponding plot of central-peak heights versus Δt for narrow strips whose long axis is parallel to the slit. A pronounced *minimum* of contrast is observed near $\Delta t \sim 150$ sec. This corresponds to a strong *anticorrelation* of velocities after such a time interval.

This anticorrelation on a Doppler sum plate is of importance in two respects. First, the fact that it is possible in this way *strongly to minimize* the sum of two velocity fields proves conclusively that the effect is *on the sun* and not in the earth's atmosphere, the instrument, or the photographic procedures. And, second, we learn that the oscillatory motion is not merely a small effect superimposed upon other, more energetic, velocity fields but is, instead, essentially *the only vertical motion present* in the range of linear dimensions resolved on our plates.

By observing the time correlation of velocity using spectral lines formed at different elevations, we may hope to study the properties of the oscillatory motions as a function of height. As mentioned previously, we have found the oscillation to be present in all lines so far studied except Hα and Ca^+ 8542. Figure 17 shows Doppler difference and sum fields for the line Ba^+ 4554, in which the oscillatory motions are especially marked.

The presence of the oscillatory motions at the rather high elevations represented by the lines Na 5896 and Mg 5173 suggests that the oscillation might manifest itself in the chromosphere in other ways than through the Doppler effect. Accordingly, we have sought and apparently found a corresponding periodicity *in the small-scale bright elements characteristic of the lower chromosphere.* Figure 18 (*top*) shows an ordinary spectroheliogram taken in the center of the D_1 line Na 5896, and Figure 18 (*bottom*) shows a "Brightness difference" plate obtained by photographic subtraction of two such plates scanned in opposite directions in quick succession. This plate and the corresponding A-C function peak-height curve of Figure 19 exhibit the same characteristic maxima and minima of contrast, with about the same period, as are seen in the Doppler difference plates of Figures 14 and 17. If the effect seen on this single plate is confirmed by further observation and if the intensity fluctuations seen at the center of the mean profile are not significantly contributed to by the Doppler effect, then it would appear that the oscillatory motions may play a significant role, and perhaps a dominant one, in the transport of energy from the granulating layer into the chromosphere.

Further observation of the periodic time changes in these and other spectral lines are under way to determine whether there may be a dependence of period upon altitude in the atmosphere or upon horizontal wavelength, and to elucidate further the physical processes involved.

The velocity amplitude of the oscillation, as measured from A-C curves, appears to be greatest near the center of the disk and to fall off near the limb, which suggests that the associated motions are primary *vertical.*

In order to measure the *average duration* of a particular oscillation before it dies out, Doppler difference plates were obtained for which the time delay was held *constant* over the entire area at one of the values $\Delta t = T/2, T, 3T/2, \ldots$, where $T = 296$ sec. This was done by scanning the spectroheliograph in the same direction at two times separated by the required interval, as explained earlier. Thus A-C curves for these plates may utilize nearly the whole area of the plate instead of only a small area, so that much im-

THE FIVE-MINUTE OSCILLATIONS ON THE SOLAR SURFACE*

ROGER K. ULRICH
University of California, Los Angeles
Received 1970 March 5; revised 1970 June 3

ABSTRACT

The acoustic properties of the subphotospheric layers are examined. It is shown that standing acoustic waves may be trapped in a layer below the photosphere. These standing waves may exist only along discrete lines in the diagnostic diagram of horizontal wavenumber versus frequency. The positions of these lines are derived from a modal analysis of the solar envelope. The lines for the fundamental mode and the first-overtone mode pass through the centers of the two peaks observed by Frazier. An examination of the energy balance of the oscillations shows that they are overstable. When they are assigned an amplitude of 0.2 km sec^{-1}, they generate about (5–7) $\times 10^6$ ergs cm^{-2} sec^{-1}. This power output suggests that the dissipation of the 5-minute oscillations above the temperature minimum is responsible for heating the chromosphere and corona.

I. INTRODUCTION

The 5-minute oscillations in the solar photosphere have been studied intensively since their discovery by Leighton and his co-workers (Leighton, Noyes, and Simon 1962; Noyes and Leighton 1963). Athay (1966) has suggested that these oscillations are involved in the process which heats the chromosphere and corona. Unfortunately, the mechanism which generates the oscillatory motion has not been well understood. In particular, the power spectrum of acoustic energy predicted by a theory of generation must be compatible with the observed spectrum. The spectrum derived by Stein (1968) from Lighthill's (1952) turbulence-generation mechanism has a peak of power near periods of 30–60 sec and falls very steeply for periods different from these values. In contrast, most of the power is observed near 300 sec (Leighton *et al.* 1962; Tanenbaum *et al.* 1969). This paper describes a process which may generate the observed 300-sec oscillations and which is essentially different from Lighthill's mechanism.

Moore and Spiegel (1966) pointed out that acoustic waves are overstable in the presence of a superadiabatic temperature gradient and radiative exchange of energy. This paper shows that the 5-minute oscillations are trapped standing acoustic waves, and gives eigenfrequencies for the fundamental and first three overtone modes. Examination of the energy balance of these modes shows that the first three overtones are overstable. In addition, the dispersion relation between frequency and horizontal wavelength may explain the apparently random location of peaks in the power spectrum observed by Howard (1967), Frazier (1968), and Gonczi and Roddier (1969). Tanenbaum *et al.* (1969) showed that the amplitude of the oscillations increases as the wavelength decreases. Consequently, it is reasonable to expect the shortest observable wavelength to dominate the power spectrum. The random locations of the power-spectrum peaks found by Howard (1967) may be interpreted as a result of random variations in the quality of the seeing.

The vertical wavelength of the oscillations is comparable to the horizontal wavelength and is roughly 1000–5000 km. As a result, a correct treatment of the problem must necessarily involve a substantial region below the photosphere. Also, the temperature and rate of radiative cooling change substantially over a distance of 1000 km. Clearly the approximations that the gas is perfect and the atmosphere is isothermal may not be

* Supported in part by the National Science Foundation [GP-9433, GP-9114] and the Office of Naval Research [Nonr-220(47)].

used. Indeed, the presence of a superadiabatic temperature gradient is critically important in the discussion of the energy balance. Another consequence of the long vertical wavelength is that the concepts of ray acoustics are useful only as general guides. Finally, although there may be some coupling between the oscillations and the convective motion, this coupling cannot be dominant since Gonczi and Roddier (1969) have found that the oscillations remain in phase for a least 1 hour whereas typical convective cells live 7–8 minutes. This lifetime of 1 hour or more differs considerably from the usually quoted lifetime of 8–10 minutes. This shorter lifetime has prompted many workers to postulate that the oscillations are generated directly by the motions of convective cells. The observations by Gonzci and Roddier show that the shorter lifetime should be interpreted as a beat period rather than a true lifetime. Such a beat period requires there to be two or more natural frequencies for the motion. We shall show in the next two sections that such multiple frequencies are the result of the variable acoustic properties below the surface of the Sun.

II. TRAPPED WAVES

A simplified form of the local dispersion relation is that given by Whitaker (1963). We let k_z, k_h, c, ω_0, and N be the vertical and horizontal wavenumbers, the adiabatic sound velocity, long acoustic cutoff frequency, and the Väisälä-Brunt gravity-wave frequency. Whitaker's dispersion relation may then be written

$$k_z^2 = \frac{\omega^2 - \omega_0^2}{c^2} - k_h^2\left(1 - \frac{N^2}{\omega^2}\right). \tag{1}$$

In a nonisothermal atmosphere ω_0 and N are given by

$$\omega_0 = \frac{c}{2H} \quad \text{and} \quad N^2 = -\frac{g}{\rho}\left(\frac{\partial\rho}{\partial S}\right)_P \frac{dS}{dz}, \tag{2}$$

where H, g, ρ, S, and z are the *density* scale height, acceleration of gravity, density, entropy, and altitude. In a stratified atmosphere ω_0, N, and c are functions of altitude so that k_z does not remain constant. Since we are ignoring the coupling with convective motions, the atmosphere is uniform in time and on horizontal planes. Consequently, a mode of oscillation may be assigned a value for ω and k_h. For such a mode, altitudes where $k_z = 0$ are boundaries between regions where waves propagate and where they are attenuated. Ray acoustics suggests that a wave packet moving from the propagating region toward the attenuating region will be reflected from the boundary surface. Although equation (1) does not include the effect of radiative exchange of energy, it shows that for some values of ω and k_h there are two altitudes where $k_z = 0$. Near the photosphere ω_0 increases by about a factor of 1.6 due to the increase in mean molecular weight. For frequencies between the minimum and maximum of ω_0 there is a reflecting surface in the photosphere. As long as this condition on ω is satisfied, it is always possible to choose k_h small enough that $k_z > 0$ below the photosphere. For finite values of k_h a second reflecting layer is present well below the photosphere as a result of the increase in sound velocity. For particular values of ω and k_h between these limits, trapped standing oscillations may be set up. The fact that the oscillations are observed in a region where they are not permitted does not constitute a problem since the decay distance for the energy density $\frac{1}{2}\rho v^2$ is quite long and in fact the velocity amplitude increases with altitude.

In the presence of radiative exchange of energy, k_z is complex rather than purely real or purely imaginary. For this case the analogue of equation (1) with $k_z = 0$ was obtained by Souffrin (1966) by equating the real and imaginary parts of k_z. In a nonisothermal atmosphere we may obtain a similar formula if we neglect the difference between the pressure and density scale heights and the gradients of c^2, ω_R, and N^2. The resulting formula for the critical horizontal wavenumber is

$$k_h = \frac{\omega^2}{c^2}\frac{\omega^2 + \gamma\omega_R^2 - \omega_0^2(1 + \omega_R^2/\omega^2)}{\omega^2 + \omega_R^2 - N^2}, \tag{3}$$

where γ is the ratio of specific heats,

$$\omega_R = \frac{16\sigma T^3\kappa}{C_P}(1 - \tau_e \cot^{-1}\tau_e), \tag{4}$$

and C_P, κ, and τ_e are, respectively, the heat capacity at constant pressure, the opacity, and the effective optical thickness of the perturbation. The formula for ω_R was derived by Spiegel (1957) for an infinite uniform medium. For this uniform case Spiegel found $\tau_e = \rho\kappa/k$ for a perturbation whose total wavenumber is k. Near the photosphere, however, the optical thickness of a perturbation cannot be greater than the optical depth of the point. The effective vertical wavenumber for any perturbation cannot be less than $\rho\kappa/\tau$. The actual vertical wavenumber for the perturbation is considerably less than the lower limit, so a good estimate for τ_e is

$$\tau_e = \left[\left(\frac{k_h}{\rho\kappa}\right)^2 + \frac{1}{\tau^2}\right]^{-1/2}. \tag{5}$$

While the general characteristics discussed above must be true for any model of the solar envelope, the particular relation between k_h and ω must depend on the structure of the photosphere. The present analysis was performed on a model calculated with the nonlocal mixing-length theory described by Ulrich (1970). The important properties of this model are given in Table 1.[1] The entropy of the adiabatic region may not be correct in this model, but the general nature of the relation between k_h and ω is independent of this uncertainty. In fact, we may hope that improved observations of the 5-minute oscillations will allow an accurate determination of the envelope adiabat.

Figure 1 shows the altitude of the reflecting layer as a function of the horizontal wavelength for several periods of oscillation. Altitude zero is at optical depth unity. The shortest-period oscillation which is reflected in the photosphere is about 180 sec. Oscillations with shorter periods will penetrate to the chromosphere before reflection and have not been studied. They may be reflected at that level by the increase in sound velocity; however, the relevant physical phenomena may be quite different. The cutoff frequency is the value of ω_0 at the temperature minimum. In fact, the power observed by Tanenbaum *et al.* (1969) drops sharply as the cutoff period is approached. Evidently, reflection above the temperature minimum is much less effective than reflection from the photosphere.

The principal conclusions of this section are that the 5-minute oscillations are acoustic waves trapped below the solar photosphere and that power in the (k_h, ω)-diagram should be observed only along discrete lines.

III. MODAL ANALYSIS

The arbitrary variation of the atmospheric parameters requires the calculation to be performed numerically. Some simplification of the momentum and continuity equations results from using the mass flux

$$j = \rho v \tag{6}$$

[1] The presence of negative values for ω_0 in this table is due to a density inversion just below the photosphere. This density inversion is unavoidable in models computed with a nonlocal convective theory. An extensive discussion of this problem is not appropriate here since the density inversion plays only a minor role in determining the acoustic properties of the atmosphere.

TABLE 1

IMPORTANT PROPERTIES OF THE AVERAGE MODEL

$-z$ (km)	T (° K)	c (km sec^{-1})	$\omega_0 \times 10^2$ (sec^{-1})	$\omega_g \times 10^2$ (sec^{-1})	$N^2 \times 10^4$ (sec^{-2})
239	4640	7.23	+3.23	0.33	+10.1
387	4680	7.23	+3.17	1.10	+ 9.5
474	4820	7.30	+3.00	2.11	+ 8.7
570	5130	7.56	+2.71	3.57	+ 6.2
685	5940	8.16	+2.40	6.36	+ 4.5
719	6330	8.41	+1.78	9.14	+ 2.3
750	6830	8.70	+1.13	7.61	− 3.9
793	7960	9.12	−1.76	4.47	−18.6
821	9280	9.39	−3.69	1.94	−30.4
837	10000	9.61	−1.67	0.72	−15.0
863	10700	9.92	+0.08	0.20	− 8.0
908	11300	10.3	+0.77	0.04	− 3.1
1040	12300	10.9	+1.06	0.00	− 1.0
1220	13200	11.6	+1.10	0.00	− 0.47
1410	14100	12.2	+1.07	0.00	− 0.27
2730	19300	15.8	+0.87	0.00	− 0.03
4640	27600	20.9	+0.67	0.00	0.00
7770	45000	29.6	+0.48	0.00	0.00
17900	119000	50.8	+0.29	0.00	0.00
49800	755000	95.2	+0.17	0.00	0.00

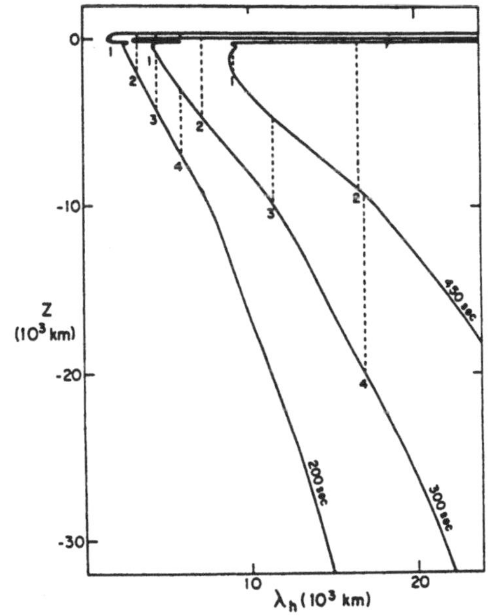

FIG. 1.—Altitude z of the reflecting layer versus horizontal wave length λ_h for three periods of oscillation. Permitted oscillations are found to the right of each line. Dashed lines indicate positions of the eigensolutions found in § III. These solutions are labeled by their modal numbers defined as the number of nodes between the reflecting layers plus one.

in place of the velocity. In terms of this variable, only the coefficients in the energy equation change significantly with depth. The second-order term dropped, $\nabla\cdot(\rho v v)$, is somewhat different from the usual term dropped, $\rho(v\cdot\nabla)v$; however, in a linearized analysis the results should not be affected. We denote fluctuations in the thermodynamic variables with primes, and subtract the hydrostatic equation from the momentum equation. The linearized equations of motion are

$$\frac{\partial j}{\partial t} = -\nabla P' + g\rho' , \tag{7}$$

$$\frac{\partial \rho'}{\partial t} = -\nabla\cdot j , \tag{8}$$

$$\frac{T\partial S'}{\partial t} = -\frac{T}{\rho} j_z \frac{dS}{dz} - C_P\omega_R T' , \tag{9}$$

where $g = (0, 0, -g)$. We may express S' and T' in terms of P' and ρ' by use of the equation of state.

The modal equations are obtained by setting $\partial/\partial t = i\omega$ and $\partial^2/\partial x^2 + \partial^2/\partial y = -k_h^2$. The z-dependence of all quantities may then be obtained from a numerical integration. A common procedure in problems such as this is to allow ω to be complex. The imaginary part of ω then indicates whether the mode grows or decays. Unfortunately, this method cannot be used at present because of the difficulties with the outer boundary condition discussed below. Consequently, these calculations have been done with real ω, and the growth or decay of the oscillations has been determined from energy balance in § IV.

The main result of the arbitrary temperature stratification is that $\partial/\partial z$ may not be replaced by $ik_z - 1/(2H)$. We may eliminate the horizontal component of j by dividing equation (7) by $i\omega$ and multiplying the x- and y-components by $\partial/\partial x$ and $\partial/\partial y$, respectively. After the results are used in equation (8), we obtain

$$\frac{\partial P'}{\partial z} = -i\omega j_z - g\rho' , \quad \frac{\partial j_z}{\partial z} = -i\omega\rho' - \frac{k_h^2 P'}{i\omega} , \tag{10}$$

$$\rho' - \frac{P'}{c^2} = \frac{N^2 j_z}{i\omega g} - \frac{\omega_R}{i\omega}\left(\rho' - \gamma\frac{P'}{c^2}\right) . \tag{11}$$

These equations give the altitude dependence of the complex amplitude of the oscillations. They must be supplemented by boundary conditions. The interior boundary condition is given by equation (1). In that equation k_z gives the z-dependence of $j_z/\rho^{1/2}$, and we may set $N^2 \sim 0$. The boundary condition is then

$$\frac{1}{j_z}\frac{dj_z}{dz} = \left(\frac{\omega_0^2 - \omega^2}{c^2} + k_h^2\right)^{1/2} - \frac{\omega_0}{c} . \tag{12}$$

The outer boundary condition is less simple. For frequencies less than the maximum value of ω_0 at the temperature minimum there is a layer about the temperature minimum through which the amplitude decreases. The simultaneous decrease in density maintains the velocity amplitude roughly constant. Beyond the temperature minimum at some point the oscillations are again permitted. Since the thickness of the attenuating layer is not great, the two propagating regions are coupled together. Ultimately, the temperature rise in the chromosphere will cause the oscillations to be attenuated. For modes whose horizontal wavelength is roughly 10^4 km, this second reflecting layer is located where the temperature is roughly 10^5 ° K. For most chromosphere models the upper permitted region is about 2000 km thick. Since the density scale height is about 150 km through this region, the velocity amplitude must increase by about a factor of exp

(2000/300) ≈ 1000. The observed velocity amplitude of 0.3 km sec^{-1} in the photosphere would be increased to 300 km sec^{-1}. Clearly some dissipation mechanism must operate to keep the velocity small.

The mechanism discussed by Osterbrock (1961) involves the conversion of acoustic waves into magnetoacoustic waves. These waves then are converted to shocks by the decreasing density, and the energy is dissipated rapidly. One formulation of the outer boundary condition would be to couple the mode to an outgoing wave where propagation becomes permitted beyond the temperature minimum. This is not an acceptable solution, however, since the vertical wavelength is about 1000–2000 km—about the distance between the temperature minimum and the point where shock formation is supposed to take place. The procedure used in the present calculation has been to find that

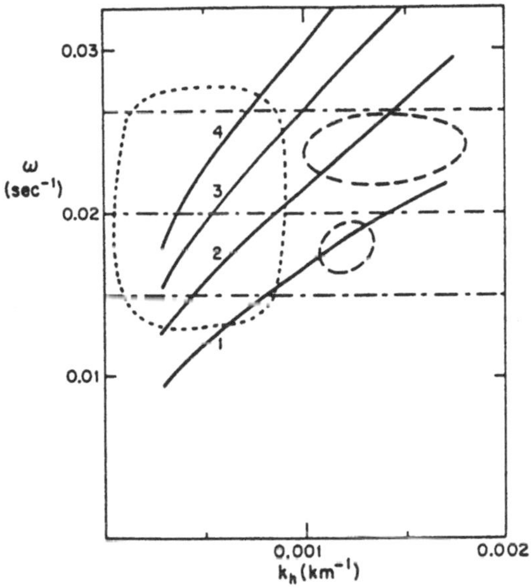

FIG. 2.—Diagnostic (k_h, ω)-diagram. Loci of the eigensolutions are shown as solid lines labeled by their modal numbers. Short dashed line indicates region where the oscillations were found by Tanenbaum *et al.* (1969). The two long dashed curves indicate the frequencies found by Frazier (1968). The three long-short dashed lines indicate the frequencies found by Gonczi and Roddier (1969). These workers did not determine the spatial wavelengths of the observed oscillations, but their results may be interpreted as modes 1–3 at about the resolution limit of Tanenbaum *et al.* (1969).

mode which has the smallest velocity amplitude above the temperature minimum. This mode may be expected to be damped least by the shock formation. The simple condition of smallest velocity amplitude does not provide any phase relation between the pressure perturbation and the velocity. Consequently, the growth or damping of the oscillations has been determined indirectly by the method described in § IV. The simple boundary condition does determine the eigenfrequencies to an accuracy of about 1 percent.

The dispersion relation between k_h and ω is shown in Figure 2, together with a rough representation of the relevant observational material. It is clear from this diagram that poor resolution in horizontal wave number will smear out the distinct frequencies. In particular, observational efforts which do not attempt to identify the wavelength of the oscillations are bound to lead to more or less random frequencies in the range $0.015 < \omega < 0.032$. Consistent observations may be obtained only in cases where a narrow range of wavelengths is selected. Such selection appears to have been achieved by Frazier (1968) by virtue of this high spatial resolution and small field of view. The work by Tanenbaum *et al.* (1969) also included a resolution of the power in one spatial dimension. Their results do not show the distinct lines predicted here, although there is a hint of diagonal ridges roughly parallel to the dispersion lines.

IV. ENERGY BALANCE

a) The Fundamental Equation of Conservation

The present incomplete understanding of the outer boundary condition eliminates the possibility of determining the growth or decay of the oscillations from the imaginary part of the frequency. A physically meaningful alternative is to compare the energy change in a fixed volume of space with the energy of the oscillation. Three processes act to change the energy of the oscillations. First, there is the work done on the oscillating fluid during one cycle. This is

$$\text{work} = -\int_0^{2\pi/\omega} v_z P' dt\,, \tag{13}$$

where v_z, P', and all quantities below associated with the oscillation are the real parts of the complex modal solutions found in § III. The second process is the flow of thermal energy in through the surface. This is

$$\text{thermal flux} = -\int_0^{2\pi/\omega} \rho v_z E dt\,, \tag{14}$$

where E is the internal energy of the fluid. Finally, there is the second-order part of the radiative flux which escapes into space. As long as the optical-depth scale does not chance significantly during the cycle, we may evaluate the radiative flux from

$$\text{radiative flux} = -\int_0^{2\pi/\omega}\int_\tau^\infty 12\sigma T^2 T'^2 E_2(\tau')d\tau' dt\,, \tag{15}$$

where σ is the Stefan-Boltzmann constant and E_2 is the second exponential integral. The minus sign appears in all these formulae because we wish to determine the rate of change of the energy in the volume of space. We shall neglect the flux of kinetic energy $\frac{1}{2}\rho v^2 v_z$ which is third order.

The energy density of the oscillations is also composed of two parts—thermal and kinetic. These are

$$\text{thermal-energy density} = \int_0^{2\pi/\omega} [\rho E - (\rho E)_0] dt\,, \tag{16}$$

$$\text{kinetic-energy density} = \int_0^{2\pi/\omega} \tfrac{1}{2}\rho v^2 dt\,. \tag{17}$$

Landau and Lifshitz (1959, § 49) show that

$$\frac{d}{dt}\int_{z_1}^{z_2} (\text{energy density})\, dz = \text{flux}\,(z_1) - \text{flux}\,(z_2) \tag{18}$$

when the energy density and flux (flux in this context includes work) are the sums of the parts given in equations (13)–(17). It is worth emphasizing that equation (18) is a rigorous result to second order.

The energy flux is virtually zero in the deep interior where radiative exchange is negligible. This is a result of our boundary-condition equation (12), which sets the velocity and thermodynamic fluctuations 90° out of phase. Equation (11) clearly shows that this phase relationship is altered when $\omega_R \sim \omega$. For most of the eigensolutions found in § III the flux at the temperature minimum was negative, a result which indicates that the oscillations are overstable. The cause for the overstability is the same as the mechanism discussed by Souffrin and Spiegel (1967) in connection with gravity modes. A comparison of the energy of the oscillations to the flux showed that the amplitude should increase by a factor of e in from twenty to thirty periods. This is a rather significant result since it eliminates the necessity of coupling the convective velocity field to the acoustic modes. It is likely that interaction between the convective and oscillatory velocity fields will take place and that this interaction may even contribute to the energy of the oscillatory field. Present observational evidence does not indicate that this interaction is dominant. Frazier (1968) has concluded from a direct inspection of his observations that the two velocity fields are uncorrelated.

b) Dissipation above the Photosphere

The discussion of the energy balance to this point has been incomplete in that the region above the temperature minimum has been ignored. As indicated in § III, the motion above the temperature minimum is coupled to that below in a rather direct fashion. In the upper region the motion is generally thought to involve conversion of mechanical energy to heat through some hydromagnetic interaction (see, for example, the discussions by Osterbrock 1962 and Athay 1966). Since the oscillations are not observed to grow in time, we may equate the nonradiative heat input to the chromosphere to the energy produced below the photosphere by the overstable oscillation. Clearly the energy production is proportional to the square of the amplitude of the oscillation. Without an explicit treatment of the dissipation, we must assign the amplitude on the basis of the observations. The summary by Tanenbaum *et al.* (1969) suggests $v_{\rm rms} = 0.2$ km sec^{-1} for wavelengths between 500 and 4000 km. Table 2 gives the required dissipation per unit area per unit time if the entire observed velocity is assigned to each of the modes and wavelengths indicated. The modal number is defined as one plus the number of nodes between the reflecting layers. The actual nonradiative heat input to the chromosphere and corona is a weighted average of these numbers and depends on the detailed distribution of the velocity amplitude on the (k_h, ω)-plane. The available evidence suggests that the modes which produce the most energy (those with periods about 300 sec) are also the modes with the largest amplitudes. Clearly this energy-production mechanism is in good agreement with the rate of energy loss by radiation of 5.6×10^6 ergs cm^{-2} sec^{-1} found by Athay (1966) for all layers above the temperature minimum (Table 2).

c) Uncertainty in ω_R

An important uncertainty in these calculations is the radiative-interaction rate ω_R. The formula we have used may be quite inaccurate in optically thick regions. In general, we may determine ω_R from

$$q = \int_0^\infty \kappa_\nu (J_\nu - B_\nu) d\nu \,, \tag{19}$$

where J_ν is the mean intensity. We expect that $q \neq 0$ in the unperturbed atmosphere because the convective flux varies with depth. This radiative exchange of heat, q_0, must be statistically balanced by the convective motions. The heat exchange due to the oscillatory motion is then the fluctuation in q. Unfortunately, we must know q_0 before we can determine the fluctuation in q since variations in κ interact directly with q_0. Nonetheless, in principle for a particular mode of oscillation we could determine ω_R from

$$\omega_R = -\frac{q - q_0}{C_P T'} \,. \tag{20}$$

For optical depths smaller than unity, the mean intensity should be unaffected by the perturbation and ω_R should reduce to the form given by Spiegel. At large depths, ω_R is quite small and may be neglected. The crucial superadiabatic layer is thus the only region where ω_R is difficult to determine. Since the superadiabatic layer is quite thin, we may explore the importance of this uncertainty by multiplying ω_R by arbitrary scale factor f in optically thick regions. Because of phase differences in the temperature fluctuation at adjacent layers we must also check complex values of f. Calculations were done with f between 0.1 and 10.0 and with phase angles between $-90°$ and $+90°$. The results showed that the power output is roughly proportional to $f^{-0.3}$ and is virtually unaffected for phase angles between $\pm 45°$. At phase angles of $\pm 90°$ the oscillations were no longer overstable. The result that the power output is inversely proportional to f is due to the fact that when T' is kept very small, the loops in the (T, S)- and (P, V)-diagrams are also very small. The dissipating loops in the optically thin region are unaffected and eventually dominate. Nonetheless, we would have to increase f by about a factor of 50 to eliminate the overstability. It seems unlikely that Spiegel's formula could be wrong by such a large factor. Also, the phase difference between the temperature fluctuations at different layers generally is less than 45°.

TABLE 2

REQUIRED DISSIPATION AS A FUNCTION OF HORIZONTAL WAVELENGTH

Modal Number	$\lambda_h = 4.83\times10^3$ km		$\lambda_h = 6.98\times10^3$ km		$\lambda_h = 12.57\times10^3$ km	
	Flux	Period	Flux	Period	Flux	Period
1.	−0.3	311	0.0	399	−3.5	532
2.	+7.4	256	6.5	313	−2.0	412
3.	+7.4	211	6.3	256	+1.2	335
4.	. . .	. . .	5.0	219	+3.7	282

NOTE.—Flux is in units of 10^6 ergs cm^{-2} sec^{-1}; period is in seconds.

V. CONCLUSIONS

The most important finding of this study has been that the 5-minute oscillations are overstable and are capable of supplying the energy lost through radiation in the chromosphere and corona. Also important is the result that the oscillations should be confined to distinct lines on the diagnostic (k_h, ω)-plane. These lines have not yet been found because of poor resolution in k_h and ω. The double peak observed by Frazier (1968) and the multiple peaks found by Gonczi and Roddier (1969) are compatible with the positions of the dispersion lines. More precise observations at longer wavelengths should permit an accurate determination of the entropy of the convective envelope.

We may set minimal conditions which will permit the observation of the dispersion lines. First, the spatial analysis must be two-dimensional. Unless very high wavenumbers are observed in one dimension, the total wavenumber is not well established by one-dimensional observations such as are made with a magnetograph. Second, the observations must be long enough, and of a large enough region, to resolve adjacent lines. At a wavelength of 8000 km the region must be roughly 60000 km in diameter and must be observed for roughly 1 hour. Third, these periods and wavelengths must be detectable. This condition requires velocity differences at points separated by 3000 km (4″ of arc) to be measurable and successive observations to be no more than 1 minute apart. Although the shorter-wavelength oscillations should be easiest to resolve, the longer wavelengths will provide the most information about the structure of the solar envelope.

I would like to thank Dr. E. N. Frazier for suggesting that the 5-minute oscillations might be dependent on the subphotospheric layers. I am grateful to Professor R. F. Christy for suggesting that my unexpected result that the oscillations are self-exciting might be more than a numerical error. I wish to thank the Kellogg Laboratory at Caltech for their hospitality and support during the initial phases of this work.

REFERENCES

Athay, R. G. 1966, *Ap. J.*, **146**, 223.
Frazier, E. 1968, *Zs. f. Ap.*, **68**, 345.
Gonczi, G., and Roddier, F. 1969, *Solar Phys.*, **8**, 225.
Howard, R. 1967, *Solar Phys.*, **2**, 3.
Landau, L. D., and Lifshitz, E. M. 1959, *Fluid Mechanics* (London: Pergamon Press).
Leighton, R. B., Noyes, R. W., and Simon, G. W. 1962, *Ap. J.*, **135**, 474.
Lighthill, M. J. 1952, *Proc. Roy. Soc. London, A*, **211**, 564.
Moore, D. W., and Spiegel, E. A. 1966, *Ap. J.*, **143**, 871.
Noyes, R. W., and Leighton, R. B. 1963, *Ap. J.*, **138**, 631.
Osterbrock, D. E. 1961, *Ap. J.*, **134**, 347.
Souffrin, P. 1966, *Ann. d'ap.*, **29**, 55.
Souffrin, P., and Spiegel, E. A. 1967, *Ann. d'ap.*, **30**, 985.
Spiegel, E. A. 1957, *Ap. J.*, **126**, 202.
Stein, R. F. 1968, *Ap. J.*, **154**, 297.
Tanenbaum, A. S., Wilcox, J. M., Frazier, E. N., and Howard, R. 1969, *Solar Phys.*, **9**, 328.
Ulrich, R. K. 1970, *Ap. and Space Sci.*, **7**, 183.
Whitaker, W. A. 1963, *Ap. J.*, **137**, 914.

A New Description of the Solar Five-Minute Oscillation

J. LEIBACHER *Harvard College Observatory, Cambridge, Massachusetts, USA*
R. F. STEIN *Brandeis University, Waltham, Massachusetts, USA*

A model is proposed for producing oscillations in the solar photosphere; it is suggested that the upper portion of the convection zone acts as a resonant cavity and drives the photosphere as a whole, at frequencies for which waves are non-propagating. The existence of such a cavity depends on a reflecting upper boundary that results from decreasing temperature and on the postulated existence of a reflecting lower boundary which, for waves of 300-sec period, lies 1000–2000 km below the top of the convection zone.

INTRODUCTION

During the last decade, a large body of evidence has emerged showing that a general feature of the solar surface is a very regular vertical oscillation (Leighton *et al.* 1962, Evans and Michard 1962). Observations of the vertical phase velocity and of the phase relations between intensity and velocity indicate that the resonance has nearly the character of a standing wave, where only a small fraction of the energy density is propagated (Frazier 1968, Tanenbaum *et al.* 1969). The observed long temporal coherence of the oscillations strengthens this interpretation (Gonczi and Roddier 1969). Theoretical descriptions of these phenomena have been founded upon the inability of an atmosphere to propagate low frequency sound waves (Lamb 1908). Thus, for example, Souffrin (1966) has argued that the high-pass filter nature of the atmosphere, combined with an incident spectrum monotonically decreasing at high frequencies, would give rise to a peak in the emergent power spectrum. Bahng and Schwarzschild (1963), Kahn (1961), and Uchida (1967) have proposed that the oscillation is a normal mode resonance arising from the reflection of compression or buoyancy waves in the temperature minimum between the increasing temperatures of the convection zone and the corona. This approach succeeds in accounting for the observed standing wave aspect of the oscillation, but encounters difficulty from the inability of sound waves at the observed frequency to propagate—and hence interfere—near the temperature minimum. Moore and Spiegel (1964) and Kato (1966a) pointed out the possibility that the photosphere experiences turbulent excitation throughout a broad range of frequencies and that the power at propagating frequencies runs out through the top of the atmosphere, leaving the photosphere oscillating at a locally non-propagating frequency.

We wish to draw attention to another mechanism for producing the observed oscillations. This mechanism came to our attention during an investigation of the non-linear impulse response of a model solar atmosphere.

THE DRIVING MECHANISM

The rapid temperature drop (40 K km^{-1}) just below the top of the convection zone acts as a good reflector for waves incident from below. If a second reflecting layer farther down in the convection zone is postulated, a resonant cavity is formed at frequencies for which waves are non-propagating near the temperature minimum. Lamb (1908) has shown that an isothermal atmosphere can propagate energy only at frequencies higher than $c/(2H)$ where c is the sound speed and H is the scale height. This cut-off frequency varies inversely as the square root of the temperature. At lower frequencies, displacements vary exponentially—rather than sinusoidally—with height. A running wave propagating into a lower temperature region becomes evanescent, and hence a pulse of energy with a distribution of frequencies will suffer reflection of its non-propagating components when it encounters a cooler region, as occurs near the top of the solar hydrogen convection zone (Vitense 1953). When a cavity is formed by imposing a reflecting lower boundary (1600 km below the photosphere), it resonates at a frequency below the critical frequency of the photosphere. The photosphere is therefore driven up and down as a whole. This produces a well-defined non-propagating oscillation at the observed frequency, below the photospheric critical frequency. With increasing height, the higher-frequency, local resonant oscillation at the critical frequency of the photosphere (about 200 sec) becomes dominant and provides the high frequency tail in the observed power spectra.

BOUNDARY CONDITIONS

The existence of trapped running waves in the hydrogen convection zone depends crucially upon the existence of some source of reflection near 1000 km below the visible surface. The precise location of the boundary determines the resonant period of the cavity. The results do not depend qualitatively upon the particular type of boundary condition; we have considered rigid boundaries, constant pressure boundaries, and distributed sources of reflection, all of which produce resonance. With a transmitting lower boundary, relaxation to the initial state is rapid. The position and nature of the upper boundary condition have virtually no influence on the nature of the photospheric oscillation, because of the rapidly decreasing density.

DISCUSSION

In addition to the source of resonance which this model provides, several other aspects of the dynamics of the visible layers emerge in qualitative agreement with observations. Besides the general property of resonance, observations show that the decay of the oscillation is very slow, much slower in fact than the calculated response to an impulse (Kato 1966b, Meyer and Schmidt 1967, Stix 1970). The acoustic cut-off frequency in the photosphere and low chromosphere exhibits a very broad maximum approximately 1.5 times the observed frequency and nearly 17 pressure scale heights thick. This extended non-propagating zone effectively isolates the resonance in our model from the propagating region in the upper chromosphere and corona. Observations have indicated that the vertical phase velocities in the photosphere are much greater than the sound speed, arguing strongly against interpretating the oscillations as the output of a high pass filter acting on a low frequency signal (Souffrin 1966). Large phase velocities may result from running waves of nearly equal amplitude propagating in opposite directions, or from evanescent waves, in which case the phase velocity becomes infinite. From our results the latter description emerges most naturally. The rapidly decreasing density inevitably leads to a rapidly increasing velocity. The non-linear interaction and dissipation included in this calculation limit the velocity growth to values comparable with observation in the uppermost observed layers. The observed phase lag between velocity and temperature does not distinguish between trapped oscillatory waves and evanescent waves.

The response of the chromosphere and the chromospheric-coronal interface should be quite different for our evanescent representation and for previous trapped-oscillatory descriptions. If the upper reflection giving rise to the resonance were to occur in the chromospheric-coronal temperature rise, then this region would display the same oscillatory motion observed in the photosphere. In our proposal, the resonance arises in the hydrogen convection zone, drives an evanescent oscillation in the photosphere, and gives rise to propagating waves in the chromosphere where the acoustic cut-off decreases to the resonant frequency of the hydrogen convection zone. In this case, we find that the chromosphere is the site of rapidly steepening, upward propagating waves, which because of non-linear interactions change from regularly spaced sinusoids, to irregularly spaced sawtooths.

Although these results depend on the arbitrary assumption of a reflecting lower boundary about 1600 km below the visible solar surface, with this one assumption many heretofore unexplained aspects of the observed solar oscillations are natural consequences. Therefore, future attention should be directed to investigating possible physical mechanisms for producing this reflection.

The computations upon which this letter is based were made possible by a grant from the National Center for Atmospheric Research (NCAR), Boulder, Colorado, which is supported by the National Science Foundation. The authors are indebted to the personnel of the NCAR Facilities Laboratory and to members of the NCAR High Altitude Observatory for many helpful conversations.

REFERENCES

Bahng, J., and Schwarzschild, M., 1963, *Astrophys. J.*, **137**, 901.
Evans, J., and Michard, R., 1962, *Astrophys. J.*, **136**, 493.
Frazier, E., 1968, *Astrophys. J.*, **152**, 557.
Gonczi, G., and Roddier, F., 1969, *Solar Phys.*, **8**, 255.
Kahn, F. D., 1961, *Astrophys. J.*, **134**, 343.
Kato, S., 1966a, *Astrophys. J.*, **143**, 372.
Kato, S., 1966b, *Astrophys. J.*, **144**, 326.
Lamb, H., 1908, *Proc. London Math. Soc.* (2), **VII**, 122.
Leighton, R., Noyes, R., and Simon, G., 1962, *Astrophys. J.*, **135**, 474.
Meyer, F., and Schmidt, H., 1967, *Z. Astrophys.*, **65**, 274.
Moore, D., and Spiegel, E., 1964, *Astrophys. J.*, **139**, 48.
Souffrin, P., 1966, *Ann. Astrophys.*, **29**, 55.
Stix, M., 1970, *Astron. Astrophys.*, **4**, 189.
Tanenbaum, A., Wilcox, J., Frazier, E., and Howard, R., 1969, *Solar Phys.*, **9**, 328.
Uchida, Y., 1967, *Astrophys. J.*, **147**, 181.
Vitense, E., 1953, *Z. Astrophys.*, **32**, 135.

Observations of Low Wavenumber Nonradial Eigenmodes of the Sun*

Franz-Ludwig Deubner
Fraunhofer-Institut, Freiburg i. Br., Fed. Rep. Germany

Received April 10, 1975

Summary. New photoelectric observations of the photospheric velocity field with high resolution in horizontal wavenumber and frequency have been carried out in the CI 5380 line. Three or four discrete stable modes of the five-minute oscillations are resolved in the k, ω diagram, which agree in many respects with the predicted fundamental modes of subphotospheric standing acoustic waves.
Some interesting aspects of the confirmation of a "global" excitation mechanism of the five-minute oscillations are pointed out. Several apparently discordant details of previous observations are rediscussed, and it is shown that they can be reconciled with the present results.

Key words: solar atmosphere — photospheric oscillations — nonradial eigenmodes

Introduction

Recent investigations of the horizontal scale of the five-minute oscillations, including both high resolution photographic and moderate or low resolution photoelectric techniques (cf. Stix and Wöhl, 1974; Deubner, 1974b; Fossat and Ricort, 1975), have unanimously and unambiguously confirmed the presence or rather concentration of oscillatory power at very low horizontal wave numbers $k = 2\pi/\lambda < 1\ \mathrm{Mm}^{-1}$. This finding, in accordance with the fact that spatial coincidence between granule appearance and the onset of an oscillation was not observed, (Frazier, 1968; Musman, 1974) and that almost no horizontal sound waves were found in the diagnostic diagrams of the photospheric velocity field (Stix and Wöhl, 1974), definitely rules out models of local excitation of the 5-min oscillations by granular pistons.
A different class of models, based on trapping of acoustic waves below the photosphere, had already been suggested and worked out in some detail by Ulrich (1970), Leibacher and Stein (1971) and Wolff (1972). These models are able to account for the existence of a pervasive wavemotion in a layer of the quiet solar atmosphere and in a frequency band where waves are essentially non progressive.
However, the "modal" character of these overstable subphotospheric oscillations, as predicted by theory, was not evident from the observations despite the considerable earlier efforts. This is not surprising, because the accumulation of data required for the computation of k, ω spectra with adequate statistical stability is enormous, and, moreover, the resolution in frequency *and* wavenumber achieved by previous writers was hardly sufficient for detection of even the lowest modes, as already pointed out by Ulrich (1970). We planned therefore a series of new observations, particularly designed to yield reliable information on the existence of any fine structure in the diagnostic spectra related to a modal character of the wavepattern. Our first attempts in this direction have proved successful, and the results will be described in this paper.

Observations

The low photospheric CI 5380 line ($EP = 7.68$ eV) was used to measure Doppler velocities with the magnetograph of the Fraunhofer Institute attached to the domeless Coudé refractor in Anacapri. The total band pass of the exit slit was 20 mÅ. A straight line extending 300″ on the solar disk was scanned periodically at 110 s intervals with a 2.0 by 2.5 (arc s)^2 aperture. The scanning step was 1.0 arc s. At the beginning of each observation the scan line was centered on the disk centre, and then displaced continuously to correct for solar rotation.
We obtained two sets of observations, one lasting 2 h 50 m with the scan line running parallel to the solar axis, and a second one on the same day of September 20, 1974 lasting 3 h 55 m with the scan line parallel to the equator. With the exception of the small intersection area of the two scan lines, the two sets of data can be regarded as statistically independent.
The seeing conditions during this day remained moderate to poor and did not permit any high resolution work. The DC circuitry of the Doppler compensator, however, was perfectly stable, which enabled us to

* Mitteilung aus dem Fraunhofer-Institut Nr. 139

perform the observations of this extremely low contrast spectral line without difficulties.

Analysis and Results

Spatial and temporal averages of the measured velocity fluctuations were subtracted from the data before they were subjected to a straight forward FFT analysis followed by moderate smoothing (Hamming) of both frequency and wavenumber, and a second transform

$$P(k) = -k \cdot \int_{k_x=k}^{\infty} \frac{dP(k_x)/dk_x}{(k_x^2 - k^2)^{1/2}}\, dk_x$$

which converts the Fourier spectra, under the assumption of two dimensional isotropy in space, into "diagnostic diagrams", yielding the correct power distribution per unit wave number. Final smoothing with 5-point running means in wavenumber k completes the numerical analysis.

Both sets of observations were evaluated separately, but finally plotted on the same scale in frequency ω, in order to facilitate the comparison between them (Fig. 1a and b). In the five-minute domain of the diagrams, the power distribution is given by contour lines outlining the increase of power in a quadratic progression. The striking similarity of some of the basic features in the two spectra and the appearance of discrete *stable* ridges in particular prompted us to combine (by adding) the results in one single spectrum which is plotted in the same way in Fig. 2.

This diagram makes it convincingly clear that the power distribution in the resonant range is not smooth in the sense that it decreases monotonically from a central maximum in all directions. A number of more or less parallel ridges running in a roughly diagonal direction at regularly spaced intervals can be easily seen. The gradient of these ridges (where they are well resolved, i.e. at $k \approx 1\ \mathrm{Mm}^{-1}$) corresponds to a horizontal group velocity of approximately $13\ \mathrm{km\ s}^{-1}$! This makes us aware of the potential importance of subphotospheric layers for the velocity field under consideration. The gradient further steepens as we approach lower wavenumbers (well visible in Fig. 1a).—At this point, we are prepared to compare our results with the theoretical predictions of trapped acoustic wave modes given by Ulrich (1970) and solutions for overstable modes of the entire solar body computed by Wolff (1974, unpublished results), which are also drawn in Fig. 2, *as given by the authors.*

In the "high wave number" domain we find that the solutions of Ulrich agree with the observed ridges in all detail to an embarassing extent. As well as a general family resemblance the observed and the predicted curves have the same curvature, appear at the same intervals and even, at almost precisely the same absolute position in k, ω space. The fundamental mode of Ulrich's also appears to be the first detectable mode in our diagram.

Obviously, our spectral resolution in k, ω space is insufficient, to resolve any of Wolff's radial eigenmodes which fall into the low wave number range of the diagram. However, it is interesting to note that 1. Wolff's modes appear to be the natural extension of Ulrich's (we have already pointed out the continuous increase of $d\omega/dk$ towards lower wavenumber in our data, and 2. the maximum of power of the combined observed spectrum clearly falls within the range of solutions

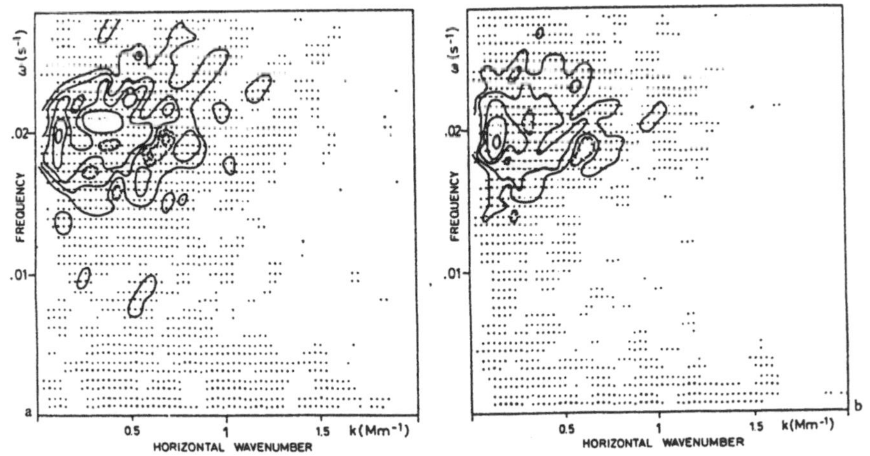

Fig. 1. (a) Diagnostic diagram of the photospheric velocity field computed from Doppler shifts of the CI 5380 line. Six different printer symbols bordered by solid contour lines indicate quadratically increasing levels of kinetic power per unit wavenumber, the lowest level printed (not outlined) being 2.8% of the maximum. Valleys are bordered by broken lines. (b) Same as 1a, obtained with different oriantation of the scan line

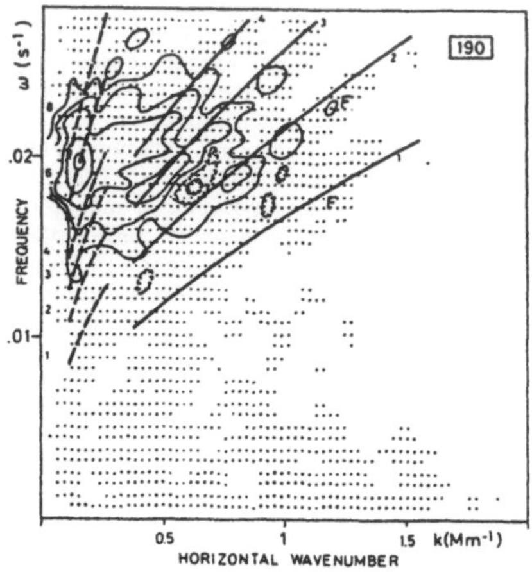

Fig. 2. Diagnostic diagram of the photospheric velocity field obtained by combining the statistically independent power spectra of Fig. 1a and 1b. The spectral resolution element with the number of "degrees of freedom" is given in the upper right corner. Predicted normal modes of trapped acoustic waves are drawn and their radial eigenvalues are given according to Ulrich (solid lines) and Wolff (dashed lines). The two power maxima observed by Frazier (1968) are indicated by a capital *F*

involving the entire solar body ($\lambda \approx 60''$). It is conceivable that, given a better resolution in k, we might be able to trace the resonant modes beyond the present limit of $k \approx 0.3\ \mathrm{Mm}^{-1}$. Observations exploring this part of the power spectrum are planned in the near future.

Discussion

We are now in the position to discuss a few of the apparent discrepancies found among previous observations of the five-minute oscillations without having to take one or the other side.

Since the oscillations evidently are a global solar phenomenon, we do not expect the oscillatory motion to be correlated to rising granules. Whenever such a coincidence was in fact observed (cf. e.g. Evans and Michard, 1962), it was probably due to the excitation of short period (<180 s) acoustic modes by powerful bright granules. The occurance of such modes has recently been demonstrated by Deubner (1974a). However, their contribution to the total power of the oscillatory velocity field is negligible.

It can be easily seen from the diagrams in Fig. 1, that the shape of one dimensional ω-spectra will depend on the choice of a particular spatial "window" just as much as on the amount of time spent for the observation. Applying a somewhat rigorous mathematical treatment to one-point records of the velocity field obtained by B. Howard, White and Cha (1973) have been able to produce a smooth "stable" ω-spectrum similar to the one obtained by Deubner (1972) by averaging a large number of statistically independent raw frequency spectra. But it is now becoming clear that physically the width of the resonance spectrum reflects the dispersion law and the growth rates of certain oscillatory modes in the deep solar photosphere rather than the randomness of the process.

Horizontal progression of the oscillatory motion at sound velocity has never been observed. The waves are standing also in horizontal direction. The resulting picture, comparable to the surface of an agitated pond, is easily confused with a closely packed field of independently oscillating cells, frequently put forward in the past. But even with standing waves prevailing, interference of the different modes visible in our diagrams is inevitable in space and time. It is not surprising that the beat period (20–25 min) and spatial "coherence length" (15–30 Mm) derived from the observed differences in frequency and wave number among the oscillatory modes is in good agreement with the values obtained from previous one dimensional measurements [see White and Cha (1973), Wolff (1973), Fossat and Ricort (1975) for discussion of previous results]. An illustrative example of spatio-temporal interference patterns is discussed by Reif and Musman (1971).

Apart from interference phenomena particularly deceptive in a one dimensional treatment of small samples, it can be seen from the diagrams of Fig. 1, that even with a record length of ca. 3 h the distribution of power is neither uniform nor constant along the dispersion curves, and maxima appear at quite different horizontal eigenvalues at different times. This also applies to the principal maximum which changes drastically its position and structure from one record to the other. At present we are not able to distinguish between real temporal fluctuations of the distribution of power among different eigenmodes and a possible anisotropy of the wave field, seen with different orientations of the scan line.

Regarding the wide range in wavenumber k covered by the power spectrum, the whole dispute about the horizontal scale of the oscillation becomes void. Considerable amount of power is still present at $k \approx 1\ \mathrm{Mm}^{-1}$, corresponding to $\lambda = 9''$ or—in terms of diameters as discussed by Deubner and Hayashi (1973) and Fossat and Ricort (1975)—to a cell size of ≈ 2000 km, which equals the lower number given by Sheeley and Bhatnagar (1971). The fact that the small elements appear to be the most powerful in the spectro-heliograms is a consequence of the presentation in λ-space with $d\lambda = -\lambda^2 dk$. The maximum of the power distribution per unit wavenumber clearly is at the limit of our spectral resolution close to $\lambda = 60''$, confirming the large numbers derived from most of the photoelectric measurements.

In their resent study of the photospheric oscillations,

Fossat and Ricort (1975) reported on a secondary power maximum at periods of 10 min, which they occasionally found in the frequency spectra deduced from recordings with apertures larger than 3 arc min. In Fig. 1b of the present paper two "modes" appear which are probably related to this feature, one at $k=0.25\ \mathrm{Mm}^{-1}$ on the extension of Ulrich's fundamental mode (or a zero mode of Wolff's) and another one at $k=0.55$ occupying the locus of a hypothetical zero mode of Ulrich's solutions. Since such a zero mode is physically meaningless in the context of Ulrich's work (it requires one open boundary) the presence of such a mode in the data would indicate that the model atmosphere used by Ulrich still needs some corrections. Of course, more observations are desirable to verify this particular mode.

It is interesting to note, that neither of the two 10 min modes appears in the other diagram 1b, thus confirming the spasmodical character of the 10 min oscillations, reported by Fossat and Ricort. The large spatial coherence length (100 Mm) estimated by these authors is not evident from our spectra.

Is there horizontal phase propagation at high velocities? Deubner and Hayashi (1973) have argued that phase velocities of the order of $140\ \mathrm{km\ s^{-1}}$ reported by Musman and Rust (1970) can be easily interpreted as random phase relations of adjacent independent oscillators. Now that we have confirmed the global character of the wave field we are entitled to compare this value measured in λ-space with the ratio of ω_{MAX}/k_{MAX} taken from the combined spectrum in Fig. 2. With $\omega_{MAX}= 2.1\times10^{-2}\ \mathrm{s}^{-1}$ and $k_{MAX}=0.144\ \mathrm{Mm}^{-1}$ we obtain $v_{phase}=145$ km in excellent agreement with the previous number.

Frazier (1968) had already observed a decomposition of the oscillatory power distribution into two separate maxima at periods of 270 s and 360 s. Since the spatial as well as the temporal window of his observations were rather small, serious doubts were raised about the statistical reliability of his results, and it remained an open question, whether the two maxima could be identified with eigenmodes of the solar atmosphere. If we believe that randomness in the velocity field is restricted to fluctuations of power along a given set of discrete "modes" in the spectrum, there is no reason to discredit Fraziers results as a first indication of the modal character of the five-minute oscillations. The maxima taken from his Fig. 3c are indicated by capital F in our Fig. 2. They can be seen to coincide closely with the ridges belonging to the first two eigensolutions given by Ulrich (1970). Again, it is mainly the correct *spacing* of Fraziers modes which gives us the confidence of not having stretched the argument too far.

Conclusions

Our observations have shown that the five-minute oscillations may in fact be interpreted as low wavenumber nonradial acoustic eigenmodes of the subphotospheric layers of the solar atmosphere.

Many of the hitherto seemingly discordant observational aspects of the five-minute oscillation can be reconciled in this interpretation quite naturally. The concentration

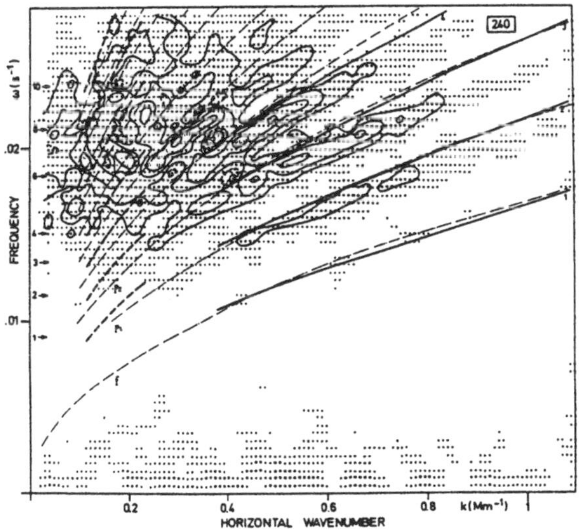

Fig. 3. Diagnostic diagram as in Fig. 2, obtained from observations with a 880 arc s scan line. The dash dotted curves delineate the solutions obtained by Ando and Osaki (1975)

of power at very low wavenumbers indicates that substantial parts of the solar interior participate in the oscillation. A direct comparison of our observations with the spectra of free (without artificial boundary) normal modes of oscillation of the whole sun has not yet been possible because of lack of spectral resolution. But the necessary resolution can certainly be achieved in the near future, putting at our disposal an excellent diagnostic tool which allows us to probe the internal structure of our sun as we can the structure of pulsating stellar variables.

Since the five-minute oscillations very likely contribute significantly to the heating of the upper chromosphere and corona (for a review of literature see Stein and Leibacher, 1974), it seems desirable that the production of this energy flux which has to be supplied continuously and over the whole surface of the sun, does not depend on localized events of a shallow surface layer. The observations provide sufficient evidence that this basic process of the solar envelope is indeed deeply and firmly rooted in the solar body.

Note: Another set of four-hour recordings of the solar eigenmodes was obtained recently (June 1975) with a larger scan line extending across half the solar diameter. The average k, ω diagram resulting from three such recordings is plotted in Fig. 3. Due to the high spectral resolution achieved, more than five modes may be traced down to wavenumbers as low as 0 s Mm^{-1}, and can be compared with the theoretical predictions.

We also included in this graph the new results of Ando and Osaki (1975, to be published) who solved the equations of linear *nonadiabatic* nonradial oscillations of the solar envelope. We note that with increasing modal index their solutions (as well as Ulrich's) tend to deviate progressively from the observed positions of the individual modes. We suppose that the model of the outer layers of the convection zone used in the calculations still needs further improvement.

Acknowledgements. I wish to express my gratitude to C.L.Wolff, who put at my disposal some of his unpublished results. The recording and preliminary reduction of the observational data was carried out with the DESO computer facilities supported by the Deutsche Forschungsgemeinschaft.

References

Deubner, F.-L. 1972, *Solar Phys.* **23**, 304

Deubner, F.-L. 1974a, in R. Grant Athay (ed.), Chromospheric Fine Structure, IAU *Symp.* **56**, D. Reidel Publ. Co., Dordrecht-Holland, p. 263

Deubner, F.-L. 1974b, *Solar Phys.* **39**, 31

Deubner, F.-L., Hayashi, N. 1973, *Solar Phys.* **30**, 39

Evans, J. W., Michard, R. 1962, *Astrophys. J.* **136**, 493

Fossat, E., Ricort, G. 1975, in press

Frazier, E. N. 1968, *Z. Astrophys.* **68**, 345

Leibacher, J. W., Stein, R. F. 1971, *Astrophys. Letters* **7**, 191

Musman, S. 1974, *Solar Phys.* **36**, 313

Musman, S., Rust, D. M. 1970, *Solar Phys.* **13**, 261

Reif, R., Musman, S. 1971, *Solar Phys.* **20**, 257

Sheeley, N., Bhatnagar, A. 1971, *Solar Phys.* **18**, 379

Stein, R. F., Leibacher, J. W. 1974, *Ann. Rev. Astron. & Astrophys.* **12**, 407

Stix, M., Wöhl, H. 1974, *Solar Phys.* **37**, 63

Ulrich, R. K. 1970, *Astrophys. J.* **162**, 993

White, O. R., Cha, M. Y. 1973, *Solar Phys.* **31**, 23

Wolff, C. L. 1972, *Astrophys. J. Letters* **177**, L 87

Wolff, C. L. 1973, *Solar Phys.* **32**, 31

F. L. Deubner
Fraunhofer Institut
D-7800 Freiburg i. Br.
Schöneckstr. 6
Federal Republic of Germany

Solar structure from global studies of the 5-minute oscillation

A. Claverie, G. R. Isaak, C. P. McLeod & H. B. van der Raay

Department of Physics, University of Birmingham, Birmingham, UK

T. Roca Cortes

Instituto de Astrofisica de Canarias, La Laguna, Tenerife

High sensitivity Doppler spectroscopy of integral sunlight in the 769.9 nm line of neutral potassium reveals several equally spaced lines of $Q > 1,000$ centred on the well known period of 5 min. These lines are interpreted as low *l*, high *n* overtones of the entire Sun and, if viewed in terms of current solar models, appear to imply a low heavy element abundance and a consequent low solar neutrino flux. Previous workers[1-3] have used spatial resolution to study the 5-min oscillation of the solar photosphere and have shown that power is concentrated along narrow ridges in the *k*, *ω* diagram. These oscillations are mainly low *n*, high *l* value acoustic modes of the entire Sun. In contrast to this, the present work (for a summary see ref. 4) is aimed at an overall view of the solar surface and consequently is biased to a study of low *l* value oscillations. These modes penetrate deeply[5] and thus probe the structure as well as the interior angular velocity of the Sun[6].

Doppler shift measurements of integral solar light show several discrete lines of amplitude 0.1–0.3 m s^{-1} within the main peak of the power spectrum of the 5-min oscillation. These lines are found to have, on average, a uniform spacing of 67.8 μHz.

If the lines are interpreted as being due to low *l* value oscillations of the entire Sun, then the observed spacing taken in the context of current solar models[7,8] implies a mean heavy element abundance of $Z \sim 0.004$ and a consequent neutrino flux of ~2.3 SNU, within two standard deviations of the experimental result[9].

Line of sight velocity measurements of the whole solar disk were made using optical resonance spectroscopy, a technique which has been described in detail previously[10]. Briefly, the velocity of the solar photosphere with respect to the laboratory is measured by comparing the position of the Fraunhofer absorption line of neutral potassium at 769.9 nm with that of the same line in the laboratory. Observations were made during 1976, 1977 and 1978 at Izana, Tenerife and in 1978 simultaneously at Pic du Midi in the Pyrenees. The two sites are ~2,300 km apart and are expected to be meteorologically uncorrelated, thus allowing solar effects to be distinguished from atmospheric or instrumental perturbations. In the 1977 and 1978 seasons the spectrometer incorporated substantial improvements with respect to guidance, optical components, thermal stability and statistical accuracy, resulting in the collection of data of improved quality.

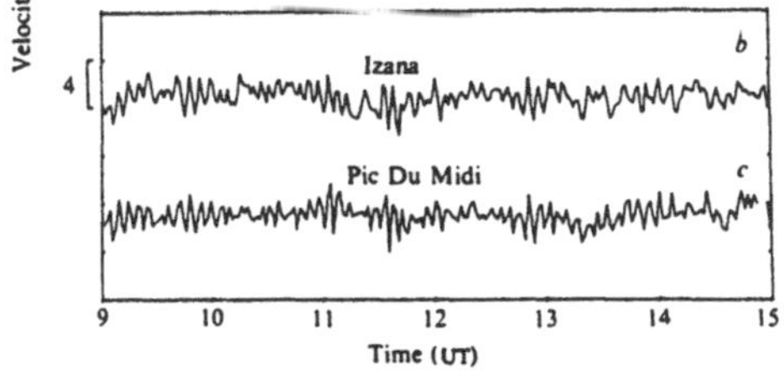

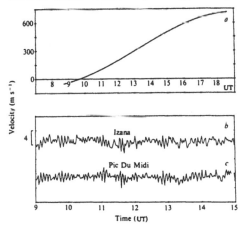

Fig. 1 Mean line of sight velocity observed on 6 August 1978 at Izana (*a*). Comparison of residuals obtained simultaneously at Izana (*b*) and Pic du Midi (*c*).

The Sun was observed for about 8 h on each of 33 days in 1976 (a total of 269 h) and 35 days in 1977 (total 280 h). These data as well as those from 7 of the days obtained in 1978 have been analysed. The data for any one day consist of a mean line of sight velocity determined every 42 s (100 s on Pic du Midi) for the whole observing day (Fig. 1*a*). Residuals corresponding to velocity amplitudes of less than 3 ms^{-1} (Fig. 1*b*) show up after subtraction of the observer's velocity relative to the centroid of the Sun due to the spin and orbital velocities of the Earth. A small allowance for residual curvature effects has been made.

Residuals of data obtained simultaneously at the other observing site are shown for comparison in Fig. 1*c*. To compare these data directly, those from Izana have been smoothed by a moving mean over three points (126 s) and then summed over two points (84 s) to simulate the 100-s integration time of the Pic du Midi data. The correlation of these two data strings clearly demonstrates that the observed signals can only be attributed to a solar origin.

These residuals form a time series which may be processed using standard power spectrum techniques[11]. The power spectra obtained for the data of Fig. 1*b* and *c* are shown in Fig. 2. The agreement of these two spectra, obtained simultaneously at two sites separated by 2,300 km, clearly reinforces the statement that the observed oscillations are of solar origin. A series of well defined peaks can also be seen.

To investigate these peaks further the power spectra for individual days were averaged for the 1976, 1977 and 1978 data. These spectra are shown in Fig. 3. The structure is clearly present and an apparent constancy in the spacing of these peaks can be seen. To verify that this spacing is not an artefact of the analysis, different lengths of data strings were used but this had no effect on the observed spacing.

The constancy of the peak spacing may be demonstrated in two ways. First, the observed peaks in the power spectra are numbered sequentially and plots of the order of the peak against the observed frequency are shown in Fig. 4*a* for the 3 yr of data. Where the line spacing clearly suggests that a particular line is absent in the experimental data, the order of the line has been appropriately increased. The data are well fitted by straight lines and yield mean spacings of 67.6, 67.4 and 67.6 μHz for the 3 yr. Second, the mean power spectra data are screened by a high pass filter by subtracting a moving mean over three points, and then subjecting the resulting points to an autocorrelation and power spectrum analysis. In this way mean line spacings of 67.8, 68.0 and 68.0 μHz are found for the 3 yr. A cross-correlation analysis of the 1976 and 1977 power spectra yields a correlation coefficient of 0.89 and has a maximum at zero frequency lag. This firmly establishes the consistency of the lines over the 2 yr.

As an alternative form of treatment, the data of 1978, which are of higher quality due to improvements in experimental technique, were subjected to a superimposed epoch analysis. Four consecutive days in early August and three in September were used for this study, the frequency range being restricted to that covered by the power spectrum analysis. The structure found in the power spectrum analysis was again evident and a plot of the frequency against peak order is shown in Fig. 4*b*. Straight line fits to these plots yielded mean line spacings of

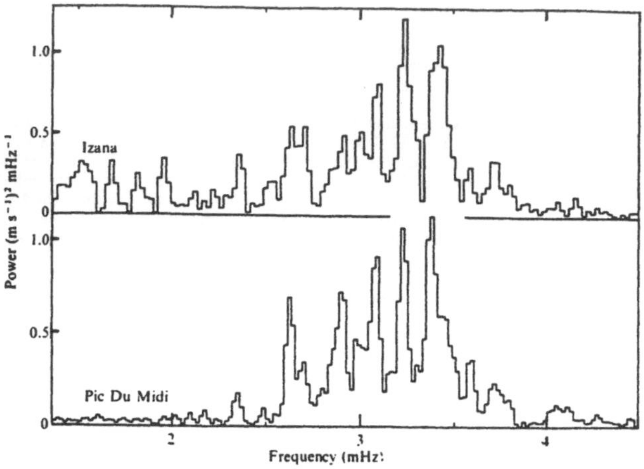

Fig. 2 Power spectra of the residuals obtained from data taken on 6 August at Izana and Pic du Midi.

67.4 ± 0.5 and 67.9 ± 0.2 μHz, respectively. The results of the superimposed epoch analysis over the four consecutive days seem to imply that the Q value of these oscillations is $>1{,}000$.

To investigate further the coherence of the oscillation, a power spectrum analysis of two consecutive days of data was undertaken with zeros in the data string corresponding to the times when actual data were not available. This resulted in a 32-h data string whose power spectrum is shown in Fig. 5*a*. The usual peaks in the power spectrum are clearly evident. To test for the coherence of these peaks sine waves of the indicated frequencies were subtracted from the original data by using a least squares fitting procedure to optimise the frequency, phase and amplitude of the fitted wave. The new power spectrum obtained after subtraction of these coherent fitted sine waves from the data is shown in Fig. 5*b*. This clearly demonstrates that the oscillations are coherent over the 32-h data string and thus have a $Q > 400$.

It seems likely that these lines correspond to normal modes of vibration of the whole Sun. To identify these modes with certainty, information on their spatial structure is required. An integral light spectrometer tends to average out modes of high l value with many peaks and troughs across the visible disk of the Sun[6], thus strongly biasing the observations towards low l value modes. An approximate calculation of the sensitivity of a velocity spectrometer to various spherical harmonic modes can be found elsewhere[12] but a more precise evaluation must include the spectral weighting function across the solar disk[13]. The above suggests that the observed modes are of low l value, possibly $l = 0$ and $l = 1$, with even and odd l values alternating as the value of n (number of radial nodes) increases. The observed mean spacing of 135.6 ± 0.3 μHz of next nearest lines falls in the range of spacings (130–142 μHz) predicted by solar models with uniform, as well as surface enhanced, heavy element abundances. An extrapolation of the uniform model calculations[7] suggests that the observed spacing is consistent with a heavy element abundance of $Z \sim 0.002$. An interpolation of the non-uniform model predictions[8], with a surface heavy element abundance of 0.02, indicates that a mean heavy element abundance of $Z \sim 0.004$ closely fits the observed spacing. Assignment of the lines to l values ≥ 2 is either not consistent with the uniform model[7] or leads to even lower Z values for the non-uniform model[8]. This suggests that any interpretation in terms of these models implies a Z value ≤ 0.004 and a solar neutrino flux of ≤ 2.3 SNU, consistent with the observed value of 1.75 ± 0.4 SNU[9].

Values of n can be tentatively assigned to the lines using the uniform model for which frequencies of particular modes in this region[7] have been calculated. Using the low l value assignment and the approximate Z value limit quoted above, the observed frequencies of 2.41 and 4.04 mHz are tentatively identified as $l = 0$ modes with $n = 17$ and 29, respectively. Such an identification is consistent with the observed slight inequality in spacing between alternate lines. Thus, if the frequencies ν of two

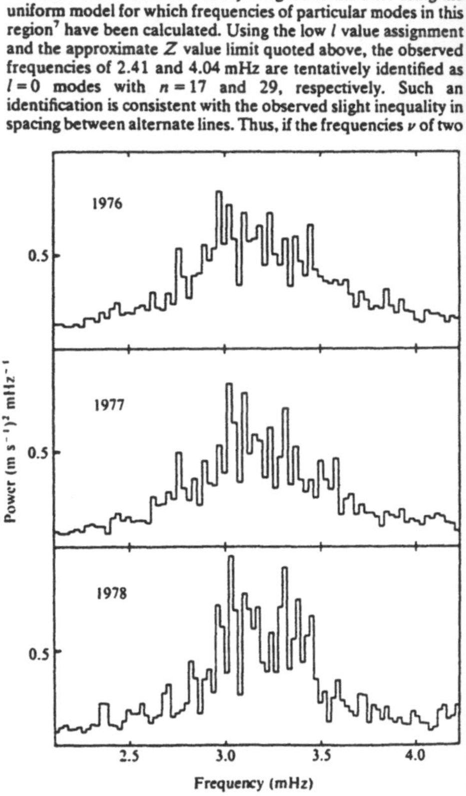

Fig. 3 Mean power spectra obtained from data collected over 33 days in 1976, 35 days in 1977 and 7 days in 1978 at Izana.

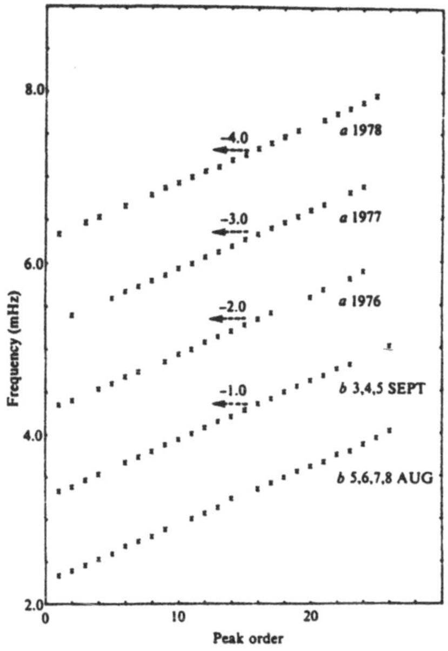

Fig. 4 *a*, Peak frequency against order plotted for the mean power spectra of 1976, 1977 and 1978. Successive plots are displaced by 1 mHz on the vertical scale as indicated. *b*, Similar plots for data determined by a superimposed epoch analysis for 4 days in August and 3 days in September 1978.

neighbouring $l = 0$ lines are compared with that of an $l = 1$ line, as expected[7], $\frac{1}{2}(\nu_{n+1,0} + \nu_{n,0}) > \nu_{odd}$. The heavy element abundance corresponding to this identification is near $Z = 0.004$.

The discrete lines are expected to be hitherto unresolved multiplets due to incomplete degeneracy of different n, l contributions. Rotation of the Sun is further expected to split the modes into $(2l + 1)$ components of spacing characterised by the mean angular velocity of the interior[6,14]. This will produce a long period amplitude modulation of the lines, some evidence for which already exists. It should be possible to resolve this fine structure by using long data strings, provided that the coherence time is adequate.

The definite assignment of the various modes is being attempted by a two-dimensional study of the solar photosphere. The amplitude variation of particular lines, as well as phase information, is being analysed further to study the excitation and damping mechanisms.

We thank B. A. Hail, J. W. Litherland, P. Morris and C. P. Sutton for technical assistance, and Professors J. Rösch and F. Sánchez Martinez for their hospitality at the two observing sites. This work was supported by the SRC and A.C. acknowledges a research associateship. T.R.C. thanks the Spanish Government for a maintenance grant during this work.

Received 23 July; accepted 5 October 1979.

1. Deubner, F.-L. *Solar Phys.* **44**, 371 (1975).
2. Rhodes, Jr, E. J., Ulrich, R. K. & Simon, G. W. *Astrophys. J.* **218**, 521 (1977).
3. Deubner, F.-L., Ulrich, R. K. & Rhodes, E. J. Jr *Astr. Astrophys.* **72**, 177–185 (1979).
4. Gough, D. O. *Nature* **278**, 685–686 (1979).
5. Wolff, C. L. *Solar Phys.* **32**, 31 (1973).
6. Brookes, J. R., Isaak, G. R. & van der Raay, H. B. *Nature* **259**, 92–95 (1976).
7. Iben, I. & Mahaffy, J. *Astrophys. J.* **209**, L39–L43 (1976).
8. Christensen-Dalsgaard, J., Gough, D. O. & Morgan, J. G. *Astr. Astrophys.* **73**, 121–128 (1979).
9. Davis, R. Jr, Evans, J. C. & Cleveland, B. T. *Purdue Univ. Conf. Proc.*, 53–65 (1978).
10. Brookes, J. R., Isaak, G. R. & van der Raay, H. B. *Mon. Not. R. astr. Soc.* **185**, 1–17 (1978).
11. Blackman, R. B. & Tukey, J. W. *The Measurement of Power Spectra* (Dover, New York, 1958).
12. Hill, H. A. *The New Solar Physics* (ed. Eddy, J. A.) 162 (Westview, Boulder, Colorado, 1978).
13. Brookes, J. R., Isaak, G. R. & van der Raay, H. B. *Mon. Not. R. astr. Soc.* **185**, 19–22 (1978).
14. Ledoux, P. *Astrophys. J.* **114**, 373–384 (1951).

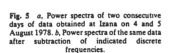

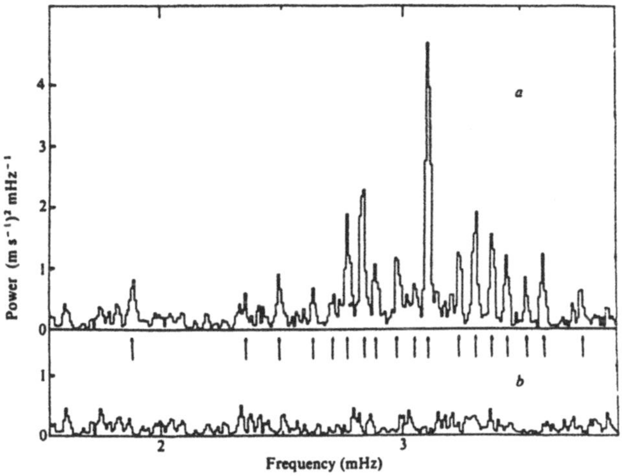

Fig. 5 *a*, Power spectra of two consecutive days of data obtained at Izana on 4 and 5 August 1978. *b*, Power spectra of the same data after subtraction of indicated discrete frequencies.

Solar oscillations: full disk observations from the geographic South Pole

Gérard Grec*, Eric Fossat† & Martin Pomerantz‡

* Département d'Astrophysique de l'IMSP, ERA 669 du CNRS, Université de Nice, Parc Valrose, F-06034 Nice Cedex, France
† Observatoire de Nice, BP 252, F-06007 Nice Cedex 2, France
‡ Bartol Research Foundation of the Franklin Institute, University of Delaware, Newark, Delaware 19711

Observing conditions at the geographic South Pole enable modes of global solar oscillations and theoretical models of the internal solar structure to be identified.

THERE are only two methods available for probing the interior of the Sun to test the theoretical models of the internal solar structure. Such calculations have been based on externally observable parameters ($M_{\odot}$, $L_{\odot}$, $R_{\odot}$, $t_{\odot}$) plus some speculations regarding the relative abundances of certain elements, such as metals and helium[1].

The first technique requires the exceedingly difficult measurement of neutrinos emanating from thermonuclear reactions in the solar core. After much effort, Davis[2] established an upper limit of the solar neutrino flux (1.75 ± 0.4 SNU) that is somewhat below original expectations. These neutrinos convey information only about the central region of the Sun, where the time scale of temporal variations is enormous ($\sim 3 \times 10^7$ yr).

The second approach attempts to detect and identify normal modes of global solar oscillations arising from internal processes. These pulsations characterize phenomena in the layers in which dynamical processes can cause effects that produce visible manifestations on the solar surface[3].

Efforts to detect global solar oscillations through observation of luminosity variations[4–7] were unsuccessful because the amplitudes are far smaller than the transparency fluctuations of the Earth's atmosphere.

Another technique utilizes continuous determinations of the apparent solar diameter[8–12]. These experiments have produced positive results which have been attributed to underestimates of the effects of atmospheric fluctuations[13–16].

Finally, various Doppler shift measurements of large-scale velocity fields have been carried out either with spatial resolution or with integral light from the entire visible disk. These have usually involved continuous observations of various spectral lines by optical resonance techniques.

Fifteen years after the discovery of 5-min oscillations by Leighton[17] came the first incontrovertible observations of global solar pulsations by Deubner[18] who succeeded in resolving the diagnostic κ–ω diagram (or spatiotemporal power spectrum) into discrete lines. The periods were ~5 min and the horizontal wavelengths typically $1\text{–}3 \times 10^4$ km corresponding to spherical harmonics with high l values (typically a few hundred). These results subsequently proved that solar seismology is an exceedingly powerful diagnostic tool for investigating the solar interior[19]. The modes discovered by Deubner have provided information on layers located several thousand kilometres below the photosphere. The detection of lower-order modes is required for probing greater depths.

Thus, there is general agreement over the occurrence of solar pulsations with periods in the range of ~5 min. In fact, a recent study has resolved some low-angular (l), high-radial (n) overtones into discrete lines of amplitudes 0.1–0.3 m s^{-1} separated by an average uniform spacing of 67.8 μHz (ref. 20). However, for various reasons (including possible solar noise, interference by the Earth's atmosphere and other local considerations and the severe problems in conducting these difficult observations at normal mid-latitude locations), the validity of reports of longer period global solar pulsations, especially one at 2 h 40 min (refs 21, 22), has not been universally accepted[23–25].

The polar solar observatory

The advantages of undertaking certain types of astronomical observations at the geographic South Pole became apparent[26] during a program of cosmic ray research in Antarctica[27]. More recently, we recognized that the South Pole enabled us to pursue the solar oscillation problem in conditions that cannot be duplicated anywhere else on Earth:

(1) during the austral summer, the Sun remains essentially at a constant angular altitude, thereby eliminating the difficulties introduced by the day–night cycle;

(2) uninterrupted data runs appreciably longer than those attainable elsewhere are feasible, thus significantly improving both the ω-resolution and the potential for detecting long-period oscillations;

(3) the Amundsen–Scott Station at the South Pole is, in fact, at an elevation corresponding to a pressure altitude of ~3,400 m;

(4) the atmosphere is colder than elsewhere, hence the precipitable water vapour content is exceedingly low, and the transparency is good, providing extended periods of coronal seeing;

(5) the polar plateau constitutes an absolutely uniform terrain (a desert of snow) out to great distances in all directions;

(6) the large trend due to the Earth's rotation, which can never be removed as perfectly as is required, is absent at the planet's rotational axis.

Consequently, an American–French programme was organized to conduct this experiment during the austral summer 1979–80. Only the main characteristics of the apparatus constructed for obtaining the required observations in the hostile polar environment ($T \leq -30$ °C) will be summarized here.

A sodium optical resonance photoelectric spectrophotometer developed at Nice (France) for full disk measurements of the Doppler shifts in the 5,896 Na D1 line[28] was modified to operate in the rigorous conditions at the South Pole. A vertical telescope (8 cm, f/18), specifically designed and constructed for this experiment at Bartol, provided two nonrotating solar images, one of which was focused on the guiding sensor array, the other on the spectrophotometer.

Extensive photography in HLyα during the preceding austral summer[29] revealed that the telescope completely satisfied the requirements for the present programme.

We attempted to create the best conditions possible by eliminating all potential man-made sources of spurious effects such as the local source of thermal turbulence arising from the

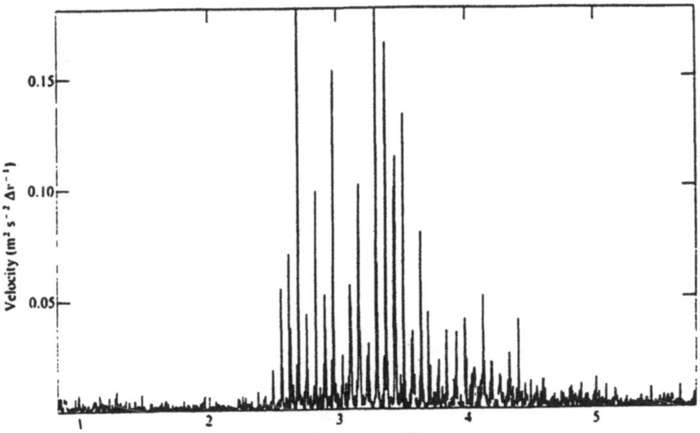

Fig. 1 Power spectrum of the continuous 5-day full-disk Doppler shift measurements recorded at the South Pole from 31 December, 1979 to 5 January, 1980. The resolution of the power in the 3-mHz range into many discrete equidistant lines separated by 68 μHz indicates that global *p*-modes corresponding at least to *l* values of 0 and 1 are observed. Note that the small peaks around 2.4 mHz represent global oscillations with an amplitude $<10\ \mathrm{cm\ s^{-1}}$, corresponding to motion of the solar radius <5 m, or 7×10^{-6} arc s.

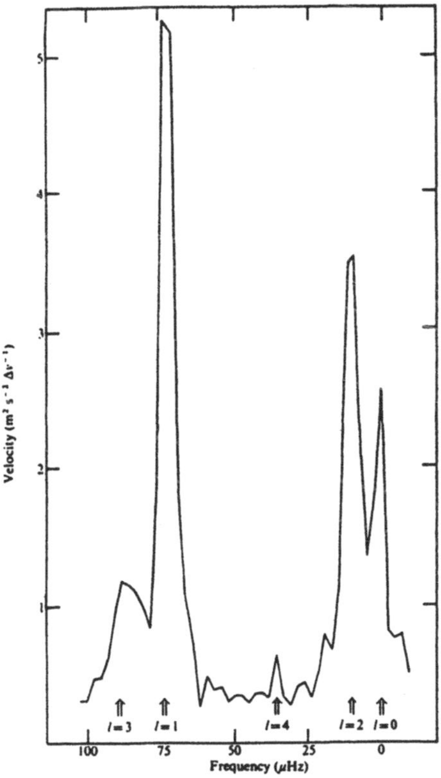

Fig. 2 A superposed frequency analysis of the frequency range between 2.4 and 4.8 mHz reveals the average shape of spectral lines displayed in the power spectrum of Fig. 1. The horizontal axis indicates where the theory predicts the positions of *l* = 1, 2, 3 and 4 modes if the one on the extreme right is assumed to be *l* = 0. The good agreement leaves no room for doubt. Note that the natural width of each separate mode indicates a *Q* value of the order of 600.

Amundsen–Scott station complex which would be seen at the same time every day, a site 8.1-km upwind from the base was selected. There, the well insulated 3.5 m × 2.5 m laboratory building mounted on a sled was installed in a deep trench. The wanigan was then enclosed by a plywood roof and vestibule walls fore and aft, and was buried in the snow. The telescope and instrument package were mounted on the surface 30 m away. A 300-m cable connected the observatory to the diesel-powered electrical generator positioned downwind with respect to the prevailing wind direction. The exhaust emerged practically at ambient temperature, and neither the generator nor the track vehicle utilized for transport produced any detectable atmospheric disturbance.

Observations

Historical meteorological records and experience gained during 1978–79 made us hope for one 5-day period of uninterrupted optimal seeing during December–January. In fact, this did materialize, and the 5-day interval was extended to ~7 days by remaining on the air during two brief cloudy periods, which did not interrupt the tracking. In addition, clean data (cloud free) were recorded during intervals of additional hours ranging from 5 to 10, yielding high-quality tape recordings covering more than 200 h.

Data analysis and results

The definitive results of the analyses so far completed use part of the available data—basically the 5-day continuous run. The results mostly concern: (1) whether the widely discussed oscillation of period 160 min could be confirmed in the ideal conditions that prevail at South Pole; and (2) the investigation of the theoretically expected line spectrum of the 5-min oscillation.

Note that the noise level at all frequencies was significantly lower than has ever been attained previously, by about an order of magnitude.

The 5-min range

Figure 1 shows the power spectrum of the 5-day data sample. It reveals that the power in the 3-mHz range is resolved into many equidistant peaks, separated by 68.0 μHz. This confirms the result published by Isaak's group[20]. This constant displacement of 68 μHz clearly indicates that modes of low degree with both odd and even values of *l* are observed, because the frequency separation between modes of high order with the same value of *l* is expected to be around 130–145 μHz (depending on the theoretical model)[31].

Modes with values of *l* between 0 and 3 may possibly contribute substantially to the observed line spectrum. Each line

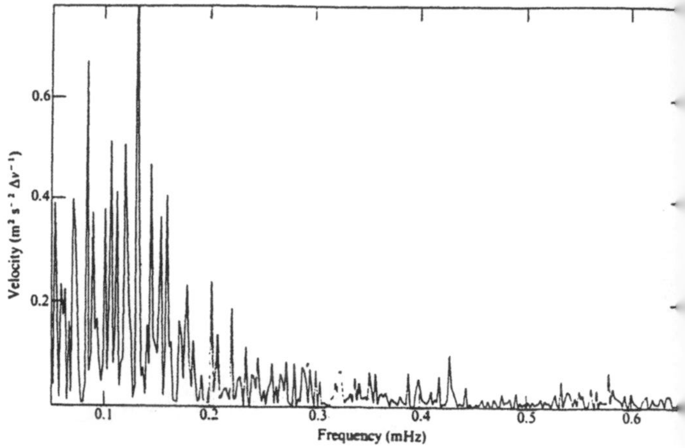

Fig. 3 The low-frequency part of the power spectrum shown in Fig. 1. Its spiked structure is consistent with statistical departure around a broad, continuous spectrum. A spiked spectrum of oscillatory *g*-modes, a broad-band solar convective spectrum as well as a broad-band telluric spectrum could separately be responsible for this result.

would then contain contributions from modes with even l or odd l with almost coincident frequencies. The frequency separation between neighbouring modes with $l=0$ and $l=2$, and with $l=1$ and $l=3$ is expected to be $\sim 10\ \mu$Hz and $16\ \mu$Hz respectively[31]. Indeed, many peaks in Fig. 1 seem to be split into two or more components. To verify the significance of this splitting, we have used the equidistance to obtain the average shape of odd and even peaks by a 'superposed frequency analysis' of the spectrum between 2.4 and 4.8 μHz for an interval equal to 136.0 μHz. Figure 2 shows the result of this analysis, and clearly indicates that all modes between $l=0$ and $l=3$ and possibly even $l=4$ contribute to the power spectrum. It is quite reassuring that the $l=3$ modes display a significantly lower amplitude due to the smearing effect of the full disk integration.

In addition to the identification of the spherical harmonics associated with the spectral peaks, another important result emerges from Fig. 2. The width of the four main peaks indicates a Q value of about 600, or a typical damping time of 2 days. Indeed, if shorter data samples, of, say, 12 h, are Fourier analysed, the energy present in a given spectral peak does not seem to be constant in time or regularly modulated as would be expected from overstable beating pulsations. Rather, an individual peak seems to be suddenly excited and then slowly damped over a 2-day interval, unless it is re-excited before this time. A more complete statistical analysis of this temporal behaviour, embracing the total amount of data, will be necessary, but it is already clear that these modes around the 5-min period have a damping time of about 2 days and are spasmodically excited.

The intermediate frequency range

Figures 1 and 2 show the asymptotic behaviour of low degree global modes that must be predicted by solar models. Between 2.4 and 5 mHz the order n (the number of nodes along a solar radius in the vertical component of the displacement eigenfunction) must typically be between ~ 15 and 35 for these peaks[31]. To be able to identify each peak with one given order would require the detection of p modes with longer periods, and if possible all modes until the fundamental, the period of which is expected to be of the order of 1 h. Figure 1 shows that peaks with approximately uniform separation can be followed, with the limit of noise, towards low frequencies below 1 mHz. The same type of superposed frequency analysis, described above, over the range 0.8–2.4 mHz, demonstrates that $l=0$, $l=1$ and $l=2$ normal modes may be present in this frequency interval, with a mean amplitude of less than 3–5 cm s^{-1}. Comparing this information with theoretical model predictions, one can be assured that, for example, the first peak which emerges significantly above the noise level near 2.23 mHz must be an $l=0$–2 pair with $n=15\pm 1$. To suppress the indicated uncertainty of the n-number definitely, the first four or five spherical harmonics must be identified, and this is more difficult because the asymptotic frequency equidistance prediction is no longer valid in the frequency range 0.3–0.8 mHz. We hope to be able to eliminate this ambiguity eventually by including the totality of recorded data.

The long period range

For periods longer than 1 h, we obtain an increasing power spectrum with decreasing frequency. Figure 3 shows this long period power spectrum, for periods between roughly 30 and 300 min. Its spiked structure is consistent with pure statistical fluctuations around a broad, continuous curve. Although it may contain some solar information from oscillatory gravity wave or broad-band convective spectrum, this result alone does not allow us to conclude that any spectral peak is characteristic of a long-period solar oscillation.

The 160-min period

The presence of the 160-min solar oscillation originally discovered in the Crimea and observed in phase coherence during the past few years both in the Crimea and at Stanford was looked for using a superposed epoch analysis of the data processed thus far (5 days). Figure 4 shows the result of this analysis, together with the sine wave which would be expected by extrapolating the phase and amplitude according to the average of the Crimean

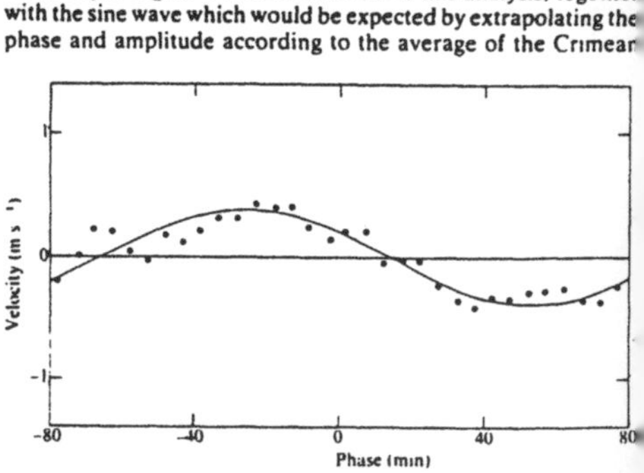

Fig. 4 The superposed epoch analysis of a data sample extending over 5 days (45 periods of 160 min). The points represent the South Pole data, and the solid line is the average based upon the observations obtained at the Crimean Observatory and Stanford.

and Californian results. The present result seems to be consistent with the latter. However, note that following a recent study of error analysis of superposed epoch procedure by Forbush *et al.*[30], standard methods of assigning statistical uncertainties may be not valid in this case.

Discussion

As we have analysed only part of the data processed so far, it would be premature to undertake a comprehensive discussion of theoretical implications concerning, for example, the depth of the convective zone, helium abundance, or other solar parameters. We have, therefore, only presented observational results which can be regarded as definitive:

(1) Although the 160-min oscillation, in phase with the wave expected from the latest Russian and American results, has been detected, further analysis is required to determine its significance.

(2) The power in the 5-min range is resolved into many equidistant peaks separated by 68 μHz. The splitting of these peaks, which is different for even and odd lines, indicates that global modes with l values of 0, 1, 2 and 3 have been identified.

(3) These global solar oscillations are not overstable, but are spasmodically excited, with a damping time of about 2 days ($Q = 600$).

(4) Comparison between theoretical and observed frequencies of a large number of different modes has made it possible to reduce the uncertainty in the specification of the order of the spherical harmonics to unity.

These results are very encouraging for future astronomical research at the South Pole.

This work was supported by the NSF's Division of Polar Programs under grant DPP-7822267. Assistance in the field was provided by Lyman Page Jr and Jon Towle. Numerous individuals associated with the US Antarctic Research Program in different capacities provided logistical help. We especially thank Robert Pfeiffer, Arthur Smith, Don McCauley, Max Azouit, Jean-Francois Manigault and Alex Robini for technical assistance and Dermott Mullan for helpful discussions.

Received 30 September; accepted 6 November 1980.

1. Mazzitelli, J. *Astr. Astrophys.* **79**, 251 (1979).
2. Davis, R. Jr, Evans, J. C. & Cleveland, B. T. *Proc. Conf. Neutrinos* **53** Lafavette (1978).
3. Gough, D. O. *2nd Assemblée Génerale Europeene de Physique Solaire*, 81 (CNRS, Paris, 1978).
4. Musman, S. & Nye, A. N. *Astrophys. J. Lett.* **212**, L95 (1977).
5. Livingston, W. D., Milkey, R. & Slaughter, C. *Astrophys. J.* **211**, 281 (1977).
6. Beckers, J. M. & Ayres, T. R. *Astrophys. J. Lett.* **217**, L69 (1977).
7. Deubner, F. L. *Astr. Astrophys.* **57**, 317 (1977).
8. Hill, H. A. & Stebbins, R. T. *Astrophys. J.* **200**, 471 (1975).
9. Hill, H. A., Stebbins, R. T. & Brown, T. M. *Atomic Masses and Fundamental Constants* Vol. 5, 622 (Plenum, New York, 1976).
10. Brown, T. M. thesis, Univ. Colorado (1977).
11. Brown, T. M., Stebbins, R. T. & Hill, H. A. *Astrophys. J.* **223**, 325 (1978).
12. Hill, H. A. & Caudell, T. P. *Mon. Not. R. astr. Soc.* **186**, 327 (1979).
13. Fossat, E., Harvey, J. W., Hausman, M. & Slaughter, C. *Astr. Astrophys.* **59**, 279 (1977).
14. Kenknight, C., Gatewood, G. D., Kipp, S. L. & Black, D. *Astr. Astrophys.* **59**, L27 (1977).
15. Clarke, D. *Nature* **274**, 670 (1978).
16. Fossat, E., Grec, G. & Harvey, J. W. *Astr. Astrophys.* (in the press).
17. Leighton, R. B. *Nuovo Cimento Suppl.* **22** (1961).
18. Deubner, F. L. *Astr. Astrophys.* **44**, 371 (1975).
19. Deubner, F. L., Ulrich, R. K. & Rhodes, E. D. *Astr. Astrophys.* **72**, 177 (1979).
20. Claverie, A., Isaak, G. R., McLeod, C. P., Van der Raay, H. B. & Roca Cortes, T. *Nature* **282**, 591 (1979).
21. Severny, A. B., Kotov, V. A. & Tsap, T. T. *Nature* **259**, 8 (1976).
22. Scherrer, P. M., Wilcox, J. J., Kotov, V. A., Severny, A. B. & Tsap, T. T. *Nature* **277**, 635 (1979).
23. Fossat, E. & Grec, G. *2nd Assemblée Génerale Europeenne de Physique Solaire*, 151 (CNRS, Paris, 1978).
24. Grec, G. & Fossat, E. *Astr. Astrophys.* **77**, 351 (1979).
25. Worden, S. D. & Simon, G. W. *Astrophys. J. Lett.* **210**, L1 (1976).
26. Willer, A. A. *Polar Research, A Survey*, 170 (National Academy of Sciences, Washington DC, 1970).
27. Pomerantz, M. A. *Antarctic Research Series* Vol. 29, 12 (American Geophysical Union, Washington DC, 1978).
28. Grec, G., Fossat, E. & Vernin, J. *Astr. Astrophys.* **50**, 221 (1976).
29. Kusoffsky, U. & Pomerantz, M. A. *Proc. IAU General Assembly*, Montreal (1979).
30. Forbush, S. E., Pomerantz, M. A. & Duggel, S. P. (in preparation).
31. Christensen-Dalsgaard, J. & Gough, D. O. *Nature* **288**, 544–547 (1980).

Internal rotation of the Sun

T. L. Duvall Jr[*], W. A. Dziembowski[†], P. R. Goode[‡], D. O. Gough[§], J. W. Harvey[‖] & J. W. Leibacher[‖]

[*] Laboratory for Astronomy and Solar Physics, NASA/Goddard Space Flight Center, Greenbelt, Maryland 20771, USA
[†] N. Copernicus Astronomical Centre, ul. Bartycka 18, 00-716 Warszawa, Poland
[‡] Arizona Research Laboratories and Department of Physics, University of Arizona, Tucson, Arizona 85721, USA
[§] Institute of Astronomy, and Department of Applied Mathematics and Theoretical Physics, University of Cambridge, Madingley Road, Cambridge CB3 0HA, UK
[‖] National Solar Observatory, Tucson, Arizona 85726, USA

The frequency difference between prograde and retrograde sectoral solar oscillations is analysed to determine the rotation rate of the solar interior, assuming no latitudinal dependence. Much of the solar interior rotates slightly less rapidly than the surface, while the innermost part apparently rotates more rapidly. The resulting solar gravitational quadrupole moment is $J_2 = (1.7 \pm 0.4) \times 10^{-7}$ and provides a negligible contribution to current planetary tests of Einstein's theory of general relativity.

SOLAR oscillations can be used to probe the structure and dynamics of the Sun. In particular, rotation produces, in the spectrum of acoustic (p-mode) oscillations, fine structure that can be used to determine the internal rotation[1-3]. Knowledge of the variation of rotation rate with depth and latitude would allow us to determine the gravitational quadrupole moment of the Sun, and would provide valuable information about the solar dynamo. Furthermore, we would gain new insight into the internal structure and history of the Sun.

The individual normal modes of oscillation of the Sun can be characterized by a radial order n, and a spherical harmonic degree l and azimuthal order m. In the absence of rotation and magnetic fields, each (n, l) set is $(2l+1)$-fold degenerate with respect to m. Rotation breaks this degeneracy, producing rotational splitting in the frequency spectrum: modes propagating in the sense of the rotation are shifted to higher frequency, while modes propagating against the rotation are shifted to lower frequency. Acoustic oscillation modes of low degree penetrate deeply into the interior, while high-degree modes are confined near the surface. Measurement of rotational splitting for modes of different degrees provides a means of determining the variation of rotation rate with depth.

Fine structure in the spectrum of low-degree solar oscillations, attributed to rotation, has been observed. Claverie *et al.*[4] reported a fine structure consisting of $2l+1$ peaks in their spectra of observations of low-degree $(l = 1, 2)$ 5-min oscillations in integrated sunlight. If, however, the fine structure they saw were due to rotation about a single axis, their spectra should have shown only $l+1$ peaks. Isaak[5] suggested that the fine structure was actually due to an inclined, rotating magnetic field much like that postulated earlier by Dicke[6]. Woodard and Hudson[7] have established an upper limit of twice the surface rotational splitting for $l = 1, 2$ modes. This limit is consistent with the result found by Claverie *et al.*[4] Duvall and Harvey[1] reported a consistent value for $l = 1$, but their value for $l = 2$ differs from that of Claverie *et al.*[4]. The finite lifetimes of these modes has been suggested[1] as a possible source of the discrepancy.

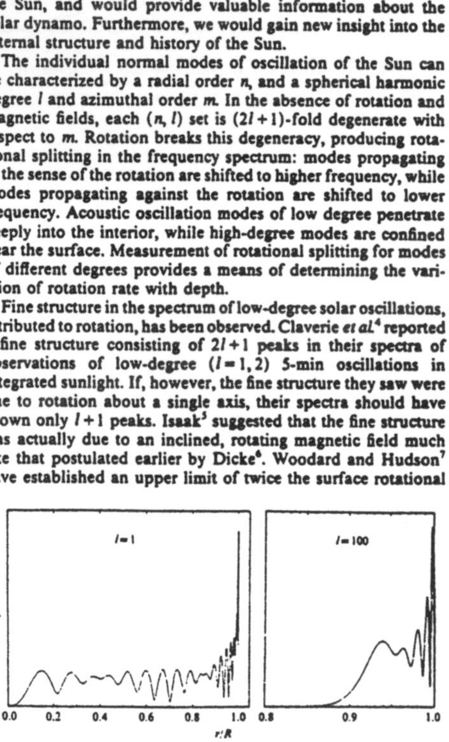

Fig. 1 The square roots of the averaged splitting kernels, $\sqrt{K_l}$, for $l = 1$ and 100 as functions of fractional radius r/R. The $l = 1$ kernel is an average of the two kernels corresponding with the two observed values of n. The ordinate scales are arbitrary.

Bos and Hill[8] observed low-frequency oscillations of limb darkening. Hill *et al.*[9] and Gough[10] used these results to infer a steep gradient in the rotation rate in the solar convection zone and an interior rotating about six times more rapidly than the surface. On the other hand, low-frequency Doppler oscillations analysed by Delache and Scherrer[11] suggested that the region below the convection zone rotates approximately only half that rapidly.

Duvall and Harvey[1] present values for rotational splitting, about the surface axis, of prograde $(m = -l)$ and retrograde $(m = +l)$ sectoral oscillation modes in the frequency range 2.1–3.7 mHz and with l between 1 and 100. Frequency differences between 180 prograde and retrograde mode pairs were computed according to

$$\nu'_{nl} = \frac{\nu_{n,l,m=-l} - \nu_{n,l,m=+l}}{2l} \tag{1}$$

These were then averaged over observed values of n, with equal weight to increase significance without appreciably degrading the resolution in radius, for each l to produce 37 values of rotational splitting, ν'_l. We use these values here to determine the internal rotation rate of the Sun.

Rotational splitting

In general, the rotation rate may be a function of radius and latitude. However, we have little information on the dependence of the rate on latitude from sectoral modes alone. These modes are concentrated near the equator, between latitudes $\approx \pm l^{-1/2}$ when l is large, therefore, we really measure only an equatorially-concentrated average of the rotation rate over a varying, restricted range of latitude. Because of this concentration, sectoral modes are well suited for investigation of the radial variation of rotation near the equatorial plane. Assuming that the rotation frequency ν_{rot}, depends only on radius r within an appropriate latitude range, the frequency splitting, ν'_{nl}, is given[12,13] by

$$\nu'_{nl} = \int_0^R K_{nl}(r)\nu_{rot}(r)\,dr \tag{2}$$

where R is the solar radius, and K_{nl} is the splitting kernel, which we have calculated from two essentially equivalent solar models (ref. 14 and H. Saio, personal communication). The averaged rotational splitting ν'_l satisfies a similar equation with K_{nl} replaced by the uniformly weighted average of appropriate kernels, K_l. If ν'_l were known for a sufficient variety of oscillations, we could obtain $\nu_{rot}(r)$ by solving the averaged integral equation (2). We present the averaged splitting kernels, K, for the lowest ($l=1$) and highest ($l=100$) degree oscillations in Fig. 1 to illustrate the range of depth penetration for the acoustic modes that we used. Slightly more than half the contribution to the $l=1$ kernel comes from beneath the convection zone, whereas the $l=100$ kernel does not penetrate even to the base of the convection zone. Although the lowest-degree oscillations do penetrate into the deep core, most of the information contained in these modes comes from the outer regions, and the calculated rotational frequency of the core must be the most uncertain.

Inversion of data

We present three different inversions of the integral equation (2): simple fitting of a polynomial, an optimal averaging procedure suggested by the work of Backus and Gilbert[15], and a fitting of a piecewise constant representation of $\nu_{rot}(r)$. Figure 2 shows a least-squares fit of an eighth-degree polynomial representation of $\nu_{rot}(r)$ to the data. The solution is essentially flat for $r>0.4R$, with only modest deviations from the equatorial surface rate. For $r<0.4R$, where the results are less certain, the curve shows a dip and then a sharp rise with decreasing r. The general features of this fit are reproduced by the other two techniques. The curve is cut off for $r<0.1R$, because only 1% of the deepest kernel ($l=1$) penetrates this region. All low-degree polynomial fits have structure similar to that shown in Fig. 2, although the positions of the maxima and minima vary with degree, and the rise at small values of r becomes more dramatic with increasing degree. Higher-degree polynomials show finer scale structure than the data can resolve.

The optimal averaging technique has the advantages that the depth resolution and error magnification are readily estimated and that a tradeoff can be made between these two quantities—increasing resolution leading to larger error magnification. The method seeks weighted averages, D, of the splitting kernels that are concentrated at different radii, r_0 and that have unit integral with respect to r:

$$D(r, r_0) = \sum_l \alpha_{l,r_0} K_l(r) \tag{3}$$

If we were successful in concentrating D about r_0, then D resembles the delta function, $\delta(r-r_0)$, and we might set

$$\nu_{rot}(r_0) = \int_0^R D(r, r_0)\nu_{rot}(r)\,dr = \sum_l \alpha_{l,r_0}\nu_l' \tag{4}$$

In our calculations, we have adjusted the parameter affecting the width of D so that the relative error in $\nu_{rot}(r)$ is small. The procedure was successful within the convection zone, $r>0.7R$, with the results shown in Fig. 3. Unfortunately, the straightforward optimal averaging procedure fails to generate D's that are well-concentrated deep in the Sun, because all of the K's have large amplitudes near the surface. With only 37 different K's there is no linear combination that is large in the core and that adequately cancels near the surface. To overcome this difficulty, we truncated the kernels by setting $K=0$ throughout the convection zone and subtracting from the splitting data the contribution from that zone. The latter was computed using a smooth curve drawn through the points in Fig. 3 at radii $>0.7R$. The optimal averaging procedure was then applied to the truncated kernels and the adjusted data to provide the solution for $r<0.7R$ shown in Fig. 3. Of course, we could have continued by successively truncating the kernels at smaller and smaller radii, but we did not for fear that the results would be unreliable.

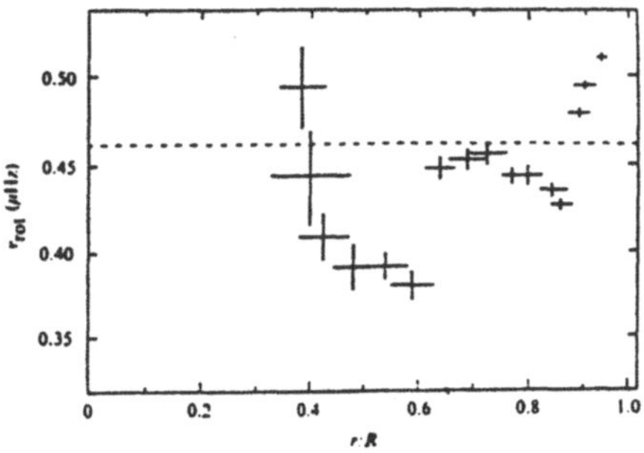

Fig. 3 An optimal-averaging-procedure solution of $\nu_{rot}(r)$ as a function of fractional radius r/R. The solution below $r=0.7R$ used a modified procedure discussed in the text. Bar widths indicate the halfwidth of the D kernels and bar heights reflect the standard errors of the observations.

Finally, we represented $\nu_{rot}(r)$ by a simple, piecewise constant function—a sequence of bins with ν_{rot} constant in each. The inversion of equation (2) then reduces to a set of linear, algebraic equations for the unknown rotation frequencies. We solved the problem by minimizing χ^2 where χ^2 is the sum of the 37 ratios of the square of the difference between each fit and the measured splitting to its variance. Ten bins were used, and the widths of the optimal averaging kernels, D, obtained from the Backus and Gilbert procedure, were used to estimate their relative widths, except for the inner two where the method fails; for these the widths were assigned arbitrarily. Figure 4a shows the rotational frequency by bin and the width of each bin. The bars reflect the uncertainties in the observed splittings, attributing no error to the solar model. Only the outer part of the innermost bin is plotted to emphasize that the rotational splitting kernel is significant only in the outer part of that bin. Figure 4a is quite similar to Fig. 3, except at small radii where the optimal averaging procedure fails to give a reliable result. The largest differences between Fig. 4a and the polynomial fit of Fig. 2 occur in the deep core where the calculations are most uncertain. All three solutions (Figs 2–4a) show much of the interior rotating at or below the equatorial surface rate.

To explore the dependence of the inferred internal rotation shown in Fig. 4a on the observed splittings, we minimized χ^2 ignoring the observed $l=90$ and 100 splittings which differ markedly from the other high-degree data. The only appreciable changes shown in Fig. 4b are an increase in the difference between the rotation rates in the two bins nearest the surface and an increase in the uncertainty of those two rates. This emphasizes the important role of the $l=90$ and 100 splittings in determining the rotation rate near the solar surface. We note that there were no noticeable changes in the deeper rotation rates. Similarly, removal of $l=1$, 2 and 3 data successively from the solution gave results for $r<0.4R$ very similar to Fig. 4a

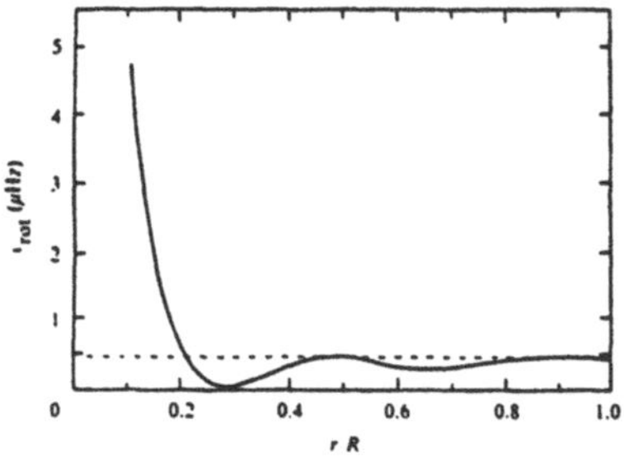

Fig. 2 An eighth-degree polynomial representation of rotation frequency $\nu_{rot}(r)$ as a function of the fractional radius r/R. For comparison, the surface rotation frequency at the equator is the dashed line at ~0.46 μHz.

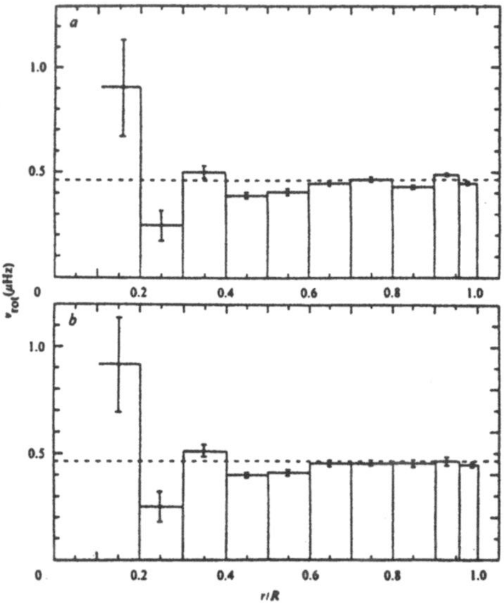

Fig. 4 *a*, A piecewise constant solution of $\nu_{rot}(r)$ as a function of fractional radius r/R. The fit to the observations was done by minimizing χ^2. The average error bars indicate the observational uncertainties. *b*, As *a*, except observations of $l = 90$ and 100 were ignored.

except with increasing uncertainty. Thus, the largest departures from a rigid rotation rate in Fig. 4*a* arise from the low and high degree splittings; unfortunately, these are the most difficult to measure. Without them, the observations are consistent with rigid rotation, at a statistically significantly slower rate than the equatorial surface value. Thus the three methods yield similar internal rotation. Figure 4*a* is a good representation of the radial variation of rotation that can be inferred from the available data. Measurements of non-sectoral modes would enable us to explore the latitude dependence. Improved measurements of low- and high-degree rotational splitting may significantly change our results for $r < 0.3R$ and $r > 0.9R$.

Discussion

We now consider some of the theoretical implications of our results. Note that some of the structure in our $\nu_{rot}(r)$ could be due to latitude variations of rotation. However, a latitudinal variation alone cannot explain both the observed surface rotation and low-l splitting measurements that are most sensitive to latitude effects. Many models (reviewed in ref. 16) of the convection zone suggest a relative maximum in $\nu_{rot}(r)$ near the surface that is consistent with our results. Between 0.9 and 0.6R there is little structure in the rotation rate on the resolution scale implied by the data, which implies that there is no significant change in rotation rate at the base of the convection zone. The general flatness of the rotation rate in the convection zone is consistent with the Solberg–Høiland criterion for dynamical stability[17]. The nearly constant rotation just beneath the convective envelope is stable to the Goldreich–Schubert–Fricke instabilities[18,19]. For $r < 0.4R$ our $\nu_{rot}(r)$ is significantly below the minimum suggested in ref. 20.

The dimensionless measure, J_2, of the solar gravitational quadrupole moment obtained from the rotation curve in Fig. 4*a* is $(1.7 \pm 0.2) \times 10^{-7}$. The quoted error follows from the observational uncertainties. An additional uncertainty of $\pm 0.3 \times 10^{-7}$ arises from the range of plausible interior models, so the total uncertainty associated with our value is $\pm 0.4 \times 10^{-7}$. The value of J_2 is slightly less than the value for rigid rotation, 1.8×10^{-7}, reflecting that much of the interior is rotating slightly slower than the surface. J_2 was calculated using

$$J_2 = \int_0^R F(r)\nu_{rot}^2(r)\,dr \qquad (5)$$

where $F(r)$ is the J_2 kernel plotted in ref. 10. Our rotation rate is most uncertain for $r < 0.3R$ and $r > 0.9R$, where $F(r)$ is quite small. By using equation (5), we have assumed that ν_{rot} is independent of latitude throughout the Sun. Thus although our quoted error might be optimistic, we point out, however, that the dominant contribution to J_2 comes from low latitudes, the region where we have measured ν_{rot}.

According to parametrized, post-newtonian theories of gravitation, the predicted advance of the perihelion of Mercury is

$$\dot{\omega} = 42.95[(2+2\gamma-\beta)/3 + 2.9 \times 10^3 J_2] \text{ arc s per 100 yr} \qquad (6)$$

where γ and β are parameters which are both unity in general relativity. Combining our value of J_2 with the planetary data of Shapiro *et al.*[21] and that of Anderson *et al.*[22], we have

$$(2+2\gamma-\beta)/3 = 1.002 \pm 0.005 \text{ and } 1.006 \pm 0.005 \qquad (7)$$

Our value of J_2 makes an essentially negligible contribution to equation (7). In particular, the contribution of the J_2 term from equation (6) to equation (7) is a factor of 10 less than the errors quoted in equation (7), which are due to the planetary data. Therefore, at the present level of accuracy of the planetary data our value of J_2 is not in conflict with general relativity.

As noted above, there is an apparent discrepancy between observational estimates of the internal rotation. We calculate a markedly slower internal rotation rate than Hill *et al.*[9]. The discrepancies between the two results are much larger than the measurement uncertainties appear to allow. Unfortunately, there are no modes common to both data sets, so a direct comparison of splittings is not possible. The theory by which the interior rotation rate is calculated from a set of rotational splittings is sufficiently well understood that the two different sets of splittings should yield comparable rotation rates, at least if ν_{rot} is a slowly varying function of r.

It is mathematically possible (see ref. 23) to find a function ν_{rot} that is consistent with all of the data, but we find it implausible that the details of the rapid rotation variation that would then be required would be such as to impart high rotational splitting just to the modes that happen to have been identified in one set of observations. This suggests that there are systematic errors in at least one set of results. The most likely source of systematic errors is mode misidentification, that is, either a feature in a spectrum that is not a mode is wrongly identified as a mode or a feature is assigned incorrect values of n, l, m. The relative simplicity and high resolution in l in the spectra of Duvall and Harvey[1] compared with Bos and Hill[8], plus the unexpected appearance (Fig. 1 of ref. 9) of zonal modes in the latter analysis, lends confidence to the present work.

This work was done while T.L.D. was a visiting astronomer at the National Solar Observatory in Tucson. The National Solar Observatory is operated by the Association of Universities for Research In Astronomy, Inc. under contract with the NSF. The work of P.R.G. was partially supported by the NSF Astronomy Division and the Air Force Office of Scientific Research.

Received 16 March; accepted 30 April 1984.

1. Duvall, T. L. Jr & Harvey, J. W. *Nature* 310, 19–22 (1984).
2. Cowling, T. G. & Newing, R. A. *Astrophys. J.* 109, 149–158 (1949).
3. Ledoux, P. *Astrophys. J.* 144, 373–384 (1951).
4. Claverie, A., Isaak, G. R., McLeod, C. P., van der Raay, H. B. & Roca Cortes, T. *Nature* 293, 443–445 (1981).
5. Isaak, G. R. *Nature* 296, 130–131 (1982).
6. Dicke, R. H. *Astrophys. J.* 228, 898–902 (1979).
7. Woodard, M. & Hudson, H. *Bull. Am. astr. Soc.* 15, 951–952 (1983).
8. Bos, R. J. & Hill, H. A. *Sol. Phys.* 82, 89–102 (1983).

9. Hill, H. A., Bos, R. J. & Goode, P. R. *Phys. Rev. Lett.* **49**, 1794–1797 (1982).
10. Gough, D. O. *Nature* **298**, 334–339 (1982).
11. Delache, P. & Scherrer, P. *Nature* **306**, 651–653 (1983).
12. Hansen, C. J., Cox, J. P. & Van Horn, J. M. *Astrophys. J.* **217**, 151–159 (1977).
13. Gough, D. O. *Mon. Not. R. astr. Soc.* **196**, 731–745 (1981).
14. Christensen-Dalsgaard, J., Gough, D. O. & Morgan, J. G. *Astr. Astrophys.* **73**, 121–128 (erratum **79**, 260) (1979).
15. Backus, G. & Gilbert, F. *Phil. Trans. R. Soc.* **A266**, 123–192 (1970).
16. Gilman, P. A. in *Physics of the Sun* (Reidel, Dordrecht, in the press).
17. Zahn, J.-P. *IAU Symp.* **59**, 187–194 (1974).
18. Goldreich, P. & Schubert, G. *Astrophys. J.* **150**, 571–587 (1967).
19. Fricke, K. *Z. Astrophys.* **68**, 317–344 (1968).
20. Spruit, H. C., Knobloch, E. & Roxburgh, I. W. *Nature* **304**, 520–522 (1983).
21. Shapiro, I. I., Counselman, C. C. & King, R. W. *Phys. Rev. Lett.* **36**, 555–558 (1976).
22. Anderson, J. D., Keesey, M. S. W., Lau, E. L., Standish, E. M. & Newhall, X. X. *Acta astronaut.* **5**, 43–61 (1978).
23. Gough, D. O. *Mem. Soc. astr. Ital.* (in the press).

Seismology of the Sun

Jørgen Christensen-Dalsgaard, Douglas Gough, Juri Toomre

Summary. Oscillations of the sun make it possible to probe the inside of a star. The frequencies of the oscillations have already provided measures of the sound speed and the rate of rotation throughout much of the solar interior. These quantities are important for understanding the dynamics of the magnetic cycle and have a bearing on testing general relativity by planetary precession. The oscillation frequencies yield a helium abundance that is consistent with cosmology, but they reinforce the severity of the neutrino problem. They should soon provide an important standard by which to calibrate the theory of stellar evolution.

The sun is an oscillating star. The oscillations are imperceptible to the naked eye, but they have been revealed by Doppler shifts in spectral lines that are formed in the solar atmosphere. The pattern of motion is complicated and for a long time was not understood, but now we know that it is a result of the interference between about 10^7 resonant modes of vibration, many of which are coherent over the entire surface of the sun. The modes have wavelengths greater than a few thousand kilometers and periods from a few minutes to several hours. The frequencies carry information about the structure and dynamics of the region where the modes have appreciable amplitudes, which in many cases spans much of the solar interior. The unraveling of this information, which is analogous to seismological studies of Earth, is rapidly evolving into a new branch of solar physics known as helioseismology.

Solar physicists grapple with a complex dynamical problem. The fluid in the outer layers of the sun is highly turbulent and carries magnetic fields, which it distorts into intricate configurations. In some places the field is concentrated into small regions that darken into sunspots; elsewhere the field erupts, causing a temporary local brightening or flare. Charged particles are ejected that subsequently disturb Earth's ionosphere, causing magnetic storms and inhibiting radio communication in particularly severe instances. The rotation of the sun is not uniform, and this feature is commonly believed to be an essential ingredient of a dynamo that maintains and modulates the magnetic activity and is responsible for the 11-year cycle of sunspots. Such magnetic activity may influence the terrestrial climate. Solar variability with longer characteristic time scales is also suspected. Yet despite this complexity of behavior, the sun is believed to be among the simplest of stars.

One of the major assumptions of stellar physics is that these complicated phenomena are not relevant to our understanding of the large-scale structure of the solar interior. Therefore relatively simple theoretical models suffice to describe the structure and evolution of a star such as the sun. In these models, the sun is treated as a perfect sphere powered by thermonuclear reactions that gradually convert hydrogen into helium in the high-temperature core. The energy is liberated in the inner 30 percent (by radius) and is transported outward by radiative diffusion. In the outer layers the energy is carried predominantly by turbulent convection, because the stellar material is too opaque to permit the radiation to pass. The thickness of the convecting region is poorly determined by theory, for it depends on the uncertain ratio of the abundances of hydrogen and helium at the time the sun was formed; values of the thickness ranging from less than 20 percent to about 30 percent of the radius of the sun have been commonly believed. But this uncertainty is not all: theoretical predictions of the rate of production of neutrinos by the nuclear reactions in the core are three times greater than the measured value. Is this a minor error, or is it a symptom of a fundamental flaw in the theory of stellar evolution? Helioseismological studies may play a major role in answering that question.

In the past few years we have started to consolidate our picture of the inside of the sun. Initially, seismological inferences were used to calibrate theoretical solar models. Thus it became possible to estimate the solar helium abundance, a quantity of considerable cosmological interest. It also led to an estimate of the depth of the base of the convection zone. This is important for our understanding of the dynamics of the outer envelope of the sun and is pertinent to theories of both the solar cycle and the long-term variability. More recently, seismological deductions have been made by means of so-called inverse methods, which are free from the uncertain details of the theory of stellar structure. In particular, it has been possible to measure the angular velocity throughout much of the solar interior and so to evaluate the oblateness of the sun's gravitational field. A knowledge of the latter is required for testing theories of gravity from observations of planetary orbits. It has also been possible to infer the sound speed throughout much of the sun.

It is the purpose of this article to describe the seismological inferences

J. Christensen-Dalsgaard is a lecturer in astronomy at the Astronomisk Institut, Aarhus Universitet, DK 8000 Aarhus C, Denmark, and a research associate at NORDITA, Copenhagen, Denmark. D. Gough is a lecturer in astronomy and applied mathematics at the Institute of Astronomy and the Department of Applied Mathematics and Theoretical Physics, University of Cambridge, Cambridge CB3 OHA, England. J. Toomre is a professor in the Department of Astrophysical, Planetary, and Atmospheric Sciences and a fellow of the Joint Institute for Laboratory Astrophysics, University of Colorado, Boulder, Colorado 80309.

that have already been made and to provide a basis for anticipating what is likely to be learned in the foreseeable future. The knowledge is important not merely for establishing a more accurate picture of the inside of the sun but to provide a standard with which to compare our theoretical models of other stars.

First Helioseismological Inferences

Solar oscillations were discovered in 1960 by Leighton, Noyes, and Simon (*1*), who found that typically about half the surface of the sun is occupied by patches that oscillate intermittently with periods near 5 minutes and with amplitudes of about 1 km sec^{-1}. The oscillations persist for six or seven periods with a spatial coherence of about 30,000 km, or 2 percent of the diameter of the sun; consequently they were for a long time regarded as a local phenomenon, triggered possibly by eruptions in the convection zone. Some 10 years later Ulrich, and independently Leibacher and Stein, proposed that these oscillations are a superposition of coherent acoustic modes (*2*). What excites the modes is an issue of unresolved debate (*3*).

The major restoring forces responsible for solar oscillations are pressure and buoyancy. Which is the more important force depends largely on frequency and to a lesser extent on wavelength. Pressure fluctuations dominate at high frequency, producing acoustic waves; at low frequency buoyancy dominates, producing internal gravity waves. Waves propagating through the sun may undergo reflection where appropriate conditions are satisfied and can thus be confined within a cavity. A resonant mode of oscillation is the result of constructive interference between such internally reflected waves. Thus it is analogous to a standing wave in, say, an organ pipe. And, as in an organ, resonance can occur only at particular frequencies, which depend on conditions in the cavity. But in an organ, as in almost all musical instruments, the oscillations take place in only one dimension; the sun oscillates in three dimensions, and this clearly permits a far richer spectrum of tones.

Standing acoustic waves are known as *p* modes. In the sun, they have periods between about 3 minutes and 1 hour. It is these modes that have been the most widely observed and have provided nearly all the useful diagnostic information. Consequently they are the principal subject of our discussion. Standing internal gravity waves are called *g* modes. Their periods exceed about 40 minutes; they depend on buoyancy, whose magnitude is determined by the stratification of density and pressure and is sensitive to the distribution of chemical elements within the sun.

Two typical ray paths of acoustic waves are shown in Fig. 1. They lie in a plane that passes through the center of the sun (*4*). The surface layers of the sun, where the density of the solar material varies rapidly with depth, act as a reflecting boundary. Downward propagating waves are refracted back upward, unless the motion is precisely vertical. This comes about because temperature, and consequently sound speed, increase with depth, causing deeper parts of the wave fronts to travel more quickly (*5*). Thus the waves are confined within a cavity that excludes a central region of the sun. The more nearly vertical a wave is as it is reflected at the surface, the greater is the depth of the acoustic cavity. On the whole, if the wavelength λ of the disturbance on the surface of the sun is shorter, the waves travel more nearly horizontally, and the cavity within which they are confined is shallower.

For each value of λ there is a fundamental acoustic frequency and a sequence of overtones. They are identified by an integer n, which increases with increasing frequency. Theoretical calculations by Ulrich (*2*), and subsequently by Ando and Osaki (*6*), showed how the frequencies ν were expected to depend on n and λ.

The theory was confirmed observationally by Deubner (*7*), who in 1975 made continuous measurements lasting many hours of Doppler velocity in an equatorial strip on the solar disk (*8*). Fourier transforms in longitude and time resulted in the determination of ν as a function of the horizontal wave number $2\pi\lambda^{-1}$. The latter is related to an integer l, called the degree of the mode, which measures the number of wavelengths that make up the circumference of the sun (*9*). A power spectrum of subsequent, more accurate observations is displayed in Fig. 2.

The frequencies depend also on the sound speed in the acoustic cavity. This is illustrated in Fig. 2, where the eigenfrequencies of two theoretical models of the sun are superposed on the power spectrum of the observations. Evidently the data select one of the models and suggest that the initial helium abundance Y of the sun is about 25 percent (by mass).

It is clear from Fig. 2 that only the frequencies of modes with quite large values of l can be discerned. These modes have small wavelengths λ and are therefore confined to only a shallow cavity; the depth to which the *p* modes

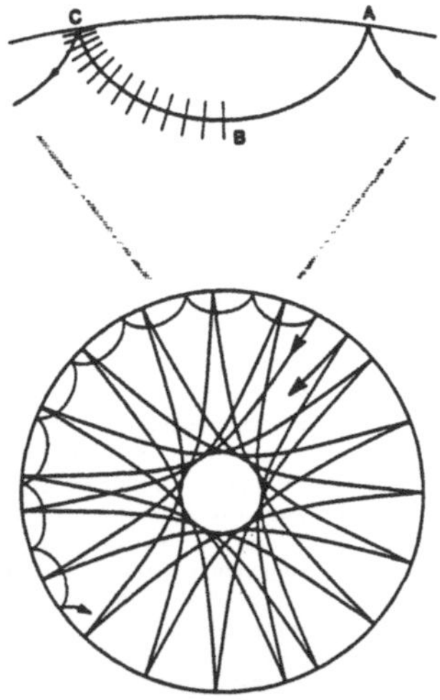

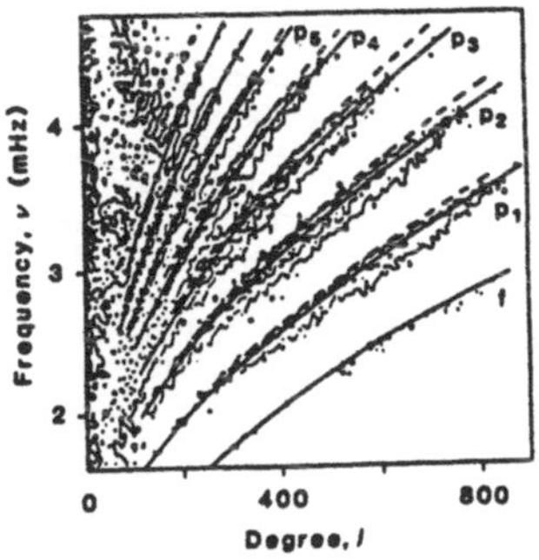

Fig. 1 (left). Ray paths of acoustic waves in the sun. The more deeply penetrating rays refer to waves with n/l = 5; the waves confined to the shallower cavity have n/l = 1/20. The circle represents the surface of the sun. In the enlargement, the lines intersecting the ray path represent wave fronts at equal intervals of phase; they are most closely spaced near the surface where the sound speed is least. At the level B the wave propagates horizontally and travels at the same angular phase speed $\lambda\nu$ about the center of the sun, as does the disturbance on the surface. Fig. 2 (right). Spectrum of Doppler data (*43*) showing contours of constant $l^{1/2}$ times power. Superposed on the ridges are the eigenfrequencies of two theoretical solar models. The continuous lines are for a model of the present sun whose helium abundance Y was 0.25 at the time the sun was formed; the dashed lines are for Y = 0.19. The theoretical curves are labeled p_n for p modes of order n and f for the f modes. The theoretical values of the latter for the two solar models are indistinguishable at the high values of l for which they are plotted. The continuous lines pass through the regions of maximum power, suggesting that Y = 0.25. [After (*44*)]

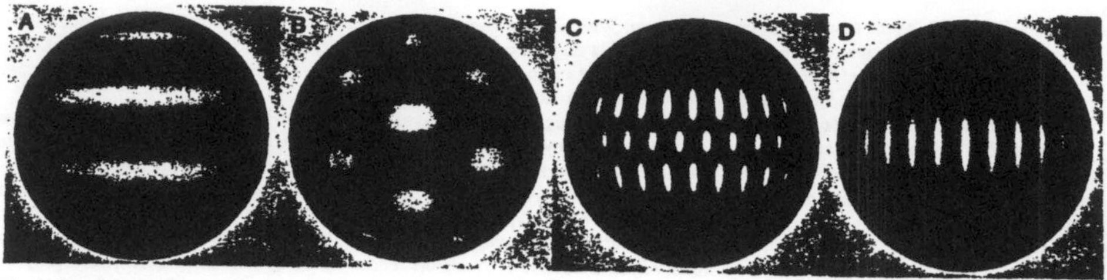

Fig. 3. Doppler velocities of solar p modes. The shading represents the line-of-sight component of velocity: dark regions are approaching the observer and light regions are receding (or vice versa). The motion is almost radial, so that the mid-grey at the edges of the sun's image represents zero velocity. (A) The zonal mode ($m = 0$) of degree $l = 10$; (B and C) tesseral modes with $(l, m) = (20, 10)$ and $(32, 30)$; (D) a sectoral mode ($m = l$) of degree 30.

penetrate is about $(2n + 3)/l$ of the radius of the sun. Therefore the determination of Y is indirect, resting heavily on the theory of stellar evolution from which the models of the sun were computed. What is measured directly is the upper part of the convection zone, and from that it has been possible to estimate that the thickness of the convection zone is 30 percent of the solar radius (*10, 11*). This was the first diagnosis from helioseismology.

Interest in the solar helium abundance had been stimulated partly by the neutrino problem. Only if Y were very low could solar models be constructed with neutrino fluxes that agreed with observation, unless the assumptions of stellar evolution theory were modified. But such low values are in conflict with many theories of cosmology, which predict Y to be in excess of 20 percent immediately after the Big Bang and to increase slowly afterward. The helioseismological calibration illustrated in Fig. 2, and a calibration from low-degree modes that will be discussed later, are consistent with cosmology. But they reinforce the severity of the neutrino problem and the doubt that is cast on the premises of stellar evolution theory (*12*).

Classification of Solar Oscillations

Because the amplitudes of solar oscillations are small, they can safely be treated as small perturbations about a spherical equilibrium state. This permits the separation of the oscillations into normal modes, which simplifies the theoretical analysis. In spherical polar coordinates (r, θ, ϕ), the radial component of the velocity of a mode with cyclic frequency ν can be expressed as

$$V_{nl}(r)\, P_l^m(\cos\theta)\, \cos(m\phi - 2\pi\nu t) \qquad (1)$$

where t is time and P_l^m is the associated Legendre function of degree l and order m. The coordinate r is the distance from the center of the sun, θ is colatitude, and ϕ is longitude.

Figure 3 shows how several individual modes might appear to an observer. At any given instant of time, the pattern of Doppler velocities of an individual mode is one of alternating regions of approaching (shown as light tones) and receding (dark tones) flow. For zonal modes ($m = 0$), all the nodal lines of the spherical harmonics are lines of latitude; for sectoral modes ($m = l$) they are lines of longitude. More common are the tesseral modes such as those in Fig. 3, B and C. Of course in reality the motion of the sun is a superposition of many such modes, and the observer's signal is much more complicated.

The functions V_{nl} are the eigenfunctions of a system of linear ordinary differential equations, the associated eigenvalues ν_{nl} being the frequencies ν of the modes. There are two discrete spectra of modes for given values of l and m: the p modes and the g modes (when $l = 0$ only the p sequence exists). The modes in each spectrum can be arranged in ascending order of frequency and period, respectively, and labeled with the integer n, which is the order of the mode. Typically the eigenfunctions $V_{nl}(r)$ possess n zeros.

In addition there is a mode whose frequency lies between those of the p modes and the g modes; it is called the f mode and exists only for $l \geq 2$. Except when l is low, it has no nodes. It is essentially a surface gravity wave, and when l is large its frequency is independent of the stratification of the equilibrium state of the sun. It can therefore be easily identified in a power spectrum such as that depicted in Fig. 2. Were the sun to be perfectly spherical, ν_{nl} and V_{nl} would be independent of m. This comes about because m depends on the axis of coordinates, and with spherical symmetry all choices must be equivalent.

The computation of theoretical eigenfrequencies must be performed numerically. For a given solar model, frequencies can now be calculated without undue computational effort to a precision as high as the observational accuracy. Results from a typical solar model are illustrated in Fig. 4; frequencies with like values of n are connected with straight lines. In all cases ν increases with l at fixed n; for the f and p modes ν varies roughly as a positive power of l, but for the g modes ν approaches a finite limit. That limit is the maximum value of the buoyancy frequency (*13*) beneath the convection zone, which in the model illustrated here is 0.47 mHz. This general behavior continues to higher values of l, as can be seen for the f and p modes in Fig. 2.

Acoustic Modes of Low Degree

Although quantitative assessment of theoretical solar models requires the numerical solution of the full eigenvalue problem, it is helpful also to carry out simpler approximate analyses. The physical understanding that is derived provides a valuable guide to designing useful comparisons between theory and observation. Moreover, it also enables us to appreciate the physical significance of any discrepancies that are found. Thus, after Deubner announced that the solar frequencies were lower than theoretical predictions, it was from a simple analysis of the oscillations of a polytrope that the principal implication was first recognized: the convection zones in the theoretical models were too shallow (*10*). The repercussions concerned not only the large-scale structure of the sun but also the dynamics of the convection zone. Of course more accurate numerical computations were required to estimate by how much the depth of the convection zone needed to be augmented (*11*).

After obtaining the high-degree oscillation spectrum, the next step forward came in 1979 when Isaak and colleagues detected a uniformly spaced sequence of frequencies in the range of 2 to 4 mHz in Doppler shifts in light integrated from the entire disk of the sun (*14*). The observations employed resonance scattering from gaseous sodium or potassium, which has high spectral stability and a low noise level. Since there was no spatial resolution, a direct determination of l and m was not possible. Nevertheless, it could be calculated that only modes with $l \leq 3$ were visible (*15*), which implies that n is quite large (Fig. 4). For such modes it was known that (*16*)

$$\nu_{nl} = \left[n + \frac{1}{2}\left(l + \frac{1}{2}\right) + \alpha\right]\nu_0 + \epsilon(n,l) \qquad (2)$$

where

$$\nu_0 = \left(2\int_0^R \frac{dr}{c}\right)^{-1} \qquad (3)$$

and α is a constant of order unity, c is the speed of sound, R is the radius of the sun, and ϵ is a small correction term. The functional form of ϵ had not yet been derived, but it was known from numerical computations to be too small to have been resolved by the observations. Thus it was deduced that modes of degrees 0 and 2 and modes of degrees 1 and 3 contributed alternately to the peaks in the power spectrum of the data (*15*). The spacing between the peaks must therefore be $\frac{1}{2}\nu_0$, which measures the sound travel time from the center to the surface of the sun.

Not long after, Fossat and colleagues (*17*) improved the temporal resolution by making similar observations from the South Pole. They obtained an uninterrupted record lasting 120 hours; it was possible to distinguish between different modes with like values of $n + \frac{1}{2}l$ and so to measure ϵ. Fortuitously at about the same time, Tassoul (*18*) presented an asymptotic analysis that gave not only the functional dependence of ϵ on n and l but also implied that ϵ is a measure of conditions close to the center of the sun. Her results predicted that the separation Δ between visible modes of degrees 0 and 2 with like $n + \frac{1}{2}l$ is three-fifths of the separation of the corresponding modes of degrees 1 and 3, whatever the structure of the sun. The 3:5 ratio was evident in the South Pole data (*19*) and, together with a comparison between power in the spectrum and the relative sensitivity of the measurements to modes of different l, provided the determination of the degrees of the modes (*20*). Subsequently, even more highly resolved data were obtained by Isaak and colleagues (*21*), who combined observations from Hawaii and Tenerife over a period of 3 months. Figure 5 is a power spectrum of their data.

The low-degree modes are commonly considered to provide a more robust means of calibrating solar models than the high-degree modes, for they depend on integrated properties of the entire star. Initially there was some ambiguity in fitting the data, for although l had been identified, a direct determination of n had not been possible without the observation of intermediate-degree modes. Moreover there was a systematic discrepancy between theory and observation (*20*). Therefore a determination of the initial solar helium abundance Y by fitting theory to observation, for example, must be considered somewhat uncertain. We judge from the comparisons, however, that $Y = 0.25 \pm 0.02$, which agrees with the high-degree calibration illustrated in Fig. 2.

Another conclusion that can be drawn from the low-degree data concerns the structure of the core. As a possible way

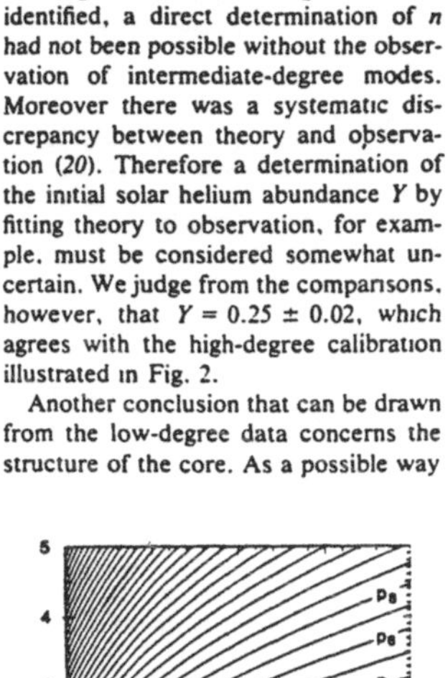

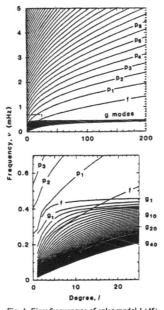

Fig. 4. Eigenfrequencies of solar model 1 (*45*) plotted against degree l; modes of like order n are connected by straight lines. The p and g modes are labeled p_n and g_n. The lower panel is an enlargement of the bottom left-hand corner of the upper panel; the connecting lines never cross, although the scale of the diagram is too small to resolve that in some cases. Plotting of g modes was stopped arbitrarily at $n = 40$; formally the order of the g modes is unbounded, and the eigenfrequencies accumulate at $\nu = 0$.

out of the neutrino problem, some have advocated that the processed material in the core has been mixed with its surroundings (*22*). This would have increased the abundance of unburned hydrogen in the core and raised the sound speed. However, the standard unmixed solar model with $Y = 0.25$ reproduces the observed properties of ϵ, whereas models with a lesser variation of chemical composition do not. In particular, a chemically homogeneous solar model predicts a separation Δ that is 50 percent greater than the value observed. Thus a substantial degree of homogenization appears to be ruled out.

Acoustic Modes of Intermediate Degree

Much of the uncertainty in the calibration of solar models was removed by the observation of modes of intermediate degree. By averaging light in the east-west direction and appropriately projecting Doppler velocities onto Legendre functions, Duvall and Harvey (*23*) were able to identify zonal models such as that illustrated in Fig. 3A. The resulting power spectrum (Fig. 6) is markedly similar to the theoretical eigenfrequencies illustrated in Fig. 4A. At higher l it overlaps Fig. 2, and the order of each ridge can therefore be determined. After careful analysis it was possible to identify modes with degree as low as unity and to confirm the identifications that had previously been made by model-fitting to the low-degree data.

The data can be interpreted by considering the properties of the rays illustrated in Fig. 1. A wave can be regarded as being locally plane, with wave number **k** whose horizontal component is $k_h = L/r$, where $L = [l(l+1)]^{1/2}$. The magnitude k of the wave number is related to ν and c by the simple acoustic dispersion relation $2\pi\nu = kc$. Since c increases with depth, k must decrease, until the radius r_t is reached where $k = k_h$. This is the radius of a caustic surface, which bounds a sphere into which the waves cannot propagate (Fig. 1). It is given by

$$\frac{c(r_t)}{r_t} = \frac{2\pi\nu}{rk_h} = \frac{2\pi\nu}{L} \qquad (4)$$

Thus the depth of penetration depends on the angular phase speed $2\pi\nu/L$ of the disturbance. At a given frequency, waves of smaller degree penetrate more deeply, because c/r is a decreasing function of r. Spherically symmetric waves, with $l = 0$, are not refracted at all according to this simple description; they propagate vertically, passing right through the center of the sun. An oscillation mode results from the interference between many such traveling waves, giving the structure exhibited in Eq. 1.

If Fig. 4A were superposed on the observations of Fig. 6, the two would be barely distinguishable. Therefore in Fig. 7 we display the differences between the observed and calculated frequencies. Although they are small, they are far from random; furthermore they are substantially larger than the errors in the observations. Thus we must conclude that there are significant errors in this theoretical model. To discover what they are it is expedient to examine the eigenfunctions.

The vertical component of velocity in several oscillations is illustrated in Fig. 8. Two sets of modes have been chosen, each with roughly the same frequency but with different degrees. For the p modes the variation in the penetration depth is in accordance with Eq. 4. Just below the surface, the structures of p modes of different degree are quite similar. Here propagation is predominantly vertical, and the dynamics is insensitive to degree. Thus modes of the same frequency sample the outer layers of the sun in much the same way; only near the caustic surface, where the waves propagate nearly horizontally, do the eigenfunctions sense the degree.

With these properties in mind we notice three salient features of the discrepancies plotted in Fig. 7. The first is that there is a significant discrepancy at the largest values of l; from this we infer that the very surface layers, where these modes are confined, cannot be adequately represented by the theoretical model. The second is that at high l there is a systematic variation of the discrepancy with l that ceases at $l \simeq 20$, beneath which the discrepancy is independent of l to within the observational accuracy. From this we infer that there is no substantial error deeper in the model than the turning points r_t of the $l = 20$ modes; modes of lower degree, penetrating more deeply, sense the erroneous regions in an l-independent fashion. The $l = 20$ turning point varies somewhat with frequency but is near $r \simeq 0.45\ R$. The third feature is that the discrepancy is least at the lowest values of l. Thus the error immediately above $r \simeq 0.45\ R$ tends to compensate the error in the surface layers, contributing negatively to the theoretical frequencies and implying that here the theoretical model underestimates the sound speed.

The behavior seen in Fig. 7 is specific to the solar model considered here for illustration and is not shared in detail by all other models. However, qualitatively

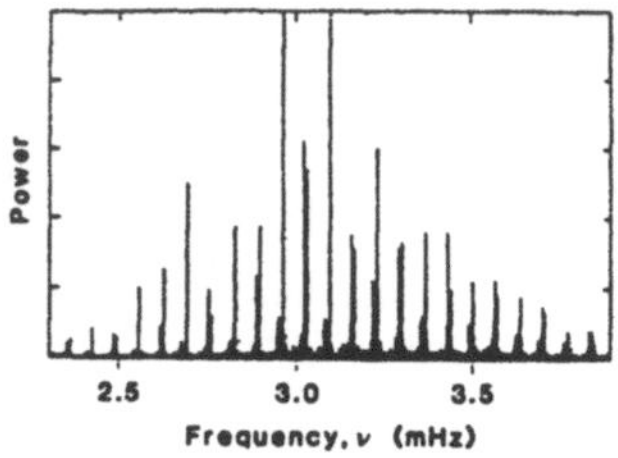

Fig. 5. Power spectrum of low-degree whole-disk Doppler measurements carried out over a period of 3 months at Tenerife and Hawaii with up to 22 hours of coverage per day (*21*). The greatest amplitude is 15 cm sec^{-1}. The separations Δ between modes with like $n + \frac{1}{2}l$ are just discernible.

similar results were obtained from the frequencies computed by Ulrich and Rhodes (*24*) for a model with a more complicated treatment of the physical properties of matter within the sun. Yet there are detailed differences in frequencies predicted by various models that exceed the uncertainties of the observed frequencies. Such differences should allow helioseismology to test and refine the physics used in the calculation of the evolution of stars.

The Internal Sound Speed

Fitting theoretical models by adjusting uncertain parameters is a simple and sometimes enlightening procedure for estimating the structure of the sun. Nevertheless it suffers the severe deficiency that the range of possibilities considered may have been too tightly constrained by the untested assumptions of the theory. This is exemplified by the attempts to determine Y; the frequencies of the optimal model deviate significantly from the observations (*20*). Attempts have been made to improve the fit by adjusting additional parameters, but with only limited success (*24*). Our discussion of Fig. 7 provides some guidance for estimating the changes that are required, but it is not adequate. It is preferable to employ a so-called inverse method, whereby the structure of the sun is estimated directly as a function of the data.

The frequency of a p mode is determined by the condition that acoustic waves interfere constructively. This requires that the phase difference between the end points A and C of a ray (see Fig. 1) should be the same, up to an integral multiple n of 2π, whether it is evaluated along the ray or along the surface. When buoyancy is neglected, this condition is

$$2\pi(n + \alpha) \simeq 2L\int_{r_t(\nu/L)}^{R}\left(\frac{4\pi^2r^2\nu^2}{c^2L^2} - 1\right)^{1/2}\frac{dr}{r}$$

$$\equiv 2\pi L\,F\!\left(\frac{\nu}{L}\right) \qquad (5)$$

which defines the function F; α is a constant that accounts for the phase changes at the caustic and on reflection at the surface (*25*). This condition exhibits a remarkably simple property of the spectrum of acoustic modes: namely, that $(n + \alpha)/L$ depends on l and ν only in the combination ν/L, a property first noticed by Duvall (*26*) in the solar data. As is evident in Fig. 9, with a suitable choice of α the many ridges in Figs. 2 and 6 collapse onto a single curve with perhaps surprising accuracy. Then the observed quantity $(n + \alpha)/L$ measures the function F.

When $l \ll n$, the radius r_t of the lower turning point, where the integrand in Eq.

Fig. 6. Power spectrum of zonal modes observed by Duvall and Harvey (*23*). Lighter tones represent greater power. The orders n of the modes contributing to the ridges were identified by comparison with Fig. 2 in which power from f modes, whose frequencies at high l are independent of the detailed structure of the sun, is clearly visible.

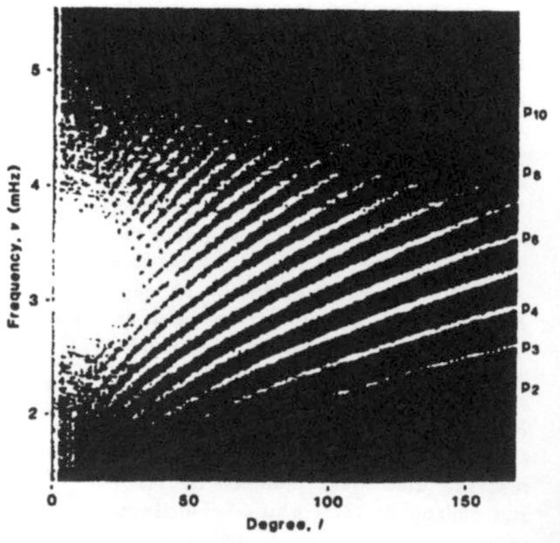

5 vanishes, is small. Moreover, the integrand is dominated by the first term in the square root throughout most of the star. If the integral is expanded about the value obtained by ignoring the second term in the square root and setting $r_t = 0$, Eqs. 2 and 3 are recovered, except that $l + \frac{1}{2}$ is replaced by L.

Equation 5 can be regarded as an integral equation for $c(r)$ in terms of F. It can be inverted analytically to yield $c(r)$ without recourse to a solar model. However, the simple acoustic dispersion relation on which the analysis depends is of questionable applicability; hence the procedure must be tested. When applied to computed eigenfrequencies of a solar model, the inversion has been shown to be accurate to within about 1 percent at radii between 0.4 and 0.9 R. In this part of the model, at least, it is possible to determine the sound speed from the set of modes available (*27*).

Figure 10 illustrates the result of the inversion. The base of the convection zone occurs where the second derivative of temperature, and hence of sound speed, changes almost discontinuously, producing a visible kink in the curve. It is located at $r/R \simeq 0.7$, close to the position of the bottom of the convection zone in typical solar models. Indeed, the sound speed deviates little from the model used to compute the frequencies in Fig. 4. Only in a region below the convection zone in the vicinity of $r/R = 0.4$ does the difference exceed the estimated error in the inversion. Here the sun appears to be hotter than the model by about 2 percent, which is in accordance with our discussion of Fig. 7. Unfortunately, the inversion fails in the core of the sun, where information relating to solar evolution and to the neutrino problem is to be found.

Gravity Modes

The frequencies of the g modes also have a distinctive pattern. The modes can be regarded as a superposition of locally plane gravity waves, just as the p modes can be decomposed into plane acoustic waves. Gravity waves can propagate only in convectively stable regions, namely in the solar atmosphere and in the interior beneath the convection zone. We confine attention to the interior modes. Except when l is low their amplitudes decrease rapidly through the convection zone, and they cannot be observed in the photosphere. Therefore the g-mode data will never be as extensive as the data from p modes. However, if the core of the sun is stably

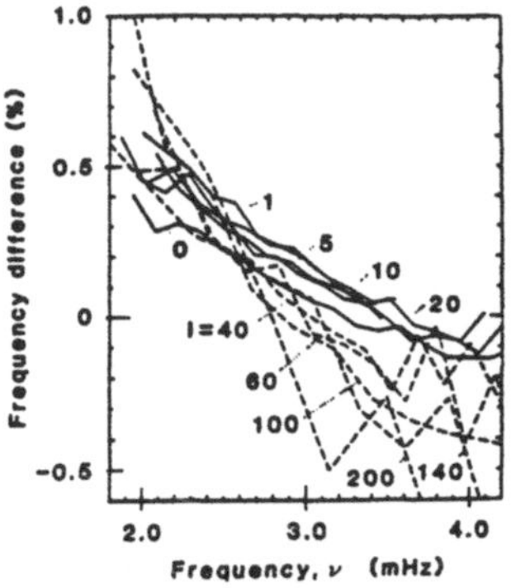

Fig. 7. Percentage differences, 100 × $(\nu_o - \nu_t)/\nu_t$, between corresponding observed frequencies ν_o deduced from Fig. 6 and the theoretical frequencies ν_t of Fig. 4, plotted against ν_t. Values corresponding to modes with like degree l are joined by continuous straight lines if $l \leq 20$ and dashed lines if $l \geq 40$. Observational uncertainties are about ±0.07 percent if $l \leq 20$ and ± 0.3 percent if $l \geq 40$. [After (*46*)]

stratified, as most theoretical models predict, the g modes approach the center more closely than do the p modes (Fig. 8). Therefore low-degree g modes are potentially a more powerful diagnostic of the innermost regions.

High-order g modes satisfy a dispersion relation analogous to Eq. 5 for p modes, namely (*28*)

$$\frac{n+\beta}{L} = G(\nu) \equiv \int_{r_1}^{r_2} \left(\frac{N^2}{4\pi^2\nu^2} - 1\right)^{1/2} \frac{dr}{r} \qquad (6)$$

which defines the function G; here N is the buoyancy frequency, the limits of integration r_1 and r_2 are the levels at which the integrand vanishes, and β is a constant of order unity (*29*). When $n \gg l$, the frequency $2\pi\nu$ is much less than N, and the integrand is dominated by its first term. The turning points r_1 and r_2 are then roughly the radii at which N vanishes (that is, the center of the sun and the base of the convection zone, $r = r_c$, in typical solar models). It may further be shown that

$$(n + \tfrac{1}{2}l + \bar{\beta})\, \nu_{n,l} \simeq L/P_0$$

where

$$P_0 = 2\pi^2 \left(\int_0^{r_c} N \frac{dr}{r}\right)^{-1} \qquad (8)$$

and $\bar{\beta}$ is a constant related to β and to the nature of the variation of N near $r = r_c$. The presence of $\frac{1}{2}l$ in this formula is a manifestation of the radiative core; if the core were convective this term would be absent and ν would depend on l only through L on the right-hand side of Eq. 7. The analysis predicts that periods are uniformly spaced as the order n varies; the spacing is P_0/L and hence decreases with increasing degree. For typical solar models, the value of P_0 is 33 to 36 minutes.

In contrast to the relations for the p modes, the asymptotic Eq. 7 is not obviously confirmed by observation. The power spectra obtained from g-mode data are quite complicated. The periods are substantial fractions of a day, and hence the sidelobes produced by interruptions in the data cause greater confusion than they do for the p modes; in addition, the splitting induced by rotation is comparable with the separation between modes of different degree and order, which further complicates mode identification. Nonetheless there have been tentative identifications of sequences of peaks in power that are uniformly spaced in period (*30, 31*), with separations proportional to L^{-1} in accordance with Eq. 7. The values inferred for P_0 are somewhat higher than those obtained for normal theoretical solar models. If the gradients in chemical composition were to be reduced in the theoretical models, so would N, and according to Eq. 8 this would increase P_0. Only a modest degree of diffusive mixing is required to bring the computed value of P_0 into agreement with the values suggested by the observations (*32*). However, as noted above, mixing may not be consistent with the observed frequency separations between 5-minute modes of low degree. Moreover, with the available data the presence of approximately uniformly spaced sequences of peaks may be purely fortuitous; indeed, alternative identifications of the observed frequencies can be made that are consistent with normal solar models (*33*). Nevertheless the sensitivity of P_0 to mixing is an indication of the potential for diagnosing the solar core from the periods of g modes.

Solar Rotation

Judging from the angular velocities of young stars, we believe that the surface layers of the sun were once rotating much more rapidly than they are today. This belief is strengthened by the observation that the sun is now losing angular momentum via the solar wind. But by how much has the center of the sun spun down? Here opinions differ, for it is not known how strong the coupling is between the surface and the core. Some have argued that the coupling is weak and that, although the surface rotates with a period of about four weeks, the core is rotating much faster, with a period of only a few days. Others believe that large-scale circulations, rotationally induced instabilities, or wave motion provide effective coupling and that the core is really rotating not much faster than the surface. There are also those who maintain that the sun is pervaded by a magnetic field that gives it sufficient rigidity to be rotating almost uniformly. Among this diversity of opinion, a recent seismological measurement of the internal angular velocity is of considerable interest.

Rotation breaks the spherical symmetry of the sun. If the entire sun rotates about a common axis, it can still oscillate in essentially normal modes with structure close to that given by Eq. 1, provided that the coordinate axis is chosen as the axis of rotation. But now ν depends on m. The principal reason is that rotation causes the patterns illustrated in Fig. 3 to rotate, which modifies the apparent frequencies of especially the high-m modes when viewed from Earth. The extent to which the m degeneracy is split provides a measure of the angular velocity $\Omega(r)$ in the cavity within which the modes are confined (*34*).

Most of the observations from which Ω has been estimated are of sectoral ($m = l$) acoustic modes (*35*) with degrees up to 100. These modes are concentrated near the equator (9), particularly when l is large (Fig. 3), and cannot provide information about the latitudinal dependence of Ω. But they can reveal how Ω varies with radius r near the equatorial plane. The magnitude of the degeneracy splitting of the acoustic modes is an average of $\Omega(r)$ weighted approximately as the square of the velocity eigenfunctions scaled as in Fig. 8. With data from a sufficient variety of modes, those averages can be inverted to estimate $\Omega(r)$.

The results of such an inversion are shown in Fig. 11. Perhaps the most startling feature is the inward decline of Ω in the solar envelope. As we have already mentioned, the solar wind braking of the outermost layers has led most theorists to infer that Ω increases everywhere with depth. To be sure, the core appears to be rotating more rapidly, but why is it surrounded by a region of such slow rotation?

The abrupt increase of Ω at the edge of the energy-generating core appears to occur just where the variation of chemical composition produced by the nuclear reactions is predicted to give the strongest stabilization to vertical motion. However, the resolution of the splitting data is poorest in this region, so that the details of the variation of Ω are uncertain.

A secure determination of the angular velocity of the core appears to require knowledge of the splitting of g modes that penetrate almost to the center of the sun. There is a report of rotational splitting of dipole g modes (*30*), but there is some doubt about its interpretation. Were we to accept the result and to combine it with the p mode splitting data, we would deduce that the mean angular velocity of the core is between about three and six times the angular velocity at the surface. This is consistent with a recent report of rotational splitting of dipole p modes by Isaak and colleagues (*36*).

The rotational flattening of the gravitational potential outside the sun depends predominantly on the value of Ω near the equatorial plane at radii between 0.3 and 0.9 R. This is the very region where Ω has been most securely determined by the splitting data. The values of the quadrupole moment J_2 inferred from the two curves in Fig. 11 are both 1.7×10^{-7} and are too small to influence significantly the interpretation of current measurements of the precession of the perihelion of Mercury (*35*). Unless the oscillation data have been misinterpreted, which seems unlikely, these measurements are consistent with Einstein's theory of general relativity.

As in any rapidly developing field, there is some controversy over the interpretation of data. For example, Hill and colleagues (*37*) have interpreted mea-

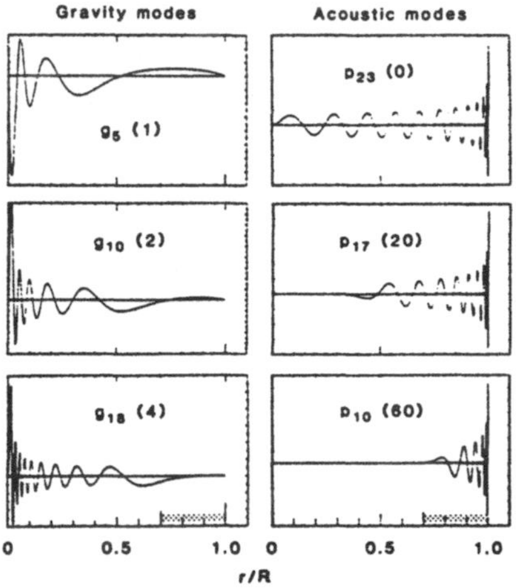

Fig. 8. Vertical component of velocity of several modes of oscillation, scaled by $r\rho^{1/2}$ where $\rho(r)$ is density. The ordinate scales are arbitrary; the horizontal line is at zero. For each value of l (given in parentheses), the order n was chosen such that the g-mode frequencies are approximately 0.10 mHz (periods of 165 minutes) and the p-mode frequencies are approximately 3.3 mHz (periods of 5 minutes). The shaded bands indicate the extent of the convection zone.

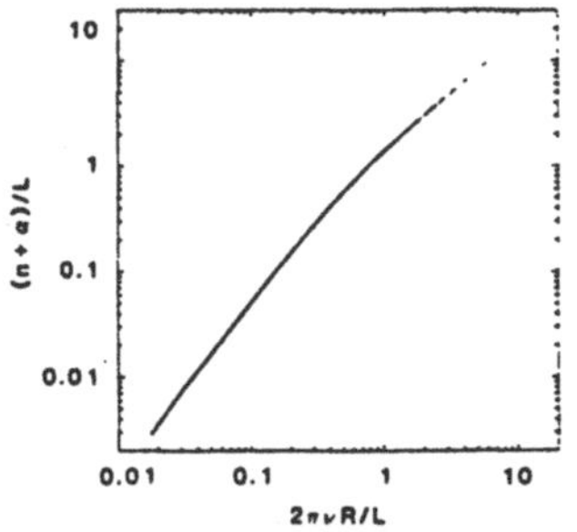

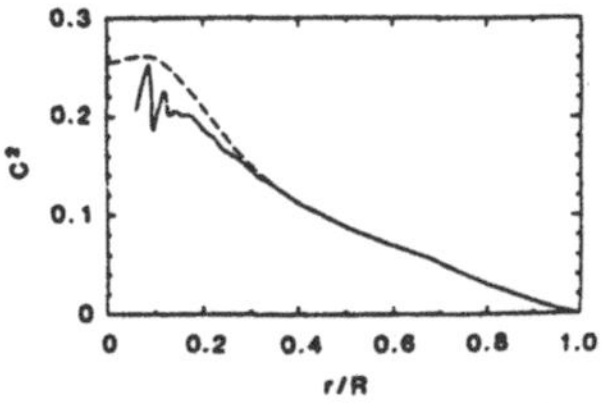

Fig. 9 (left). The function $(n + \alpha)/L$ plotted against the surface phase velocity $2\pi\nu R/L$ (which is in units of Mm sec^{-1}). The data include the high-degree modes considered by Duvall (*26*), which are similar to those of Fig. 2, and the modes of low and intermediate degree discussed by Duvall and Harvey (*23*) and illustrated in Fig. 6. The value $\alpha = 1.57$ was chosen to minimize the scatter of the data about a single curve. Fig. 10 (right). The square of the sound speed c obtained by inverting Eq. 5, using a smooth curve through the points in Fig. 9 to determine the function F, is shown as a continuous line. The actual value of c^2 of solar model 1 (*45*) is shown as a dashed line. The large discrepancy at $r \lesssim 0.3\ R$ results mainly from the inaccuracy of the simple asymptotic theory. The unit of speed is Mm sec^{-1}. [After (*27*)]

surements of fluctuations at the periphery of the solar image to be the result of acoustic and gravity modes with high rotational splitting. Thus they have concluded that most of the sun is rotating substantially more rapidly than the surface. When considered in relation to the precession of the orbit of Mercury, this raises difficulties with Einstein's theory of general relativity, but it is apparently consistent with some other covariant theories of gravity. From earlier observations of the oblateness of the solar image, Dicke had postulated that the interior of the sun is rotating rapidly and is distorted by an intense magnetic field (*38*). We must await more detailed seismological probing before these issues are resolved.

The analysis used to obtain Fig. 11 did not permit the determination of the rotation close to the solar surface. However observations of modes of higher degree (*39*), which provide better depth resolution near the surface, have suggested that the rotation rate increases with depth in the outer part of the convection zone. Such observations have also shown indications of day-to-day variations in the subphotospheric velocity field. These could be caused by the passage through the field of view of giant convective cells, with associated velocities of order 100 m sec^{-1}. Although these results are tentative, they hint at the potential of helioseismology for probing the dynamics of the solar convection zone.

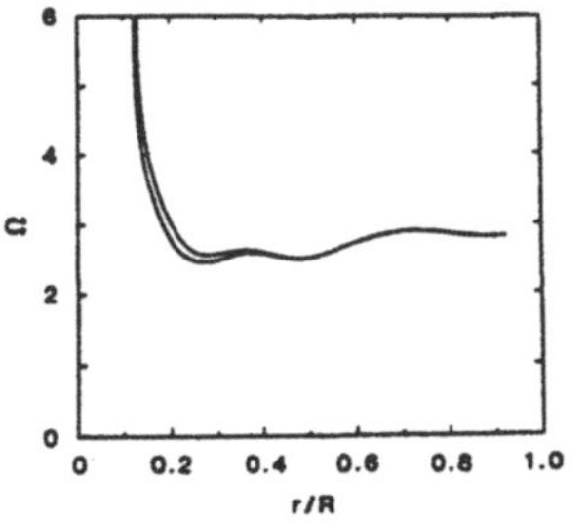

Fig. 11. Examples of the equatorial angular velocity $\Omega(r)$ of the sun, in units of 10^{-6} sec^{-1}, inferred from the rotational splitting of 5-minute *p* modes reported by Duvall and Harvey (*35*), using an asymptotic inversion procedure (*47*). The lower curve was computed under the assumption that Ω is independent of colatitude θ except very close to the surface; the upper curve results from the assumption that the θ-dependence of Ω is at all values of *r* the same as the observed surface variation. The curves are uncertain for $r \lesssim 0.3$ *R* because the data from the deeply penetrating low-degree modes are imprecise (*35*); yet the difference between them represents well the sensitivity to the assumptions about the θ-dependence. The *p*-mode data provide no information about the rotation of the very center of the sun. The photospheric equatorial angular velocity $\Omega(R)$ is 2.86×10^{-6} sec^{-1}.

Conclusions

Ten years ago there was little empirical information about the solar interior. What was available, the neutrino flux, disagreed with the theoretical models. The neutrino problem is still with us, but the observational data from the solar interior have expanded tremendously. As a result we can now study the sun at a level of detail that would have seemed hopelessly out of reach only a decade ago.

Helioseismology has provided, through the observations of 5-minute oscillations, an estimate of the sound speed throughout much of the solar interior. It has also provided a test of the structure of the deepest layers of the sun predicted by the theory of stellar evolution. There are certainly discrepancies between our present best models and the observations; these must be studied and eliminated. Nevertheless it is perhaps surprising how well normal solar models fit the data. In contrast, models with a low helium abundance seem to be excluded, as do models with substantial mixing of the core with its environment. Among standard solar models the seismological data appear to favor those with a high neutrino flux. This may strengthen the case for searching for a physical, as opposed to an astrophysical, solution to the neutrino problem, such as the possibility that neutrinos have mass and undergo transitions from one form to another. Nevertheless, it is essential to explore the astrophysical options, including that of the sun being in a transient interlude of thermal imbalance after a sudden overturning in the core. Such processes would almost certainly not be restricted to the sun.

It may have been thought, perhaps with some justification, that theory has provided a reasonably reliable estimate of the structure of the solar interior. The same can certainly not be said about the internal rotation, where all theoretical models must be highly uncertain. Indeed, none of them has predicted a rotation law such as the one that has recently been measured.

What are the requirements for the future? It behooves us to develop more subtle theory, particularly for extracting more detailed information about the core of the sun. And as always there is a need for better and more extensive data. Modes of low frequency, particularly the *g* modes, are sensitive to conditions in the core and would provide valuable information. But we also need better data from high-degree *p* modes to measure the region immediately beneath the solar surface. These will permit us to study the giant cells in the convection zone, which may control much of the magnetic activity in the solar atmosphere. The region sampled by high-degree modes also influences the frequencies of the *p* modes that penetrate more deeply, for all acoustic modes have their upper reflection point just below the solar surface. Reliable information about the interior requires accounting for the influence of this region on the frequencies of the lower-degree modes.

It is of great interest to study temporal variations in the stratification of the sun. A change in the solar radius causes a change in the *p*-mode frequencies of roughly the same relative magnitude. Given the high accuracy of some of the observed frequencies of the 5-minute oscillations (*40*), this may provide a sensitive method for detecting structural changes associated with solar activity and the solar cycle. We must also look for variations with time in the rotation. Could it be that the curious rotation law shown in Fig. 11 is a time-dependent phenomenon associated with torsional oscillations of the solar interior, perhaps under the influence of a magnetic field?

We may be getting near the limit of what can be attained from single ground-based observatories. The presence of night-time gaps in the data produces confusing aliases in the power spectra. Such gaps might be eliminated by observing from the South Pole during austral summer, but it is rare for continuous periods of clear weather to last more than a week. Therefore observations are being combined from widely separated sites (*21*); it has been estimated that nearly continuous coverage could be ensured with six stations suitably placed around the world (*41*). Such networks are now being established by investigators from Birmingham and Nice using Doppler measurements in integrated sunlight to study the low-degree modes. More ambitious efforts, to be carried out by the Global Oscillations Network Group and coordinated by the National Solar Observatory, are aimed at studying solar oscillations of higher degree by making spatially resolved Doppler measurements with identical instruments from a network of observatories.

Harder to eliminate are the effects of the terrestrial atmosphere. They can introduce daily modulations of the observed signal that hinder the combining of data from several observatories, particularly at the long periods of *g* modes. Modes with periods of 30 to 60 minutes may be overpowered by atmospheric noise; these modes have considerable diagnostic potential, but their amplitudes are so low that they have not been detected. Finally, atmospheric seeing distortions have particularly serious ramifications for measuring high-degree modes whose length scales are comparable with the scale of the seeing. Yet to study dynamics within the convection zone, we must observe these modes with a high signal-to-noise ratio, and this must be achieved within a time sufficiently short that the underlying convection cells do not change or move out of the field of view. It may not be possible to do this from the ground. Therefore several helioseismology instruments are now under study by the European Space Agency and by the National Aeronautics and Space Administration to fly on the satellite SOHO. They would acquire uninterrupted observations of extremely high quality, limited only by the intrinsic solar "noise" in the form of random surface velocity and intensity fluctuations.

Stars similar to the sun might be expected to exhibit a similar spectrum of oscillations. Analogs of the solar 5-minute oscillations may already have been detected in the stars HR 1217, α Centauri A, and ε Eridani (*42*). Rich spectra of oscillations with substantial diagnostic value have also been found in other classes of stars, notably the white dwarfs. Thus we are now seeing the birth of asteroseismology. Continued developments in the acquisition of better data and improvements in theoretical understanding will undoubtedly lead to the establishment of a firm empirical basis for the study of solar and stellar evolution and activity.

References and Notes

1. R. B. Leighton, *Proc. IAU Symp.* **12**, 321 (1960); ———, R. W. Noyes, G. W. Simon, *Astrophys. J.* **135**, 474 (1962).
2. R. K. Ulrich, *Astrophys. J.* **162**, 993 (1970); J. W. Leibacher and R. F. Stein, *Astrophys. Lett.* **7**, 191 (1971).
3. It is likely that either the modes are excited by their interaction with the flux of radiative and convective energy passing through the sun or that they are driven stochastically by the vigorous turbulent motion in the convection zone.
4. A mode of oscillation is a standing wave produced by the constructive interference of progressive waves such as those whose ray paths are illustrated in Fig. 1. Each wave propagates essentially in a plane and produces an interference pattern in two dimensions. Three-dimensional patterns are formed by the interference of waves in differently orientated planes.
5. The rapid increase with depth of molecular weight (that is, mean mass per particle) brought about by thermonuclear transformations in the core causes the sound speed to decrease with depth close to the center of the sun (Fig. 10). However, c/r is monotonic in normal solar models, so that the solution to Eq. 4, which determines the bottom of the acoustic cavity, is unique.
6. H. Ando and Y. Osaki, *Publ. Astron. Soc. Jpn.* **27**, 581 (1975).
7. F.-L. Deubner, *Astron. Astrophys.* **44**, 371 (1975).
8. The solar disk is the projection of the visible surface of the sun onto the plane of the sky. By averaging in the north-south direction on the disk, Deubner filtered out essentially all but the sectoral modes (an example of which is shown in Fig. 3D). These are formed solely by waves that propagate close to the equatorial plane. Consequently a Fourier transform in longitude alone is sufficient to determine the horizontal component of the wave number.
9. If the sun were cylindrical, exactly l wavelengths of a sectoral mode such as that in Fig. 3D would fit around the circumference $2\pi R$. But because the sun is a sphere, the mode varies away from the equatorial plane between latitudes of $\pm \cos^{-1} [l/(l + 1)]^{1/2}$. This effectively reduces the wavelength of the surface disturbance to $2\pi R/L$, where $L^2 = l(l + 1)$. More complicated modes, such as those in Fig. 3, A, B, and C, can be regarded as superpositions of sectoral modes in different planes with the same value of l.
10. D. O. Gough, in *The Energy Balance and Hydrodynamics of the Solar Chromosphere and Corona*, R. M. Bonnet and P. Delache, Eds. (de Bussac, Clermont-Ferrand, 1977), p. 3.
11. R. K. Ulrich and E. J. Rhodes, Jr., *Astrophys. J.* **218**, 521 (1977); G. Berthomieu *et al.*, in *Nonradial and Nonlinear Stellar Pulsation*, H. A. Hill and W. A. Dziembowski, Eds. (Springer, Heidelberg, 1980), p. 307.
12. Neutrinos are usually considered to be massless. If that were not the case (a possibility that has received considerable attention in the last few years), some of the neutrinos produced by the nuclear reactions in the sun could have been transformed during their passage to Earth into a form that would not have been detected. The neutrino observations may then be consistent with the seismological calibration.
13. Buoyancy frequency is the frequency with which a fictitious fluid element would oscillate adiabatically under the action of buoyancy forces alone.
14. A. Claverie, G. R. Isaak, C. P. McLeod, H. B. van der Raay, T. Roca Cortes, *Nature (London)* **282**, 591 (1979).
15. J. Christensen-Dalsgaard and D. O. Gough, in *Nonradial and Nonlinear Stellar Pulsation*, H. A. Hill and W. A. Dziembowski, Eds. (Springer, Heidelberg, 1980), p. 184; *Mon. Not. R. Astron. Soc.* **198**, 141 (1982).
16. Yu. V. Vandakurov, *Astron. Zh.* **44**, 786 (1967).
17. G. Grec, E. Fossat, M. Pomerantz, *Nature (London)* **288**, 541 (1980); *Solar Phys.* **82**, 55 (1983).
18. M. Tassoul, *Astrophys. J. Suppl. Ser.* **43**, 469 (1980).
19. The l-dependence of Tassoul's formula for ε was subsequently confirmed up to $l = 5$ [P. H. Scherrer, J. M. Wilcox, J. Christensen-Dalsgaard, D. O. Gough, *Solar Phys.* **82**, 75 (1983)].
20. J. Christensen-Dalsgaard and D. O. Gough, *Nature (London)* **288**, 544 (1980); *Astron. Astrophys.* **104**, 173 (1981).
21. A. Claverie *et al.*, *Mem. Soc. Astron. Ital.* **55**, 63 (1984).
22. Most recently, E. Schatzman, A. Maeder, F. Angrand, R. Glowinski, *Astron. Astrophys.* **96**, 1 (1981).
23. T. L. Duvall, Jr., and J. W. Harvey [*Nature (London)* **302**, 24 (1983)] include a list of frequencies and the power spectrum shown in Fig. 6. Additional frequencies have been listed [J. W. Harvey and T. L. Duvall, Jr., in *Solar Seismology from Space*, R. K. Ulrich, J. W. Harvey, E. J. Rhodes, Jr., J. Toomre, Eds. (Jet Propulsion Laboratory publication 84-84, Pasadena, 1984), p. 165] and discussed [J. W. Harvey and T. L. Duvall, Jr., in *Theoretical Problems in Stellar Stability and Oscillations*, M. Gabriel and A. Noels, Eds. (Institut d'Astrophysique, Liège, 1984), p. 209].
24. R. K. Ulrich and E. J. Rhodes, Jr., *Astrophys. J.* **265**, 551 (1983); in *Solar Seismology from Space*, R. K. Ulrich, J. W. Harvey, E. J. Rhodes, Jr., J. Toomre, Eds. (Jet Propulsion Laboratory publication 84-84, Pasadena, 1984), p. 371. Other models have been described recently [A. Noels, R. Scuflaire, M. Gabriel, *Astron. Astrophys.* **130**, 389 (1984); H. Shibahashi, A. Noels, M. Gabriel, *ibid.* **123**, 283 (1983); *Mem. Soc. Astron. Ital.* **55**, 163 (1984); E. J. Rhodes, Jr., R. K. Ulrich, W. M. Brunish, *ibid.*, p. 37].
25. F.-L. Deubner and D. O. Gough, *Annu. Rev. Astron. Astrophys.* **22**, 593 (1984); J. Christensen-Dalsgaard, in *Theoretical Problems in Stellar Stability and Oscillations*, M. Gabriel and A. Noels, Eds. (Institut d'Astrophysique, Liège, 1984), p. 155.
26. T. L. Duvall, Jr., *Nature (London)* **300**, 242 (1982).
27. J. Christensen-Dalsgaard, T. L. Duvall, Jr., D. O. Gough, J. W. Harvey, E. J. Rhodes, Jr., *ibid.* **315**, 378 (1985).
28. D. O. Gough [in *Solar Physics and Interplanetary Traveling Phenomena*, B. Chen and C. de Jager, Eds. (Yunnan Observatory, Kunming, 1984)] obtains Eq. 6 and discusses the difficulties of identifying *g* modes in currently available observations. The theoretical low-degree limit was previously discussed more precisely and in greater detail [Yu V. Vandakurov (*16*); J.-P. Zahn, *Astron. Astrophys.* **4**, 452 (1970); M. Tassoul (*18*); P. Ledoux and J. Perdang, *Bull. Soc. Math. Belg.* **32**, 133 (1980)].
29. It may be the case that the integrand in Eq. 6 can vanish more than twice and that the star therefore contains several resonant cavities. Theoretical solar models suggest that this will not be so when $n \gg l$ and consequently $\nu^2 \ll N^2$, which is the regime in which it appears that reliable observations will most imminently be available.
30. P. Delache and P. H. Scherrer, *Nature (London)* **306**, 651 (1983).
31. A. B. Severny *et al.*, *ibid.* **307**, 247 (1984); C. Frölich and P. Delache, *Mem. Soc. Astron. Ital.* **55**, 99 (1984); G. R. Isaak, H. B. van der Raay, P. L. Palle, T. Roca Cortes, P. Delache, *ibid.*, p. 91; P. H. Sherrer, *ibid.*, p. 83.
32. G. Berthomieu, J. Provost, E. Schatzman, *Nature (London)* **308**, 254 (1984); *Mem. Soc. Astron. Ital.* **55**, 107 (1984).
33. M. Gabriel, *Astron. Astrophys.* **134**, 387 (1984).
34. T. G. Cowling and R. A. Newing, *Astrophys. J.* **109**, 149 (1949); P. Ledoux, *Mém. Soc. Roy. Sci. Liège Collect.* 4° **9**, 263 (1949). Rotational splitting appears to have been first mentioned in connection with the sun by J. R. Brookes, G. R. Isaak, and H. B. van der Raay [*Nature (London)* **259**, 92 (1976)].
35. T. L. Duvall, Jr., and J. W. Harvey [*Nature (London)* **310**, 19 (1984)] present the mean rotational splitting of modes of various orders for 33 values of l. Various estimates of the sun's equatorial angular velocity, and an evaluation of the induced gravitational quadrupole moment J_2, have been reported [T. L. Duvall, Jr., *et al.*, *ibid.*, p. 22; J. W. Leibacher, in *Theoretical Problems in Stellar Stability and Oscillations*, M. Gabriel and A. Noels, Eds. (Institut d'Astrophysique, Liège, 1984), p. 298]. Recently, T. M. Brown [in *Seismology of the Sun and the Distant Stars*, D. O. Gough, Ed. (Reidel, Dordrecht, in press)] has presented more extensive data that give some indication of Ω away from the equatorial plane.
36. G. R. Isaak, in *Seismology of the Sun and the Distant Stars*, D. O. Gough, Ed. (Reidel, Dordrecht, in press).
37. H. A. Hill, R. J. Bos, P. R. Goode, *Phys. Rev. Lett.* **49**, 1794 (1982).
38. R. H. Dicke, *Annu. Rev. Astron. Astrophys.* **8**, 297 (1970); *Solar Phys.* **78**, 3 (1982).
39. F. Hill, D. O. Gough, J. Toomre, *Mem. Soc. Astron. Ital.* **55**, 153 (1984).
40. The precision with which some oscillation frequencies can be measured is higher than that of the gravitational constant, a quantity upon which those frequencies depend.
41. F. Hill and G. Newkirk, Jr., *Solar Phys.* **95**, 201 (1985).
42. D. W. Kurtz and J. Seeman, *Mon. Not. R. Astron. Soc.* **205**, 11 (1983); E. Fossat, G. Grec, B. Gelly, Y. Decanini, *C. R. Acad. Sci. Paris Ser. B* **299**, 17 (1984); R. W. Noyes *et al.*, *Astrophys. J. Lett.* **285**, L23 (1984).
43. F.-L. Deubner, R. K. Ulrich, E. J. Rhodes, Jr., *Astron. Astrophys.* **72**, 177 (1979).
44. D. O. Gough, *Phys. Bull.* **34**, 502 (1983).
45. J. Christensen-Dalsgaard, *Mon. Not. R. Astron. Soc.* **199**, 735 (1982).
46. ——— and D. O. Gough, in *Solar Seismology from Space*, R. K. Ulrich, J. W. Harvey, E. J. Rhodes, Jr., J. Toomre, Eds. (Jet Propulsion Laboratory publication 84-84, Pasadena, 1984), p. 199.
47. D. O. Gough, *Phil. Trans. R. Soc. London Ser. A* **313**, 27 (1984).

48. We thank J. Harvey and G. Isaak for making available illustration material from their observations. Supported in part by grants from NASA and the Air Force Geophysics Laboratory.

THE DEPTH OF THE SOLAR CONVECTION ZONE

J. Christensen-Dalsgaard
Astronomisk Institut, Aarhus Universitet, DK-8000 Aarhus C, Denmark; and
High Altitude Observatory, National Center for Atmospheric Research[1]
D. O. Gough[2]
Institute of Astronomy, and Department of Applied Mathematics and Theoretical Physics, University of Cambridge, England; and
Joint Institute for Laboratory Astrophysics, University of Colorado
AND
M. J. Thompson[3]
High Altitude Observatory, National Center for Atmospheric Research[1]
Received 1990 February 28; accepted 1991 January 4

ABSTRACT

The transition of the temperature gradient between being subadiabatic and adiabatic at the base of the solar convection zone gives rise to a clear signature in the sound speed. Helioseismic measurements of the sound speed therefore permit a determination of the location of the base of the convection zone. We have tested two techniques by applying them to artificial data, obtained by adding simulated noise to frequencies computed from two different solar models. The determinations appear to be relatively insensitive to uncertainties of the physics of the solar interior, although, if present, a large-scale megagauss magnetic field or some other phenomenon not normally incorporated into theoretical models at the base of the convection zone might have a significant effect on the results. On the assumption that this is not the case, we conclude, from an analysis of observed frequencies of solar oscillation, that the depth of the solar convection zone is (0.287 ± 0.003) solar radii.

Subject headings: Sun: abundances — Sun: interior — Sun: oscillations

1. INTRODUCTION

The properties of the Sun's convection zone play a major role in many aspects of solar physics. In particular, it is presumed that the abundance of lithium in the convection zone, and consequently in the photosphere, has been depleted by nuclear transmutations to the observed photospheric level, which is considerably below the general cosmic value and the abundance in meteorites (e.g., Boesgaard & Steigman 1985; Anders & Grevesse 1989). The fact that lithium has not been completely destroyed indicates that vigorous mixing cannot today extend to regions where the temperature exceeds about 2.8×10^6 K, because in that case the time scale for lithium destruction would be sufficiently short for all the lithium in the convection zone to have been destroyed while the Sun has been on the main sequence. On the other hand, beryllium, for which substantial destruction takes place only at temperatures exceeding 3.4×10^6 K, appears to have been at most slightly depleted (Anders & Grevesse 1989). It is evident that the rate of lithium and beryllium destruction, and hence the expected solar photospheric abundances, must depend strongly on the temperature at the base of the convection zone. Another issue is atomic diffusion, which could have a significant effect on the chemical composition of the solar material within and beneath the convection zone (e.g., Noerdlinger 1977; Gabriel, Noels, & Scuflaire 1984; Wambsganss 1988; Cox, Guzik, & Kidman 1989); this affects not only the observed surface abundances but also the overall structure of the Sun. The diffusion time scale is a sensitive function of conditions at the point where macroscopic mixing ceases to be effective.

There are several other issues upon which the extent of the convection zone bears. For example, it is widely believed that the solar magnetic activity originates through some form of dynamo process, taking place either in the convection zone (see, for example, the review by Parker 1987) or in a thin layer at its base (Parker 1975; also, for example, DeLuca & Gilman 1986 and references therein). Numerical simulations (e.g., Gilman & Miller 1986) have revealed that the depth of the convection zone has an important controlling influence on the form of the differential rotation. Through the differential rotation, the depth is therefore likely to affect the details of the dynamo process and the surface manifestation of magnetic activity. Thus it is evidently of considerable importance to determine the radius, and other properties, at the base of the convection zone.

Conditions near the base of the convection zone are uncertain. In most computations of solar models, it is assumed that motion extends only over the region where the star is convectively unstable according to the Schwarzschild criterion, namely where

$$\gamma^{-1} - \Gamma^{-1} > 0 \, . \qquad (1.1)$$

Here $\gamma = (\partial \ln p/\partial \ln \rho)_s$, p being pressure and ρ density, the partial derivative being taken at constant specific entropy s, and $\Gamma = d \ln p/d \ln \rho$ characterizes the actual stratification of the equilibrium state. In a region of uniform chemical composition, the criterion (1.1) can be rewritten in terms of gradients of temperature T as

$$\nabla > \nabla_{ad} \, , \qquad (1.2)$$

where $\nabla_{ad} = (\partial \ln T/\partial \ln p)_s$ and $\nabla \equiv d \ln T/d \ln p$. This is the form of the instability condition that is most commonly used in

[1] The National Center for Atmospheric Research is sponsored by the National Science Foundation.

[2] Postal address: Institute of Astronomy, University of Cambridge, Madingley Road, Cambridge CB3 0HA, England.

[3] Postal address: Astronomy Unit, School of Mathematical Sciences, Queen Mary and Westfield College, University of London, Mile End Road, London E1 4NS, England.

Advances in Helioseismology

K. G. LIBBRECHT AND M. F. WOODARD

Globally coherent oscillation modes were discovered in the sun about a decade ago, providing a unique seismological probe of the solar interior. Current observations detect modes that are phase-coherent for up to 1 year, with surface velocity amplitudes as low as 2 millimeters per second, and thousands of mode frequencies have been measured to accuracies as high as 1 part in 10^5. This article discusses the properties of these oscillation modes and the ways in which they are adding to our understanding of the structure and dynamics of the sun.

HELIOSEISMOLOGY, THE STUDY OF SOLAR OSCILLATIONS and their use as a probe of the sun's interior, is still a relatively young field. High-degree solar oscillations were first observed in the early 1960s (*1, 2*), and it was just over a decade ago that globally coherent oscillation modes were observed in the sun (*3*). In this short amount of time, however, the quality of the observations has improved greatly, and there have been substantial advances in our theoretical understanding of these oscillations [see the review articles on helioseismology (*4*)]. The observations have allowed us to accurately measure the depth of the solar convection zone as well as the sun's internal rotation profile, and to infer the temperature of the solar core (important for unraveling the solar neutrino problem).

Starting with a model for the overall structure of the sun, one can show with linear adiabatic perturbation theory that small-amplitude oscillations of the model about its equilibrium state can be classified into three types (*4*): (i) p-modes, which have pressure as the dominant restoring force; (ii) g-modes, for which gravity, or buoyancy, is the primary restoring force; and (iii) f-modes, which are nearly compressionless surface waves. Our discussion here will focus on p- and f-modes, for which there is clear and abundant observational data. g-Modes, according to calculation, are predominantly trapped deep in the solar interior, and it seems likely that these modes are not excited to observable amplitudes at the sun's surface. Some possible g-mode detections have been reported, but the observations at this point are still not convincing.

p-Modes are essentially acoustic (sound) waves propagating through the solar interior. For frequencies below a maximum acoustic cutoff frequency, $\nu < \nu_{a,\,max} \approx 5.3$ mHz, acoustic waves in the interior are reflected near the solar surface, forming an acoustic cavity inside the sun (*4*); for $\nu > \nu_{a,\,max}$, waves propagate through the surface and their energy is quickly dissipated. Because the wave damping is small inside the sun, interference organizes the reflected waves into the normal modes of the acoustic cavity, which are the p-modes. For our discussion these modes can be thought of as a set of harmonic oscillators, each uncoupled to all the rest, independently interacting with weak driving and damping forces. [Nonlinear coupling between modes has been found theoretically to be quite small (*5*).]

The perturbation of a scalar quantity in the solar interior, such as pressure p, resulting from an oscillation mode, can be written as an eigenfunction

$$\delta p(\mathbf{r},t) = \mathrm{Re}[\delta p_{n\ell}(r) Y_\ell^m(\theta,\phi) \exp(i 2\pi \nu_{n\ell m} t)] \qquad (1)$$

with frequency eigenvalue $\nu_{n\ell m}$, where $Y_\ell^m(\theta,\phi)$ is a spherical harmonic. Each mode is labeled with three integer "quantum"

K. G. Libbrecht is in the Department of Astrophysics and M. F. Woodard is a senior research fellow at California Institute of Technology, Mail Stop 264-33, Pasadena, CA 91125.

numbers n, ℓ, and m; n counts the number of radial nodes in the wavefunction, and ℓ and m describe the nodes in θ and ϕ. Sample radial wavefunctions have been plotted by several investigators (*4*). Figure 1 shows some p-mode frequencies measured by Libbrecht *et al.* as a function of degree ℓ (*6*). Note the very high accuracy of the observed frequencies, possible because the modes have measured quality factors as high as $Q \approx 20{,}000$. The f-modes, which have no radial nodes, can be thought of as an extension of the p-modes, forming another lower frequency ridge with $n = 0$. Although f-modes have not yet been observed at the ℓ values shown in Fig. 1, they are frequently observed at higher ℓ, where their surface amplitudes are greater.

In order to understand some of the observed trends in the p-mode data, one must be aware of the basic properties of the mode eigenfunctions, particularly near the upper and lower reflection points. Close to the upper reflection point, always near the solar surface, the vertical wavelength of the eigenfunction is much smaller than the horizontal wavelength (for $\ell \ll 1000$). Thus the structure of the eigenfunction is nearly ℓ-independent near the surface. On the other hand, an acoustic wave ceases to propagate once the frequency is below the acoustic cutoff frequency, that is, $\nu < \nu_a \approx c/4\pi H$, where c is the speed of sound and H is the atmospheric scale height. Because $c/H \propto T^{-1/2}$, and temperature T increases rapidly with depth below the solar surface, we find that low-frequency modes are trapped farther below the solar surface than higher frequency modes. Above the upper reflection point the remaining solar atmosphere oscillates up and down like a rigid body. The structure of the *p*-mode eigenfunctions near the solar surface, which in general depends on n and ℓ, is to first order dependent only on $\nu_{n\ell}$.

The lower reflection point r_t, deep in the solar interior, is given by $c(r_t)/r_t \approx 2\pi\nu/L$, where $L^2 = \ell(\ell + 1)$, and thus depends on both ν and ℓ; for fixed ν, lower ℓ modes propagate deeper into the solar interior than those with higher ℓ. At $\nu = 3$ mHz, near the peak in p-mode amplitudes, the reflection point is at the base of the convection zone, $r_t/R_\odot = 0.71$, for $\ell \approx 40$, where $R_\odot$ is the radius of the sun.

A particularly useful concept is that of the mode "mass," defined by $M = E/\langle v^2_{surf}\rangle$, where E is the mode energy and $\langle v^2_{surf}\rangle$ is its mean squared surface velocity; typical mode masses are shown in Fig. 2. The mass increases sharply for $\nu < 2$ mHz because these lower frequency modes are trapped deeper inside the sun, leading to a smaller $\langle v^2_{surf}\rangle$ for a given E. Similarly, at constant ν, a deeper lower reflection point (lower ℓ) leads to a higher mode mass.

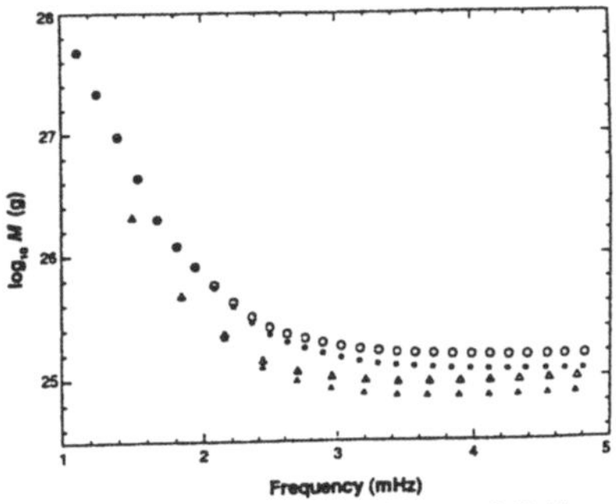

Fig. 2. Mode mass $M_{n\ell m}$, defined as the mode energy divided by its mean squared surface velocity. A range in n is shown for $\ell = 0$ (circles) and $\ell = 100$ (triangles); the open characters use the surface velocity at $\tau_{5000} = 0.5$, and the filled characters are for $\tau_{5000} = 0.1$. For higher ℓ, the mode mass at fixed frequency is roughly proportional to $\ell^{-1/2}$ [Adapted from (*24*)]

Structure of the Solar Interior

By comparing measured p-mode frequencies with solar model calculations, we can in principle infer some properties of the solar interior; recent attempts to determine the solar helium abundance are a good example (7). In practice this process has been a difficult one, primarily because of the complexity of the physics that goes into the solar interior models. If we use state-of-the-art solar models, we find that the calculated frequencies match the measured frequencies to better than 1% (*4*), which is an indication of the overall accuracy of our current solar models. Of the remaining discrepancy, most can be shown to come from inaccuracies in the model near the solar surface (*8*); this is because the mode frequencies are most sensitive to the surface structure (because the sound speed is lowest there), while at the same time the surface layers are the most difficult to model accurately.

By comparing modes with different ℓ and ν, we find that it is possible to peel away the surface uncertainties and infer in a roughly

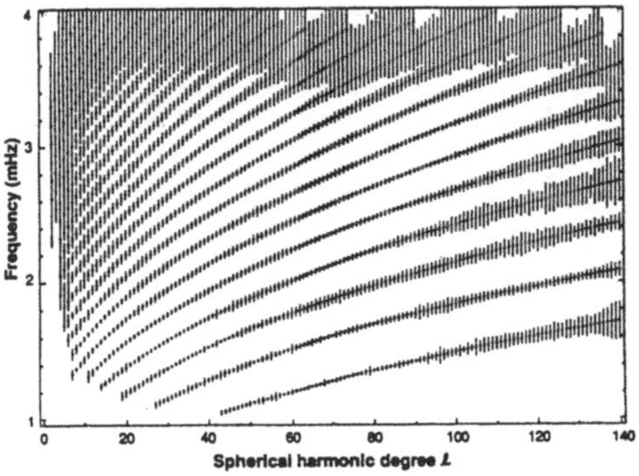

Fig. 1. Schematic plot showing p-mode frequencies measured by Libbrecht and Woodard in 1986 from Big Bear Solar Observatory, California Institute of Technology. The usual 1σ error bars have been magnified by a factor of 1000 to make them visible. Each point shows a mode multiplet frequency $\bar{\nu}_{n\ell}$; the different ridges correspond to fixed values of radial order n (equal to the number of radial nodes in the wavefunction), where the lowest frequency ridge has $n = 1$. [Adapted from (*15*)]

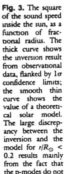

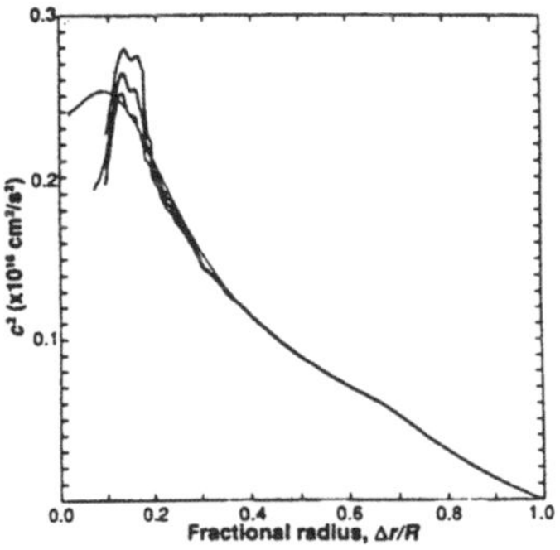

Fig. 3. The square of the sound speed inside the sun, as a function of fractional radius. The thick curve shows the inversion result from observational data, flanked by 1σ confidence limits; the smooth thin curve shows the value of a theoretical solar model. The large discrepancy between the inversion and the model for $r/R_\odot < 0.2$ results mainly from the fact that the p-modes do not probe the deep interior very well. [Adapted from (*10*)]

model-independent way the sound speed as a function of depth deep inside the sun (*9–11*); Fig. 3 shows one example of such an inversion. The difference between inferred and model sound speeds, $\Delta c/c$, is less than 1 to 2% for $r/R_\odot > 0.2$, and the difference appears to be mostly dependent on the details of the solar model used, for instance, the detailed formulation of the equation of state for the solar plasma. Known uncertainties in the calculated opacity tables used to generate the solar models are sufficient to lead to $\Delta c/c$ values of this order.

Of particular interest is the structure of the deepest core of the sun, where nuclear burning and neutrino production are taking place. The sound speed inversions are most uncertain below $r/R_\odot = 0.2$ (Fig. 3); this is primarily because the p-modes are the most sensitive to surface structure and therefore do not probe the deep interior very well. Nevertheless, data from the lowest ℓ modes now indicate that standard solar models give a fairly accurate picture of the structure of the solar core, suggesting that neutrino physics, not solar physics, is the most likely explanation of the solar neutrino problem (*12*). Additional data and analysis are needed to confirm this result, however.

One spectacular success of the sound speed inversions is a very accurate determination of the depth of the convection zone (Fig. 4). Defining the base of the convection zone as that position r_b where the temperature gradient makes a transition from subadiabatic to adiabatic, Christensen-Dalsgaard *et al.* (*13*) find $r_b/R_\odot = 0.713 \pm 0.003$, and $c(r_b) = 0.223 \pm 0.002$ Mm/s. Such an inference shows how a relatively simple bit of information can be extracted to high accuracy from a complex spectrum of p-mode frequencies.

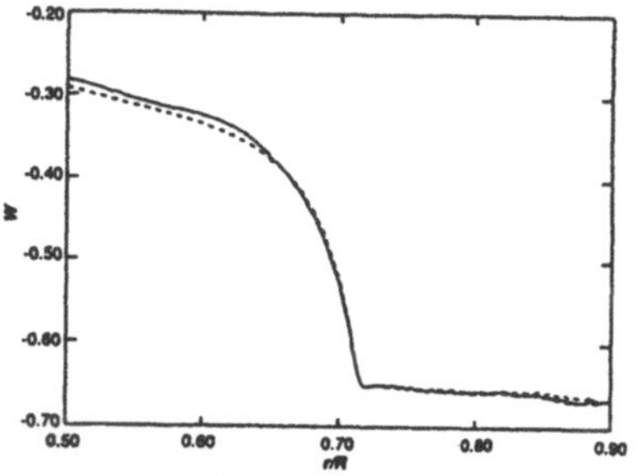

Fig. 4. The function $W = (r^2/GM_\odot)(dc^2/dr)$ as a function of fractional radius near the base of the convection zone, where G is the gravitational constant and $M_\odot$ is the solar mass; in the convection zone we should have $W = 1 - \gamma$, where γ is the adiabatic index. The solid curve is from an inversion of solar data, and the dashed curve is from a reference model. The break at $r/R = 0.713$ shows the change in solar temperature gradient from nearly adiabatic inside the convection zone to subadiabatic below. [Adapted from (*13*)]

The Solar Rotation

Another area in which significant advances have been made recently is that of the p-mode frequency splittings, which have been used to determine the solar rotation rate through most of the solar interior. Because the sun is not spherically symmetric, the mode frequencies are not completely degenerate in m, and the frequencies $\nu_{n\ell m}$ in an $(n\ell)$ multiplet are said to be split, analogous to the Zeeman splitting of degenerate atomic energy levels. Observers tend to fit the measured frequencies for each multiplet to a sum of Legendre polynomials in m/L:

$$\nu_{n\ell m} \to \bar{\nu}_{n\ell} + L \sum_{i=1} a_i P_i(m/L) \qquad (2)$$

where the a_i have been measured to be significantly nonzero up to as high as $i = 12$ (*14*). In such an expansion, the odd-coefficient terms, a_1, a_3, a_5, and so forth, are nonzero because of the latitude- and depth-dependent solar rotation velocity; waves propagating along the equator in the direction of the solar rotation appear, to a fixed observer, to have a higher frequency than waves propagating in the opposite direction. The terms with even coefficients, however, are by symmetry insensitive to the solar rotation (a small but measurable a_2 does arise from the solar oblateness) and instead measure predominantly the latitude- and depth-dependent perturbation of the solar surface structure resulting from solar magnetic activity (*15*).

Figure 5 shows some recent measurements and indicates how the calculated a_i depend on the assumed rotation profile inside the sun (*16*). Figure 6 shows an inversion of the data to determine the best-fit

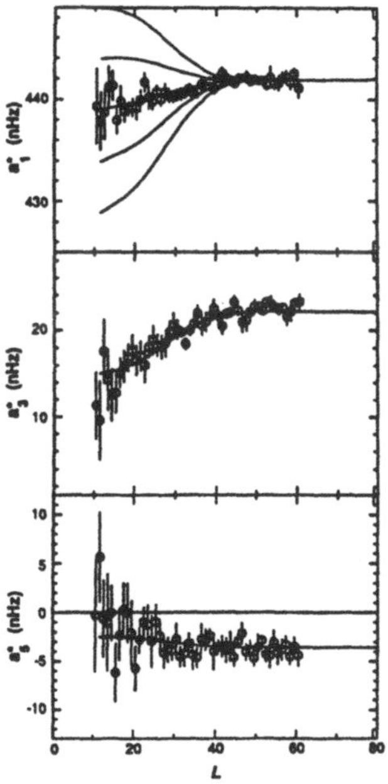

Fig. 5. p-Mode splitting data, where $a_i^*(\ell)$ represents the splitting coefficient a_i (ℓ, n) interpolated to a frequency of 2.5 mHz, together with splittings calculated from a family of solar rotation models. The models have surface-like rotation throughout the convection zone, $\Omega(r, \theta)/2\pi = 461 - 60\cos^2\theta - 75\cos^4\theta$ nHz for $r/R_\odot > 0.71$, where θ is solar colatitude, followed by solid-body rotation for $r/R_\odot < 0.71$. From bottom to top, the a_1^* curves correspond to models with interior rotation rates from 404 to 468 nHz, in evenly spaced 16-nHz intervals. Models like this demonstrate that the radiative interior of the sun, at least for $r/R_\odot > 0.4$, cannot be rotating much faster than the surface rate. [Adapted from (*16*)]

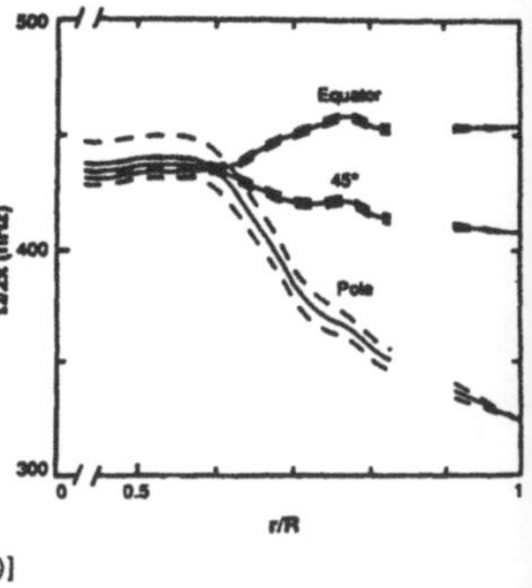

Fig. 6. Solar interior rotation profile, as inferred from an inversion of p-mode splitting data. The rotation rate depends on latitude but not radius throughout the convection zone, with a transition to solid-body rotation below. This inversion does not extend much above 0.8 $R_\odot$ or below 0.4 $R_\odot$; the observed surface rotation is indicated, but the rotation rate of the deep interior remains largely unknown. [Adapted from (*17*)]

rotation profile (*17*). These results reveal that the latitude-dependent rotation profile seen at the solar surface extends down through the convection zone; that is, the angular velocity is $\Omega(r, \theta) \approx \Omega(R_\odot, \theta)$ for $r/R_\odot > 0.7$, where θ is the solar colatitude, whereas in the radiative interior the rotation assumes a solid-body profile, $\Omega(r, \theta)/2\pi \approx 430$ nHz for $0.4 < r/R_\odot < 0.6$. Computer simulations of the solar convective zone suggest a qualitatively different picture of the sun's interior rotation in that region (*18*). At present there is no obvious theoretical explanation for the seismologically inferred rotation profile.

For $r/R_\odot < 0.4$ the existing data do not provide a clear picture. Solar model calculations including rotation suggest that a rapidly rotating core may be present (*19*); however, even a small magnetic field in the radiative interior would halt any such a differential rotation (*20*). Further data for the lowest ℓ values should decide this point in the not-too-distant future.

Time-Dependent Mode Frequencies

p-Mode frequency measurements with accuracies like those in Fig. 1 have been measured for several years, allowing the detection of frequency shifts associated with the changing level of solar magnetic activity. Frequency shifts were first detected in low-ℓ data by Woodard and Noyes (*21*), and more recent observations show that the frequency shifts depend strongly on p-mode frequency [Fig. 7 (*15*)]. The more obvious trend in these data, that the high-ν modes experience a greater shift than low-ν modes, is understood as a time-dependent perturbation in the near-surface layers of the sun, presumably related to the surface magnetic fields associated with the solar magnetic cycle. Because the low-ν modes are trapped deeper below the surface, these modes do not "see" the perturbation as strongly and thus show a smaller frequency shift. The ℓ dependence of the frequency shift (Fig. 7B) is also consistent with this interpretation. Goldreich *et al.* (*22*) have found that a flux-tube model of the surface magnetic fields can fit these data quite well, provided that the magnetic field strength B increases with depth approximately as $B^2 \sim \rho^{1/3}$, where ρ is density, at least at shallow depths.

A particularly surprising aspect of the measurements (Fig. 7A) is that at the highest frequencies accurately measured, $\nu \approx 4$ mHz, the frequency shift turns sharply down. Perhaps the most probable explanation for this behavior (*22*) involves a second acoustic cavity that is theoretically expected to be present in the solar chromosphere. Modes trapped in the solar interior can "tunnel" through the photosphere and interact with this chromospheric cavity; the greatest interaction would naturally be for modes with frequencies in resonance with the chromospheric cavity, $\nu \approx 4$ mHz. As the chromosphere changes with solar cycle, it affects the frequencies of modes that propagate in it. The sudden change in $\Delta\nu$ near $\nu \approx 4$ mHz is explained by the fact that only modes very near resonance have large amplitudes in the chromosphere, so only these modes are significantly affected by the chromospheric changes.

Excitation and Damping

Given a model of the structure of the sun, it is straightforward to calculate the various mode eigenfrequencies $\nu_{n\ell m}$, if we assume small-amplitude adiabatic oscillations. And, as we have discussed, the calculated frequencies are in excellent agreement with those measured, the remaining discrepancies being due mainly to surface uncertainties. It is a much more challenging task to calculate the expected mode amplitudes, because these depend in detail on the dynamics of the excitation and damping mechanisms, which cannot be approximated in an adiabatic limit. A solution to this problem will be welcomed by asteroseismologists, because an understanding of mode excitation in the sun should allow the prediction of acoustic mode amplitudes on other stars.

Figures 8 and 9 show the observed mode amplitudes (converted to average solar surface velocity per mode) and linewidths at low ℓ values. The lowest frequency modes observed have surface amplitudes of only a few millimeters per second and are phase-coherent for nearly 1 year. Kaufman (*23*) found that at higher ℓ the mode amplitudes first increase

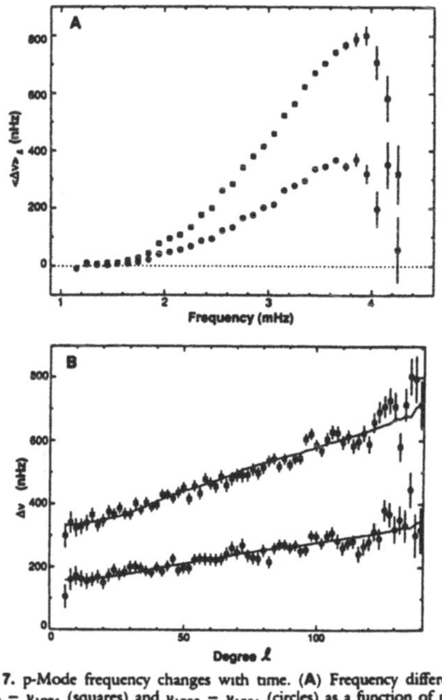

Fig. 7. p-Mode frequency changes with time. (**A**) Frequency differences $\nu_{1989} - \nu_{1986}$ (squares) and $\nu_{1988} - \nu_{1986}$ (circles) as a function of mode frequency ν, averaged over degree with $4 \le \ell \le 140$. The solar minimum was in 1986, and the solar maximum was near 1989. (**B**) p-Mode frequency differences at $\nu = 3$ mHz, as a function of degree ℓ. The solid lines show the inverse mode mass $M^{-1}(\nu = 3 \text{ mHz}, \ell)$ scaled to fit the data (*15*).

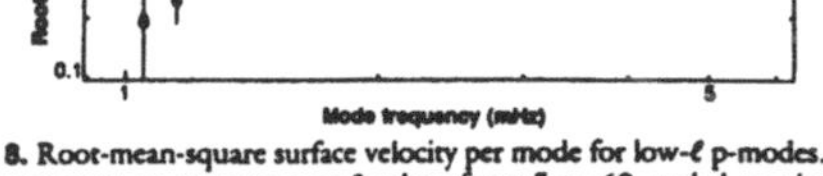

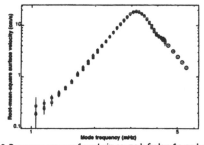

Fig. 8. Root-mean-square surface velocity per mode for low-ℓ p-modes. The data represent an average over ℓ values from 5 to 60, scaled to give the amplitude of $\ell = 0$ modes. Squares, circles, and diamonds show data from 1986, 1988, and 1989, respectively; the different years were scaled to match the 1986 measurements. The large circles at high frequency are from a separate analysis of $\ell = 60$ data (*34*). Between solar minimum (near 1986) and maximum (near 1989) the functional form of the p-mode amplitudes did not change significantly. The lowest frequency modes have surface amplitudes of only a few millimeters per second.

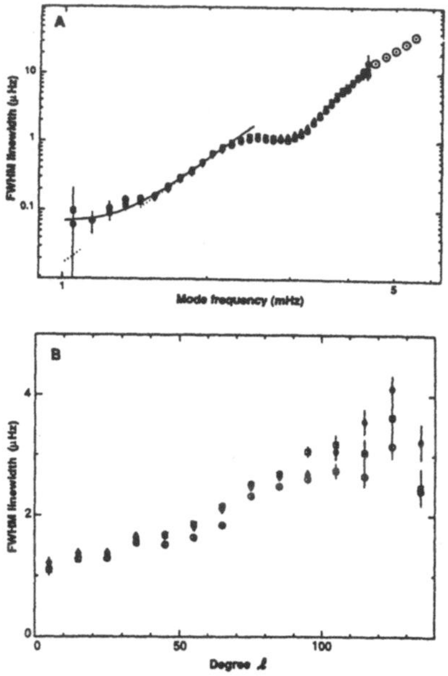

Fig. 9. (**A**) Full-width at half-maximum (FWHM) linewidths of low-ℓ solar p-modes, assuming Lorentzian line profiles; the notation is similar to that in Fig. 8. Here the different years of data were not independently scaled, but all were extrapolated [based on the results in (B)] to the expected $\ell = 0$ linewidths. The dotted line shows a power law behavior, with $\Gamma \sim \nu^5$; the solid line is the expected measured linewidth from this power law, with the finite time span of the observations (about 5 months) taken into account. The lowest frequency modes measured are probably phase-coherent for of order 1 year. (**B**) p-Mode linewidths averaged over $2.4 < \nu < 3.0$ mHz, as a function of spherical harmonic degree ℓ. Note from these plots the apparent lack of any obvious solar cycle changes in linewidths.

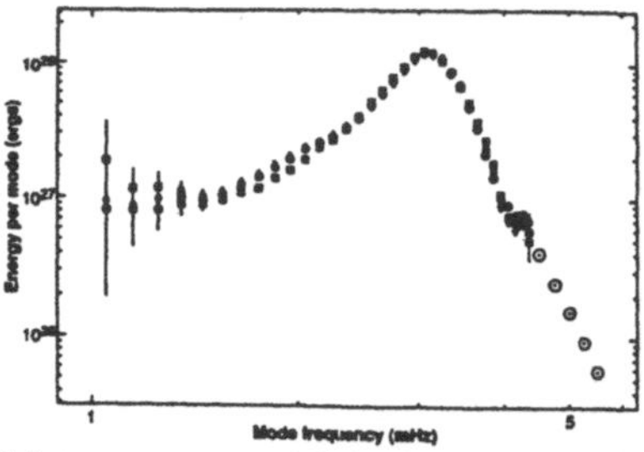

Fig. 10. Average energy per p-mode for low-ℓ modes, where the notation is the same as in Fig. 8. Here a solar model was used to generate p-mode masses, which with the amplitude measurements were used to estimate mode energies.

with ℓ out to $\ell \approx 200$, then decrease at still higher ℓ; the mode linewidths almost certainly increase at higher ℓ, but at present there are no accurate linewidth data at ℓ values above those shown in Fig. 9. Extrapolating the measured linewidths to higher ℓ, we find that the linewidth $\Gamma > d\nu/d\ell$ for all modes with $\ell > 400$; thus for these larger ℓ values no p-modes are globally coherent.

Although the details of the excitation and damping mechanisms are still uncertain, it is clear that all these mechanisms operate very near the solar surface. For example, excitation of modes by the radiation field, the kappa mechanism, is occurring primarily in the hydrogen ionization zone, which is located a few scale heights below the photosphere. The competing process of radiative damping is significant near the photosphere but is negligible in the solar interior, because below the photosphere the time scale for photons to diffuse a distance of one oscillation wavelength is much greater than the period of the modes. Also, energy exchange with convection is greatest near the solar surface, because it depends strongly on the Mach number of the convective motion, which approaches unity near the photosphere and is small below.

This fact, that the excitation and damping processes all occur near the solar surface, qualitatively explains many of the observed trends in the mode properties. For example, for $\ell < 100$ the observed linewidths approximately separate into a function of frequency $\Gamma_0(\nu)$ times a weak function of ℓ, as shown in Fig. 9. The constant frequency dependence results from the fact that at these ℓ values the eigenfunctions near the surface, the most important region, depend nearly exclusively on ν, as discussed above, independent of ℓ. The remaining ℓ dependence in the linewidth, approximately given by $M^{-1}(\ell)$, arises from the fact that modes with higher ℓ have higher surface amplitudes than lower ℓ modes (for fixed ν and mode energy). The lowest ν modes, which are trapped deepest below the surface, have the smallest linewidths, owing to their weaker interaction with the surface driving and damping forces.

The most important process responsible for the observed p-mode amplitudes appears to be stochastic excitation by turbulent convection (*24, 25*), first proposed for p-modes by Goldreich and Keeley over a decade ago (*26*) and recently refined by Goldreich and Kumar (*27, 28*). The basic idea is that turbulence in the sun's convection zone generates acoustic noise, and this acoustic noise trapped inside the sun excites the cavity's resonant modes, the p-modes. Recent calculations predict p-mode energies that are given by $E \approx mv^2$, where E is the energy per mode, m is the mass of a resonant turbulent convective eddy, and v is the convective velocity. In the photosphere a convective eddy is essentially a granulation cell, and we can write $E \approx \rho H L^2 v^2$, where $\rho \approx 10^{-7}$ g/cm^3 is the density at τ_{5000} (optical depth at 5000 Å) = 1, $H \approx 100$ km is the scale height, $L \approx 1000$ km is the horizontal size of a granule, and $v \approx 0.1\,c \approx 1$ km/s, giving $E \sim 10^{26}$ ergs, not too far from the observations (Fig. 10). The turbulent excitation model not only predicts nearly the correct order of magnitude for the p-mode energies, but it also naturally explains the observation that millions of modes are simultaneously excited. However, a model reproducing the correct energy spectrum does not yet exist.

An interesting quantity to consider is the product of mode energy and linewidth as a function of frequency, $E\Gamma(\nu)$, for low-ℓ modes (*24*), because this is easier to describe theoretically. For a stochastically driven p-mode, $E\Gamma = \dot{E}$ is a measure of the power being pumped into the mode, presumably by convection. If one were to leave this input power fixed and arbitrarily increase, say, the radiative damping, then the mode energy would decrease and the linewidth would increase, but $E\Gamma$ would remain unchanged. The observations indicate that $E\Gamma \sim \nu^8$ for $\nu < 3$ mHz, $E\Gamma \sim \nu^{-5}$ for $\nu > 4$ mHz, and $E\Gamma \approx 10^{22}$ ergs/s at $\nu = 3.5$ mHz (*24*). Goldreich and Kumar (*28*) found that they can reproduce quite well the observed trends in $E\Gamma(\nu)$, both above and below the peak at $\nu \approx 3.5$ mHz, supporting the hypothesis that p-modes are stochastically excited by turbulent convection. The observation of p-mode structure for $\nu > \nu_{a,max}$ also points toward turbulent convection as the most likely excitation mechanism (*29*).

The Future

Work is under way on a new generation of helioseismology instruments that promise a tenfold or greater improvement in the accuracy of the measurements shown above. The Global Oscillation Network Group (GONG) (*30*), scheduled to begin operation in 1993, will consist of a network of six telescopes, spaced in longitude to provide continuous Doppler measurements of the solar surface for several years. Other networks of integrated sunlight instruments (*31*) will provide more accurate measurements of the properties of modes with $\ell \leq 3$. Meanwhile, the Solar Heliospheric Observatory (SOHO) spacecraft will contain several helioseismology instruments, including the GOLF (Global Oscillations at Low Frequencies) (*32*), and MDI (Michelson Doppler Imager) (*33*) instruments, which will provide precision measurements of low-frequency and high-ℓ oscillations, respectively.

These instruments, combined with theoretical advances yet to come, should provide an extremely detailed picture of many properties of the solar interior. Going beyond a basic understanding of the structure of the sun, these new helioseismological measurements are expected to turn the sun into a precision laboratory for learning about the physics of high-temperature plasmas and magnetohydrodynamics, neutrino oscillations, radiative transfer, and the dynamics of large-scale stratified convection and rotation.

REFERENCES AND NOTES

1. R. B. Leighton, R. W. Noyes, G. W. Simon, *Astrophys. J.* 135, 474 (1962).
2. F.-L. Deubner, *Astron. Astrophys.* 44, 371 (1975).
3. A. Claverie *et al.*, *Nature* 282, 591 (1979); G. Grec, E. Fossat, M. Pomerantz, *ibid.* 288, 541 (1980); *Solar Phys.* 82, 55 (1983).
4. F.-L. Deubner and D. Gough, *Annu. Rev. Astron. Astrophys.* 22, 593 (1984); J. Christensen-Dalsgaard, D. Gough, J. Toomre, *Science* 229, 923 (1985); K. G. Libbrecht, *Space Sci. Rev.* 47, 275 (1988); S. V. Vorontsov and V. N. Zharkov, *Sov. Sci. Rev. Sect. E Astrophys. Space Phys.* 7, 1 (1989); H. Shibahashi, in *Progress of Seismology of the Sun and Stars*, Y. Osaki and H. Shibahashi, Eds. (Springer-Verlag, Berlin, 1990), p. 3.
5. P. Kumar and P. Goldreich, *Astrophys. J.* 342, 558 (1989).
6. K. G. Libbrecht, M. F. Woodard, J. M. Kaufman, *Astrophys. J. Suppl. Ser.* 74, 1129 (1990).
7. S. V. Vorontsov, V. A. Baturin, A. A. Pamyatnykh, *Nature* 349, 49 (1991); other solar helium determinations are cited in this paper.
8. J. Christensen-Dalsgaard, in *Proceedings of the Symposium on the Seismology of the Sun and Sun-Like Stars*, E. J. Rolfe, Ed. [European Space Agency (ESA), Noordwijk, Netherlands, 1988], p. 431.
9. J. Christensen-Dalsgaard *et al.*, *Nature* 315, 378 (1985).
10. H. Shibahashi and T. Sekii, in *Proceedings of the Symposium on the Seismology of the Sun and Sun-Like Stars*, E. J. Rolfe, Ed. (ESA, Noordwijk, Netherlands, 1988), p. 471.
11. S. V. Vorontsov, *ibid.*, p. 475.
12. Y. Elsworth *et al.*, *Nature* 347, 536 (1990).
13. J. Christensen-Dalsgaard, D. O. Gough, M. J. Thompson, in preparation.
14. K. G. Libbrecht and M. F. Woodard, in preparation.
15. ———, *Nature* 345, 779 (1990). Low ℓ results are also described on p. 768 of that issue.
16. K. G. Libbrecht and C. A. Morrow, in *The Solar Interior and Atmosphere* (Univ. of Arizona Press, Tucson, in press).
17. J. Christensen-Dalsgaard and J. Schou, in *Proceedings of the Symposium on the Seismology of the Sun and Sun-Like Stars*, E. J. Rolfe, Ed. (ESA, Noordwijk, Netherlands, 1988), p. 149.
18. G. A. Glatzmaier, in *The Internal Solar Angular Velocity*, B. R. Durney and S. Sofia, Eds. (Reidel, Dordrecht, 1987), p. 263; P. A. Gilman and J. Miller, *Astrophys. J. Suppl. Ser.* 61, 585 (1986).
19. M. H. Pinsonneault *et al.*, *Astrophys. J.* 338, 424 (1989).
20. H. C. Spruit, in *The Internal Solar Angular Velocity*, B. R. Durney and S. Sofia, Eds. (Reidel, Dordrecht, 1987), p. 185.
21. M. Woodard and R. W. Noyes, *Nature* 318, 449 (1985).
22. P. Goldreich *et al.*, *Astrophys. J.*, in press.
23. J. Kaufman, thesis, California Institute of Technology (1990).
24. K. G. Libbrecht, in *Proceedings of the Symposium on the Seismology of the Sun and Sun-Like Stars*, E. J. Rolfe, Ed. (ESA, Noordwijk, Netherlands, 1988), p. 3.
25. Y. Osaki, *Progress of Seismology of the Sun and Stars*, Y. Osaki and H. Shibahashi, Eds. (Springer-Verlag, Berlin, 1990), p. 75.
26. P. Goldreich and D. A. Keeley, *Astrophys. J.* 211, 934 (1977); *ibid.* 212, 243 (1977).
27. P. Goldreich and P. Kumar, *ibid.* 326, 462 (1988).
28. ———, *ibid.* 363, 694 (1990).
29. P. Kumar and E. Lu, in preparation.
30. J. Harvey *et al.*, in *Proceedings of the Symposium on the Seismology of the Sun and Sun-Like Stars*, E. J. Rolfe, Ed. (ESA, Noordwijk, Netherlands, 1988), p. 203.
31. A. Aindow *et al.*, *ibid.*, p. 157; E. Fossat, *ibid.*, p. 161.
32. L. Dame, *ibid.*, p. 367.
33. P. H. Scherrer, J. T. Hoeksema, R. S. Bogart, *ibid.*, p. 375.
34. K. G. Libbrecht, *Astrophys. J.* 334, 510 (1988).

35. We thank N. Murray and P. Goldreich for enlightening conversations. This work was supported by NSF grants ATM-8907012 and AST-8657393.

VI. Transition from "Problem" to "Opportunity"

Introduction: Bahcall, Davis, Parker, Smirnov, Ulrich

The three papers included in this section have been described in the Preface to the paperback edition of 2002. We only need to add here that the announcement of the epochal results from the SNO experiment (VI. 3) on June 18, 2001 immediately led to a torrent of papers that used the SNO results, together with those of the Super-Kamiokande collaboration (VI. 2) and the previously announced results from the chlorine, Kamiokande, SAGE, GALLEX, and GNO experiments, to elucidate the importance of the experimental results for physics and astronomy.

We list directly below the three reprinted articles. Following the list of reprinted papers, we cite (in chronological order of appearance) a few of the articles that appeared in the two months following the SNO announcement in order to give the interested reader a feeling for the jubilation and the enthusiasm of the immediate response to the new results.

VI. Reprinted papers

1. J.N. Bahcall, S. Basu, M.H. Pinsonneault, "How uncertain are solar neutrino predictions?" *Phys. Lett. B* **433** (1998).
2. S. Fukuda et al. (Super-Kamiokande Collaboration), "Solar ^{8}B and hep neutrino measurements from 1258 days of Super-Kamiokande data, *Phys. Rev. Lett.* **86** (2001) 5651.
3. Q.R. Ahmad et al., "Measurement of charged current interactions produced by ^{8}B solar neutrinos at the Sudbury Neutrino Observatory," *Phys. Rev. Lett.* **87** (2001) 071301.

VI. References.

(1.) "Unknowns after the SNO charged-current measurement," V. Barger, D. Marfatia, and K. Whisnant, *Phys. Rev. Lett.* **88**, 011302 (2002).

(2.) "Model-dependent and independent implications of the first Sudbury Neutrino Observatory results," G.L. Fogli, E. Lisi, D. Montanino, and A. Palazzo, *Phys. Rev. D* **64**, 093007 (2001).

(3.) "Global analysis of solar neutrino oscillations including SNO CC measurement," J.N. Bahcall, M.C. Gonzalez-Garcia, Carlos Pe\~na-Garay, *JHEP* 08(2001)014.

(4.) "Impact of the first SNO results on neutrino mass and mixing," A. Bandyopadhyay, S. Choubey, S. Goswami, and K. Kar, *Phys. Lett. B* **519**, 83 (2001).

(5). "Frequentist analyses of solar neutrino data", P. Creminelli, G. Signorelli, and A. Strumia, *JHEP* 0105(2001)052.

(6.) "Are there í$_1$ or í$_6$ in the flux of solar neutrinos on Earth?", C. Giunti, *Phys. Rev.D* **65**, 033006 (2002).
(7.) "How Many Sigmas is the Solar Neutrino Effect?," John N. Bahcall, *Phys. Rev. C* **65**, 015802 (2002).
(8.) "Exact analysis of the combined data of SNO and Super-Kamiokande," V. Berezinsky [hep-ph/0108166].
(9) "Global Analysis with SNO: Toward a Solution of the Solar Neutrino Problem", P. I. Krastev and A. Yu. Smirnov, [hep-ph/0108177]
(10.) "Bayesian view of solar neutrino oscillations," M.V. Garzelli and C. Giunti, *JHEP* 12(2001)017.

ELSEVIER

6 August 1998

PHYSICS LETTERS B

Physics Letters B 433 (1998) 1–8

How uncertain are solar neutrino predictions?

John N. Bahcall [a,1], Sarbani Basu [a,2], M.H. Pinsonneault [b,3]

[a] *School of Natural Sciences, Institute for Advanced Study, Princeton, NJ 08540, USA*
[b] *Department of Astronomy, Ohio State University, Columbus, OH 43210, USA*

Received 30 March 1998; revised 8 May 1998
Editor: W. Haxton

Abstract

Solar neutrino fluxes and sound speeds are calculated using a systematic reevaluation of nuclear fusion rates. The largest uncertainties are identified and their effects on the solar neutrino fluxes are estimated. © 1998 Elsevier Science B.V. All rights reserved.

Five solar neutrino experiments (chlorine, Kamiokande, GALLEX, SAGE, and Super-Kamiokande) have measured solar neutrinos with approximately the fluxes and energies predicted by standard solar models, confirming empirically the basic picture of stellar energy generation. However, robust quantitative differences exist between the neutrino experiments and the combined predictions of minimal standard electroweak theory and stellar evolution models. Many authors have suggested that these results provide the first evidence of physics beyond the minimal standard electroweak model.

What are the principal uncertainties in the standard model predictions? In this paper, we determine the uncertainties in the solar neutrino calculations that arise from errors in the nuclear fusion cross sections and show that these uncertainties, while relatively small, are currently the largest sources of recognized errors in the neutrino predictions.

In January, 1997, the Institute for Nuclear Theory (INT) hosted a workshop devoted to determining the best estimates and the uncertainties in the most important solar fusion reactions. Thirty-nine experts in low energy nuclear experiments and theory, representing many different research groups and points of view, participated in the workshop and evaluated the existing experimental data and theoretical calculations. Their conclusions have been summarized in a detailed article authored jointly by the participants and to be published by the Reviews of Modern Physics [1]. In general outline, the conclusions of the INT workshop paper confirmed and strengthened previous standard analyses of nuclear fusion rates, although in a few important cases (for the $^{3}He(\alpha,\gamma)^{7}Be$, $^{7}Be(p,\gamma)^{8}B$, and $^{14}N(p,\gamma)^{15}O$ reactions) the estimated uncertainties were determined to be larger than previously believed.

The purpose of this article is to present calculations of solar neutrino fluxes and solar sound veloci-

[1] E-mail address: jnb@sns.ias.edu.
[2] E-mail address: basu@sns.ias.edu.
[3] E-mail address: pinsono@payne.mps.ohio.state.edu.

0370-2693/98/$ – see frontmatter
PII: S0370-2693(98)00657-1

ties, with special attention to their uncertainties, that were made using the recommended INT nuclear reaction rates and the best available other input data.

Our results can be compared directly with the observed rates in solar neutrino experiments and be used as input for detailed analyses of the particle physics implications of the measured solar neutrino rates. We identify the most important nuclear parameters that need to be measured more accurately in laboratory experiments and determine the precision that is required. By comparing our solar models with five recent, precise helioseismological determinations of the sound velocities, we estimate the size of the remaining errors in the model calculations.

Table 1 gives the neutrino fluxes and their uncertainties for our best standard solar model (hereafter BP98). The solar model makes use of the INT nuclear reaction rates [1], recent (1996) Livermore OPAL opacities [2], the OPAL equation of state [3], and electron and ion screening as indicated by recent calculations [4]. The adopted uncertainties in input parameters are given in Table 2 and the associated text. We have also made small improvements in our energy generation code, which will be described in detail in a future publication.

The theoretical predictions in Table 1 disagree with the observed neutrino event rates, which are [6]: 2.55 ± 0.25 SNU (chlorine), 73.4 ± 5.7 SNU (GALLEX and SAGE gallium experiments), and $(2.80 \pm 0.19(\text{stat}) \pm 0.33(\text{syst})) \times 10^{6}\text{cm}^{-2}\text{s}^{-1}$ (^{8}B flux from Kamiokande).

Table 1
Standard Model Predictions (BP98): solar neutrino fluxes and neutrino capture rates, with 1σ uncertainties from all sources (combined quadratically).

Source	Flux ($10^{10}\ \text{cm}^{-2}\text{s}^{-1}$)	Cl (SNU)	Ga (SNU)
pp	$5.94(1.00^{+0.01}_{-0.01})$	0.0	69.6
pep	$1.39\times10^{-2}(1.00^{+0.01}_{-0.01})$	0.2	2.8
hep	2.10×10^{-7}	0.0	0.0
^{7}Be	$4.80\times10^{-1}(1.00^{+0.09}_{-0.09})$	1.15	34.4
^{8}B	$5.15\times10^{-4}(1.00^{+0.19}_{-0.14})$	5.9	12.4
^{13}N	$6.05\times10^{-2}(1.00^{+0.19}_{-0.13})$	0.1	3.7
^{15}O	$5.32\times10^{-2}(1.00^{+0.22}_{-0.15})$	0.4	6.0
^{17}F	$6.33\times10^{-4}(1.00^{+0.12}_{-0.11})$	0.0	0.1
Total		$7.7^{+1.2}_{-1.0}$	129^{+8}_{-6}

The principal differences between the results shown in Table 1 and the results presented in our last systematic publication of calculated solar neutrino fluxes [7] is a 1.3σ decrease in the ^{8}B neutrino flux and 1.1σ decreases in the ^{37}Cl and ^{71}Ga capture rates. These decreases are due principally to the lower $^{7}\text{Be}(p,\gamma)^{8}\text{B}$ cross section adopted by Adelberger et al. [1]. If we use, as in our recent previous publications, the Caltech (CIT) value for the ^{8}B production cross section [8], then the ^{8}B flux is $\phi(^{8}\text{B},\text{CIT}) = 6.1^{+1.1}_{-0.9} \times 10^{6}\ \text{cm}^{-2}\text{s}^{-1}$, $\Sigma(\phi\sigma)_{i}|_{\text{Cl,CIT}} = 8.8^{+1.4}_{-1.1}$ SNU, and $\Sigma(\phi\sigma)_{i}|_{\text{Gallium}} = 131^{+9}_{-7}$ SNU, all of which are within ten percent of the Bahcall-Pinsonneault 1995 best-estimates. The difference between the INT and the CIT estimates of the ^{8}B production cross section is due almost entirely to the decision by the INT group to base their estimate on only one (the best documented) of the six experiments analyzed by the CIT collaboration.

Table 2 summarizes the uncertainties in the most important solar neutrino fluxes and in the Cl and Ga event rates due to different nuclear fusion reactions (the first four entries), the heavy element to hydrogen mass ratio (Z/X), the radiative opacity, the solar luminosity, the assumed solar age, and the helium and heavy element diffusion coefficients. The $^{14}\text{N} + p$ reaction causes a 0.2% uncertainty in the predicted pp flux and a 0.1 SNU uncertainty in the Cl (Ga) event rates.

The predicted event rates for the chlorine and gallium experiments use recent improved calculations of neutrino absorption cross sections [5]. The uncertainty in the prediction for the gallium rate is dominated by uncertainties in the neutrino absorption cross sections, $+6.7$ SNU (7% of the predicted rate) and -3.8 SNU (3% of the predicted rate). The uncertainties in the chlorine absorption cross sections cause an error, ± 0.2 SNU (3% of the predicted rate), that is relatively small compared to other uncertainties in predicting the rate for this experiment. For non-standard neutrino energy spectra that result from new neutrino physics, the uncertainties in the predictions for currently favored solutions (which reduce the contributions from the least well-determined ^{8}B neutrinos) will in general be less than the values quoted here for standard spectra and must be calculated using the appropriate cross section uncertainty for each neutrino energy [5].

Table 2
Average uncertainties in neutrino fluxes and event rates due to different input data. The flux uncertainties are expressed in fractions of the total flux and the event rate uncertainties are expressed in SNU. The ^{7}Be electron capture rate causes an uncertainty of $\pm 2\%$ [9] that affects only the ^{7}Be neutrino flux. The average fractional uncertainties for individual parameters are shown. See text for discussion of asymmetric uncertainties and uncertainties due to radiative opacity or diffusion.

⟨Fractional uncertainty⟩	pp 0.017	^{3}He^{3}He 0.060	^{3}He^{4}He 0.094	^{7}Be + p 0.106	Z/X 0.033	opac	lum 0.004	age 0.004	diffuse
Flux									
pp	0.002	0.002	0.005	0.000	0.002	0.003	0.003	0.0	0.003
^{7}Be	0.0155	0.023	0.080	0.000	0.019	0.028	0.014	0.003	0.018
^{8}B	0.040	0.021	0.075	0.105	0.042	0.052	0.028	0.006	0.040
SNUs									
Cl	0.3	0.2	0.5	0.6	0.3	0.4	0.2	0.04	0.3
Ga	1.3	0.9	3.3	1.3	1.6	1.8	1.3	0.20	1.5

The nuclear fusion uncertainties in Table 2 were taken from Adelberger et al. [1], the neutrino cross section uncertainties from [5], the heavy element uncertainty was taken from helioseismological measurements [10], the luminosity and age uncertainties were adopted from BP95 [7], the 1σ fractional uncertainty in the diffusion rate was taken to be 15% [11], which is supported by helioseismological evidence [12], and the opacity uncertainty was determined by comparing the results of fluxes computed using the older Los Alamos opacities with fluxes computed using the modern Livermore opacities [13]. To include the effects of asymmetric errors, the code exportrates.f (see below) was run with different input uncertainties and the results averaged.

Many authors have used the results of solar neutrino experiments and the calculated best-estimates and uncertainties in the standard solar model fluxes to determine the allowed ranges of neutrino parameters in different particle physics models. To systematize the calculations in the solar neutrino predictions, we have constructed an exportable computer code, exportrates.f, which evaluates the uncertainties in the predicted neutrino fluxes and capture rates from all the recognized sources of errors in the input data. This code is available at www.sns.ias.edu/~jnb (see Solar Neutrino Software and Data); it contains a description of how each of the uncertainties listed in Table 2 was determined and used.

The low energy cross section of the ^{7}Be + p reaction is the most important quantity that must be determined more accurately in order to decrease the error in the predicted event rates in solar neutrino experiments. The ^{8}B neutrino flux that is measured by the Kamiokande [6], Super-Kamiokande [14], and SNO [15] experiments is, in all standard solar model calculations, directly proportional to the ^{7}Be + p cross section. If the 1σ uncertainty in this cross section can be reduced by a factor of two to 5%, then it will no longer be the limiting uncertainty in predicting the crucial ^{8}B neutrino flux (cf. Table 2).

The ^{7}Be neutrino flux will be measured by BOREXINO [16]. Table 2 shows that the theoretical uncertainty in this flux is dominated by the uncertainty in the measured laboratory rate for the ^{3}He–^{4}He reaction. In order that the uncertainty from the ^{3}He–^{4}He cross section be reduced to a level, 3%,that is comparable to the uncertainties in calculating the ^{7}Be flux that arise from other sources, the fractional uncertainty in the ^{3}He–^{4}He low energy cross section factor must be reduced to 3.5%. This goal is not easy, but it appears to be achievable. The six published determinations of the low energy cross section factor using measurements of the capture γ-rays currently have a 1σ uncertainty of 3.2%, but they differ by 2.5σ (14%) from the value determined by counting the ^{7}Be activity [1].

Could the solar model calculations be wrong by enough to explain the discrepancies between predictions and measurements for solar neutrino experiments? Helioseismology, which confirms predictions of the standard solar model to high precision, suggests that the answer is probably "No."

Fig. 1 shows the fractional differences between the most accurate available sound speeds measured by helioseismology [17] and sound speeds calculated

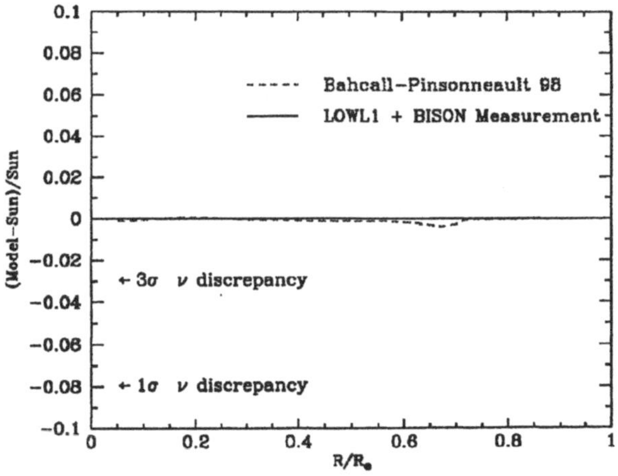

Fig. 1. Predicted versus Measured Sound Speeds. This figure shows the excellent agreement between the calculated (solar model BP98, Model) and the measured (Sun) sound speeds, a fractional difference of 0.001 rms for all speeds measured between $0.05R_\odot$ and $0.95R_\odot$. The vertical scale is chosen so as to emphasize that the fractional error is much smaller than generic changes in the model, 0.03 to 0.08, that might significantly affect the solar neutrino predictions.

with our best solar model (with no free parameters). The horizontal line corresponds to the hypothetical case in which the model predictions exactly match the observed values. The rms fractional difference between the calculated and the measured sound speeds is 1.1×10^{-3} for the entire region over which the sound speeds are measured, $0.05R_\odot < R < 0.95R_\odot$. In the solar core, $0.05R_\odot < R < 0.25R_\odot$ (in which about 95% of the solar energy and neutrino flux is produced in a standard model), the rms fractional difference between measured and calculated sound speeds is 0.7×10^{-3}.

Helioseismological measurements also determine two other parameters that help characterize the outer part of the sun (far from the inner region in which neutrinos are produced): the depth of the solar convective zone (CZ), the region in the outer part of the sun that is fully convective, and the present-day surface abundance by mass of helium ($Y_{\rm surf}$). The measured values, $R_{\rm CZ} = (0.713 \pm 0.001)R_\odot$ [18], and $Y_{\rm surf} = 0.249 \pm 0.003$ [10], are in satisfactory agreement with the values predicted by the solar model BP98, namely, $R_{\rm CZ} = 0.714R_\odot$, and $Y_{\rm surf} = 0.243$. However, we shall see below that precision measurements of the sound speed near the transition between the radiative interior (in which energy is transported by radiation) and the outer convective zone (in which energy is transported by convection) reveal small discrepancies between the model predictions and the observations in this region.

If solar physics were responsible for the solar neutrino problems, how large would one expect the discrepancies to be between solar model predictions and helioseismological observations? The characteristic size of the discrepancies can be estimated using the results of the neutrino experiments and scaling laws for neutrino fluxes and sound speeds.

All recently published solar models predict essentially the same fluxes from the fundamental pp and pep reactions (amounting to 72.4 SNU in gallium experiments, cf. Table 1), which are closely related to the solar luminosity. Comparing the measured and the standard predicted rate for the gallium experiments, the ^{7}Be flux must be reduced by a factor N if the disagreement is not to exceed n standard deviations, where N and n satisfy $72.4 + (34.4)/N = 73.4 + n\sigma$. For a 1σ (3σ) disagreement, $N = 5.1(1.9)$. Sound speeds scale like the square root of the local temperature divided by the mean molecular weight and the ^{7}Be neutrino flux scales approximately as the

10th power of the temperature [19]. Assuming that the temperature changes are dominant, agreement to within 1σ would require fractional changes of order 0.08 in sound speeds (3σ could be reached with 0.03 changes), if all model changes were in the temperature. This argument is conservative because it ignores the contributions from the ^{8}B and CNO neutrinos which contribute to the observed counting rate (cf. Table 1) and which, if included, would require an even larger reduction of the ^{7}Be flux.

We have chosen the vertical scale in Fig. 1 to be appropriate for fractional differences between measured and predicted sound speeds that are of order 0.03 to 0.08 and that might therefore affect solar neutrino calculations. Fig. 1 shows that the characteristic agreement between solar model predictions and helioseismological measurements is more than a factor of 30 better than would be expected if there were a solar model explanation of the solar neutrino problems.

Given the helioseismological measurements, how uncertain are the solar neutrino predictions? We provide a tenative answer to this question by examining more closely the small discrepancies shown in Fig. 1 between the observed and calculated sound speeds.

There has been an explosion of precise helioseismological data in the last year. Fig. 2 compares the results of five different observational determinations of the sound speeds in the sun with the results of our best solar model; the helioseismological discussion in Ref. [12] made use of only the LOWL1 data. The small features discrepancies shown in Fig. 2 are robust; they occur when comparisons are made with all the data sets. The vertical scale for Fig. 2 has been expanded by a factor of 20 with respect to the scale of Fig. 1 in order to show the small but robust discrepancies. References to the different helioseismological measurements are given in the caption to Fig. 2.

All five of the precise helioseismological measurements show essentially the same difference between the model and the solar sound speeds near the base of the convective zone. The sharp edge to this feature occurs near the present base of the convective zone at $R_{CZ} = 0.713 R_{\odot}$.

What could be the cause of the broad feature shown in Fig. 2 that stretches from about $0.3 R_{\odot}$ to about $0.7 R_{\odot}$? This feature may be due to some combination of small errors in the adopted radiative opacities or equation of state and the oversimplification we use in our stellar evolution code of a sharp

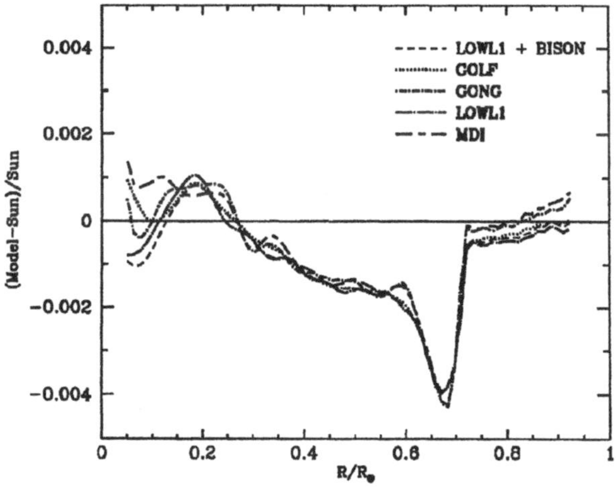

Fig. 2. Five Precise Helioseismological Measurements. The predicted BP98 sound speeds are compared with five different helioseismological measurements [20]. The vertical scale has been expanded by a factor of 20 relative to Fig. 1.

boundary between the radiative and the convective zones. We can make an estimate of the likely effect of hypothetical improved physics on the calculated neutrino fluxes by considering what happens if we change the adopted opacity.

Fig. 3 shows the fractional difference in the computed sound speeds obtained from two solar models that are identical except for the adopted radiative opacity [2], either the version from OPAL92 or the later version from OPAL95. It is apparent from Fig. 3 that the difference in sound speeds caused by using the OPAL95 opacity rather than OPAL92 opacity produces a feature that is similar in shape and in magnitude to the broad feature in Fig. 2. For $0.3 < R/R_\odot < 0.6$, the OPAL95 opacity is about 2% less than the OPAL92 opacity for the conditions in the BP98 model. The opacity difference increases to about 6% near the base of the convective zone. A change in opacity of about the same size but opposite in sign to that which occurred between OPAL92 and OPAL95 would remove most of the broad discrepancy, but at the price of producing a slightly deeper convective zone, $0.7105 R_\odot$, than is observed.

What are the implications for solar neutrino predictions of the existence of the broad 0.2% discrepancy highlighted in Fig. 2? Table 3 shows the neutrino fluxes predicted by the BP98 model constructed using OPAL92 opacities. The chlorine rate is increased relative to the standard BP98 model (cf. Table 1) by 5% and the ^{7}Be and ^{8}B fluxes are increased by 3% and 5.5%, respectively. These changes are plausible estimates of the changes in the neutrino rates that may be anticipated from further precision improvements in solar models in the outer radiative zone between $0.3 R_\odot$ and $0.7 R_\odot$.

The narrow feature near the base of the convective zone may be caused by mixing that is related to the observed [21] depletion of Li and Be. Several independent calculations [22] of the required amount of mixing indicate that the effect on the neutrino fluxes is small, less than a 1% change in the pp flux, and a 2% (4%) decrease in the ^{7}Be (^{8}B) neutrino flux.

There is a smaller difference between the models and the observations centered near $R = 0.2 R_\odot$. This feature occurs using all available helioseismological data and is robust against changes in the method of inversion and inversion parameters. This feature could result, for example, from a 2% inaccuracy in the Livermore opacity at 9×10^6 K [23] or a 0.1% inaccuracy in the OPAL evaluation of the adiabatic index Γ_1. Detailed calculations [24] of the sensitivity

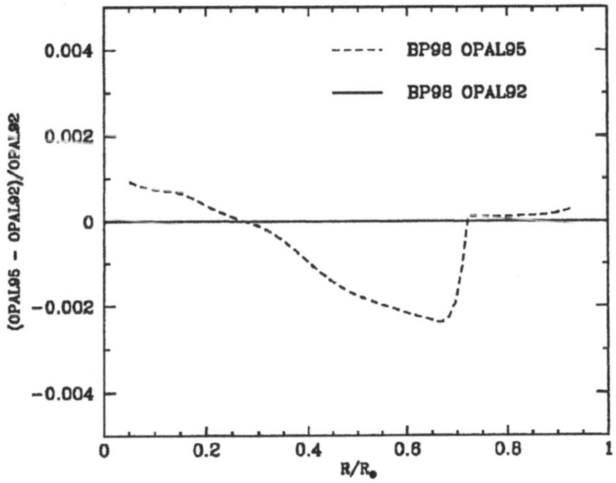

Fig. 3. The effect of opacity on the calculated sound speeds. The figure shows the difference between the calculated sound speeds for two solar models that differ only in the version used of the OPAL opacities [2], 1992 or 1995.

Table 3
Standard solar model with OPAL92 opacities: solar neutrino fluxes. The predicted chlorine capture rate is 8.1 SNU and the gallium rate is 131 SNU.

Source	Flux (10^{10} cm^{-2}s^{-1})
pp	5.92
pep	1.39×10^{-2}
hep	2.08×10^{-7}
^{7}Be	4.94×10^{-1}
^{8}B	5.44×10^{-4}
^{13}N	6.25×10^{-2}
^{15}O	5.52×10^{-2}
^{17}F	6.59×10^{-4}

of the neutrino fluxes to changes in opacity or equation of state suggest that either of the changes mentioned above would affect the most sensitive calculated neutrino fluxes by about 2%.

In conclusion, we note that three decades of refining the input data and the solar model calculations has led to a predicted standard model event rate for the chlorine experiment, 7.7 SNU, which is very close to the best-estimate value obtained in 1968 [25], which was 7.5 SNU. The situation regarding solar neutrinos is, however, completely different now, thirty years later. Four experiments have confirmed the detection of solar neutrinos. Helioseismological measurements show (cf. Fig. 1) that hypothetical deviations from the standard solar model that seem to be required by simple scaling laws to fit just the gallium solar neutrino results are at least a factor of 30 larger than the rms disagreement between the standard solar model predictions and the helioseismological observations. This conclusion does not make use of the additional evidence which points in the same direction from the chlorine, Kamiokande, and SuperKamiokande experiments. The comparison between observed and calculated helioseismological sound speeds is now so precise ($\sim 0.1\%$ rms) that Fig. 2 indicates the need for an improved physical description of the broad region between $0.3R_{\odot}$ and $0.7R_{\odot}$. The indicated improvement may increase the ^{7}Be and ^{8}B neutrino fluxes by $\sim 5\%$ (cf. Tables 1 and 3). The narrow deep feature near $0.7R_{\odot}$ suggests that mixing, possibly associated with Li depletion, might reduce the ^{7}Be and ^{8}B neutrino fluxes by somewhat less than 5% [22]. Measurements of the low energy cross sections for the ^{3}He$(\alpha,\gamma)^{7}$Be reaction to a 1σ accuracy of 3% and of the ^{7}Be$(p,\gamma)^{8}$B reaction to an accuracy of 5% are required in order that uncertainties in these laboratory experiments not limit the information that can be obtained from solar neutrino experiments about the solar interior and about fundamental neutrino physics.

Acknowledgements

J.N.B. acknowledges support from NSF grant #PHY95-13835.

References

[1] E. Adelberger et al., Rev. Mod. Phys. (accepted, Oct. 1998) astro-ph/9805121.
[2] C.A. Iglesias, F.J. Rogers, Astrophys. J. 464 (1996) 943; D.R. Alexander, J.W. Ferguson, Astrophys. J. 437 (1994) 879. These references describe the different versions of the OPAL opacities.
[3] F.J. Rogers, F.J. Swenson, C.A. Iglesias, Astrophys. J. 456 (1996) 902.
[4] A.V. Gruzinov, J.N. Bahcall, Astrophys. J. (in press), astro-ph/9801028 (1998).
[5] J.N. Bahcall, Phys Rev C 56 (1997) 3391; J.N. Bahcall et al., Phys Rev C 54 (1996) 411.
[6] R. Davis Jr., Prog. Part. Nucl. Phys. 32 (1994) 13; B.T. Cleveland, T. Daily, R. Davis Jr., J.R. Distel, K. Lande, C.K. Lee, P.S. Wildenhain, J. Ullman, Astrophys. J. 495 (March 10, 1998); GALLEX Collaboration, P. Anselmann et al., Phys. Lett. B 342 (1995) 440; GALLEX Collaboration, W. Hampel et al., Phys. Lett. B 388 (1996) 364; SAGE Collaboration, V. Gavrin et al., in: Neutrino'96, Proceedings of the 17th International Conference on Neutrino Physics and Astrophysics (Helsinki) eds. K. Huitu, K. Enqvist, J. Maalampi (World Scientific, Singapore, 1997), p. 14; KAMIOKANDE Collaboration, Y. Fukuda et al., Phys. Rev. Lett. 77 (1996) 1683.
[7] J.N. Bahcall, M.H. Pinsonneault, Rev. Mod. Phys. 67 (1995) 781.
[8] C.W. Johnson, E. Kolbe, S.E. Koonin, K. Langanke, Astrophys. J. 392 (1992) 320.
[9] A.V. Gruzinov, J.N. Bahcall, Astrophys. J. 490 (1997) 437.
[10] S. Basu, H.M. Antia, Mon. Not. R. Astron. Soc. 287 (1997) 189.
[11] A.A. Thoul, J.N. Bahcall, A. Loeb, Astrophys. J. 421 (1994) 828.
[12] J.N. Bahcall, M.H. Pinsonneault, S. Basu, J. Christensen-Dalsgaard, Phys. Rev. Lett. 78 (1997) 171.
[13] J.N. Bahcall, M.H. Pinsonneault, Rev. Mod. Phys. 64 (1992) 885.
[14] Y. Totsuka, to appear in the proceedings of the 18th Texas

Symposium on Relativistic Astrophysics, December 15–20, 1996, Chicago, Illinois, eds. A. Olinto, J. Frieman, D. Schramm (World Scientific, Singapore).
[15] A.B. McDonald, Proceedings of the 9th Lake Louise Winter Institute, eds. A. Astbury et al. (World Scientific, Singapore, 1994), p. 1.
[16] C. Arpesella et al., BOREXINO proposal, Vols. 1 and 2, eds. G. Bellini, R. Raghavan et al. (Univ. of Milano, Milano, 1992).
[17] S. Basu et al., Mon. Not. R. Astron. Soc. 292 (1997) 234.
[18] S. Basu, H.M. Antia, Mon. Not. R. Astron. Soc. 276 (1995) 1402.
[19] J.N. Bahcall, A. Ulmer, Phys. Rev. D 53 (1996) 4202.
[20] LOWL1 + BISON: S. Basu et al. Mon. Not. R. Astron. Soc. 292 (1997) 234; GOLF: S. Turck Chiéze et al., Solar Phys. 175 (1997) 247; GONG, MDI: S. Basu, Mon. Not. R. Astron. Soc., in press, astro-ph/9712733; LOWL1: S. Basu et al., Astrophys. J. 460 (1996) 1064.
[21] M.H. Pinsonneault, Annu. Rev. Astron. Astrophys. 35 (1997) 557.
[22] C.R. Proffitt, G. Michaud, Astrophys. J. 371 (1991) 584; B.C. Chaboyer, P. Demarque, M.H. Pinsonneault, Astrophys. J. 441 (1995) 865.
[23] S.C. Tripathy, S. Basu, J. Christensen-Dalsgaard, Proc. IAU Symp. 181: Sounding Solar and Stellar Interiors, eds. J. Provost, F.X. Schmider (Nice Observatory).
[24] J.N. Bahcall, N.A. Bahcall, R.K. Ulrich, Astrophys. J. 156 (1969) 559.
[25] J.N. Bahcall, N.A. Bahcall, G. Shaviv, Phys. Rev. Lett. 20 (1968) 1209.

VOLUME 86, NUMBER 25 PHYSICAL REVIEW LETTERS 18 JUNE 2001

Solar ^{8}B and hep Neutrino Measurements from 1258 Days of Super-Kamiokande Data

S. Fukuda,[1] Y. Fukuda,[1] M. Ishitsuka,[1] Y. Itow,[1] T. Kajita,[1] J. Kameda,[1] K. Kaneyuki,[1] K. Kobayashi,[1] Y. Koshio,[1] M. Miura,[1] S. Moriyama,[1] M. Nakahata,[1] S. Nakayama,[1] A. Okada,[1] N. Sakurai,[1] M. Shiozawa,[1] Y. Suzuki,[1] H. Takeuchi,[1] Y. Takeuchi,[1] T. Toshito,[1] Y. Totsuka,[1] S. Yamada,[1] S. Desai,[2] M. Earl,[2] E. Kearns,[2] M.D. Messier,[2] K. Scholberg,[2,*] J.L. Stone,[2] L.R. Sulak,[2] C.W. Walter,[2] M. Goldhaber,[3] T. Barszczak,[4] D. Casper,[4] W. Gajewski,[4] W.R. Kropp,[4] S. Mine,[4] D.W. Liu,[4] L.R. Price,[4] M.B. Smy,[4] H.W. Sobel,[4] M.R. Vagins,[4] K.S. Ganezer,[5] W.E. Keig,[5] R.W. Ellsworth,[6] S. Tasaka,[7] A. Kibayashi,[8] J.G. Learned,[8] S. Matsuno,[8] D. Takemori,[8] Y. Hayato,[9] T. Ishii,[9] T. Kobayashi,[9] K. Nakamura,[9] Y. Obayashi,[9] Y. Oyama,[9] A. Sakai,[9] M. Sakuda,[9] M. Kohama,[10] A.T. Suzuki,[10] T. Inagaki,[11] T. Nakaya,[11] K. Nishikawa,[11] T.J. Haines,[12,4] E. Blaufuss,[13,14] S. Dazeley,[13] K.B. Lee,[13,†] R. Svoboda,[13] J.A. Goodman,[14] G. Guillian,[14] G.W. Sullivan,[14] D. Turcan,[14] A. Habig,[15] J. Hill,[16] C.K. Jung,[16] K. Martens,[16,‡] M. Malek,[16] C. Mauger,[16] C. McGrew,[16] E. Sharkey,[16] B. Viren,[16] C. Yanagisawa,[16] C. Mitsuda,[17] K. Miyano,[17] C. Saji,[17] T. Shibata,[17] Y. Kajiyama,[18] Y. Nagashima,[18] K. Nitta,[18] M. Takita,[18] M. Yoshida,[18] H.I. Kim,[19] S.B. Kim,[19] J. Yoo,[19] H. Okazawa,[20] T. Ishizuka,[21] M. Etoh,[22] Y. Gando,[22] T. Hasegawa,[22] K. Inoue,[22] K. Ishihara,[22] T. Maruyama,[22] J. Shirai,[22] A. Suzuki,[22] M. Koshiba,[23] Y. Hatakeyama,[24] Y. Ichikawa,[24] M. Koike,[24] K. Nishijima,[24] H. Fujiyasu,[25] H. Ishino,[25] M. Morii,[25] Y. Watanabe,[25] U. Golebiewska,[26] D. Kielczewska,[26,4] S.C. Boyd,[27] A.L. Stachyra,[27] R.J. Wilkes,[27] and K.K. Young[27,§]

(Super-Kamiokande Collaboration)

[1]*Institute for Cosmic Ray Research, University of Tokyo, Kashiwa, Chiba 277-8582, Japan*
[2]*Department of Physics, Boston University, Boston, Massachusetts 02215*
[3]*Physics Department, Brookhaven National Laboratory, Upton, New York 11973*
[4]*Department of Physics and Astronomy, University of California, Irvine, Irvine, California 92697-4575*
[5]*Department of Physics, California State University, Dominguez Hills, Carson, California 90747*
[6]*Department of Physics, George Mason University, Fairfax, Virginia 22030*
[7]*Department of Physics, Gifu University, Gifu, Gifu 501-1193, Japan*
[8]*Department of Physics and Astronomy, University of Hawaii, Honolulu, Hawaii 96822*
[9]*Institute of Particle and Nuclear Studies, High Energy Accelerator Research Organization (KEK), Tsukuba, Ibaraki 305-0801, Japan*
[10]*Department of Physics, Kobe University, Kobe, Hyogo 657-8501 Japan*
[11]*Department of Physics, Kyoto University, Kyoto 606-8502, Japan*
[12]*Physics Division, P-23, Los Alamos National Laboratory, Los Alamos, New Mexico 87544*
[13]*Department of Physics and Astronomy, Louisiana State University, Baton Rouge, Louisiana 70803*
[14]*Department of Physics, University of Maryland, College Park, Maryland 20742*
[15]*Department of Physics, University of Minnesota, Duluth, Minnesota 55812-2496*
[16]*Department of Physics and Astronomy, State University of New York, Stony Brook, New York 11794-3800*
[17]*Department of Physics, Niigata University, Niigata, Niigata 950-2181, Japan*
[18]*Department of Physics, Osaka University, Toyonaka, Osaka 560-0043, Japan*
[19]*Department of Physics, Seoul National University, Seoul 151-742, Korea*
[20]*International and Cultural Studies, Shizuoka Seika College, Yaizu, Shizuoka 425-8611, Japan*
[21]*Department of Systems Engineering, Shizuoka University, Hamamatsu, Shizuoka 432-8561, Japan*
[22]*Research Center for Neutrino Science, Tohoku University, Sendai, Miyagi 980-8578, Japan*
[23]*The University of Tokyo, Tokyo 113-0033, Japan*
[24]*Department of Physics, Tokai University, Hiratsuka, Kanagawa 259-1292, Japan*
[25]*Department of Physics, Tokyo Institute for Technology, Meguro, Tokyo 152-8551, Japan*
[26]*Institute of Experimental Physics, Warsaw University, 00-681 Warsaw, Poland*
[27]*Department of Physics, University of Washington, Seattle, Washington 98195-1560*

(Received 19 March 2001)

Solar neutrino measurements from 1258 days of data from the Super-Kamiokande detector are presented. The measurements are based on recoil electrons in the energy range 5.0–20.0 MeV. The measured solar neutrino flux is $2.32 \pm 0.03(\text{stat})^{+0.08}_{-0.07}(\text{syst}) \times 10^6\ \text{cm}^{-2}\,\text{s}^{-1}$, which is $45.1 \pm 0.5(\text{stat})^{+1.6}_{-1.4}(\text{syst})\%$ of that predicted by the BP2000 SSM. The day vs night flux asymmetry $(\Phi_n - \Phi_d)/\Phi_{\text{average}}$ is $0.033 \pm 0.022(\text{stat})^{+0.013}_{-0.012}(\text{syst})$. The recoil electron energy spectrum is consistent with no spectral distortion. For the hep neutrino flux, we set a 90% C.L. upper limit of $40 \times 10^3\ \text{cm}^{-2}\,\text{s}^{-1}$, which is 4.3 times the BP2000 SSM prediction.

DOI: 10.1103/PhysRevLett.86.5651 PACS numbers: 26.65.+t, 95.85.Ry, 96.40.Tv

0031-9007/01/86(25)/5651(5)$15.00

Solar neutrinos have been detected using chlorine-, gallium-, and water-based detectors [1–5]; all have measured significantly lower solar neutrino fluxes than predicted by standard solar models (SSMs) [6–8]. This disagreement between the measured and expected solar neutrino flux, known as the "solar neutrino problem," is generally believed to be due to neutrino flavor oscillations. Signatures of neutrino oscillations in Super-Kamiokande (SK) might include distortion of the recoil electron energy (E_{recoil}) spectrum, difference between the nighttime solar neutrino flux relative to the daytime flux, or a seasonal variation in the neutrino flux. Observation of these effects would be strong evidence in support of solar neutrino oscillations independent of absolute flux calculations. Conversely, nonobservation would constrain oscillation solutions to the solar neutrino problem. We describe here solar neutrino measurements from 1258 days of SK data.

SK, located at Kamioka Observatory, Institute for Cosmic Ray Research, University of Tokyo, is a 22.5 kton fiducial volume water Cherenkov detector that detects solar neutrinos via the elastic scattering of neutrinos off atomic electrons. The scattered recoil electron is detected via Cherenkov light production, allowing both the direction and total energy to be measured. These quantities are related to the original neutrino direction and energy. Detailed descriptions of SK can be found elsewhere [5,9–11].

The 1258-day solar neutrino data were collected in four periods with different trigger thresholds between 31 May 1996 and 6 October 2000 (Table I). The analysis threshold has been at 5.0 MeV except for the first 280 days where the data were analyzed with a threshold of 6.5 MeV. The analysis threshold is determined by the level of irreducible background events and the event trigger threshold. An event is triggered when the sum of the photomultiplier tubes (PMTs) registering a hit in a 200 nsec time window (N_{hit}) is above a threshold (Table I). This threshold should be sufficiently low that the trigger efficiency at the analysis threshold is nearly 100%. The lowering of the trigger threshold in periods 2–4 was made possible by the addition of a software filter to the data acquisition system that removes a large portion of background events. This removal is accomplished by reconstructing the event vertex and rejecting events with vertices within 2 m of the inner detector wall, most of which are due to external radioactivity. Each lowering of the trigger threshold in the course of the experiment was made possible by increasing the computing power for the filter program.

There are 2.0×10^9 events in the raw data sample before background reduction. After removing cosmic ray muon events, the sample in the 22.5 kton fiducial volume with energy between 5.0–20.0 MeV contains 3.0×10^7 events. The dominant background sources in the low-energy region ($E \lesssim 6.5$ MeV) are ^{222}Rn in the water and external radioactivity; in the high-energy region ($E \gtrsim 6.5$ MeV), radioactive decay of muon-induced spallation products accounts for most of the background. Background reduction takes place in the following steps: first reduction, spallation cut, second reduction, and external gamma-ray cut. The first reduction includes cuts that remove events due to electronic noise and arcing PMTs. In addition, a cut on the goodness of the reconstructed vertex is used to remove obvious background events originating from various nonphysical sources. The number of remaining events after the first reduction is 1.5×10^7. The spallation cut has been improved compared to that used in earlier publications [5,10,11]. We have improved the likelihood functions used in removing spallation events and introduced a new cut for ^{16}N events that originate from absorption of cosmic ray stopped μ^- on ^{16}O. The number of events in the high-energy region (6.5–20 MeV) before and after the spallation cut is 1.6×10^6 and 3.3×10^5, respectively. The spallation cut is 79% efficient for solar neutrino events. The second reduction removes events with poor vertex fit quality or with blurred Cherenkov ring patterns, characteristics of low-energy background events, and external gamma rays. This newly introduced reduction step has improved the signal-to-noise ratio in the low-energy region by almost an order of magnitude. The number of events before and after the second reduction in the 5.0–6.5 MeV region are 1.0×10^7 and 1.4×10^6 events, respectively. In addition, the gamma-ray cut, which removes external events, has been tightened for those events with $E < 6.5$ MeV. The combined efficiency of the first reduction, second reduction, and the external gamma-ray cut for solar neutrino events is ~73% for $E \geq 6.5$ MeV and ~52% for $E < 6.5$ MeV. After these reduction steps, 236 140 events remain in the fiducial volume above 5 MeV, with $S/N \approx 1$ in the solar direction.

The SK detector simulation is based on GEANT 3.21 [12]. The energy scale was measured using a larger sample of data from an *in situ* electron linear accelerator [9] (LINAC) compared to that used in earlier results. The detector simulation's reliability was tested using the well known β decay of ^{16}N, which is produced *in situ* by an (n, p) reaction on ^{16}O. Fast neutrons for this reaction are produced using a portable deuterium-tritium neutron generator (DTG) [13]. The energy scale measured by the DTG agrees with that from the LINAC within ±0.3%. The total systematic uncertainty in the absolute energy scale, including

TABLE I. The trigger and analysis thresholds and live times during which they were used. The third column shows the recoil electron energy at which the trigger is 50% and 95% efficient. The software filter was added starting in May 1997.

Run period	N_{hit} threshold	50%/95% efficiency (MeV)	Analysis threshold (MeV)	Live time (days)
(1) May 1996~	40.6	5.7/6.2	6.5	280
(2) May 1997~	34.5	4.7/5.2	5.0	650
(3) Sep. 1999~	30.4	4.2/4.6	5.0	320
(4) Sep. 2000~	27.7	3.7/4.2	5.0	8

possible long term variation and direction dependence, is ±0.6%.

We compare our solar neutrino measurements against reference fluxes and neutrino spectra in order to search for signatures of neutrino oscillations. For $E_{recoil} \geq 5.0$ MeV, solar neutrinos are expected to come almost exclusively from the β decay of ^{8}B, with a slight admixture of neutrinos from ^{3}He-proton (hep) fusion. For the absolute flux of ^{8}B and hep neutrinos, we take the BP2000 [6] SSM as our reference [14]. The β decay spectrum of the ^{8}B neutrinos is dominated by the transition to a broad excited state of ^{8}Be, which decays immediately to two α particles. Bahcall *et al.* [15] use a neutrino spectrum deduced from a comparison of world data on ^{8}Be α decay [16–18] with the direct measurement of the positron spectrum from ^{8}B decay measured by Napolitano, Freedman, and Camp [19]. Energy-dependent systematic errors are deduced from a combination of experimental uncertainties and the theoretical uncertainties in radiative and other corrections that must be made to convert the charged particle data into a neutrino spectrum [15]. Recently, Ortiz *et al.* [20] have made an improved measurement of the ^{8}B spectrum based on ^{8}Be α decay in which some of the major sources of systematic errors present in previous measurements were reduced or eliminated. We have adopted the neutrino spectral shape and experimental uncertainties from this measurement. These experimental uncertainties were then added in quadrature with the theoretical uncertainties given by Bahcall *et al.* [15].

The solar neutrino signal is extracted from the data using the $\cos\theta_{sun}$ distribution [5]. The angle θ_{sun} is that between the recoil electron momentum and the vector from the sun to the earth. The solar neutrino flux is obtained by a likelihood fit of the signal and background shapes to the $\cos\theta_{sun}$ distribution in data. The signal shape is obtained from the known angular distribution and detector simulation, while the background shape is nearly flat in $\cos\theta_{sun}$. In the ^{8}B flux measurement, the data are subdivided into 19 energy bins in the range 5.0–20.0 MeV (binning as in Fig. 2). The likelihood function is defined as follows:

$$\mathcal{L} = \prod_{j=1}^{19} \frac{e^{-(Y_j \cdot S + B_j)}}{N_j!} \prod_{i=1}^{N_j} [B_j F_b(\cos\theta_i, E_i) + Y_j S F_s(\cos\theta_i, E_i)]. \quad (1)$$

S is the total number of signal events, while N_j, B_j, and Y_j represent the number of observed events, the number of background events, and the expected fraction of signal events in the jth bin, respectively. F_b and F_s are the probability for the background and signal events as a function of $\cos\theta_{sun}$ and energy (E_i) of each event. The likelihood function is maximized with respect to S and B_j. For the energy spectrum measurement, each term in the product over bins is maximized separately.

The best-fit value of S is $18464 \pm 204(\text{stat})^{+646}_{-554}$(syst which is $45.1 \pm 0.5(\text{stat})^{+1.6}_{-1.4}(\text{syst})$% of the reference pre diction. The corresponding ^{8}B flux at 1 AU is

$$2.32 \pm 0.03(\text{stat})^{+0.08}_{-0.07}(\text{syst}) \times 10^6 \text{ cm}^{-2}\text{s}^{-1}.$$

The total systematic error is $^{+3.5\%}_{-3.0\%}$, with the largest source coming from the reduction cut efficiency ($^{+2.2\%}_{-1.7\%}$), energ scale and resolution (±1.4%), systematic shifts in the even vertex (±1.3%), and the angular resolution of the recoi electron momentum (±1.2%).

Figure 1 shows the solar neutrino flux as a function o the solar zenith angle θ_z (the angle between the vertica axis at SK and the vector from the sun to the earth). Th daytime solar neutrino flux Φ_d is defined as the flux o events when $\cos\theta_z \leq 0$, while the nighttime flux Φ_n i that when $\cos\theta_z > 0$. The measured fluxes are

$$\Phi_d = 2.28 \pm 0.04(\text{stat})^{+0.08}_{-0.07}(\text{syst}) \times 10^6 \text{ cm}^{-2}\text{s}^{-1},$$
$$\Phi_n = 2.36 \pm 0.04(\text{stat})^{+0.08}_{-0.07}(\text{syst}) \times 10^6 \text{ cm}^{-2}\text{s}^{-1}.$$

Some neutrino oscillation parameters predict a nonzer difference between Φ_n and Φ_d due to the matter effec in the earth's mantle and core [21]. The degree of thi difference is measured by the day-night asymmetry defined as $\mathcal{A} = (\Phi_n - \Phi_d)/\Phi_{average}$, where $\Phi_{average} = \frac{1}{2}(\Phi_n + \Phi_d)$. We find

$$\mathcal{A} = 0.033 \pm 0.022(\text{stat})^{+0.013}_{-0.012}(\text{syst}).$$

Including systematic errors, this is 1.3σ from zero asym metry. Many sources of systematic errors cancel out in the day-night asymmetry measurement. The largest sources o

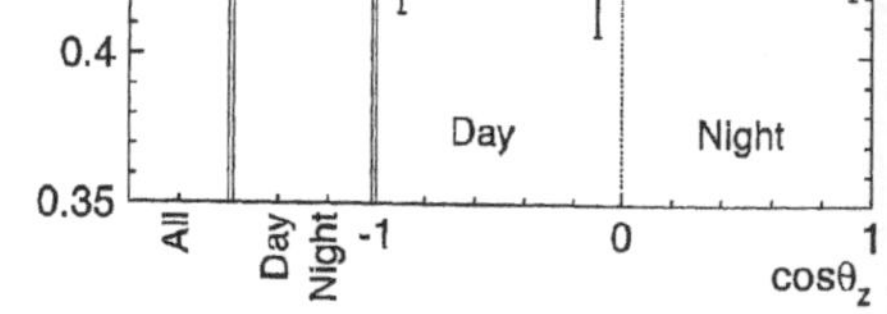

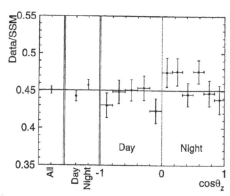

FIG. 1. The solar zenith angle (θ_z) dependence of the solar neutrino flux (error bars show statistical error). The width of the nighttime bins was chosen to separate solar neutrinos that pass through the earth's dense core ($\cos\theta_z \geq 0.84$) from those that pass through the mantle ($0 < \cos\theta_z < 0.84$). The horizontal line shows the flux for all data.

error in the asymmetry are the energy scale and resolution ($^{+0.012}_{-0.011}$) and the nonflat background shape of the $\cos\theta_{sun}$ distribution (± 0.004).

Figure 2 shows the measured recoil electron energy spectrum relative to the Ortiz *et al.* spectrum normalized to BP2000. A fit to an undistorted energy spectrum gives χ^2/d.o.f. = 19.1/18. Energy-correlated systematic errors are considered in the definition of χ^2 [10]. The energy-correlated systematic error (shown in Fig. 2 as a band around the total flux) is due to uncertainties that could cause a systematic shift in the energy spectrum. The sources of this error are uncertainties in the energy scale, resolution, and the reference ^{8}B spectrum against which the data are compared.

The seasonal dependence of the solar neutrino flux is shown in Fig. 3. The points represent the measured flux, and the curve shows the expected variation due to the orbital eccentricity of the earth (assuming no neutrino oscillations, and normalized to the measured total flux). The data are consistent with the expected annual variation (χ^2/d.o.f. = 3.9/7). A fit to a flat distribution gives χ^2/d.o.f. = 8.1/7. Systematic errors are included in the calculation of χ^2. The total systematic error on the relative flux values in each seasonal bin is $\pm 1.3\%$, the largest sources coming from energy scale and resolution ($^{+1.2\%}_{-1.1\%}$) and reduction cut efficiency ($\pm 0.6\%$).

The hep neutrino flux given by BP2000 is 9.3×10^3 cm^{-2} s^{-1} [6,22], which is 3 orders of magnitude smaller than the ^{8}B neutrino flux. Since the theoretically calculated hep flux is highly uncertain because of many delicate cancellations in calculating the astrophysical S factor, the uncertainty of the flux is not given in BP2000. The effect of hep neutrinos on solar neutrino measurements at SK is expected to be small. However, since the end point of the hep neutrino spectrum is 18.77 MeV compared to about 16 MeV for the ^{8}B spectrum, the high energy end of the E_{recoil} spectrum should be relatively enriched with hep neutrinos. An unexpectedly large hep flux may distort the E_{recoil} spectrum. In our measurement of the hep flux, we extract the number of events in the window E_{recoil} = 18–21 MeV from the $\cos\theta_{sun}$ distribution. This window was chosen because it optimizes the significance of the hep flux measurement in the Monte Carlo simulation assuming BP2000 ^{8}B and hep fluxes. We find 1.3 ± 2.0 events in the chosen window. Assuming that all of these events are due to hep neutrinos, the 90% confidence level upper limit of the hep neutrino flux is 40×10^3 cm^{-2} s^{-1} (4.3 times the BP2000 prediction for the unoscillated assumption). Figure 4 shows the expected energy spectra with various hep contributions.

In summary, SK has lowered the analysis energy threshold to 5.0 MeV, collected more than twice the data previously reported, and reduced systematic errors through refinements in data analysis and extensive detector calibrations. With those improvements, and with the 18 464 observed solar neutrino events, SK provides very precise measurements of the recoil electron energy spectrum, day-night flux asymmetry, and the absolute solar neutrino flux. The measured flux is $45.1 \pm 0.5(\text{stat})^{+1.6}_{-1.4}(\text{syst})\%$ of the BP2000 prediction. We found no statistically significant energy spectrum distortion (χ^2/d.o.f. = 19.1/18 relative to the predicted ^{8}B spectrum), and the day-night flux difference of 3.3% of the average flux is 1.3σ from zero. However, the precision of these measurements should provide strong and important constraints on the neutrino oscillation parameters. The seasonal dependence of the flux shows the expected 7% annual variation due to the eccentricity of the earth's orbit. This is the

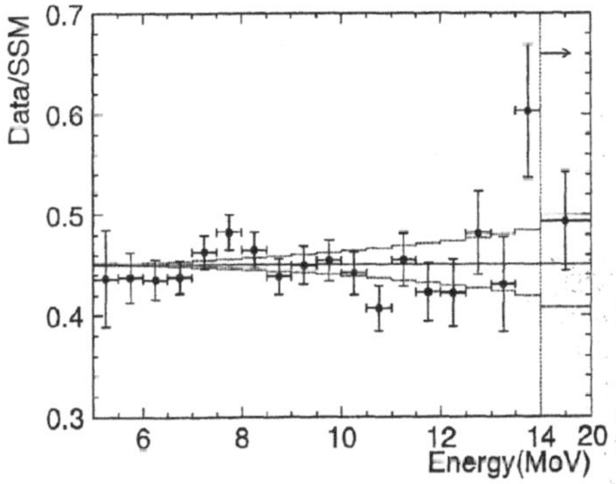

FIG. 2. The measured ^{8}B + hep solar neutrino spectrum relative to that of Ortiz *et al.* [20] normalized to BP2000 [6]. The data from 14 to 20 MeV are combined into a single bin. The horizontal solid line shows the measured total flux, while the dotted band around this line indicates the energy correlated uncertainty. Error bars show statistical and energy-uncorrelated errors added in quadrature.

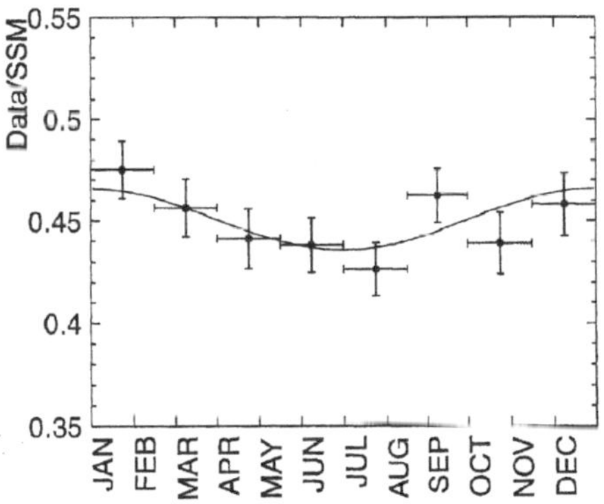

FIG. 3. Seasonal variation of the solar neutrino flux. The curve shows the expected seasonal variation of the flux introduced by the eccentricity of the earth's orbit. Error bars show statistical errors only.

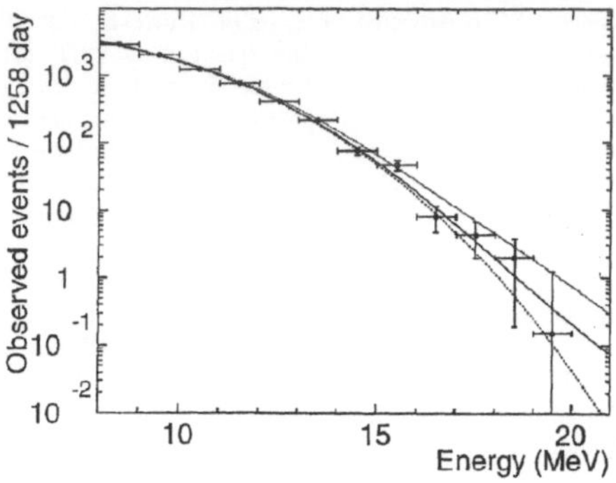

FIG. 4. Energy spectrum of recoil electrons produced by ^8B and hep neutrinos, in 1 MeV bins. The points show data with statistical error bars. The curves show expected spectra with various hep contributions to the best-fit ^8B spectrum. The solid, dotted, and dashed curves show the spectrum with 1, 4.3, and 0 times the BP2000 hep flux, respectively.

first neutrino-based observation of the earth's orbital eccentricity. A stringent limit on the hep neutrino flux ($\Phi_{hep} < 40 \times 10^3$ cm^{-2} s^{-1}) was obtained, which corresponds to 4.3 times the predicted value from BP2000.

The authors acknowledge the cooperation of the Kamioka Mining and Smelting Company. The Super-Kamiokande detector has been built and operated from funding by the Japanese Ministry of Education, Culture, Sports, Science and Technology, the U.S. Department of Energy, and the U.S. National Science Foundation. This work was partially supported by the Korean Research Foundation (BK21) and the Korea Ministry of Science and Technology.

*Present address: Department of Physics, Massachusetts Institute of Technology, Cambridge, MA 02139.
†Present address: Korea Institute of Standards and Science, Yusong, P.O. Box 102, Taejon, 305-600, Korea.
‡Present address: Department of Physics, University of Utah, Salt Lake City, UT 84112.
§Deceased.

[1] B.T. Cleveland *et al.*, Astrophys. J. **496**, 505 (1998).
[2] Y. Fukuda *et al.*, Phys. Rev. Lett. **77**, 1683 (1996).
[3] J.N. Abdurashitov *et al.*, Phys. Rev. C **60**, 055801 (1999).
[4] P. Anselmann *et al.*, Phys. Lett. B **327**, 377 (1994).
[5] Y. Fukuda *et al.*, Phys. Rev. Lett. **81**, 1158 (1998).
[6] J.N. Bahcall *et al.*, astro-ph/0010346.
[7] J.N. Bahcall, S. Basu, and M.H. Pinsonneault, Phys. Lett. B **433**, 1 (1998).
[8] S. Turck-Chièze and I. Lopes, Astrophys. J. **408**, 347 (1993).
[9] M. Nakahata *et al.*, Nucl. Instrum. Methods Phys. Res., Sect. A **421**, 113 (1999).
[10] Y. Fukuda *et al.*, Phys. Rev. Lett. **82**, 2430 (1999).
[11] Y. Fukuda *et al.*, Phys. Rev. Lett. **82**, 1810 (1999).
[12] GEANT Detector Description and Simulation Tool, Cern Programming Library W5013, 1994.
[13] E. Blaufuss *et al.*, Nucl. Instrum. Methods Phys. Res., Sect. A **458**, 636 (2001).
[14] The ^8B flux we have used is 5.15×10^6 cm^{-2} s^{-1}. After this Letter was prepared we were informed that the ^8B flux has been updated to 5.05×10^6 cm^{-2} s^{-1} in BP2000 [6].
[15] J.N. Bahcall *et al.*, Phys. Rev. C **54**, 411 (1996).
[16] D.H. Wilkinson and D.E. Alburger, Phys. Rev. Lett. **26**, 1127 (1971).
[17] B.J. Farmer and C.M. Class, Nucl. Phys. **15**, 626 (1960).
[18] α energy spectrum measurement by L. De Braeckeleer and D. Wright (unpublished); quoted in L. De Braeckeleer *et al.*, Phys. Rev. C **51**, 2778 (1995).
[19] J. Napolitano, S.J. Freedman, and J. Camp, Phys. Rev. C **36**, 298 (1987).
[20] C.E. Ortiz *et al.*, Phys. Rev. Lett. **85**, 2909 (2000).
[21] E.D. Carlson, Phys. Rev. D **34**, 1454 (1986); J. Bouchez *et al.*, Z. Phys. C **32**, 499 (1986).
[22] L.E. Marcucci *et al.*, Phys. Rev. Lett. **84**, 5959 (2000); L.E. Marcucci *et al.*, Phys. Rev. C **63**, 015801 (2001).

OLUME 87, NUMBER 7 PHYSICAL REVIEW LETTERS 13 AUGUST 2001

Measurement of the Rate of $\nu_e + d \rightarrow p + p + e^-$ Interactions Produced by ^{8}B Solar Neutrinos at the Sudbury Neutrino Observatory

Q. R. Ahmad,[15] R. C. Allen,[11] T. C. Andersen,[12] J. D. Anglin,[7] G. Bühler,[11] J. C. Barton,[13,*] E. W. Beier,[14] M. Bercovitch,[7] J. Bigu,[4] S. Biller,[13] R. A. Black,[13] I. Blevis,[3] R. J. Boardman,[13] J. Boger,[2] E. Bonvin,[9] M. G. Boulay,[9] M. G. Bowler,[13] T. J. Bowles,[6] S. J. Brice,[6,13] M. C. Browne,[15] T. V. Bullard,[15] T. H. Burritt,[15,6] K. Cameron,[12] . Cameron,[13] Y. D. Chan,[5] M. Chen,[9] H. H. Chen,[11,†] X. Chen,[5,13] M. C. Chon,[12] B. T. Cleveland,[13] E. T. H. Clifford,[9,1] J. H. M. Cowan,[4] D. F. Cowen,[14] G. A. Cox,[15] Y. Dai,[9] X. Dai,[13] F. Dalnoki-Veress,[3] W. F. Davidson,[7] P. J. Doe,[15,11,6] G. Doucas,[13] M. R. Dragowsky,[6,5] C. A. Duba,[15] F. A. Duncan,[9] J. Dunmore,[13] E. D. Earle,[9,1] S. R. Elliott,[15,6] H. C. Evans,[9] G. T. Ewan,[9] J. Farine,[3] H. Fergani,[13] A. P. Ferraris,[13] R. J. Ford,[9] M. M. Fowler,[6] K. Frame,[13] E. D. Frank,[14] W. Frati,[14] J. V. Germani,[15,6] S. Gil,[10] A. Goldschmidt,[6] D. R. Grant,[3] R. L. Hahn,[2] A. L. Hallin,[9] E. D. Hallman,[4] A. Hamer,[6,9] A. A. Hamian,[15] R. U. Haq,[4] C. K. Hargrove,[3] P. J. Harvey,[9] R. Hazama,[15] R. Heaton,[9] K. M. Heeger,[15] W. J. Heintzelman,[14] J. Heise,[10] R. L. Helmer,[10,‡] J. D. Hepburn,[9,1] H. Heron,[13] J. Hewett,[4] A. Hime,[6] M. Howe,[15] J. G. Hykawy,[4] M. C. P. Isaac,[5] P. Jagam,[12] N. A. Jelley,[13] C. Jillings,[9] G. Jonkmans,[4,1] J. Karn,[12] P. T. Keener,[14] K. Kirch,[6] J. R. Klein,[14] A. B. Knox,[13] R. J. Komar,[10,9] R. Kouzes,[8] T. Kutter,[10] C. C. M. Kyba,[14] J. Law,[12] I. T. Lawson,[12] M. Lay,[13] H. W. Lee,[9] K. T. Lesko,[5] J. R. Leslie,[9] I. Levine,[3] W. Locke,[13] M. M. Lowry,[8] S. Luoma,[4] J. Lyon,[13] S. Majerus,[13] H. B. Mak,[9] A. D. Marino,[5] N. McCauley,[13] A. B. McDonald,[9,8] D. S. McDonald,[14] K. McFarlane,[3] G. McGregor,[13] W. McLatchie,[9] R. Meijer Drees,[15] H. Mes,[3] C. Mifflin,[3] G. G. Miller,[6] G. Milton,[1] B. A. Moffat,[9] M. Moorhead,[13,5] C. W. Nally,[10] M. S. Neubauer,[14] F. M. Newcomer,[14] H. S. Ng,[10] A. J. Noble,[3,‡] E. B. Norman,[5] V. M. Novikov,[3] M. O'Neill,[3] C. E. Okada,[5] R. W. Ollerhead,[12] M. Omori,[13] J. L. Orrell,[15] S. M. Oser,[14] A. W. P. Poon,[5,6,10,15] T. J. Radcliffe,[9] A. Roberge,[4] B. C. Robertson,[9] R. G. H. Robertson,[15,6] J. K. Rowley,[2] V. L. Rusu,[14] E. Saettler,[4] K. K. Schaffer,[15] A. Schuelke,[5] M. H. Schwendener,[4] H. Seifert,[4,6,15] M. Shatkay,[3] J. J. Simpson,[12] D. Sinclair,[3] P. Skensved,[9] A. R. Smith,[5] M. W. E. Smith,[15] N. Starinsky,[3] T. D. Steiger,[15] R. G. Stokstad,[5] R. S. Storey,[7,†] B. Sur,[1,9] R. Tafirout,[4] N. Tagg,[12] N. W. Tanner,[13] R. K. Taplin,[13] M. Thorman,[13] P. Thornewell,[6,13,15] P. T. Trent,[13,*] Y. I. Tserkovnyak,[10] R. Van Berg,[14] R. G. Van de Water,[14,6] C. J. Virtue,[4] C. E. Waltham,[10] J.-X. Wang,[12] D. L. Wark,[13,6,§] N. West,[13] J. B. Wilhelmy,[6] J. F. Wilkerson,[15,6] J. Wilson,[13] P. Wittich,[14] J. M. Wouters,[6] and M. Yeh[2]

(SNO Collaboration)

[1]*Atomic Energy of Canada Limited, Chalk River Laboratories, Chalk River, Ontario K0J 1J0 Canada*
[2]*Chemistry Department, Brookhaven National Laboratory, Upton, New York 11973-5000*
[3]*Carleton University, Ottawa, Ontario K1S 5B6 Canada*
[4]*Department of Physics and Astronomy, Laurentian University, Sudbury, Ontario P3E 2C6 Canada*
[5]*Institute for Nuclear and Particle Astrophysics and Nuclear Science Division, Lawrence Berkeley National Laboratory, Berkeley, California 94720*
[6]*Los Alamos National Laboratory, Los Alamos, New Mexico 87545*
[7]*National Research Council of Canada, Ottawa, Ontario K1A 0R6 Canada*
[8]*Department of Physics, Princeton University, Princeton, New Jersey 08544*
[9]*Department of Physics, Queen's University, Kingston, Ontario K7L 3N6 Canada*
[10]*Department of Physics and Astronomy, University of British Columbia, Vancouver, BC V6T 1Z1 Canada*
[11]*Department of Physics, University of California, Irvine, California 92717*
[12]*Physics Department, University of Guelph, Guelph, Ontario N1G 2W1 Canada*
[13]*Nuclear and Astrophysics Laboratory, University of Oxford, Keble Road, Oxford, OX1 3RH, United Kingdom*
[14]*Department of Physics and Astronomy, University of Pennsylvania, Philadelphia, Pennsylvania 19104-6396*
[15]*Center for Experimental Nuclear Physics and Astrophysics, and Department of Physics, University of Washington, Seattle, Washington 98195*

(Received 18 June 2001; published 25 July 2001)

Solar neutrinos from ^{8}B decay have been detected at the Sudbury Neutrino Observatory via the charged current (CC) reaction on deuterium and the elastic scattering (ES) of electrons. The flux of ν_e's is measured by the CC reaction rate to be $\phi^{CC}(\nu_e) = 1.75 \pm 0.07(\mathrm{stat})^{+0.12}_{-0.11}(\mathrm{syst}) \pm 0.05(\mathrm{theor}) \times 10^6\ \mathrm{cm}^{-2}\,\mathrm{s}^{-1}$. Comparison of $\phi^{CC}(\nu_e)$ to the Super-Kamiokande Collaboration's precision value of the flux inferred from the ES reaction yields a 3.3σ difference, assuming the systematic uncertainties are normally distributed, providing evidence of an active non-ν_e component in the solar flux. The total flux of active ^{8}B neutrinos is determined to be $5.44 \pm 0.99 \times 10^6\ \mathrm{cm}^{-2}\,\mathrm{s}^{-1}$.

DOI: 10.1103/PhysRevLett.87.071301 PACS numbers: 26.65.+t, 14.60.Pq, 95.85.Ry

VOLUME 87, NUMBER 7 PHYSICAL REVIEW LETTERS 13 AUGUST 2001

Solar neutrino experiments over the past 30 years [1–6] have measured fewer neutrinos than are predicted by models of the Sun [7,8]. One explanation for the deficit is the transformation of the Sun's electron-type neutrinos into other active flavors. The Sudbury Neutrino Observatory (SNO) measures the ^{8}B solar neutrinos through the reactions

$$\nu_e + d \rightarrow p + p + e^- \quad \text{(CC)},$$
$$\nu_x + d \rightarrow p + n + \nu_x \quad \text{(NC)},$$
$$\nu_x + e^- \rightarrow \nu_x + e^- \quad \text{(ES)}.$$

The charged current (CC) reaction is sensitive exclusively to electron-type neutrinos, while the neutral current (NC) is sensitive to all active neutrino flavors ($x = e, \mu, \tau$). The elastic scattering (ES) reaction is sensitive to all flavors as well, but with reduced sensitivity to ν_μ and ν_τ. By itself, the ES reaction cannot provide a measure of the total ^{8}B flux or its flavor content. Comparison of the ^{8}B flux deduced from the ES reaction, assuming no neutrino oscillations [$\phi^{\mathrm{ES}}(\nu_x)$], to that measured by the CC reaction [$\phi^{\mathrm{CC}}(\nu_e)$] can provide clear evidence of flavor transformation without reference to solar model flux calculations. If neutrinos from the Sun change into other active flavors, then $\phi^{\mathrm{CC}}(\nu_e) < \phi^{\mathrm{ES}}(\nu_x)$.

This Letter presents the first results from SNO on the ES and CC reactions. SNO's measurement of $\phi^{\mathrm{ES}}(\nu_x)$ is consistent with previous measurements described in Ref. [5]. The measurement of $\phi^{\mathrm{CC}}(\nu_e)$, however, is significantly smaller and is therefore inconsistent with the null hypothesis that all observed solar neutrinos are ν_e. A measurement using the NC reaction, which has equal sensitivity to all neutrino flavors, will be reported in a future publication.

SNO [9] is an imaging water Čerenkov detector located at a depth of 6010 m of water equivalent in the INCO, Ltd. Creighton mine near Sudbury, Ontario. It features 1000 metric tons of ultrapure D_2O contained in a 12-m diameter spherical acrylic vessel. This sphere is surrounded by a shield of ultrapure H_2O contained in a 34-m-high barrel-shaped cavity of maximum diameter 22 m. A stainless steel structure 17.8 m in diameter supports 9456 20-cm photomultiplier tubes (PMTs) with light concentrators. Approximately 55% of the light produced within 7 m of the center of the detector will strike a PMT if it is not absorbed by intervening media.

The data reported here were recorded between November 2, 1999 and January 15, 2001 and correspond to a live time of 240.95 days. Events are defined by a multiplicity trigger of 18 or more PMTs exceeding a threshold of ~0.25 photoelectrons within a time window of 93 ns. The trigger reaches 100% efficiency at 23 PMTs. The total instantaneous trigger rate is 15–18 Hz, of which 6–8 Hz is the data trigger. For every event trigger, the time and charge responses of each participating PMT are recorded.

The data were partitioned into two sets, with approximately 70% used to establish the data analysis procedures and 30% reserved for a blind test of statistical bias in the analysis. The analysis procedures were frozen before the blind data set was analyzed, and no statistically significant differences in the data sets were found. We present here the analysis of the combined data sets.

Calibration of the PMT time and charge pedestals, slopes, offsets, charge vs time dependencies, and second order rate dependencies are performed using electronic pulsers and pulsed light sources. Optical calibration is obtained by using a diffuse source of pulsed laser light at 337, 365, 386, 420, 500, and 620 nm. The absolute energy scale and uncertainties are established with a triggered ^{16}N source (predominantly 6.13-MeV γ's) deployed over two planar grids within the D_2O and a linear grid in the H_2O. The resulting Monte Carlo predictions of detector response are tested using a ^{252}Cf neutron source, which provides an extended distribution of 6.25-MeV γ rays from neutron capture, and a $^3\mathrm{H}(p,\gamma)^4\mathrm{He}$ [10] source providing 19.8-MeV γ rays. The volume-weighted mean response is approximately nine PMT hits per MeV of electron energy.

Table I details the steps in data reduction. The first of these is the elimination of instrumental backgrounds. Electrical pickup may produce false PMT hits, while electrical discharges in the PMTs or insulating detector materials produce light. These backgrounds have characteristics very different from Čerenkov light, and are eliminated by using cuts based only on the PMT positions, the PMT time and charge data, event-to-event time correlations, and veto PMTs. This step in the data reduction is verified by comparing results from two independent background rejection analyses.

For events passing the first stage, the calibrated times and positions of the hit PMTs are used to reconstruct the vertex position and the direction of the particle. The reconstruction accuracy and resolution are measured using Compton electrons from the ^{16}N source, and the energy and source variation of reconstruction are checked with a ^{8}Li β source. Angular resolution is measured using Compton electrons produced more than 150 cm from the ^{16}N source. At these energies, the vertex resolution is 16 cm and the angular resolution is 26.7°.

An effective kinetic energy, T_{eff}, is assigned to each event passing the reconstruction stage. T_{eff} is calculated

TABLE I. Data reduction steps.

Analysis step	Number of events
Total event triggers	355 320 964
Neutrino data triggers	143 756 178
$N_{\mathrm{hit}} \geq 30$	6 372 899
Instrumental background cuts	1 842 491
Muon followers	1 809 979
High level cuts[a]	923 717
Fiducial volume cut	17 884
Threshold cut	1169
Total events	1169

[a]Reconstruction figures of merit, prompt light, and $\langle\theta_{ij}\rangle$.

using prompt (unscattered) Čerenkov photons and the position and direction of the event. The derived energy response of the detector can be characterized by a Gaussian:

$$R(E_{\rm eff}, E_e) = \frac{1}{\sqrt{2\pi}\,\sigma_E(E_e)} \exp\left[-\frac{1}{2}\left(\frac{E_{\rm eff} - E_e}{\sigma_E(E_e)}\right)^2\right],$$

where E_e is the total electron energy, $E_{\rm eff} = T_{\rm eff} + m_e$, and $\sigma_E(E_e) = (-0.4620 + 0.5470\sqrt{E_e} + 0.008722 E_e)$ MeV is the energy resolution. The uncertainty on the energy scale is found to be $\pm 1.4\%$, which results in a flux uncertainty nearly 4 times larger. For validation, a second energy estimator counts all PMTs hit in each event, $N_{\rm hit}$, without position and direction corrections.

Further instrumental background rejection is obtained by using reconstruction figures of merit, PMT time residuals, and the average angle between hit PMTs ($\langle\theta_{ij}\rangle$), measured from the reconstructed vertex. These cuts test the hypothesis that each event has the characteristics of single electron Čerenkov light. The effects of these and the rest of the instrumental background removal cuts on neutrino signals are quantified using the ^{8}Li and ^{16}N sources deployed throughout the detector. The volume-weighted neutrino signal loss is measured to be $1.4^{+0.7}_{-0.6}\%$ and the residual instrumental contamination for the data set within the D_2O is $<0.2\%$. Lastly, cosmic ray induced neutrons and spallation products are removed using a 20 s coincidence window with the parent muon.

Figure 1 shows the radial distribution of all remaining events above a threshold of $T_{\rm eff} \geq 6.75$ MeV. The distribution is expressed as a function of the volume-weighted radial variable $(R/R_{\rm AV})^3$, where $R_{\rm AV} = 6.00$ m is the radius of the acrylic vessel. Above this energy threshold, there are contributions from CC events in the D_2O, ES events in the D_2O and H_2O, a residual tail of neutron capture events, and high energy γ rays from radioactivity in the outer detector. The data show a clear signal within the D_2O volume. For $(R/R_{\rm AV})^3 > 1.0$ the distribution rises into the H_2O region until it is cut off by the acceptance of the PMT light collectors at $R \sim 7.0$ m. A fiducial volume cut is applied at $R = 5.50$ m to reduce backgrounds from regions exterior to the D_2O, and to minimize systematic uncertainties associated with optics and reconstruction near the acrylic vessel.

Possible backgrounds from radioactivity in the D_2O and H_2O are measured by regular low level radio assays of U and Th decay chain products in these regions. The Čerenkov light character of D_2O and H_2O radioactivity backgrounds is used *in situ* to monitor backgrounds between radio assays. Low energy radioactivity backgrounds are removed by the high threshold imposed, as are most neutron capture events. Monte Carlo calculations predict that the H_2O shield effectively reduces contributions of low energy (<4 MeV) γ rays from the PMT array, and these predictions are verified by deploying an encapsulated Th source in the vicinity of the PMT support sphere. High energy γ rays from the cavity are also attenuated by the

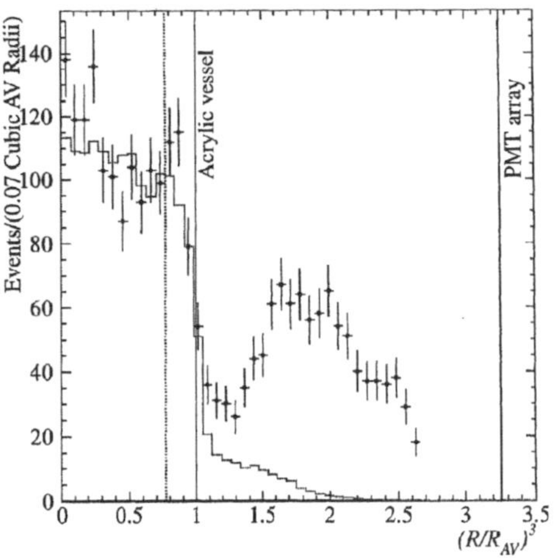

FIG. 1. Distribution of event candidates with $T_{\rm eff} \geq 6.75$ MeV as a function of the volume-weighted radial variable $(R/R_{\rm AV})^3$. The Monte Carlo simulation of the signals, weighted by the results from the signal extraction, is shown as a histogram. The dotted line indicates the fiducial volume cut used in this analysis.

H_2O shield. A limit on their leakage into the fiducial volume is estimated by deploying the ^{16}N source near the edge of the detector's active volume. The total contribution from all radioactivity in the detector is found to be $<0.2\%$ for low energy backgrounds and $<0.8\%$ for high energy backgrounds.

The final data set contains 1169 events after the fiducial volume and kinetic energy threshold cuts. Figure 2(a) displays the distribution of $\cos\theta_\odot$, the angle between the reconstructed direction of the event and the instantaneous direction from the Sun to the Earth. The forward peak in this distribution arises from the kinematics of the ES reaction, while CC electrons are expected to have a distribution which is $(1 - 0.340\cos\theta_\odot)$ [11], before accounting for detector response.

The data are resolved into contributions from CC, ES, and neutron events above threshold using probability density functions (pdf's) in $T_{\rm eff}$, $\cos\theta_\odot$, and $(R/R_{\rm AV})^3$, generated from Monte Carlo simulations assuming no flavor transformation and the shape of the standard ^{8}B spectrum [12] (hep neutrinos are not included in the fit). The extended maximum likelihood method used in the signal extraction yields 975.4 ± 39.7 CC events, 106.1 ± 15.2 ES events, and 87.5 ± 24.7 neutron events for the fiducial volume and the threshold chosen, where the uncertainties given are statistical only. The dominant sources of systematic uncertainty in this signal extraction are the energy scale uncertainty and reconstruction accuracy, as shown in Table II. The CC and ES signal decomposition gives consistent results when used with the $N_{\rm hit}$ energy estimator, as well as with different choices of the analysis threshold

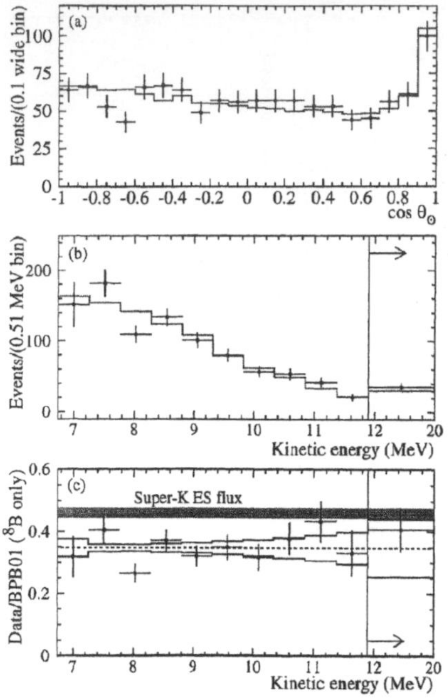

FIG. 2. Distributions of (a) $\cos\theta_\odot$ and (b) extracted kinetic energy spectrum for CC events with $R \leq 5.50$ m and $T_{\rm eff} \geq 6.75$ MeV. The Monte Carlo simulations for an undistorted ^{8}B spectrum are shown as histograms. The ratio of the data to the expected kinetic energy distribution with correlated systematic errors is shown in (c). The uncertainties in the ^{8}B spectrum [12] have not been included.

and the fiducial volume up to 6.20 m with backgrounds characterized by pdf's.

The CC spectrum can be extracted from the data by removing the constraint on the shape of the CC pdf and repeating the signal extraction.

Figure 2(b) shows the kinetic energy spectrum with statistical error bars, with the ^{8}B spectrum of Ortiz *et al.* [12] scaled to the data. The ratio of the data to the prediction [7] is shown in Fig. 2(c). The bands represent the 1σ uncertainties derived from the most significant energy-dependent systematic errors. There is no evidence for a deviation of the spectral shape from the predicted shape under the nonoscillation hypothesis.

Normalized to the integrated rates above the kinetic energy threshold of $T_{\rm eff} = 6.75$ MeV, the measured ^{8}B neutrino fluxes assuming the standard spectrum shape [12] are

$$\phi_{\rm SNO}^{\rm CC}(\nu_e) = 1.75 \pm 0.07({\rm stat})^{+0.12}_{-0.11}({\rm syst}) \pm 0.05({\rm theor}) \times 10^6\ {\rm cm}^{-2}\,{\rm s}^{-1}$$

$$\phi_{\rm SNO}^{\rm ES}(\nu_x) = 2.39 \pm 0.34({\rm stat})^{+0.16}_{-0.14}({\rm syst}) \times 10^6\ {\rm cm}^{-2}\,{\rm s}^{-1},$$

TABLE II. Systematic error on fluxes.

Error source	CC error (percent)	ES error (percent)
Energy scale	−5.2, +6.1	−3.5, +5.
Energy resolution	±0.5	±0.3
Energy scale nonlinearity	±0.5	±0.4
Vertex accuracy	±3.1	±3.3
Vertex resolution	±0.7	±0.4
Angular resolution	±0.5	±2.2
High energy γ's	−0.8, +0.0	−1.9, +0.
Low energy background	−0.2, +0.0	−0.2, +0.
Instrumental background	−0.2, +0.0	−0.6, +0.
Trigger efficiency	0.0	0.0
Live time	±0.1	±0.1
Cut acceptance	−0.6, +0.7	−0.6, +0.
Earth orbit eccentricity	±0.1	±0.1
^{17}O, ^{18}O	0.0	0.0
Experimental uncertainty	−6.2, +7.0	−5.7, +6.
Cross section	3.0	0.5
Solar Model	−16, +20	−16, +20

where the theoretical uncertainty is the CC cross section uncertainty [13]. Radiative corrections have not been applied to the CC cross section, but they are expected to decrease the measured $\phi^{\rm CC}(\nu_e)$ flux [14] by up to a few percent. The difference between the ^{8}B flux deduced from the ES rate and that deduced from the CC rate in SNO is $0.64 \pm 0.40 \times 10^6$ cm^{-2} s^{-1}, or 1.6σ. The SNO's ES rate measurement is consistent with the precision measurement by Super-Kamiokande Collaboration of the ^{8}B flux using the same ES reaction [5]:

$$\phi_{\rm SK}^{\rm ES}(\nu_x) = 2.32 \pm 0.03({\rm stat})^{+0.08}_{-0.07}({\rm syst}) \times 10^6\ {\rm cm}^{-2}\,{\rm s}^{-1}.$$

The difference between the flux $\phi^{\rm ES}(\nu_x)$ measured by Super-Kamiokande via the ES reaction and the $\phi^{\rm CC}(\nu_e)$ flux measured by SNO via the CC reaction is $0.57 \pm 0.17 \times 10^6$ cm^{-2} s^{-1}, or 3.3σ [15], assuming that the systematic errors are normally distributed. The probability that a downward fluctuation of the Super-Kamiokande result would produce a SNO result $\geq 3.3\sigma$ is 0.04%. For reference, the ratio of the SNO CC ^{8}B flux to that of the BPB01 solar model [7] is 0.347 ± 0.029, where all uncertainties are added in quadrature.

If oscillation solely to a sterile neutrino is occurring, the SNO CC-derived ^{8}B flux above a threshold of 6.75 MeV will be consistent with the integrated Super-Kamiokande ES-derived ^{8}B flux above a threshold of 8.5 MeV [16]. By adjusting the ES threshold [5], this derived flux difference is $0.53 \pm 0.17 \times 10^6$ cm^{-2} s^{-1}, or 3.1σ. The probability of a downward fluctuation $\geq 3.1\sigma$ is 0.13%. These data are therefore evidence of a nonelectron active flavor component in the solar neutrino flux. These data are also inconsistent with the "Just-So^2" parameters for neutrino oscillation [17].

VOLUME 87, NUMBER 7 PHYSICAL REVIEW LETTERS 13 AUGUST 2001

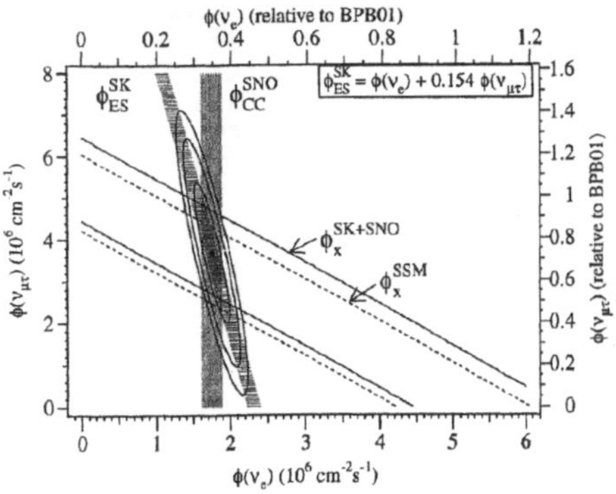

FIG. 3. Flux of ^{8}B solar neutrinos which are μ or τ flavor vs the flux of electron neutrinos as deduced from the SNO and Super-Kamiokande data. The diagonal bands show the total ^{8}B flux $\phi(\nu_x)$ as predicted by BPB01 (dashed lines) and that derived from the SNO and Super-Kamiokande measurements (solid lines). The intercepts of these bands with the axes represent the $\pm 1\sigma$ errors.

Figure 3 displays the inferred flux of nonelectron flavor active neutrinos $[\phi(\nu_{\mu\tau})]$ against the flux of electron neutrinos. The two data bands represent the one standard deviation measurements of the SNO CC rate and the Super-Kamiokande ES rate. The error ellipses represent the 68%, 95%, and 99% joint probability contours for $\phi(\nu_e)$ and $\phi(\nu_{\mu\tau})$. The best fit to $\phi(\nu_{\mu\tau})$ is

$$\phi(\nu_{\mu\tau}) = 3.69 \pm 1.13 \times 10^6 \text{ cm}^{-2}\text{ s}^{-1}.$$

The total flux of active ^{8}B neutrinos is determined to be

$$\phi(\nu_x) = 5.44 \pm 0.99 \times 10^6 \text{ cm}^{-2}\text{ s}^{-1}.$$

This result is displayed as a diagonal band in Fig. 3, and is in excellent agreement with predictions of standard solar models [7,8].

Assuming that the oscillation of massive neutrinos explains both the evidence for the electron neutrino flavor change presented here and the atmospheric neutrino data of the Super-Kamiokande collaboration [18], two separate splittings of the squares of the neutrino mass eigenvalues are indicated: $<10^{-3}$ eV^2 for the solar sector [19,17] and $\simeq 3.5 \times 10^{-3}$ eV^2 for atmospheric neutrinos. These results, together with the beta spectrum of tritium [20], limit the sum of mass eigenvalues of active neutrinos to be between 0.05 and 8.4 eV, corresponding to a constraint of $0.001 < \Omega_\nu < 0.18$ for the contribution to the critical density of the Universe [21,22].

In summary, the results presented here are the first direct indication of a nonelectron flavor component in the solar neutrino flux, and enable the first determination of the total flux of ^{8}B neutrinos generated by the Sun.

This research was supported by the Natural Sciences and Engineering Research Council of Canada, Industry Canada, National Research Council of Canada, Northern Ontario Heritage Fund Corporation, the Province of Ontario, the United States Department of Energy, and in the United Kingdom by the Science and Engineering Research Council and the Particle Physics and Astronomy Research Council. Further support was provided by INCO, Ltd., Atomic Energy of Canada Limited (AECL), Agra-Monenco, Canatom, Canadian Microelectronics Corporation, AT&T Microelectronics, Northern Telecom, and British Nuclear Fuels, Ltd. The heavy water was loaned by AECL with the cooperation of Ontario Power Generation.

---

*Permanent address: Birkbeck College, University of London, Malet Road, London WC1E 7HX, UK.

†Deceased.

‡Permanent address: TRIUMF, 4004 Wesbrook Mall, Vancouver, BC V6T 2A3, Canada.

§Permanent address: Rutherford Appleton Laboratory, Chilton, Didcot, Oxon, OX11 0QX, and University of Sussex, Physics and Astronomy Department, Brighton BN1 9QH, United Kingdom.

[1] B.T. Cleveland *et al.*, Astrophys. J. **496**, 505 (1998).

[2] K.S. Hirata *et al.*, Phys. Rev. Lett. **65**, 1297 (1990); K.S. Hirata *et al.*, Phys. Rev. D **44**, 2241 (1991); **45**, 2170(E) (1992); Y. Fukuda *et al.*, Phys. Rev. Lett. **77**, 1683 (1996).

[3] J.N. Abdurashitov *et al.*, Phys. Rev. C **60**, 055801 (1999).

[4] W. Hampel *et al.*, Phys. Lett. B **447**, 127 (1999).

[5] S. Fukuda *et al.*, Phys. Rev. Lett. **86**, 5651 (2001).

[6] M. Altmann *et al.*, Phys. Lett. B **490**, 16 (2000).

[7] J.N. Bahcall, M.H. Pinsonneault, and S. Basu, Astrophys. J. **555**, 990 (2001). The reference ^{8}B neutrino flux is 5.05×10^6 cm^{-2} s^{-1}.

[8] A.S. Brun, S. Turck-Chièze, and J.P. Zahn, Astrophys. J. **525**, 1032 (1999); S. Turck-Chièze *et al.*, Astrophys. J. Lett. **555**, L69 (2001).

[9] The SNO Collaboration, J. Boger *et al.*, Nucl. Instrum. Methods Phys. Res., Sect. A **449**, 172 (2000).

[10] A.W.P. Poon *et al.*, Nucl. Instrum. Methods Phys. Res., Sect. A **452**, 115 (2000).

[11] P. Vogel and J.F. Beacom, Phys. Rev. D **60**, 053003 (1999).

[12] C.E. Ortiz *et al.*, Phys. Rev. Lett. **85**, 2909 (2000).

[13] S. Nakamura, T. Sato, V. Gudkov, and K. Kubodera, Phys. Rev. C **63**, 034617 (2001); M. Butler, J.-W. Chen, and X. Kong, Phys. Rev. C **63**, 035501 (2001); G. 't Hooft, Phys. Lett. **37B**, 195 (1971). The Butler *et al.* cross section with $L_{1A} = 5.6$ fm^3 is used.

[14] I.S. Towner, J. Beacom, and S. Parke (private communication); I.S. Towner, Phys. Rev. C **58**, 1288 (1998); J. Beacom and S. Parke, hep-ph/0106128; J.N. Bahcall, M. Kamionkowski, and A. Sirlin, Phys. Rev. D **51**, 6146 (1995).

[15] Given the limit set for the hep flux by Ref. [5], the effects of the hep contribution may increase this difference by a few percent.

[16] G.L. Fogli, E. Lisi, A. Palazzo, and F.L. Villante, Phys. Rev. D **63**, 113016 (2001); F.L. Villante, G. Fiorentini, and E. Lisi, Phys. Rev. D **59**, 013006 (1999).
[17] J.N. Bahcall, P.I. Krastev, and A. Yu. Smirnov, J. High Energy Phys. **05**, 015 (2001).
[18] T. Toshito *et al.*, hep-ex/0105023.
[19] M. Apollonio *et al.*, Phys. Lett. B **466**, 415 (1999).
[20] J. Bonn *et al.*, Nucl. Phys. (Proc. B Suppl.) **91**, 273 (2001).
[21] *Allen's Astrophysical Quantities*, edited by Arthur Cox (Springer-Verlag, New York, 2000), 4th ed.; D.E. Groom *et al.*, Eur. Phys. J. C **15**, 1 (2000).
[22] H. Pas and T.J. Weiler, Phys. Rev. D **63**, 113015 (2001).

Permissions and Acknowledgments

I.A.1. Reprinted with permission from *Physical Review Letters*, Vol. **12**, pp. 300-302; ©1964 American Physical Society.

I.A.2. Reprinted with permission from *Physical Review Letters*, Vol. **13**, pp. 764-766; ©1964 American Physical Society.

I.A.3. Reprinted with permission from *Physical Review Letters*, Vol. **13**, pp. 767-769; ©1964 American Physical Society.

I.A.4. Reprinted with permission from *Physical Review Letters*, Vol. **136**, pp. B1164, B1170; ©1964 American Physical Society.

I.A.5. Reprinted with permission from *Physical Review Letters*, Vol. **16**, pp. 145-147; ©1966 American Physical Society.

I.A.6. Reprinted with permission from *Physical Review Letters*, Vol. **17**, pp. 398-401; ©1966 American Physical Society.

I.A.7. Reprinted with permission from *Physics Letters*, Vol. **37B**, pp. 195-196; ©1971 Elsevier Science Publishers.

I.A.8. Reprinted with permission from *Review Modern Physics*, Vol. **50**, p. 881; ©1978 American Physical Society.

I.B.1. Reprinted from *Astrophysical Journal*, Vol. **137**, pp. L344-L346 (1963).

I.B.2. Reprinted from *Astrophysical Journal*, Vol. **140**, pp. 477-484 (1964).

I.B.3. Reprinted with permission from *Canadian Journal of Physics*, Vol. **43**, p. 1497; ©1965 National Research Council Canada.

I.B.4. Reprinted with permission from *Stellar Evolution*, pp. 241-243; ©1966 Plenum Press, NY.

I.C.1 Reprinted with permission from *Physical Review Letters*, Vol. **20**, pp. 1209-1212; ©1968 American Physical Society.

I.C.2. Reprinted with permission from *Physical Review Letters*, Vol. **21**, pp. 1208-1212; ©1968 American Physical Society.

I.C.3. Reprinted from *Astrophysical Journal*, Vol. **156**, pp. 559, 564 (1969).

I.C.4. Reprinted from *Astrophysical Journal*, Vol. **170**, p. 157 (1971).

I.D.1. Reprinted with permission from *Review Modern Physics*, Vol. **54**, p. 767; ©1982 American Physical Society.

I.D.2. Reprinted with permission from *Review Modern Physics*, Vol. **60**, p. 297; ©1988 American Physical Society.

I.D.3. Reprinted from *Astrophysical Journal*, Vol. **360**, p. 727 (1990).

I.D.4. Reprinted with permission from *Review Modern Physics*, Vol. **64**, p. 885; ©1992 American Physical Society.

I.D.5. Reprinted with permission from *Astronomy & Astrophysics*, Vol. **264**, p. 673 ©1992.

I.D.6. Reprinted with permission from *Astronomy & Astrophysics*, Vol. **268**, p. 275 ©1993.

I.D.7. Reprinted from *Astrophysical Journal*, Vol. **408**, p. 347 (1993).

I.D.8. Reprinted with permission from *Astronomy & Astrophysics*, Vol. **271**, p. 601 ©1993.

I.D.9. Reprinted with permission from *Physical Review*, Vol. **D47**, pp. 1298-1301 ©1993 American Physical Society.

I.D.10. Reprinted from *Astrophysical Journal*, Vol. **425**, pp. 849-855 (1994).

II.A.1. Chalk River Laboratory Report PD-205 (1948), reprinted with permission of Atomic Energy of Canada, Ltd.

II.A.2. Reprinted with permission from *Physical Review Letters*, Vol. **12**, pp. 303-305; ©1964 American Physical Society.

II.A.3. Reprinted with permission from *Physical Review Letters*, Vol. **12**, pp. 457-459; ©1964 American Physical Society.

II.A.4. Reprinted with permission from *Physics Letters*, Vol. **13**, pp. 332-333; ©1964 Elsevier Science Publishers.

II.A.5. Reprinted with permission from *Soviet Physics JETP*, Vol. **22**, p. 1051; ©1966 American Institute of Physics.

II.A.6. Reprinted with permission from *Physical Review Letters*, Vol. **40**, pp. 1351-1354; ©1978 American Physical Society.

II.A.7. Reprinted with permission from *Physical Review Letters*, Vol. **55**, pp. 1534-1536; ©1985 American Physical Society.

II.A.8. Reprinted with permission from *Physics Letters B*, Vol. **178**, p. 324-328; ©1986 Elsevier Science Publishers.

II.A.9. Reprinted with permission from *Physical Review Letters*, Vol. **57**, pp. 1801; ©1986 American Physical Society.

II.A.10. Reprinted with permission from *Physical Review Letters*, Vol. **60**, p. 768; ©1988 American Physical Society.

II.A.11. Reprinted with permission from *Frontiers of Neutrino Astrophysics*, eds. Y. Suzuki and K. Nakamura, pp. 135-137; ©1993 Universal Academy Press, Inc.

II.A.12. Reprinted with permission from *Frontiers of Neutrino Astrophysics*, eds. Y. Suzuki and K. Nakamura, pp. 147-149; ©1993 Universal Academy Press, Inc.

II.B.1. Reprinted with permission from *Physical Review Letters*, Vol. **20**, pp. 1205-1209; ©1968 American Physical Society.

II.B.2. Reprinted from *Science*, Vol. **191**, pp. 264-267, Bahcall, J.N. and Davis, R. Jr.; ©1976, with permission of the AAAS.

II.B.3. Reprinted with permission from ***Solar Neutrinos and Neutrino Astronomy***, AIP Conference Proceedings No. 126, Eds. Cherry, M.L., Fowler, W. A., and Lande, K., p. 1; ©1985 American Institute of Physics.

II.B.4. Reprinted with permission from *Physical Review Letters*, Vol. **65**, pp. 1297-1300; ©1990 American Physical Society.

II.B.5. Reprinted with permission from *Physical Review Letters*, Vol. **67**, p. 332; ©1991 American Physical Society.

II.B.6. Reprinted with permission from *Physics Letters*, Vol. **B285**, pp. 376-389; ©1992 Elsevier Science Publishers.

II.B.7. Reprinted with permission from *Physics Letters*, Vol. **B327**, pp. 377-378; ©1994 Elsevier Science Publishers.

II.B.8. Reprinted with permission from *Physics Letters*, Vol. **B328**, p. 234; ©1994 Elsevier Science Publishers.

II.C.1. Reprinted from *Science*, Vol. **193**, pp. 1117-1118, Freedman, M.S. et al.; ©1976, with permission of the AAAS.

II.C.2 Reprinted with permission from *Physical Review Letters*, Vol. **37**, p. 259; ©1976 American Physical Society.

II.C.3. Reprinted from *Science*, Vol. **216**, p. 51, Cowan, G.A. and Haxton, W.C.; ©1982, with permission of the AAAS.

II.C.4 Reprinted with permission from *Physical Review Letters*, Vol. **53**, p. 1116; ©1984 American Physical Society.

II.C.5 Reprinted with permission from *Physical Review Letters*, Vol. **58**, p. 2498; ©1987 American Physical Society.

III.A.1 Reprinted with permission from *Physical Review*, Vol. **113**, pp. 1556-1559; ©1959 American Physical Society.

III.A.2 Reprinted from *Astrophysical Journal*, Vol. **127**, pp. 551-556 (1958).

III.A.3 Reprinted with permission from *BAPS*, Vol. **3**, p. 227; ©1958 American Physical Society.

III.A.4 Reprinted with permission from *Nuclear Physics*, Vol. **24**, pp. 89-90, ©1961 Elsevier Science Publishers.

III.B.1 Reprinted with permission from *Physical Review*, Vol. **128**, pp. 1297-1301; ©1962 American Physical Society.

III.B.2 Reprinted with permission from *Canadian Journal of Physics*, Vol. **41**, pp. 724, 731-733; ©1963 National Research Council Canada.

III.B.3 Reprinted with permission from *Physical Review*, Vol. **131**, pp. 2578-2582; ©1963 American Physical Society.

III.A.4 Reprinted from *Astrophysical Journal*, Vol. **139**, pp. 602, 605 (1964).

III.A.5 Reprinted from *Astrophysical Journal*, Vol. **150**, pp. 1001-1004 (1967).

III.A.6 Reprinted from *Astrophysical Journal*, Vol. **152**, pp. L17-L20 (1968).

III.A.7 Reprinted from *Astrophysical Journal*, Vol. **155**, p. 511 (1969).

III.B.8 Reprinted with permission from *Physical Review*, Vol. **C4**, pp. 2578-2582; ©1971 American Physical Society.

III.B.9 Reprinted from *Cosmology, Fusion and Other Matters: George Gamow Memorial Volume*," ed. F. Reines (Colorado Associated University Press, ©1972), pp. 169, 180-182, with permission of the University Press of Colorado.

III.C.1 Reprinted with permission from *Physical Review Letters*, Vol. **48**, pp. 1664-1666; ©1982 American Physical Society.

III.C.2. Reprinted with permission from *Physical Review*, Vol. **C28**, pp. 2222-2229; ©1983 American Physical Society.

III.C.3. Reprinted with permission from *Nuclear Physics*, Vol. **A467**, pp. 273-275, 285, 288-290, ©1987 Elsevier Science Publishers.

III.C.4 Reprinted from *Astrophysical Journal*, Vol. **392**, pp. 320-327 (1992).

IV.A.1. Reprinted with permission from *Physics Letters*, Vol. **B28**, pp. 493-496; ©1969 Elsevier Science Publishers.

IV.A.2. Reprinted with permission from *Physics Letters*, Vol. **B29**, pp. 623-625; ©1969 Elsevier Science Publishers.

IV.A.3. Reprinted with permission from *Physical Review*, Vol. **D24**, pp. 538-541; ©1981 American Physical Society.

IV.A.4. Reprinted with permission from *Physics Letters*, Vol. **B190**, p. 199; ©1987 Elsevier Science Publishers.

IV.B.1 Reprinted with permission from *Physical Review Letters*, Vol. **28**, pp. 316-318; ©1972 American Physical Society.

IV.B.2. Reprinted with permission from *Physics Letters*, Vol. **B199**, pp. 281-285; ©1987 Elsevier Science Publishers.

IV.B.3. Reprinted with permission from *Physics Letters*, Vol. **B200**, p. 115; ©1988 Elsevier Science Publishers.

IV.C.1. Reprinted with permission from *Physical Review*, Vol. **D17**, pp. 2369-2374; ©1978 American Physical Society.

IV.C.2. Reprinted with permission from *Physical Review*, Vol. **D22**, pp. 2718-2720; ©1980 American Physical Society.

IV.C.3. Reprinted with permission from *Physical Review*, Vol. **D27**, pp. 1228, 1231; ©1983 American Physical Society.

IV.C.4. Reprinted with permission from *Soviet Journal of Nuclear Physics*, Vol. **42** pp. 913-917; ©1985 American Institute of Physics.

IV.C.5. Reprinted with permission from *Nuovo Cimento*, **C9**, p. 24 © 1986 Societá Italiana di Fisica.

IV.C.6 Reprinted with permission from *Soviet Physics JETP*, Vol. **64**, pp. 4-7; ©1986 American Institute of Physics.

IV.C.7 Reprinted from *Proceedings of the 6th Moriond Workshop on Massive Neutrinos in Particle Physics and Astrophysics*, ed. O. Fackler and J. Tran Thanh Van, Tignes, pp. 373, 376, 378 ©1986, by permission of Editions Frontieres.

IV.C.8 Reprinted with permission from *Physical Review*, Vol. **D34**, pp. 969-974; ©1986 American Physical Society.

IV.C.9. Reprinted with permission from *Physical Review Letters*, Vol. **56**, pp. 1305-1308; ©1986 American Physical Society.

IV.C.10. Reprinted with permission from *Physical Review Letters*, Vol. **57**, pp. 1275-1278; ©1986 American Physical Society.

IV.C.11. Reprinted with permission from *Physical Review Letters*, Vol. **57**, pp. 1271-1274; ©1986 American Physical Society.

IV.C.12. Reprinted with permission from *International Journal of Modern Physics*, **A2**, pp. 304-305; ©1987 World Scientific Publishing Co.

IV.D.1. Reprinted with permission from *Astrophysics & Space Science*, Vol. **10**, pp. 87-92; ©1971 Kluwer Academic Publishers.

IV.D.2. Reprinted with permission from *Soviet Physics JETP*, Vol. **64**, pp. 446-452; ©1986 American Institute of Physics.

IV.D.3. Reprinted with permission from *Physical Review*, Vol. **D37**, pp. 1368-1373; ©1988 American Physical Society.

IV.D.4. Reprinted with permission from *Physics Letters*, Vol. **B213**, pp. 64-68; ©1988 Elsevier Science Publishers.

IV.E.1. Reprinted with permission from *Physical Review*, Vol. **D38**, pp. 2635-2637; ©1988 American Physical Society.

IV.E.2. Reprinted with permission from *Physical Review*, Vol. **D44**, pp. R935-938; ©1991 American Physical Society.

IV.E.3. Reprinted with permission from *Physics Letters*, Vol. **B260**, pp. 154-156; ©1991 Elsevier Science Publishers.

V.A.1. Reprinted from *Astrophysical Journal*, Vol. **135**, pp. 474-477. 488-490 (1962).

V.A.2. Reprinted from *Astrophysical Journal*, Vol. **162**, pp. 993-1002 (1970).

V.A.3. Reprinted with permission from *Astrophysics Letters*, Vol. **7**, pp. 191-192 © 1971 Gordon & Breach Science Publishers.

V.A.4. Reprinted with permission from *Astronomy & Astrophysics*, Vol. **44**, pp. 371-375 ©1975.

V.A.5. Reprinted with permission from *Nature*, Vol. **282**, pp. 591-594, © 1979 Macmill Magazines Ltd.

V.A.6. Reprinted with permission from *Nature*, Vol. **288**, pp. 541-544, © 1980 Macmill Magazines Ltd.

V.A.7. Reprinted with permission from *Nature*, Vol. **310**, pp. 22-25, © 1984 Macmill Magazines Ltd.

II.A.8. Reprinted from *Science*, Vol. **229**, pp. 923-931, Christensen-Dalsgaard, J., Goug D.O. and Toomre, J.; ©1985, with permission of the AAAS.

V.A.9. Reprinted from *Astrophysical Journal*, Vol. **378**, p. 413 (1991).

V.A.10. Reprinted from *Science*, Vol. **253**, pp. 152-157, Libbrecht, K.G. and Woodard, M. ©1991, with permission of the AAAS.

VI.1 Reprinted with permission from PHYSICS LETTERS B, Vol 433, No 1-2, 1998, pp 1-8, Bahca et al: "How uncertain are solar neutrino predictions?", copyright 1998, with permission fron Elsevier Science.

VI.2 Reprinted with permission from Phys. Rev. Lett. 86 (2001) 5651. S. Fukuda et al. (Supe Kamiokande Collaboration). "Solar and hep neutrino measurements from 1258 days c Super-Kamiokande data," copyright 2001 by the American Physical Society.

VI.3 Reprinted with permission from Phys. Rev. Lett. 87 (2001) 071301. Q. R. Ahmad et al., "Measurement of charged current interactions produced by solar neutrinos at the Sudbury Neutrino Observatory," copyright 2001 by the American Physical Society.

Milton Keynes UK
Ingram Content Group UK Ltd.
UKHW040712141024
449569UK00012B/606

9 780813 340371